Some Useful Constants and Physical Measurements*

astronomical unit	1 A.U. = 1.496×10^8 km (1.5×10^8 km)
light year	1 ly = 9.46×10^{12} km (10^{13} km; 6 trillion miles)
parsec	1 pc = 3.09×10^{13} km = 3.3 ly
speed of light	c = 299,792.458 km/s (3×10^5 km/s)
Stefan-Boltzmann constant	σ [Greek sigma] = 5.67×10^{-8} W/m$^2 \cdot$ K^4
Planck's constant	h = 6.63×10^{-34} Js
gravitational constant	G = 6.67×10^{-11} Nm^2/kg^2
mass of the Earth	$M_\oplus$ = 5.97×10^{24} kg (6×10^{24} kg; about 6000 billion billion tons)
radius of the Earth	$R_\oplus$ = 6378 km (6500 km)
mass of the Sun	$M_\odot$ = 1.99×10^{30} kg (2×10^{30} kg)
radius of the Sun	$R_\odot$ = 6.96×10^5 km (7×10^5 km)
luminosity of the Sun	$L_\odot$ = 3.90×10^{26} W
effective temperature of the Sun	$T_\odot$ = 5778 K (5800 K)
Hubble constant	$H_0 \approx$ 75 km/s/Mpc
mass of the electron	m_e = 9.11×10^{-31} kg
mass of the proton	m_p = 1.67×10^{-27} kg

The rounded-off values used in the text are shown in parentheses.

Conversions between English and Metric Units

1 inch	=	2.54 centimeters (cm)	1 mile	=	1.609 kilometers (km)
1 foot (ft)	=	0.3048 meters (m)	1 pound (lb)	=	453.6 grams (g) or .4536 kilograms (kg) (on Earth)

The Entire Electromagnetic Spectrum

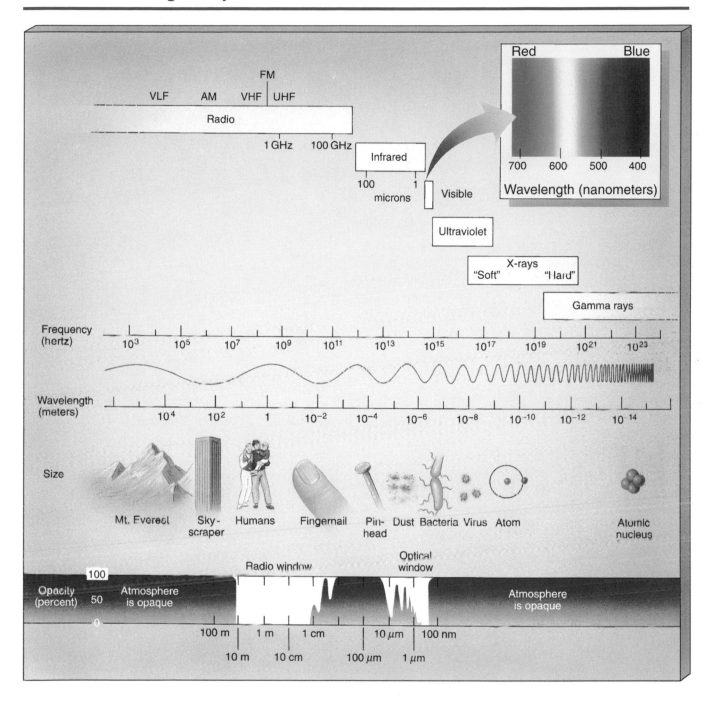

Astronomy

About the Authors

Eric Chaisson Eric holds a doctorate in Astrophysics from Harvard University, where he spent ten years on the faculty of Arts and Sciences. For five years, Eric was a Senior Scientist and Director of Educational Programs at the Space Telescope Science Institute and Adjunct Professor of Physics at Johns Hopkins University. He recently joined Tufts University, where he is now Professor of Physics, Professor of Education, and Director of the Wright Center for Innovative Science Education. He has written nine books on astronomy, which have received such literary awards as the Phi Beta Kappa Prize, two American Institute of Physics Awards, and Harvard's Smith Prize for Literary Merit. He has published more than 100 scientific papers in professional journals, and has also received Harvard's Bok Prize for original contributions to astrophysics.

Steve McMillan Steve holds a bachelor's and master's degree in Mathematics from Cambridge University and a doctorate in Astronomy from Harvard University. He held post-doctoral positions at the University of Illinois and Northwestern University, where he continued his research in theoretical astrophysics, star clusters, and numerical modeling. Steve is currently Distinguished Professor of Physics at Drexel University and a frequent visiting researcher at Princeton's Institute for Advanced Study and the University of Tokyo. He has published over 40 scientific papers in professional journals.

Astronomy

A Beginner's Guide to the Universe

Second Edition

Eric Chaisson
Tufts University

Steve McMillan
Drexel University

Prentice Hall
Upper Saddle River, New Jersey 07458

Chaisson, Eric.
    Astronomy : a beginner's guide to the universe / Eric Chaisson,
Steve McMillan. — 2nd ed.
      p.   cm.
    Includes index.
    ISBN 0-13-085848-X
    1. Astronomy.   I. McMillan, S. (Stephen).   II. Title
QB43.2.C43   1998
520—dc21   97-38104
                                                      CIP

Executive Editor: Alison Reeves
Editor in Chief: Paul F. Corey
Development Editor: Irene Nunes
Editorial Director: Tim Bozik
Assistant Vice President of Production and Manufacturing:
   David W. Riccardi
Executive Managing Editor: Kathleen Schiaparelli
Assistant Managing Editor: Shari Toron
Production Editor: Alison Aquino
Director of Marketing: John Tweedale
Marketing Manager: Kelly McDonald
Senior Marketing Manager: Leslie Cavaliere
Creative Director: Paula Maylahn
Art Manager: Gus Vibal
Art Editor: Karen Branson
Art Director: Amy Rosen

Interior and Cover Designer: Carole Anson
Manufacturing Manager: Trudy Pisciotti
Assistant Editor: Wendy Rivers
Editorial Assistant: Gillian Kieff
Image Resource Center Director: Lori Morris-Nantz
Image Permission Supervisor: Kay Dellosa
Image Coordinator: Charles Morris
Copy Editor: Sally Ann Bailey
CD Rom Project Manager: Cynthia Dunn
CD Rom Assistant Project Manager: Barbara Booth
CD Rom Production: Jeff Henn, Patty Gutierrez, Ray
   Caramanna, Karen Stephens, Charles Pelletreau, Molly
   Pike Riccardi, Michael D'Angelo
New Media Project Manager: Cindy Harford
Testing Supervisor: David Moles

Printed in the United States of America

10 9 8 7 6 5 4

ISBN 0-13-085848-X

Prentice Hall International (UK) Limited, *London*
Prentice-Hall of Australia Pty. Limited, *Sydney*
Prentice-Hall Canada Inc., *Toronto*
Prentice-Hall Hispanoamericana, S.A., *Mexico*
Prentice-Hall of India Private Limited, *New Delhi*
Prentice-Hall of Japan, Inc., *Tokyo*
Pearson Education Asia Pte. Ltd., *Singapore*
Editora Prentice-Hall do Brasil, Ltda., *Rio de Janeiro*

Brief Contents

Contents

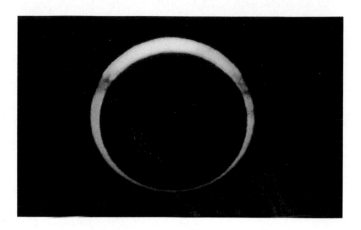

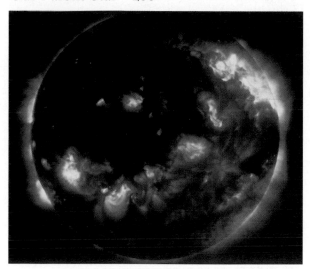
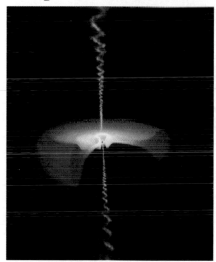

Animations, Videos, and Extensions on the CD

The enclosed CD contains a free, hyperlinked version of the text with integrated links to animations, videos, and extensions, as well as links to the text's WWW site, which provides on-line archives, on-line destinations, and on-line exercises for each chapter.

Following is a list of animations, videos, and extensions for each chapter.

Preface

Astronomy continues to enjoy a golden age of exploration and discovery. Fueled by new technologies and novel theoretical insights, the study of the cosmos has never been more exciting. We are pleased to have the opportunity to present in this book a representative sample of the known facts, evolving ideas, and frontier discoveries in astronomy today.

Astronomy: A Beginner's Guide to the Universe has been written for students who have taken no previous college science courses and who will likely not major in physics or astronomy. It is intended primarily for use in a one-semester, nontechnical astronomy course. We present a broad view of astronomy, straightforwardly descriptive and without complex mathematics. The absence of sophisticated mathematics, however, in no way prevents discussion of important concepts. Rather, we rely on qualitative reasoning as well as analogies with objects and phenomena familiar to the student to explain the complexities of the subject without over-simplification. We have tried to impart the enthusiasm that we feel about astronomy, and to awaken students to the marvelous universe around us.

We are very gratified that the first edition of this text has been received so well by many in the astronomy education community. In using that earlier text, many of you—teachers and students alike—have sent us your helpful feedback and constructive criticisms. From these, we have learned to communicate better both the fundamentals and the excitement of astronomy. Many improvements inspired by your comments, as well as numerous innovations and popular new features from our companion hardback text *Astronomy Today, Media Edition*, have been incorporated into this new edition.

Organization and Approach

As in the first edition, our organization follows the popular and effective "Earth-out" progression. We have found that most students, especially those with little scientific background, are much more comfortable studying the (relatively familiar) solar system before tackling stars and galaxies. Thus, Earth is the first object we discuss in detail. With Earth and Moon as our initial planetary models, we move through the solar system. Integral to our coverage of the solar system is a discussion of its formation. This line of investigation leads directly into a study of our Sun. With the Sun as our model star, we broaden the scope of our discussion to include stars in general—their properties, their evolutionary histories, and their varied fates. This journey naturally leads us to coverage of the Milky Way Galaxy, which in turn serves as an introduction to our treatment of other galaxies, both normal and active. Finally, we reach the subject of cosmology and the large-scale structure and dynamics of the universe as a whole. Throughout, we strive to emphasize the dynamic nature of the cosmos—virtually every major topic, from planets to quasars, includes a discussion of how those objects formed and how (we think) they evolve.

The second edition of *Astronomy: A Beginner's Guide to the Universe* contains two fewer chapters, and is almost 20 percent shorter (in terms of total text), than its predecessor. All chapters have been updated in content and several have seen significant internal reorganization. Specifically, the first two chapters have been restructured to create a brief Prologue, which presents some basic introductory material—the essentials of the celestial sphere and angular measure, and some elementary geometry—and a new Chapter 1, which discusses the motions of the Sun, the Moon, and the planets in a single chapter, resulting in a more concise and effective presentation. The solar system section has been reduced from 6 chapters to 5 by merging the overview of our planetary system with the discussion of its formation. Instructors presenting this material in a 1-quarter course, who wish to (or have time to) cover only the essentials of the solar system before proceeding on to the study of stars and the rest of the universe, may want to teach only Chapter 4, and then move directly to Chapter 9 (the Sun). Finally, our discussion of stellar evolution, which was spread over two chapters in the first edition, has been reworked into a more succinct single-chapter format.

We continue to place much of the needed physics in the early chapters—an approach derived from years of experience teaching thousands of students. Additional physical principles are developed as needed later, both in the text narrative and in the boxed *More Precisely* features (described below). We feel strongly that this is the most economical and efficient means of presentation. However, we acknowledge that not all instructors feel the same way. Accordingly, we have made the treatment of physics, as well as the more quantitative discussions, as modular as possible, so that these topics can be deferred to later stages of an astronomy course if desired.

The Illustration Program

Visualization plays an important role in both the teaching and the practice of astronomy, and we continue to place strong emphasis on this aspect of our book. We have tried to combine aesthetic beauty with scientific accuracy in the artist's conceptions that adorn the text, and we have sought to present the best and latest imagery of a wide range of cosmic objects. Each illustration has been carefully crafted to enhance student learning; each is pedagogically sound and tightly tied to nearby discussion of important scientific facts and ideas.

Compound Art. ▶

It is rare that a single image, be it a photograph or an artist's conception, can capture all aspects of a complex subject. Wherever possible, multiple-part figures are used in an attempt to convey the greatest amount of information in the most vivid way:

- Visible images are often presented along with their counterparts captured at other wavelengths.

- Interpretive line drawings are often superimposed on or juxtaposed with real astronomical photographs, helping students to really "see" what the photographs reveal.

- Breakouts often multiple ones—are used to zoom in from wide-field shots to closeups, so that detailed images can be understood in their larger context.

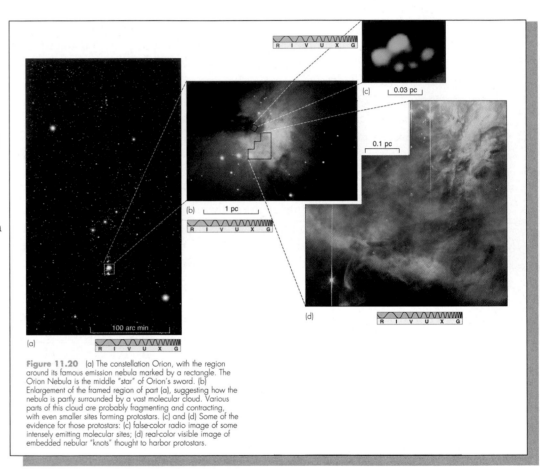

Figure 11.20 (a) The constellation Orion, with the region around its famous emission nebula marked by a rectangle. The Orion Nebula is the middle "star" of Orion's sword. (b) Enlargement of the framed region of part (a), suggesting how the nebula is partly surrounded by a vast molecular cloud. Various parts of this cloud are probably fragmenting and contracting, with even smaller sites forming protostars. (c) and (d) Some of the evidence for those protostars: (c) false-color radio image of some intensely emitting molecular sites; (d) real-color visible image of embedded nebular "knots" thought to harbor protostars.

▲

Explanatory Captions. Students often review a chapter by "looking at the pictures." For this reason, the captions in this book are often a bit longer and more detailed than those in other texts.

Full-Spectrum Coverage and Spectrum Icons. 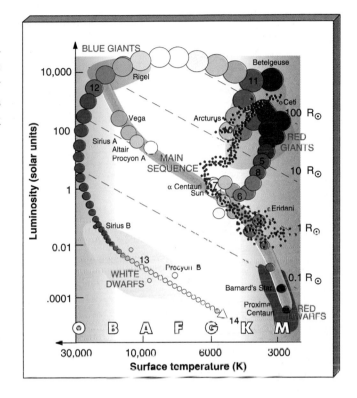 Increasingly, astronomers are exploiting the full range of the electromagnetic spectrum to gather information about the cosmos. Throughout this book, images taken at radio, infrared, ultraviolet, X ray, or gamma ray wavelengths are used to supplement visible-light images. As

it is sometimes difficult (even for a professional) to tell at a glance which images are visible-light photographs and which are false-color images created with other wavelengths, each photo in the text is provided with an icon that identifies the wavelength of electromagnetic radiation used to capture the image.

H–R Diagrams and Acetate Overlays. ▶

All of the book's H-R diagrams have been redrawn in a uniform format, using real data. In addition, a unique set of transparent acetate overlays dramatically demonstrate to students how the H-R diagram helps us to organize our information about the stars and track their evolutionary histories.

Pedagogy

As with many other parts of our text, adopting instructors have helped guide us toward what is most helpful for effective student learning. With their assistance, we have revised both our in-chapter and end-of-chapter pedagogical apparatus to increase its utility to students.

Learning Goals.   Studies indicate that beginning students often have trouble prioritizing textual material. For this reason, a few (typically 5 or 6) well-defined Learning Goals are provided at the start of each chapter. These help students to structure their reading of the chapter and then test their mastery of key facts and concepts. The Goals are numbered, and cross-referenced to key sections in the body of each chapter. This

in-text highlighting of the most important aspects of the chapter also helps students to review. The Goals have also been reorganized and rephrased to make them more objectively testable, affording students better means of gauging their own progress.

Key Terms.   Like all subjects, astronomy has its own specialized vocabulary. To aid student learning, the most important astronomical terms are boldfaced at their first appearance in the text. Each boldfaced Key Term is also incorporated in the appropriate chapter summary, together with the page number where it was defined. In addition, a full alphabetical glossary, defining each Key Term and locating its first use in the text, appears at the end of the book.

Interludes. These explore a ▶ variety of interesting supplementary topics.

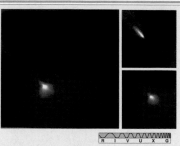
◀ **More Precisely** boxes. New to the second edition these provide more quantitative treatments of subjects discussed qualitatively in the text. Removing these more challenging topics from the main flow of the narrative and placing them within a separate modular element of the chapter design (so that they can be covered in class, assigned as supplementary material, or simply left as optional reading for those students who find them of interest) will allow instructors greater flexibility in setting the level of their coverage.

∞ **Cross-Links.** In astronomy, as in many ▶ scientific disciplines, almost every topic seems to have some bearing on almost every other. In particular, the connection between the specifically astronomical material and the physical principles set forth early in the text is crucial. It is important that students, when they encounter, say, Hubble's Law in Chapter 16, recall what they learned about spectral lines and the Doppler shift in Chapter 2. Similarly, the discussions of the masses of binary star components (Chapter 10) and of galactic rotation (Chapter 14) both depend on the discussion of Kepler's and Newton's laws in

Chapter 1. To remind students of these links, so that the reader can recall the principles on which later discussions rest, and if necessary, review them, we have inserted "cross-links" throughout the text—symbols that mark key intellectual bridges between material in different chapters. The links, denoted by the symbol ∞, together with a section reference, signal to students that the topic under discussion is related in some significant way to ideas developed earlier, and direct them to material that they might need to review before proceeding.

Chapter Summaries. The Chapter Summaries, a primary review tool for the student, have been expanded and improved for the second edition. All Key Terms introduced in each chapter are listed again, in context and in boldface, in these Summaries, along with page references to the text discussion.

Questions, Problems, and Projects. Many elements of the end-of-chapter material have seen substantial reorganization:

- Each chapter now incorporates some 20–30 Self-Test Questions, roughly equally divided between "true/false" and "fill-in-the-blank" formats, designed to allow students to assess their understanding of the chapter material. Answers to all these questions appear at the end of the book.

- Each chapter has about 15 Review and Discussion Questions, which may be used for in-class review or for assignment. As with the Self-Test Questions, the material needed to answer Review Questions may be found within the chapter. The Discussion Questions explore particular topics more deeply, often asking for opinions, not just facts. As with all discussions, these questions usually have no single "correct" answer.

- Several Problems in each chapter entail some numerical calculation; their answers are not contained verbatim within the chapter, but the information necessary to solve them has been presented in the text.

- Each chapter ends with a few Projects meant to get the student out of the classroom and looking at the sky, although some entail research in libraries or other extracurricular activities.

CD-ROM. Following the enthusiastic response to the *Astronomy Today, 2E 1997 Media Edition*, this sec-

ond edition of the *Beginner's Guide* comes complete with a fully integrated CD-ROM. *Astronomy: A Beginner's Guide 2/e* includes a free CD in the back of the text which includes a fully hyperlinked electronic version of the text to help the reader quickly find related information and assist in review, integrated animations and videos to bring text figures to life, links to our companion website, which is organized by text chapter and updated monthly, and a separate executable multimedia study guide program.

A special feature of the CD for this text is a series of "Extensions"—1–2 page sections that expand on discussions in the printed text. In this way, we present essential material in the print-based text, but still provide additional material for those students who want to delve deeper into some topics, without making the text itself too long, detailed, or overwhelming. We are excited about the innovative use of media to complement the text and look forward to your response to it.

The CD-ROM material can be used on Macintosh and PC computers using any standard browser (such as Netscape Navigator or Microsoft Explorer). For those students who do not already have a browser, Microsoft's Internet Explorer is included on the CD. A UNIX script for using the CD or a UNIX system is available at

ftp://ftp.prenhall.com/pub/esm/physics.s-085/chaissonbg/

Content Updates in the Second Edition

This second edition of the Beginner's Guide has been updated throughout, both in the text itself and in the CD-ROM extensions, with new and late-breaking information, including

- the latest developments in telescope technology, covering both ground-based adaptive optics and interferometry, and the present status of the *Hubble Space Telescope* and other orbiting instruments (Chapter 3)

- coverage of the recent widely viewed Comets Hyakutake and Hale-Bopp (Chapter 4)

- the continuing story of the search for, and the apparent discovery of, planets orbiting stars other than our Sun (Chapter 4)

- the search for life on Mars, including the possibility that fossilized bacteria may have been detected in a meteorite believed to have originated on the Martian surface (Chapter 6)

- the Mars *Pathfinder* mission and plans for future visits to the red planet (Chapter 6)

- the *Galileo* mission to Jupiter, and its main findings so far (Chapters 7 and 8)

- revision of all distance scales throughout the text in light of the recent findings by the *Hipparcos* satellite. The *Hipparcos* mission is described in Extension 10.1

- Hubble Space Telescope observations of the Eagle Nebula, the Orion Nebula, and other star-forming regions (Chapter 11)

- the ongoing mystery of the cosmic gamma-ray bursts, and the recent strong evidence that at least one lies at cosmological distances (Chapter 13)

- the Hubble Deep Field, and its significance to studies of galaxy formation and cosmic evolution (Chapter 15)

- observations of quasar host galaxies, and how they relate to current theories of active galaxy evolution (Chapter 16)

- a review of how the *Hipparcos* data impact the long-standing debate on the age of the universe (Chapter 17)

Supplementary Material

This edition is accompanied by an outstanding set of instructional aids.

World Wide Web Site. For both teachers and students we have a companion website specifically for *Astronomy: A Beginner's Guide 2/e* at

http://www.prenhall.com/chaisson/bg

This powerful resource organizes material from a variety of sources on the web on a chapter-by-chapter basis, is updated monthly, and provides interactive on-line exercises for each chapter.

Each chapter of the website for *Astronomy: A Beginner's Guide 2/e* has the following three categories of materials:

- Online Archives—annotated images, videos, animations and free downloadable software

- Online Destinations—annotated links to relevant websites

- Online Exercises—interactive questions for students to answer on-line; scoring and feedback are provided immediately.

Comets. This is an annual update kit for Astronomy containing videos, slides, and *New York Times* articles. The VHS tape in the Fall 1997 Comets kit includes 27 custom animations prepared by the Wright Center for Science Visualization to accompany *Astronomy: A Beginner's Guide to the Universe 2/e* and *Astronomy Today, 2/e* 1997 *Media Edition* as well as three videos from the Space Telescope Science Institute, six from the Jet Propulsion Laboratories, a simulation on Galaxy Formation by Edward Bertschinger and a simulation from Roeland Van Der Marel called "A Black Hole in Galaxy M32." The slides, videos and animations can be shown in class; the collection of *New York Times* articles, called *Themes of the Times*, is published twice yearly and available free in quantity for your students using either text. A newsletter in the *Comets* kit provides descriptions of everything on the VHS tape and the slides and cross-references them, as well as the *Times* articles, with appropriate chapters of the Chaisson/McMillan texts. (ISBN: 0-13-754169-4)

Instructor's Manual, by Leo Connolly (California State University at San Bernardino). This manual provides an overview of each chapter; pedagogical tips, useful analogies, and suggestions for classroom demonstrations; answers to the end-of-chapter review and discussion questions and problems; and a list of selected readings. (ISBN: 0-13-754177-5)

Presentation Manager CD for Astronomy: A Beginner's Guide 2E.   This flexible, easy-to-use tool contains a wealth of photographs, line art, animations, and videos to use in class lectures. With the *Presentation Manager* system, instructors can easily search, access, and organize the materials according to their lecture outlines and add their own visuals and lecture notes. The CD contains all the art and tables from *Astronomy: A Beginner's Guide 2E* as well as all the animations and videos that are on the CD in the back of the student text. In addition, the *Presentation Manager* incorporates over 80 slides from the past four editions of *Comets*. [(Mac) ISBN: 0-13-080420-7; (Win) ISBN: 0-13-754151-1]

Acetates and Slides.   An extensive set of color acetates and a comprehensive 35-mm slide set are available free to qualified adopters. [(Slide set) ISBN: 0-13-754144-9; (Transparency pack) ISBN: 0-13-754136-8]

Test Item File.   An extensive file of test questions, newly compiled for the second edition is offered free upon adoption. Available in both printed and electronic form (Macintosh or IBM-compatible formats). (ISBN: 0-13-754094-9)

Prentice Hall Custom Test. *Prentice Hall Custom Test* is based on the powerful testing technology developed by Engineering Software Associates, Inc. (ESA). Available for Windows, Macintosh, and DOS, *Prentice Hall Custom Test* allows educators to create and tailor the exam to their own needs. With the Online Testing option, exams can also be administered online and data can then be automatically transferred for evaluation. A comprehensive desk reference guide is included, along with on-line assistance. [(Mac) ISBN: 0-13-754128-7; (Win) ISBN: 0-13-754110-4]

Student Observation Guide with Laboratory Exercises, by Michael Seeds and Joseph Holzinger (Franklin and Marshall College). The second edition of this useful supplement contains 42 classic labs and observational activities, along with cardboard cutout instruments that students can build and use for observations. Available for sale to students. (ISBN: 0-13-644196-3)

Basic Astronomy Labs, by Jay Huebner and Terry Smith (University of North Florida, Jacksonville), and Michael Reynolds (Chabot Observatory and Science Center). A collection of 40 laboratory exercises, including a wide range of both traditional and innovative topics (such as the nature of human vision, radioactivity and time, and astronomy on the Internet). Detailed introductions provide a fully-developed context for each exercise. Available for sale to students. (ISBN: 0-13-376336-6)

Astronomy on the Internet, by Andrew Stull and Alan Sill (Texas Tech University). A guide to general astronomy resources on the Internet. Everything you need to know to get yourself online and browsing the World Wide Web! (ISBN: 0-13-89011-2)

Acknowledgments

Throughout the many drafts that have led to this book, we have relied on the critical analysis of many colleagues. Their suggestions ranged from the macroscopic issue of the book's overall organization to the minutiae of the technical accuracy of each and every sentence. We have also benefited from much good advice and feedback from users of the first edition of the text and our longer book, *Astronomy Today*, *Media Edition*. To these many helpful colleagues, we offer our sincerest thanks.

Stephen G. Alexander, *Miami University*
Martin Goodson, *Delta College*
David J. Griffiths, *Oregon State University*
Andrew R. Lazarewicz, *Boston College*
Richard Nolthenius, *Cabrillo College*
Robert S. Patterson, *Southwest Missouri State University*
John C. Schneider, *Catonsville Community College*
Don Sparks, *Los Angeles Pierce College*
Jack W. Sulentic, *University of Alabama*

We would also like to acknowledge our gratitude to Leo Connolly for preparing the end-of-chapter questions and problems; to Ray Villard of the Space Telescope Science Institute for compiling the *Comets* supplement; and to Alan Sill of Texas Tech University for creating and maintaining our World Wide Web site.

The publishing team at Prentice Hall has assisted us at every step along the way in creating this text. Much of the credit for getting the project completed on time goes to our Executive Editor, Alison Reeves, who has successfully navigated us through the twists, turns, and "absolute final deadlines" of the publishing world, all the while managing the many variables that go into a multifaceted publication such as this. Irene Nunes, our Development Editor, has skillfully helped us revise the manuscript and has been a constant source of insight and strength in making some very difficult content decisions. Production Editor Alison Aquino has done a remarkable job of tying together the threads of this very complex project, made all the more complex by the necessity of combining text, art, and electronic media into a coherent whole. Cindy Dunn has shown much technical leadership in this book's accompanying CD-ROM project. Finally, we would like to express our gratitude to renowned space artist Dana Berry for allowing us to use many of his beautiful renditions of astronomical scenes, and to Lola Judith Chaisson for assembling and drawing all the H–R diagrams (including the acetate overlays) for this edition.

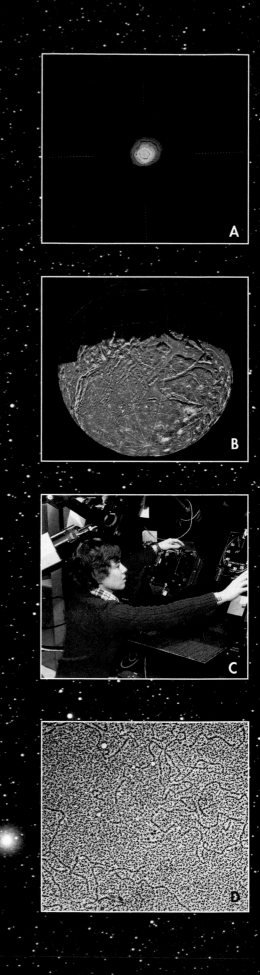

Prologue
Charting the Heavens *www*

◀ (Opposite page, background) One of the most easily recognizable star fields in the winter nighttime sky, the familiar constellation Orion. This field of view spans roughly 100 light years, or 10^{15} kilometers. (See also Figure P.6.)

(Inset A) If we magnify the wide view of the Orion constellation, shown at left, by a million times, we enter into the realm of the largest stars, with sizes of about a billion kilometers. Such a star is seen in this false-color image of the red-giant star Betelgeuse (which is actually the bright star at the upper left of the Orion constellation).

(Inset B) Another magnification of a million brings us to the scale of typical moons—roughly 1,000 kilometers—represented here by Ariel, one of the many moons of Uranus.

(Inset C) With yet another million-times magnification, we reach scales of meters, represented here by an astronomer at the controls of her telescope.

(Inset D) At a final magnification of an additional million, we reach the scale of molecules (about 10^{-6} meter), represented by this coiled DNA molecule of a rat's liver.

LEARNING GOALS

Studying this prologue will enable you to:

1   Describe the concept of the celestial sphere and the conventions of angular measurement that enable us to locate objects in the sky.

2   Explain the simple geometric reasoning that allows astronomers to measure distances to faraway objects.

Nature offers no greater splendor than the starry sky on a clear, dark night. Jeweled with the constellations of ancient myth and legend, the night sky has inspired wonder throughout the ages—a wonder that leads our imaginations far from the confines of Earth and out into the distant reaches of space and time. Astronomy, born in response to that wonder, is built on two basic traits of human nature: the need to explore and the need to understand. Through the interplay of curiosity, discovery, and analysis, people have sought answers to questions about the universe since the earliest times. Astronomy is the oldest of all the sciences, yet never has it been more exciting than it is today.

P.1 Our Place in Space

Of all the scientific insights achieved to date, one stands out boldly: Earth is neither central nor special. We inhabit no unique place in the universe. Astronomical research, especially within the past few decades, strongly suggests that we live on an ordinary rocky *planet* called Earth, one of nine known planets orbiting an average *star* called the Sun, a middle-aged star near the edge of a huge collection of stars called the Milky Way *galaxy*, one galaxy among countless billions of others spread throughout the observable *universe*. To get a feel for these relationships, consult Figures P.1 through P.4, and put these objects in perspective by studying Figure P.5.

We are connected to these distant realms of space not just by our imaginations but also through a common cosmic heritage: most of the chemical elements in our bodies were created billions of years ago in the hot centers of long-vanished stars. Their fuel supply spent, these gigantic stars died in huge explosions, scattering afar the elements created deep within their cores. Eventually, this matter collected into clouds of gas that slowly collapsed to give birth to a new generation of stars. In this way, the Sun and its family of planets formed nearly 5 billion years ago. Everything on Earth embodies atoms from other parts of the universe and from a past far more remote than the beginning of human evolution. Elsewhere, other beings, perhaps with an intelligence much greater than our own, might at this very moment be gazing in wonder at their own night sky. Our Sun might be nothing more than an insignificant point of light to them, if it is visible at all. Yet, if such beings exist, they must share our cosmic origin.

Simply put, the **universe** is the totality of all space, time, matter, and energy. **Astronomy** is the study of the universe. It is a subject unlike any other, for it requires us to change profoundly our view of the cosmos and to consider matter on scales totally unfamiliar from everyday experience. Look again at the galaxy shown in Figure P.3. It is a swarm of about a hundred billion stars—more stars than the number of people who have ever lived on Earth. The entire assemblage is spread across a vast expanse of space some 100,000 **light years** across. What is a light year? It is the *distance* traveled by light, moving at a speed of about 300,000 kilometers per second, in a year. One light year equals about 10 trillion kilometers (around 6 trillion miles)—astronomical systems are truly "astronomical" in size! The light year is

15,000 kilometers

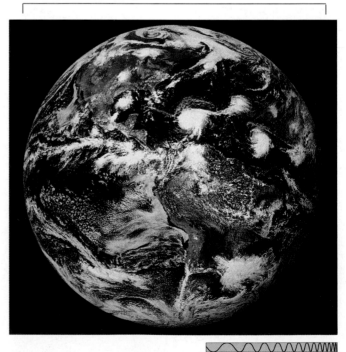

R I V U X G

Figure P.1   Earth is a planet, a mostly solid object, though it has some liquid in its oceans and its core, and gas in its atmosphere. (In this view, you can clearly see the North and South American continents.)

1,500,000 kilometers

Figure P.2 Much bigger than Earth, the Sun is a very hot ball of gas held together by its own gravity.

About 10,000,000 light years

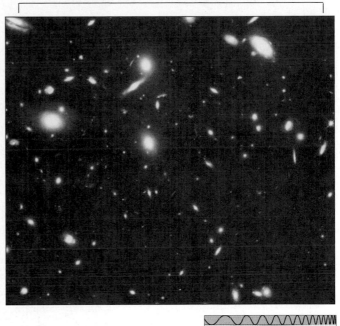

Figure P.4 A portion of the Coma cluster of galaxies, some 300 million light years from Earth. Each galaxy contains hundreds of billions of stars, probably a comparable number of planets, and possibly living creatures.

About 100,000 light years

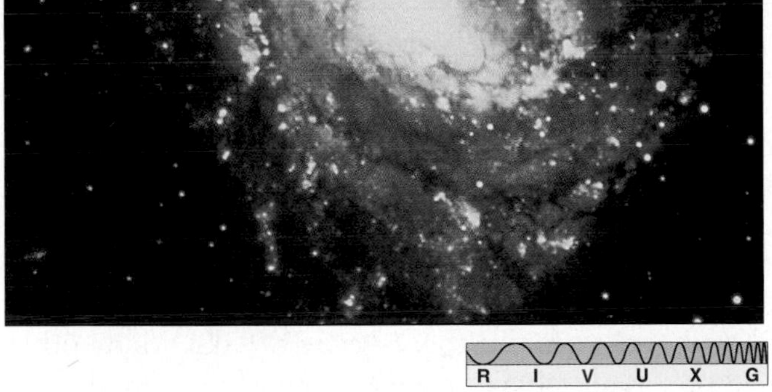

Figure P.3 A typical galaxy is a collection of a few hundred billion stars, each separated by vast regions of nearly empty space. This galaxy is M83, the 83rd entry in the catalog compiled by the eighteenth-century French astronomer Charles Messier. Our Sun is an undistinguished star near the edge of another galaxy, the Milky Way.

a convenient unit introduced by astronomers to help them describe immense distances. We will encounter many more such "custom" units in our studies (see Appendix 2).

A thousand (1000), a million (1,000,000), a billion (1,000,000,000), a trillion (1,000,000,000,000) — let's take a moment to understand the magnitude of these numbers. One thousand is easy enough to understand: at the rate of one number per second, you could count to a thousand in about 16 minutes. However, if you wanted to count to a million, you would need more than 2 weeks of counting at the rate of one number per second, 16 hours per day (allowing 8 hours per day for sleep). To count from 1 to a billion at the same rate of one number per second for 16 hours per day would take nearly 50 years. In this text we consider spatial domains spanning not just billions of kilometers but billions of light years, objects containing not just trillions of atoms but trillions of stars, time intervals of not just billions of seconds or hours but billions of years. You will need to become familiar with—and comfortable with—such enormous numbers. A good way to start is to recognize just how much larger than a thousand is a million and how much larger still is a billion.

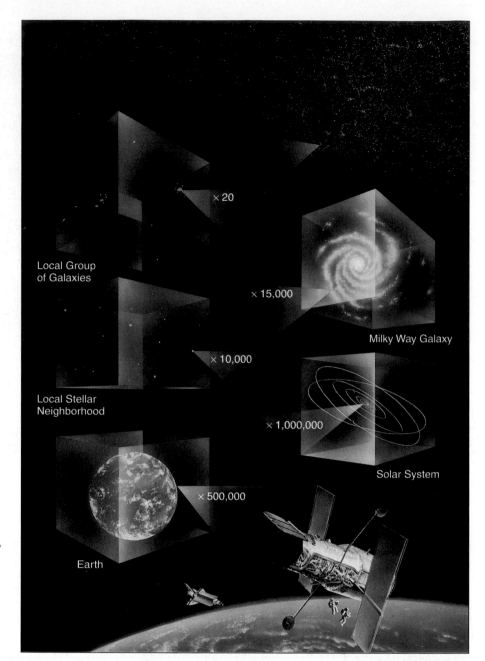

Figure P.5 This artist's conception puts the previous four illustrations in perspective. The bottom shows spacecraft (and astronauts) in Earth orbit, a view that widens progressively from bottom to top in the next five cubes: Earth, the Solar System, the Local Stellar Neighborhood, the Milky Way Galaxy, and the Local Group of Galaxies. The numbers indicate approximately the increase in scale between successive images.

P.2 The Obvious View

Constellations in the Sky

Between sunset and sunrise on a clear night, we can see some 3000 points of light. Include the view from the opposite side of Earth, and nearly 6000 stars are visible to the unaided eye. A natural human tendency is to see patterns and relationships between objects even when no true connection exists, and people long ago connected the brightest stars into configurations called **constellations,** which ancient astronomers named after mythological heroes and animals. Figure P.6 shows a constellation especially prominent in the northern

night sky from October through March: the hunter Orion, named for a mythical Greek hero famed, among other things, for his amorous pursuit of the Pleiades, the seven daughters of the giant Atlas. According to Greek mythology, in order to protect the Pleiades from Orion, the gods placed them among the stars, where Orion nightly stalks them across the sky. Many other constellations have similarly fabulous connections with ancient cultures.

Generally speaking, the stars making up a particular constellation are not close together in space. They merely are bright enough to observe with the naked eye and happen to lie in the same direction in the sky

(a)

R I V U X G

(b)

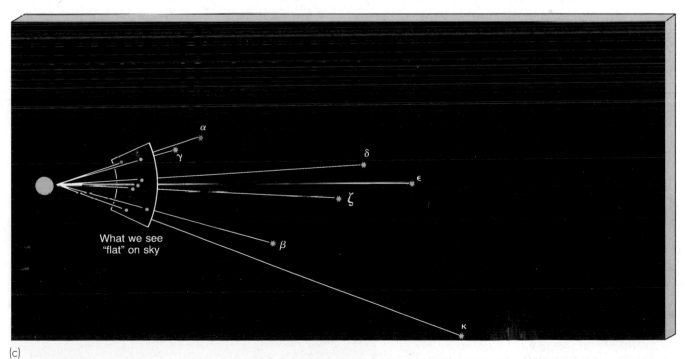

(c)

Figure P.6 (a) A photograph of the group of bright stars that make up the constellation Orion. (b) The stars connected to show the pattern visualized by the Greeks: the outline of a hunter. You can easily find this constellation in the northern winter sky by identifying the line of three bright stars in the hunter's belt. (c) The three-dimensional relationships for the prominent stars in Orion. The Greek letters are astronomical notations indicating brightness.

as seen from Earth. Figure P.6(c) illustrates this point for the constellation Orion, showing the true relationships between that constellation's brightest stars. Despite the fact that constellation patterns have no physical significance, the terminology is still used today. The constellations provide a convenient means for astronomers to specify large areas of the sky, much as geologists use continents or politicians use voting precincts to identify certain localities on Earth. In all, there are 88 constellations, most of them visible from North America at some time during the year.

The Celestial Sphere

Over the course of a night, the constellations seem to move across the sky from east to west. However, ancient sky-watchers noted that the *relative* positions of the constellations seemed unchanging as this nightly march took place. It was natural for those first astronomers to conclude that the stars must be firmly attached to a **celestial sphere** surrounding Earth—a canopy of stars resembling an astronomical painting on a heavenly ceiling. Figure P.7 shows how early astronomers pictured the stars as moving with this celestial sphere as it turned around a fixed, unmoving Earth. Figure P.8 shows how all stars appear to move in circles around a point in the sky very close to the star Polaris (better known as the Pole Star or the North Star). To the ancients, this point represented the axis around which the celestial sphere turned.

From our modern standpoint, the apparent motion of the stars is the result of the spin, or **rotation**, not of the celestial sphere but of Earth. Polaris indicates the direction—due north—in which Earth's

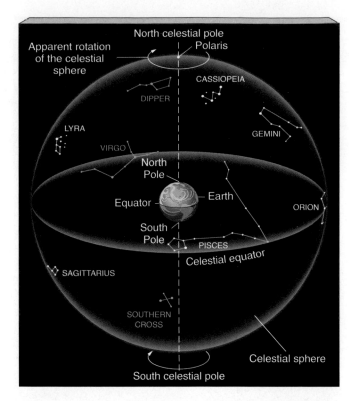

Figure P.7 Planet Earth sits fixed at the hub of the celestial sphere, which contains all the stars. This is one of the simplest possible models of the universe, but it doesn't agree with all the facts that astronomers know about the universe.

rotation axis points. Even though we now know that a revolving celestial sphere is an incorrect description of the heavens, we still use the idea as a convenient fiction that helps us visualize the positions of stars in the sky. The point where Earth's axis intersects the celestial sphere in the Northern Hemisphere is known as the **north celestial pole**, and it is directly above Earth's North Pole. In the Southern Hemisphere, the extension

Figure P.8 A time-lapse photograph of the northern sky. Each trail is the path of a single star across the night sky. The duration of the exposure is about 5 hours. (How can you tell that this is so?) The center of the concentric circles is near the North Star, Polaris, whose short, bright arc is prominently visible.

of Earth's axis in the opposite direction defines the **south celestial pole**, directly above Earth's South Pole. Midway between the north and south celestial poles lies the **celestial equator**, representing the intersection of Earth's equatorial plane with the celestial sphere.

Celestial Coordinates

The simplest method of locating stars in the sky is to specify their constellation and then rank the stars in it in order of brightness. The brightest star is denoted by the Greek letter α (alpha), the second brightest by β (beta), and so on. Thus, the two brightest stars in the constellation Orion—Betelgeuse and Rigel—are also known as α Orionis and β Orionis, respectively (Figure P.6). (Precise recent observations show that Rigel is actually brighter than Betelgeuse, but the names are now permanent.) Because there are many more stars in any given constellation than there are letters in the Greek alphabet, this method is of limited utility. However, for naked-eye astronomy, where only bright stars are involved, it is quite satisfactory.

For more precise measurements, astronomers find it helpful to lay down a system of **celestial coordinates** on the sky. If we think of the stars as being attached to the celestial sphere centered on Earth, then the familiar system of latitude and longitude on Earth's surface extends quite naturally to the sky. The celestial analogs of latitude and longitude are called **declination** and **right ascension**, respectively (Figure P.9). Just as latitude and longitude are tied to Earth, right ascension and declination are fixed on the celestial sphere. Although the stars appear to move across the sky because of Earth's rotation, their celestial coordinates remain constant over the course of a night.

Declination (dec) is measured in *degrees* (°) north or south of the celestial equator, just as latitude is measured in degrees north or south of Earth's equator. (See *More Precisely P-1*.) Thus, the celestial equator is at a declination of 0°, the north celestial pole is at +90°, and the south celestial pole is at −90° (the minus sign here just means "south of the celestial equator"). Right ascension (RA) is measured in angular units called *hours*, *minutes*, and *seconds*, and it increases in the eastward direction. Like the choice of the Greenwich Meridian as the zero-point of longitude on Earth, the choice of zero right ascension is quite arbitrary—it is conventionally taken to be the position of the Sun in the sky at the instant of the vernal equinox (to be discussed in Chapter 1).

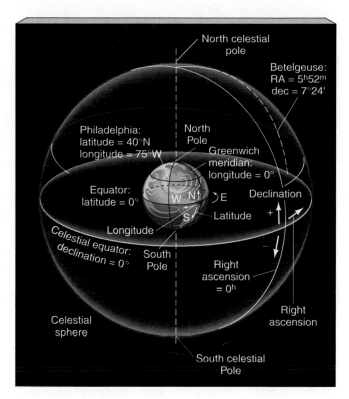

Figure P.9 Just as longitude and latitude allow us to locate a point on the surface of Earth, right ascension and declination specify locations on the sky. For example, to find Philadelphia on Earth, look 75° west of the Greenwich Meridian and 40° north of the Equator. Similarly, to locate the star Betelgeuse on the celestial sphere, look 5ʰ52ᵐ east of the vernal equinox (the line on the sky with a right ascension of zero) and 7°24′ north of the celestial equator.

P.3 The Measurement of Distance

2 Knowing the direction in which an object lies is only part of the information needed to locate it in space. Before we can make a systematic study of the heavens, we must find a way of measuring *distances*, too. One distance-measurement method, called **triangulation**, is based on the principles of Euclidean geometry and finds widespread application today in both terrestrial and astronomical settings. It forms the foundation of the family of distance-measurement techniques that together make up the **cosmic distance scale**.

Imagine trying to measure the distance to a tree on the other side of a river. The most direct method is to lay a tape across the river, but that's not the simplest way. A smart surveyor would make the measurement by visualizing an imaginary triangle (hence the term *triangulation*), sighting the tree on the far side of the river from two positions on the near side, as illustrated in Figure P.10. The simplest possible triangle is a right triangle, in which one of the angles is exactly 90°, so it

P-1 MORE PRECISELY

ANGULAR MEASURE

The size and scale of things are often specified by measuring lengths and angles. The concept of length measurement is fairly intuitive. The concept of angular measurement may be less familiar, but it too can become second nature if you remember a few simple facts.

A full circle contains 360 arc degrees (or just 360°). Therefore, the half-circle that stretches from horizon to horizon, passing directly overhead and spanning the portion of the sky visible to one person at any one time, contains 180°.

Each 1° increment can be further subdivided into fractions of an arc degree, called arc minutes; there are 60 arc minutes (60′) in 1 arc degree. Both the Sun and the Moon project an angular size of 30 arc minutes on the sky. Your little finger, held at arm's length, does about the same, covering about a 40-arc-minutes slice of the 180° horizon-to-horizon arc.

An arc minute can be divided into 60 arc seconds (60"). Put another way, an arc minute is 1/60 of an arc degree, and an arc second is $1/60 \times 1/60 = 1/3600$ of an arc degree. An arc second is an extremely small unit of angular measure. It is, in fact, the angle subtended (projected) by a centimeter-sized object at a distance of about 2 kilometers, or approximately the angle subtended by a dime when seen from a mile away. The accompanying figure illustrates this subdivision of the circle into progressively smaller units.

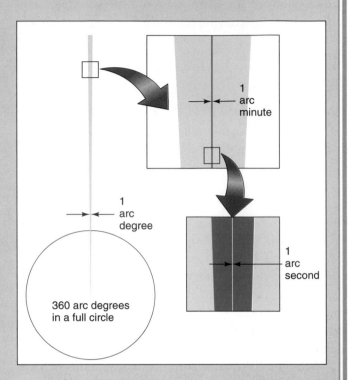

1 arc minute

1 arc degree

1 arc second

360 arc degrees in a full circle

One final note. Arc degrees have nothing to do with temperature, and arc minutes and arc seconds have nothing to do with time. However, the angular units used to measure right ascension (and only right ascension)—hours (^{h}), minutes (^{m}), and seconds (^{s})—*are* constructed to parallel the units of time, the connection being provided by Earth's rotation. In 24 hours, Earth rotates once on its axis, or through 360°. Thus, in a time period of 1 hour, Earth rotates through 360°/24 = 15°, or 1^h. In 1 minute of time, Earth rotates through 1^m; in 1 second, Earth rotates through 1^s. Just bear in mind that these units refer only to right ascension and you should avoid undue confusion.

is usually convenient to set up one observation position directly opposite the object, as at point A. The surveyor then moves to another observation position at point B, noting the distance covered between A and B. This distance is called the **baseline** of the imaginary triangle. Finally, the surveyor, standing at point B, sights toward the tree and notes the angle formed at point B by the intersection of this sightline and the baseline. No further observations are required.

Knowing the length of one side (AB) and two angles (the right angle at A and the angle at B) of the triangle, the surveyor can geometrically construct the remaining sides and angles and so establish the distance from A to the tree.

To use triangulation to measure distances, a surveyor must be familiar with trigonometry, the mathematics of geometrical angles. However, even if we knew no trigonometry at all, we could still solve the

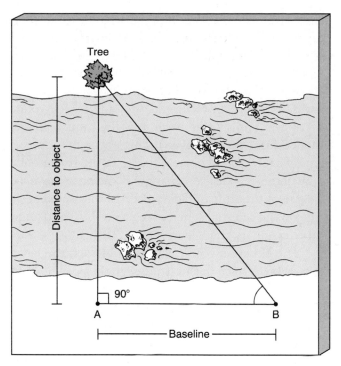

Figure P.10 Surveyors often use simple geometry and trigonometry to estimate the distance to a faraway object by triangulation.

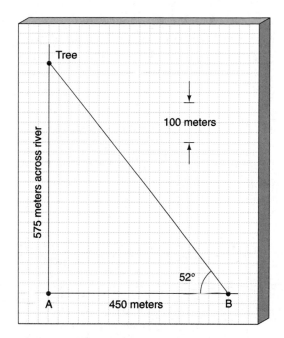

Figure P.11 We don't need trigonometry to estimate distances indirectly. Scaled estimates, like this one on a piece of graph paper, often suffice.

problem by graphical means, as shown in Figure P.11. Suppose we pace off the baseline AB, measuring it to be 450 m and measure the angle between the baseline and the line from B to the tree to be 52°, as illustrated in the figure. Each division on our graph represents 25 m on the ground. Drawing the line AB, completing the other two sides of the triangle, at angles of 90° (at A) and 52° (at B), we measure the distance from A to the tree to be 23 divisions—that is, 575 m. We have solved the real problem by *modeling* it on paper. The point to remember here is this: Nothing more complex than basic geometry is needed to infer the distance, the size, and even the shape of an object too far away or too inaccessible for direct measurement.

Triangles with larger baselines are generally needed to measure greater distances. Figure P.12 shows a triangle having a fixed baseline between two observation positions at points A and B. Note how the triangle becomes narrower as an object's distance becomes progressively greater. Narrow triangles cause problems because the angles at points A and B become hard to measure accurately. The measurements can be made easier by "fattening" the triangle—in other words, by lengthening the baseline.

Now consider an imaginary triangle extending from Earth to a nearby object in space, perhaps a neighboring planet. The triangle is now extremely long

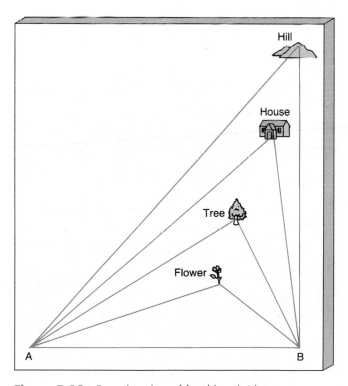

Figure P.12 For a baseline of fixed length (distance between points A and B), a triangle becomes narrower as the distance from the baseline to the third angle increases. As shown here, the imaginary triangle used to estimate the distance to a remote hill is much narrower than the triangle used to estimate the distance to a nearby flower.

and narrow, even for this relatively nearby cosmic object. Figure P.13(a) illustrates a case in which the longest baseline on Earth—Earth's diameter, measured from point A to point B—is used. In principle, two observers could sight the planet from opposite sides of Earth, measuring the triangle's angles at A and B. In practice, however, it is easier to measure the third angle of the imaginary triangle. Here's how.

The observers each sight toward the planet, taking note of its position *relative to some distant stars* seen on the plane of the sky. The observer at A sees the

planet at apparent location A′ relative to those stars, as indicated in Figure P.13(a). The observer at B sees the planet at location B′. If each observer takes a photograph of the appropriate region of the sky, the planet will appear at slightly different places in the two images, as shown in Figure P.13(b). (The background stars appear undisplaced because of their much greater distance from the observer.) This apparent displacement of a foreground object relative to the background as the observer's location changes is known as **parallax**. The size of the shift in Figure P.13(b), measured as an angle on the celestial sphere, is equal to the third angle of the imaginary triangle in Figure P.13(a).

The closer an object is to the observer, the larger the parallax. To see this for yourself, hold a pencil vertically just in front of your nose, as sketched in Figure P.14. Look at some far-off object—a distant wall, say. Close one eye, then open it while closing the other. By blinking in this way, you should be able to see a large shift of the apparent position of the pencil relative to the

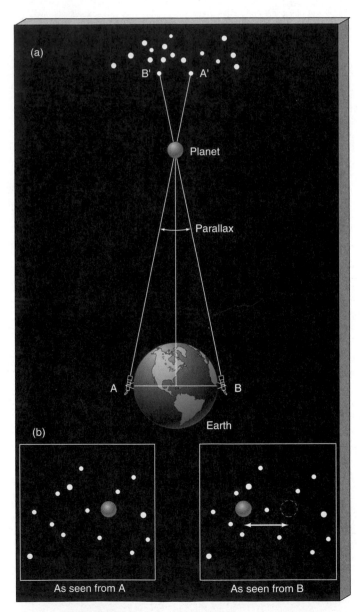

Figure P.13   (a) This imaginary triangle extends from Earth to a nearby object in space (such as a planet). The group of stars at the top represents a background field of very distant stars.
(b) Hypothetical photographs of the same star field showing the nearby object's apparent displacement, or shift, relative to the distant undisplaced stars.

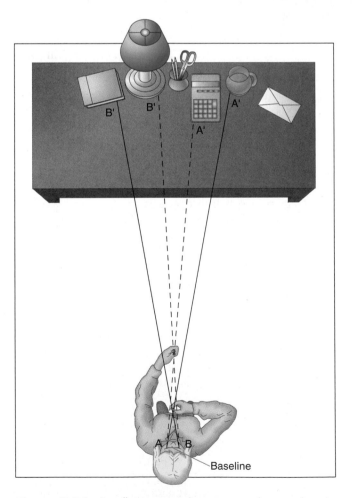

Figure P.14   Parallax is inversely proportional to an object's distance from the observer. An object near your nose has a much larger parallax than an object held at arm's length.

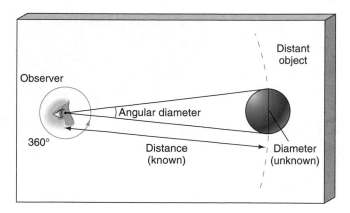

Figure P.15 If the angular diameter of an observed object can be measured and its distance is known, its true diameter can be calculated by simple geometry.

wall—a large parallax. In this example, one eye corresponds to point A in our previous example, the other eye to point B, the distance between your eyeballs to the baseline, the pencil to the planet, and the distant wall to the remote field of stars. Now hold the pencil at arm's length, corresponding to a more distant object (but still not as far away as the distant stars). The apparent shift of the pencil will be less. By moving the pencil farther away, you are narrowing the triangle and decreasing the parallax. If you were to paste the pencil to the wall, corresponding to the case where the object of interest is as far away as the background star field, blinking would produce no apparent shift of the pencil at all.

The amount of parallax is thus inversely proportional to an object's distance. Small parallax implies large distance. Conversely, large parallax implies small distance. Knowing the amount of parallax (as an angle) and the length of the baseline, we can easily derive the distance through triangulation.

Once we know the distance to an object, we can determine many other properties. For example, by measuring the object's *angular diameter*, we can calculate its size. Figure P.15 illustrates the geometry involved. Notice that this is basically the same picture as Figure P.13(a), except that now the angle (the angular diameter) and the distance are known, instead of the angle (parallax) and the baseline. We compute the object's diameter by noting that the ratio of the diameter to the circumference of the circle centered on the observer and passing through the object (2π times the distance) must be equal to the ratio of its angular diameter to one full revolution, $360°$:

$$\frac{\text{diameter}}{2\pi \times \text{distance}} = \frac{\text{angular diameter}}{360°}.$$

Surveyors of the land routinely use such simple geometric techniques to map out our planet Earth. As surveyors of the sky, astronomers use the same basic principles to chart the universe.

Summary *www*

The **universe** (p. 2) is the totality of all space, time, matter, and energy. **Astronomy** (p. 2) is the study of the universe. A widely used unit of distance in astronomy is the **light year** (p. 2), the distance traveled by light in one year.

Early observers grouped the thousands of stars visible to the naked eye into patterns called **constellations** (p. 4). These patterns have no physical significance, although they are a useful means of labeling regions of the sky.

The nightly motion of the stars across the sky is the result of Earth's **rotation** (p. 6) on its axis. Early astronomers, however, imagined that the stars were attached to a vast **celestial sphere** (p. 6) centered on Earth, and that the motions of the heavens were caused by the rotation of the celestial sphere about a fixed Earth. The points where Earth's rotation axis intersects the celestial sphere are called the **north** and **south celestial poles** (p. 6, 7). The line where Earth's equatorial

plane cuts the celestial sphere is the **celestial equator** (p. 7). We can locate a star by specifying its constellation and its brightness relative to other stars in that constellation. A more precise method is to use **celestial coordinates** (p. 7) on the celestial sphere. **Declination** and **right ascension** (p. 7) are the celestial equivalents of latitude and longitude on Earth's surface.

Surveyors on Earth use **triangulation** (p. 7) to determine the distances to distant objects. Astronomers use the same technique to measure the distances to planets and stars. The **cosmic distance scale** (p. 7) is the family of distance-measurement techniques by which astronomers chart the universe. **Parallax** (p. 10) is the apparent motion of a foreground object relative to a distant background as the observer's position changes. The larger the **baseline** (p. 8), the distance between the two observation points, the greater the parallax. Astronomers use parallax to find the distances to planets and stars by triangulation.

1 The Copernican Revolution

The Birth of Modern Science ✦www✦

LEARNING GOALS

Studying this chapter will enable you to:

1 Account for the apparent motions of the Sun, Moon, and stars in terms of the actual motions of Earth and the Moon.

2 Show how the relative motions of Earth, the Sun, and the Moon lead to eclipses.

3 Explain how the observed motions of the planets led to our modern view of a Sun-centered solar system.

4 Sketch the major contributions of Galileo and Kepler to the development of our understanding of the solar system.

5 State Kepler's laws of planetary motion.

6 State Newton's laws of motion and his law of universal gravitation, and explain how the latter permits us to measure the masses of astronomical bodies.

*L*iving in the Space Age, we have become accustomed to the modern view of our place in the universe. Images of our planet taken from space leave little doubt that Earth is round, and no one seriously questions the idea that we orbit the Sun. Yet there was a time, not so long ago, when our ancestors maintained that Earth was flat and lay at the center of all things. Our view of the universe—and of ourselves—has undergone a radical transformation since those early days. In this transformed view, Earth is a planet like many others. Humankind has been torn from its throne at the center of the cosmos and relegated to an unremarkable position on the periphery of the Milky Way Galaxy. But we have been amply compensated for our loss of prominence, for we have gained a wealth of scientific knowledge in the process. The story of how all this came about is the story of the rise of science and the genesis of modern astronomy.

1.1 The Motion of the Sun and the Stars

Many ancient cultures took a keen interest in the changing night sky. The records and artifacts that have survived make that abundantly clear. Unlike today, however, the major driving force behind the development of astronomy in those early societies was probably neither science nor religion. Instead, it was decidedly practical and very down to earth: seafarers needed to navigate their vessels, and farmers needed to know when to plant their crops. In a very real sense, then, human survival depended on knowledge of the heavens.

Day-to-Day Changes

1 As we saw in the Prologue the apparent daily movement of the Sun and other stars across the sky (known as *diurnal motion*) is a consequence of Earth's rotation on its axis. ∞ (Sec. P-2) When we say "one day" conversationally, we usually mean the 24-hour period of time from one sunrise (or noon, or sunset) to the next; in astronomy, this length of time is defined as one **solar day**. But there is another kind of day used by astronomers, one based on the fact that the stars' positions in the sky do not repeat themselves exactly from one night to the next. Each night, the whole celestial sphere appears shifted a little compared with the night before—you can confirm this for yourself by noting over the course of a week or two which stars are visible near the horizon just after sunset or just before dawn. Because of this shift, a day measured by the stars—called a **sidereal day** after the Latin word *sidus*, meaning "star"—differs in length from a solar day.

The reason for the difference in length between a solar day and a sidereal day is sketched in Figure 1.1. It is a result of the fact that Earth moves in two ways simultaneously: it rotates on its central axis while at the same time **revolving** around the Sun. Each time Earth rotates once on its axis, it also moves a small

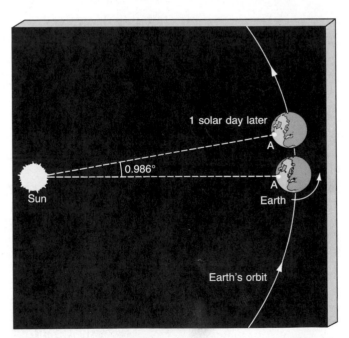

Figure 1.1 The difference in length between a solar and a sidereal day can be easily explained once we understand that Earth revolves around the Sun at the same time as it rotates on its axis. A solar day is the time from one noon to the next. In that time, Earth also moves a little in its solar orbit. Because Earth completes one circuit (360°) around the Sun in 1 year (365 days), it moves through nearly 1° in 1 day. Thus, between noon at point A on one day and noon at the same point the next day, Earth rotates through about 361°. Consequently, the solar day exceeds the sidereal day (360° rotation) by about 4 minutes. Note that the diagram is not drawn to scale; the true 1° angle is in reality much smaller than shown here.

distance along its orbit about the Sun. Each day, therefore, Earth has to rotate through slightly more than 360° in order for the Sun to return to the same apparent location in the sky. Thus, the interval of time between noon one day and noon the next (a solar day) is slightly greater than the true rotation period (one sidereal day). Our planet takes 365 days to orbit the Sun, so the additional angle is 360°/365 = 0.986°. Because Earth takes about 3.9 minutes to rotate through this angle, the solar day is 3.9 minutes longer than the sidereal day.

Seasonal Changes

Because Earth orbits the Sun, the Sun appears, to an observer on Earth, to move relative to the background stars. This apparent motion of the Sun on the sky over the course of a year traces out a path on the celestial sphere known as the **ecliptic**. As illustrated in Figure 1.2(a), the ecliptic forms a great circle on the celestial sphere, inclined at an angle of about 23.5° to the celestial equator. In reality, as illustrated in Figure 1.2(b), the plane defined by the ecliptic is *the plane of Earth's orbit around the Sun*. Its tilt is a consequence of the *inclination* of our planet's rotation axis to its orbital plane.

The point on the ecliptic where the Sun is at its northernmost point above the celestial equator is known as the **summer solstice** (from the Latin words *sol*, meaning "sun," and *stare*, "to stand"). As shown in Figure 1.2, it represents the point in Earth's orbit where our planet's North Pole points closest to the Sun. This occurs on or near June 21—the exact date varies slightly from year to year because the actual length of a year is not a whole number of days. As Earth rotates on

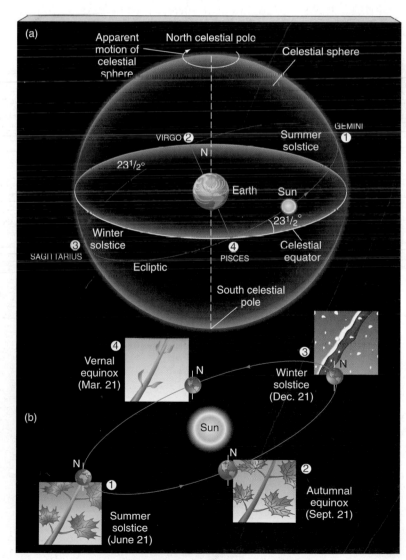

Figure 1.2 (a) The apparent path of the Sun on the celestial sphere and (b) the Sun's actual relationship to Earth's rotation and revolution. The seasons result from the changing height of the Sun above the horizon. At the summer solstice (the points marked 1), the Sun is highest in the sky, as seen from Earth's northern hemisphere, and the days are longest. In the "celestial sphere" picture (a), the Sun at this time is at its northernmost point on its path around the ecliptic; in reality (b), the summer solstice corresponds to the point on Earth's orbit where our planet's North Pole points most nearly toward the Sun. The reverse is true at the winter solstice (points 3). At the vernal and autumnal equinoxes, day and night are of equal length. These are the times when, as seen from Earth (a), the Sun crosses the celestial equator. These times correspond to the points in Earth's orbit when our planet's axis is perpendicular to the line joining Earth and the Sun (b).

that date, points north of the equator spend the greatest fraction of their time in sunlight, so the summer solstice corresponds to the longest day of the year in Earth's Northern Hemisphere and the shortest day in Earth's Southern Hemisphere. Six months later, the Sun is at its southernmost point below the celestial equator, and we have reached the **winter solstice** (December 21)—the shortest day in Earth's Northern Hemisphere and the longest in the Southern Hemisphere. These two effects—the height of the Sun above the celestial equator and the length of the day—combine to account for the **seasons** we experience. In northern summer, the Sun is high in the sky and the days are long, with the result that temperatures are generally much higher than in winter, when the Sun is low and the days are short.

The two points where the ecliptic intersects the celestial equator are known as **equinoxes**. On those dates, day and night are of equal duration. (The word *equinox* derives from the Latin for "equal night.") In the fall (in Earth's northern hemisphere), as the Sun crosses from the northern into the southern celestial hemisphere, we have the **autumnal equinox** (on September 21). The **vernal equinox** occurs in spring, on or near March 21, as the Sun crosses the celestial equator moving north. The vernal equinox pays an important role in human time keeping. The interval of time from one vernal equinox to the next—365.242 solar days—is known as one **tropical year**.

Because Earth revolves around the Sun, our planet's darkened hemisphere faces in a slightly different direction each night. The change in direction is only about 1° per night (Figure 1.1)—too small to be easily noticed with the naked eye from one evening to the next but clearly noticeable over the course of weeks and months, as illustrated in Figure 1.3. In 6 months, Earth moves to the opposite side of its orbit, and we face an entirely different group of stars and

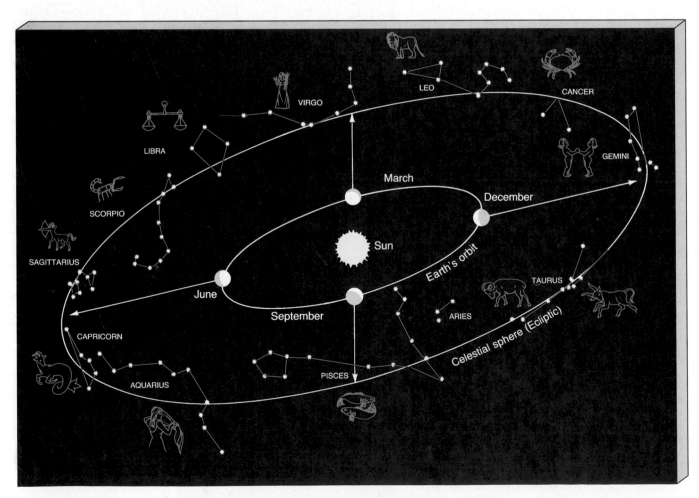

Figure 1.3 The view of the night sky changes as Earth moves in its orbit about the Sun. As drawn here, the night side of Earth faces a different set of constellations at different times of the year. The 12 constellations named here make up the astrological zodiac.

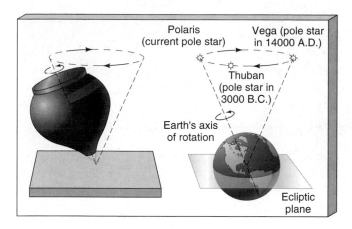

Figure 1.4 Earth's axis currently points nearly toward the star Polaris. Some 12,000 years from now—nearly half-way through one cycle of precession—Earth's axis will point toward a star called Vega, which will then be the "North Star." Five thousand years ago, the North Star was a star named Thuban in the constellation Draco.

constellations at night. The 12 constellations through which the Sun passes as it moves along the ecliptic—that is, the constellations we would see looking in the direction of the Sun if they weren't overwhelmed by the Sun's light—had special significance for astrologers of old. These constellations are collectively known as the **zodiac**.

Long-Term Changes

The time Earth takes to complete one orbit around the Sun is called a **sidereal year**. One sidereal year is 365.256 solar days long, about 20 minutes longer than a tropical year. The reason for this slight discrepancy is a phenomenon known as **precession**. Like a spinning top that rotates rapidly on its own axis while that axis slowly revolves about the vertical, Earth's axis changes its direction over the course of time (although the angle between the axis and a line perpendicular to the plane of the ecliptic remains close to 23.5°). Figure 1.4 illustrates Earth's precession, which is caused mostly by the gravitational pulls of the Moon and the Sun. During a complete cycle of precession, which takes about 26,000 years, Earth's axis traces out a cone. Because of this slow shift in the orientation of Earth's rotation axis, the vernal equinox drifts slowly around the zodiac over the course of the precession cycle.

The tropical year is the year our calendars measure. If our timekeeping were tied to the sidereal year, the seasons would slowly march around the calendar as

Earth precessed—13,000 years from now, summer in the northern hemisphere would be at its height in mid February! By using the tropical year instead, we ensure that July and August will always be (northern) summer months. However, in 13,000 years' time, Orion, now a prominent feature of the northern winter sky, will be a summer constellation.

1.2 The Motion of the Moon

The Moon is our nearest neighbor in space. Apart from the Sun, it is by far the brightest object in the sky. Like the Sun, the Moon appears to move relative to the background stars. Unlike the Sun, however, the explanation for this motion is the obvious one—the Moon really does revolve around the Earth.

The Moon's appearance undergoes a regular cycle of changes, or **phases**, taking a little more than 29 days to complete. (The word *month* is derived from the word *Moon*.) Figure 1.5 illustrates the appearance of the Moon at different times in this monthly cycle. Starting from the so-called **new Moon**, which is all but invisible in the sky, the Moon appears to *wax* (grow) a little each night and is visible as a growing *crescent* (panel 1 of Figure 1.5). One week after new Moon, half of the lunar disk can be seen (panel 2). This phase is known as a **quarter Moon**. During the next week, the Moon continues to wax, passing through the *gibbous* phase (more than half of the lunar disk visible, panel 3) until, two weeks after new Moon, the **full Moon** (panel 4) is visible. During the next 2 weeks, the Moon *wanes* (shrinks), passing in turn through the gibbous, quarter, and crescent phases (panels 5–7), eventually becoming new again.

The Moon doesn't actually change its size and shape on a monthly basis, of course; its full circular disk is present at all times. Why then don't we always see a full Moon? The answer to this question lies in the fact that, unlike the Sun and the other stars, the Moon emits no light of its own. Instead, it shines by reflecting sunlight. As illustrated in Figure 1.5, half of the Moon's surface is illuminated by the Sun at any instant. However, not all of the Moon's sunlit face can be seen because of the Moon's position with respect to Earth and the Sun. When the Moon is full, we see the entire "daylit" face because the Sun and the Moon are in opposite directions from Earth in the sky. In the case of a new Moon, the Moon and the Sun are in almost

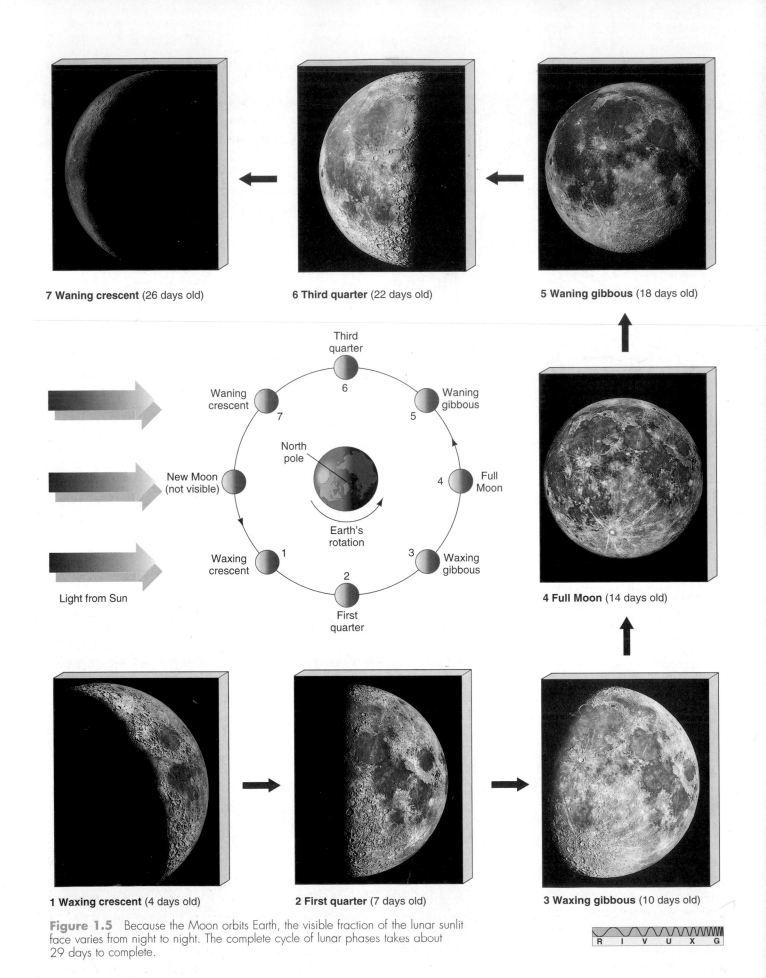

7 Waning crescent (26 days old)

6 Third quarter (22 days old)

5 Waning gibbous (18 days old)

Third
quarter

6

Waning
crescent

7

Waning
gibbous

5

New Moon
(not visible)

North
pole

4

Full
Moon

Earth's
rotation

Light from Sun

Waxing
crescent

1

3

Waxing
gibbous

2

First
quarter

4 Full Moon (14 days old)

1 Waxing crescent (4 days old)

2 First quarter (7 days old)

3 Waxing gibbous (10 days old)

Figure 1.5 Because the Moon orbits Earth, the visible fraction of the lunar sunlit face varies from night to night. The complete cycle of lunar phases takes about 29 days to complete.

R I V U X G

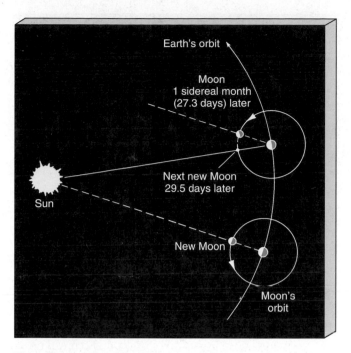

Figure 1.6 The difference in length between a synodic month and a sidereal month stems from Earth's motion relative to the Sun. Because Earth orbits the Sun in 365 days, in the 29.5 days from one new Moon to the next (one synodic month), Earth moves through an angle of approximately 29°. Thus, the Moon must revolve more than 360° between new moons. The sidereal month, which is the time taken for the Moon to revolve through exactly 360°, relative to the stars, is about 2 days shorter.

the same part of the sky, and the sunlit side of the Moon is oriented away from us. At new Moon, the Sun must be almost behind the Moon, from our perspective.

As it revolves around Earth, the Moon's position in the sky changes with respect to the stars. In one **sidereal month** (27.3 days), the Moon completes one revolution and returns to its starting point on the celestial sphere, having traced out a great circle in the sky. The time required for the Moon to complete a full cycle of phases, one **synodic month**, is a little longer—about 29.5 days. The synodic month is a little longer than the sidereal month for the same reason that a solar day is slightly longer than a sidereal day: Because of Earth's motion around the Sun, the Moon must complete slightly more than one full revolution to return to the same phase in its orbit (Figure 1.6).

1.3 Eclipses

2 From time to time—but only at new or full Moon—the Sun, Earth, and the Moon line up precisely and we observe the spectacular phenomenon known as an **eclipse**. When the Sun and the Moon are in exactly *opposite* directions as seen from Earth, Earth's shadow sweeps across the Moon, temporarily blocking the Sun's light and darkening the Moon in a **lunar eclipse**, as illustrated in Figure 1.7. From Earth, we see the curved edge of Earth's shadow begin to cut

Figure 1.7 A lunar eclipse occurs when the Moon passes through Earth's shadow. At these times we see a darkened, copper-colored Moon, as shown in the inset photograph. The coloration is caused by sunlight deflected by Earth's atmosphere onto the Moon's surface. (Note that this figure is not drawn to scale.)

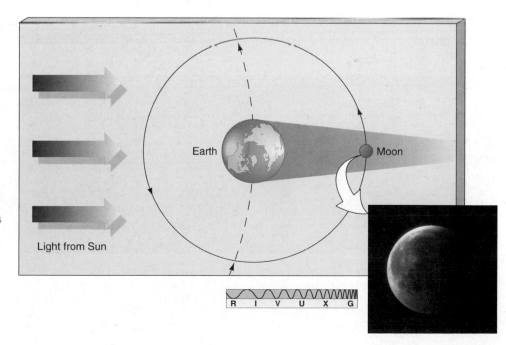

across the face of the full Moon and slowly eat its way into the lunar disk. Usually, the alignment of the Sun, Earth, and the Moon is imperfect, so the shadow never completely covers the Moon. Such an occurrence is known as a **partial eclipse**. Occasionally, however, the entire lunar surface is obscured in a **total eclipse** (such as that shown in the inset in Figure 1.7). Total lunar eclipses last only as long as is needed for the Moon to pass through Earth's shadow—no more than about 100 minutes. During that time, the Moon often acquires an eerie, deep red coloration, the result of a small amount of sunlight being refracted (bent) by Earth's atmosphere onto the lunar surface, preventing the shadow from being completely black.

When the Moon and the Sun are in exactly the *same* direction as seen from Earth, an even more awe-inspiring event occurs. The Moon passes directly in front of the Sun, briefly turning day into night in a **solar eclipse**. In a total solar eclipse, when the alignment is perfect, planets and some stars become visible in the daytime as the Sun's light is reduced to nearly nothing. We can also see the Sun's ghostly outer atmosphere, or *corona* (Figure 1.8). In a partial solar eclipse, the Moon's path is slightly "off center," and only a portion of the Sun's face is covered.

Unlike a lunar eclipse, which is simultaneously visible from all locations on Earth's night side, a total solar eclipse can be seen from only a small portion of the daytime side. The Moon's shadow on Earth's surface is about 7000 kilometers wide—roughly twice the diameter of the Moon (Figure 1.9). Outside that shadow, no eclipse is seen. However, only within the central region of the shadow, the **umbra**, is the eclipse

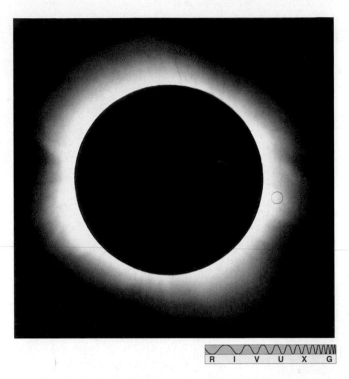

Figure 1.8 During a total solar eclipse, the Sun's corona becomes visible as an irregularly shaped halo surrounding the blotted-out disk of the Sun. This was the July 1991 eclipse as seen from the Baja Peninsula.

total. Within the shadow but outside the umbra, in the **penumbra**, the eclipse is partial, with less and less of the Sun being obscured the farther one travels from the shadow's center. The connections between the umbra, the penumbra, and the relative locations of Earth, Sun, and Moon during eclipses of various types are illustrated in Figure 1.10. The umbra is always very small—even under the most favorable circumstances, its diam-

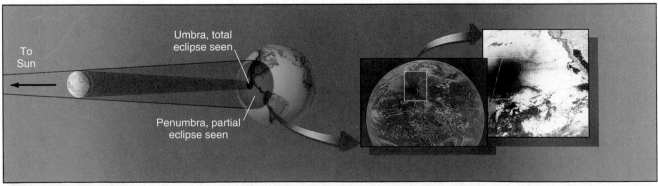

Figure 1.9 The Moon's shadow on Earth during a solar eclipse consists of the umbra, where the eclipse is total, and the penumbra, where the Sun is only partially obscured. The insets show photographs of the Moon's shadow projected onto Earth's surface (near the Baja Peninsula) during the total solar eclipse of July 11, 1991. The photographs were taken by an Earth-orbiting weather satellite.

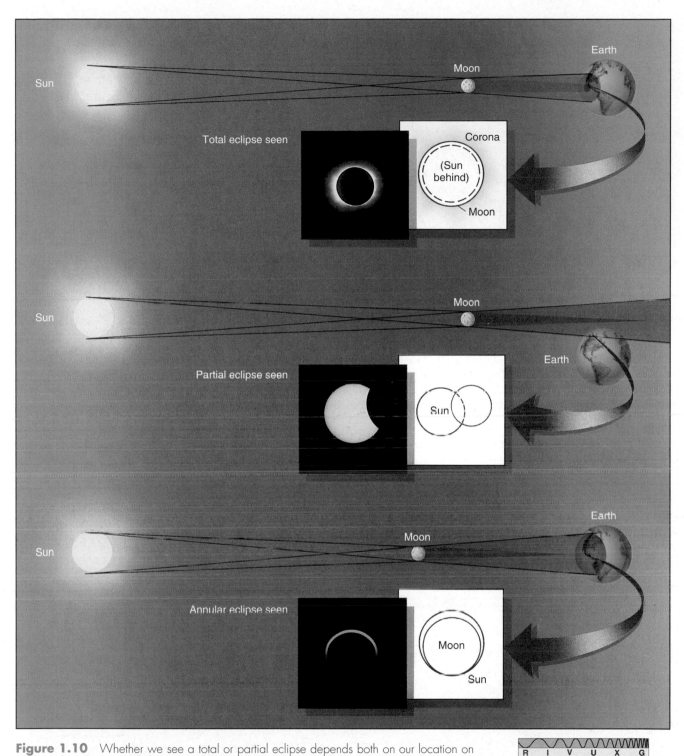

Figure 1.10 Whether we see a total or partial eclipse depends both on our location on Earth and on the precise alignment of Earth, Moon, and Sun. If the Moon is too far from Earth at the moment of the eclipse, there is no region of totality; instead, an annular eclipse is seen.

R   I   V   U   X   G

eter never exceeds 270 kilometers. Because the Moon's shadow sweeps across Earth's surface at a speed of more than 1700 kilometers per hour, the duration of a total solar eclipse at any given point can never exceed 7.5 minutes.

The Moon's orbit around Earth is not exactly circular. Thus, the Moon may be far enough from Earth at the moment of an eclipse that its disk fails to cover the disk of the Sun completely, even though their centers coincide. In that case, there is no

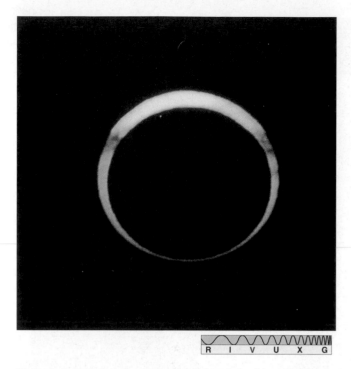

Figure 1.11 During an annular eclipse, the Moon fails to completely hide the Sun, with the result that a thin ring of light remains. No corona is seen in this case because even the small amount of the Sun still visible completely overwhelms the corona's faint glow. This was the December 1973 eclipse as seen from Algiers. (The gray fuzzy areas at top left and right of the ring are clouds in Earth's atmosphere.)

region of totality—the umbra never reaches Earth at all, and a thin ring of sunlight can still be seen surrounding the Moon. Such an occurrence, called an **annular eclipse** (from the word *annulus*, meaning "ring"), is depicted at the bottom of Figure 1.10 and in Figure 1.11. Roughly half of all solar eclipses are annular.

Why isn't there a solar eclipse at every new Moon and a lunar eclipse at every full Moon? The answer is that the Moon's orbit is slightly inclined to the ecliptic (at an angle of 5.2°) so the chance of a new (or full) Moon occurring just as the Moon happens to cross the ecliptic plane (so Earth, Moon, and Sun are perfectly aligned) is quite low. For this reason, eclipses, especially solar eclipses, are relatively rare events. Nevertheless, because we know the orbits of both Earth and the Moon to great accuracy, we can predict eclipses far into the future. Figure 1.12 shows the location and duration of all total and annular eclipses of the Sun from 1995 to 2005.

Figure 1.12 Regions of Earth that will see total or annular solar eclipses between the years 1995 and 2005. Each track represents the path of the Moon's umbra across Earth's surface during an eclipse.

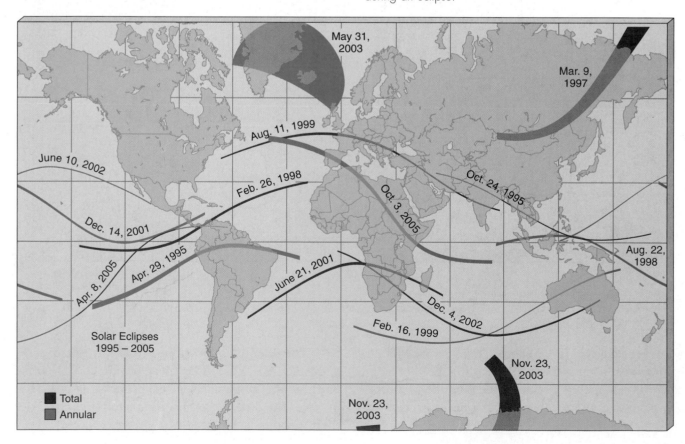

1.4 Planetary Motion

3 Over the course of a night, the stars slide smoothly across the sky. Over the course of a month, the Moon moves smoothly and steadily along its path on the sky relative to the stars, passing through its familiar cycle of phases. Over the course of a year, the Sun progresses along the ecliptic at an almost constant rate, varying little in brightness from day to day. In short, the behavior of the Sun, the Moon, and the stars seems fairly simple and orderly. But ancient astronomers were also aware of five other bodies in the sky—the planets Mercury, Venus, Mars, Jupiter, and Saturn—whose behavior was not so easy to grasp.

Planets do not behave in as regular a fashion as do the Sun, Moon, and stars. They vary in brightness and don't maintain a fixed position in the sky. Unlike the Sun and the Moon, they seem to wander around the celestial sphere—indeed, the word *planet* derives from the Greek word *planetes*, meaning "wanderer." Planets never stray far from the ecliptic and generally traverse the celestial sphere from west to east, as the Sun does. However, they seem to speed up and slow down during their journeys, and at times they even appear to loop back and forth relative to the stars, as shown in Figure 1.13. Motion in the eastward sense is usually referred to as *direct*, or *prograde*, motion; the backward (westward) loops are known as **retrograde motion**.

Like the Moon, the planets produce no light of their own; instead, they shine by reflected sunlight. Ancient astronomers correctly reasoned that the apparent brightness of a planet in the night sky is

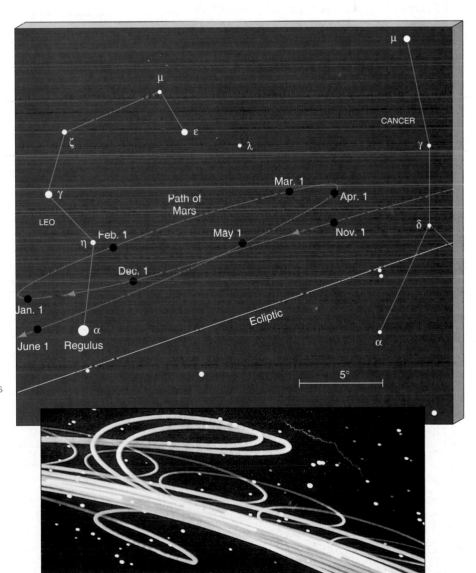

Figure 1.13 Most of the time, planets move from west to east relative to the background stars. Occasionally, however, they change direction and temporarily undergo retrograde motion (east to west) before looping back. The image at right shows an actual retrograde loop in the motion of the planet Mars. The inset depicts the movements of several planets over the course of several years, as reproduced on the inside dome of a planetarium. The motion of the planets relative to the stars (represented as unmoving points) produces continuous streaks on the planetarium "sky."

related to its distance from Earth—planets appear brightest when closest to us. However, the planets Mars, Jupiter, and Saturn are always brightest during the retrograde portions of their orbits. The challenge facing astronomers was to explain the observed motions of the planets and to relate those motions to the variations in planetary brightness.

The Geocentric Universe

The earliest models of the solar system followed the teachings of the Greek philosopher Aristotle (384–322 B.C.) and were **geocentric**, meaning that Earth lay at the center of the universe and all other bodies moved around it. (Figures P.7 and 1.2a illustrate the basic geocentric view.) ∞ (Sec. P.2) These models employed what Aristotle had taught was the perfect form: the circle. The simplest possible description—uniform motion around a circle having Earth at its center—provided a fairly good approximation to the orbits of the Sun and the Moon, but it could not account for the observed variations in planetary brightness or their retrograde motion. A more complex model was needed to describe the planets.

In the first step toward this new model, each planet was taken to move uniformly around a small circle, called an **epicycle**, whose *center* moved uniformly around Earth on a second and larger circle, known as the **deferent** (Figure 1.14). The motion was now composed of two separate circular orbits, creating the possibility that, at some times, the planet's apparent motion could be retrograde. Also, the distance from the planet to Earth would vary, accounting for changes in brightness. By tinkering with the relative sizes of epicycle and deferent, with the planet's speed on the epicycle, and with the epicycle's speed along the deferent, early astronomers were able to bring this "epicyclic" motion into fairly good agreement with the observed paths of the planets in the sky. Moreover, this model had good predictive power, at least to the accuracy of observations at the time.

However, as the number and the quality of observations increased, small corrections had to be introduced into the simple epicyclic model to bring it into line with new observations. The center of the deferents had to be shifted slightly from the center of Earth, and the motion of the epicycles had to be imagined uniform with respect not to Earth but to yet another point

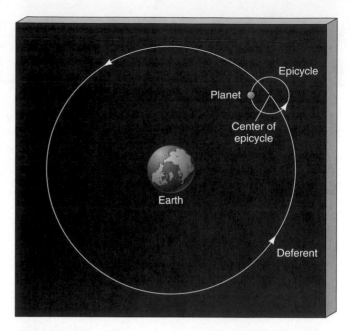

Figure 1.14 In the geocentric model of the solar system, the observed motions of the planets made it impossible to assume that they moved on simple circular paths around Earth. Instead, each planet was thought to follow a small circular orbit (the epicycle) about an imaginary point that itself traveled in a large, circular orbit (the deferent) about Earth.

in space. Around A.D. 140, a Greek astronomer named Ptolemy constructed perhaps the best geocentric model of all time. Illustrated in simplified form in Figure 1.15, it explained remarkably well the observed paths of the five planets then known, as well as the paths of the Sun and the Moon. However, to achieve its explanatory and predictive power, the full **Ptolemaic model** required a series of no fewer than 80 circles. To account for the paths of the Sun, the Moon, and all the nine planets (and their moons) that we know today would require a vastly more complex set.

Today, our scientific training leads us to seek simplicity because simplicity in the physical sciences has so often proved to be an indicator of truth. We would regard the intricacy of a model as complicated as the Ptolemaic system as a clear sign of a fundamentally flawed theory. With the benefit of hindsight, we now recognize that the major error lay in the assumption of a geocentric universe, compounded by the insistence on uniform circular motion, the basis of which was largely philosophical rather than scientific in nature.

Actually, history records that some ancient Greek astronomers reasoned differently about the motions of heavenly bodies. Foremost among them was Aristarchus of Samos (310–230 B.C.), who proposed

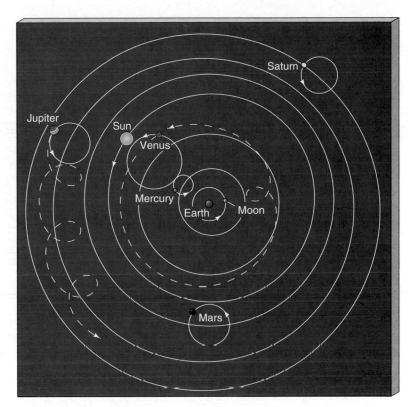

Figure 1.15 The basic features, drawn roughly to scale, of Ptolemy's geocentric model of the inner solar system, a model that enjoyed widespread popularity prior to the Renaissance. To avoid confusion, we have drawn partial paths (dashed) of only two planets, Venus and Jupiter.

that all the planets, including Earth, revolve around the Sun and, furthermore, that Earth rotates on its axis once each day. This, he argued, would create an apparent motion of the sky—a simple idea that is familiar to anyone who has ridden on a merry-go-round and watched the landscape appear to move past in the opposite direction. However, Aristarchus's description of the heavens, though essentially correct, did not gain widespread acceptance during his lifetime. Aristotle's influence was too strong, his followers too numerous, his writings too comprehensive.

The Heliocentric Model of the Solar System

The Ptolemaic picture of the universe survived, more or less intact, for almost 13 centuries, until a sixteenth-century Polish cleric, Nicholas Copernicus (Figure 1.16), rediscovered Aristarchus's **heliocentric** (Sun-centered) model. Copernicus asserted that Earth spins on its axis and, like all other planets, orbits the Sun. Not only does this model explain the observed daily and seasonal changes in the heavens, as we have seen, but it also naturally accounts for planetary retrograde motion and brightness variations. The critical realiza-

tion that Earth is not at the center of the universe is now known as the **Copernican revolution**.

Figure 1.17 shows how the Copernican view explains both the changing brightness of a planet (in this case, Mars) and its apparent looping motions. If we suppose that Earth moves faster than Mars, then every so often Earth "overtakes" that planet. Each time this

Figure 1.16 Nicholas Copernicus (1473–1543).

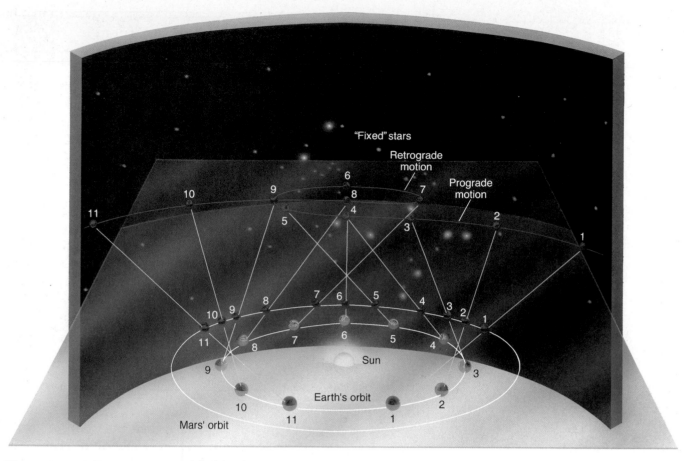

Figure 1.17   The Copernican model of the solar system explains both the varying brightnesses of the planets and the phenomenon of retrograde motion. Here, for example, when Earth and Mars are relatively close to one another in their respective orbits (as at position 6), Mars seems brighter; when they are farther apart (as at position 1), Mars seems dimmer. Also, because the line of sight from Earth to Mars changes as the two planets orbit the Sun, Mars appears to loop back and forth in retrograde motion. The line of sight changes because Earth, on the inside track, moves faster in its orbit than Mars moves in its orbit.

happens, Mars appears to move backwards in the sky, in much the same way as a car we overtake on the highway seems to slip backwards relative to us. Furthermore, at these times, Earth is closest to Mars, so Mars appears brightest, in agreement with observations. Notice that in the Copernican picture the planet's looping motions are only apparent; in the Ptolemaic view, they were real.

Copernicus's major motivation for introducing the heliocentric model was simplicity. Even so, he was still influenced by Greek thinking and clung to the idea of circles to model the planets' motions. To bring his theory into agreement with observations, he was forced to retain the idea of epicyclic motion, though with the deferent centered on the Sun rather than on Earth, and with the epicycles being smaller than in the Ptolemaic picture. Thus, he retained unnecessary complexity and actually gained little in accuracy over the

geocentric model. The heliocentric model did rectify some small discrepancies and inconsistencies in the Ptolemaic system, but for Copernicus, the primary attraction of heliocentricity was its simplicity, its being "more pleasing to the mind." To this day, scientists still are guided by simplicity, symmetry, and beauty in modeling all aspects of the universe.

1.5 The Birth of Modern Astronomy

4   In the century following the death of Copernicus and the publication of his theory of the solar system, two scientists—Galileo Galilei and Johannes Kepler—made indelible imprints on the study of astronomy. Each achieved fame for his discoveries and made great strides in popularizing the Copernican viewpoint.

Figure 1.18   Galileo Galilei (1564–1642).

Galileo Galilei (Figure 1.18) was an Italian mathematician and philosopher. By his willingness to perform experiments to test his ideas—a radical approach in those days (see *Interlude 1-1*)—and by embracing the brand-new technology of the telescope, he revolutionized the way science was done, so much so that he is now widely regarded as the father of experimental science. The telescope was invented in Holland in the early seventeenth century. Having heard of the invention (but without having seen one), Galileo built a telescope for himself in 1609 and aimed it at the sky. What he saw conflicted greatly with the philosophy of Aristotle and provided much new data to support the ideas of Copernicus.

Using his telescope, Galileo discovered that the Moon had mountains, valleys, and craters—terrain in many ways reminiscent of that on Earth. He found that the Sun had imperfections—dark blemishes now known as *sunspots*. By noting the changing appearance of these sunspots from day to day, he inferred that the Sun *rotates*, approximately once per month, around an axis roughly perpendicular to the ecliptic plane. These observations ran directly counter to the orthodox wisdom of the day. Galileo also saw four small points of light, invisible to the naked eye, orbiting the planet Jupiter and realized that they were moons. To Galileo, the fact that another planet had moons provided the strongest support for the Copernican model; clearly, Earth was not the center of all things. He also found that Venus showed a complete cycle of phases, like those of our Moon (Figure 1.19), a finding that could

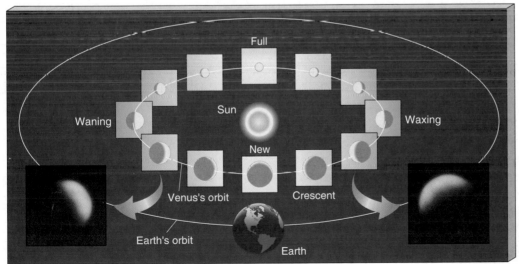

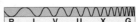

Figure 1.19   The phases of Venus, rendered at different points in the planet's orbit. If Venus orbits the Sun and is closer to the Sun than is Earth, as Copernicus maintained, then Venus should display phases, much as our Moon does. As shown here, when directly between Earth and the Sun, Venus's unlit side faces us and the planet is invisible to us. As Venus moves in its orbit (at a faster speed than Earth moves in its orbit), progressively more of its illuminated face is visible from Earth. Note also the connection between orbital phase and the apparent size of the planet. Venus seems much larger in its crescent phase than when it is full because it is much closer to us during its crescent phase. (The insets at bottom left and right are photographs of Venus at two of its crescent phases.)

be explained only by the planet's motion around the Sun. These observations were further strong evidence that Earth is not the center of things and that at least one planet orbited the Sun.

Galileo published his findings and his controversial conclusions supporting the Copernican theory in 1610, challenging the scientific establishment and religious dogma of the time. In 1616, his ideas were judged heretical, both his and Copernicus's works were banned by the Roman Church, and Galileo was instructed to abandon his astronomical pursuits. This he refused to do. In 1632, he raised the stakes by publishing a side-by-side comparison of the Ptolemaic and Copernican models, in the form of a discussion among three people, one of them a dull-witted Aristotelian whose views time and again were roundly defeated by the arguments of one of his two companions, an articulate proponent of the heliocentric system.

To make the book accessible to a wide popular audience, Galileo wrote it in Italian rather than Latin (the standard language of academic discourse at the time). These actions brought Galileo into direct conflict with the Church. The Inquisition forced him to retract his claim that Earth orbits the Sun, and he was placed under house arrest in 1633; he remained imprisoned for the rest of his life. Not until 1992 were Galileo's "crimes" publicly forgiven by the Roman Catholic Church. But the damage to the orthodox view of the universe was done, and the Copernican genie was out of the bottle once and for all.

1.6 Kepler's Laws of Planetary Motion

At about the time Galileo was becoming famous for his telescopic observations, Johannes Kepler (Figure 1.20), a German mathematician and astronomer, announced his discovery of a set of simple empirical (that is, based on observation) "laws" that accurately described the motions of the planets. While Galileo was the first "modern" observer, Kepler was a pure theorist; he based his work almost entirely on the observations of another scientist (in part because of his own poor eyesight). Those observations, which predated the telescope by several decades, had been made by Kepler's employer, Tycho Brahe (1546–1601), arguably one of the greatest observational astronomers who ever lived.

Figure 1.20 Johannes Kepler (1571–1630).

Brahe's Complex Data

Tycho, as he is often called, was born in Denmark and studied astrology, alchemy, and medicine at some of the best universities in Europe. In 1597 he moved to Prague, which happens to be fairly close to Graz, in Austria, where Kepler lived. Kepler joined Tycho in Prague in 1600. There Kepler was put to work trying to find a theory that could explain Brahe's planetary data. When Tycho died a year later, Kepler inherited not only Brahe's position as Imperial Mathematician of the Holy Roman Empire (then located in Eastern Europe) but also his most priceless possession: the accumulated observations of the planets, spanning several decades. Tycho's observations, though made with the naked eye, were nevertheless of very high quality. Kepler set to work seeking a unifying principle to explain the motions of the planets without the need for epicycles. The effort was to occupy much of the remaining 29 years of his life.

Kepler's goal was to find a simple description of the solar system, within the basic framework of the Copernican model, that fit Tycho's complex mass of detailed observations. In the end, he had to abandon Copernicus's original notion of circular planetary orbits, but even greater simplicity emerged as a result. Kepler determined the shapes and relative sizes of each planet's orbit by triangulation ∞ (Sec. P.3) not from different points on Earth but from different points on Earth's orbit, using observations made at many different times of the year. Noting where the planets were on successive nights, he was able to infer the speeds at which they moved. After long years working with Brahe's planetary data and after many false starts and blind alleys, Kepler succeeded in summarizing the motions of all the known planets, including Earth, in the three **laws of planetary motion** that now bear his name.

Kepler's Simple Laws

5 *Kepler's first law* has to do with the *shapes* of the planetary orbits:

The orbital paths of the planets are elliptical (*not* circular), with the Sun at one focus.

An **ellipse** is simply a flattened circle. Figure 1.21 illustrates a means of constructing an ellipse using a piece of string and two thumbtacks. Each point at which the string is pinned is called a **focus** (plural: *foci*) of the

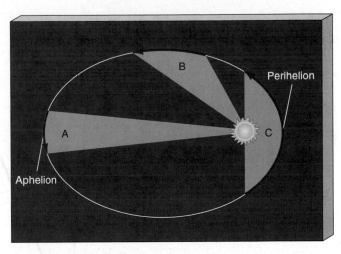

Figure 1.22 Kepler's second law: equal areas are swept out in equal intervals of time. The three shaded areas A, B, and C are equal. Any object traveling along the elliptical path would take the same amount of time to cover the distance indicated by the three red arrows. Therefore, planets move faster when closer to the Sun.

ellipse. The long axis of the ellipse, containing the two foci, is known as the *major axis*. Half the length of this long axis is referred to as the **semi-major axis**; it is a measure of the ellipse's size. The **eccentricity** of the ellipse is the ratio of the distance between the foci to the length of the major axis. The length of the semi-major axis and the eccentricity are all we need to describe the size and shape of a planet's orbital path. (A circle is a special kind of ellipse in which the two foci happen to coincide, so the eccentricity is zero. The semi-major axis of a circle is simply its radius.)

In fact, no planet's elliptical orbit is nearly as elongated as the one shown in Figure 1.21. With two exceptions (the paths of Mercury and Pluto), planetary orbits have such small eccentricities that our eyes would have trouble distinguishing them from true circles. Only because the orbits are so nearly circular were the Ptolemaic and Copernican models able to come as close as they did to describing reality.

Kepler's second law, illustrated in Figure 1.22, addresses the *speed* at which a planet traverses different parts of its orbit:

An imaginary line connecting the Sun to any planet sweeps out equal areas of the ellipse in equal intervals of time.

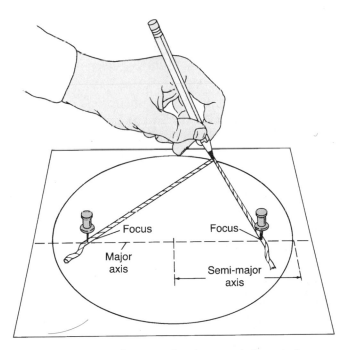

Figure 1.21 An ellipse can be drawn with the aid of a string, a pencil, and two thumbtacks. The wider the separation of the foci, the more elongated, or eccentric, the ellipse. In the special case where the two foci are at the same place, the drawn curve is a circle.

While orbiting the Sun, a planet traces the arcs labeled A, B, and C in Figure 1.22 in equal times. Notice, how-

1-1 INTERLUDE

THE SCIENTIFIC METHOD

Most ancient philosophers held firmly to the belief that, whatever the reasons for the motions of the heavens, Earth in general and humankind in particular were absolutely central to the workings of the universe. Modern science, by contrast, has arrived at a diametrically opposite view. Our present-day outlook is that Earth, the solar system, and (some would argue) humanity are ordinary in every way. This idea is often (and only half-jokingly) called the "principle of mediocrity," and it is deeply embedded in modern scientific thought It is a natural extension of the Copernican principle discussed in Chapter 1. Nowadays, any theory or observation that singles out Earth, the solar system, or the Milky Way Galaxy as being in any way special is regarded with great suspicion in scientific circles.

The principle of mediocrity extends far beyond mere philosophical preference. Simply put, without it we could not make much headway in science, and we could not do astronomy at all. Virtually every statement made in this text rests squarely on the premise that the laws of physics, as we know them here on Earth, apply everywhere else too, without modification and without exception.

This transformation in the perception of humanity's place in the universe went hand in hand with a gradual but radical shift in the way scientists conducted their investigation of the cosmos. The earliest known models of the universe were based largely on imagination and mythology, with little attempt to explain the workings of the heavens in terms of known earthly experience. However, history shows that some early scientists did come to realize the importance of careful observation and testing to the formulation of their theories. The success of their approach changed, slowly but surely, the way science was done and opened the door to a fuller understanding of nature.

As knowledge from all sources was sought and embraced for its own sake, the influence of logic and reasoned argument grew and the power of myth diminished. People began to inquire more critically about themselves and the universe. They realized that thinking about nature was no longer sufficient, that looking at it was also necessary. Experiments and observations became a central part of the process of inquiry. To be effective, a *theory*—the framework of ideas and assumptions used to explain some set of observations and make predictions about the real world—must be continually tested. If experiments and observations favor it, a theory can be further developed and refined. If they do not, the theory must be rejected, no matter how appealing it originally seemed.

The process is illustrated schematically in the accompanying figure. This new approach to investigation, combining thinking and doing—that is, theory and experiment—is known as the *scientific method*. It lies at the heart of modern science.

Notice that there is no "end-point" to the process depicted in the figure. A theory can be invalidated by a single wrong prediction, but no

ever, that the distance traveled along arc C is greater than the distance traveled along arc A or arc B. Because the time is the same and the distance is different, the speed must vary: when a planet is close to the Sun, as in sector C, it moves much faster than when farther away, as in sector A.

Note that these laws are not restricted to planets. They apply to *any* orbiting object. Spy satellites, for example, move very rapidly as they swoop close to Earth's surface not because they are propelled by powerful on-board rockets but because their highly eccentric orbits are governed by Kepler's laws.

Kepler published his first two laws in 1609, stating that he had proved them only for the orbit of Mars. Ten years later, he extended them to all the known planets (Mercury, Venus, Earth, Mars, Jupiter, and Saturn) and added a third law relating the size of a planet's orbit to its sidereal orbital **period**, defined as

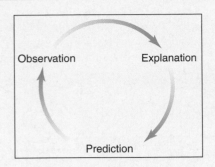

amount of observation or experimentation can ever prove it correct. Theories simply become more and more widely accepted as their predictions are repeatedly confirmed.

In astronomy, we are rarely afforded the luxury of performing experiments to test our theories, so observation is vitally important. One of the first documented uses of the scientific method in an astronomical context was by Aristotle (384–322 B.C.) nearly 25 centuries ago. He noted that during a lunar eclipse, Earth casts a curved shadow onto the surface of the Moon, as shown in the accompanying figure. Earth's shadow, projected onto the Moon's surface, is indeed slightly curved. This is what Aristotle must have seen and recorded so long ago.

Because the shadow seemed always to be an arc of the same circle, Aristotle theorized that Earth,

the cause of the shadow, must be round. On the basis of this *hypothesis*—this possible explanation of the observed facts—he then predicted that any and all future lunar eclipses would show Earth's shadow to be curved, regardless of our planet's orientation. That prediction has been tested every time a lunar eclipse has occurred. It has yet to be proved wrong. Aristotle was not the first person to argue that Earth is round, but he was apparently the first to offer proof using the lunar-eclipse method.

The reasoning Aristotle used forms the basis of all scientific inquiry today. He first made an observation. He then formulated a hypothesis to explain that observation. Then he tested the validity of his hypothesis by making predictions that could be confirmed or refuted by further observations. Observation, theory, and testing—these are the cornerstones of the scientific method, a technique whose power is demonstrated again and again throughout our text.

Used properly over a period of time, this rational, methodical approach enables us to arrive at conclusions that are mostly free of the personal bias and human values of any one scientist. The scientific method is designed to yield an objective view of the universe we inhabit.

    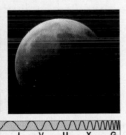

R   I   V   U   X   G

the time needed for the planet to complete one circuit around the Sun. *Kepler's third law* states that

The square of a planet's orbital period is proportional to the cube of its semi-major axis.

Kepler's third law becomes particularly simple when we choose the (Earth) year as our unit of time and the *astronomical unit* as our unit of length. One **astro-nomical unit** (A.U.) is the semi-major axis of Earth's orbit around the Sun—essentially the average distance between Earth and the Sun. Like the light year, the astronomical unit is a unit custom-made for the vast distances encountered in astronomy. Using these units for time and distance, we can write Kepler's third law for any planet in the form

$$P^2 \text{ (in Earth years)} = a^3 \text{ (in astronomical units)},$$

Table 1.1 Some Planetary Properties

Planet	Orbital Semi-Major Axis, a (astronomical units)	Orbital Period, P (Earth years)	Orbital Eccentricity	P^2/a^3
Mercury	0.387	0.241	0.206	1.002
Venus	0.723	0.615	0.007	1.001
Earth	1.000	1.000	0.017	1.000
Mars	1.524	1.881	0.093	1.000
Jupiter	5.203	11.86	0.048	0.999
Saturn	9.539	29.46	0.056	1.000
Uranus	19.19	84.01	0.046	0.999
Neptune	30.06	164.8	0.010	1.000
Pluto	39.53	248.6	0.248	1.001

where P is the planet's sidereal orbital period and a is the length of its semi-major axis.

Table 1.1 presents some basic data describing the orbits of the nine planets now known. Renaissance astronomers knew these properties for the innermost six planets and used them to construct the heliocentric model of the solar system. For purposes of verifying Kepler's third law, the rightmost column lists the ratio P^2/a^3. As we have just seen, in the units used in the table, the third law implies that this number should equal 1 in all cases.

The main points to be grasped from Table 1.1 are these: (1) with the exception of Mercury and Pluto, the planets' orbits are very nearly circular (that is, their eccentricities are close to zero), and (2) the farther a planet is from the Sun, the greater is its orbital period, in precise agreement with Kepler's third law, to within the four-digit accuracy of the numbers in the table. For example, in the case of Pluto, verify for yourself that $39.53^3 = 248.6^2$ (at least, to three significant figures). Most important, note that Kepler's laws are exactly obeyed by *all* the known planets, *not just by the six on which he based his conclusions*.

The Dimensions of the Solar System

Kepler's laws allow us to construct a scale model of the solar system, with the correct shapes and *relative* sizes of all the planetary orbits, but they do not tell us the *actual* size of any orbit. We can express the distance to each planet only in terms of the distance from Earth to the Sun—in other words, only in astronomical units. Why is this? Because Kepler's triangulation measurements all used a portion of Earth's orbit as a baseline, so his distances could be expressed only relative to the size of that orbit, which was not itself determined. Our model of the solar system would be analogous to a road map of the United States showing the *relative* positions of cities and towns but lacking the all-important scale marker indicating distances in kilometers or miles. For example, we would know that Kansas City is about three times more distant from New York than it is from Chicago, but we would not know the actual mileage between any two points on the map. If we could somehow determine the value of the astronomical unit—in kilometers, say—we would be able to add the vital scale marker to our map of the solar system and compute the exact distances between the Sun and each of the planets.

The modern method for deriving the absolute scale of the solar system uses a technique called *radar ranging*. The word **radar** is an acronym for **ra**dio detec-tion **and r**anging. Radio waves are transmitted toward an astronomical body, such as a planet. Their returning echo indicates the body's direction and range, or dis-tance, in absolute terms—in other words, in kilometers rather than in astronomical units. Multiplying the round-trip travel time of the radar signal (the time

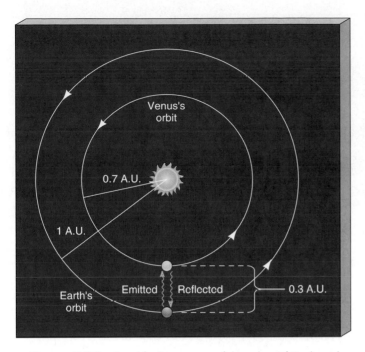

Figure 1.23 Simplified geometry of the orbits of Earth and Venus as they move around the Sun. The wavy lines represent the paths along which radar signals might be transmitted toward Venus and received back at Earth at the moment when Venus is at its minimum distance from Earth. Because the radius of Earth's orbit is 1 A.U. and that of Venus is about 0.7 A.U., we know that the one-way distance covered by the signal is 0.3 A.U. Thus, radar measurements allow us to determine the astronomical unit in kilometers.

elapsed between transmission of the signal and reception of the echo) by the speed of light (300,000 km/s, which is also the speed of radio waves), we obtain twice the distance to the target planet.

We cannot use radar ranging to measure the distance to the Sun directly because radio signals are absorbed at the solar surface and do not reflect back to Earth. Instead, the planet Venus, whose orbit periodically brings it closest to Earth, is the most common target for this technique. Figure 1.23 is an idealized diagram of the Sun–Earth–Venus orbital geometry. Neglecting for the sake of simplicity the small eccentricities of the planets' orbits, we see from the figure that the distance from Earth to Venus at the point of closest approach is approximately 0.3 A.U. Radar signals bounced off Venus at that instant return to Earth in about 300 seconds, indicating that Venus lies 45,000,000 km from Earth. With this information, we can compute the magnitude of the astronomical unit: 0.3 A.U. is 45,000,000 km, so 1 A.U. is 45,000,000 km/0.3, or 150,000,000 km.

Through precise radar ranging, the astronomical unit is now known to be 149,597,870 km. In this text,

we will use the rounded-off value of 1.5×10^8 km. (For more on the use of scientific notation to represent very large or very small numbers, see Appendix 1.) Having determined the value of the astronomical unit, we can reexpress the sizes of the other planetary orbits in terms of more familiar units, such as miles or kilometers. The entire scale of the solar system can then be calibrated to high precision.

1.7 Newton's Laws

6 What prevents the planets from flying off into space or falling into the Sun? What causes them to revolve about the Sun, apparently endlessly? To be sure, the motions of the planets obey Kepler's three laws, but only by considering something more fundamental than those laws can we understand these motions.

The Laws of Motion

The heliocentric system was secured when, in the seventeenth century, the British mathematician Isaac Newton (Figure 1.24) developed a deeper understanding of the way *all* objects move and interact with one another. Newton's theories form the basis for what today is known as **Newtonian mechanics**. Three basic laws of motion, the law of universal gravitation, and a

Figure 1.24 Isaac Newton (1642–1727).

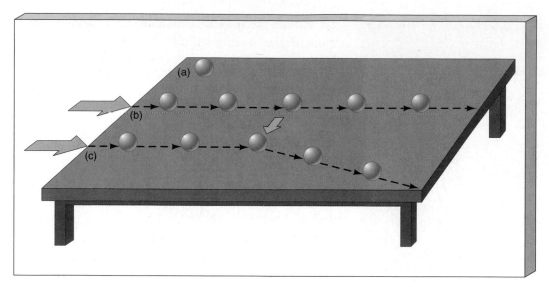

Figure 1.25 (a) A body at rest remains at rest if no force acts on it. (b) Once set in motion by the application of a force (represented by the pink arrow), the body remains in that state of uniform motion (constant speed and direction) until another force acts on it. (c) Application of a second force (green arrow) causes the body to change its state of motion. Put another way, its acceleration changes.

little calculus (which Newton invented) are sufficient to explain and quantify virtually all of the complex dynamic behavior we see on Earth and throughout the universe.

Figure 1.25 illustrates *Newton's first law of motion*, which states that, unless some external **force** changes its state of motion, an object at rest remains at rest and a moving object continues to move forever in a straight line with constant speed. An example of an external force would be the force exerted by, say, a brick wall when a rolling ball glances off it, or the force exerted on a pitched ball by a baseball bat. In either case, a force changes the original motion of the object.

The tendency of an object to keep moving at the same speed and in the same direction unless acted upon by a force is known as **inertia**. A familiar measure of an object's inertia is its **mass**—loosely speaking, the total amount of matter the object contains. The greater an object's mass, the more inertia it has and the greater is the force needed to change its state of motion.

Newton's first law contrasts sharply with the view of Aristotle, who maintained (incorrectly) that the natural state of an object was to be *at rest*—most probably an opinion based on Aristotle's observations of the effect of friction. To simplify our discussion, we will neglect friction—the force that slows balls rolling along the ground, blocks sliding across tabletops, and baseballs moving through the air. In any case, this is

not an issue for the planets because there is no appreciable friction in outer space. The fallacy in Aristotle's argument was first realized and exposed by Galileo, who conceived of the notion of inertia long before Newton formalized it into a law.

The rate of change of the velocity of an object—speeding up, slowing down, or simply changing direction—is called its **acceleration**. *Newton's second law* states that the acceleration of an object is directly proportional to the applied force and inversely proportional to the object's mass—that is, the greater the force acting on the object, or the smaller the mass of the object, the greater its acceleration. Thus, if two objects are pulled with the same force, the more massive one will accelerate less; if two identical objects are pulled with different forces, the one experiencing the greater force will accelerate more.

Finally, *Newton's third law* tells us that forces cannot occur in isolation—if body A exerts a force on body B, then body B necessarily exerts a force on body A that is equal in magnitude but oppositely directed.

Gravity

Forces can act either *instantaneously* or *continuously*. To a good approximation, the force from a baseball bat that hits a home run can be thought of as instantaneous in nature. A good example of a continuous force is the one that prevents the baseball from zooming off into space—**gravity**, the phenomenon that started Newton on the path to the discovery of his laws. Newton hypothesized that any object having mass exerts an attractive *gravitational force* on all other massive objects. The more massive an object, the stronger its gravitational pull.

Consider a baseball thrown upward from Earth's surface, as illustrated in Figure 1.26. In accordance with Newton's first law, the downward force of Earth's gravity steadily modifies the baseball's velocity, slowing its initial upward motion and eventually causing the ball to fall back to the ground. Of course, the baseball, having some mass of its own, also exerts a gravitational pull on Earth. By Newton's third law, this force is equal in magnitude to the weight of the ball (the weight of any object is a measure of the force with which Earth attracts that object), but oppositely directed. By Newton's second law, however, Earth has a much greater effect on the light baseball than the baseball has on the much more massive Earth. The ball and Earth each feel the same gravitational force, but Earth's *acceleration* as a result of this force is much smaller.

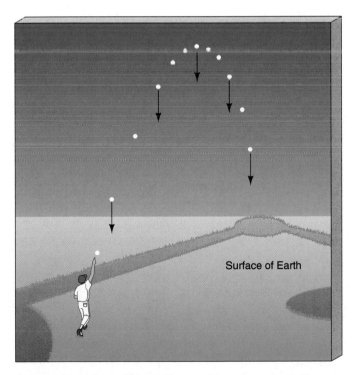

Figure 1.26 A ball thrown upward from the surface of a massive object, such as a planet, is pulled continuously back toward the surface by the gravity of that planet (and, conversely, the gravity of the ball continuously pulls the planet toward the ball).

Now consider the trajectory of a baseball batted from the surface of the Moon, which has much less mass than Earth has. Because the pull of gravity is about one-sixth as great on the Moon as on Earth, a baseball's path changes more slowly near the Moon. A typical home run in a ballpark on Earth would travel nearly half a mile on the Moon. The Moon, less massive than Earth, has less gravitational influence on the baseball. The magnitude of the gravitational force, then, depends on the *masses* of the attracting bodies. Theoretical insight, as well as detailed experiments, tells us that the force is in fact directly proportional to the product of the two masses.

Studying the motions of the planets around the Sun reveals a second aspect of the gravitational force. At all locations equidistant from the Sun's center, the gravitational force has the same strength, and it is always directed toward the Sun. Furthermore, the force of gravity weakens with distance from any object. Forces that decrease with distance from their source are encountered throughout all of science. Many of them, including gravity, decrease in proportion to the *square* of the distance. They are said to obey an **inverse-square law**. As shown in Figure 1.27, inverse-square forces decrease rapidly with distance from their source. For example, tripling the distance makes the force $3^2 = 9$ times weaker, while multiplying the distance by 5 results in a force that is $5^2 = 25$ times weaker. Despite this rapid decrease, the force never quite reaches zero. The gravitational pull of any object having some mass, however small, can never be completely extinguished.

We can combine the preceding statements about mass and distance to form a law of gravity that dictates the way in which *all* material objects attract one another. As a proportionality, Newton's law of gravity is

$$\frac{\text{gravitational}}{\text{force}} \propto \frac{\text{mass of object \#1} \times \text{mass of object \#2}}{\text{distance}^2}$$

This relationship is a compact way of stating that the gravitational pull between two objects is directly proportional to the product of their masses and inversely proportional to the square of the distance separating them. (See *More Precisely 1-1*.)

To Newton, gravity was a force that acted at a distance, with no obvious way in which it was transmitted from place to place. Newton was not satisfied with this explanation, but he had none better. To

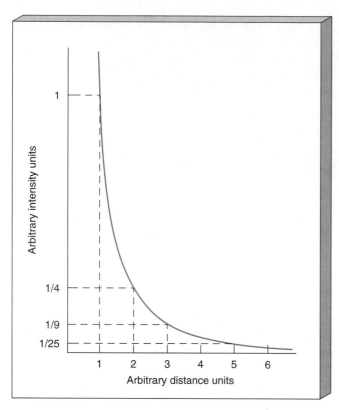

Figure 1.27 Inverse-square forces rapidly weaken with distance from their source. The strength of the Sun's gravitational force decreases with the square of the distance from the Sun. The force never quite diminishes to zero, however, no matter how far away from the Sun we go.

appreciate the modern view of gravity, consider any piece of matter having some mass—it could be smaller than an atom or larger than a galaxy. Extending outward from this object in all directions is a **gravitational field** produced by the matter. We now regard such a field as a property of space itself—a property that determines the influence of one massive object on another. All other matter "feels" the field as a gravitational force.

Planetary Motion

The mutual gravitational attraction of the Sun and the planets, as expressed by Newton's law of gravity, produces the observed planetary orbits. As depicted in Figure 1.28, this gravitational force continuously pulls each planet toward the Sun, deflecting its forward motion into a curved orbital path. Because the Sun is much more massive than any of the planets, it dominates the interaction. We might say that the Sun "controls" the planets, not the other way around. The Sun–planet interaction sketched here is analogous to

what occurs when you whirl a rock at the end of a string above your head. The Sun's gravitational field is your hand and the string, and the planet is the rock at the end of that string. The tension in the string provides the force necessary for the rock to move in a circular path. If you were suddenly to release the string—which would be like eliminating the Sun's gravity—the rock would fly away along a straight line tangent to the circle, in accordance with Newton's first law.

We can use Newtonian mechanics to calculate the relationship between the distance and the orbital speed of a planet moving in a circular orbit around the Sun. By calculating the force required to keep the planet moving in a circle and comparing that force with the gravitational force due to the Sun, it can be shown that

$$\text{orbital speed}^2 = \frac{G \times \text{mass}}{\text{distance}}$$

where the gravitational constant G is defined in *More Precisely 1-1*. Because we have measured G in the laboratory on Earth and because we know the length of a year and the size of the astronomical unit, we can use Newtonian mechanics to determine the mass of the Sun. Inserting the known values of orbital

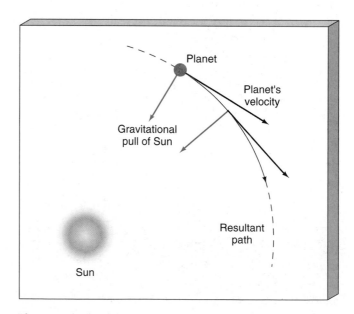

Figure 1.28 The Sun's inward pull of gravity on a planet competes with the planet's tendency to continue moving in a straight line. These two effects combine, causing the planet to move smoothly along an intermediate path, which continuously "falls around" the Sun. This unending tug-of-war between the Sun's gravity and the planet's inertia results in a stable orbit.

1-1 MORE PRECISELY

NEWTON'S LAWS OF MOTION AND GRAVITATION

The Three Laws of Motion

1. Every body continues in a state of rest or in a state of uniform motion in a straight line unless it is compelled to change that state by a force acting on it.

It requires no force to maintain motion in a straight line with constant velocity—that is, motion with constant speed and constant direction in space. The tendency of a body to remain in a state of uniform motion is usually called *inertia*. When velocity does vary (the speed increases or decreases or the direction of motion changes), its rate of change is called *acceleration*. The relation of acceleration to any forces acting on a body is the subject of the second law of motion:

2. A force F acting on a body of mass m produces in that body an acceleration a equal to the force divided by the mass. Thus, $a = F/m$, or $F = ma$.

In honor of Newton, the SI unit of force is named after him. By definition, 1 newton (N) is the force required to cause a mass of 1 kilogram to accelerate at a rate of 1 meter per second every second.

Newton's third law relates the forces acting between separate bodies:

3. To every action, there is an equal and opposite reaction.

This law means, for example, that you attract Earth with exactly the same force as it attracts you (a force known as your *weight*). This attraction is governed by one final law:

THE LAW OF UNIVERSAL GRAVITATION ("NEWTON'S LAW OF GRAVITY")
Every particle of matter in the universe attracts every other particle with a force that is directly proportional to the product of the masses of the particles and inversely proportional to the square of the distance between them.

In other words, two bodies of masses m_1 and m_2, separated by a distance R, attract each other with a force F that is proportional to $(m_1 \times m_2)/R^2$. The constant of proportionality is known as the *gravitational constant*, or often simply as *Newton's constant*, and is always denoted by the letter G. We can then express the law of gravity as

$$F = \frac{Gm_1m_2}{R^2}.$$

The value of G has been measured in extremely delicate laboratory experiments. In SI units, its value is 6.67×10^{-11} newton meter^2/kilogram^2 (N m^2/kg^2).

speed = 30 km/s and distance = 1 A.U. = 1.5×10^{11} m, we calculate the mass of the Sun to be 2×10^{30} kg. Similarly, knowing the distance to the Moon (Chapter 5) and the length of the (sidereal) month, we can measure the mass of Earth to be 6×10^{24} kg.

In fact, this is how basically *all* masses are measured in astronomy. When we need to know the mass of an astronomical object, we must look for its gravitational influence on something else. This principle applies to planets, stars, galaxies, and even clusters of galaxies—very different objects, but all subject to the same physical laws.

Kepler's Laws Reconsidered

Newton's laws of motion and his law of universal gravitation provided a theoretical explanation for Kepler's empirical laws of planetary motion. Just as Kepler modified Copernicus's model by introducing ellipses in place of circles, so too did Newton make corrections to Kepler's first and third laws. It turns out that a planet does not orbit the exact center of the Sun. Instead, both the planet and the Sun orbit their common **center of mass**. Because the Sun and the planet feel equal and opposite gravitational forces (by Newton's third law), the Sun must also move (by Newton's first law), driven by the gravitational influence of the planet. The Sun is so much more massive than any planet that the center of mass of any planet–Sun system is very close to the center of the Sun, which is why Kepler's laws are so accurate. Thus, Kepler's first law becomes

> The orbit of a planet around the Sun is an ellipse having the center of mass of the planet–Sun system at one focus.

As shown in Figure 1.29, the center of mass for a system consisting of two objects of comparable mass does not lie within either object. For identical masses (Figure 1.29a), the orbits are identical ellipses, with a common focus located midway between the two objects. For unequal masses (as in Figure 1.29b), the elliptical orbits still share a focus and both have the same eccentricity, but the more massive object moves more slowly and on a tighter orbit. (Note that Kepler's second law, as stated earlier, continues to apply without modification to each orbit separately, but the rates at which the two orbits sweep out area are different.) In the extreme case of a planet orbiting the much more massive Sun (Figure 1.29c), the path traced out by the Sun's center lies entirely within the Sun.

The change to Kepler's third law is also small in the case of a planet orbiting the Sun but very important in other circumstances, such as the orbital motion of two stars that are gravitationally bound to one another. Following through the mathematics of Newton's theory, we find that the true relationship between the semi-major axis a (measured in astronomical units) of the planet's orbit relative to the Sun and its orbital period P (in Earth years) is

$$P^2 \text{ (in Earth years)} = \frac{a^3 \text{ (in astronomical units)}}{M_{\text{total}} \text{ (in solar masses)}},$$

where M_{total} is the combined mass of the two objects expressed in terms of the mass of the Sun. Notice that Newton's restatement of Kepler's third law preserves the proportionality between P^2 and a^3, but now the proportionality includes M_{total}, so it is *not* quite the same for all the planets. The Sun's mass is so great, however, that the differences in M_{total} among the various combinations of the Sun and the other planets are almost unnoticeable. Therefore Kepler's third law, as originally stated, is a very good approximation. This modified form of Kepler's third law is true in *all* circumstances, inside or outside the solar system.

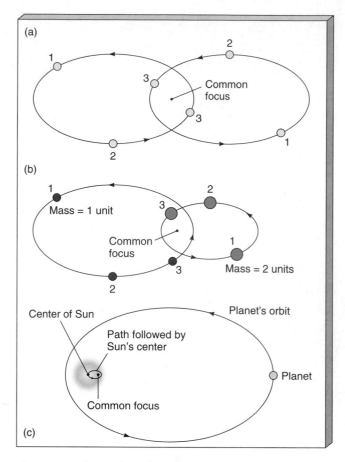

Figure 1.29 (a) The orbits of two bodies (stars, for example) with equal masses, under the influence of their mutual gravity, are identical ellipses with a common focus. That focus is not at the center of either star but instead at the center of mass of the two-star system, located midway between them. The positions of the two bodies at three times are indicated by the pairs of numbers. (Notice that a line joining the bodies at any given time always passes through the common focus.) (b) The orbits of two bodies when one body is twice as massive as the other. Here again the elliptical orbits have a common focus (at the center of mass of the two-body system), and the two ellipses have the same eccentricity. However, in accordance with Newton's laws of motion, the more massive body moves more slowly and in a smaller orbit, staying closer to the system's center of mass. The larger ellipse is twice the size of the smaller one. (c) In the extreme case of a hypothetical planet orbiting the Sun, the common focus of the two orbits lies inside the Sun.

Newton's laws explain the paths of objects moving at any point in space near any gravitating body. These laws provide a firm physical and mathematical foundation for Copernicus's heliocentric model of the solar system and for Kepler's laws of planetary motions, but they also do much more than that. Newtonian gravitation governs not only the planets, moons, and satellites in their elliptical orbits but also the stars and galaxies in their motion throughout our universe.

Chapter Review

Summary

The time from one sunrise to the next is a **solar day** (p. 14), and that between successive risings of any given star is one **sidereal day** (p. 14). Because of Earth's **revolution** (p. 14) around the Sun, the solar day is a few minutes longer than the sidereal day. The Sun's yearly path around the celestial sphere or, equivalently, the plane of Earth's orbit around the Sun is called the **ecliptic** (p. 15). The constellations lying along the ecliptic are collectively called the **zodiac** (p. 17).

Because Earth's axis is inclined at an angle of 23.5° to the ecliptic plane, we experience seasons (p. 15). At the **summer solstice** (p. 15), the Sun is highest in the sky and the length of the day is greatest. At the **winter solstice** (p. 16), the Sun is lowest and the day is shortest. At the **vernal** and **autumnal equinoxes** (p. 16), Earth's rotation axis is perpendicular to the line joining Earth to the Sun, so day and night are of equal length. The ime interval between one vernal equinox and the next is one **tropical year** (p. 16). The time required for the same zodiac constellations to reappear at the same location in the sky, as viewed from a given point on Earth, is one **sidereal year** (p. 17).

In addition to its rotation about its axis and its revolution around the Sun, Earth undergoes a motion called **precession** (p. 17), where the influence of the Moon causes Earth's axis to wobble slightly. As a result, the particular constellations that happen to be visible on any given night change slowly over the course of many years.

As the Moon orbits Earth, we see lunar **phases** (p. 17) as the fraction of the Moon's sunlit face visible to us varies. At **full Moon** (p. 17), we see the entire illuminated side. At **quarter Moon** (p. 17), only half the sunlit side can be seen. At **new Moon** (p. 17), the sunlit face points away from us, and the Moon is all but invisible from Earth. The time between successive full Moons is one **synodic month** (p. 19). The time taken for the Moon to return to the same position in the sky relative to the stars is one **sidereal month** (p. 19). Because of Earth's motion around the Sun, the synodic month is about two days longer than the sidereal month.

A **lunar eclipse** (p. 19) occurs when the Moon enters Earth's shadow. The eclipse may be **total** (p. 20)—the entire Moon is (temporarily) darkened—or **partial** (p. 20)—only a portion of the Moon's surface is affected. A **solar eclipse** (p. 20) occurs when the Moon passes between Earth and the Sun and a small part of Earth's surface is plunged into shadow. For observers in the **umbra** (p. 20), the entire Sun is obscured and the solar eclipse is total. In the **penumbra** (p. 20), a **partial solar eclipse** (p. 20) is seen. If the Moon happens to be too far from Earth for its disk to completely hide the Sun, an **annular eclipse** (p. 22) occurs.

Planets sometimes appear to temporarily reverse their direction of motion relative to the stars; this "backward" motion is called **retrograde motion** (p. 23). **Geocentric** (p. 24) models of the universe, such as the **Ptolemaic model** (p. 24) have the Sun, the Moon, and all the other planets orbiting Earth. To account for retrograde motion within the geocentric picture, it was necessary to suppose that planets moved on small circles called **epicycles** (p. 24), whose centers orbited Earth on larger circles called **deferents** (p. 24). The **heliocentric** (p. 25) view of the solar system holds that Earth, like all the other planets, orbits the Sun. This model accounts for retrograde motion and the observed brightness variations of the planets in a much more natural way than the geocentric model. The widespread realization that the solar system is Sun-centered and not Earth-centered is known as the **Copernican revolution** (p. 25), in honor of Nicholas Copernicus, who laid the foundations of the modern heliocentric model.

Kepler's three **laws of planetary motion** (p. 29) state that (1) planetary orbits are **ellipses** (p. 29) having the Sun as one **focus** (p. 29), (2) a planet moves faster as its orbit takes it closer to the Sun, and (3) the **semi-major axis** (p. 29) of the orbit is related in a simple way to the planet's orbit **period** (p. 30). Most planets move on orbits whose **eccentricities** (p. 29) are quite small, so their paths differ only slightly from perfect circles.

The average distance from Earth to the Sun is one **astronomical unit** (p. 31) today determined by bouncing **radar** (p. 32) signals off the planet Venus.

According to the principles of **Newtonian mechanics** (p. 33), the tendency of a body to keep moving at constant velocity is the body's **inertia** (p. 34). The greater the body's **mass** (p. 34), the greater its inertia. To change the velocity, a **force** (p. 34) must be applied to the body. The rate of change of velocity, called **acceleration** (p. 34), is equal to the applied force divided by the body's mass. To explain planetary orbits,

Newton postulated that **gravity** (p. 35) attracts the planets to the Sun. Every object with any mass is surrounded by a **gravitational field** (p. 36), the strength of which decreases with distance according to an **inverse-square law** (p. 35). This field determines the gravitational force exerted by the object on any other body in the universe. Newton's laws imply that a planet does not orbit the precise center of the Sun; instead the planet and the Sun orbit the **center of mass** (p. 37) of the planet–Sun system.

Self-Test: True or False?

_____ 1. The solar day is longer than the sidereal day.

_____ 2. The constellations lying immediately adjacent to the north celestial pole are collectively referred to as the zodiac.

_____ 3. The seasons are caused by the precession of Earth's axis.

_____ 4. The vernal equinox marks the beginning of fall.

_____ 5. The new phase of the Moon cannot be seen because it always occurs during the daytime.

_____ 6. A lunar eclipse can occur only during the full phase.

_____ 7. Solar eclipses are possible during any phase of the Moon.

_____ 8. An annular eclipse is a type of eclipse that occurs every year.

_____ 9. Aristotle was first to propose that all planets revolve around the Sun.

_____ 10. Ptolemy was responsible for a geocentric model that was successful at predicting the positions of the planets, Moon, and the Sun.

_____ 11. The heliocentric model of the universe holds that Earth is at the center of the universe and everything else moves around it.

_____ 12. Kepler's discoveries regarding the orbital motion of the planets were based on his own observations.

_____ 13. The Sun's location in a planet's orbit is at the center.

_____ 14. A circle has an eccentricity of zero.

_____ 15. The astronomical unit is a distance equal to the semi-major axis of Earth's orbit around the Sun.

_____ 16. The speed of a planet orbiting the Sun is independent of the planet's position in its orbit.

_____ 17. Kepler's laws hold only for the six planets known in his time.

_____ 18. Kepler never knew the true distances between the planets and the Sun, only their relative distances.

_____ 19. Using his laws of motion and gravity, Newton was able to prove Kepler's laws.

_____ 20. You throw a baseball to someone. Before the ball is caught, it is temporarily in orbit around Earth's center.

Self-Test: Fill in the Blanks

1. The solar day is measured relative to the Sun; the sidereal day is measured relative to the _____.

2. The apparent path of the Sun across the sky is known as the _____.

3. On December 21, known as the _____, the Sun is at its _____ point on the celestial sphere.

4. When the Sun, Earth, and the Moon are positioned to form a right angle at Earth, the Moon is seen in the _____ phase.

5. A _____ eclipse can be seen by about half of Earth at once.

6. When the planets Mars, Jupiter, and Saturn appear to move "backwards" (westward) in the sky relative to the stars, this is known as _____ motion.

7. The heliocentric model was reinvented by _____.

8. Central to the heliocentric model is the assertion that the observed motions of the other planets and of the Sun, all viewed from Earth, are the result of _____ motion around the Sun.

9. Galileo discovered _____ of Jupiter, the _____ of Venus, and the Sun's rotation from observations of _____.

10. Kepler discovered that the shape of an orbit is an _____, not a _____ as had previously been believed.

11. Kepler's third law relates the _____ of the orbital period to the _____ of the semi-major axis.

12. The modern method of measuring the astronomical unit uses _____ measurements of Venus.

13. Newton's first law states that a moving object will continue to move in a straight line with constant speed unless acted upon by a _____.

14. Newton's law of gravity states that the gravitational force between two objects depends on the _____ of their masses and inversely on the _____ of their separation.

15. Newton discovered that, in Kepler's third law, the orbital period depends on the semi-major axis and on the sum of the _____ of the two objects involved.

Review and Discussion

1. Why does the Sun rise in the east and set in the west? Does the Moon also rise in the east and set in the west? Why? Do stars do the same? Why?

2. How many times in your life have you orbited the Sun?

3. Why are there seasons on Earth?

4. Why do we see different stars in summer and in winter?

5. If one complete hemisphere of the Moon is always lit by the Sun, why do we see different phases of the Moon?

6. What causes a lunar eclipse? A solar eclipse?

7. Why aren't there lunar and solar eclipses every month?

8. Describe the geocentric model of the universe.

9. The benefit of our current knowledge lets us see flaws in the Ptolemaic model of the universe. What is its basic flaw?

10. What was the great contribution of Copernicus to our knowledge of the solar system? What was a flaw in the Copernican model?

11. Describe the Copernican Revolution.

12. What discoveries of Galileo helped confirm the views of Copernicus?

13. State Kepler's three laws of orbital motion.

14. If radio waves cannot be reflected from the Sun, how can radar be used to find the distance from Earth to the Sun?

15. List the two modifications made by Newton to Kepler's laws.

16. Why would a baseball thrown upward from the surface of the Moon go higher than one thrown with the same velocity from the surface of Earth?

17. Do you think the climate for new ideas is better today than it was during the time of Copernicus? Why or why not?

Problems

1. How long would an Earth–Mars radar signal take to complete its round trip when Earth and Mars are closest to one another (approximately 0.5 A.U. apart)?

2. Jupiter's moon Callisto orbits it at a distance of 1.88 million km. Its orbital period about the planet is 16.7 days. What is the mass of Jupiter? (Assume that Callisto's mass is negligible compared with that of Jupiter.) Use the modified version of Kepler's third law (Section 1.7).

3. Use Newton's law of gravity to calculate the speed at which Earth orbits the Sun. (Assume a circular orbit.)

Projects

1. Look in an almanac for the date of opposition of one or all of these bright planets: Mars, Jupiter, and Saturn. At opposition, these planets are at their closest points to Earth and therefore are at their largest and brightest in the night sky. Observe these planets. How long before opposition does each planet's retrograde motion begin? How long afterwards does it end?

2. Following the setup sketched in Figure 1.21, draw an ellipse. What is the eccentricity of the ellipse you have drawn?

3. Use a small telescope to replicate Galileo's observations of Jupiter's four largest moons. Note the moons' brightnesses and their locations with respect to Jupiter. If you watch over a period of several nights, draw what you see; you'll notice that these moons change their positions as they orbit the gigantic planet. Check the charts given monthly in *Astronomy* or *Sky & Telescope* magazines to identify each moon you see.

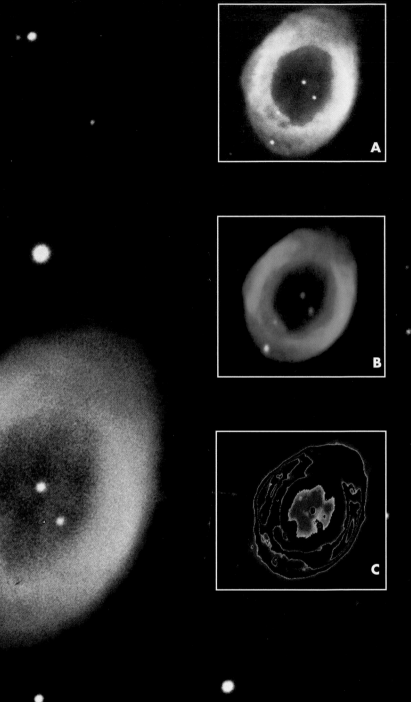

2

Light and Matter
The Inner Workings
of the Cosmos *www*

◄ (Opposite page, back ground) The Ring Nebula in the constellation Lyra is one of the most magnificent sights in the nighttime sky. Seen here glowing in the light of its own emitted radiation, the nebula is actually the expanding outer atmosphere of a nearly dead star.

(Inset A) This nearly true color view clearly shows the dying dwarf star at the center of the expanding gas cloud, which is really a three-dimensional shell and not a ring.

(Inset B) False-color images can sometimes enhance certain features. Here some of the fine structure in the shell of gas can be seen more clearly.

(Inset C) Contour images, like this one derived from red light emitted by hydrogen gas in the nebula's shell and star's lingering atmosphere, can be used to map regions of relative brightness.

LEARNING GOALS

Studying this chapter will enable you to:

1 Discuss the nature of electromagnetic radiation and how that radiation transfers energy and information through interstellar space.

2 Name the major regions of the electromagnetic spectrum.

3 Tell how we can determine an object's temperature by observing the radiation it emits.

4 Describe the characteristics of continuous, emission, and absorption spectra, and the conditions under which each is produced.

5 Specify the basic components of the atom and describe our modern conception of its structure.

6 Explain how electron transitions within atoms produce unique emission and absorption spectra.

Astronomical objects are more than just things of beauty in the night sky. Planets, stars, and galaxies are of vital significance if we are to understand fully the big picture—the grand design of the universe. Each object is a source of information about the universe—its temperature, its chemical composition, its state of motion, its past history. The starlight we see tonight began its journey to Earth decades, centuries—even millennia—ago. The faint rays from the most distant galaxies have taken billions of years to reach us. The stars and galaxies in the night sky show us not just the far away but also the long ago. In this chapter, we begin our study of how astronomers extract information from the light emitted by astronomical objects. The observational and theoretical techniques that enable researchers to determine the nature of distant atoms by the way they emit and absorb light are the indispensable foundation of modern astronomy.

2.1 Information from the Skies

Figure 2.1 shows a galaxy in the constellation Andromeda. On a dark, clear night, far from cities or other sources of light, the Andromeda Galaxy, as it is generally called, can be seen with the naked eye as a faint, fuzzy patch on the sky, comparable in diameter to the full Moon. Yet the fact that it is visible from Earth belies this galaxy's enormous distance from us. It lies roughly 3 million light years away. An object at such a distance is truly inaccessible in any realistic human sense. Even if a space probe could miraculously travel at the speed of light, it would need 3 million years to reach this galaxy and 3 million more to return with its findings. Considering that civilization has existed on Earth for fewer than 10,000 years (and its prospects for

the next 10,000 are far from certain), even this unattainable technological feat would not provide us with a practical means of exploring other galaxies—or even the farthest reaches of our own galaxy, several tens of thousands of light years away.

Light and Radiation

Given the impossibility of traveling to such remote parts of the universe, how do astronomers know anything about objects far from Earth? How can we obtain detailed information about any planet, star, or galaxy too distant for a personal visit or any kind of controlled experiment? The answer is that we use the laws of physics, as we know them here on Earth, to interpret

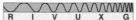

Figure 2.1 The pancake-shaped Andromeda Galaxy is about 3 million light years away and contains a few hundred billion stars.

the **electromagnetic radiation** emitted by these objects. *Radiation* is any way in which energy is transmitted through space from one point to another without the need for any physical connection between those two locations. The term *electromagnetic* means that the energy is carried in the form of rapidly fluctuating *electric* and *magnetic fields* (Section 2.2). Virtually all we know about the universe beyond Earth's atmosphere has been gleaned from analysis of electromagnetic radiation received from afar.

Visible light is the particular type of electromagnetic radiation to which the human eye happens to be sensitive, but there is also *invisible* electromagnetic radiation, which goes completely undetected by our eyes. **Radio**, **infrared**, and **ultraviolet** waves, as well as **X rays** and **gamma rays**, all fall into this category. Recognize that, despite the different names, the words *light, rays, electromagnetic radiation,* and *waves* really all refer to the same thing. The names are just historical accidents, reflecting the fact that it took many years for scientists to realize that these apparently very different types of radiation are in reality one and the same physical phenomenon. Throughout this text, we use the general terms "light" and "electromagnetic radiation" more or less interchangeably.

Wave Motion

Scientists now know that all types of electromagnetic radiation travel through space in the form of *waves*. To understand the behavior of light, then, we must know a little about this kind of motion. Simply stated, a **wave** is a way in which energy is transferred from place to place without physical movement of material from one location to another. In wave motion, the energy is carried by a *disturbance* of some sort that occurs in a distinctive, repeating pattern. Ripples on the surface of a pond, sound waves in air, and electromagnetic waves in space, despite their many obvious differences, all share this basic defining property.

As a familiar example, imagine a twig floating in a pond (Figure 2.2). A pebble thrown into the pond at some distance from the twig disturbs the surface of the water, setting it into up-and-down motion. This disturbance propagates outward from the point of impact in the form of waves. When the waves reach the twig,

Figure 2.2 The passage of a wave across a pond causes the surface of the water to bob up and down, but there is no movement of water from one part of the pond to another. Here waves ripple out from the point where a pebble hits the water to the point where a twig is floating.

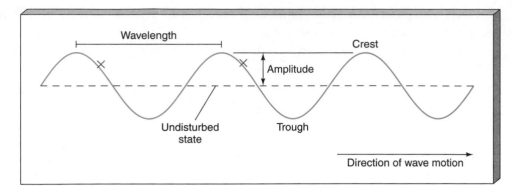

Figure 2.3 Representation of a typical wave, showing its direction of motion, wavelength, and amplitude.

some of the pebble's energy is imparted to it, causing the twig to bob up and down. In this way, both energy and *information*—the fact that the pebble entered the water—are transferred from the place where the pebble landed to the location of the twig. We could tell just by observing the twig that a pebble (or some object) had entered the water. With a little additional physics, we could even estimate the pebble's energy.

A wave is *not* a physical object. No water traveled from the point of impact of the pebble to the twig—at any location on the surface, the water surface simply moved up and down as the wave passed. What, then, *does* move across the pond surface? The answer is that the wave is the *pattern* of up-and-down motion, and it is this pattern that is transmitted from one point to the next as the disturbance moves across the water. Figure 2.3 shows how wave properties are quantified and establishes some standard terminology. We characterize a wave not only by the speed with which it moves, but also by the length of its cycle. How many seconds does it take for a wave to repeat itself at some point in space? This is its **wave period**. How many meters does it take for the wave to repeat itself at a given moment in time? This is the **wavelength**. The wavelength can be measured as the distance between two adjacent wave *crests*, two adjacent wave *troughs*, or any other two similar points on adjacent wave cycles (for example, the points marked "X" in Figure 2.3). The maximum departure of the wave from the undisturbed state—still air, say, or a flat pond surface—is called its **amplitude**.

If a wave moves at high speed, then the number of wave crests passing any given point per unit time—the wave's **frequency**—is high. Conversely, if a wave moves slowly, with only a few crests passing a point per unit time, it has a low frequency. The frequency of a wave is just 1 divided by the wave's period:

$$\text{wave frequency} = \frac{1}{\text{wave period}}.$$

Frequency is expressed in units of inverse time, cycles per second, termed hertz (Hz) in honor of the nineteenth-century German scientist Heinrich Hertz, who studied the properties of radio waves. Thus, a wave having a period of 5 s, meaning that one wave crest passes a given point every 5 seconds, has a frequency of (1/5) cycles/s = (1/5) Hz = 0.2 Hz.

Wavelength and wave frequency are *inversely* related. Doubling the frequency halves the wavelength, halving the frequency doubles the wavelength, and so on. The product of wavelength and frequency equals the *wave velocity*:

$$\text{wavelength} \times \text{frequency} = \text{velocity}.$$

Put another way, a wave crest moves a distance equal to one wavelength in one wave period. For example, if the wave in our earlier example has a wavelength of 0.5 m, we can calculate its velocity to be (0.5 m) × (0.2 Hz) = 0.1 m/s.

2.2 Waves in What?

1 Waves of radiation differ in one fundamental respect from water waves, sound waves, or any other waves that travel through a material medium: radiation needs *no* such medium. When light travels from a distant galaxy, or from any other cosmic object, it moves through the virtual vacuum of space. Sound waves, by contrast, cannot do this; if we were to remove all the air from a room, conversation would be impossible because sound waves cannot exist without air or some other physical medium to support them.

Communication by flashlight or radio, however, would be entirely feasible.

The ability of light to travel through empty space was once a great mystery. The idea that light, or any other kind of radiation, could move as a wave through nothing at all seemed to violate common sense; yet it is now a cornerstone of modern physics.

Interactions Between Charged Particles

To understand more about the nature of light, consider an *electrically charged* particle, such as an **electron** or a **proton**. Electrons and protons are elementary particles—fundamental components of atoms and all matter—that carry the basic unit of charge. Electrons are said to carry a *negative* charge, while protons carry an equal and opposite *positive* charge. Just as a massive object exerts a gravitational force on any other massive object (as we saw in Chapter 1), an electrically charged particle exerts an *electrical* force on every other charged particle in the universe. Unlike the gravitational force, however, which is always attractive, electrical forces can be either attractive or repulsive. Particles having like charges (that is, both negative or both positive) repel one another; particles having unlike charges attract (Figure 2.4a).

Extending outward in all directions from our charged particle is an **electric field**, which determines the electric force exerted by the particle on other charged particles (Figure 2.4b). The strength of the electric field, like the strength of the gravitational field, decreases with increasing distance from the source charge according to an inverse-square law. ∞ (Sec. 1.7) By means of the electric field, the particle's presence is "felt" by other charged particles, near and far.

Now suppose our particle begins to vibrate, perhaps because it becomes heated or collides with some other particle. Its changing position causes its associated electric field to change, and this changing field in turn causes the electrical force exerted on other charges to vary (Figure 2.4c). If we measure the change in the force on these other charges, we learn about our original particle. Thus, *information about our particle's state of motion is transmitted through space via a changing electric field*. This *disturbance* in the particle's electric field travels through space as a wave.

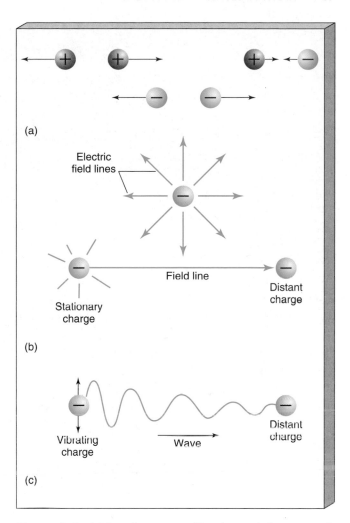

Figure 2.4 (a) Particles carrying like electrical charges repel one another, whereas particles carrying unlike charges attract. (b) A charged particle is surrounded by an electric field, which determines the particle's influence on other charged particles. We represent the field by a series of field lines. (c) If a charged particle begins to vibrate back and forth, its electric field changes. The resulting disturbance travels through space as a wave.

Electromagnetic Waves

The laws of physics tell us that a **magnetic** field must accompany every changing electric field. Magnetic fields govern the influence of *magnetized* objects on one another, much as electric fields govern interactions between charged particles. The fact that a compass needle always points to magnetic north is the result of the interaction between the magnetized needle and Earth's magnetic field (Figure 2.5). Magnetic fields also exert forces on moving electric charges (that is, electric currents)—electric meters and motors rely on this basic fact. Conversely, moving charges create magnetic fields (electromagnets are a familiar example).

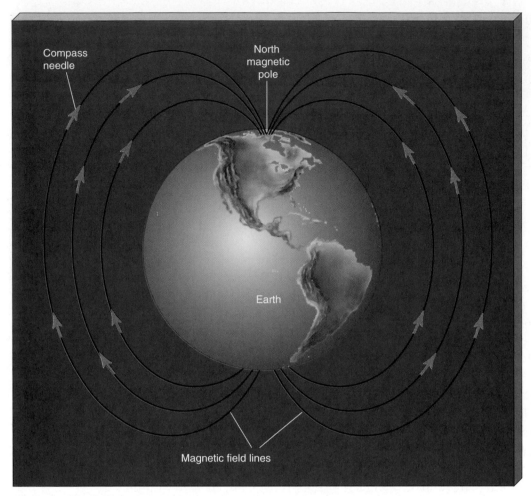

Figure 2.5 Earth's magnetic field interacts with a magnetic compass needle, causing the needle to become aligned with the field.

Electric and magnetic fields are inextricably linked to one another: a change in either one *necessarily* creates the other. For this reason, the disturbance produced by our moving charge actually consists of vibrating electric *and* magnetic fields, always oriented perpendicular to one another and moving together through space (Figure 2.6). These fields do not exist as independent entities; rather, they are different aspects of a single physical phenomenon: **electromagnetism**. Together, they constitute an *electromagnetic wave* that carries energy and information from one part of the universe to another.

Now consider a cosmic object—a star, say. It is made up of charged particles, mainly protons and electrons, in constant motion. As these charged contents move around, their electric fields change, and electromagnetic waves are produced. These waves travel outward into space, and eventually some reach Earth. Other charged particles, either in our eyes or in our

experimental equipment, respond to the electromagnetic field changes by vibrating in tune with the received radiation. This response is how we detect the radiation—with our eyes or with our detectors.

How *quickly* does one charge feel the change in the electromagnetic field when another charge begins to move? This is an important question because it is equivalent to asking how fast an electromagnetic wave travels. The answer is that all electromagnetic waves move at a very specific speed—the **speed of light** (always denoted by the letter c). Its value is 299,792.458 km/s in a vacuum (and somewhat less in material substances, such as air or water). In this text, we round this value off to $c = 3.00 \times 10^5$ km/s. This is an extremely high speed. In the time needed to snap a finger—about a tenth of a second—light can travel three quarters of the way around our planet! If the currently known laws of physics are correct, then the speed of light is the fastest speed possible.

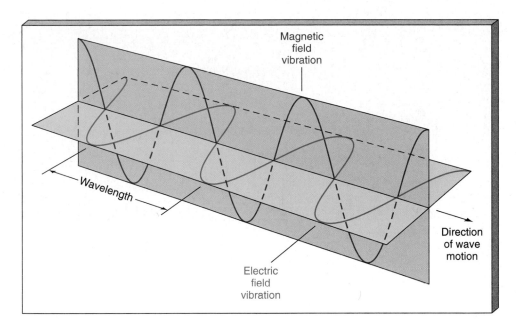

Magnetic
field
vibration

Wavelength

Direction
of wave
motion

Electric
field
vibration

Figure 2.6 Electric and magnetic fields vibrate perpendicular to each other. Together they form an electromagnetic wave that moves through space at the speed of light.

The speed of light is very large, but it is still finite. That is, light does not travel instantaneously from place to place. This fact has some interesting consequences for our study of distant objects. It takes time—often lots of time—for light to travel through space. The light we see from the nearest large galaxy—the Andromeda Galaxy, shown in Figure 2.1—left that object about 3 million years ago, around the time our first human ancestors appeared on Earth. We can know nothing about this galaxy as it exists today. For all we know, it might no longer even exist! Only our descendants 3 million years from now will know if it exists now. So, as we study objects in the cosmos, remember that the light now seen left those objects long ago. We can never observe the universe as it is—only as it was.

2.3 The Electromagnetic Spectrum

2 White light is a mixture of colors, which we conventionally divide into six major hues—red, orange, yellow, green, blue, and violet. As shown in Figure 2.7, we can separate a beam of white light into a rainbow of these basic colors by passing it through a prism. In principle, the original beam of white light could be produced once again by passing the entire red-to-violet range of colors—called a *spectrum* (plural, *spectra*)—through a second prism to recombine the colored beams. This experiment was first reported by Isaac Newton over 300 years ago.

The Components of Visible Light

What determines the color of a beam of light? The answer is its wavelength (or, equivalently, its frequency)—we see different colors because our eyes react differently to electromagnetic waves of different wavelengths. Red light has a frequency of roughly 4.3×10^{14} Hz, corresponding to a wavelength of about 7.0×10^{-7} m. Violet light, at the other end of the visible range, has nearly double the frequency—7.5×10^{14} Hz—and (since the speed of light is the same in either case) just over half the wavelength—4.0×10^{-7} m. The other colors we see have frequencies and wavelengths intermediate between these two extremes.

Astronomers often use a unit called the *nanometer* (nm) when describing the wavelength of light (see Appendix 2). There are 10^9 nanometers in 1 meter. An older unit called the *angstrom* ($1\mathring{A} = 10^{-10}$ m = 0.1 nm) is also widely used. (The unit is named after the nineteenth-century Swedish physicist Anders Ångstrom—pronounced "ongstrem.") However, the nanometer is preferred. Thus, the visible spectrum covers the wavelength range from 400 to 700 nm (4000 to 7000 Å). The radiation to which our eyes are most sensitive has a wavelength near the middle of this range, at about 550 nm (5500 Å), in the yellow-green region of the spectrum. It is no coincidence that this wavelength falls within the range of wavelengths at which the Sun emits most of its electromagnetic energy—our eyes have evolved to take greatest advantage of the available light.

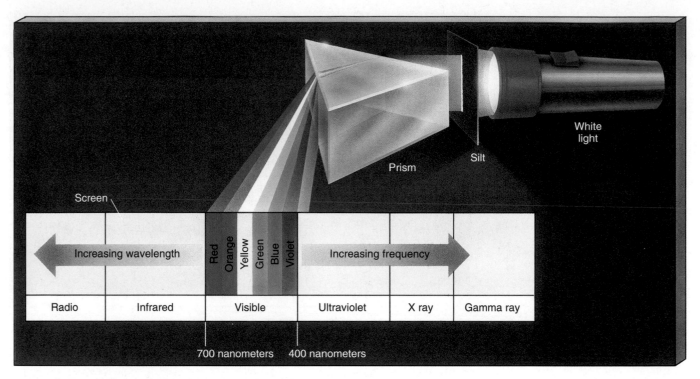

Figure 2.7 When passed through a prism, white light splits into its component colors, spanning red to violet in the visible part of the electromagnetic spectrum. The slit immediately in front of the flashlight narrows the beam of radiation. The image on the screen is a series of colored images of the slit. Human eyes are insensitive to radiation of wavelength shorter than 400 nm or longer than 700 nm.

The Full Range of Radiation

Figure 2.8 plots the entire range of electromagnetic radiation. The only characteristic that distinguishes one type of radiation from another is frequency—or, equivalently, wavelength. To the low-frequency, long-wavelength side of visible light lies radio and infrared radiation. Radio frequencies include radar, microwave radiation, and the familiar AM, FM, and TV bands. We perceive infrared radiation as heat. To the high-frequency, short-wavelength side of visible light lies ultraviolet, X-ray, and gamma-ray radiation. Ultraviolet radiation, lying just beyond the violet end of the visible spectrum, is responsible for suntans and sunburns. X rays are perhaps best known for their ability to penetrate human tissue and reveal the state of our insides without our resorting to surgery. Gamma rays are the shortest-wavelength radiation. They are often associated with radioactivity and are invariably damaging to any living cells they encounter.

All these spectral regions, including the visible, collectively make up the **electromagnetic spectrum**. Remember that, despite their greatly differing wave-lengths and the very different roles they play in everyday life on Earth, all types of radiation are basically the same phenomenon, and all move at the same speed—the speed of light c.

Figure 2.8 is worth studying carefully, as it contains a great deal of information. Note that wave frequency (in hertz) increases from left to right and wavelength (in meters) increases from right to left. These wave properties behave in opposite ways because, as noted earlier, they are inversely related. When picturing wavelengths and frequencies, this book adheres to the convention that frequency increases toward the right. Notice that the wavelength and frequency scales in Figure 2.8 do not increase by equal increments of 10. Instead, successive values marked on the horizontal axis differ by factors of 10—each successive value is 10 times greater than its neighbor. This type of scale, called a *logarithmic* scale, is often used in science in order to condense a very large range of some quantity into a manageable size. Had we used a linear scale for the wavelength range shown in Figure 2.8, the figure would have been many light years long! Throughout the text we will often find it convenient to use a logarithmic

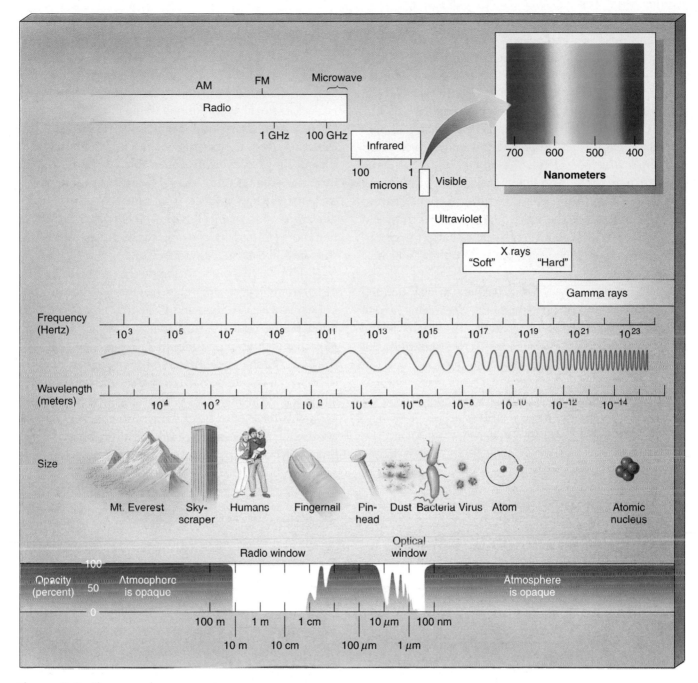

Figure 2.8   The entire electromagnetic spectrum.

scale in order to compress a wide range of some quantity onto a single easy-to-view plot.

Figure 2.8 shows wavelengths extending from the height of mountains for radio radiation to the diameter of an atomic nucleus for gamma-ray radiation. The box at the upper right emphasizes how small the visible portion of the electromagnetic spectrum is. Most objects in the universe emit large amounts of invisible radiation. Indeed, many objects emit only a tiny fraction of their total energy in the visible range. A wealth of extra knowledge can be gained by studying the invisible regions of the electromagnetic spectrum. To remind you of this important fact and to identify the region of the electromagnetic spectrum in which a particular observation was made, we have attached a spectrum icon—an idealized version of the wavelength scale in Figure 2.8—to every astronomical image presented in this text.

Only a small fraction of the radiation arriving at our planet actually reaches Earth's surface because of the *opacity* of Earth's atmosphere. **Opacity** is the extent to which radiation is blocked by the material through which it is passing—in this case, air. The more opaque an object is, the less radiation gets through. Earth's atmospheric opacity is plotted along the wavelength and frequency scales at the bottom of Figure 2.8. The extent of shading is proportional to the opacity. Where the shading is greatest, no radiation can get in or out—the energy is completely absorbed by atmospheric gases. Where there is no shading at all, our atmosphere is almost totally transparent. Extraterrestrial radiation in these regions of the electromagnetic spectrum can reach Earth's surface and terrestrial radiation from human transmissions can pass virtually unhindered into space.

There are only a few *windows*, at well-defined locations in the electromagnetic spectrum, where Earth's atmosphere is transparent. In much of the radio and in the visible portions of the spectrum, the opacity is low, so we can study the universe at those wavelengths from ground level. In parts of the infrared range, the atmosphere is partially transparent, so we can make certain infrared observations from the ground. In the rest of the spectrum, however, the atmosphere is opaque. As a result, ultraviolet, X-ray, and gamma-ray observations can be made only from above the atmosphere, from orbiting satellites.

2.4 The Distribution of Radiation

All macroscopic objects—fires, ice cubes, people, stars—emit radiation at all times. They radiate because the microscopic charged particles in them are in constant random motion, and whenever charges change their state of motion, electromagnetic radiation is emitted. The temperature of an object is a direct measure of the amount of microscopic motion within it (see *More Precisely 2-1*). The hotter the object—that is, the higher its temperature—the faster its constituent particles move and the more energy they radiate.

The Black-Body Spectrum

Intensity is a term often used to specify the amount or strength of radiation at any point in space. Like frequency and wavelength, intensity is a basic property of radiation. No natural object emits all of its radiation at just one frequency. Instead, the energy is often spread out over a range of frequencies. By studying the way in which the intensity of this radiation is distributed across the electromagnetic spectrum, we can learn much about the object's properties.

Figure 2.9 illustrates schematically the distribution of radiation emitted by any object. The curve peaks at a single, well-defined frequency and falls off to lesser values above and below that frequency. Note that the curve is not shaped like a symmetrical bell that declines evenly on either side of the peak. The intensity falls off more slowly from the peak to lower frequencies than it does from the peak to high frequencies. This overall shape is characteristic of the radiation emitted by *any* object, regardless of its size, shape, composition, or temperature.

The curve drawn in Figure 2.9 is the radiation-distribution curve for a mathematical idealization known as a *black body*—an object that absorbs all radiation falling upon it. In a steady state, a black body must reemit the same amount of energy as it absorbs; the **black-body curve** shown in the figure describes the distribution of that reemitted radiation. No real object absorbs and radiates as a perfect black body. In many cases, however, the black-body curve is a very good approximation to reality, and the properties of black bodies provide important insights into the behavior of real objects.

The Radiation Laws

3 The black-body curve shifts toward higher frequencies (shorter wavelengths) and greater intensities as an object's temperature increases. Even so,

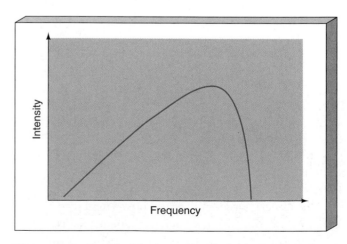

Figure 2.9 The black-body, or Planck, curve represents the distribution of the intensity of the radiation emitted by any object.

the *shape* of the curve remains the same. This shifting of radiation's peak frequency with temperature is familiar to us all: very hot glowing objects, such as toaster filaments or stars, emit visible light. Cooler objects, such as warm rocks or household radiators, produce invisible radiation—warm to the touch but not glowing hot to the eye. These latter objects emit most of their radiation in the lower-frequency infrared part of the electromagnetic spectrum.

As a further example, imagine a piece of metal placed in a hot furnace. At first, the metal becomes warm, although its appearance doesn't change. As it heats up, it begins to glow dull red, then orange, brilliant yellow, and finally white. How do we explain this? As illustrated in Figure 2.10, when the metal is at room temperature (300 K), it emits only invisible infrared radiation. As the metal becomes hotter, the peak of its black-body curve shifts toward higher frequencies. At 1000 K, for instance, most of the emitted radiation is still infrared, but now there is also a small amount of visible (dull red) radiation being emitted (note in Figure 2.10 that the high-frequency portion of the 1000 K curve just overlaps the visible region of the graph). As the temperature continues to rise, the peak of the metal's black-body curve moves through the visible spectrum, from red (the 4000 K curve) through yellow. The metal eventually becomes

white hot because when its black-body curve peaks in the blue or violet part of the spectrum (the 7000 K curve), the low-frequency tail of the curve extends through the entire visible spectrum (to the left in Figure 2.10), meaning that substantial amounts of green, yellow, orange, and red light are also emitted. Together, all these colors combine to produce white.

From detailed studies of the precise form of the black-body curve, we obtain a very simple connection between the wavelength at which most radiation is emitted and the absolute temperature (that is, temperature measured in kelvins—see *More Precisely 2-1*) of the emitting object:

$$\text{wavelength of peak emission} \propto \frac{1}{\text{temperature}}.$$

This relationship is known as **Wien's law.** Simply put, it tells us that the hotter the object, the bluer its radiation.

Finally, it is also a matter of everyday experience that as the temperature of an object increases, the *total* amount of energy it radiates (summed over all frequencies) increases rapidly. For example, the heat given off by an electric heater increases sharply as the heater warms up and begins to emit visible light. In fact, the total amount of energy radiated per unit time is pro-

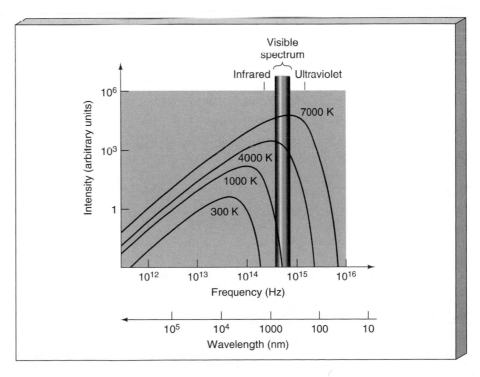

Figure 2.10 As an object is heated, the radiation it emits peaks at higher and higher frequencies. Shown here are curves corresponding to temperatures of 300 K (room temperature), 1000 K (beginning to glow deep red), 4000 K (red hot), and 7000 K (white hot).

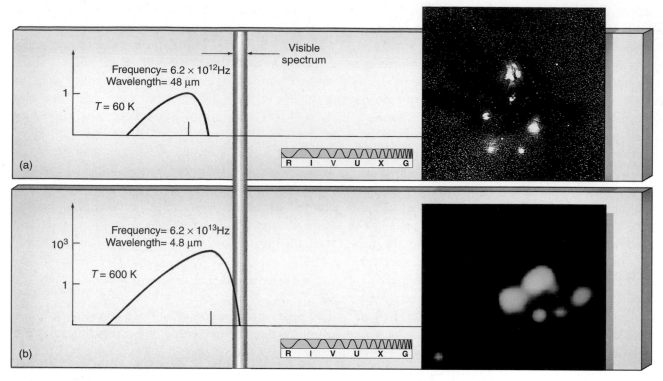

Figure 2.11 Comparison of black-body curves for four cosmic objects. (a) A cool, invisible galactic gas cloud called Rho Ophiuchi. At a temperature of 60 K, it emits mostly low-frequency radio radiation. (b) A dim, young star (shown red in the inset photograph) near the center of the Orion Nebula. The star's atmosphere, at 600 K, radiates primarily in the infrared. (c) The Sun's surface, at approximately 6000 K, is brightest in the visible region of the electromagnetic spectrum. (d) A cluster of very bright stars, called Omega Centauri, as observed by a telescope aboard the space shuttle. At a temperature of 60,000 K, these stars radiate strongly in the ultraviolet.

portional to the fourth power of an object's temperature:

$$\text{total energy radiated} \propto \text{temperature}^4.$$

This relationship is called **Stefan's law**. It implies that the energy emitted by a body rises dramatically as the body's temperature increases. Doubling the temperature, for example, causes the total energy radiated to increase by a factor of 16.

Astronomical Applications

Astronomers often use black-body curves as thermometers to determine the temperatures of distant objects. For example, study of the solar spectrum makes it possible to measure the temperature of the Sun's surface. Observations of the radiation from the Sun at many frequencies yield a curve shaped somewhat like

that shown in Figure 2.9. The Sun's curve peaks in the visible part of the electromagnetic spectrum; the Sun also emits a lot of infrared and a little ultraviolet radiation. Using Wien's law, we find that the temperature of the Sun's surface is approximately 6000 K. (A more precise measurement, applying Wein's law to the black-body curve that best fits the solar spectrum, yields a temperature of 5800 K.)

Other cosmic objects have surfaces very much cooler or hotter than the Sun's, emitting most of their radiation in invisible parts of the spectrum (Figure 2.11). For example, the relatively cool surface of a very young star might measure 600 K and emit mostly infrared radiation. Cooler still is the interstellar gas cloud from which the star formed; at a temperature of 60 K, such a cloud would emit mainly long-wavelength radiation in the radio and infrared parts of the spectrum. The brightest stars, by contrast, have surface temperatures as high as 60,000 K and hence emit mostly ultraviolet radiation.

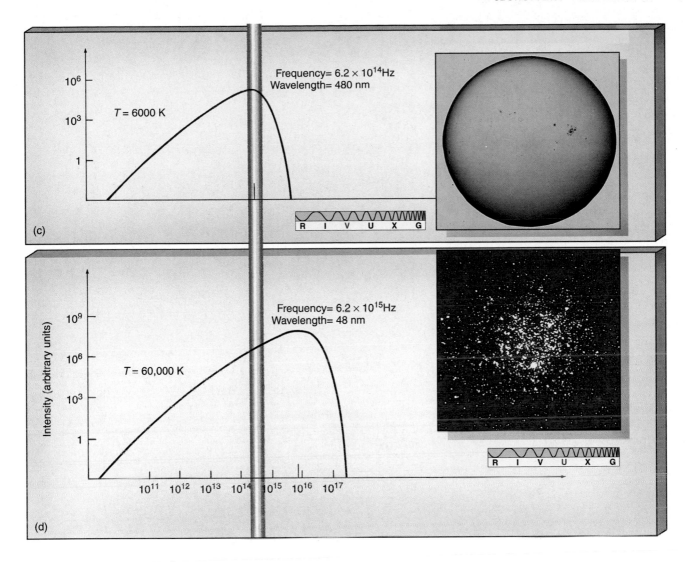

2.5 Spectral Lines

4 Radiation can be analyzed with an instrument known as a **spectroscope**. In its most basic form, this device consists of an opaque barrier with a slit in it (to form a narrow beam of light), a prism (to split the beam into its component colors), and either a detector or a screen (to allow the user to view the resulting spectrum). Figure 2.12 shows such an arrangement.

Emission Lines

The spectra encountered in the previous section are examples of **continuous spectra**. A light bulb, for instance, emits radiation of all wavelengths (mostly in the visible range), with an intensity distribution that is well described by the black-body curve corresponding to the bulb's temperature. Viewed through a spectroscope, the spectrum of the light from the bulb would show the familiar rainbow running from red to violet without interruption, as presented in Figure 2.13a.

Not all spectra are continuous. For instance, if we took a glass jar containing pure hydrogen gas and passed an electrical discharge through it (a little like a lightning bolt arcing through Earth's atmosphere), the gas would begin to glow—that is, it would emit radiation. If we were to examine that radiation with our spectroscope, we would find that its spectrum consisted of only a few bright lines on an otherwise dark background, quite unlike the continuous spectrum described for the light bulb. Figure 2.13(b)

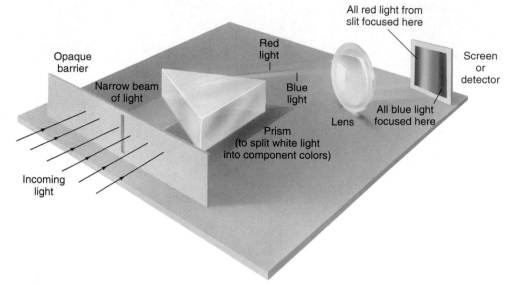

Figure 2.12 Diagram of a simple spectroscope. A slit in the barrier allows a narrow beam of light to pass. The beam passes through a prism and is split into its component colors. A lens then focuses the light into a sharp image that is either projected onto a screen, as shown here, or analyzed as it is passed through a detector.

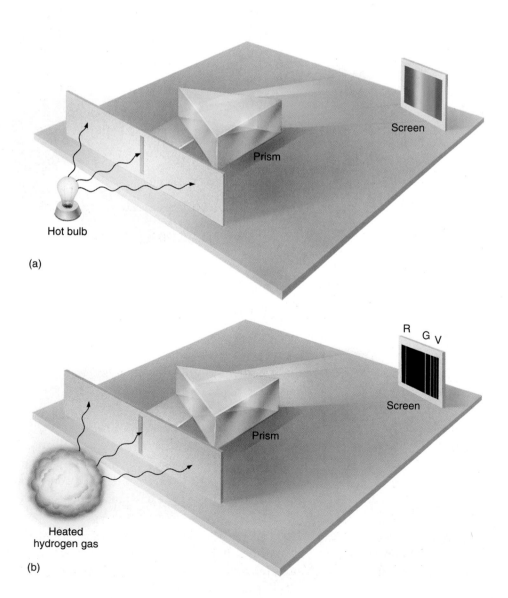

Figure 2.13 (a) When passed through a slit and split up by a prism, light from a source of continuous radiation gives rise to the familiar rainbow of colors. (b) By contrast, the light from excited (heated) hydrogen gas consists of a series of distinct spectral lines. (For simplicity, the focusing lens has been omitted from these drawings.)

2-1 MORE PRECISELY

THE KELVIN TEMPERATURE SCALE

The atoms and molecules that make up any piece of matter are in constant random motion. This motion represents a form of energy known as thermal energy or, more commonly, heat. The quantity we call temperature is a direct measure of this internal motion: the higher an object's temperature, the faster the random motion of its constituent particles. More precisely, the temperature of a piece of matter specifies the average thermal energy of the particles it contains.

The temperature scale probably most familiar to you, the Fahrenheit scale, is now a peculiarity of American society. Most of the rest of the world uses the Celsius temperature scale, in which water freezes at 0 degrees (0°C) and boils at 100 degrees (100°C), as illustrated in the accompanying figure.

There are, of course, temperatures below the freezing point of water. Although we know of no matter anywhere in the universe that is actually this cold, temperatures can in theory reach as low as −273.15°C. This is the temperature at which all atomic and molecular motion virtually ceases. It is convenient to construct a temperature scale based on this lowest possible temperature, which is called *absolute zero*. Scientists commonly use such a scale, called the Kelvin scale in honor of the nineteenth-century British physicist Lord Kelvin. Since it takes absolute zero as its starting point, the Kelvin

scale differs from the Celsius scale by 273.15°. In this book, we round off the decimal places and simply use the relationship

$$\text{kelvins} = \text{degrees Celsius} + 273.$$

Thus,

- Translational motion in atoms and molecules ceases at 0 kelvins (0 K).
- Water freezes at 273 kelvins (273 K).
- Water boils at 373 kelvins (373 K).

Note that the unit is "kelvins," or "K," not "degrees kelvin" or "°K."

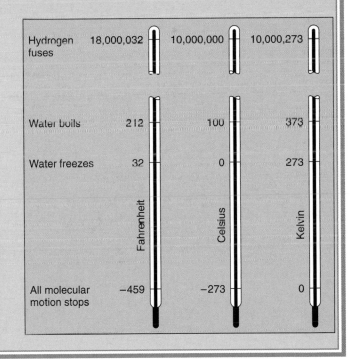

	Fahrenheit	Celsius	Kelvin
Hydrogen fuses	18,000,032	10,000,000	10,000,273
Water boils	212	100	373
Water freezes	32	0	273
All molecular motion stops	−459	−273	0

shows this schematically, and more detailed rendering of the spectrum of hydrogen appears in the top panel of Figure 2.14. The light produced by the hydrogen in this experiment does *not* consist of all possible colors, but instead includes only a few narrow, well-defined **emission lines**, narrow "slices" of the continuous spectrum. The black background represents all the wavelengths *not* emitted by hydrogen.

After further experimentation, we would also find that although we could alter the intensity of the lines (for example, by changing the amount of hydrogen in the jar or the strength of the electrical discharge), we could not alter their color (in other words, their frequency or wavelength). This particular pattern of spectral emission lines is a property of the element hydrogen—whenever we perform this experiment, the same

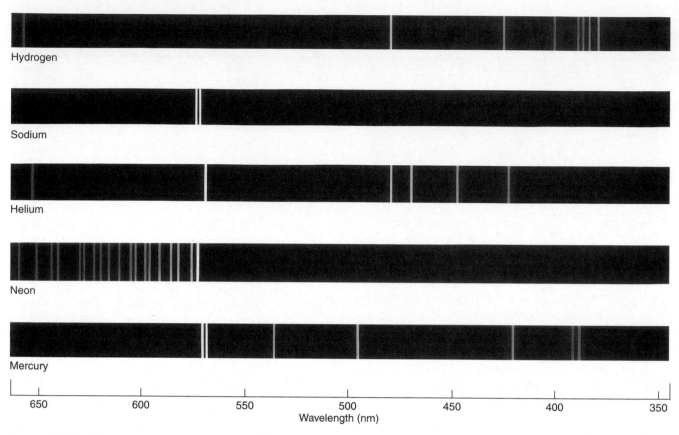

Hydrogen

Sodium

Helium

Neon

Mercury

650   600   550   500   450   400   350

Wavelength (nm)

Figure 2.14 The emission spectra of some well-known elements.

characteristic emission spectrum is the result. Other elements yield different emission spectra. Depending on which element is involved, the pattern of lines can be fairly simple or very complex. Always, though, it is *unique* to that element.

Scientists have accumulated extensive catalogs of the specific wavelengths at which many different hot gases emit radiation. For gas of a given chemical composition, the particular pattern of the light it emits is known as its **emission spectrum**. The emission spectrum of a gas provides a kind of "fingerprint" that allows scientists to deduce its presence by spectroscopic means. Examples of the emission spectra of some common substances are shown in Figure 2.14.

Absorption Lines

When sunlight is split by a prism, at first glance it appears to produce a continuous spectrum. However, closer scrutiny shows that the solar spectrum is interrupted by a large number of narrow dark lines, as shown in Figure 2.15. We now know that many of these lines represent wavelengths of light that have been removed (absorbed) by gases present either in the outer layers of the Sun or in Earth's atmosphere. These gaps in the spectrum are called **absorption lines**. The absorption lines in the solar spectrum are referred to collectively as *Fraunhofer lines*, after the nineteenth-century German physicist Joseph Fraunhofer, who measured and cataloged more than 600 of them.

At around the time solar absorption lines were discovered, scientists found that absorption lines could also be produced in the laboratory by passing a beam of light from a continuous source through a cool gas, as shown in Figure 2.16. They quickly observed an intriguing connection between emission and absorption lines: the absorption lines associated with a given gas occur at precisely the *same* wavelengths as the emission lines produced when the gas is heated.

As an example, consider the element sodium, whose emission spectrum appears in Figure 2.14. When heated to high temperatures, a sample of sodium vapor emits visible light strongly at just two wavelengths—

Figure 2.15 This visible spectrum of the Sun shows hundreds of dark absorption lines superimposed on a bright continuous spectrum. Here, the scale extends from long wavelengths (red) at the upper left to short wavelengths (blue) at the lower right.

589.9 nm and 589.6 nm—lying in the yellow part of the spectrum. When a continuous spectrum is passed through some relatively cool sodium vapor, two sharp, dark absorption lines appear at precisely the same wavelengths. The emission and absorption spectra of sodium are compared in Figure 2.17.

The analysis of the ways in which matter emits and absorbs radiation is called **spectroscopy**. The observed relationships between the three types of spectra—continuous, emission line, and absorption line—may be summarized as follows:

1. A luminous solid or liquid, or a sufficiently dense gas, emits light of all wavelengths and so produces a continuous spectrum of radiation (Figure 2.13a).

2. A low-density hot gas emits light whose spectrum consists of a series of bright emission lines. These lines are characteristic of the chemical composition of the gas (Figure 2.13b).

3. A low-density cool gas absorbs certain wavelengths from a continuous spectrum, leaving dark absorption lines in their place, superimposed on the continuous spectrum. These lines are characteristic of the composition of the intervening gas; they occur at precisely the same wavelengths as the emission lines produced by that gas at higher temperatures (Figure 2.16).

These rules are known as **Kirchhoff's laws**, after the German physicist Gustav Kirchhoff, who published them in 1859.

Figure 2.16 When a cool gas is placed between a source of continuous radiation (a light bulb) and the detector/screen of a spectroscope, the resulting spectrum is crossed by a series of dark absorption lines. These lines are formed when the cool gas absorbs certain wavelengths (colors) from the light radiating from the light bulb. The absorption lines appear at precisely the same wavelengths as the emission lines that would be produced if the gas were heated to high temperatures (see Figure 2.13b). (For simplicity, the focusing lens has been omitted from this drawing.)

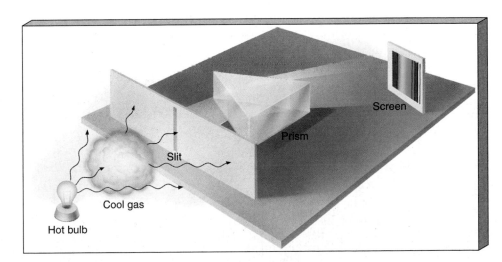

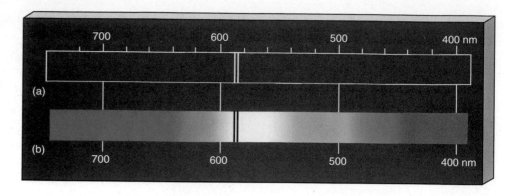

Figure 2.17 (a) The emission spectrum of sodium. The two bright lines in the center appear in the yellow part of the visible spectrum. (b) The absorption spectrum of sodium. The two dark lines appear at exactly the same wavelengths as the bright lines in the sodium emission spectrum.

Astronomical Applications

Once astronomers realized that spectral lines are indicators of chemical composition, they set about identifying the observed lines in the Sun's spectrum. Almost all of the lines in light from extraterrestrial sources could be attributed to known elements (for example, many of the Fraunhofer lines in sunlight are associated with the element iron). However, some new lines also appeared in the solar spectrum. In 1868, astronomers realized that those lines must correspond to a previously unknown element. It was given the name helium, after the Greek word *helios*, meaning "sun." Only in 1895, almost three decades after its detection in sunlight, was helium discovered on Earth.

For all the information that nineteenth-century astronomers could extract from observations of stellar spectra, they still lacked a theory explaining *how* those spectra arose. Despite their sophisticated spectroscopic equipment, they knew scarcely any more about the physics of stars than did Galileo or Newton. To understand how spectroscopy can be used to extract detailed information about astronomical objects from the light they emit, we must delve more deeply into the processes that produce line spectra.

2.6 The Formation of Spectral Lines

By the start of the twentieth century, physicists had accumulated evidence that light sometimes behaves in a manner that cannot be explained by the wave theory of radiation. As we have just seen, the production of absorption and emission lines involves only certain very specific wavelengths of light. This result would not be expected if light behaved only as a continuous wave and matter always obeyed the laws of Newtonian mechanics. Other experiments conducted around the same time strengthened the conclusion that the notion of radiation as a wave was incomplete. It became clear that, when light interacts with matter on very small scales, it does so not in a smooth, continuous way but in a discontinuous, stepwise manner. The challenge was to find an explanation for this unexpected behavior. The solution revolutionized our view of nature and now forms the foundation not just for physics and astronomy, but for virtually all of modern science.

Atomic Structure

5 To explain the formation of spectral lines, we must understand not just the nature of light but also something of the structure of **atoms**—the microscopic building blocks from which all matter is constructed. Let us start with the simplest atom, hydrogen, which consists of an electron, with a negative electrical charge, orbiting a proton, which carries a positive charge. The proton forms the central **nucleus** (plural: nuclei) of the atom. Because the positive charge on the proton exactly cancels the negative charge on the electron, the hydrogen atom as a whole is electrically neutral.

How does this picture of the hydrogen atom relate to the characteristic emission and absorption lines associated with hydrogen gas? If an atom emits some energy in the form of radiation, that energy has to come from somewhere within the atom. Similarly, if energy is absorbed by the atom, that energy must cause some internal change. It is reasonable (and correct) to suppose that the energy emitted or absorbed by the atom is associated with changes in the motion of the orbiting electron.

The first theory of the atom to provide an explanation of hydrogen's observed spectral lines was propounded by the Danish physicist Niels Bohr. This the-

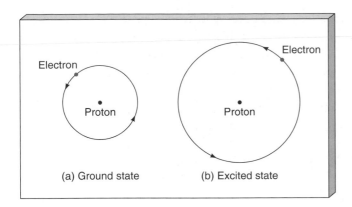

Figure 2.18 An early conception of the hydrogen atom—the Bohr model—pictured its electron orbiting the central proton in a well-defined orbital, like a planet orbiting the Sun. Two electron orbitals of different energies are shown: (a) the ground state and (b) an excited state.

ory is now known simply as the **Bohr model** of the atom. Its essential features are as follows. First, there is a state of lowest energy—the **ground state**—which represents the "normal" condition of the electron as it orbits the nucleus. Second, there is a maximum energy that the electron can have and still be part of the atom. Once the electron acquires more than that maximum energy, it is no longer bound to the nucleus, and the atom is said to be *ionized;* an atom missing one or more of its electrons is called an **ion**. Third, and most important (and also least intuitive), between those two energy levels, the electron can exist only in certain sharply defined energy states, often referred to as *orbitals.*

This description of the atom contrasts sharply with the predictions of Newtonian mechanics, which would permit orbits with *any* energy, not just at certain specific values. In the atomic realm, such discontinuous behavior is the norm. In the jargon of the field, the orbital energies are said to be **quantized**. The rules of *quantum mechanics,* the branch of physics governing the behavior of atoms and subatomic particles, are far removed from everyday experience.

In Bohr's model, each electron orbital was pictured as having a specific radius, much like a planetary orbit in the solar system, as shown in Figure 2.18. However, the modern view is not so simple. Although each orbital *does* have a precise energy, the electron is now envisioned as being smeared out in an *electron cloud* surrounding the nucleus, as illustrated in Figure 2.19. It is common to speak of the average distance from the cloud to the nucleus as the radius of the electron's orbital. When a hydrogen atom is in its ground

state, the radius of the orbital is about 0.05 nm (0.5 Å). As the orbital energy increases, the radius increases, too. For the sake of clarity in the diagrams that follow, we represent electron orbitals by solid lines, but bear in mind always that the fuzziness shown in Figure 2.19 is a more accurate depiction of reality.

Atoms do not always remain in their ground state. An atom is said to be in an **excited state** when an electron occupies an orbital at a greater than normal distance from its parent nucleus. An atom in such an excited state has a greater than normal amount of energy. The excited state with the lowest energy (that is, the one closest in energy to the ground state) is called the *first excited state,* that with the second-lowest energy the *second excited state,* and so on. An atom can become excited in one of two ways: by absorbing some light energy from a source of electromagnetic radiation or by colliding with some other particle—another atom, for example. However, the electron cannot stay in a higher orbital forever; the ground state is the only level where it can remain indefinitely. After about 10^{-8} s, an excited atom returns to its ground state.

The Particle Nature of Radiation

Here now is the crucial point that links atoms to radiation and allows us to interpret atomic spectra. Because electrons may exist only in orbitals having specific energies, atoms can absorb only specific amounts of energy as their electrons are boosted into excited states. Likewise, atoms can emit only specific amounts of energy as their electrons fall back to lower energy

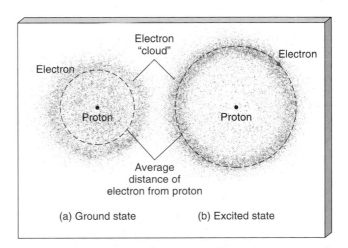

Figure 2.19 The modern view of the hydrogen atom sees the electron as a "cloud" surrounding the nucleus. The same two energy states are shown as in Figure 2.18.

states. Thus, the amount of light energy absorbed or emitted in these processes *must correspond precisely to the energy difference between two orbitals*. The quantized nature of the atom's energy levels requires that light must be absorbed and emitted in the form of little "packets" of electromagnetic radiation, each carrying a very specific amount of energy. We call these packets **photons**. A photon is, in effect, a "particle" of electromagnetic radiation.

The idea that light sometimes behaves not as a continuous wave but as a stream of particles was first proposed by Albert Einstein in 1905. In order to explain all the experimental results he knew, Einstein found that the energy contained within a photon must be proportional to the frequency of the radiation:

$$\text{photon energy} \propto \text{radiation frequency}.$$

Thus, for example, a "red" photon having a frequency of 4×10^{14} Hz (corresponding to a wavelength of about 750 nm, or 7500 Å) has 4/7 the energy of a "blue" photon having a frequency of 7×10^{14} Hz. Because it connects the *energy* of a photon with the *color* of the light it represents, this relationship is the final piece in the puzzle of how to understand the spectra we see.

Many people are confused by the idea that light can behave in two such different ways. To be truthful, modern physicists don't yet fully understand why nature displays this wave–particle duality. Nevertheless, there is irrefutable experimental evidence for both of these aspects of radiation. Environmental conditions ultimately determine which description—wave or stream of particles—better fits the behavior of electromagnetic radiation. As a general rule of thumb, in the macroscopic realm of everyday experience, radiation is more usefully described as a wave, and in the microscopic domain of atoms, it is best characterized as a stream of particles.

Absorption and emission of photons by a hydrogen atom are illustrated in Figure 2.20. Figure 2.20(a) shows the atom absorbing a photon of radiation and making a transition from the ground state to the first excited state, then emitting a photon of precisely the same energy and dropping back to the ground state. The energy difference between the two states corresponds to an ultraviolet photon, of wavelength 121.6 nm (1216 Å).

Absorption can also boost an electron into an excited state higher than the first excited state. Figure 2.20(b) depicts the absorption of a more energetic (higher-frequency, shorter-wavelength) ultraviolet photon, this one having a wavelength of 102.6 nm (1026 Å). Absorption of this photon causes the atom to jump to the *second* excited state. From that state, the electron may return to the ground state via either one of two alternate paths. It can proceed directly back to the ground state, in the process emitting an ultraviolet photon identical to the one that excited the atom in the first place, or it can *cascade* down one orbital at a time. If the latter possibility occurs, the atom will emit *two* photons: one having an energy equal to the difference between the second and first excited states, and the other having an energy equal to the difference between the first excited state and the ground state.

The second step of this cascade produces a 121.6-nm ultraviolet photon, just as in Figure 2.20a. However, the first transition of the cascade—the one from the second to the first excited state—produces a photon of wavelength 656.3 nm (6563 Å), which is in the visible part of the electromagnetic spectrum. This photon is seen as red light. An individual atom—if one could be isolated—would emit a momentary red flash. This is the origin of the red line in the hydrogen spectrum shown in Figure 2.14. The inset in Figure 2.20 shows an astronomical object whose red coloration is the result of this process. As ultraviolet photons from a hot star pass through surrounding hydrogen gas, some are absorbed by the gas, boosting its atoms into excited states. Transitions from the second to the first excited state as the atoms cascade back to their ground states produce the 656.3-nm red glow characteristic of excited hydrogen gas. (In fact, *all* of hydrogen's visible lines are the result of transitions between the first and higher excited states.)

Kirchhoff's Laws Explained

6 Let's reconsider our earlier discussion of emission and absorption lines in terms of the model just presented. In Figure 2.16, a beam of continuous radiation shines through a cloud of cool gas. The beam contains photons of all energies, but most of them cannot interact with the gas because the gas can absorb only photons having precisely the right energy to cause an electron to jump from one orbital to another. Photons having energies that cannot produce such a jump do not interact with the gas at all; they pass through it unhindered. Photons having the right energies are absorbed, excite the gas, and are removed from the beam. This is the cause of the dark absorption lines

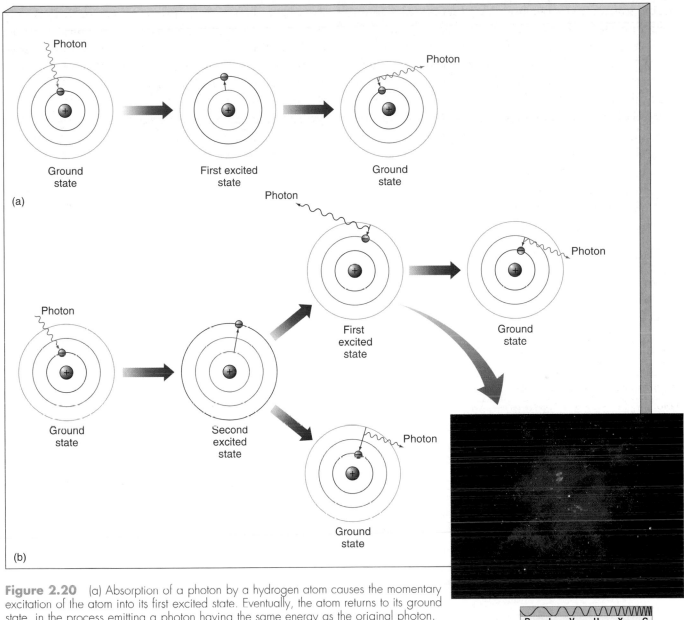

Figure 2.20   (a) Absorption of a photon by a hydrogen atom causes the momentary excitation of the atom into its first excited state. Eventually, the atom returns to its ground state, in the process emitting a photon having the same energy as the original photon. (b) Absorption of a photon might also boost the atom into a higher excited state, from which there are several possible paths back to the ground state. The object shown in the inset, designated NGC 2440, is an emission nebula: an interstellar cloud consisting largely of hydrogen gas excited by an extremely hot star (the white dot in the center).

in the spectrum of Figure 2.16. These lines are direct indicators of the energy differences between orbitals in the atoms making up the gas.

The excited gas atoms return rapidly to their original states, each emitting one or more photons in the process. We might think, then, that although some photons from the beam are absorbed by the gas, they are quickly replaced by reemitted photons, with the result that we could never observe the effects of absorption. This is not the case, however, for two important reasons. First, while the photons not absorbed by

the gas continue on directly to the detector, the reemitted photons can leave in *any* direction. Most of the reemitted photons leave at angles that do not take them through the slit and on to the detector, and so they are effectively lost from the original beam. Second, as we have just seen, electrons can cascade back to the ground state, emitting several lower-energy photons instead of a single photon equal in energy to the one originally absorbed. A second detector looking at the cloud from the side would record the reemitted energy as an emission spectrum. (This is what we are seeing in the inset to Figure 2.20.) Like the absorption spectrum, the emission spectrum is characteristic of the gas, not of the original beam.

Absorption and emission spectra are created by the same atomic processes. They correspond to the same atomic transitions. They contain the same information about the composition of the gas cloud. In the laboratory, we can move our detector and can measure both. In astronomy, we are not able to change our vantage point, so the type of spectrum we see depends on our chance location with respect to both the source and the intervening gas cloud.

More Complex Atoms

All hydrogen atoms have the same structure—a single electron orbiting a single proton—but, of course, there are many other kinds of atoms, each having a unique internal structure. The number of protons in the nucleus of an atom determines the **element** that the atom represents. That is, just as all hydrogen atoms have a single proton, all oxygen atoms have 8 protons, all iron atoms have 26 protons, and so on.

The next simplest element after hydrogen is helium. The central nucleus of the most common form of helium is made up of two protons and two **neutrons** (another kind of elementary particle having a mass slightly larger than that of a proton but carrying no

electrical charge). About this nucleus orbit two electrons. As with hydrogen and all other atoms, the "normal" condition for helium is to be electrically neutral, with the negative charge of the orbiting electrons exactly canceling the positive charge of the nucleus (Figure 2.21a).

More complex atoms contain more protons (and neutrons) in the nucleus and have correspondingly more orbiting electrons. For example, an atom of carbon (Figure 2.21b) consists of six electrons orbiting a nucleus containing six protons and six neutrons. As we progress to heavier and heavier elements, the number of orbiting electrons increases, and consequently the number of possible electronic transitions rises rapidly. The result is that very complicated spectra can be produced. The complexity of atomic spectra generally reflects the complexity of the source atoms. A good example is the element iron, which contributes several hundred of the Fraunhofer absorption lines seen in the solar spectrum. The many possible transitions of its 26 orbiting electrons yield an extremely rich line spectrum.

Molecular Spectra

A **molecule** is a tightly bound group of atoms held together by interactions among their orbiting electrons—interactions called *chemical bonds*. Much like

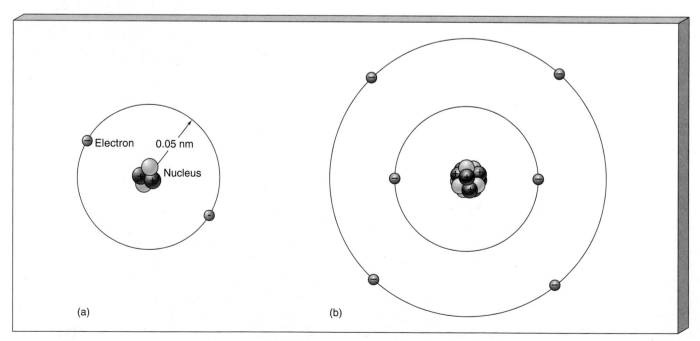

Figure 2.21   (a) A helium atom in its ground state. Two electrons occupy the lowest-energy orbital around a nucleus containing two protons and two neutrons. (b) A carbon atom in its ground state. Six electrons orbit a six-proton, six-neutron nucleus; two of the electrons are in an inner orbital, the other four are at a greater distance from the center.

atoms, molecules can exist only in certain well-defined energy states, and again like atoms, molecules produce emission or absorption spectral lines when they make a transition from one state to another. Because molecules are more complex than atoms, the rules of molecular physics are also much more complex. Nevertheless, as with atomic spectral lines, painstaking experimental work over many decades has determined the precise frequencies (or wavelengths) at which millions of molecules emit and absorb radiation.

In addition to the lines resulting from electron transitions, molecular lines result from two other kinds of changes not possible in atoms: molecules can *rotate*, and they can *vibrate*. Figure 2.22 illustrates these basic molecular motions. Just as with atomic states, only certain rotations and vibrations are allowed by the rules of molecular physics. Whenever a molecule changes its rotational state or its vibrational state, a photon is emitted or absorbed. Spectral lines characteristic of the specific kind of molecule result. These lines are molecular fingerprints, just like their atomic counterparts, enabling researchers to identify and study one kind of molecule to the exclusion of all others.

Molecular lines usually bear little resemblance to the spectral lines associated with their component atoms. For example, Figure 2.23(a) shows the emission spectrum of the simplest molecule known—molecular hydrogen. Notice how different it is from the spectrum of atomic hydrogen shown in part (b).

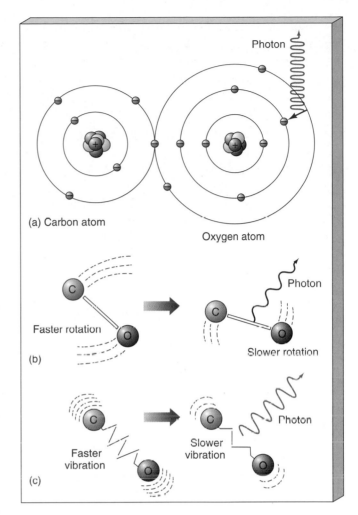

Figure 2.22   Molecules can change in three ways while emitting or absorbing electromagnetic radiation. Sketched here is the molecule carbon monoxide (CO) experiencing (a) a change in electron arrangement, in which an electron in the outermost orbital of the oxygen atom drops to a lower-energy state, (b) a change in rotational state, and (c) a change in vibrational state.

2.7 Spectral-Line Analysis

Astronomers apply the laws of spectroscopy in analyzing radiation from beyond Earth. A nearby star or a distant galaxy takes the place of the light bulb in our previous examples. A galactic cloud or a stellar (or even planetary) atmosphere plays the role of the intervening cool gas. And a spectrograph attached to a telescope replaces our simple prism and detector.

We began this chapter by stating that virtually all we know about planets, stars, and galaxies is gleaned from studies of the light we receive from them, and we have presented some of the ways in which that knowl-

Figure 2.23   The emission spectra of (a) molecular hydrogen and (b) atomic hydrogen.

2-2 MORE PRECISELY

THE DOPPLER EFFECT

Imagine a rocket ship launched from Earth with enough fuel to accelerate to speeds comparable to the speed of light. As the ship's speed increased, a remarkable thing would happen. Passengers looking out from the front of the spacecraft would notice that the light emitted from stars ahead of them would seem to be getting *bluer*, and the greater the ship's speed, the greater the color change would be. Furthermore, stars behind the vessel would seem *redder* than normal. As the spacecraft slowed down and came to rest relative to Earth, all stars would resume their normal appearance. The travelers would conclude that the stars had changed their colors not because of any real change in their physical properties, but because of the spacecraft's own *motion*—movement toward or away from a source of radiation changes the way we perceive that radiation.

Such changes are not restricted to electromagnetic radiation and fast-moving spacecraft. Waiting at a railroad crossing for a train to pass, most of us have had the experience of hearing the pitch of the engine's horn change from high shrill to low blare as the train approaches and then recedes. High-pitched sound has shorter wavelength than low-pitched sound; this change in wavelength is analogous to the color change observed by our spacecraft passengers. A motion-induced change in the observed wavelength of a wave—be it an electromagentic (light) wave or an acoustical (sound) wave—is known as the *Doppler effect*, in honor of Christian Doppler, the nineteenth-century Austrian physicist who first explained it.

To understand how the Doppler effect occurs, imagine a wave moving from the place where it is generated toward an observer who is not moving with respect to the wave source, as shown in the figure below. By counting the number of wave crests passing per unit time, the observer could determine the wavelength of the emitted wave. Suppose now that the wave source begins to move. Because the source moves between the times of emission of one wave crest and the next, successive wave crests in the direction of motion of the source will be seen to be *closer together* than normal, while crests behind the source will be more widely spaced. Thus, an observer in front of the source will measure a *shorter* wavelength than normal, while one behind will see a *longer* wavelength. (The numbers indicate successive wave crests emitted by the source and the location of the source at the instant each wave crest was emitted.)

The greater the relative speed of source and observer, the greater the observed shift. In terms of the net velocity of *recession* between source and observer (so a positive velocity means that the two are moving apart, a negative value that they are approaching), the apparent wavelength (measured by the observer) is always related to the true wavelength (measured by an observer with no motion relative to the source) by

$$\frac{\text{apparent wavelength}}{\text{true wavelength}} = 1 + \frac{\text{recession velocity}}{\text{wave speed}},$$

where the wave speed is the speed of light in the case of electromagnetic radiation. In the above figure, the source is shown in motion. However, the same general statements hold whenever there is any *relative* motion between source and observer.

In the case of electromagnetic radiation, the wave measured by an observer situated in front of

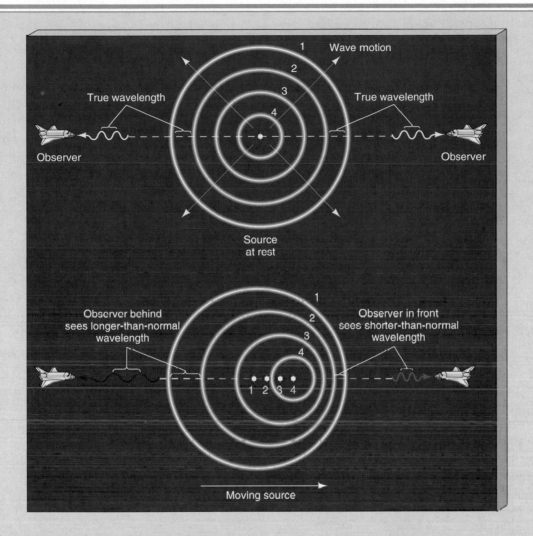

the moving source is said to be *blueshifted*, because blue light has a shorter wavelength than red light. Similarly, an observer situated behind the source will measure a longer-than-normal wavelength—the radiation is said to be *redshifted*. Note that this terminology also holds even for invisible radiation, for which "red" and "blue" have no meaning. Any shift toward shorter wavelengths is called a blueshift, and any shift toward longer wavelengths is called a redshift.

Astronomers can use the Doppler effect to measure the speed of any cosmic object along the line of sight, simply by determining the extent to which its light is red- or blueshifted. The motions of nearby stars and distant galaxies—even the expansion of the universe itself—have all been measured in this way. Motorists stopped for speeding on the highway have experienced a much more down-to-earth application: Police radar measures speed by means of the Doppler effect, as do the radar guns used to clock the velocity of a pitcher's fastball or a tennis player's serve.

edge is obtained. We will not describe in detail the myriad ways in which astronomers extract information from spectroscopic studies of starlight—such a description could fill an entire library of textbooks. Here, we simply list a few of the properties of emitters and absorbers that can be determined by careful analysis of radiation received on (or near) Earth. We will encounter other important examples as our study of the cosmos unfolds.

- The *composition* of an object is determined by matching its spectral lines with the laboratory spectra of known atoms and molecules.

- The *temperature* of an object emitting a continuous spectrum can be measured by matching the overall distribution of radiation—specifically, the wavelength at which the continuous energy emission peaks—with a black-body curve. This measurement is then refined by detailed studies of the intensities of individual spectral lines, the width and strength of which are affected by the temperature of the object producing them.

- The *magnetic field* of an object can be inferred from a characteristic splitting it produces in many spectral lines, when a single line divides into two. Generally speaking, the degree of splitting increases as the magnetic field strengthens.

- The *pressure* of the gas in the emitting region of an object can be measured by its tendency to broaden spectral lines. The greater the pressure, the broader the line.

- The *line-of-sight velocity* of an object is measured by determining the Doppler shift of its spectral lines (see *More Precisely 2-2*). In other words, a set of spectral lines might be recognized as belonging to a particular element, except that they are all offset—blueshifted or redshifted—by the same amount from the expected wavelengths. Interpreting that offset as the result of the Doppler effect yields the emitter's radial velocity relative to the observer.

Given sufficiently sensitive equipment, there is almost no end to the wealth of data contained in starlight. However, deciphering the extent to which each of the foregoing factors influences a spectrum can be a very difficult task. Typically, the spectra of many elements are superimposed on one another, and several competing physical effects are occurring simultaneously, each modifying the spectrum in its own way. The challenge facing astronomers is to unravel the extent to which each mechanism contributes to spectral-line profiles and so obtain meaningful information about the source of the lines.

Chapter Review *www*

Summary

Visible light (p. 45) is a particular type of **electromagnetic radiation** (p. 45). It travels through space in the form of a **wave** (p. 45). A wave is characterized by the **wave period** (p. 46), the length of time taken for one complete cycle; the **wavelength** (p. 46), the distance between successive wave crests; and the wave **amplitude** (p. 46), which measures the size of the disturbance associated with the wave. The wave **frequency** (p. 46) is simply 1 over the wave period—it counts the number of wave crests that pass a given point in one second.

Electrons (p. 47) and **protons** (p. 47) are elementary particles that carry equal and opposite electrical charges. Any electrically charged object is surrounded by an **electric field** (p. 47) that determines the force it exerts on other charged objects. Like gravitational fields, electric fields decrease as the square of the distance from their source. When a charged particle moves, information about that motion is transmitted throughout the universe by the particle's changing electric and **magnetic** (p. 47) fields. The information travels at the **speed of light** (p. 48) in the form of a wave. The phenomenon is known as **electromagnetism** (p. 48).

The **electromagnetic spectrum** (p. 50) consists of (in order of increasing frequency) **radio waves**, **infrared radiation**, **visible light**, **ultraviolet radiation**, **X rays**, and **gamma rays** (p. 45). The **opacity** (p. 52) of Earth's atmosphere—the extent to which it absorbs radiation—varies greatly with the wavelength of the radiation. Only radio waves, some infrared wavelengths, and visible light can penetrate the atmosphere and reach the ground.

The **intensity** (p. 52) of radiation emitted by an object has a characteristic distribution, called a **black-body curve** (p. 52), that depends only on the temperature of the object. **Wien's law** (p. 53) tells us that the wavelength at which the object radiates most energy is directly proportional to its temperature. **Stefan's law** (p. 54) states that the total amount of energy radiated is proportional to the fourth power of the temperature.

Many hot objects emit a **continuous spectrum** (p. 55) of radiation, containing light of all wavelengths. A hot gas may instead produce an **emission spectrum** (p. 58), consisting only of a few well-defined **emission lines** (p. 57) of specific frequencies, or colors. Passing a continuous beam of radiation through cool gas will produce **absorption lines** (p. 58) at precisely the same frequencies as would be present in the gas's emission spectrum. **Kirchhoff's laws** (p. 59) describe the relationships between these different types of spectra. The emission and absorption lines produced by each element are unique—they provide a "fingerprint" of that element in the light it emits or absorbs. A **spectroscope** (p. 55) splits a beam of radiation into its component frequencies and delivers them to a detector as a series of spectral lines; **spectroscopy** (p. 59) is the study of these spectral lines.

Atoms (p. 60) are made up of negatively charged electrons orbiting a positively charged **nucleus** (p. 60) consisting of positively charged **protons** and, with the exception of the hydrogen nucleus, electrically neutral **neutrons** (p. 64). In normal circumstances, the number of orbiting electrons equals the number of protons in the nucleus, and the atom is electrically neutral. The number of protons in the nucleus determines the type of **element** (p. 64) the atom represents.

The **Bohr model** (p. 61) of the hydrogen atom was an early attempt to explain how atoms produce emission and absorption line spectra. An atom has a minimum-energy **ground state** (p. 61), representing its "normal" condition; in this state, the orbiting electron is at its normal distance from the nucleus. An atom is in an **excited state** (p. 61) when the electron occupies an orbital located at a greater than normal distance from the nucleus. When an electron in an atom acquires sufficient energy, it is no longer bound to the atom; an atom which has lost one or more electrons is called an **ion** (p. 61). In the modern view, the electron is envisaged as being spread out in a "cloud" around the nucleus but still having a sharply defined energy.

As an electron moves from one energy level to another in an atom, the difference in the energy between the states is emitted or absorbed in the form of "packets" of electromagnetic radiation—**photons** (p. 62). Because the energy levels have definite energies, the photons also have definite energies that are characteristic of the type of atom involved. For this reason, the energy is said to be **quantized** (p. 61). The energy of a photon determines the frequency, and hence the color, of the light emitted or absorbed.

Molecules (p. 64) are groups of two or more atoms bound together by interactions known as chemical bonds. Molecules can exist only in certain well-defined energy states that obey rules similar to those governing the internal structure of atoms. As with atoms, molecules moving from one energy state to another emit or absorb a characteristic spectrum of radiation that identifies them uniquely.

Self-Test: True or False?

_____ **1.** Light, radio, ultraviolet, and gamma rays are all forms of electromagnetic radiation.

_____ **2.** Sound is a familiar form of electromagnetic wave.

_____ **3.** Electromagnetic waves cannot travel through a perfect vacuum.

_____ **4.** Electromagnetic waves all travel at the same speed, the speed of light.

_____ **5.** Visible light makes up the greatest part of the electromagnetic spectrum.

_____ **6.** Ultraviolet light has the shortest wavelength of any electromagnetic wave.

_____ **7.** A black body emits all its radiation at a single wavelength or frequency.

_____ **8.** A perfect black-body emits exactly as much radiation as it absorbs from outside.

_____ **9.** Emission spectra are characterized by narrow, bright lines of different colors.

_____ **10.** For an emission spectrum produced by a container of hydrogen gas, changing the amount of hydrogen in the container will change the color of the lines in the spectrum.

_____ **11.** In the previous question, changing the gas

in the container from hydrogen to helium will change the color of the lines in the spectrum.

_____ 12. An absorption spectrum appears as a continuous spectrum interrupted by a series of dark lines.

_____ 13. The wavelengths of the emission lines produced by an element are different from the wavelengths of the absorption lines produced by the same element.

_____ 14. The energy of a photon is inversely proportional to the wavelength of the radiation.

_____ 15. An atom is in its ground state when its electrons are in the lowest energy levels (orbitals).

_____ 16. An electron can have any energy within an atom, so long as that energy is above the ground state energy.

_____ 17. An atom can remain in an excited state indefinitely.

_____ 18. Emission and absorption lines correspond to the specific energy differences between orbitals in an atom.

_____ 19. The number of electrons in an atom determines the element it represents.

_____ 20. More than one element or molecule can have the same emission or absorption spectrum.

Self-Test: Fill in the Blank

1. The speed of light is _____ km/s.

2. The _____ of a wave is the distance between any two adjacent wave crests.

3. The hertz (Hz) is a unit used to measure the _____ of a wave.

4. When a charged particle moves, information about this motion is transmitted through space by means of the particle's changing _____ and _____ fields.

5. The visible spectrum ranges from _____ nm to _____ nm in wavelength.

6. Light having a wavelength of 700 nm is perceived to be _____ in color.

7. Earth's atmosphere has low opacity for three forms of electromagnetic radiation. They are _____, _____, and _____.

8. The peak of an object's emitted radiation occurs at a frequency or wavelength determined by the object's _____.

9. Two identical objects have temperatures of 1000 K and 1200 K. It is observed that one of the objects emits roughly twice as much radiation as the other. Which one is it? _____.

10. If an astronomical object is observed to emit X rays, it is reasonable to assume its temperature is very _____.

11. A _____ is a glass wedge that splits light into a spectrum.

12. Black-body radiation is an example of a _____ spectrum.

13. Fraunhofer discovered absorption lines in the spectrum of _____.

14. A continuous spectrum can be produced by a luminous solid, a liquid, or a _____ gas.

15. An absorption spectrum is produced when a _____ gas lies in front of a continuous source.

16. Light behaves both as a wave and as a _____.

17. Protons carry a _____ charge; electrons carry a _____ charge.

18. When moving to a higher energy level in an atom, an electron _____ a photon of a specific energy.

19. When moving to a lower energy level in an atom, an electron _____ a photon of a specific energy.

20. The "specific energy" referred to in the preceding two questions is exactly equal to the energy _____ between the two energy levels the electron moves.

Review and Discussion

1. Define the following wave properties: period, wavelength, amplitude, frequency.

2. What is the relationship between wavelength, wave frequency, and wave velocity?

3. What's so special about *c*?
4. Compare the gravitational force with the electric force.
5. Describe the way in which light radiation leaves a star, travels through the vacuum of space, and finally is seen by someone on Earth.
6. Name the colors that combine to make white light. What is it about the various colors that cause us to perceive them differently?
7. What do radio waves, infrared radiation, visible light, ultraviolet radiation, X rays, and gamma rays have in common? How do they differ?
8. In what regions of the electromagnetic spectrum is the atmosphere transparent enough to allow observations from the ground?
9. What is a black body? What are the characteristics of the radiation it emits?
10. If Earth were completely blanketed with clouds and we couldn't see the sky, could we learn about the realm beyond the clouds? What forms of radiation might penetrate the clouds and reach the ground?
11. Describe how its black-body curve changes as a red-hot glowing coal cools off.
12. What is spectroscopy? Why is it so important to astronomers?
13. Describe the basic components of a spectroscope.
14. What is a continuous spectrum? An absorption spectrum?
15. Why are gamma rays generally harmful to life forms but radio waves generally harmless?
16. In the particle description of light, what is color?
17. Give a brief description of a hydrogen atom.
18. What is the normal condition for atoms? What is an excited atom? What are orbitals?
19. Why do excited atoms absorb and reemit radiation at characteristic frequencies?
20. Suppose a luminous cloud of gas is discovered emitting an emission spectrum. What can be learned about this cloud from this observation?

Problems

1. What is the wavelength of a 100-MHz ("FM 100") radio signal?
2. The black-body emission spectrum of object A peaks in the ultraviolet region of the electromagnetic spectrum, at a wavelength of 200 nm. That of object B peaks in the red region, at 650 nm. According to Wien's law, A is how many times hotter than B? According to Stefan's law, A radiates how much more energy per unit area per second?
3. The Sun has a temperature of almost 6000 K, and its black-body emission peaks at a wavelength of approximately 550 nm. At what wavelength does a star-forming cloud having a temperature of 1000 K radiate most strongly?
4. A 1-nm gamma ray has how many times more energy than a 10-MHz radio photon?
5. How many different photons (that is, photons of different frequencies) can be emitted as a hydrogen atom in the second excited state falls back, directly or indirectly, to the ground state? What about a hydrogen atom in the third excited state?

Projects

1. Locate the constellation Orion. Its two brightest stars are Betelgeuse and Rigel. Which is hotter? Which is cooler? How can you tell? Which of the other stars scattered across the night sky are hot, and which are cool?
2. Use a hand-held spectroscope, available through Learning Technologies, Inc. While in the shade, point the spectroscope at a white cloud or white piece of paper that is in direct sunlight. Look for the absorption lines in the Sun's spectrum. Note their wavelength from the scale inside the spectroscope. Compare your list to the Fraunhofer lines given in many physics, astronomy, or chemistry reference books.

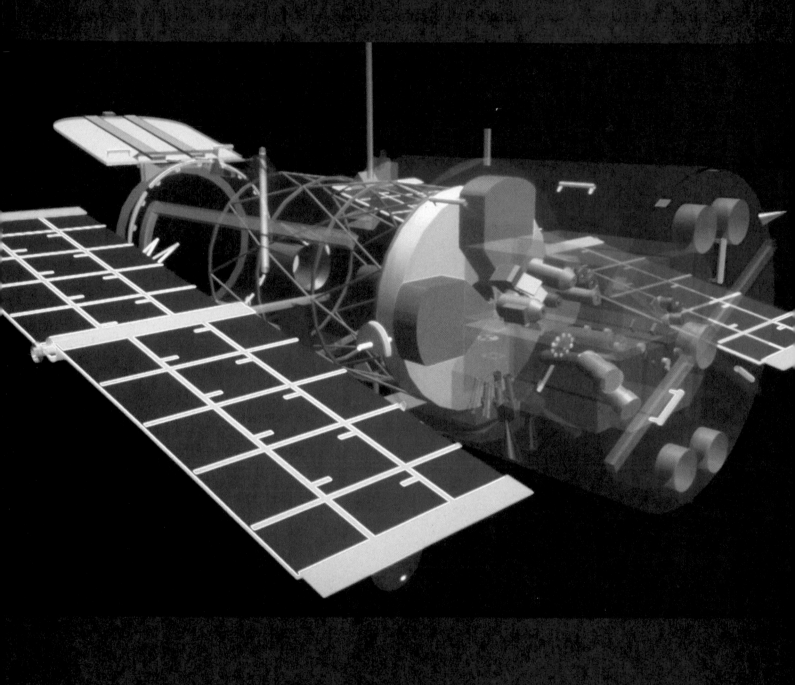

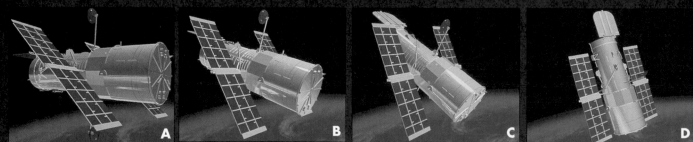

3 Telescopes
The Tools of Astronomy ✦www✦

◀ (Opposite page, background) This semi-transparent illustration shows some of the main features of the Hubble Space Telescope. The large blue disk at center of the spacecraft is the primary mirror, and the red gadgets to its rear are the sensors that guide the pointing of the telescope. The open aperture door is at upper left. The huge solar panels are shown in yellow-checkered blue at left and partly obscured at right. Looking inside the aft bay of the vehicle, we can see key components of each of the science instruments—the spectrometers are shown in copper and blue (in the foreground), the cameras in pink, green and lavender (mostly in the background).

(Insets A, B, C, and D) The four small insets are computer-rendered views of Hubble in orbit. Despite having the size of a city bus, Hubble is designed to move in space with the grace of a prima ballerina. All these illustrations are taken from video animations made by the astronomy artist and animator Dana Berry.

LEARNING GOALS

Studying this chapter will enable you to:

1 Sketch and describe the basic designs of the major types of optical telescopes.

2 Explain why very large telescopes are needed for most astronomical studies and specify the advantages of reflecting telescopes for astronomical use.

3 Describe how Earth's atmosphere affects astronomical observations and discuss some of the current efforts to improve ground-based astronomy.

4 Discuss the advantages and disadvantages of radio astronomy.

5 Explain how interferometry can enhance the usefulness of radio observations.

6 Discuss the advantages, limitations, and chief uses of infrared, ultraviolet, high-energy, and full-spectrum astronomies.

7 Explain why it is important to make astronomical observations in many different regions of the electromagnetic spectrum.

At its heart, astronomy is an observational science. More often than not, observations of cosmic phenomena precede any clear theoretical understanding of their nature. As a result, our detecting instruments—our telescopes—have evolved to observe as broad a range of wavelengths as possible. Until the middle of the twentieth century, telescopes were limited to visible light. Since then, however, technological advances have expanded our view of the universe to all regions of the electromagnetic spectrum. Some telescopes are sited on Earth, others must be placed in space, and design considerations vary widely from one part of the spectrum to another. Whatever the details of its construction, however, a telescope is a device the basic purpose of which is to collect electromagnetic radiation and deliver it to a detector for detailed study.

3.1 Optical Telescopes

1 In essence, a **telescope** is a "light bucket"—a device whose primary function is to capture as much radiation as possible from a given region of the sky and concentrate it into a focused beam for analysis. An *optical* telescope is one designed to collect wavelengths visible to the human eye. Optical telescopes have a long history, reaching back to the early seventeenth century. They are probably also the best-known type of telescope, so it is fitting that we begin our study of astronomical hardware with these devices. Later we will turn our attention to telescopes designed to capture and analyze radiation in other, *invisible* regions of the electromagnetic spectrum.

Reflecting and Refracting Telescopes

Optical telescopes fall into two basic categories—*reflectors* and *refractors*. Figure 3.1 shows how a **reflecting** telescope uses a curved mirror to gather and concentrate a beam of light. The mirror, usually called the *primary mirror* because telescopes sometimes contain more than one mirror, is constructed so that all light rays arriving parallel to its axis (the imaginary line through the center of and perpendicular to the mirror), regardless of their distance from that axis, are reflected to pass through a single point, called the *focus*. The distance between the primary mirror and the focus is the *focal length*. In astronomical contexts, the focus of the primary mirror is referred to as the **prime focus**.

A **refracting telescope** uses a lens to focus the incoming light. **Refraction** is the bending of a beam of light as it passes from one transparent medium (for example, air) into another (such as glass). For example, consider how a pencil half immersed in a glass of water looks bent. The pencil is straight, of course, but the light by which we see it is bent—refracted—as that light leaves the water and enters the air. When that light then enters our eyes, we perceive the pencil as

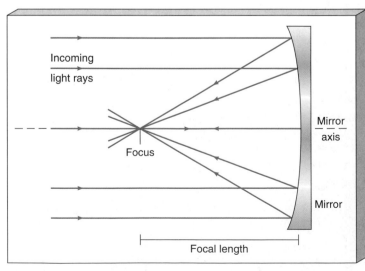

Figure 3.1 A curved mirror can be used to focus to a single point all rays of light arriving parallel to the mirror axis. Light rays traveling along the axis are reflected back along the axis, as indicated by the arrowheads pointing in both directions. Off-axis rays are reflected through greater and greater angles the farther they are from the axis, so that they all pass through the same point—the focus.

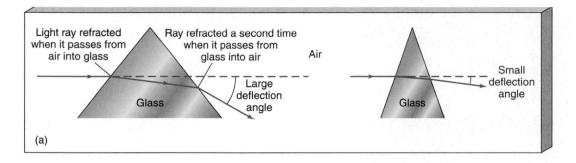

Figure 3.2 (a) Refraction by a prism changes the direction of a light ray by an amount that depends on the angle between the prism's faces. (b) A lens can be thought of as a series of prisms. A light ray traveling along the axis of a lens is unrefracted as it passes through the lens. Parallel rays arriving at progressively greater distances from the axis are refracted by increasing amounts, in such a way that all are focused to a single point.

being bent. Figure 3.2(a) illustrates the process and shows how a prism can be used to change the direction of a beam of light. As illustrated in Figure 3.2(b), we can think of a lens as a series of prisms combined in such a way that all light rays striking the lens parallel to the axis are refracted to pass through the focus.

Optical telescopes are often used to make images of their fields of view. Figure 3.3 illustrates how this is accomplished, in this case by the primary mirror in a reflecting telescope. Light from a distant object reaches the telescope as nearly parallel rays. A ray of light entering the instrument parallel to the mirror axis is reflected through the focus. Light from a slightly different direction—that is, at a slight angle to the axis—is focused to a slightly different point. In this way, an image is formed near the focus. Each point in the image corresponds to a different angle in the field of view.

The images produced by large telescopes are quite small—the image of the entire field of view may be as little as 1 centimeter across. Often, the image is magnified with a lens known as an *eyepiece*. Figure 3.4(a) shows the basic design of a simple reflecting telescope, illustrating how a small secondary mirror and an eyepiece are used to view the image. Figure 3.4(b) shows

how a refracting telescope accomplishes the same function.

The two telescope designs shown in Figure 3.4 achieve the same result—light from a distant object is captured and focused to form an image. However, as telescope *size* has steadily increased over the years (for reasons to be discussed in Section 3.2), a number of important factors have tended to favor reflecting instruments over refractors:

- The fact that light must pass through the lens is a major disadvantage of refracting telescopes. Just as a prism separates white light into its component colors, the lens in a refracting telescope focuses red and blue light differently. This deficiency is known as *chromatic aberration*. Careful design and choice of materials can largely correct this problem, but it is very difficult to eliminate entirely. Mirrors do not suffer from this defect.

- As light passes through the lens, some of it is absorbed by the glass. This absorption is a relatively minor problem for visible radiation, but it can be severe for infrared and ultraviolet observations because glass blocks most of the radiation

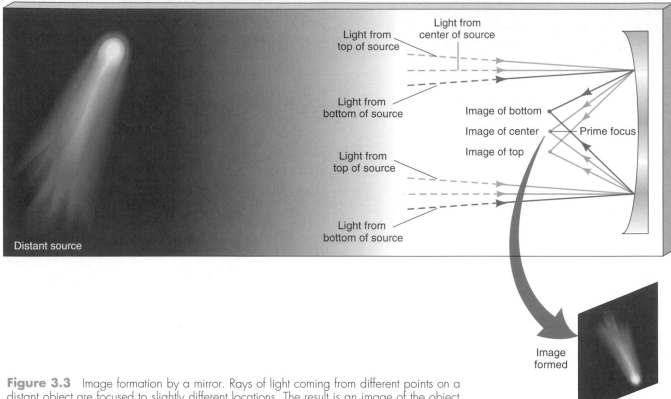

Figure 3.3 Image formation by a mirror. Rays of light coming from different points on a distant object are focused to slightly different locations. The result is an image of the object formed around the focus. Notice that the image is inverted (that is, upside down).

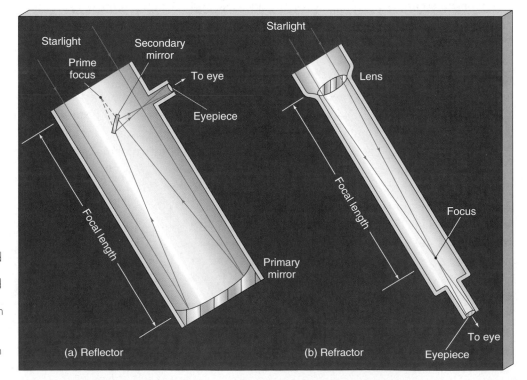

Figure 3.4 Comparison of (a) reflecting and (b) refracting telescopes. Both types are used to gather and focus cosmic radiation that is either observed by human eyes or recorded on photographs or in computers. In both types, the image formed at the focus is viewed with a small magnifying lens called an eyepiece.

coming from those regions of the electromagnetic spectrum. This problem obviously does not affect mirrors.

- A large lens can be quite heavy. Because it can be supported only around its edge (so as not to block the incoming radiation), the lens tends to deform under its own weight. A mirror does not have this drawback because it can be supported over its entire back surface.

- A lens has two surfaces that must be accurately machined and polished, which can be a difficult task. A mirror has only one surface to be dealt with.

For these reasons, all large modern optical telescopes are reflecting telescopes. The world's largest refracting telescope, installed in 1897 at the Yerkes Observatory in Wisconsin and still in use today, has a lens diameter of 1 m. By contrast, the largest modern reflecting telescopes have mirror diameters in the 10-m range, and even larger instruments are on the way.

Telescope Design

Figure 3.5 shows some basic reflecting telescope designs. Radiation from a star enters the instrument, passes down the main tube, strikes the primary mirror, and is reflected back toward the prime focus, near the top of the tube. Sometimes astronomers place their recording instruments at the prime focus. However, it can be very inconvenient, or even impossible, to suspend bulky pieces of equipment there. More often, the light is intercepted on its path to the focus by a secondary mirror and redirected to a more convenient location, as in Figures 3.5(b) through (d).

In a **Newtonian telescope** (named after Isaac Newton, who invented this design), the light is intercepted by a secondary mirror before it reaches the prime focus and deflected 90°, usually to an eyepiece at the side of the instrument (Figure 3.5b). This is a popular design for smaller reflecting telescopes, such as those used by amateur astronomers.

Alternatively, astronomers may choose to work on a rear platform where they can use detecting equipment too heavy or delicate to hoist to the prime focus. In this case, the light reflected by the primary mirror toward the prime focus is intercepted by a secondary mirror, which reflects the light back down the tube and through a small hole at the center of the primary mirror (Figure 3.5c). This arrangement is known as a **Cassegrain telescope** (after Guillaume Cassegrain, a French lensmaker). The point behind the primary mir-

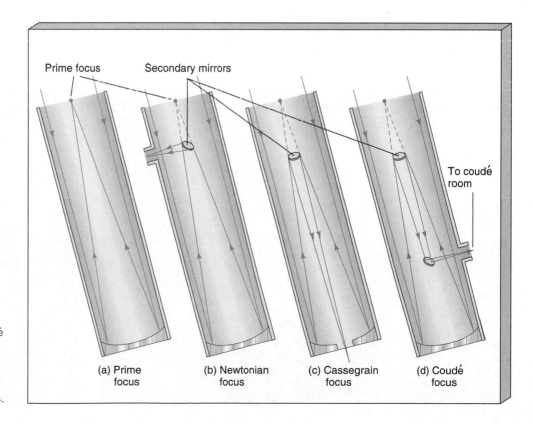

Figure 3.5 Four reflecting telescope designs: (a) prime focus, (b) Newtonian focus, (c) Cassegrain focus, and (d) coudé focus. Each design uses a primary mirror at the bottom of the telescope to capture radiation, which is then directed along different paths for analysis.

ror where the light from the star finally converges is called the *Cassegrain focus.*

In a variant on the Cassegrain design, light is first reflected by the primary mirror toward the prime focus and reflected back down the tube by a secondary mirror. A third, much smaller mirror then reflects the light into an environmentally controlled laboratory (Figure 3.5d). Known as the *coudé* room (from the French word for "bent"), this laboratory is separate from the telescope, enabling astronomers to use very heavy and finely tuned equipment that could not possibly be

lifted to either the prime focus or the Cassegrain focus. The light path to the coudé room lies along the axis of the telescope's mount—that is, the axis around which the telescope rotates as it tracks objects across the sky—so that the light path does not change as the telescope moves.

To illustrate some of these points, Figure 3.6 depicts the Hale 5-m-diameter optical telescope on California's Palomar Mountain. As the size of the person drawn in the observer's cage at the prime focus indicates, this is indeed a very large telescope. In fact,

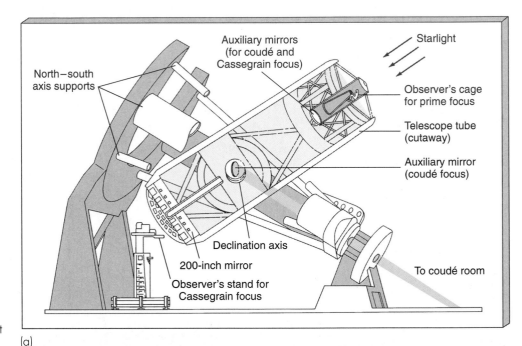

Figure 3.6 (a) An artist's illustration of the 5-m-diameter Hale reflecting telescope on Palomar Mountain in California. (b) A photograph of the telescope. (c) Astronomer Edwin Hubble in the observer's cage at the Hale prime focus.

(a)

(b)

(c)

for almost three decades after its dedication in 1948, the Hale telescope was the largest in the world. It has been at or near the forefront of astronomical research for much of the last half century. Observations can be made at the prime, the Cassegrain, or the coudé focus, depending on the needs of the user.

3.2 Telescope Size

2 Astronomers generally prefer large telescopes over small ones, for two main reasons. The first has to do with the amount of light a telescope can gather—its *light-gathering power*. The second is related to the amount of detail that can be seen—the telescope's *resolving power*.

Light-Gathering Power

One important reason for using a larger telescope is that it has a greater **collecting area**, which is the area capable of gathering radiation. The larger the telescope's reflecting mirror (or refracting lens), the more light it collects and the easier it is to measure and study an object's radiative properties. Astronomers spend a large fraction of their time observing very distant—and hence very *faint*—cosmic sources. In order to make detailed observations of such objects, very large telescopes are essential. Figure 3.7 illustrates the effect of increasing telescope size by comparing images of the Andromeda Galaxy taken with two different instruments.

The observed brightness of an astronomical object is directly proportional to the area of the telescope's mirror and hence to the *square* of the mirror diameter. Thus, a 5-m telescope produces an image 25 times brighter than a 1-m instrument because a 5-m mirror has $5^2 = 25$ times the collecting area of a 1-m mirror. We can also think of this relationship in terms of the time required for a telescope to collect enough energy to create a recognizable image on a photographic plate. A 5-m telescope produces an image 25 times faster than a 1-m device because it gathers energy at a rate 25 times greater. Put another way, a 1-hour time exposure with a 1-m telescope is roughly equivalent to a 2.4-minute time exposure with a 5-m instrument.

Currently, the largest optical telescopes are the twin Keck instruments atop Mauna Kea in Hawaii (Figure 3.8), operated jointly by the California Institute of Technology and the University of California. Each telescope combines 36 hexagonal 1.8-m mirrors

R I V U X G

Figure 3.7 Effect of increasing light-gathering power on an image of the Andromeda Galaxy. Both photographs had the same exposure time, but the bottom image was taken with a telescope twice the size of that used to make the top image. Fainter detail can be seen as the diameter of the telescope mirror increases because larger telescopes are able to collect more photons per unit time.

into the equivalent collecting area of a single 10-m reflector. The first Keck telescope became fully operational in 1992; the second was completed in 1996.

Resolving Power

A second advantage of large telescopes is their finer **angular resolution**. In general, *resolution* refers to the ability of any device, such as a camera or a telescope, to form distinct, separate images of objects lying close

(a)

(b)

Figure 3.8 (a) The world's highest ground-based observatory, at Mauna Kea, Hawaii, is perched atop an extinct volcano more than 4 km above sea level. Among the domes visible in the picture are those that house the Canada–France–Hawaii 3.6-m telescope, the 2.2-m telescope of the University of Hawaii, Britain's 3.8-m infrared facility, and the twin Keck telescopes. The thin air at this high-altitude site guarantees less atmospheric absorption of incoming radiation and hence a clearer view than at sea level, but the air is so thin that astronomers must occasionally wear oxygen masks while working. (b) The 10-m mirror in the first Keck telescope. Note the technician in orange coveralls at center.

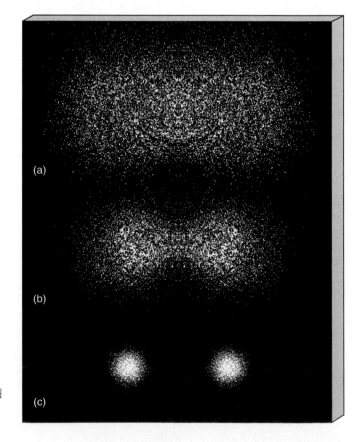

(a)

(b)

(c)

Figure 3.9 Two comparably bright light sources become progressively clearer when viewed at finer and finer angular resolution. (a) When the angular resolution is much poorer than the separation of the objects, the objects appear as a single fused image. (b) and (c) As the resolution improves, the two sources become discernible as separate objects.

together in the field of view. The finer the resolution, the better we can distinguish the objects and the more detail we can see. In astronomy, where we are always concerned with angular measurement (see Section P.2), "close together" means "separated by a small angle on the sky," so angular resolution is the factor that determines our ability to see fine structure. Figure 3.9 illustrates how the appearance of two objects— stars, say—might change as the angular resolution of our telescope varies. Figure 3.10 illustrates the result of increasing resolving power with views of the Andromeda Galaxy at several different resolutions.

One important factor limiting a telescope's resolution is **diffraction**, the tendency of light to "bend" around corners (Figure 3.11). Because of diffraction, when a parallel beam of light enters a telescope, the rays spread out slightly, making it impossible to focus the beam to a sharp point, even with a perfectly constructed mirror. Diffraction introduces a certain "fuzziness," or loss of resolution, into the system. The degree of fuzziness—the minimum angular separation that can be distinguished—determines the angular resolution of the telescope.

The amount of diffraction is proportional to the wavelength of the radiation divided by the diameter of the telescope mirror. For a given telescope size, therefore, the blurring effects of diffraction increase in proportion to the wavelength used. Observations in the infrared or radio range are often severely limited by diffraction. For example, in an otherwise perfect observing environment, the best possible angular resolution of blue light (wavelength 400 nm) using a 1-m telescope is about 0.1″. This quantity is known as the *diffraction-limited resolution* of the telescope. For the same

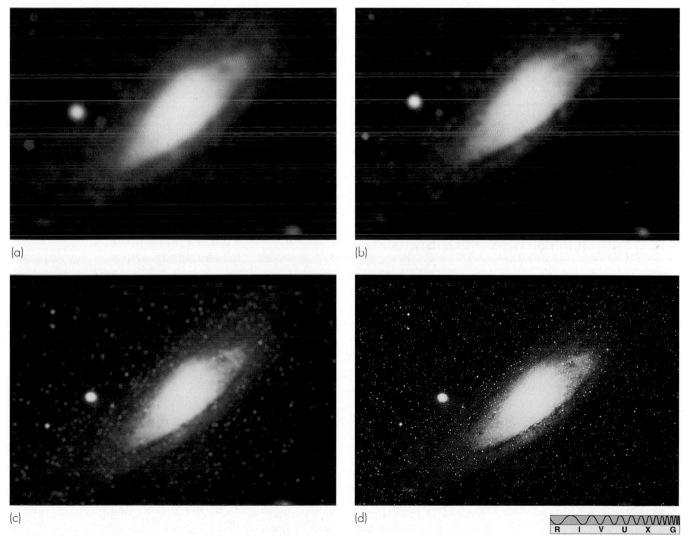

(a)

(b)

(c)

(d)

R I V U X G

Figure 3.10 Detail becomes clearer in the Andromeda Galaxy as the angular resolution is improved some 600 times, from (a) 10′, to (b) 1′, (c) 5″, and (d) 1″.

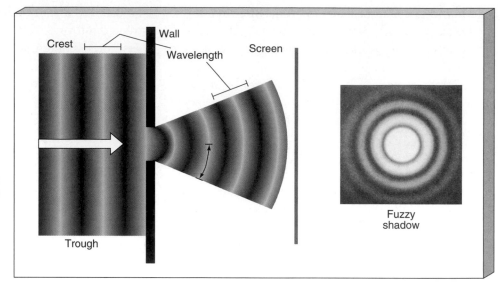

Figure 3.11 A wave passing through a gap is diffracted through an angle that depends on the ratio of the wavelength of the wave to the size of the gap. The longer the wavelength and/or the smaller the gap, the greater the angle through which the wave is diffracted.

telescope used in the near infrared range, at a wavelength of 10,000 nm, the best resolution obtainable is only 2.5″.

For light of a given wavelength, large telescopes produce less diffraction than small ones. A 5-m telescope observing in blue light has a diffraction-limited resolution of 0.02″, five times finer than that of the 1-m telescope just discussed. A 0.1-m telescope has a diffraction limit of 1, and so on. For comparison, the angular resolution of the human eye in blue light is about 0.5′.

3.3 High-Resolution Astronomy

3 Even large telescopes have their limitations. For example, according to the discussion in the preceding section, the 5-m Hale telescope should have an angular resolution of around 0.02″. In practice, however, it cannot do better than about 1″. In fact, apart from instruments using special techniques developed to examine some particularly bright stars, *no ground-based optical telescope built before 1990 can resolve astronomical objects to much better than 1″.* The reason is Earth's turbulent atmosphere, which blurs the image even before the light reaches our instruments. In recent years great strides have been made in overcoming this obstacle. Telescopes have been placed above the atmosphere, and computers are playing an increasingly important role in both telescope operation and image processing.

Atmospheric Blurring

As we observe a star, atmospheric turbulence produces continuous small changes in the optical properties of the air between the star and our telescope (or eye). The light from the star is refracted slightly, again and again, and the stellar image dances around on the detector (or retina). This is the cause of the well-known "twinkling" of stars. It occurs for the same basic reason that objects appear to shimmer when viewed across a hot roadway on a summer day.

On a good night at the best observing sites, the maximum deflection produced by the atmosphere is slightly less than 1″. Consider taking a photograph of a star under such conditions. After a few minutes' exposure time (long enough for the intervening atmosphere to have undergone many small, random changes), the image of the star has been smeared out over a roughly circular region 1″ or so in diameter. Astronomers use the term **seeing** to describe the effects of atmospheric turbulence. The circle over which a star's light (or the light from any other astronomical source) is spread is called the seeing disk. Figure 3.12 illustrates the formation of the **seeing disk** for a small telescope.

To achieve the best possible seeing, telescopes are sited on mountaintops (to get above as much of the atmosphere as possible) in locations where the atmosphere is known to be fairly stable and relatively free of dust, moisture, and light pollution from cities. In the continental United States, these sites tend to be in the

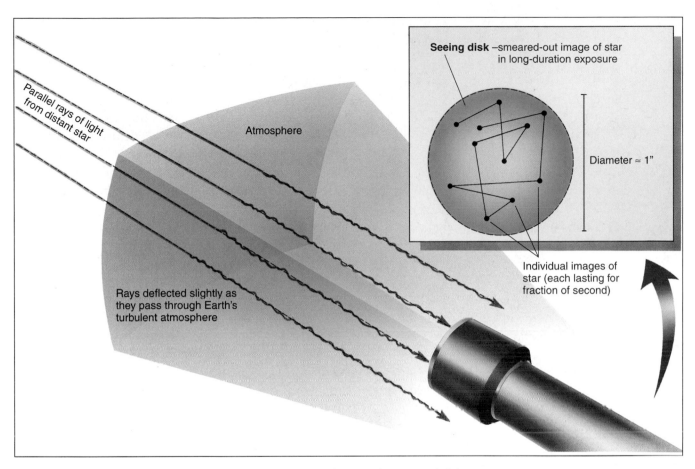

Figure 3.12   Individual photons from a distant star strike a telescope detector at slightly different locations because of turbulence in Earth's atmosphere. Over time, the photons cover a roughly circular region on the detector, and even the pointlike image of a star is recorded as a small disk, called the seeing disk.

desert Southwest. The U.S. National Observatory for optical astronomy in the Northern Hemisphere, completed in 1973, is located high on Kitt Peak near Tucson, Arizona. The site was chosen because of its many dry, clear nights. Seeing of 1″ from such a location is regarded as good, and seeing of a few arc seconds is tolerable for many purposes. Even better conditions are found on Mauna Kea in Hawaii (Figure 3.8) and at Cerro Tololo and La Silla in the Andes Mountains of Chile (Figure 3.13), which is why many large telescopes have recently been constructed at those two exceptionally clear sites.

Figure 3.13   Located in the Andes Mountains of Chile, the European Southern Observatory at La Silla is run by a consortium of European nations. Numerous domes house optical telescopes of different sizes, each carrying various pieces of support equipment, making this one of the most versatile observatories south of the equator.

3-1 INTERLUDE

THE HUBBLE SPACE TELESCOPE

The *Hubble Space Telescope* (HST) is the largest, most complex, most sensitive observatory ever deployed in space. At over $3 billion (including the cost of two missions to service and refurbish the system), it is also the most expensive scientific instrument ever constructed. Built jointly by NASA and the European Space Agency, *HST* is designed to allow astronomers to probe the universe with at least 10 times finer resolution and with some 30 times greater sensitivity to light than existing Earth-based devices. At the heart of *HST* is a 2.4-m-diameter mirror designed to capture optical, ultraviolet, and infrared radiation before it reaches Earth's murky atmosphere. The accompanying figure shows the telescope being lifted out of the cargo bay of the space shuttle *Discovery* in the spring of 1990.

The telescope reflects light from its primary mirror back to a smaller, 0.3-m secondary mirror, which sends the light through a hole in the primary mirror and into the aft bay of the spacecraft (see the chapter opening illustration). There, any of five major scientific instruments wait to analyze the incoming radiation. Most of these instruments are about the size of a telephone booth, including two cameras to image various regions of the sky, two spectrographs to split the radiation into its component wavelengths, and a group of fine-guidance sensors to measure the positions of stars in the sky. Some of these instruments have since been upgraded or replaced by NASA astronauts.

Soon after launch, astronomers discovered that the telescope's primary mirror had been polished to the wrong shape. The mirror is too flat by 2 μm, about 1/50 the width of a human hair. Even though it is the smoothest mirror ever made, this imperfection makes it impossible to focus light as well as expected. This optical flaw (known as *spherical aberration*) meant that *HST* was not as sensitive as designed, although it could still see many objects in the universe with unprecedented resolution. In late 1993, astronauts aboard the space shuttle *Endeavour* visited *HST* and succeeded in repairing some of its ailing equipment. They replaced *Hubble*'s gyroscopes to help the telescope point more accurately and installed sturdier versions of the solar arrays that power the telescope's electronics. They also partly corrected Hubble's flawed vision by inserting an intricate set of small mirrors (each about the size of a coin) to compensate for the faulty primary mirror—in much the same way we use eyeglasses or contact lenses to help humans see better.

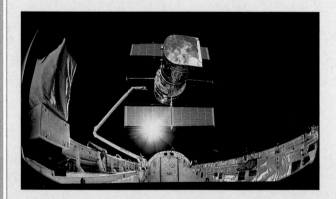

Following the 1993 repair mission, *Hubble*'s resolution is close to the original design specifications, and the telescope has regained much of its lost sensitivity, enabling it to see very faint objects. In early 1997, a second servicing mission replaced several instruments with more sensitive models, performed maintenance on the satellite's fine-guidance system, and upgraded some of the telescope's data systems. The next service mission is scheduled for 1999.

A good example of *Hubble*'s scientific capabilities today can be seen by comparing the two images of the spiral galaxy M100, shown below. On the left is perhaps the best ground-based photograph of this beautiful galaxy, showing rich detail and color in its spiral arms. On the right is an *HST* image showing improvement in both resolution and sensitivity. (The chevron-shaped field of view is caused by the corrective optics inserted into the telescope; an

additional trade-off is that *Hubble*'s field of view is smaller than those of ground-based telescopes.)

Given that *Hubble* was so expensive to build, was meant to be the flagship of a whole new generation of NASA spacecraft, and was greeted with such public fanfare, it is perhaps understandable that the news media sensationalized so many aspects of the telescope's problems and of the repair mission. The bottom line, however, is that *HST* was never really "broken," nor is it now completely "fixed." Despite the fact that it still does not operate as originally designed, *Hubble* has been a superb telescope since its deployment in space in 1990, and it remains so today. In fact, especially due to the heroic efforts of the astronaut repair crew, *Hubble* is probably the best telescope built by humans to date. Many spectacular examples of its remarkable data appear throughout this book.

R I V U X G

One way to overcome the limitations the atmosphere imposes on ground-based observations is to place a telescope above the atmosphere, in Earth orbit. Without atmospheric blurring, resolution close to the diffraction limit can be achieved, subject only to the engineering restrictions of building or placing large structures in space. The *Hubble Space Telescope* (*HST*; named for one of America's most notable astronomers, Edwin Hubble) was placed in orbit by NASA's space shuttle *Discovery* in 1990. This telescope has a 2.4-m mirror that has a diffraction limit of only 0.05″, giving astronomers a view of the universe as much as 20 times sharper than that normally available from even much larger ground-based instruments. (See *Interlude 3-1*.)

Image Processing

Computers play an important role in observational astronomy. Most large telescopes today are controlled either by computers or by operators who rely heavily on computer assistance, and images and data are recorded in a form that can be easily read and manipulated by computer programs.

It is becoming rare for photographic equipment to be used as the primary means of data acquisition at large observatories. Instead, electronic detectors known as **charge-coupled devices**, or **CCDs**, are in widespread use. Their output goes directly to a computer. A CCD (Figure 3.14) consists of a wafer of silicon divided into a two-dimensional array of many tiny picture elements, known as *pixels*. When light strikes a pixel, an electric charge builds up on the device. The amount of charge is directly proportional to the number of photons striking each pixel—in other words, to the intensity of the light at that point. The charge buildup is monitored electronically, and a two-dimensional image is obtained. A CCD is typically a few square centimeters in area and may contain several million pixels, generally arranged on a square grid. As the technology improves, both the areas of CCDs and the number of pixels they contain are steadily increasing. Incidentally, the technology is not limited to astronomy—many home video cameras contain CCD chips similar in basic design to those in use at the great astronomical observatories of the world.

CCDs have two important advantages over photographic plates, which were the staple of astronomers for over a century. First, CCDs are much more *efficient* than photographic plates, recording as many as 75 percent of the photons striking them, compared with less than 5 percent for photographic methods. This means that a CCD image can show objects ten to twenty times fainter than can a photograph made using the same telescope and the same exposure time. The high efficiency of CCDs also means they can record the same level of detail in less than a tenth of the time required by photographs, or record that detail with a much smaller telescope. Second, CCDs produce a faithful representation of an image in a digital format that can be placed directly on magnetic tape or disk, or even sent across a computer network to an observer's home institution.

Computers are also widely used to reduce *background noise* in astronomical images. Noise is anything that corrupts the integrity of a message, such as static on an AM radio or "snow" on a television screen. With the aid of high-speed computers, the background noise in the raw image from a telescope can be greatly

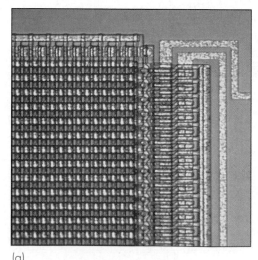

Figure 3.14 A charge-coupled device consists of hundreds of thousands, or even millions, of tiny light-sensitive cells called pixels, usually arranged in a square array. Light striking a pixel causes an electrical charge to build up on it. By electronically reading out the charge on each pixel, a computer can reconstruct the pattern of light—the image—falling on the chip. (a) Detail of a CCD array. (b) A CCD chip mounted for use at the focus of a telescope.

(a)

(b)

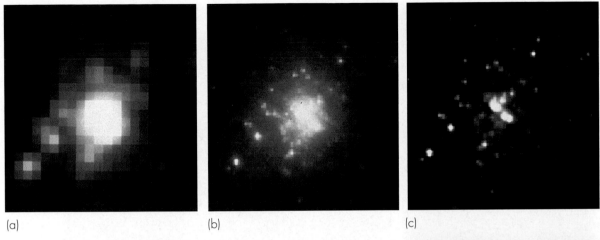

(a)   (b)   (c)

Figure 3.15 (a) A ground-based view of the star cluster R136, a group of stars in the Large Magellenic Cloud (a nearby galaxy). (b) The raw image of R136 as seen by the *Hubble Space Telescope* in 1990, before the repair mission. (c) The same image after computer processing that partly compensated for imperfections in the mirror. (d) The same region as seen by the repaired *HST* in 1994.

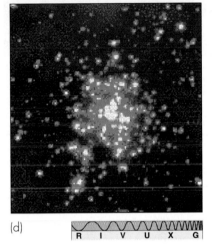

(d)   R I V U X G

reduced, allowing astronomers to see features that would otherwise remain hidden.

The noise corrupting telescopic images has many causes. In part, it results from faint, unresolved sources in the telescope's field of view and from light scattered into the line of sight by Earth's atmosphere. It can also be caused by electronic "hiss" within the detector. Whatever the origin of noise, its characteristics can be determined (for example, by observing a part of the sky where there are no known sources of radiation) and its effects partially removed.

Using computer processing, astronomers can compensate for known instrumental defects and even correct some effects of bad seeing. In addition, the computer can often carry out many of the tedious and time-consuming chores that must be performed before an image or spectrum reaches its final form. Figure 3.15 illustrates how computerized image-processing techniques were used to correct for known instrumental problems in the *Hubble Space Telescope*, allowing much of the planned resolution of the telescope to be recovered even before its repair in 1993.

New Telescope Design

An exciting development promising to bring about striking improvements in the resolution of ground-based optical telescopes takes these ideas of computer control and image processing one stage further. If an image can be analyzed while the light is still being collected, a process that can take many minutes (and even hours in some cases), it is possible to adjust the telescope from moment to moment to reduce the effects of mirror distortion, temperature changes, and bad seeing. In principle, the telescope might even come close to its theoretical (diffraction-limited) resolution.

Some of these techniques, collectively known as **active optics**, are in use in the New Technology Telescope (NTT) at the European Southern Observatory in Chile (indicated on Figure 3.13). This 3.5-m instrument, employing the latest in real-time telescope controls, achieves resolution of about 0.5" by making minute modifications to the tilt of the mirror as its temperature and orientation change, thus maintaining the best possible focus at all times. From its very first observing run, NTT became the highest-resolution optical telescope on Earth (Figure 3.16). The Keck 10-m instruments employ similar methods and may ultimately achieve resolution as fine as 0.25".

An even more ambitious undertaking is known as **adaptive optics**. This technique deforms the shape of

(a)

(b)

R I V U X G

Figure 3.16 Infrared images of part of the star cluster R136 contrast the resolution obtained (a) without and (b) with an active optics system.

(a)

Figure 3.17 (a) Until mid-1991, the Starfire Optical Range at Kirtland Air Force Base in New Mexico was one of the U.S. Air Force's most closely guarded secrets. Here, beams of laser light probe the atmosphere above the system's 1.5-m telescope, allowing minute computer-controlled changes to be made to the mirror surface thousands of times each second. (b) The improvement in seeing produced by such systems can be dramatic, as seen in these images acquired at another military observatory, atop Mt. Haleakala in Maui, Hawaii, employing similar technology. The uncorrected image (left) of the bright star Procyon is a blur spread over several arc seconds. With adaptive optics applied (right), the resolution is improved to 0.2".

R I V U X G

(b)

the mirror's surface, under computer control, while the image is being exposed. The intent is to undo the effects of atmospheric turbulence. In the experimental system shown in Figure 3.17(a), lasers probe the atmosphere above the telescope, returning information about the air's swirling motion to a computer that modifies the mirror thousands of times per second to compensate for poor seeing. Adaptive optics presents formidable theoretical and technological problems, but the rewards are so great that the field is presently the subject of intense research. Recently declassified SDI ("Star Wars") technology has provided an enormous boost to this effort. Already, impressive improvements in image quality have been obtained (Figure 3.17b). In the next decade, it may well be possible to achieve with a large ground-based telescope the kind of resolution presently attainable only from space.

3.4 Radio Astronomy

4 In addition to the visible radiation that penetrates Earth's atmosphere on a clear day, radio radiation also reaches the ground. Indeed, as indicated in Figure 2.8, the radio window in the electromagnetic spectrum is far wider than the optical window). ∞ (Sec. 2.3) Because the atmosphere is no hindrance to long-wavelength radiation, astronomers have built many ground-based **radio telescopes** capable of detecting cosmic radio waves. These devices have all been constructed since the 1950s—radio astronomy is a much younger subject than optical astronomy.

Essentials of Radio Telescopes

Figure 3.18 shows a typical radio telescope, the large 43-m-diameter telescope located at the National Radio Astronomy Observatory in West Virginia. Most radio telescopes are built in basically the same way as reflecting optical telescopes, although radio telescopes are generally much bigger. They have a large, horseshoe-shaped mount supporting a huge, curved metal dish that serves as the collecting area. The dish captures cosmic radio waves and reflects them to the focus, where a receiver detects the signals and channels them to a computer. Conceptually, the operation of a radio telescope is similar to the operation of an optical reflector with the detecting instruments placed at the prime focus (Figure 3.5a). However, unlike optical

Figure 3.18 The 43-m-diameter radio telescope at the National Radio Astronomy Observatory in Green Bank, West Virginia.

instruments, which can detect all visible wavelengths simultaneously, radio detectors normally register only a narrow band of wavelengths at any one time. To observe radiation at another radio frequency, we must retune the equipment, much as we tune a television set to a different channel.

Large radio telescopes are very sensitive and so can detect extremely faint radio sources. Their angular resolution is generally poor compared with that of their optical counterparts, however, despite the enormous size of many radio dishes. It is not Earth's atmosphere that is to blame—the radio wavelengths normally studied pass through air without any significant distortion. The problem is that the typical wavelengths of radio waves are about a million times longer than those of visible light, and these longer wavelengths impose a corresponding crudeness in angular resolution because of the effects of diffraction. Recall from Section 3.2 that the longer the wavelength, the greater the diffraction.

The best angular resolution obtainable with a single radio telescope is about 10″ (for the largest instruments operating at millimeter wavelengths), which is at least 10 times coarser than the capabilities of the largest optical mirrors. The resolution in radio instruments varies widely, depending on the wavelength being observed. For example, the 43-m telescope shown in Figure 3.18 can achieve resolution of about

1′ when receiving 1-cm radio waves. However, it was designed to operate most efficiently (that is, it is most sensitive to radio signals) at wavelengths closer to 5 cm, where the resolution is only about 6′.

Radio telescopes are large in part because that is the only way they can achieve good resolution. But, as with optical telescopes, here is another reason: light-gathering power. The amount of energy arriving at Earth in the form of radio radiation is extremely small. In fact, the total amount of radio energy received by Earth's entire surface is less than a trillionth of a watt. Compare this with the roughly 10 *million* watts our planet's surface receives in the form of infrared and visible light from any of the bright stars visible in the night sky.

Figure 3.19 shows the world's largest radio telescope, built in 1963 in Arecibo, Puerto Rico. Approximately 300 m in diameter, the telescope's reflecting surface lies in a natural depression in a hillside and spans nearly 20 acres. The receiver is strung among several limestone hills. Its enormous collecting area makes this telescope the most sensitive on Earth. The huge size of the dish creates one distinct disadvantage, however: the Arecibo telescope cannot be pointed very well to follow cosmic objects across the sky, limiting its observations to those objects that happen to pass within about 20° of overhead as Earth rotates.

The Value of Radio Astronomy

Despite the inherent disadvantage of relatively poor angular resolution, radio astronomy enjoys many advantages. Radio telescopes can observe 24 hours a day. Darkness is not needed for receiving radio signals because the Sun is a relatively weak source of radio energy, with the result that its emission does not swamp radio signals arriving at Earth from elsewhere in the sky. In addition, radio observations can often be made through cloudy skies, and radio telescopes can detect the longest-wavelength radio waves even during rain or snowstorms.

However, perhaps the greatest value of radio astronomy (and, in fact, of all invisible astronomies) is that it opens up a whole new window on the universe. There are two main reasons for this. First, just as objects that are bright in the visible part of the spectrum (the Sun, for example) are not necessarily strong radio emitters, many of the strongest radio sources in the universe emit little or no visible light. Second, visible light may be strongly absorbed by interstellar dust along the line of sight to a source. Radio waves, on the other hand, are generally unaffected by intervening matter. Many parts of the universe cannot be seen at all by optical means but are easily detectable at radio wavelengths. The center of the Milky Way Galaxy is a

Figure 3.19 The 300-m-diameter dish at the National Astronomy and Ionospheric Center near Arecibo, Puerto Rico. The receivers that detect the focused radiation are suspended nearly 300 m (about 80 stories) above the center of the dish.

prime example of such a totally invisible region—our knowledge of the Galactic Center is based almost entirely on radio and infrared observations. Thus, radio observations do not just afford us the opportunity of studying the same objects at different wavelengths. They allow us to see whole new classes of objects that would otherwise be completely unknown.

Figure 3.20 shows an optical photograph of the Orion Nebula (a huge cloud of interstellar gas) taken with the 4-m telescope on Kitt Peak. Superimposed on the optical image is a radio contour map of the same region, obtained by scanning a radio telescope back and forth across the nebula and taking many measurements of radio intensity. The map is drawn as a series of contour lines connecting locations of equal radio brightness, similar to pressure contours drawn by meteorologists on weather maps or height contours drawn by cartographers on topographic maps. The inner contours represent stronger radio signals, the outside contours weaker signals.

The radio map in Figure 3.20 has many similarities to the visible-light image of the nebula. For instance, the radio emission is strongest near the cen-

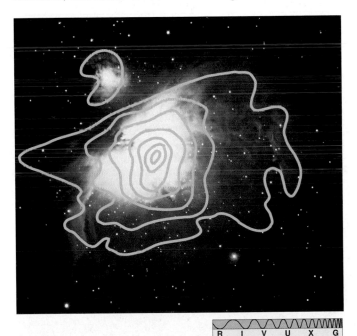

ter of the optical image and declines toward the nebular edge. But there are also subtle differences between the radio and optical images. The two differ mainly toward the upper left of the main cloud, where visible light seems to be absent, despite the existence of radio waves. How can radio waves be detected from locations not showing any light emission? The answer is that this particular nebular region is known to be especially dusty in its top left quadrant. The dust obscures the short-wavelength visible light but not the long-wavelength radio radiation. Thus, our radio map allows us to see the true extent of this cosmic source.

Interferometry

5 Radio astronomers can sometimes overcome the problem of poor angular resolution by using a technique known as **interferometry**. This technique makes it possible to produce radio images of much higher angular resolution than can be achieved with even the best optical telescopes, on Earth or in space.

In interferometry, two or more radio telescopes are used in tandem to observe the same object at the same wavelength and at the same time. The combined instruments together make up an **interferometer** (Figure 3.21). By means of electronic cables or radio links, the signals received by each antenna in the array making up the interferometer are sent to a central computer that combines and stores the data as the antennas track their target. After extensive computer processing, a high-resolution image of the target results.

An interferometer is essentially a substitute for a single huge antenna. As far as resolving power is concerned, the effective diameter of an interferometer is the distance between its outermost dishes. In other words, two small dishes can act as opposite ends of an imaginary but huge single radio telescope, dramatically improving the angular resolution. For example, resolution of a few arc seconds can be achieved at typical radio wavelengths (such as 10 cm), either by using a single radio telescope 5 km in diameter (which is impossible to build) or by using two or more much smaller dishes separated by 5 km and connected electronically. The larger the distance separating the telescopes—the longer the *baseline* of the interferometer—the better the resolution attainable.

Large interferometers made up of many dishes, like the instrument shown in Figure 3.21, now rou-

Figure 3.20 The Orion Nebula is a star-forming region about 1500 light years from Earth. (The nebula is located in the constellation Orion and can be seen in Figure P.6.) The bright regions in this photograph are stars and clouds of glowing gas. The dark regions are not empty, but their visible emission is obscured by interstellar matter. Superimposed on the optical image is a radio contour map of the same region. Each curve of the contour map represents a different intensity of radio emission. The resolution of the optical image is about 1″; that of the radio map is 1′.

Figure 3.21 This large interferometer, located on the Plain of San Augustin in New Mexico, comprises 27 dishes spread along a Y-shaped pattern about 30 km across. The most sensitive radio device in the world, it is called the Very Large Array, or VLA. The dishes are mounted on railroad tracks so that they can be repositioned easily.

tinely attain radio resolution comparable to that of optical images. Figure 3.22 compares an interferometric radio map of a nearby galaxy with a photograph of that same galaxy made using a large optical telescope. Note that the radio clarity is much better than that in the radio map of Figure 3.20.

Astronomers have created radio interferometers spanning very great distances. A typical very-long-baseline interferometry (VLBI) experiment might use radio telescopes in North America, Europe, Australia, and Russia to achieve angular resolution on the order of 0.001″. It now seems that even Earth's diameter is no limit. Proposals exist to place large interferometers entirely in Earth orbit and even on the Moon.

Although the technique was originally developed by radio astronomers, interferometry is no longer restricted to the radio domain. Radio interferometry became feasible when electronic equipment and computers achieved speeds great enough to combine and analyze radio signals from separate radio detectors

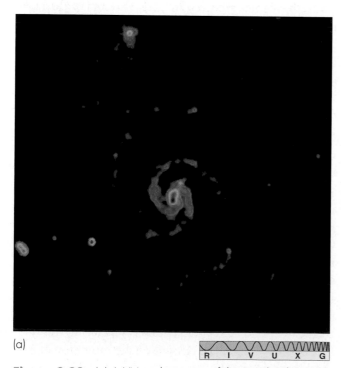

(a)

(b)

Figure 3.22 (a) A VLA radio image of the spiral galaxy M51, observed at radio frequencies having an angular resolution of a few arc seconds. (b) A visible-light image of M51, made with the 4-m Kitt Peak optical telescope.

without loss of data. As the technology has improved, it has become possible to apply the same methods to higher-frequency radiation. Millimeter-wavelength interferometry has already become an established and important observational technique, and infrared interferometry will become commonplace in the next few years. Optical interferometry is the subject of intensive research. The Keck telescopes on Mauna Kea are designed to be used for infrared—and perhaps someday for optical—interferometric work.

3.5 Other Astronomies

6 Optical and radio astronomy are the oldest and best-established branches of astronomy, but since the 1970s there has been a virtual explosion of observational techniques covering the rest of the electromagnetic spectrum. Today, all portions of the spectrum are studied, from radio waves to gamma rays, to maximize the amount of information available about astronomical objects. As noted earlier, the *types* of astronomical objects that can be observed may differ markedly from one wavelength range to another. Full-spectrum coverage is essential not only to see things more clearly, but even to see some things at all.

Because of the transmission characteristics of Earth's atmosphere, astronomers must study most wavelengths other than optical and radio from space. The rise of these "other astronomies" has therefore been closely tied to the development of the space program.

Infrared and Ultraviolet Astronomy

Infrared studies are a very important component of modern observational astronomy. Generally, **infrared telescopes** resemble optical telescopes (indeed, many optical telescopes are also used for infrared work), but the infrared detectors are sensitive to longer-wavelength radiation. Although most infrared radiation is absorbed by the atmosphere (primarily by water vapor), there are a few windows in the high-frequency part of the infrared spectrum where the opacity is low enough to allow ground-based observations (Figure 2.8). ∞ (Sec. 2.3) Indeed, some of the most useful infrared observing is done from the ground, even though the radiation is somewhat diminished in intensity by our atmosphere. Because of its 4-km altitude, Mauna Kea is one of the finest locations on Earth for both optical and infrared ground-based astronomy.

As with radio observations, the longer wavelength of infrared radiation often enables us to perceive objects partially or totally hidden from optical view. As an example of the penetrating properties of infrared radiation, Figure 3.23 shows a dusty and hazy region in California, hardly viewable optically but easily seen using infrared radiation.

Astronomers can make still better infrared observations if they can place their instruments above most or all of Earth's atmosphere. Improvements in balloon-, aircraft-, rocket-, and satellite-based telescope technologies have made infrared research a pow-

Figure 3.23 (a) An optical photograph taken near San Jose, California and (b) an infrared image of the same area taken at the same time. Longer wavelength infrared radiation can penetrate smog much better than shorter-wavelength visible light.

Figure 3.24 (a) A gondola containing a 1-m infrared telescope (lower left) is readied for its balloonborne ascent to an altitude of about 30 km, where it will capture infrared radiation that cannot penetrate the atmosphere. (b) An artist's conception of the *Infrared Astronomy Satellite*, placed in orbit in 1983. This 0.6-m telescope surveyed the infrared sky at wavelengths ranging from 10 to 100 μm. During its 10 months of operation, it greatly increased astronomers' understanding of many aspects of the universe, from the formation of stars and planets to the evolution of galaxies.

(a)

(b)

erful tool for studying the universe (Figure 3.24). However, as might be expected, the infrared telescopes that can be carried above the atmosphere are considerably smaller than the massive instruments found in ground-based observatories.

The most advanced facility to function in this part of the spectrum is the *Infrared Astronomy Satellite* (*IRAS*), shown in Figure 3.24(b). Launched into Earth orbit in 1983 but now inoperative, this British–Dutch–U.S. satellite housed a 0.6-m mirror having an angular resolution as fine as 30″ (depending on the wavelength observed). Its sensitivity was greatest for radiation in the 10- to 100-μm range. During its 10-month lifetime (and long afterwards—the data archives are still heavily used today), *IRAS* contributed greatly to our knowledge of clouds of galactic matter that seem destined to become stars, and possibly planets. These regions are composed of warm gas that cannot be seen with optical telescopes or adequately studied with radio telescopes. Throughout this text we will encounter many findings made by this satellite about comets, stars, galaxies, and the scattered dust and rocky debris found between the stars.

Figure 3.25(a) shows an *IRAS* image of the Orion Nebula. At about 1′ angular resolution, the fine details of Orion visible in the optical image of Figure 3.20 cannot be perceived. Nonetheless, astronomers can extract useful information about this object and others like it from such observations. For example, they can see clouds of warm dust and gas, believed to play a crit-

ical role in star formation, and extensive groups of bright young stars that are completely obscured at visible wavelengths.

Unfortunately, by Wien's law (Chapter 2), telescopes radiate strongly in the infrared unless they are cooled to extremely low temperatures. ∞ (Sec. 2.4) The end of *IRAS*'s mission came not because of any equipment malfunction or unexpected mishap but simply because its supply of liquid helium coolant ran out. *IRAS*'s own thermal emission then overwhelmed the radiation the instrument was built to detect.

To the short-wavelength side of the visible spectrum lies the ultraviolet domain. This region of the spectrum, extending in wavelength from 400 nm (4000 Å, blue light) down to a few nanometers ("soft" X rays), has only recently begun to be explored. Because Earth's atmosphere is partially opaque to radiation below 400 nm and is totally opaque to radiation below about 300 nm, astronomers cannot conduct any useful ultraviolet observations from the ground, not even from the highest mountaintop. Rockets, balloons, or satellites are therefore essential to any **ultraviolet telescope**—a device designed to capture and analyze this high-frequency radiation. The *Hubble Space Telescope* (*Interlude 3-1*), best known as an optical telescope, is also a superb ultraviolet instrument.

An alternative means of placing astronomical payloads into (temporary) Earth orbit is provided by NASA's space shuttle. In December 1990 and March 1995, a shuttle carried aloft the *Astro* package of three

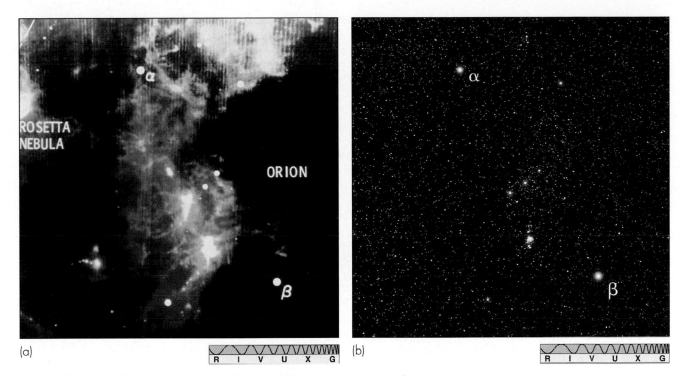

(a)   (b)

Figure 3.25 (a) This infrared image of the Orion Nebula and its surrounding environment was made by *IRAS*. The whiter regions denote greater strength of infrared radiation; the false colors denote different temperatures, descending from white to red to black. (b) The same region photographed in visible light. The labels α and β refer, respectively, to Betelgeuse and Rigel, the two brightest stars in the constellation.

ultraviolet telescopes (Figure 3.26). Astronomical shuttle missions offer a potentially very flexible way for astronomers to get instruments into space without having to deal with the long lead times and great expense of permanent satellite missions like *HST*.

High-Energy Astronomy

High-energy astronomy studies the universe as it presents itself to us in X rays and gamma rays—the types of radiation whose photons have the highest frequencies and hence the greatest energies. How do we detect radiation of such short wavelengths? First, it must be captured high above Earth's atmosphere because none of it reaches the ground. Second, its detection requires the use of equipment basically different in design from that used to capture the relatively low-energy radiation discussed up to this point.

The basic difference in the design of **high-energy telescopes** comes about because X and gamma rays cannot be reflected easily by any kind of surface. Rather, these rays tend to either pass straight through

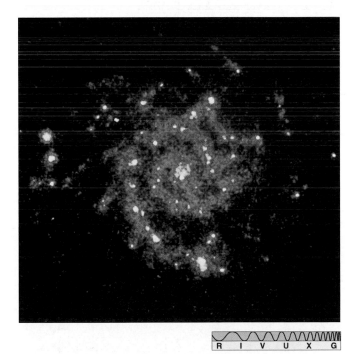

Figure 3.26 This false-color image of the spiral galaxy M74 was made by an ultraviolet telescope aboard the *Astro* payload carried by the space shuttle in 1990 and 1995.

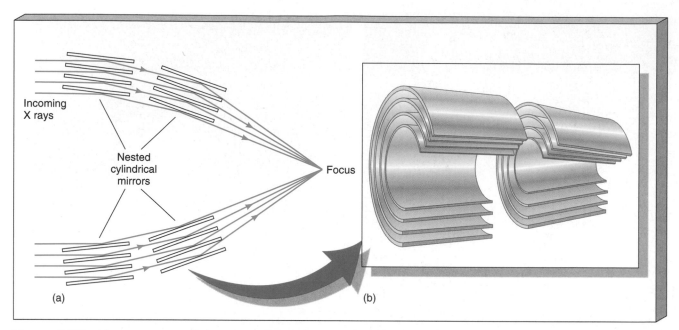

Figure 3.27 The arrangement of mirrors in an X-ray telescope allows the rays to be reflected at grazing angles and focused into an image.

Figure 3.28 *HEAO-2*, also known as the *Einstein Observatory*, the first imaging X-ray telescope. The left end of the mirror arrangement depicted in Figure 3.27 is at the bottom of the satellite as oriented here. Although *Einstein's* collecting diameter was only 0.6 m, its angular resolution was a mere 3″. Accordingly, this spacecraft could produce images of quality comparable to that of optical photographs.

or else be absorbed by any material they strike. When X rays barely graze a surface, however, they can be reflected from it in a way that yields an image, although the mirror design is fairly complex (Figure 3.27). For gamma rays, no such method of producing an image has yet been devised. Present-day gamma-ray telescopes simply point in a specified direction and count photons received.

Figure 3.28 is a photograph of the second *High-Energy Astronomy Observatory* (HEAO-2, also known as the *Einstein Observatory*). Launched in 1978, this was the first X-ray telescope capable of forming an image of its field of view. During its two-year lifetime, this spacecraft made major advances in our understanding of high-energy phenomena throughout the universe; its observational database is still heavily used. The most recent major X-ray satellite is the German *ROSAT* (short for *Röntgen Satellite*, after Wilhelm Röntgen, the discoverer of X rays), launched in 1991 by a European *Ariane* rocket. With more sensitivity, a wider field of view, and better resolution than *Einstein*, *ROSAT* is providing high-energy astronomers with new levels of observational detail (Figure 3.29).

Gamma-ray astronomy is the youngest entrant into the observational arena. As just mentioned, imaging gamma-ray telescopes do not exist, so only fairly coarse (1° resolution) observations can be made. Nevertheless, even at this resolution, there is much to be learned. Cosmic gamma rays were originally

Figure 3.29 An X-ray image of the Orion region of the sky taken by the *ROSAT* satellite. (Compare with Figures P.6, 3.20, and 3.25.) Note the three stars of Orion's belt (middle left) and the glowing nebula below them at bottom left.

(a)

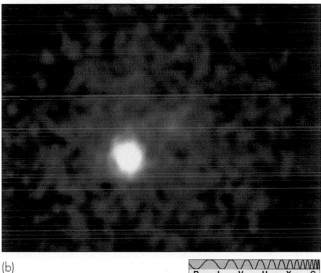

(b)

Figure 3.30 (a) This photograph of the 17-ton *Gamma-Ray Observatory* was taken by an astronaut during the satellite's deployment from the space shuttle *Atlantis* over the Pacific Coast of the United States. (b) A typical false-color gamma-ray image showing a violent event known as a *gamma-ray blazar* in the distant galaxy 3C279.

detected in the 1960s by the U.S. *Vela* series of satellites, whose primary mission was to monitor illegal nuclear detonations on Earth. Since then, several X-ray telescopes have also been equipped with gamma-ray detectors. By far the most advanced instrument is the *Gamma Ray Observatory* (GRO), launched by space shuttle in 1991. This satellite can scan the sky and study individual objects in much greater detail than previously attempted. Figure 3.30 shows GRO in low Earth orbit, along with a false-color gamma-ray image of a highly energetic outburst in the nucleus of a distant galaxy.

Full-Spectrum Coverage

7 Table 3.1 lists the basic regions of the electromagnetic spectrum and describes objects typically studied in each frequency range. Bear in mind that the list is far from exhaustive and that many astronomical objects are now routinely observed at many different electromagnetic wavelengths. As we proceed through the text, we will discuss more fully the wealth

of information that high-precision astronomical instruments can provide us.

In the twenty-first century, if all goes according to plan, it will be possible, for the first time ever, to make simultaneous high-quality measurements of any astronomical object at all wavelengths, from radio to gamma ray. The consequences of this development for our understanding of the workings of the universe may be little short of revolutionary. As an illustration of the sort of comparison that full-spectrum coverage allows,

Table 3.1 Astronomy at Many Wavelengths

Radiation	General Considerations	Common Applications (Chapter Reference)
Radio	Can penetrate dusty regions of interstellar space. Earth's atmosphere largely transparent to radio wavelengths. Can be detected in daytime as well as at night. High resolution at long wavelengths requires very large telescopes.	Radar studies of planets (1, 6) Planetary magnetic fields (7) Interstellar gas clouds (11) Center of Milky Way Galaxy (14) Galactic structure (14, 15) Active galaxies (16) Cosmic background radiation (17)
Infrared	Can penetrate dusty regions of interstellar space. Earth's atmosphere only partially transparent to IR radiation, so some observations must be made from space.	Star formation (11) Cool stars (11, 12) Center of Milky Way Galaxy (14) Active galaxies (16) Large-scale structure of universe (16, 17)
Visible	Earth's atmosphere transparent to visible light.	Planets (6, 7) Stars and stellar evolution (9, 10, 12) Galactic structure (14, 15) Large-scale structure of universe (16, 17)
Ultraviolet	Earth's atmosphere opaque to UV radiation, so observations must be made from space.	Interstellar medium (11) Hot stars (12)
X ray	Earth's atmosphere opaque to X rays, so observations must be made from space. Special mirror configurations needed to form images.	Stellar atmospheres (9) Neutron stars and black holes (13) Hot gas in galaxy clusters (15) Active galactic nuclei (16)
Gamma ray	Earth's atmosphere opaque to gamma rays, so observations must be made from space. Cannot form images.	Neutron stars (13) Active galactic nuclei (16)

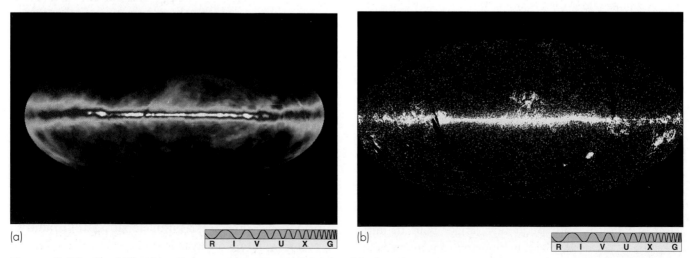

(a)   R I V U X G   (b)   R I V U X G

Figure 3.31 The Milky Way Galaxy as it appears at (a) radio, (b) infrared.

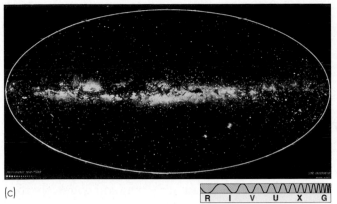

(c)

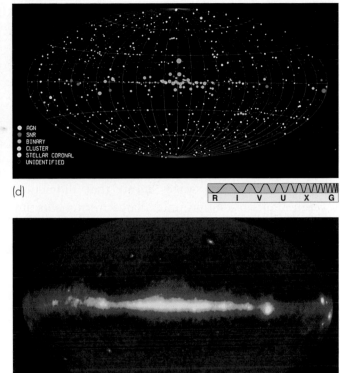

(d)

Figure 3.31(Continued)   (c) Visible, (d) X-ray, and (e) gamma-ray wavelengths.

Figure 3.31 shows a series of images of the Milky Way Galaxy. They were made by several instruments, at wavelengths ranging from radio to gamma ray, over a period of about five years. By comparing the features visible in each, we immediately see how multiwavelength observations can complement each other, greatly extending our perception of the universe around us.

(e)

Chapter Review

Summary

A **telescope** (p. 74) is a device designed to collect as much light as possible from some distant source and deliver it to a detector for detailed study. **Reflecting telescopes** (p. 74) use a mirror to concentrate and focus the light. **Refracting telescopes** (p. 74) use a lens; **refraction** (p. 74) is the bending of light as it passes from one medium to another. The **prime focus** (p. 74) of a telescope is the point where the incoming beam is focused and where analysis instruments may be placed. The **Newtonian** (p. 77) and **Cassegrain telescope** (p. 77) designs employ secondary mirrors to avoid placing heavy equipment at the prime focus. All astronomical telescopes larger than about 1 m in diameter use mirrors in their design.

The light-gathering power of a telescope depends on its **collecting area** (p. 79), which is proportional to the square of the mirror diameter. To study the faintest sources of radiation, astronomers must use large telescopes.

Angular resolution (p. 79) is a telescope's ability to distinguish between light sources lying close together on the sky. One limitation on resolution is

diffraction (p. 81), which makes it impossible to focus a beam perfectly. The amount of diffraction is proportional to the wavelength of the radiation under study and inversely proportional to the size of the mirror. Thus, at any given wavelength, larger telescopes suffer least from the effects of diffraction.

The resolution of most ground-based optical telescopes is limited by **seeing** (p. 82)—the blurring effect of Earth's turbulent atmosphere, which smears the pointlike images of stars out into **seeing disks** (p. 82) a few arc seconds in diameter. Because radio and space-based telescopes do not suffer from atmospheric effects, their resolution is determined by the effects of diffraction.

Most modern telescopes use **charge-coupled devices** (p. 86) instead of photographic plates to collect data. CCDs are many times more sensitive than photographic plates, and the resultant data are easily saved directly on disk or tape for later image processing.

Using **active optics** (p. 87), in which a telescope's environment and focus are carefully monitored and controlled, and **adaptive optics** (p. 87), in which the blurring effects of atmospheric turbulence are corrected for in real time, it may soon be possible to achieve diffraction-limited resolution in ground-based optical instruments.

Radio telescopes (p. 89) are conceptually similar in construction to optical reflecting telescopes. However, radio telescopes are generally much larger than optical instruments, for two reasons. First, the amount of radio radiation reaching Earth from space is much less than the amount of visible radiation, so a large collecting area is essential. Second, the long wavelengths of radio waves mean that diffraction severely limits resolution unless large instruments are used.

In order to increase the effective area of a telescope, and hence improve its resolution, several instruments may be combined into an **interferometer** (p. 91). Using **interferometry** (p. 91), radio telescopes can produce images much sharper than those from the best optical equipment.

Infrared (p. 93) and **ultraviolet telescopes** (p. 94) are similar in basic design to optical systems. Infrared studies in some parts of the infrared range can be carried out using large ground-based systems. Ultraviolet astronomy must be carried out from space.

High-energy telescopes (p. 95) study the X-ray and gamma-ray regions of the electromagnetic spectrum. X-ray telescopes can form images of their field of view, although the mirror design is more complex than for lower-energy instruments. Gamma-ray telescopes simply point in a certain direction and count photons received. Because the atmosphere is opaque at these short wavelengths, both types of telescope must be placed in space.

Radio and other nonoptical telescopes are essential to studies of the universe because they allow astronomers to probe regions of space that are completely opaque to visible light and to study the many objects that emit little or no optical radiation.

Self-Test: True or False?

_____ 1. The primary purpose of any telescope is to collect as much radiation as possible and magnify the image.

_____ 2. A Newtonian telescope has no secondary mirror.

_____ 3. A Cassegrain telescope has a hole in the middle of the primary mirror to allow light reflected from its secondary mirror to reach a focus behind the primary mirror.

_____ 4. The term "seeing" is used to describe how faint an object can be detected by a telescope.

_____ 5. The primary advantage to using the _Hubble Space Telescope_ is the increased amount of "night" time available to it.

_____ 6. One of the primary advantages of CCDs over photograph plates is the former's high efficiency in detecting light.

_____ 7. The _Hubble Space Telescope_ can observe objects in the optical, infrared, and ultraviolet parts of the spectrum.

_____ 8. The Keck telescope has the largest single mirror ever produced.

_____ 9. Radio telescopes are large in part to improve their angular resolution, which is poor because of the long wavelengths at which they observe.

_____ 10. Radio telescopes are large in part because the sources of radio radiation they observe are very faint.

_____ 11. Infrared astronomy must be done from space.

_____ 12. Because the ozone layer absorbs ultraviolet light, astronomers must make observations in the ultraviolet from the highest mountain tops.

_____ 13. X-ray and gamma-ray telescopes employ the same basic design as optical instruments.

Self-Test: Fill in the Blank

1. A telescope that uses a lens to focus light is called a _____ telescope.

2. A telescope that uses a mirror to focus light is called a _____ telescope.

3. All large modern telescopes are of the _____ type.

4. The light-gathering power of a telescope is determined by the _____ of its mirror or lens.

5. The angular resolution of a telescope is limited by the _____ of the telescope and the _____ of the radiation being observed.

6. The angular resolution of ground-based optical telescopes is more seriously limited by Earth's _____ than by diffraction.

7. Optical telescopes on Earth can see angular detail down to about _____ arc second.

8. CCDs produce images in _____ form that can be easily transmitted, stored, and processed by computers.

9. Active optics and adaptive optics are both being used to improve the _____ of ground-based optical telescopes.

10. All radio telescopes are of the _____ design.

11. An _____ is two or more telescopes used in tandem to observe the same object, in order to improve angular resolution.

12. An object having a temperature of 300 K would be best observed with an _____ telescope.

Review and Discussion

1. Cite two reasons astronomers are continually building larger and larger telescopes.

2. What are three advantages of reflecting telescopes over refracting telescopes?

3. How does Earth's atmosphere affect what is seen through an optical telescope?

4. What advantages does the *Hubble Space Telescope* have over ground-based telescopes? List some disadvantages.

5. What are the advantages of a CCD over a photographic plate?

6. What is image processing?

7. Describe some ways in which optical astronomers can compensate for the blurring effects of Earth's atmosphere.

8. Why do radio telescopes have to be very large?

9. What kind of astronomical objects can we best study with radio techniques?

10. What is interferometry, and what problem in radio astronomy does it address?

11. Compare the highest resolution attainable with optical telescopes with the highest resolution attainable with radio telescopes (including interferometers).

12. What special conditions are required to conduct observations in the infrared?

13. What is the main advantage of studying objects at various wavelengths of radiation?

14. Our eyes can see light with an angular resolution of 1′. Suppose our eyes detected only infrared radiation, with 1° angular resolution. Would we be able to make our way around on Earth's surface? To read? To sculpt? To create technology?

Problems

1. A 2-m telescope can collect a given amount of light in 1 h. Under the same observing conditions, how much time would be required for a 6-m telescope to perform the same task?

2. A certain space-based telescope can achieve (diffraction-limited) angular resolution of 0.05″ for red light (wavelength 700 nm). What would its resolution be (a) in the infrared, at wavelength 3.5 μm, and (b) in the ultraviolet, at wavelength 140 nm?

3. The photographic equipment on a telescope is replaced by a CCD. If the photographic plate records 5 percent of the light reaching it but the CCD records 75 percent, how much time will the new system take to collect as much information as the old detector recorded in a 1-h exposure?

4. The Andromeda Galaxy lies about 2.9 million light years away. To what distances do the angular resolutions of *HST* (0.05″) and a radio interferometer (0.001″) correspond to at that distance?

Projects

1. Here's how to take some easy pictures of the night sky. You will need a location with a clear, dark sky; a 35-mm camera with a standard 50-mm lens, tripod, and cable release; a watch with a seconds display visible in the dark; and a role of high-speed color slide film. Set your camera to the "bulb" setting for the exposure and attach the cable release so that you can take a long exposure. Set the focus on infinity. Point the camera to a favored constellation, seen through your viewfinder, and take a 20-s to 30-s exposure. In order to minimize vibration, don't hold on to the cable release during the exposure. Keep a log of your shots. When finished, have the film developed in the standard way.

2. For some variations, vary your exposure times, use different films, take hours-long exposures for star trails, use a wide-angle or telephoto lens, place the camera piggyback on a telescope that is tracking and take exposures that are a few minutes long. Experiment and have fun!

The Solar System

Interplanetary Matter and the Birth of the Planets *www*

A truly great comet—one of the brightest of the twentieth century—illuminated the skies of the northern hemisphere in early 1997. Comet Hale-Bopp is shown here in this wide-angle photograph, taken at the moment of perihelion, or closest approach to the Sun, yet still some 0.9 A.U. away. Note how the gas (glowing bluish light) boiling off the comet extends straight away from the Sun, while the comet's heavier dust particles (reflected whitish light) lag behind, forming a more gently curved tail. The full extent of the tail measured nearly 40 arc degrees, sweeping a huge arc across the nighttime sky. Almost surely, Hale-Bopp was the most-photographed comet in history.

LEARNING GOALS

Studying this chapter will enable you to:

1 Describe the scale and structure of the solar system and summarize the basic differences between the terrestrial and the jovian planets.

2 Summarize the orbital and physical properties of the major groups of asteroids.

3 Describe the composition and structure of a typical comet and explain how a cometary tail forms.

4 Explain what cometary orbits tell us about the probable origin of comets.

5 Summarize the orbital and physical properties of meteoroids and explain how these bodies are related to asteroids and comets.

6 List the major facts that any theory of solar system formation must explain and indicate how the leading theory accounts for them.

7 Outline the process by which planets form as natural by-products of star formation and explain the crucial role of dust in planet formation.

In less than a generation we have learned more about our solar system—the Sun and everything that orbits it—than in all the centuries that went before. By studying the planets, their moons, and the countless fragments of material that orbit in interplanetary space, astronomers have gained a richer outlook on our own home in space. The discoveries of the past few decades have revolutionized our understanding not only of the present state of our cosmic neighborhood but also of its history, for our solar system is filled with clues to its own origin and evolution. The richest sources of information about the earliest days of the solar system are the asteroids, comets, and meteoroids. These objects may seem to be only rocky and icy "debris," but, more than the planets themselves, they hold a record of the formative stages of our planetary system and have much to teach us about the origin of our world.

4.1 An Inventory of the Solar System

Our **solar system** contains 1 star (the Sun), 9 planets orbiting that star, 63 moons (at last count) orbiting those planets, 6 *asteroids* larger than 300 km in diameter, more than 4000 smaller (but well-studied) asteroids, myriad *comets* a few kilometers in diameter, and countless *meteoroids* less than 100 meters across. This list will likely grow as we continue to explore our cosmic neighborhood. The arrangement of the major bodies in the solar system is shown in Figure 4.1. The planet closest to the Sun is Mercury. Moving outward, we encounter Venus, Earth, Mars, Jupiter, Saturn, Uranus, Neptune, and Pluto. The asteroids lie mainly in a broad belt between the orbits of Mars and Jupiter.

Planetary Properties

Table 4.1 lists some properties of the nine planets, with the Sun and Moon included for comparison. Each planet's distance from the Sun is known from Kepler's laws once the scale of the solar system is set by radar ranging on Venus (Section 1.6). ∞ (Sec. 1.6) Planetary radii are found by measuring the planets' angular sizes and then employing elementary geometry, as described in the Prologue. ∞ (Sec. P.3) Mass is determined by observing a planet's gravitational influence on some nearby object and applying Newton's laws of motion and gravity (Section 1.7). ∞ (Sec. 1.7) Prior to the Space Age, astronomers calculated planetary masses either by tracking the orbits of the

planets' moons (if any) or by measuring the small (but detectable) distortions the planets produce in each other's orbits. Today, the masses of all but one (Pluto) of the objects in Table 4.1 are most accurately known through their gravitational effects on artificial satellites and space probes launched from Earth.

Figure 4.2 shows the sizes of the planets relative to the Sun, clearly the largest object in our solar system. As indicated in Table 4.1, it is more than 1000 times more massive than the next largest object, the planet Jupiter. The Sun in fact contains about 99.9 percent of all solar system material. The planets—including our home planet—are insignificant in comparison.

By earthly standards, the solar system is immense. The distance from the Sun to Pluto is about 40 A.U., almost a million times the radius of Earth and roughly 15,000 times the distance from Earth to the Moon. Despite the solar system's vast extent, the planets all lie very close to the Sun, astronomically speaking. The diameter of the largest orbit, that of Pluto, is less than 1/1000 of a light year, while the star nearest the solar system is several light years distant.

Note in Figure 4.1 that the planetary orbits are not evenly spaced; instead, they get farther and farther apart as we move outward from the Sun. Nevertheless, there is a regularity in their spacing. Roughly speaking, the spacing between adjacent orbits doubles as we move outward from the Sun.

All the planets orbit the Sun counterclockwise as seen from above Earth's North Pole, and in nearly the same plane as Earth (the ecliptic plane; see Section 1.2). ∞ (Sec. 1.2) Mercury and Pluto deviate from this rule—their orbital planes lie at 7° and 17° to the eclip-

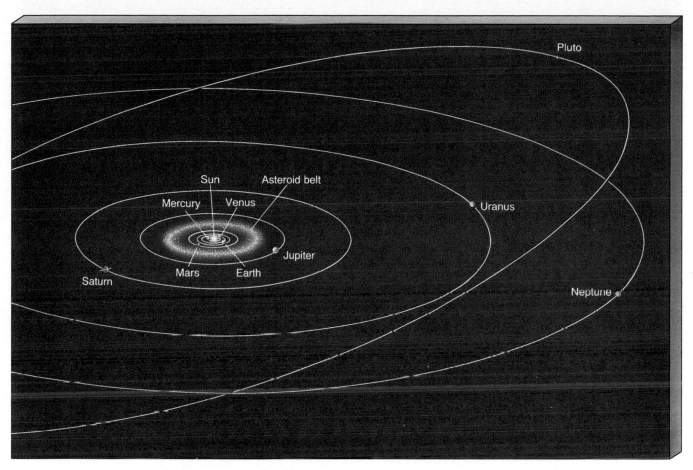

Figure 4.1 Major bodies of the solar system: Sun, planets, asteroids. Except for Mercury and Pluto, the orbits of the planets all lie nearly in the same plane. As we move outward from the Sun, the distance between adjacent orbits increases. The entire solar system spans nearly 80 A.U.

tic, respectively. Still, we can think of the solar system as being flat. Its "thickness" perpendicular to the plane of the ecliptic is less than 1/50 the diameter of Pluto's orbit. If we were to view the planets' orbits from a vantage point in the ecliptic plane about 50 A.U. from the Sun, only Pluto's orbit would be noticeably tilted. Figure 4.3 is a photograph of Mercury, Venus, Mars, and Jupiter taken during the July 1991 solar eclipse. These four planets are visible in this one photograph in large part because their orbits all lie nearly in the same plane.

The final column in Table 4.1 lists a quantity called **density**. This property is a measure of the "compactness" of an object—how much matter is packed into a given volume. It is computed by dividing the object's mass (in kilograms) by its volume (in cubic meters). Dividing Earth's mass (6×10^{24} kg, determined by observations of the Moon's orbit) by its volume (which we know because we know Earth's radius,

6400 km), we obtain an average density of 5500 kg/m^3. *On average*, then, there are about 5500 kg of Earth matter in every cubic meter of Earth volume. For comparison, the density of water is 1000 kg/m^3, rocks on Earth's surface have densities in the range 2000–3000 kg/m^3, and iron has a density of about 8000 kg/m^3. Earth's atmosphere at sea level has a density of only a few kilograms per cubic meter.

Terrestrial and Jovian Planets

A clear distinction can be drawn between the inner and outer members of our planetary system based on densities and other physical properties. The inner planets—Mercury, Venus, Earth, and Mars—are small, dense, and *rocky* in composition. The outer worlds—Jupiter, Saturn, Uranus, and Neptune (but not Pluto)—are large, of low density, and *gaseous*.

Figure 4.2 Relative sizes of the planets and our Sun. Notice that Jupiter, Saturn, Uranus, and Neptune are much larger than Pluto and much larger than Earth and the other inner planets. However, even these large planets are dwarfed by the still larger Sun.

	Table 4.1 Properties of Some Solar System Objects					
Object	**Orbit Semi-Major Axis (A.U.)**	**Orbit Period (Earth years)**	**Mass (Earth masses)**	**Radius (Earth radii)**	**Number of Known Moons**	**Average Density (kg/m³)**
Mercury	0.39	0.24	0.055	0.38	0	5400
Venus	0.72	0.62	0.81	0.95	0	5200
Earth	1.0	1.0	1.0	1.0	1	5500
Moon	1.0	—	0.012	0.27	—	3300
Mars	1.5	1.9	0.11	0.53	2	3900
Jupiter	5.2	11.9	318	11.2	16	1300
Saturn	9.5	29.5	95	9.5	20	700
Uranus	19.2	84	15	4.0	15	1200
Neptune	30.1	165	17	3.9	8	1700
Pluto	39.5	249	0.003	0.2	1	2300
Sun	—	—	332,000	109	—	1400

Figure 4.3 Taken from Hawaii during the July 1991 eclipse of the Sun, this single photograph shows Mercury, Venus, Mars, and Jupiter. Because these planets all orbit in nearly the same plane, it is possible for them all to appear (by chance) in the same region of the sky, as seen from Earth.

Because the physical and chemical properties of Mercury, Venus, and Mars are somewhat similar to Earth's, the four innermost planets are called the **terrestrial planets**. (The word *terrestrial* derives from the Latin word *terra*, meaning "land" or "earth.") The larger outer planets—Jupiter, Saturn, Uranus, and Neptune—are all similar to one another chemically and physically (and very different from the terrestrial worlds). They are labeled the **jovian planets**, after Jupiter, the largest member of the group. (The word *jovian* comes from *Jove*, another name for the Roman god Jupiter.) The jovian worlds are all much larger than the terrestrials and quite different from them in both composition and structure.

The four terrestrial planets all lie within about 1.5 A.U. of the Sun. All are small and of relatively low mass, and all have generally rocky composition and solid surfaces. Beyond that, however, the similarities end. When we take into account how the weight of overlying layers compresses the interiors of the planets to different extents (greatest for Earth, least for Mercury), we find that the average *uncompressed densities* of the terrestrial worlds—that is, the densities they would have in the absence of any compression—decrease steadily as we move farther from the Sun. This decrease in density indicates that the overall compositions of these planets differ significantly one from the other.

There are many more differences among the terrestrial worlds. All have atmospheres, but the atmospheres are about as dissimilar as we could imagine, ranging from a near-vacuum on Mercury to a hot, dense inferno on Venus. Earth alone has oxygen in its atmosphere (as well as liquid water on its surface). The present-day conditions on the surfaces of the four planets are quite distinct from one another. Earth and Mars spin at roughly the same rate—one rotation every 24 (Earth) hours—but Mercury and Venus both take months to rotate just once, and Venus rotates in the opposite sense from the others. Earth and Mars have moons, but Mercury and Venus do not. Earth and Mercury have measurable magnetic fields, of very different strengths, whereas Venus and Mars have none. Finding the common threads in the evolution of four such diverse worlds is no simple task.

Yet for all their differences, the terrestrial worlds still seem very similar when compared with the jovian planets. Perhaps the simplest way to express the major differences between the terrestrial and jovian worlds is to say that the jovian planets are everything the terrestrial planets are not. Table 4.2 compares and contrasts some key properties of these two planetary classes.

The terrestrial worlds lie close together, near the Sun; the jovian worlds are widely spaced through the

Table 4.2 Comparison of the Terrestrial and Jovian Planets	
Terrestrial	**Jovian**
close to Sun	far from Sun
closely spaced orbits	widely spaced orbits
small masses	large masses
small radii	large radii
predominantly rocky	predominantly gaseous
solid surface	no solid surface
high density	low density
slower rotation	faster rotation
weak magnetic fields	strong magnetic fields
no rings	many rings
few moons	many moons

outer solar system. The terrestrial worlds are small, dense, and rocky; the jovian worlds are large and gaseous, being made up predominantly of hydrogen and helium (the lightest elements), which are rare on the inner planets. The terrestrial worlds have solid surfaces; the jovian worlds have none (their dense atmospheres thicken with depth, eventually merging with their liquid interiors). The terrestial worlds have weak magnetic fields, if any; the jovian worlds all have strong magnetic fields. The terrestrial worlds have only three moons among them; the jovian worlds each have many moons, no two of them alike and none of them like our own. Furthermore, all the jovian planets have *rings*, a feature unknown on the terrestrial planets.

Finally, beyond the outermost jovian planet, Neptune, lies one more small world, frozen and mysterious. Pluto doesn't fit well into either planetary category. Indeed, there is debate among planetary scientists as to whether it should be classified as a planet at all. In both mass and composition, it much more in common with the icy jovian moons than with any terrestrial or jovian planet. Astronomers speculate that it may in fact be the largest member of a newly recognized class of solar system objects that reside beyond the jovian worlds.

Interplanetary Matter

In the vast space among the nine known planets move countless small chunks of matter ranging in size from a few hundred kilometers in diameter down to tiny grains of interplanetary dust. The three major constituents of this cosmic "debris" are asteroids, comets, and meteoroids. **Asteroids** and **meteoroids** are fragments of rocky material, somewhat similar in composition to the outer layers of the terrestrial planets. The distinction between the two is simply a matter of size—anything larger than 100 m in diameter (corresponding to a mass of about 10,000 tons) is conventionally termed an asteroid, anything smaller is a meteoroid. **Comets** are predominantly icy, rather than rocky in composition (although they do contain some rocky material), and have typical diameters in the 1–10 km range.

Taken together, these small bodies account for a negligible fraction of the total mass of the solar system. They play no important role in the present-day workings of the planets or their moons. Yet they are of crucial importance to our studies, for they are the keys to

answering some very fundamental questions about our planetary environment.

One of the goals of planetary science is to understand how the solar system formed and to explain the physical conditions now found on Earth and elsewhere in our planetary system. Ironically, studies of Earth itself do not help us much in this quest because information about our planet's early stages was obliterated long ago by atmospheric erosion and geological activity. Much the same holds true for the other planets (with the possible exception of Mercury, where a near-airless and geologically inactive surface still retains some imprint of the distant past). The bottom line is that the large bodies of the solar system have all *evolved* significantly since they formed, making it very difficult to decipher the circumstances of their births.

A better place to look for hints of conditions in the early solar system is on its *small* bodies—the planetary moons and the asteroids, meteoroids, and comets that make up interplanetary debris—for nearly all such fragments contain traces of solid and gaseous matter from the earliest times. Many have scarcely changed since the solar system formed billions of years ago. As a prologue to our study of the formation of the solar system, therefore, we will first examine in a little more detail the present-day contents of interplanetary space.

4.2 Solar System Debris

Asteroids

2 Astronomers have so far cataloged over 4000 asteroids with well-determined orbits. The vast majority are found in a region of the solar system known as the **asteroid belt**, located between 2.1 and 3.3 A.U. from the Sun—roughly midway between the orbits of Mars (at 1.5 A.U.) and Jupiter (at 5.2 A.U.). All but one of the known asteroids revolve about the Sun in prograde orbits (that is, in the same sense as Earth and the other planets; see Section 1.4), and, like the planets, most asteroids stay fairly close to the ecliptic plane (within 10 or 20 degrees, say). Unlike the almost circular paths of the major planets, however, asteroid orbits are generally quite eccentric. The overall layout of the asteroid belt is sketched in Figure 4.4.

In addition to the main-belt asteroids, 50 or so **Trojan asteroids** share an orbit with Jupiter, remaining a constant 60° ahead of or behind that planet as it circles the Sun. This peculiar orbital behavior is not a

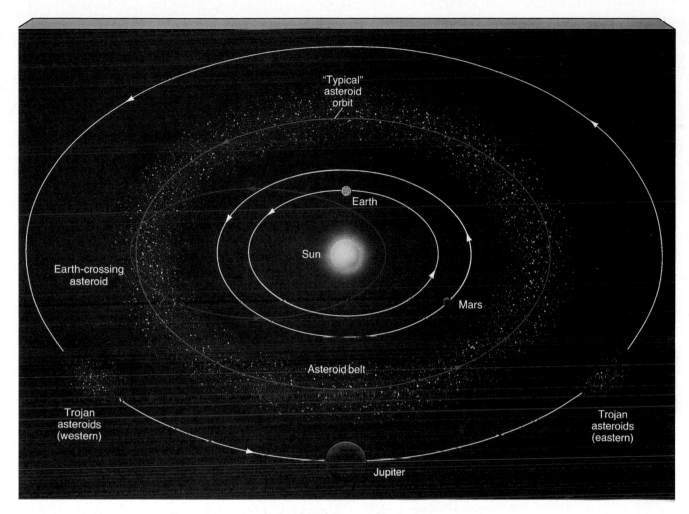

Figure 4.4 The main asteroid belt, along with the orbits of Earth, Mars, and Jupiter. Note the Trojan asteroids at two locations in Jupiter's orbit. A typical orbit for an Earth-crossing asteroid is also shown.

matter of chance—the Trojan asteroids are held in place by a delicate but stable balance between the gravitational fields of Jupiter and the Sun. Calculations first performed by the eighteenth-century French mathematician Joseph Louis Lagrange show that interplanetary matter that happens to stray into one of the two regions of space now occupied by the Trojan asteroids can remain there indefinitely, perfectly synchronized with Jupiter's motion.

The orbits of most asteroids have eccentricities lying in the range 0.05–0.3, ensuring that they always remain between the orbits of Mars and Jupiter. The few asteroids having orbital eccentricities greater than about 0.4 are of great interest to us, as their orbits may intersect that of Earth, leading to the possibility of a collision. These stray **Earth-crossing asteroids** having very elliptical orbits have probably been influenced by the gravitational fields of Mars and Jupiter, which have deflected those bodies into the inner solar system.

The potential for collision with Earth is real. For example, Figure 4.5 shows the asteroid Icarus, which passes within 0.2 A.U. of the Sun. On its way past Earth in 1968, it missed our planet by "only" 6 million km—a close call by cosmic standards. In 1989 an unnamed asteroid (designated 1989FC) came even closer, passing only 800,000 km from Earth, only twice the distance to the Moon. In 1991, asteroid 1991BA missed us by a mere 170,000 km. In fact, calculations imply that most Earth-crossing asteroids will eventually collide with Earth and that, during any given million-year period, our planet is struck by about three asteroids. Several dozen large basins and eroded craters on Earth are suspected to be sites of ancient asteroid collisions. The many large impact craters on the Moon, Venus, and Mars are direct evidence of similar events on those worlds.

Most known Earth-crossing asteroids are relatively small—about 1 km in diameter (although one

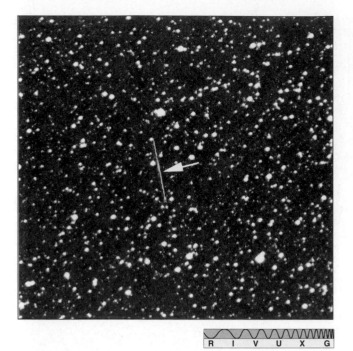

R I V U X G

Figure 4.5 The Earth-crossing asteroid Icarus has an orbit that passes within 0.2 A.U. of the Sun, well within Earth's orbit. Icarus occasionally comes close to Earth, making it one of the best-studied asteroids in the solar system. Its motion relative to the stars makes it appear as a streak (marked) in this long-exposure photograph.

10-km asteroid has been identified). Even so, the impact of even a kilometer-sized asteroid could be catastrophic by human standards. Such an object carries enough energy to devastate an area 100 km in diameter. Its explosive power would be equivalent to about a million 1-megaton nuclear bombs, a hundred times more than all the nuclear weapons currently in existence on Earth. (Figures 4.14–4.16 show the results of some recent, but much more modest, impacts.) Should an asteroid hit our planet hard enough, it might even cause the extinction of entire species. Indeed, many scientists think that the extinction of the dinosaurs was the result of just such an impact (see *Interlude 18-1*).

With few exceptions, asteroids are too small to be resolved by Earth-based telescopes. We must therefore rely on indirect methods to determine their composition, size, and shape. Consequently, only a few of their physical and chemical properties are accurately known. To the extent that astronomers can determine their composition, asteroids have been found to differ not only from the nine known planets and their moons but also from one another. The darkest, or least reflective, asteroids contain a large fraction of carbon and are

known as *carbonaceous* asteroids. The more reflective *silicate* asteroids are composed of rocky material. Generally speaking, silicate asteroids predominate in the inner portions of the asteroid belt and the fraction of carbonaceous bodies steadily increases as we move outward. Overall, about 15 percent of all asteroids are silicate, 75 percent are carbonaceous, and 10 percent are other types (such as those containing large fractions of iron). Most planetary scientists believe that the carbonaceous asteroids consist of very primitive material, representative of the earliest stages of the solar system, and have not experienced significant heating or chemical evolution since they formed.

Astronomers can estimate the sizes of asteroids from the amount of sunlight they reflect and the amount of heat they radiate, and size measurements now exist for more than 1000 of these bodies. Of these, only two dozen or so are more than 200 km across; most are far smaller. The three largest asteroids, Ceres, Pallas, and Vesta, have diameters of 940, 580, and 540 km, respectively. Occasionally, an asteroid passes directly in front of a star, as seen from Earth, an event that allows astronomers to determine both the size and the shape of the asteroid with great accuracy. Larger asteroids are roughly spherical (because, as with planets, gravity is the dominant force determining their shape), but the smaller bodies can be highly irregular. Masses are known only for the half-dozen or so largest asteroids, mainly because the gravitational effects of asteroids on their neighbors are very small and hard to measure accurately. The computed densities are generally consistent with the compositions just described. The most massive asteroid, Ceres, is just 1/10,000 the mass of Earth; the total mass of all the known asteroids probably amounts to less than one-tenth the mass of the Moon.

The first close-up views of asteroids were provided by the Jupiter probe *Galileo*, which, on its path to the giant planet, passed twice through the asteroid belt, making close encounters with asteroid Gaspra in October 1991 and asteroid Ida in August 1993 (Figure 4.6). Both Gaspra and Ida are irregularly shaped bodies having maximum diameters of about 20 and 50 km, respectively. They are pitted with craters ranging in size from a few hundred meters to 2 km across and are covered with a layer of dust of variable thickness. Both asteroids are thought to be fragments of much larger objects that broke up following violent collisions long ago. Ida is much more heavily cratered than Gaspra,

(a)

(b)

R I V U X G

Figure 4.6 (a) The asteroid Gaspra as seen from a distance of 1600 km by the space probe *Galileo*. (b) The asteroid Ida photographed by *Galileo* from a distance of 3400 km. (Ida's moon, Dactyl, is visible at the right.) The resolution in these photographs is on the order of 100 m. True-color images showed the surfaces of both bodies to be a fairly uniform gray. Spacecraft sensors indicated that the amount of infrared radiation absorbed by these surfaces varies from place to place, probably as a result of variations in the thickness of the dust layer blanketing them.

both because Ida resides in a denser part of the asteroid belt and because it is considerably older. Astronomers estimate that the collision responsible for Ida occurred nearly a billion years ago, compared to just 100 million years for the impact which created Gaspra.

To the surprise of most astronomers, close inspection of the Ida image (Figure 4.6b) revealed the presence of a tiny moon, just 1.5 km across, orbiting the asteroid. By studying *Galileo*'s images of Ida and its moon (now named Dactyl), astronomers hope to determine the asteroid's mass, allowing them to gain more insight into the composition and structure of these tiny worlds.

The compact concentration of most asteroids in a well-defined belt suggests one of two possible origins: either they are the fragments of a planet broken up long ago, or they are primal rocks that somehow never managed to form a genuine planet. On the basis of the best evidence currently available, and consistent with current theories of solar system formation, researchers strongly favor the latter view. There is far too little mass in the belt to constitute a planet, and the marked chemical differences between individual asteroids indicate that they could not all have originated in a single planet. Instead, astronomers believe that the strong gravitational field of Jupiter continuously disturbs the motions of these chunks of primitive matter, nudging

and pulling at them, preventing them from aggregating into a planet.

Comets *Extension*

Comets are usually discovered as faint, fuzzy patches of light on the sky while still several astronomical units away from the Sun. Traveling in a highly elliptical orbit with the Sun at one focus, a comet brightens and develops an extended **tail** as it nears the Sun. (The name "comet" derives from the Greek *kome*, meaning "hair.") As the comet departs from the Sun's vicinity, its brightness and its tail diminish until it once again becomes a faint point of light receding into the distance. The various parts of a typical comet are shown in Figure 4.7(a) Like the planets, comets emit no visible light of their own—they shine by reflected (or reemitted) sunlight.

The **nucleus**, or main solid body, of a comet is only a few kilometers in diameter. During most of the comet's orbit, far from the Sun, only this frozen nucleus exists. When a comet comes within a few astronomical units of the Sun, however, its icy surface becomes too warm to remain stable. Part of it becomes gaseous and expands into space, forming a diffuse **coma** ("halo") of dust and evaporated gas around the nucleus. The coma gets larger and brighter as the comet nears the Sun. At

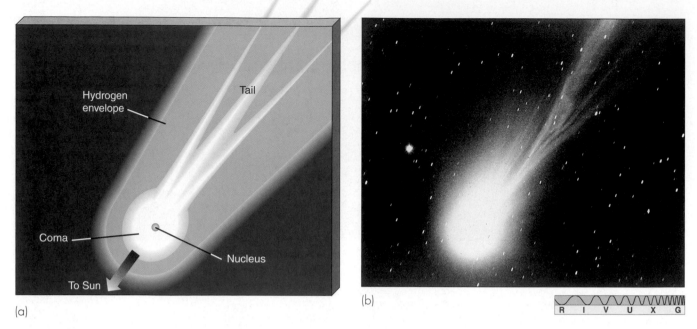

(a)

(b)

Figure 4.7 (a) Diagram of a typical comet, showing the nucleus, coma, hydrogen envelope, and tail. The tail is not a sudden streak in time across the sky, as in the case of meteors or fireworks. Instead, it travels along with the rest of the comet as long as the comet is close enough to the Sun. (b) Halley's Comet in 1986, about one month before it rounded the Sun.

maximum size, it can measure 100,000 km in diameter—almost as large as Saturn or Jupiter. Engulfing the coma, an invisible **hydrogen envelope** stretches across millions of kilometers of space. The comet's tail, most pronounced when the comet is closest to the Sun, is larger still, sometimes spanning as much as 1 A.U. From Earth, only the coma and tail of a comet are vis-

ible to the naked eye. Despite the size of the tail, most of the comet's light comes from the coma, while most of the mass resides in the nucleus.

Two types of comet tail may be distinguished. An **ion tail** is approximately straight, often made of glowing, linear streamers like those seen in Figure 4.8(a). Its emission spectrum indicates numerous ionized atoms

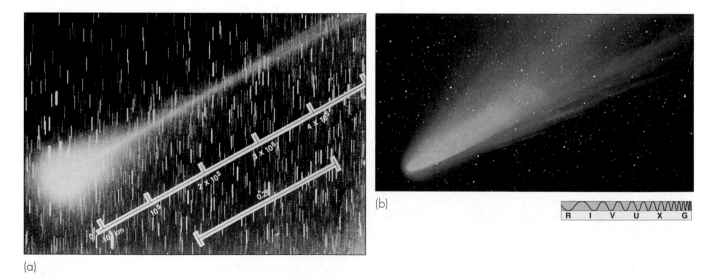

(a)

(b)

Figure 4.8 (a) A comet having a primarily ion tail—Comet Giacobini–Zinner, seen here in 1959. The coma of this comet measured 70,000 km across, and its tail was well over 500,000 km long. (b) A comet having (mostly) a dust tail. This photograph, of Comet West in 1976, shows both the gentle curvature of the dust tail and its inherent fuzziness. This tail stretched 13° across the sky. Being composed largely of ice, comets tend to be fragile. Shortly after this photograph was taken, the comet split into several fragments.

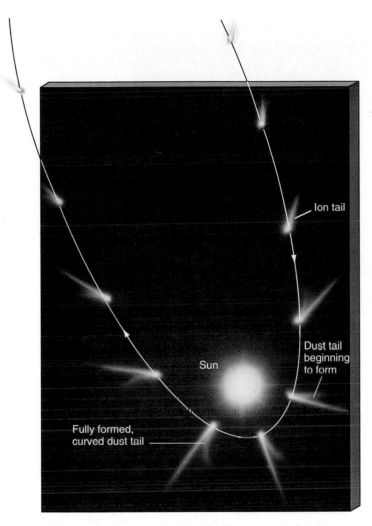

Figure 4.9 As it approaches the Sun, a comet develops an ion tail, which is always directed away from the Sun. Closer in, a curved dust tail, also directed generally away from the Sun, may appear. Notice that the ion tail always points directly away from the Sun on both the inbound and the outgoing portion of the orbit. The dust tail has a marked curvature and tends to lag behind the ion tail.

and molecules that have lost some of their normal complement of electrons. ∞ (Sec. 2.6) A **dust tail** is usually broad, diffuse, and gently curved (Figure 4.8b). It is rich in microscopic dust particles that reflect sunlight, making the tail visible from afar. Both types of tails are in all cases directed away from the Sun by the **solar wind**, an invisible stream of matter and radiation escaping from the Sun. (In fact, astronomers first inferred the existence of the solar wind from observations of comet tails.) Consequently, as depicted in Figure 4.9, the tail, be it ion or dust, always lies outside a comet's orbit and *leads* the comet during the portion of the orbit that is outbound from the Sun.

Probably the most famous comet of all is Halley's Comet (Figure 4.7b), whose appearance at 76-year in-

tervals has been documented at every passage since 240 B.C. A spectacular show, the tail of Halley's Comet can reach almost a full astronomical unit in length, stretching many tens of degrees across the sky.

When Halley's Comet rounded the Sun in 1986, a small armada of spacecraft launched by the USSR, Japan, and a group of western European countries went to meet it. The Soviet craft *Vega 2* traveled through the comet's coma, approaching to within 8000 km of the nucleus. Using positional knowledge gained from the *Vega* encounter, the European spacecraft *Giotto* was navigated to within 600 km of the nucleus—a daring trajectory since, at 70 km/s (the speed of the craft relative to the comet), a colliding dust particle becomes a devastating bullet. Debris did in fact damage *Giotto*'s camera but not before it sent back a wealth of data. Figure 4.10 shows *Giotto*'s view of Halley's nucleus, along with a sketch its structure. The comet's next visit to the inner solar system is expected in 2061. *Interlude 4-1* discusses two more recent, and very spectacular, comets whose arrival stirred the public imagination.

In seeking to understand the physical makeup of comets, astronomers are guided by the observation that comets contain dust that reflects light, as well as gas that emits spectral lines of many atoms and molecules, including hydrogen, nitrogen, carbon, and water. Based on the best available observations, experts now consider the nucleus of a comet to be composed of dust particles plus some small rocky fragments all trapped within a loosely packed mixture of methane, ammonia, and ordinary water ice, the density of the mixture being about 100 kg/m³. Even as atoms, molecules, and dust particles boil off into space, creating a comet's coma and tail, the nucleus remains frozen at a temperature of only a few tens of kelvins. Comets are often described as dirty snowballs, a term coined by Fred Whipple of Harvard University. Estimates of typical cometary masses range from 10^{12} to 10^{16} kg, comparable to the masses of small asteroids.

Their highly elliptical orbits take most comets far beyond Pluto, where, in accordance with Kepler's second law, they spend most of their time. ∞ (Sec. 1.6) The majority of comets take hundreds of thousands, even millions, of years to complete a single orbit around the Sun. However, a few *short-period* comets (conventionally defined as those having a period of less than 200 years) return for another encounter within a relatively short period of time. According to Kepler's

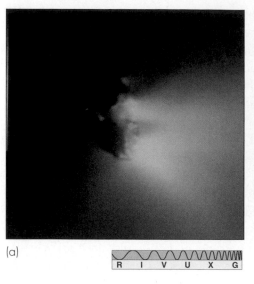

(a)

R I V U X G

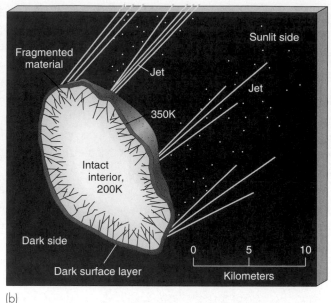

(b)

Figure 4.10 (a) The *Giotto* spacecraft resolved the nucleus of Halley's Comet, showing it to be very dark, although heavy dust in the area obscured any surface features. Resolution here is about 50 m. The Sun is toward the right in this image. The brightest areas are jets of evaporated gas and dust spewing from the comet's nucleus. (b) A diagram of the nucleus of Halley's Comet.

third law, short-period comets never venture far beyond the orbit of Pluto.

Unlike the orbits of other solar system objects, the orbits of comets are *not* confined to within a few degrees of the ecliptic plane. Short-period comets do tend to have prograde orbits lying close to the ecliptic, but long-period comets exhibit all inclinations and all orientations, both prograde and retrograde, roughly uniformly distributed in all directions from the Sun.

Astronomers believe that short-period comets originate beyond the orbit of Neptune, in a region of the outer solar system called the **Kuiper Belt** (after Gerard Kuiper, a pioneer in infrared and planetary astronomy). A little like the asteroids in the inner solar system, most Kuiper Belt comets move in roughly circular orbits between about 30 and 100 A.U. from the Sun, never venturing inside the orbits of the jovian planets. Occasionally, however, either a close encounter between two comets or (more likely) the cumulative gravitational influence of an outer planet "kicks" a Kuiper Belt comet into an eccentric orbit that brings it into the inner solar system and into our view. The observed orbits of these comets reflect the flattened structure of the Kuiper Belt.

What of long-period comets? How do we account for their apparently random orbital orientations? Only a tiny portion of a typical long-period cometary orbit lies within the inner solar system, so it follows that, for

every comet we see, there must be many more similar objects far from the Sun. On these general grounds, many astronomers reason that there must be a huge "cloud" of comets lying far beyond the orbit of Pluto, completely surrounding the Sun. It is named the **Oort Cloud**, after the Dutch astronomer Jan Oort, who first wrote (in the 1950s) of the possibility of such a vast and distant reservoir of inactive, frozen comets. The Kuiper Belt and the orbits of some typical Oort Cloud comets are sketched in Figure 4.11.

Based on the observed orbital properties of long-period comets, researchers believe that the Oort Cloud may be up to 100,000 A.U. in diameter. Like their Kuiper Belt counterparts, however, most Oort Cloud comets never come anywhere near the Sun. Indeed, Oort Cloud comets rarely approach even the orbit of Pluto, let alone that of Earth. Only when the gravitational field of a passing star happens to deflect a comet into an extremely eccentric orbit that passes through the inner solar system do we get to see it at all. Because the Oort Cloud surrounds the Sun in all directions instead of being confined to the ecliptic plane like the Kuiper Belt, the long-period comets we see can come from any direction in the sky. Despite their great distances and long orbital periods, however, Oort Cloud comets are still gravitationally bound to the Sun. Their orbits are governed by the same laws of motion that control the planets.

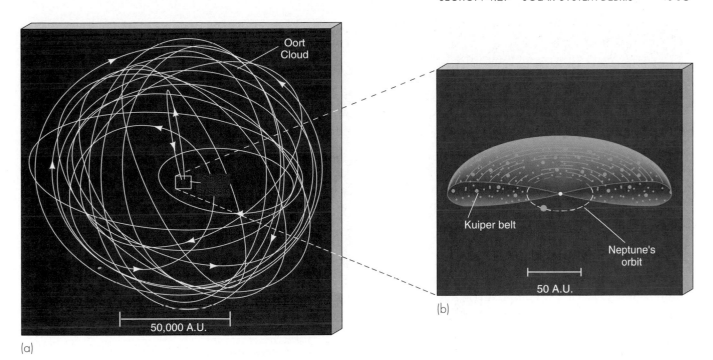

(a)

(b)

Figure 4.11 (a) Diagram of the Oort Cloud, showing a few cometary orbits. Of all the orbits shown, only the most elongated ellipse represents a comet that will enter the solar system (which, on the scale of this drawing, is much smaller than the red dot at the center of the figure) and possibly become visible from Earth. (b) The Kuiper Belt, believed to be the source of short-period comets.

Meteoroids   *Extension*

5 On a clear night, it is possible to see a few *meteors*—"shooting stars"—every hour. A **meteor** is a sudden streak of light in the night sky caused by friction between air molecules in Earth's atmosphere and an incoming piece of asteroid, meteoroid, or comet. This friction heats and excites the air molecules, which then emit light as they return to their ground states, producing the characteristic bright streak shown in Figure 4.12. Note that the brief flash that is a meteor is in no way similar to the broad, steady

(a)

(b)

R   I   V   U   X   G

Figure 4.12 A bright streak of light is produced when a fragment of interplanetary debris plunges into the atmosphere, heating the air to incandescence. (a) A small meteor photographed against a backdrop of stars. (b) The Northern Lights provide the background for a brighter meteor trail.

4-1 INTERLUDE

COMETS HYAKUTAKE AND HALE-BOPP

One of the most spectacular comets in recent years was Comet Hyakutake 1996. Named after a Japanese amateur astronomer who noticed it as "something odd and out of place" while scanning the skies with a pair of binoculars, Hyakutake grew from a small smudge while still far from the Sun into a splendid display comprising a huge coma nearly the apparent size of the Moon and a tail that eventually stretched a third of the way across the sky. The accompanying figure shows a *Hubble Space Telescope* image of Hyakutake taken in March 1996, when the comet passed closest to Earth—only 15 million km (0.1 A.U.) away. The comet's icy nucleus, the brightest point in the image, is unresolved here (the field of view is about 1000 km across), but radar pulses sent toward Hyakutake did return an echo, indicating that the diameter of the nucleus was 1 to 3 km. In the image, the Sun is out of the frame at bottom right, and the innermost part of the comet's tail is at upper left (on the side opposite the Sun, as explained in the text).

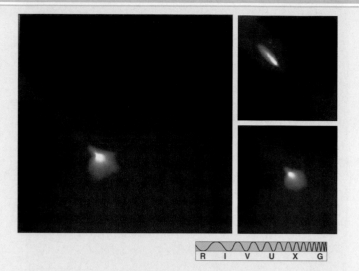

Other images showed sporadic jets pointing mostly sunward. These jets are gases gushing from the side of the comet closest to the Sun before wrapping around to become part of the graceful tail. The comet was examined at every conceivable wavelength, but perhaps the most surprising result was the intense X rays emitted from its head. Even the sunward side of the comet was far too cool to emit X rays, which are usually associated with very-high-temperature phenomena. Astronomers speculate that the X rays were produced by shock waves created as the solar wind hit the leading edge of the comet's coma.

---

swath of light associated with a comet's tail. A meteor is a fleeting event in Earth's atmosphere whereas a comet tail exists in deep space, and can be visible in the sky for weeks or even months.

Before encountering the atmosphere, the chunk of debris causing a meteor was almost certainly a meteoroid, simply because these small interplanetary fragments are far more common than either asteroids or comets. Any piece of interplanetary debris that survives its fiery passage through our atmosphere and finds its way to the ground is called a **meteorite**.

Smaller meteoroids are mainly the rocky remains of broken-up comets. Each time a comet passes near

the Sun, some fragments dislodge from the main body. The fragments initially travel in a tightly knit group of dust or pebble-sized objects called a **meteoroid swarm**, moving in nearly the same orbit as the parent comet. Over the course of time, the swarm gradually disperses along the orbit, so that eventually the **micrometeoroids**, as these small meteoroids are known, become more or less smoothly spread around the parent comet's orbit. If Earth's orbit happens to intersect the orbit of such a young cluster of meteoroids, a spectacular *meteor shower* can result. Earth's motion takes it across a given comet's orbit at most twice a year (depending on the precise orbit of each body). Because

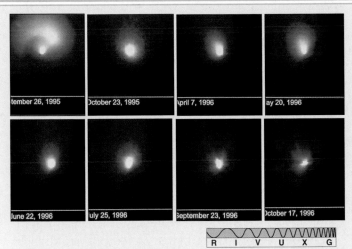

tember 26, 1995 | October 23, 1995 | April 7, 1996 | May 20, 1996

June 22, 1996 | July 25, 1996 | September 23, 1996 | October 17, 1996

R I V U X G

The comet or asteroid that struck the Earth 65 million years ago (see *Interlude 18-1*), perhaps causing the extinction of the dinosaurs, is thought to have been about 10–15 kilometers in size.

The accompanying images show a series of *Hubble* telescope observations of the inner part of Comet Hale-Bopp. The colors are artificial, with white representing the brightest parts of the comet, and red less bright; resolution is nearly 500 kilometers, far larger than the estimated size of the icy nucleus itself. These eight pictures were taken over the course of a year, from September, 1995, to October, 1996, as the comet neared the Sun.

Astronomers used multiple images like these to chronicle the evolution of the comet's nucleus. Especially interesting is the dust outburst in the first photo at top left—even when Hale-Bopp was still far beyond the orbit of Jupiter. At bottom right, as the comet neared the Sun, multiple jets are seen emanating from the surface. The impression that comets dramatically increase their activity while approaching the Sun seems amply confirmed in this time-sequence. *Hubble* was unable to follow the comet even closer to the Sun (which it rounded on April 1st, some 0.9 A.U. from the Sun), for fear of damaging its delicate optics.

Hot on the heels of Hyakutake came another interplanetary vagabond—Comet Hale-Bopp 1995. Discovered in 1995 by two American amateur astronomers, it reached its maximum brightness in the spring of 1997. Outshining everything in the night sky except the Moon and the brightest planets and stars, Hale-Bopp was probably the most widely viewed and studied comet in history.

This comet's unusual brightness and long (20°) tail was probably caused by a huge nucleus—about 30–40 kilometers in diameter. That's a very large ball of dirty ice, compared to the average comet core which measures some 3–5 kilometers across.

---

the intersection occurs at the same time each year (Figure 4.13), the appearance of certain meteor showers is a regular and (fairly) predictable event (Table 4.3).

Meteor showers are usually named for their *radiant*, the constellation from whose direction they appear to come. For example, the Perseid shower is seen to emanate from the constellation Perseus. It can last for several days, but reaches maximum every year on the morning of August 12, when upward of 50 meteors per hour can be observed. Astronomers use the speed and direction of a meteor's flight to compute its interplanetary trajectory. This is how certain meteoroid swarms

have come to be identified with well-known comet orbits.

Larger meteoroids—those more than a few centimeters in diameter—are usually *not* associated with comets. Generally regarded as small bodies that have strayed from the asteroid belt, possibly as the result of asteroid collisions, these objects have produced most of the cratering on the surfaces of the Moon, Mercury, Venus, Mars, and some of the moons of the jovian planets. When these large meteoroids enter Earth's atmosphere at typical speeds of nearly 20 km/s, they produce energetic shock waves, or "sonic booms," as well as bright sky streaks and dusty trails of discarded

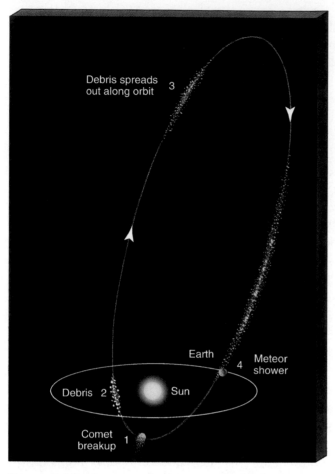

Figure 4.13 A meteoroid swarm associated with a given comet intersects Earth's orbit at specific locations, giving rise to meteor showers at specific times of the year. A portion of the comet breaks up as it rounds the Sun, at the point marked 1. The fragments continue along the comet orbit, gradually spreading out (points 2 and 3). The rate at which the debris disperses around the orbit is much slower than depicted here. It takes many orbits for the material to spread out as shown, but eventually the fragments extend all around the orbit, more or less uniformly. If the orbit happens to intersect Earth's, the result is a meteor shower each time Earth passes through the intersection (point 4).

debris. Such large meteors are sometimes known as *fireballs*.

More massive meteoroids—those at least a ton in mass and a meter across—do make it to Earth's surface, converting their kinetic energy (motion) to mechanical energy (damage), thermal energy (heat), and acoustical energy (sound) and producing a crater such as the kilometer-wide Barringer Crater shown in Figure 4.14. From the size of this crater, we can estimate that the meteoroid responsible must have had a mass of about 200,000 tons. Only 25 tons of iron meteorite

fragments have been found at the crash site. The remaining mass must have been scattered by the explosion at impact, broken down by subsequent erosion, or buried in the ground.

The orbits of large meteorites can sometimes be reconstructed in a manner similar to that used to determine the orbits of meteor showers. In most cases, their computed orbits do indeed intersect the asteroid belt, providing the strongest evidence we have that this is where they originated.

Currently, Earth is scarred with nearly 100 craters larger than 0.1 km in diameter. Most of these are so heavily eroded by weather and geological activity that they can be identified only in satellite photography, as shown in Figure 4.15. Fortunately, such major collisions between Earth and large meteoroids are thought to be rare events now. Researchers believe that, on

Table 4.3 Some Prominent Meteor Showers

Morning of Maximum Activity	Shower Name	Rough Hourly Count	Parent Comet
Jan. 3	Quadrantid	40	—
Apr. 21	Lyrid	10	1861I (Thatcher)
May 4	Eta Aquarid	20	Halley
June 30	Beta Taurid	25	Encke
July 30	Delta Aquarid	20	—
Aug. 12	Perseid	50	1862III (Swift–Tuttle)
Oct. 9	Draconid	up to 500	Giacobini–Zimmer
Oct. 20	Orionid	30	Halley
Nov. 7	Taurid	10	Encke
Nov. 16	Leonid	12[1]	1866I (Tuttle)
Dec. 13	Geminid	50	3200 Phaeton[2]

[1]Every 33 years, as Earth passes through the densest region of this meteoroid swarm, we see intense showers that can reach 1000 meteors per minute for brief periods of time. This is next expected to occur in 1999.

[2]Phaeton is actually an asteroid and shows no signs of cometary activity, but its orbit matches the meteoroid paths very well.

Figure 4.14 The Barringer Meteor Crater, near Winslow, Arizona, is 1.2 km in diameter and 0.2 km deep, the result of a meteorite impact about 25,000 years ago. The meteoroid was probably about 50 m across and likely weighed around 200,000 tons.

average, they occur only once every few hundred thousand years (see *Interlude 18-1*).

One of the most recent documented meteoritic events occurred in central Siberia on June 30, 1908 (Figure 4.16). The presence of only a shallow depression as well as a complete lack of fragments implies that this Siberian intruder exploded several kilometers

above the ground, leaving a blasted depression at ground level but no well-formed crater. Recent calculations suggest that the object in question was a rocky meteoroid about 30 m across. The explosion, estimated to have been equal in energy to a 10-megaton nuclear detonation, was heard hundreds of kilometers away and produced measurable increases in atmospheric dust levels all across the Northern Hemisphere.

One feature that distinguishes micrometeoroids, which burn up in Earth's atmosphere, from larger meteoroids, which reach the ground, is composition. The average density of meteors too small to reach the

R I V U X G

Figure 4.15 This photograph, taken from orbit by the U.S. *Skylab* space station, shows the ancient impact basin that forms Quebec's Manicouagan Reservoir. A large meteorite landed there about 200 million years ago. The central floor of the crater rebounded after the impact, forming an elevated central peak. The lake, 70 km in diameter, now fills the resulting ring-shaped depression.

Figure 4.16 The Tunguska event of 1908 leveled trees over a vast area. Although the impact of the blast was tremendous and its sound audible for hundreds of kilometers, the Siberian site is so remote that little was known about the event until scientific expeditions arrived to study it many years later.

(a)   (b)

Figure 4.17 (a) A stony meteorite often has a dark crust, created when its surface is melted by the tremendous heat generated during passage through the atmosphere. (b) Iron meteorites, much rarer than stony ones, usually contain some nickel as well. Most iron meteorites show characteristic crystalline patterns when their surfaces are cut, polished, and etched with acid.

ground (but captured by high-flying aircraft) is about 500–1000 kg/m^3. Such a low density is typical of comets, which, as we have just seen, are made of loosely packed ice and dust. By contrast, the meteorites that reach Earth's surface are often much denser—up to 5000 kg/m^3—suggesting a composition more like that of asteroids. Meteorites like the ones shown in Figure 4.17 have received close scrutiny from planetary scientists. Prior to the Space Age, meteorites were the only type of extraterrestrial matter we could touch and examine in terrestrial laboratories.

Most meteorites are rocky (Figure 4.17a) although a few are composed mainly of iron and nickel (Figure 4.17b). Their basic composition is much like that of the rocky inner planets or the Moon, except that some of their lighter elements—such as hydrogen and oxygen—are depleted. It is thought that these light elements boiled away long ago when the bodies from which the meteorites originated were molten. Some meteorites show clear evidence of strong heating at some time in their past, indicating they originated from bodies that either experienced some geological activity or were partially melted during the collision that liberated the fragments that eventually became the meteorites. Others show no such evidence and probably date back to the formation of the solar system.

Most primitive of all are carbonaceous meteorites, black or dark gray and most likely related to carbonaceous asteroids. (Similarly, the silicate-rich rocky meteorites are probably associated with silicate asteroids.) Many carbonaceous meteorites contain significant amounts of ice and other volatile substances, and they are usually rich in organic molecules.

Finally, almost all meteorites are *old*. Radioactive dating shows most of them to be between 4.4 and 4.6 billion years old—roughly the age of the oldest Moon rocks brought back to Earth by *Apollo* astronauts. Meteorites, along with asteroids, comets and some lunar rocks, provide essential clues to the original state of matter in the solar neighborhood and to the birth of our planetary system.

4.3 The Formation of the Solar System

You might be struck by the vast range of physical and chemical properties found in the solar system. Indeed, our astronomical neighborhood may seem more like a great junkyard than a smoothly running planetary system. Can we really make any sense of solar system matter? Is there some underlying principle that unifies the facts we have outlined in this chapter? Remarkably, the answer is "yes." The origin of the solar system is a complex and as yet incompletely solved puzzle, but the basic outlines are now quite well understood.

Despite the recent widely publicized discoveries of planets orbiting other stars (see *Interlude 4-2* and Chapter 18), astronomers still have little detailed information on their properties, and so far there is no firm evidence that planets like our own exist anywhere beyond our solar system. For that reason, our theories of planet formation concentrate on the planetary system in which we live. Bear in mind, however, that no part of the scenario we describe here is in any way unique to our own system. The same basic processes could have occurred—and, many astronomers believe, probably did occur—during the formative stages of most of the stars in our Galaxy.

Model Requirements

6 Based on the measured ages of the oldest meteorites, as well as Earth and lunar rocks, planetary scientists believe that the age of the solar system is 4.6 billion years. What happened 4.6 billion years ago to create the planetary system we see today? Any theory of the origin and architecture of our planetary system must adhere to these nine known facts:

1. *Each planet is relatively isolated in space.* The planets exist as independent bodies at progressively larger distances from the central Sun; they are not bunched together.

2. *The orbits of the planets are nearly circular.* In fact, with the exceptions of Mercury and Pluto, which we will argue are special cases, each planetary orbit closely describes a perfect circle.

3. *The orbits of the planets all lie in nearly the same plane.* The planes swept out by the planets' orbits are accurately aligned to within a few degrees. Again, Mercury and Pluto are slight exceptions.

4. *The direction in which the planets orbit the Sun (counterclockwise as viewed from above Earth's north pole) is the same as the direction in which the Sun's rotates on its axis.* Virtually all the large-scale motions in the solar system (other than comet orbits) are in the same plane and in the same sense. The plane is that of the Sun's equator, and the sense is that of the Sun's rotation.

5. *The direction in which most planets rotate on their axis is roughly the same as the direction in which the Sun rotates on its axis.* This property is less general than the one just described for revolution, as three planets—Venus, Uranus, and Pluto—do not share it.

6. *The direction in which most of the known moons revolve about their parent planet is the same as the direction in which the planets rotate on their axis.*

7. *Our planetary system is highly differentiated.* The terrestrial planets are characterized by high densities, moderate atmospheres, slow rotation rates, and few or no moons. The jovian planets (Pluto excepted, as usual) have low densities, thick atmospheres, rapid rotation rates, and many moons.

8. *Asteroids are very old and exhibit a range of properties not characteristic of either the terrestrial or the jovian planets or their moons.* Asteroids share, in rough terms, the bulk orbital properties of the planets. However, they appear to be made of primitive, unevolved material, and the meteorites that strike Earth are the oldest rocks known.

9. *Comets are primitive, icy fragments that do not necessarily orbit in the ecliptic plane and reside primarily at large distances from the Sun, in the Kuiper Belt and the Oort Cloud.*

All these observed facts taken together strongly suggest a high degree of order in our solar system. The system is not a random assortment of objects spinning or orbiting this way or that. It is not possible that our solar system could have formed by the slow accumulation of ready-made interstellar "planets" casually captured by our Sun over the course of billions of years. The large-scale architecture is too neat, and the ages of the components too uniform, to be the result of random chaotic events. The overall organization points toward a single formation, an ancient but one-time event, 4.6 billion years ago. A convincing theory that explains all of the nine features just listed has been a goal of astronomers for centuries.

It is important to recognize what our theory of the solar system does *not* have to explain. There is plenty of scope for planets to have evolved after their formation, and so circumstances that have developed in the eons since the solar system formed need not be included in our list. In the next few chapters, we will see many planetary properties for which a satisfactory *evolutionary* explanation exists. However, the items in the preceding list are *not* evolutionary in nature. For example, Newton's laws imply that the planets must move in elliptical orbits with the Sun at one focus, but they offer no explanation of why the observed orbits should be roughly circular, coplanar, and prograde. Knowing of no way in which the planets could have started off in random paths, then later evolved into the orbits we see today, we must conclude that the basic orbital properties were established at the outset.

Finally, in addition to its many regularities, our solar system also has many notable *irregularities*, some of which we have already mentioned. Far from threatening our theory, however, these irregularities are important facts for us to consider in shaping our explanations. For example, any theory explaining solar system formation must not insist that *all* planets rotate in

4-2 INTERLUDE

THE DISCOVERY OF PLANETS BEYOND THE SOLAR SYSTEM

Many claims of evidence for extrasolar planets have been reported (and even published) over the past few decades, but virtually none have been confirmed and most have been discredited. Either the data were inaccurate, or the analyses flawed, or the astronomers involved too eager to release their results. Only within the past few years have we seen genuine advances in this fascinating astronomical specialty. These advances have come not through any dramatic scientific or technical breakthrough but through steady improvements in both telescope and detector technology and computerized data analysis.

It is not yet possible to image any of these newly discovered planets. The techniques used to find them are indirect, based on analysis of light from the parent star and not light from the unseen planet. As a planet orbits a star, gravitationally pulling one way and then the other, the star wobbles slightly. The more massive the planet or the less massive the star or the closer the planet to the star, the greater the star's movement. If the wobble happens to occur along our line of sight to the star, then we see small fluctuations in the star's velocity (which can be measured using the Doppler effect, discussed in *More Precisely 2-2*). Alternatively, if the wobble is predominantly perpendicular to our line of sight, then the star's position in the sky changes slightly from night to night. The star's motion is very small, which is why unambiguous measurements have been hard to obtain. However, *both* types of wobble have now been seen and confirmed.

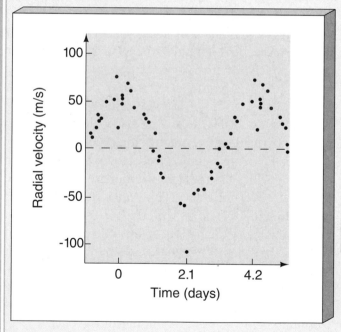

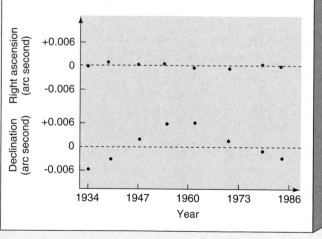

The figures on the previous page show two sets of data that betray the presence of planets. The figure on the left shows the line-of-sight velocity of the star 51 Pegasi, a near-twin to our Sun lying some 40 light years away. These data were acquired in 1994 by Swiss astronomers using the 1.9-m telescope at Haute-Provence Observatory in France. The regular 50-m/s fluctuations in the star's velocity have been confirmed by several groups of astronomers and imply that a planet at least half the mass of Jupiter orbits 51 Peg with a period of just 4.2 days.

The figure on the right plots, for the star Lalande 21185, right ascension and declination (corrected for the motion of Lalande through the Milky Way Galaxy and the motion of Earth around the Sun) over the last half century. A 30 year-period declination wobble at the level of about 0.01 arc seconds can be seen—about the amount of wobble that would be produced in our own Sun by the planets of the solar system. These observations have triggered a minirevolution in astronomy and have doubtless paved the way for a flood of new planet searches (and reinterpretation of old data) in the next few years.

Currently, these techniques can be used only to detect planets that are at least as large as Jupiter. Thus far, none of the reports of nearby extrasolar planets has claimed an object of Earth dimensions. In fact, many of the findings to date indicate "hot Jupiters"—gigantic planets surprisingly close (within 1 A.U.) to their parent star. Furthermore, only one such unseen planet has been found (and confirmed) in each system, although tentative analysis of the Lalande data suggests that there might be as many as three Jupiter-sized planets, each with a period of several years. So do not think that the newly discovered planetary systems closely resemble our own solar system. At least

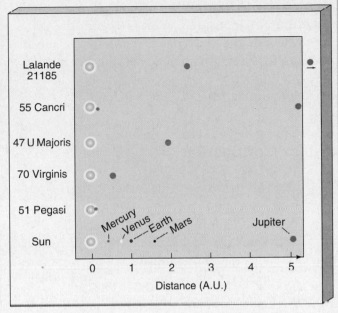

so far, they don't—and that has theorists puzzled.

The accompanying chart summarizes all the confirmed extrasolar planets found as of the end of 1996. Most appear to fall into the hot-Jupiter category, although some may well turn out to be *brown dwarfs*—"failed stars" having insufficient mass to become true stars. The dividing line between genuine Jupiter-like planets and brown dwarfs is still unclear, but it is thought to be around five Jupiter masses.

The number of extrasolar planets resembling those in our solar system still amounts to only a handful, but our awareness of whole new worlds is likely to grow quickly in the next few years as this field expands rapidly. It seems that we have now moved from the foggy realm of theoretical speculation into the clearer domain of hard observational fact.

the same sense or have only prograde moons because that is not what we observe. Instead, the theory should provide strong reasons for the observed planetary characteristics, yet be flexible enough to allow for and explain the deviations, too. And, of course, the existence of the asteroids and comets that tell us so much about our past must be an integral part of the picture. That's a tall order, yet many researchers now believe we are close to that level of understanding.

Nebular Contraction

7 One of the earliest heliocentric models of solar system formation is known as the **nebular theory**. A **nebula** is any large cloud of interstellar dust and gas. According to the nebular theory, a nebula began to collapse under the influence of its own gravity. As it contracted, it became denser and hotter, eventually forming a star—the Sun—at its center. While all this was going on, the outer, cooler, parts of the cloud formed a giant swirling region of matter, creating the planets and their moons essentially as byproducts of star formation. This swirling mass destined to become our solar system is usually referred to as the **solar nebula**.

In 1796 the French mathematician-astronomer Pierre Simon de Laplace showed mathematically that conservation of angular momentum (see *More Precisely 4-1*) demands that the hypothetical solar nebula had to spin faster as it contracted. The increase in rotation speed, in turn, caused the solar nebula's *shape* to change as it collapsed. As shown in Figure 4.18 the cloud eventually flattened into a pancake-shaped disk that was the primitive solar system. If we now suppose that planets formed out of this spinning disk, we can begin to understand the origin of some of the architecture observed in our planetary system today, such as the circularity of the planets' orbits and the fact that they move in nearly the same plane.

Astronomers are fairly confident that the solar nebula formed such a disk because similar disks have been observed (or inferred) around other stars. Figure 4.19 shows a visible-light image of the region around the star Beta Pictoris, which lies about 50 light years from the Sun. When the light from Beta Pictoris is suppressed and the resulting image computer-enhanced, a faint disk of warm matter (viewed almost edge-on here) can be seen. This particular disk is roughly 1000 A.U. across—about 10 times the diameter of Pluto's

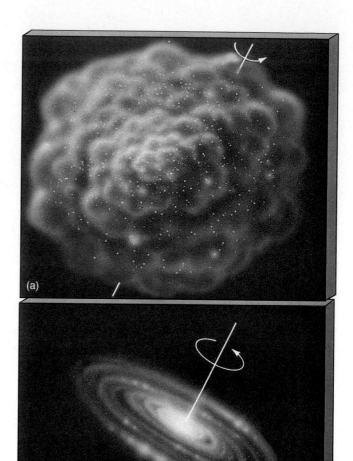

Figure 4.18 (a) Conservation of angular momentum demands that a contracting, rotating cloud must spin faster as its size decreases. (b) Eventually, the solar nebula, which is that particular nebula destined to become the solar system, came to resemble a gigantic pancake. The large blob at the center ultimately became the Sun.

orbit. Astronomers believe that Beta Pictoris is a very young star, perhaps only 100 million years old, and that we are witnessing it pass through an evolutionary stage similar to that experienced by our own Sun 4.6 billion years ago.

Laplace imagined that, as the spinning solar nebula contracted, it left behind a series of concentric rings, each of which would eventually become a planet orbiting a central **protosun**—a hot ball of gas well on its way to becoming the Sun. Each ring then clumped into a **protoplanet**—a forerunner of a genuine planet. This description of the collapse and flattening of the solar nebula is essentially correct, but when modern astronomers used computers to study the more subtle aspects of the problem, some fatal flaws were found in

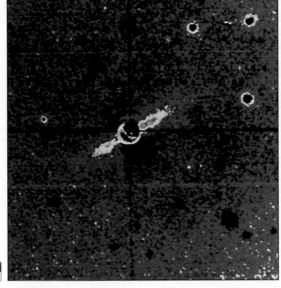

Figure 4.19 (a) A computer-enhanced photograph (taken from Las Campanas Observatory in Chile) of a disk of warm matter surrounding the star Beta Pictoris. Most of the light from the star is blocked by an instrument called a coronagraph, designed to detect faint halos around bright objects. The full extent of the disk, seen almost edge-on here, is about 1000 A.U. (b) An artist's conception of the disk of clumped matter.

R I V U X G

(a)

(b)

Laplace's nebular picture. Calculations show that a ring of the sort assumed in the theory would probably not form and, even if it did, would not condense to form a planet in any case. In fact, computer calculations predict just the opposite: the rings would tend to disperse.

The model currently favored by most astronomers is a more sophisticated version of the nebular theory. Known as the **condensation theory**, it combines the good features of the nebular theory with new information about interstellar chemistry to avoid most of the old theory's problems. The key new ingredient is *interstellar dust* in the solar nebula. Astronomers now recognize that the space between the stars is strewn with microscopic dust grains, an accumulation of the ejected matter of many long-dead stars. These dust particles probably formed in the cool atmospheres of old stars, then grew by accumulating more atoms and molecules from the interstellar gas. The end result is that interstellar space is littered with tiny chunks of icy and rocky matter having typical diameters of about 10^{-5} m. Figure 4.20 shows one of many such dusty regions in the vicinity of the Sun.

Dust grains play an important role in the evolution of any gas cloud. Dust helps to cool warm matter by efficiently radiating heat away in the form of infrared radiation, allowing the nebula to collapse more easily. Furthermore, the dust grains greatly speed up the process of collecting enough atoms to form a planet. They act as **condensation nuclei**—microscopic platforms to which other atoms can attach, forming larger and larger balls of matter. This is similar to the way that raindrops form in Earth's atmosphere; dust

Figure 4.20 Interstellar gas and dark dust lanes mark this region of star formation. The dark cloud known as Barnard 86 (left) flanks a cluster of young blue stars called NGC 6520 (right). Barnard 86 may be part of a larger interstellar cloud that gave rise to these stars.

R I V U X G

4-1 MORE PRECISELY

THE CONCEPT OF ANGULAR MOMENTUM

Most celestial objects rotate. Planets, moons, stars, galaxies all have some *angular momentum*, which we can define as the tendency of a body to either keep spinning or keep moving in a circle. Angular momentum is as important a property of an object as its mass or its energy.

Consider first a simpler notion—*linear momentum*, defined as the product of an object's mass and its speed:

linear momentum = mass × speed.

A truck and a bicycle rolling equally fast down a street each have linear momentum, but the linear momentum of the truck is much greater than that of the bicycle because the mass of the truck is so much greater than that of the bicycle. Because its momentum is less, you would find it much easier to stop the bicycle. Because linear momentum also depends on speed, if two bicycles of equal mass were rolling down the street at different speeds, the slower one would have the lower momentum and so could be stopped more easily.

Angular momentum is an analogous property of objects that are either rotating or revolving. However, in addition to mass and speed, angular momentum also depends on the way in which an object's mass is distributed. The farther the mass from the object's axis of rotation, the greater the angular momentum. For example, if you whirl a ball with constant speed at the end of a string, its angular momentum will depend directly on the length of the string—the longer the string, the greater the ball's angular momentum. We can therefore say, loosely, that

angular momentum = mass × speed × "size,"

where "size" is a quantity that depends not only on the object's dimensions but also on the distribution of its mass. For the simple case of a ball on a string, size is just the length of the string.

According to Newton's laws of motion, both types of momentum—linear and angular—must be *conserved* at all times. In other words, both linear and angular momentum must remain constant before, during, and after a physical change in any object (so long as no external forces act). For example, if a spinning sphere begins to contract (in other words, its size decreases), the equation for angular momentum demands that it spin faster so that the product mass × speed × size stays constant. The sphere's mass does not change during the contraction, but its size decreases. The circular speed of the sphere must therefore increase in order to keep the total angular momentum unchanged. Figure skaters use the principle of angular-momentum conservation. They spin faster by drawing in their arms (as shown in the accompanying figure) and slow down by extending them. Here, the mass of the human body remains the same, but its size changes, causing the body's circular speed in order to change to conserve angular momentum.

and soot in the air act as condensation nuclei around which water molecules cluster.

Planet Formation

Modern condensation theory models trace the formative stages of our solar system along the following broad lines. Imagine a dusty interstellar cloud fragment measuring about a light year across. Intermingled with the preponderance of hydrogen and helium atoms were some heavier elements, in the form of both gas and dust. Some external influence, such as the passage of another interstellar cloud or perhaps the explosion of a nearby star, started the fragment contracting, down to a size of about 100 A.U. As the cloud collapsed, it rotated faster and began to flatten (just as described in the nebular theory). By the time it had shrunk to 100 A.U., the solar nebula had formed into an extended, rotating disk (Figure 4-21a, b).

According to the condensation theory, the planets formed in three stages (Figure 4.21c–e). Early on, dust grains in the solar nebula formed condensation nuclei around which matter began to accumulate. This vital step greatly hastened the critical process of forming the first small clumps of matter. Once these clumps formed, they grew rapidly by sticking to other clumps. (Imagine a snowball thrown through a fierce snowstorm, growing bigger as it encounters more snowflakes.) As the clumps grew larger, their surface areas increased and consequently the rate at which they swept up new material accelerated. They gradually grew into objects of pebble size, baseball size, basketball size, and larger.

Eventually, this process of **accretion**—the gradual growth of small objects by collision and sticking—created objects a few hundred kilometers across. By that time, their gravity was strong enough to sweep up material that would otherwise not have collided with them, so their rate of growth became faster still. At the end of this first stage, the solar system was made up of

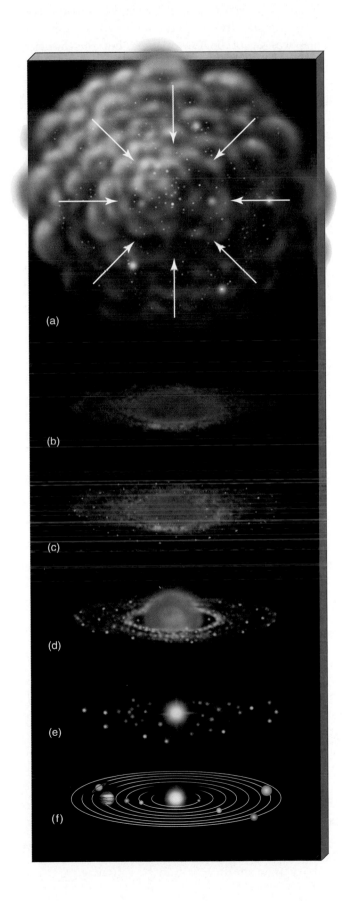

(a)

(b)

(c)

(d)

(e)

(f)

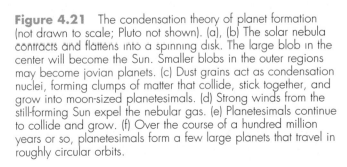

Figure 4.21 The condensation theory of planet formation (not drawn to scale; Pluto not shown). (a), (b) The solar nebula contracts and flattens into a spinning disk. The large blob in the center will become the Sun. Smaller blobs in the outer regions may become jovian planets. (c) Dust grains act as condensation nuclei, forming clumps of matter that collide, stick together, and grow into moon-sized planetesimals. (d) Strong winds from the still-forming Sun expel the nebular gas. (e) Planetesimals continue to collide and grow. (f) Over the course of a hundred million years or so, planetesimals form a few large planets that travel in roughly circular orbits.

hydrogen and helium gas and millions of **planetesimals**—objects the size of small moons, having gravitational fields just strong enough to affect their neighbors. In the second phase of accretion, gravitational forces between the planetesimals caused them to collide and merge, forming larger and larger objects. Because larger objects exert stronger gravitational pulls, the rich became richer in the early solar system, and eventually almost all the planetesimal material was swept up into a few large protoplanets—the accumulations of matter that would eventually evolve into the planets we know today.

As the protoplanets grew, another process became important. The strong gravitational fields produced many high-speed collisions between planetesimals and protoplanets. These collisions led to **fragmentation** as small objects broke into still smaller chunks that were then swept up by the protoplanets. Not only did the rich get richer, but the poor were mostly driven to destruction! Only a relatively small number of 10–100-km fragments escaped capture by a planet or a moon and became the asteroids and comets. After about 100 million years, the primitive solar system had evolved into nine protoplanets, dozens of protomoons, and a big protosolar mass at the center. Computer simulations generally reproduce the increasing spacing between the planets, although the reasons for the regularity seen in the actual planetary spacing remain unclear. Roughly a billion more years were required to sweep the system clear of interplanetary trash.

The four largest protoplanets became large enough to enter a third phase of planetary development: sweeping up large amounts of gas from the solar nebula to form what would ultimately become the jovian planets. The smaller, inner protoplanets never reached that point, and as a result their masses remained relatively low. Alternatively, the jovian planets may have formed through instabilities in the cool outer regions of the solar nebula, mimicking on small scales the collapse of the initial interstellar cloud. In this way, the jovian protoplanets might have formed directly, skipping the initial accretion stage. These first protoplanets would have had gravitational fields strong enough to scoop up more of the remaining gas and dust in the solar nebula, allowing them to grow into the gas giants we see today. Their large size reflects their head start in the accretion process.

What of the gas that made up most of the original cloud? Why don't we see it today throughout the planetary system? In the outer solar system, some of that gas was swept up into planets, but this did not occur in the inner regions, where the terrestrial protoplanets never became massive enough to accrete light material. Instead, the newly formed Sun took a hand. All young stars apparently experience a highly active evolutionary stage known as the *T-Tauri phase* (Chapter 11), during which their radiation emission and stellar winds are very intense. When our Sun entered this phase, just before nuclear burning started at its center, any gas remaining between the planets was blown away into interstellar space by the solar wind and the Sun's radiation pressure. Afterwards, all that remained were protoplanets and planetesimal fragments, ready to continue their long evolution into the solar system we know today.

4.4 The Differentiation of the Solar System

6 We can use the condensation theory to understand the basic differences in the terrestrial planets, jovian planets, and smaller bodies that constitute the solar system. Indeed, it is in this context that the term *condensation* derives its true meaning. To see why the composition of a planet, moon, or interplanetary fragment depends on its location in the solar system, it is necessary to consider the temperature structure of the solar nebula.

The Role of Heat

As the solar nebula contracted under the influence of gravity, it heated up as it flattened into a disk. The density and temperature were greatest near the central protosun and much lower in the outlying regions. In the warmer regions, dust grains broke apart into molecules, which in turn split into excited atoms. Because the extent to which the dust was destroyed depended on temperature, it also depended on location in the solar nebula. Most of the original dust in the inner solar system disappeared at this stage, but the grains in the outermost parts probably remained largely intact.

The destruction of the dust in the hot inner portion of the solar nebula introduced an important new ingredient into the theoretical mix, one that we omitted from our earlier description of accretion. With the passage of time, the gas radiated away its heat, and the temperature decreased at all locations except in the very core, where the Sun was forming. Everywhere beyond the protosun, new dust grains began to condense out, much as raindrops, snowflakes, and hailstones condense from moist, cooling air here on Earth. It may seem strange that although there was plenty of interstellar dust early on, it was mostly destroyed, only to form again later. However, a critical change had occurred. Initially, the nebular gas was uniformly peppered with dust grains. When the dust reformed later, the distribution of grains was very different.

Figure 4.22 plots the temperature in various parts of the primitive solar system just prior to the onset of accretion. At any given location, the only materials to condense out were those able to survive the temperature there. As marked on the figure, in the innermost regions, around Mercury's present orbit, only metallic

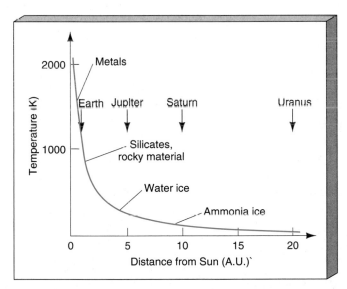

Figure 4.22 Theoretically computed variation of temperature across the primitive solar nebula. In the hot central regions, only metals could condense out of the gaseous state to form grains. At greater distances from the central protosun, the temperature was lower, so rocky and icy grains could also form. The labels indicate the minimum radii at which grains of various types could condense out of the nebula.

grains could form—it was simply too hot for anything else to exist. A little farther out, at about 1 A.U., it was possible for rocky, silicate grains to form, too. Beyond about 3 or 4 A.U., water ice could exist, and so on, with the condensation of more and more material possible at greater and greater distances from the Sun. The composition of the material that could condense out at any given distance from the Sun ultimately determined the type of planet that formed there.

The Jovian Planets

In the middle and outer regions of the primitive planetary system, beyond about 5 A.U. from the center, the temperature was low enough to allow several abundant gases to condense into solid form. After hydrogen and helium, the most common materials in the solar nebula (as they are today in the universe as a whole) were the elements carbon, nitrogen, and oxygen. The most common chemical compounds were those containing those elements—specifically, water vapor, ammonia, and methane. These compounds are still the primary constituents of jovian atmospheres. At temperatures of a few hundred kelvins or less, these gases condensed out of the nebula. Consequently, the ancestral fragments destined to become the cores of the jovian planets were formed under cold conditions out of low-density, icy material. The planetesimals that formed at these distances were predominantly composed of ice. Because more material could condense out of the solar nebula at these radii than in the inner regions near the protosun, accretion began sooner, with more resources to draw on. The outer planets grew rapidly to the point where they could accrete not just grains but nebular gas also, and the eventual result was the hydrogen-rich jovian worlds we see today.

With the formation of the four jovian planets, the remaining planetesimals were subject to those planets' strong gravitational fields. Over a period of hundreds of millions of years and after repeated gravitational "kicks" from the giant planets, many of the interplanetary fragments in the outer solar system were flung into orbits taking them far from the Sun. Astronomers believe that those fragments now make up the Oort Cloud. During this period, many icy planetesimals were also deflected into the inner solar system, where they played an important role in the evolution of the terrestrial planets. A key prediction of this model is that some of the original planetesimals should have remained behind, in the Kuiper Belt beyond the orbit of Neptune. In 1993, several such asteroid-sized objects were discovered, lying between 30 and 35 A.U. from the Sun, lending strong support to the condensation theory. Over 30 Kuiper Belt objects, having diameters ranging from 100 to 400 km, are now known.

The Terrestrial Planets

In the inner regions of the primitive solar system, condensation from gas to solid began when the average temperature was about 1000 K. The environment there was too hot for ices to survive. Many of the abundant heavier elements, such as silicon, iron, magnesium, and aluminum, combined with oxygen to produce a variety of rocky materials. Planetesimals in the inner solar system were therefore rocky in nature, as were the protoplanets and planets they ultimately formed.

These heavier materials condensed into grains in the outer solar system, too, of course. However, there they would have been vastly outnumbered by the far more abundant light elements. The more accurate way to describe the solar system is not that outer solar system is deficient in heavy elements but rather that the inner solar system is *underrepresented in light material.* Here we have another reason the jovian planets grew so much bigger than the terrestrial worlds. The inner regions of the nebula had to wait for the temperature to drop so that a few rocky grains could appear and begin the accretion process, but the outer regions did not have to wait at all. Accretion in the outer solar system began almost as soon as the solar nebula collapsed into a disk.

Very abundant light elements such as hydrogen and helium, as well as any other gases that failed to condense into solids, either escaped from the terrestrial protoplanets or, more likely, were simply never accreted from the solar nebula. The inner planets' surface temperature was too high, and their gravity too low, to capture and retain those gases. Where then did Earth's volatile gases, particularly water, come from? The answer seems to be that icy fragments—comets—

from the outer solar system, deflected into eccentric orbits by the jovian planets' gravity, participated in the meteoritic bombardment of the newly born inner planets, supplying them with water after their formation.

The myriad rocks of the asteroid belt failed to accumulate into a protoplanet, probably because nearby Jupiter's huge gravitational field prevented them from doing so. The result is a band of planetesimals, still colliding and occasionally fragmenting, but never coalescing into a larger body—surviving witnesses to the birth of the planets.

Random Encounters in the Solar Nebula

The condensation theory accounts for the nine facts listed at the start of Section 4.3. The growth of planetesimals throughout the solar nebula, with each protoplanet ultimately sweeping up the material near it, accounts for the planets' wide spacing (point 1, although the theory does not adequately explain the regularity of the spacing). That the planets' orbits are circular (2), in the same plane (3), and in the same direction as the Sun's rotation on its axis (4) is a direct consequence of the solar nebula's shape and rotation. The rotation of the planets (5) and the orbits of the moon systems (6) are due to the tendency of the smaller structures to inherit the nebula's overall sense of rotation. The heating of the nebula and the Sun's ignition resulted in the observed differentiation (7), while the debris from the accretion–fragmentation stage naturally accounts for the asteroids (8) and comets (9).

We stressed earlier that an important aspect of any theory of solar system formation is its ability to accommodate deviations. In the condensation theory, that capacity is provided by the randomness inherent in the encounters that combined planetesimals into protoplanets. As the numbers of large bodies decreased and their masses increased, individual collisions acquired greater and greater importance. The effects of these collisions can still be seen today in many parts of the solar system. A case in point is the anomalous rotation rate of Venus (mentioned earlier), which can be explained if we assume that the last major collision in the formation history of that planet just happened to involve a near-head-on encounter between two protoplanets of comparable mass. Scientists usually do not like to invoke random events to explain observations. However, as we will discover in upcoming chapters, there are many instances where pure chance has played an important role in determining the present state of the universe.

Chapter Review

Summary

The planets that make up our **solar system** (p. 104) all orbit the Sun counterclockwise, as viewed from above Earth's North Pole, on roughly circular orbits that lie close to the ecliptic plane. The orbits of the innermost planet, Mercury, and the outermost, Pluto, are the most eccentric and also have the greatest orbital inclination. The spacing between planetary orbits increases as we move outward from the Sun.

Density (p. 105) is a measure of the compactness of any object. The average density of a planet is obtained by dividing the planet's total mass by its volume. The four planets closest to the Sun—Mercury, Venus, Earth, Mars—are the **terrestrial planets** (p. 107). Jupiter, Saturn, Uranus, and Neptune are the **jovian planets** (p. 107). The terrestrial planets are all of comparable density and generally rocky, whereas all the jovian planets have much lower densities and are made up mostly of gaseous or liquid hydrogen and helium.

More than 4000 **asteroids** (p. 108) have been cataloged. Most orbit in a broad band called the **asteroid belt** (p. 108) lying between the orbits of Mars and Jupiter. Asteroids are probably primal rocks that never clumped together to form a planet. A few **Earth-crossing asteroids** (p. 109) have orbits that intersect Earth's orbit and will probably collide with our planet one day. The **Trojan asteroids** (p. 108) share Jupiter's orbit, remaining 60° ahead or behind that planet as it moves around the Sun. The largest asteroids are a few hundred kilometers across. Most are much smaller. Brighter silicate asteroids dominate the inner asteroid belt, while darker carbonaceous asteroids are more plentiful in the outer regions.

Comets (p. 108) are fragments of icy and rocky material that normally orbit far from the Sun. Unlike most other bodies in the solar system, their orbits are often highly elongated and not confined to the ecliptic plane. Most comets are thought to reside in the **Oort Cloud** (p. 114), a vast reservoir of cometary material tens of thousands of astronomical units across and completely surrounding the Sun. Comets having an orbital period of less than about 200 years are thought to originate not in the Oort Cloud but in the **Kuiper Belt** (p. 114), a broad band beyond the orbit of Neptune and lying roughly in the ecliptic plane. As a comet approaches the Sun, the surface ice of the comet begins to vaporize. We see the comet by the sunlight reflected from the dust and vapor released. The **nucleus** (p. 111), or core, of a comet may be only a few kilometers in diameter. It is surrounded by a **coma** (p. 111) of dust and gas. Surrounding this is an extensive invisible **hydrogen envelope** (p. 112). Stretching behind the comet is a long **tail** (p. 111), formed by the interaction between the cometary material and the **solar wind** (p. 113). The **ion tail** (p. 112) consists of ionized gas particles and always points directly away from the Sun. The **dust tail** (p. 113) is less affected by the solar wind and has a curved shape.

A **meteoroid** (p. 108) is a piece of interplanetary debris smaller than an asteroid and traveling in interstellar space. The dividing line between meteoroids and asteroids is usually taken to be a diameter of 100 m. Any meteoroid, asteroid, or comet fragment that enters Earth's atmosphere produces a **meteor** (p. 115), a bright streak of light across the sky. If the meteoroid, asteroid, or comet fragment causing the meteor reaches the ground, it is called a **meteorite** (p. 116). Each time a comet rounds the Sun, some cometary material becomes dislodged, forming a **meteoroid swarm** (p. 116)—a group of small **micrometeoroids** (p. 116) that travel in the comet's orbit. Larger meteoroids are pieces of material chipped off asteroids following collisions in the asteroid belt. Meteorite composition is thought to mirror the composition of the parent asteroids, and the few orbits that have been determined are consistent with an origin in the asteroid belt. Most meteorites are between 4.4 and 4.6 billion years old.

The organization of the solar system points toward formation as the product of an ancient, one-time event that occurred 4.6 billion years ago. An ideal theory of the solar system must provide strong reasons for the observed characteristics of our planetary system while at the same time be flexible enough to allow for deviations.

A **nebula** (p. 124) is any large cloud of interstellar dust and gas. In the **nebular theory** (p. 124) of solar system formation, the **solar nebula** (p. 124) began to collapse under its own gravity. As it did so, it began to spin faster, eventually forming a disk. **Protoplanets** (p. 124) formed in the disk and became planets, while the central **protosun** (p. 124) eventually evolved into the Sun. The **condensation theory** (p. 125) incorporates into the nebular theory the effects of particles of interstellar dust, which help to cool the solar nebula and act as **condensation nuclei** (p. 125), allowing the planet-building process to begin.

Small clumps of matter grew by **accretion** (p. 127), gradually sticking together and growing into moon-sized **planetesimals** (p. 128) whose gravitational fields were strong enough to accelerate accretion. Competing with accretion in the solar nebula was **fragmentation** (p. 128), the process whereby small bodies were broken up following collisions with larger ones. Eventually, only a few planet-sized objects remained. The planets in the outer solar system became so large that they could capture the hydrogen and helium gas in the solar nebula, forming the jovian worlds.

The terrestrial planets are rocky because they formed in the hot inner regions of the solar nebula, near the Sun, where only rocky and metallic materials condensed out. Farther out, the nebula was cooler, and so ices of water and ammonia could also form, leading to the observed differences in composition between the inner and outer solar system.

When the Sun became a star, its strong winds blew away any remaining gas in the solar nebula. Many leftover planetesimals were ejected into the Oort Cloud by the gravitational fields of the outer planets. Much, if not all, of Earth's water was carried to our world by comets deflected from the outer solar system. The asteroid belt is a collection of planetesimals that never managed to form a planet, probably because of Jupiter's gravitational influence.

Self-Test: True or False?

_____ 1. Most planets orbit the Sun in nearly the same plane as Earth.

_____ 2. The largest planets also have the highest densities.

_____ 3. All planets have moons.

_____ 4. Asteroids generally move on almost circular orbits.

_____ 5. Carbonaceous asteroids are most common in the inner portion of the asteroid belt.

_____ 6. Some comets travel as far away as 50,000 A.U. from the Sun.

_____ 7. Cometary orbits always lie close to the ecliptic plane.

_____ 8. The Oort Cloud is the extended cloud of gas surrounding a comet while it is near the Sun.

_____ 9. Tails of comets always lie along the comet's orbit.

_____ 10. The light emitted from a meteor is due to a meteoroid burning up in Earth's atmosphere.

_____ 11. Comets are the sources of meteor showers.

_____ 12. The direction of planetary revolution is the same as the direction of the Sun's rotation.

_____ 13. The solar system is highly differentiated.

_____ 14. Asteroids were recently formed from the collision and breakup of an object orbiting in the asteroid belt.

_____ 15. Most comets have short periods and orbit close to the ecliptic plane.

_____ 16. The terrestrial planets formed out of the original dust that made up the solar nebula.

_____ 17. Water could not have condensed out of the solar nebula any closer than 3 or 4 A.U. from the Sun.

_____ 18. Accretion occurred faster in the inner part of the solar system than in the outer regions.

Self-Test: Fill in the Blank

1. The two planets with the highest eccentricities and orbital tilts are _____ and _____.

2. Asteroids and meteoroids have a _____ composition, in contrast to comets, which have an _____ composition.

3. Most asteroids are found between the orbits of _____ and _____.

4. The largest asteroids are _____ kilometers in diameter; the smallest are only _____ meters across.

5. The Trojan asteroids share an orbit with _____.

6. Comets have orbits that are highly _____.

7. The nucleus of a comet is typically _____ kilometers across; its tail may be up to _____ long.

8. Passage of a comet near the Sun may leave a _____ moving in the comet's orbit.

9. The Kuiper Belt is the source of short-period _____.

10. When a meteoroid, asteroid, or comet fragment enters Earth's atmosphere, _____ you see a _____.

11. The oldest meteorites are _____ years old.

12. In formulating the condensation theory, astronomers realized the critical role played by _____ in the formation of small clumps of matter.

13. By the time planetesimals had formed, the accretion process was accelerated by the effect of _____.

14. In the final stage of accretion, the largest protoplanets were able to attract large quantities of _____ from the solar nebula.

15. Unlike the planetesimals that formed the terrestrial planets, those that formed the jovian planets were made up of _____ material.

16. High-speed collisions between planetesimals often led to _____ rather than accretion.

17. The large number of left-over planetesimals formed beyond 5 A.U. were destined to become _____.

18. The water now found on Earth was probably brought here by _____.

19. The reason the planetesimals of the asteroid belt did not form a larger object was probably the gravitational influence of _____.

20. Angular momentum depends on the mass, speed, and _____ of an object.

Review and Discussion

1. Name three differences between terrestrial and jovian planets.

2. Why are asteroids and meteoroids important to planetary scientists?

3. What are comets like when they are far from the Sun? What happens when they enter the inner solar system?

4. Where are most comets found?

5. Describe the various parts of a comet near the Sun.

6. What are the typical ingredients of a comet nucleus?

7. Explain the difference between a meteor, a meteoroid, and a meteorite.

8. What causes a meteor shower?

9. What do meteorites reveal about the age of the solar system?

10. Why can comets approach the Sun from any direction but asteroids generally orbit close to the ecliptic plane?

11. Why do meteorites contain information about the early solar system and yet Earth does not?

12. What might be the consequences of a 10-km meteorite striking Earth today?

13. Describe the basic features of the nebular theory of solar system formation and give three examples of how this theory explains some observed features of the present-day solar system.

14. Explain the difference between angular momentum and linear momentum.

15. What key ingredient in the modern condensation theory of solar system formation was missing from the nebular theory?

16. What are the two phases of accretion that played a role in forming the planets?

17. Why are the jovian planets so much larger than the terrestrial planets?

18. Which solar system objects, still observable today, resulted from fragmentation?

19. How did the temperature at various locations in the solar nebula determine planetary composition?

20. Describe the possible history of a single comet now visible from Earth, starting with its birth in the solar nebula somewhere near the planet Jupiter.

Problems

1. Suppose the average mass of each of 6000 asteroids (which includes a few yet-to-be-discovered ones) in the solar system is 10^{17} kg. Compare the total mass of all asteroids to the mass of Earth.

2. The largest asteroid, Ceres, has a radius 0.073 times the radius of Earth and a mass of 0.0002 Earth masses. How much would a 100-kg astronaut weigh on Ceres?

3. (a) Using the Section 1.6 version of Kepler's laws of planetary motion, calculate the orbital period of an Oort Cloud comet if the semimajor axis of the comet's orbit is 50,000 A.U. (b) A Kuiper Belt comet has an orbital period of 125 years. Calculate the semimajor axis of its orbit.

4. The orbital angular momentum of a planet in a circular orbit is the product of its mass times its orbital speed times its distance from the Sun. Compare the orbital angular momenta of Jupiter, Saturn, and Earth.

5. A typical comet contains some 10^{13} kg of water ice. How many comets would have to strike Earth in order to account for the roughly 2×10^{21} kg of water presently found on our planet? If this amount of water accumulated over a period of 0.5 billion years, how often would Earth be hit by a comet during that time?

Projects

1. The only way to tell an asteroid from a star is to watch over several nights so that you can detect the asteroid's movement in front of the star background. The astronomy magazines *Sky & Telescope* and *Astronomy* often publish charts for prominent asteroids, the three brightest of which are Ceres, Pallas, and Vesta. Use one of these charts to locate the appropriate star field for one of these asteroids. Aiming binoculars at that star field, you may be able to pick out the asteroid from its location in the chart. If you can't, make a rough drawing of the field. Come back a night or two later, and look again. The "star" that has moved is the asteroid.

2. Although a spectacular naked-eye comet comes along only about once a decade, fainter comets can be seen with binoculars and telescopes in the course of every year. *Sky & Telescope* often runs a "Comet Digest" column announcing the whereabouts of comets. In addition, a comprehensive list of periodic comets expected to return in a given year can be found in Guy Ottewell's Astronomical Calendar. This calendar contains a wealth of other sky information as well, including monthly star charts. At the time of this writing, it costs $15 a year and can be purchased from Astronomical Workshop, Furman University, Greenville, South Carolina 29613 (803-294-2208).

3. There are a number of major meteor showers every year, but if you plan to watch one, be sure to notice the phase of the Moon because bright moonlight or city lights can obliterate a meteor shower. A common misconception about meteor watching is that most meteors are seen in the direction of the shower's radiant point. It's true that if you trace the paths of the meteors backward in the sky, they all can be seen to come from the radiant. However, most meteors don't become visible until they are 20 or 30 degrees from the radiant. Meteors can appear in all parts of the sky! Just relax and let your eyes rove among the stars. You will generally see many more meteors in the hours before dawn than in the hours after sunset. Why do you suppose meteors have different brightnesses? Can you detect their variety of colors? Watch for meteors that appear to "explode" as they fall, and for vapor trails that linger after the meteor has disappeared.

5 Earth and Its Moon

Our Cosmic Backyard *www*

LEARNING GOALS

Studying this chapter will enable you to:

1 Summarize and compare the basic properties of Earth and the Moon, and explain why the two bodies differ.

2 Describe the consequences of gravitational interactions between Earth and the Moon.

3 Discuss how Earth's atmosphere helps heat us as well as protect us.

4 Explain how dynamic events early in the Moon's history formed its surface features.

5 Outline our current model of Earth's interior structure and describe some experimental techniques used to establish this model.

6 Describe the nature and origin of Earth's magnetosphere.

7 Summarize the evidence for continental drift and discuss the physical processes that drive it.

8 Discuss theories of the formation and evolution of Earth and the Moon.

If we are to appreciate the universe, we must first come to know our own home. By cataloging Earth's properties and attempting to explain them, we set the stage for a comparative study of all the other planets. From an
astronomical perspective, this is a compelling reason to study the structure and history of our own planet.

The Moon is by far our closest neighbor in space. Yet, despite its nearness, it is a world very different from
our own. It has no air, no sound, no water. Weather is nonexistent; clouds, rainfall, and blue sky are all absent;
boulders and pulverized dust litter the landscape. The Moon is magnificent in its desolation. Why then do we
study it? In part, simply because it is our nearest neighbor and dominates our night sky. Beyond that, however, we
study this body because it holds important clues to our own past. The very fact that it hasn't changed much since
its formation means the Moon is a truly primitive object, a vital key to unlocking the secrets of the solar system.

5.1 Earth and the Moon in Bulk

1 Table 5.1 presents some basic properties of
Earth and the Moon. The data in this table
will form the basis for our comparative study of our
home planet and its nearest neighbor. Figure 5.1 compares and contrasts the main structural features of these
two very dissimilar worlds. We will expand our catalog
of "planetary" characteristics and peculiarities in the
next three chapters as we study in turn the other members of the solar system.

Physical Properties

The first five columns of Table 5.1 list mass, radius, and
density for Earth and the Moon, determined as
described earlier in Section 4.1. The next two columns
present important measures of a body's gravitational
field. *Surface gravity* is the strength of the gravitational
force at the body's surface. *Escape speed* is the speed

required for any object—an atom, a baseball, or a
spaceship—to escape forever from the body's gravitational pull (see *More Precisely 5-1*). The final column
in the table lists (sidereal) rotation periods.

Because its mass is so much less than that of
Earth, the Moon's gravitational pull is also weaker. ∞
(*More Precisely 1-1*) The Moon's surface gravity is only
one-sixth that on Earth, meaning that an astronaut
weighing 180 pounds on Earth would weigh a mere 30
pounds on the lunar surface—those bulky spacesuits
and backpacks used by the *Apollo* astronauts were not
nearly as heavy as they appeared! Similarly, the
Moon's escape speed is nearly five times smaller than
the escape speed of our own planet.

The rotation periods of both Earth and the
Moon have long been accurately known from Earth-
based observations. The Moon rotates once on its
axis in 27.3 days—exactly the same time as
it takes to complete one revolution around Earth.
∞ (Sec. 1.2) As a result, the Moon presents the *same*
face toward Earth at all times; that is, the Moon has a

	Mass			Radius		Density	Surface Gravity	Escape Speed	Rotation
	(kg)	(Earth = 1)	(km)	(Earth = 1)		(kg/m^3)	(Earth = 1)	(km/s)	Period
Earth	6.0×10^{24}	1.00	6400	1.00		5500	1.00	11.2	23^h 56^m
Moon	7.3×10^{22}	0.012	1700	0.27		3300	0.17	2.4	27.3 days

Table 5.1 Some Properties of Earth and the Moon

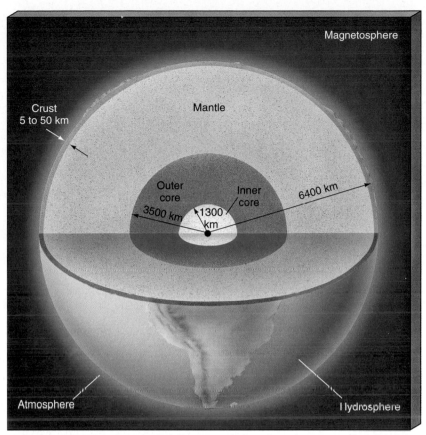

(a)

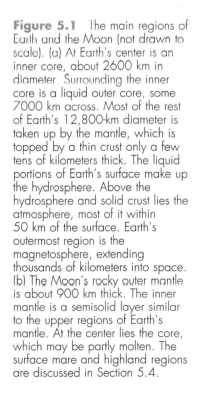

Figure 5.1 The main regions of Earth and the Moon (not drawn to scale). (a) At Earth's center is an inner core, about 2600 km in diameter. Surrounding the inner core is a liquid outer core, some 7000 km across. Most of the rest of Earth's 12,800-km diameter is taken up by the mantle, which is topped by a thin crust only a few tens of kilometers thick. The liquid portions of Earth's surface make up the hydrosphere. Above the hydrosphere and solid crust lies the atmosphere, most of it within 50 km of the surface. Earth's outermost region is the magnetosphere, extending thousands of kilometers into space. (b) The Moon's rocky outer mantle is about 900 km thick. The inner mantle is a semisolid layer similar to the upper regions of Earth's mantle. At the center lies the core, which may be partly molten. The surface mare and highland regions are discussed in Section 5.4.

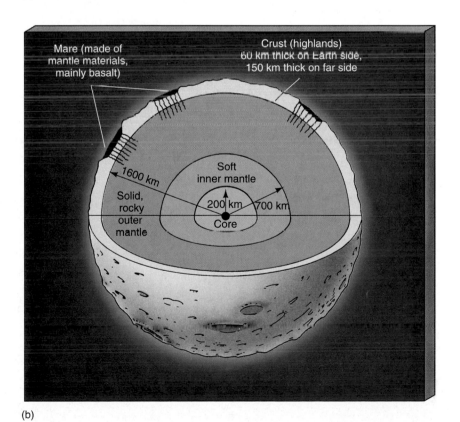

(b)

"near" side, which is always visible from Earth, and a "far" side, which never is. Until recently, no one on Earth had any idea what the Moon's hidden half looked like. It was only when spacecraft flew around the Moon that we finally saw the far side.

This condition, in which the spin of one body is precisely equal to (or *synchronized* with) its revolution around another body, is known as a **synchronous orbit**. The fact that the Moon is in a synchronous orbit around Earth is no accident. As we will see, it is an inevitable consequence of the gravitational interaction between those two bodies.

One important quantity not listed in Table 5.1 is the *distance* from Earth to the Moon. Parallax, with Earth's diameter as a baseline, has long provided astronomers with reasonably accurate measurements of the Earth–Moon distance. ∞ (Sec. P.3) However, radar ranging and laser ranging now yield far more accurate results (in fact, the Moon's precise distance from Earth at any instant is known to within a few centimeters). These high-tech methods tell us that the semi-major axis of the Moon's orbit around Earth is 384,000 km.

Overall Structure

As indicated in Figure 5.1(a), our planet may be divided into six main regions. In Earth's interior, a thick **mantle** surrounds a smaller, two-part **core**. At the surface we have a relatively thin **crust**, comprising the solid continents and the seafloor, and the **hydrosphere**, which contains the liquid oceans and accounts for some 70 percent of our planet's total surface area. Earth is unique among the planets in that it has large quantities of liquid water on its surface. There is no counterpart to the hydrosphere on any other planet in the solar system. An **atmosphere** of air lies just above the surface. At much greater altitudes, a zone of charged particles trapped by our planet's magnetic field forms Earth's **magnetosphere**.

The Moon lacks a hydrosphere, an atmosphere, and a magnetosphere. Its internal structure (Figure 5.1b) is not as well studied as that of Earth, for the very good reason that the Moon is much less accessible. Nevertheless, as we will see, the same basic interior regions as found on Earth—*crust, mantle,* and *core*—

may be discerned for the Moon, although their properties are somewhat different from those of Earth's crust, mantle, and core.

5.2 Gravitational Deformation

Tides

2 Most people are familiar with the daily fluctuation in ocean level known as **tides**. At most coastal locations on Earth, there are two low tides and two high tides each day. The *height* of the tides—the magnitude of the variation in sea level—can range from a few centimeters to many meters, depending on location and time of year. The height of a typical tide on the open ocean is about a meter. An enormous amount of energy is contained in this daily motion of the oceans, constantly eroding and reshaping our planet's coastlines.

What causes the tides? A clue comes from the fact that they exhibit daily, monthly, and yearly cycles. In fact, the tides are a direct result of the gravitational influence of the Moon and the Sun on Earth. We have already seen (in Section 1.7) how gravity keeps Earth and the Moon in orbit about one another, and both in orbit around the Sun. For simplicity, let us first consider just the interaction between Earth and the Moon.

Recall that the strength of the gravitational force depends on the distance separating any two objects. Thus, the Moon's gravitational attraction is greater on the side of Earth that faces the Moon than on the opposite side, some 12,800 km (Earth's diameter) farther away. This difference in the gravitational force is small—only about 3 percent—but it produces a noticeable effect—a **tidal bulge**. As illustrated in Figure 5.2, Earth becomes slightly elongated, with the long axis of the distortion pointing toward the Moon.

Earth's oceans undergo the greatest deformation because liquid can most easily move around on our planet's surface. Thus, the ocean becomes a little deeper in some places (along the line joining Earth to the Moon) and shallower in others (along the line perpendicular to the Earth–Moon line). The daily tides we see result as Earth rotates beneath this deformation.

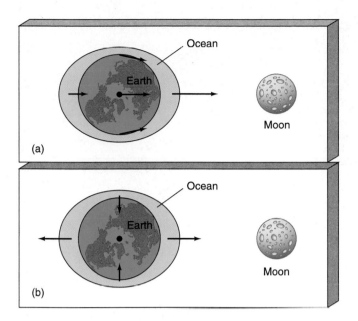

Figure 5.2 This exaggerated illustration shows how the Moon induces tides on both the near and far sides of Earth. The lengths of the arrows indicate the relative strengths of the Moon's gravitational pull on various parts of Earth. (a) The lunar gravitational forces acting on several locations on and inside Earth. The force is greatest on the side nearest the Moon and smallest on the opposite side. (b) The difference between the lunar forces experienced at the locations shown in part (a) and the force acting on Earth's center. The arrows represent the force with which the Moon tends to either pull matter away from (near side) or squeeze it toward (far side) Earth's center. Material on the side of Earth nearest the Moon tends to be pulled away from Earth's center, and material on the far side is "left behind" as Earth is tugged toward the Moon. Two ocean bulges form as a result of this pushing and pulling. High and low tides result, twice per day, as Earth rotates beneath these bulges.

Notice in Figure 5.2 that the side of Earth opposite the Moon also experiences a tidal bulge. The different gravitational pulls—greatest on that part of Earth closest to the Moon, weaker at Earth's center, and weakest of all on Earth's opposite side—cause average tides on opposite sides of our planet to be approximately equal in height. On the side nearer the Moon, the ocean water is pulled slightly toward the Moon. On the opposite side, the ocean water is left behind as Earth is pulled closer to the Moon. Thus, high tide occurs *twice*, not once, every day.

Both the Moon and the Sun create tides on our planet. Even though the Sun is roughly 375 times far-

ther away from Earth than is the Moon, the Sun's mass is so much greater (by about a factor of 27 million) that its tidal influence is still significant—about half that of the Moon. Thus, instead of one tidal force being exerted on Earth, there are two—one pointing toward the Moon, the other toward the Sun—and the interaction between them accounts for the changes in the height of the tides over the course of a month or a year. When Earth, Moon, and Sun are roughly lined up—at new or full Moon—the gravitational effects reinforce one another, and the highest tides occur (Figure 5.3a). These tides are known as *spring tides*. When the Earth–Moon line is perpendicular to the Earth–Sun line (at the first and third quarters), the daily tides are smallest (Figure 5.3b). These are termed *neap tides*.

The variation of the Moon's (or the Sun's) gravity across Earth is an example of a *differential force*, or **tidal force**. The *average* gravitational interaction between two bodies determines their orbit around one another. However, the tidal force, superimposed on that average, tends to deform the bodies. It diminishes very rapidly with increasing distance. We will see many situations in this book where tidal forces are critically important in understanding astronomical phenomena. Notice that we still use the word *tidal* in these other contexts, even though we are not discussing oceanic tides, and possibly not even planets.

Effect of Tides on Earth's Rotation

According to Table 5.1, Earth rotates once on its axis (relative to the stars) in 23^h56^m—1 sidereal day. We know from fossil measurements, however, that Earth's rotation is gradually slowing down, causing the length of the day to increase by about 2 milliseconds every century—not much on the scale of a human lifetime, but over millions of years, this steady slowing of Earth's spin adds up. Half a billion years ago, the day was just over 21 hours long and the year contained 410 days.

The tidal effect of the Moon is the main reason for Earth's slowing spin. Because of Earth's rotation, the tidal bulge raised by the Moon does not point directly at the Moon, as Figure 5.2 shows it. Instead,

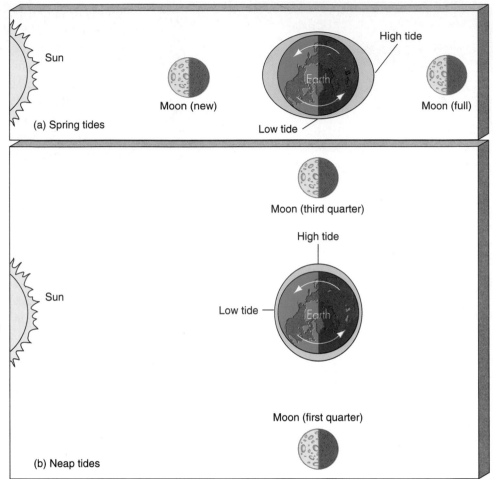

Figure 5.3 The combined effects of the Sun and the Moon produce variations in high and low tides. (a) When the Moon is either full or new, Earth, Moon, and Sun are approximately aligned, and the tidal bulges raised in Earth's oceans by the Moon and the Sun reinforce one another. (b) When the Moon is in either its first or its third quarter, the tidal effects of the Moon and the Sun partially cancel each other, and the tides are smallest. Because the Moon's tidal effect is greater than that of the Sun, the net bulge points toward the Moon.

through the effects of friction, both between the crust and the oceans and within Earth's interior, Earth's rotation tends to drag the tidal bulge around with it, causing the bulge to be displaced by a small angle from the Earth–Moon line, in the same direction as Earth's spin. The net effect of the Moon's gravitational pull on this slightly offset bulge is to reduce our planet's rotation rate. At the same time, the Moon is spiraling slowly away from Earth, increasing its average distance from our planet by about 4 cm per century.

This process will continue until Earth rotates on its axis at exactly the same rate as the Moon orbits Earth—that is, Earth's rotation will become synchronized with the Moon's motion. At that time the Moon will always be above the same point on Earth and will no longer lag behind the bulge it raises. Earth's rotation period will then be 47 of our present days, and the distance to the Moon will be 550,000 km (about 43 percent greater than at present). However, this will take a very long time—many billions of years—to occur.

Effect of Tides on the Moon

Just as the Moon raises tides on Earth, Earth also produces a tidal bulge in the Moon. In fact, because Earth is so much more massive, the tidal force it exerts on the Moon is about 20 times greater than the tidal force the Moon exerts on it, and the lunar tidal bulge is correspondingly larger. We have just seen how tidal forces are causing Earth's spin to slow and how, as a result, Earth will someday rotate on its axis at the same rate as the Moon revolves around Earth. In the case of the Moon, this process has already gone to completion. The Moon's much larger tidal deformation caused it to evolve into a synchronous orbit long ago. The Moon is said to have become *tidally locked* to Earth. Most of the moons in the solar system are similarly locked by the tidal fields of their parent planets.

5.3 Atmospheres

Earth's Atmosphere

Our planet's atmosphere is a mixture of gases, the most abundant of which are nitrogen (78 percent by volume), oxygen (21 percent), argon (0.9 percent), and carbon dioxide (0.03 percent). Water vapor is a variable constituent of the atmosphere, making up anywhere from 0.1 to 3 percent, depending on location and climate. The presence of a large amount of oxygen makes our atmosphere unique in the solar system—Earth's oxygen is a direct consequence of the emergence of life on our planet (Chapter 6).

Figure 5.4 shows a cross section of Earth's atmosphere. Compared with Earth's overall dimensions, the extent of the atmosphere is not great. Half of it lies within 5 km of the surface, and all but 1 percent of it is found below 30 km. The region below about 12 km is called the *troposphere*. Above it, extending up to an altitude of 40 to 50 km, lies the *stratosphere*. Between 50 and 80 km from the surface lies the *mesosphere*. Above about 80 km, in the *ionosphere*, the atmosphere is kept partly ionized by solar ultraviolet radiation. These various atmospheric regions are distinguished from one another by the behavior of the temperature (decreasing or increasing with altitude) in each.

Atmospheric density decreases steadily with increasing altitude, and as the right-hand vertical axis in Figure 5.4 shows, so does pressure. Climbing even a modest mountain—4 or 5 km high, say—clearly demonstrates the thinning of the air in the troposphere. Climbers must wear oxygen masks when scaling the tallest peaks on Earth.

The troposphere is the region of Earth's (or any other planet's) atmosphere where *convection* occurs. **Convection** is the constant upwelling of warm air and the concurrent downward flow of cooler air to take its place. In Figure 5.5 (p. 146), part of Earth's surface is heated by the Sun. The air immediately above the warmed surface is heated, expands a little, and becomes less dense. As a result, it becomes buoyant and starts to rise. At higher altitudes, the opposite effect occurs: the air gradually cools, grows denser, and sinks back to the ground. The cool air at the surface rushes in to replace the hot buoyant air that has risen. In this way, a circulation pattern is established. These *convection cells* of

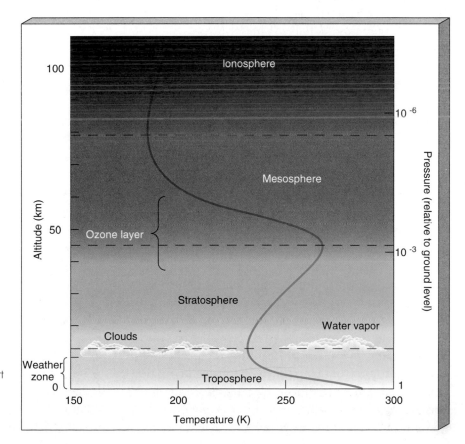

Figure 5.4 Diagram of Earth's atmosphere, showing the changes in temperature (blue line) and pressure (right-hand axis) from the surface to the lower part of the ionosphere. Pressure decreases steadily with increasing altitude, but the temperature may fall or rise, depending on height above the ground.

5-1 MORE PRECISELY

WHY AIR STICKS AROUND

Why does Earth have an atmosphere but the Moon does not? Why doesn't our atmosphere disperse into space? The answer is that *gravity* holds it down. However, gravity is not the only influence acting, for if it were, all of Earth's air would have fallen to the surface long ago. *Heat* competes with gravity to keep the atmosphere buoyant. Let's explore this competition between gravity and heat in a little more detail.

All gas molecules are in constant random motion. The temperature of any gas is a direct measure of this motion—the hotter the gas, the faster the molecules are moving (see *More Precisely 2-1*). The Sun continuously supplies heat to our planet's atmosphere, and the resulting rapid movement of heated molecules produces pressure. This pressure tends to oppose the force of gravity, exerting a net upward force and preventing our atmosphere from collapsing under its own weight.

An important measure of the strength of an object's gravity is the object's *escape speed*—the speed needed for any other object to escape forever from the surface of the parent object. This speed increases with increased mass or decreased radius of the parent object. In fact, the escape speed is proportional to the speed of a circular orbit at the parent object's surface. Mathematically, it can be expressed as

$$\text{escape speed} \propto \sqrt{\frac{\text{mass of parent object}}{\text{radius of parent object}}}.$$

Thus, if the mass of the parent object were to quadruple, the escape speed would double. If the parent object's radius were to quadruple, then the escape speed would be halved.

To determine whether or not a planet will retain an atmosphere, we must compare the planet's escape speed with the average speed of the gas particles making up the atmosphere. This speed depends not only on the temperature of the gas but also on the mass of the individual molecules—the hotter the gas or the smaller the molecular mass, the higher the average speed of the molecules:

$$\text{average molecular speed} \propto \sqrt{\frac{\text{temperature of gas}}{\text{molecular mass}}}.$$

Thus, increasing the temperature of a sample of gas by a factor of 4—from, for example, 100 K to

---

rising and falling air contribute to atmospheric heating and are responsible for surface winds and all the weather we experience. Above the troposphere, the atmosphere is stable and the air is calm.

Straddling the boundary between the stratosphere and the mesosphere is the **ozone layer**, where incoming solar ultraviolet radiation is absorbed by atmospheric oxygen, ozone, and nitrogen. (Ozone is a form of oxygen, consisting of three oxygen atoms combined into a single molecule.) The ozone layer is one of the insulating layers that protect life on Earth from the harsh realities of outer space. Not so long ago, scientists judged space to be hostile to advanced life forms because of what is missing out there—breathable air and a warm environment. Now, most scientists regard

outer space harsh also because of what is present out there—fierce radiation and energetic particles, both of which are injurious to human health. The ozone layer is one of our planet's umbrellas. Without it, advanced life (at least on Earth's surface) would be at best unlikely and at worst impossible.

The Greenhouse Effect

3 Much of the Sun's radiation manages to penetrate Earth's atmosphere, eventually reaching the ground. Most of this energy is in the form of visible radiation—ordinary sunlight. All the solar radiation not absorbed or reflected from clouds in the upper

400 K—doubles the average speed of its constituent molecules. And, at a given temperature, molecules of hydrogen in air move, on average, 4 times faster than molecules of oxygen, which are 16 times heavier.

For nitrogen and oxygen in Earth's atmosphere, where the temperature near the surface is about 300 K, the typical molecular speed is about 0.5 km/s, far smaller than the 11.2 km/s needed for a molecule to escape into space (Table 5.1). As a result, Earth is able to retain its atmosphere. On the whole, the gravity of our planet simply has more influence than the heat of our atmosphere.

In reality, the situation is a little more complicated. Atmospheric molecules can gain or lose speed either by bumping into one another or by colliding with objects near the ground. Thus, while we can characterize a gas by its average molecular speed, the molecules do not really all move at the same speed. A tiny fraction of the molecules in any gas have speeds much greater than average—one molecule in 2 million has a speed more than three times the average, while one in 10^{16} exceeds the average by more than a factor of 5. This means that, at any instant, some molecules are moving fast enough to escape, even when the average molecular speed is much less than the escape speed.

The result is that all planetary atmospheres slowly leak away into space.

Don't be alarmed—the leakage is usually very gradual! As a rule of thumb, if the escape speed from a planet or moon exceeds the average speed of a given type of molecule by a factor of 6 or more, then molecules of that type will not have escaped from the planet's atmosphere in significant quantities since the solar system formed.

For air on Earth, the mean molecular speed of oxygen and nitrogen is comfortably below one-sixth of the escape speed. However, if the Moon originally had an Earth-like atmosphere, that lunar atmosphere would have been heated by the Sun to much the same temperature as Earth's air today, and so the average molecular speed would have been about 0.5 km/s. Because the Moon's escape speed is only 2.4 km/s, any original lunar atmosphere long ago dispersed into interplanetary space.

We can use these arguments to understand atmospheric composition. For example, hydrogen molecules move, on average, at about 2 km/s in Earth's atmosphere at sea level. Consequently, they have had plenty of time to escape since our planet formed more than 4 billion years ago. As a result, we find very little hydrogen in Earth's atmosphere today.

---

atmosphere shines directly onto Earth's surface. The result is that our planet's surface and most objects on it heat up considerably during the day. Earth can't absorb this solar energy indefinitely, however. If it did, the surface would soon become hot enough to melt, and life on our planet would not exist.

As it heats up, Earth's surface reradiates much of its absorbed energy. The reradiated radiation follows the usual black-body curve studied in Chapter 2. (Sec. 2.4) As the surface temperature rises, the amount of energy radiated increases rapidly, according to Stefan's law. Eventually Earth radiates as much energy back into space as it receives from the Sun, and a stable balance is struck. In the absence of any complicating effects, this balance would be achieved at an average surface temperature of about 250 K (−23°C). At that temperature, Wien's law tells us that most of the reemitted energy is in the form of infrared (heat) radiation.

There are complications, however. Infrared radiation is partially blocked by Earth's atmosphere. The primary reason for this is the presence of molecules of carbon dioxide and water vapor, both of which absorb very efficiently in the infrared portion of the spectrum. Even though these two gases are only trace constituents of our atmosphere, they manage to absorb a large fraction of all the infrared radiation emitted from the surface. Consequently, only some of that radiation escapes back into space. The remainder is trapped within our atmosphere, causing the temperature to rise.

Figure 5.5 Convection occurs whenever cool matter overlies warm matter. The resulting circulation currents are familiar to us as the winds in Earth's atmosphere, caused by the solar-heated ground. Over and over, hot air rises, cools, and falls back to Earth. Eventually, steady circulation patterns are established and maintained, provided the source of heat remains intact.

This partial trapping of solar radiation is known as the **greenhouse effect**. The name comes about because a similar process operates in a greenhouse—sunlight passes relatively unhindered through glass panes, but much of the infrared radiation reemitted by the plants is blocked by the glass and cannot get out.[1] Consequently, the interior of the greenhouse heats up, and flowers, fruits, and vegetables can grow even on cold winter days. The radiative processes that determine the temperature of Earth's atmosphere are illustrated in Figure 5.6. Earth's greenhouse effect makes our planet about 40 K hotter than would otherwise be the case.

The magnitude of the greenhouse effect is very sensitive to the concentration of *greenhouse gases* (that is, gases that absorb infrared radiation efficiently) in the atmosphere. Of greatest importance among these is carbon dioxide, although water vapor also plays a significant role. The amount of carbon dioxide in Earth's atmosphere is increasing, largely as a result of the burning of fossil fuels (principally oil and coal) in the industrialized world. Carbon dioxide levels have increased by more than 20 percent in the last century, and they continue to rise at a present rate of 4 percent per decade. Many scientists believe that this increase, if

left unchecked, may result in global temperature increases of several kelvins over the next half-century, enough to melt much of the polar ice caps and cause dramatic, perhaps even catastrophic, changes in Earth's climate.

Lunar Air?

What about the Moon's atmosphere? That's easy—for all practical purposes, there is none! All of it escaped long ago. More massive objects have a better chance of retaining their atmospheres because the more massive an object, the greater the speed needed for atoms and molecules to escape (see *More Precisely 5-1*). The Moon's escape speed is only 2.4 km/s, compared with 11.2 km/s for Earth (Table 5.1). Simply put, the Moon has a lot less pulling power—any atmosphere it might once have had is gone forever.

With no atmosphere to help retain it, any water that ever existed on the Moon's surface has also evaporated and escaped. Not only is there no lunar hydrosphere, but all the lunar samples returned by the American and Soviet Moon programs were absolutely bone dry. Lunar rock doesn't even contain minerals containing water molecules locked within their crystal structure. Terrestrial rocks, conversely, are almost always 1 or 2 percent water. In large part because of the Moon's hostile environment, no life exists there, and no fossils were found in *Apollo* samples. Lunar rocks are

---

[1]Although this process does contribute to warming the interior of a greenhouse, it is not the most important effect. A greenhouse works mainly because its glass panes prevent convection from carrying heat up and away from the interior. Nevertheless, the name "greenhouse effect" to describe the heating effect due to Earth's atmosphere has stuck.

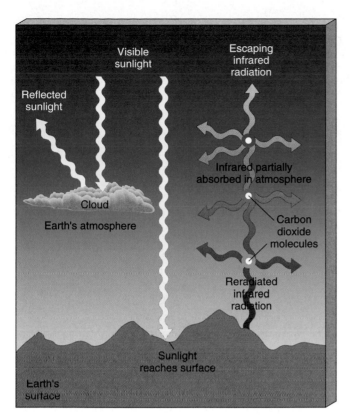

Figure 5.6 The greenhouse effect. Sunlight that is not reflected by clouds reaches Earth's surface, warming it up. Infrared radiation reradiated from the surface is partially absorbed by carbon dioxide in the atmosphere, causing the overall surface temperature to rise.

barren of life and apparently always have been. NASA was so confident of this that the astronauts were not even quarantined on their return from the last few *Apollo* landings.

Lacking the moderating influence of an atmosphere, the Moon experiences wide variations in surface temperature. Noontime temperatures can reach 400 K, well above the boiling point of water (373 K). At night (which lasts nearly 14 Earth days) or in the shade, temperatures fall to about 100 K, well below water's freezing point (273 K).

5.4 The Surface of the Moon

Large-Scale Features

The first observers to point their telescopes at the Moon noted large, roughly circular, dark areas that resemble (they thought) Earth's oceans. They called these regions **maria**, a Latin word meaning "seas" (singular: *mare*). The largest of them (Mare Imbrium, the "Sea of Showers") is about 1100 km in diameter. Today we know that the maria are not oceans but rather extensive flat plains that resulted from the spread of lava during an earlier, volcanic period of the Moon's evolution. In a sense, then, the maria *are* oceans—ancient seas of molten lava, now solidified.

Early observers also saw light-colored areas that resembled Earth's continents. Originally dubbed *terrae*, from the Latin word for "land," these regions are now known to be elevated several kilometers above the maria. Accordingly, they are usually called the lunar **highlands**. Both types of region are visible in Figure 5.7, a photographic mosaic of the full Moon. These light and dark surface features are also evident to the naked eye, creating the face of the familiar "Man-in-the-Moon."

Based on studies of lunar rock brought back to Earth by *Apollo* astronauts and unmanned Soviet landers, geologists have found that two basic kinds of surface rock exist on the Moon. The highlands, which have typical densities of about 2900 kg/m^3, represent the Moon's crust. The maria are made of denser (roughly 3300 kg/m^3) mantle material. The maria rock is quite similar to terrestrial basalt, and geologists believe that it arose on the Moon much as basalt did on Earth, through the upwelling of molten material through the crust. Radioactive dating indicates ages of

Figure 5.7 A photographic mosaic of the full Moon, north pole at the top. Some prominent maria are labeled.

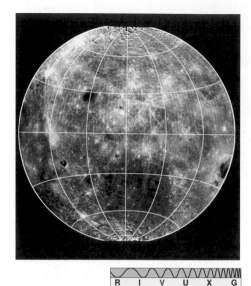

Figure 5.8 The far side of the Moon, as seen by the Clementine military spacecraft. The large, dark region at center bottom outlines the South Pacific basin. This image shows only a few small maria on the far side.

more than 4 billion years for highland rocks and from 3.2 to 3.9 billion years for those from the maria.[2]

To the surprise of most astronomers, when the far side of the Moon was mapped first by Soviet and later by American spacecraft, no major maria were found. The lunar far side (Figure 5.8) is composed almost entirely of highlands.

Cratering

4 Because the smallest lunar features we can distinguish with the naked eye are about 200 km across, we see little more than the maria and highlands when we gaze at the Moon. Through a telescope, however, we see that the lunar surface is scarred by numerous bowl-shaped depressions, or **craters** (after the Latin word for "bowl"), as Figure 5.9 shows. For an even

[2]Radioactive dating compares the rates at which different radioactive elements in a sample of rock decay into lighter elements. The "age" returned by this technique is the time since the rock solidified.

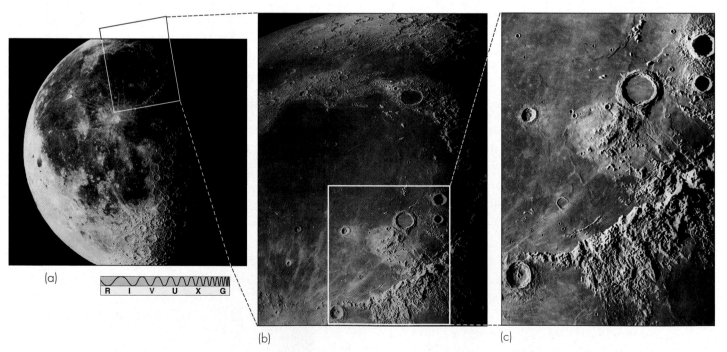

(a)   (b)   (c)

Figure 5.9 (a) The Moon near third quarter. Notice that surface features are much more visible near the terminator, the line separating light from dark, where sunlight strikes at a sharp angle and shadows highlight the topography. (b) Magnified view of a region near the terminator, as seen from Earth through a large telescope. Crater Copernicus is at bottom left, and the central dark area is Mare Imbrium, ringed at the bottom right by the Apennine mountains. (c) Enlargement of the lower right portion of (b).

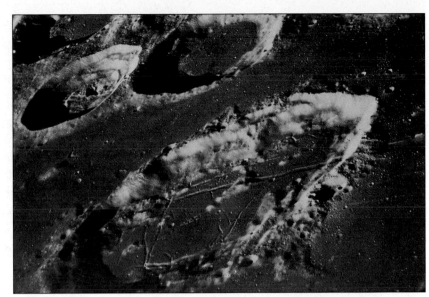

Figure 5.10 The Moon as seen from the *Apollo 8* orbiter during the first human circumnavigation in 1968. Craters of many sizes are visible, with diameters ranging from 50 km to 500 m.

R   I   V   U   X   G

closer look, Figure 5.10 shows a view of the Moon taken from a spacecraft in lunar orbit. The smallest features that can be distinguished in this image are about 0.5 km across.

Lunar craters come in all sizes. The largest are hundreds of kilometers in diameter, the smallest are microscopic. Craters are found everywhere on the Moon's surface, although they are much more prevalent in the highlands.

Most craters formed eons ago as the result of meteoritic impact. ∞ (Sec. 4.2) Meteoroids generally strike the Moon at speeds of several kilometers per second. At these speeds, even a small piece of matter carries an enormous amount of energy—for example, a 1-kg object hitting the Moon's surface at 10 km/s would release as much energy as the detonation of 10 kg of TNT. As illustrated in Figure 5.11, impact by a meteoroid causes sudden and tremendous pressures to build up on the lunar surface, heating the normally brittle rock and deforming the ground. The ensuing explosion pushes previously flat layers of rock up and out, forming a crater.

The material thrown out by the explosion surrounds the crater in a layer called an *ejecta blanket*, the ejected debris ranging in size from fine dust to large boulders. The accumulated dust from countless impacts now covers the lunar surface to an average depth of about 20 m, thinnest on the maria (about 10 m) and thickest on the highlands (more than 100 m in places). The larger pieces of ejecta may themselves form secondary craters. Figure 5.12 shows the result of one large meteoritic impact on the Moon.

The older highlands are much more heavily cratered than the younger maria. This is what we expect because craters can form only after a surface has solidified. However, the difference is not simply a matter of exposure time; the *rate* of meteoroid bombardment was also a factor. Knowing the ages of the highlands and maria from radioactive dating of lunar rocks, researchers can estimate the rate of cratering in the past. They conclude that the Moon, and presumably the entire inner solar system, experienced a sudden sharp drop in meteoritic bombardment rate about 3.9 billion years ago. The rate of cratering has been (very) roughly constant since that time.

This time—3.9 billion years in the past—is taken to represent the end of the accretion process through which planetesimals became planets. ∞ (Sec. 4.3) The highlands solidified and received most of their craters before that time. The great basins that formed the maria are thought to have been created during the final stages of heavy meteoritic bombardment between about 4.1 and 3.9 billion years ago. Subsequent volcanic activity filled the craters with lava, creating the formations we see today.

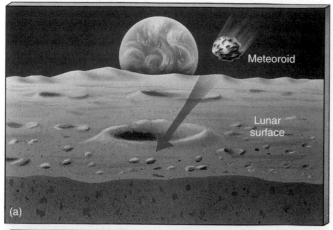

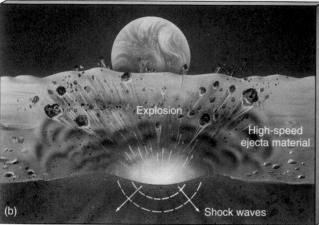

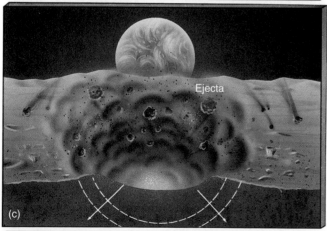

Lunar Erosion

On Earth, the combined actions of air and water erode our planet's surface and reshape its appearance almost daily. As a result, most of the ancient history of our planet's surface is lost to us. The Moon, though, has no air, no water, no ongoing volcanic or other geological activity. Consequently, features dating back almost to the formation of the Moon are still visible today. For this reason, studies of the lunar surface are of great importance to Earth geologists and have played a major role in shaping theories of the early development of our planet.

The only important source of erosion on the Moon is meteoritic impact. On Earth, most meteoroids burn up in the atmosphere, but the Moon, with no atmosphere, has no protection against this onslaught. Over billions of years, collisions with meteoroids, large and small, have scarred, cratered, and sculpted the lunar landscape. The rate of cratering decreases rapidly with crater size—fresh large craters are very few and far between, but small craters are very common. The reason for this is simple: because there aren't very many large chunks of the interplanetary debris, their collisions with the Moon are rare. At the present average rates, one new 10-km-diameter lunar crater is formed every 10 million years, one new 1-m-diameter crater is created about once a month, and 1-cm-diameter craters are formed every few minutes. In addition, a steady "rain" of micrometeoroids also eats away at the lunar surface (Figure 5.13).

Despite this barrage from space, the Moon's present-day erosion rate is still very low—about 10,000 times less than on Earth—simply because wind and water are far more effective erosive agents than meteoritic bombardment. For example, the Barringer Meteor Crater (Figure 4.14) in the Arizona desert, one of the largest meteor craters on Earth, is only 25,000 years old, but it is already decaying. It will probably disappear completely in a mere million years, quite a short time geologically. If a crater that size had formed on the Moon even a billion years ago, it would still be plainly visible today.

Figure 5.11 Stages in the formation of a crater by meteoritic impact. (a) The meteoroid strikes the surface, releasing a large amount of energy. (b, c) The resulting explosion ejects material from the impact site and sends shock waves through the underlying surface. (d) Eventually, a characteristic crater surrounded by a blanket of ejected material results.

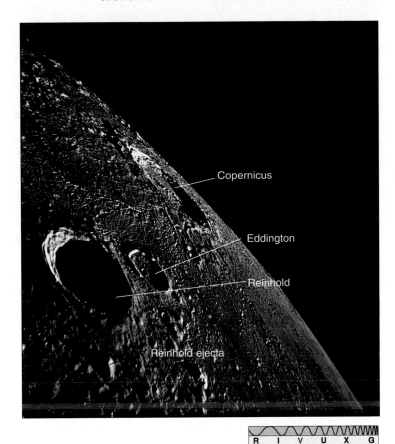

Figure 5.12   Two smaller craters called Reinhold and Eddington sit amid the secondary cratering resulting from the impact that created the 90-km-wide Copernicus Crater (near the horizon) about a billion years ago. The ejecta blanket from crater Reinhold, 40 km across, in the foreground, can be seen clearly.

Figure 5.13   The lunar surface is not entirely changeless. Despite the complete lack of wind and water on the airless Moon, the surface has still eroded a little under the constant "rain" of impacting meteoroids, especially micrometeoroids. Note the soft edges of the craters visible in the foreground of this image. In the absence of erosion, these features would be as jagged and angular today as they were when they formed. (The twin tracks were made by the *Apollo* lunar rover.)

5.5 Interiors

Seismology

5 Although we reside on Earth, we cannot easily explore our planet's interior because drilling gear can penetrate rock only so far before breaking. No substance—not even diamond, the hardest known material—can withstand the conditions below a depth of about 10 km. That's rather shallow compared with Earth's 6400-km radius. Fortunately, geologists have developed techniques that can probe the deep recesses of our planet *indirectly*.

A sudden dislocation of rocky material near Earth's surface—an **earthquake**—causes the entire planet to vibrate a little. It literally rings like a bell (but one tuned so low that human ears cannot detect the sound). These vibrations are not random. They are systematic waves, called **seismic waves** (after the Greek word for "earthquake"), that move outward from the site of the quake. Like all waves, they carry information. This information can be detected and recorded using sensitive equipment—a *seismograph*—designed to monitor Earth tremors.

Decades of earthquake research have demonstrated several types of seismic waves. The speed of each type depends on the density and physical state of the matter through which it is traveling. Consequently, if we can measure the time taken for waves to move from the site of an earthquake to one or more monitoring stations on Earth's surface, we can infer the density of matter in the interior.

Modeling Earth's Interior

Because earthquakes occur often and at widespread places across the globe, geologists have accumulated a large amount of data about seismic-wave properties. They have used these data, along with direct knowledge of surface rocks, to model Earth's interior.

Figure 5.14 illustrates some paths followed by seismic waves from the site of an earthquake. Some of the waves (those arbitrarily colored red in the figure)

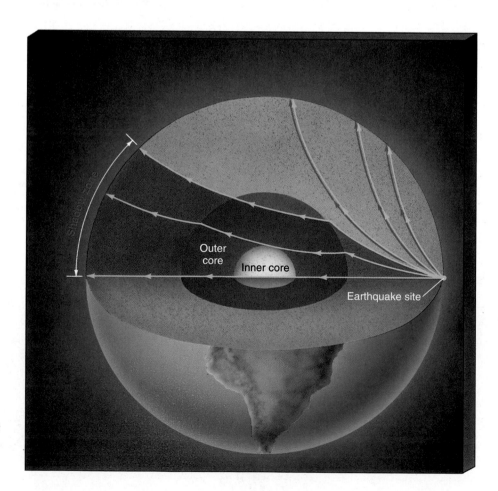

Figure 5.14 Earthquakes generate seismic waves that can be detected at seismographic stations around the world. (The waves bend as they move through Earth's interior because of the variation in density and temperature within our planet.) Some waves are not detected by stations "shadowed" by the outer core of Earth, indicating that the outer core is liquid.

are blocked by Earth's **outer core.** The result is the *shadow zone* shown in Figure 5.14. The explanation for this behavior is that these particular waves cannot pass through *liquid*. The observation that every earthquake exhibits these shadow zones is the best evidence we have that the outer core of our planet is liquid. The radius of the outer core, as determined from seismic data, is about 3500 km. There is also evidence that some waves are reflected off the surface of a solid **inner core,** of radius 1300 km, lying at the center of the liquid outer core.

As mentioned at the beginning of this chapter, Earth's core is surrounded by a thick mantle and topped with a thin crust. The mantle is about 3000 km thick and accounts for 80 percent of our planet's volume. The crust has an average thickness of only 15 km—around 8 km under the oceans and 20 to 50 km under the continents. The average density of the crust material is around 3000 kg/m³.

Earth's average density is about 5500 kg/m³ (Table 5.1). We can immediately conclude that because the surface layers have densities much less than the average, much denser material must lie deeper in. In fact, studies of Earth's interior now indicate that the central density is more than 12,000 kg/m³. Figure 5.15 presents a model that most scientists accept. As the graphs show, both density and temperature increase sharply with depth. Earth's high central density suggests to geologists that the inner parts of our planet must be rich in nickel and iron. Under the heavy pressure of the overlying layers, these metals (whose densities under normal surface conditions are around 8000 kg/m³) are compressed to the high densities predicted by the model.

The model suggests that much of the mantle has a density midway between the densities of the core and crust—about 5000 kg/m³. The sharp density increase at the mantle–core boundary results from the difference in composition between the two regions. The mantle is composed of dense but *rocky* material. The core consists primarily of even denser *metallic* elements. There is no similar jump in density or temperature at the inner core boundary. At the high pressures found near the center—about 4 million times the atmospheric pressure at Earth's surface—the material simply changes from liquid to solid.

Because geologists have been unable to drill deeper than about 10 km, no experiment has yet recov-

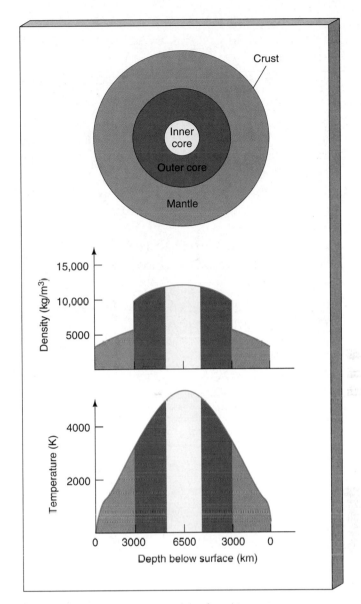

Figure 5.15 Computer models of Earth's interior imply that the density and temperature vary considerably through the mantle and the core. Note the sharp density change at the mantle–outer core boundary.

ered a sample of the mantle. However, we are not entirely ignorant of the mantle's properties. In a **volcano** (Figure 5.16), molten rock upwells from below the crust, bringing a little of the mantle to us in the form of lava and providing some inkling of the composition of Earth's interior. The chemical makeup and physical state of the newly emerged lava are generally consistent with predictions based on the model just described.

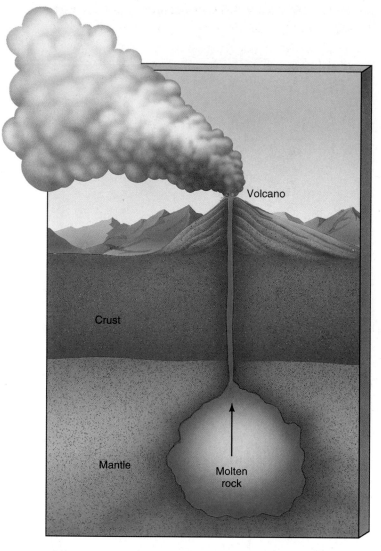

Figure 5.16 The material that erupts from a volcano originates deep below Earth's surface, in the upper mantle.

Differentiation

Earth, then, is not a homogeneous ball of rock. Instead, it has a layered structure, with a low-density rocky crust at the surface, intermediate-density rocky material in the mantle, and a high-density metallic core. Such variation in density and composition is known as **differentiation**.

Why isn't our planet just one big, rocky ball of uniform density? The answer is that, at some time in the distant past, much of Earth was *molten*, allowing the higher-density matter to sink to the core, displacing lower-density material toward the surface. A remnant of this ancient heating exists today: Earth's central temperature is nearly equal to that of the Sun's surface.

Two processes played important roles in heating Earth to the point where differentiation could occur. Very early in its history, our planet experienced a violent bombardment by interplanetary debris, an essential part of the process by which Earth and the other planets formed and grew (Section 4.3). This bombardment generated enough heat to melt much of our planet. As Earth began to soften and differentiate and heavy material sank to the center, gravitational energy was released, increasing the interior temperature still further.

The second process that heated Earth after its formation and so contributed to differentiation was **radioactivity**—the release of energy by certain elements, such as uranium, thorium, and plutonium. These elements emit energy as their complex, heavy nuclei break up into simpler, lighter ones. The energy produced by the breakup of a single radioactive atom is tiny, but young Earth contained a lot of radioactive atoms, and a lot of time was available. Rock is such a poor conductor of heat that the energy released through radioactivity would have taken a very long time to reach the surface and leak away into space. For this reason, the heat built up in the interior, adding to the energy left there by Earth's formation. Provided that enough radioactive elements were originally spread throughout the primitive planet, like raisins in a cake, the entire planet—from crust to core—could have melted and remained molten, or at least semisolid, for about a billion years.

The Lunar Interior

The Moon's average density, about 3300 kg/m^3, is quite similar to the density of the lunar surface rock obtained by U.S. and Soviet missions. This similarity all but eliminates any chance that the Moon has a large, massive, dense nickel–iron core like that within Earth. The low average lunar density suggests that the Moon contains substantially fewer heavy elements (such as iron) than does Earth.

Most of our detailed information on the Moon's interior comes from seismic data obtained from equipment left on the lunar surface by astronauts. These measurements indicate only very weak *moonquakes* deep within the lunar interior. Even if you stood directly above one of these quakes, you would not feel the vibrations. The average moonquake releases about as much energy as a firecracker, and no large quakes have ever been detected. This barely perceptible seismic activity confirms the idea that the Moon is geologically dead. Even so, researchers can use these weak lunar vibrations to learn about the Moon's interior.

Modeling all the available data indicates that the Moon's interior is of uniform density but chemically differentiated—that is, the chemical properties change from core to surface. As illustrated in Figure 5.1(b), the models suggest a central core perhaps 200 km in radius, surrounded by a 500-km-thick inner mantle of semisolid rock having properties similar to those of Earth's upper mantle. Above these regions lies a 900-km-thick outer mantle of solid rock, topped by a 60- to 150-km-thick crust. The core is probably somewhat more iron-rich than the rest of the Moon, although it is still iron-poor relative to Earth's core. The models imply that, near the center, the current temperature may be as low as 1500 K, too cool to melt rock. However, some of the seismic data suggest that the inner core may be at least partially molten, implying a temperature higher than 1500 K. Our knowledge of the details of the Moon's central regions is very limited.

The crust on the Moon's far side is thicker (150 km) than the crust on the side nearer Earth (60 km). Why is this? The answer is probably connected to Earth's gravitational pull. Just as heavier material tries to sink to the center of Earth, the denser far-side lunar mantle tended to sink below the lighter far-side crust in the presence of Earth's gravitational field. In other words, while the Moon was cooling and solidifying, the far-side lunar mantle was pulled a little closer to Earth than the far-side crust. In this way the crust and mantle became slightly off-center with respect to one another. The result was the thicker crust on the Moon's far side.

5.6 Magnetospheres *Animation*

6 Simply put, the magnetosphere is the region around a planet that is influenced by that planet's magnetic field. It forms a buffer zone between the planet and the high-energy particles of the solar wind. As we will see in the next two chapters, it can also provide important insights into the planet's interior structure.

Earth's Magnetosphere

Sketched in Figure 5.17, Earth's magnetic field extends far above the atmosphere and is similar in overall structure to the field of a gigantic bar magnet. The field has a north and a south pole, and it completely surrounds

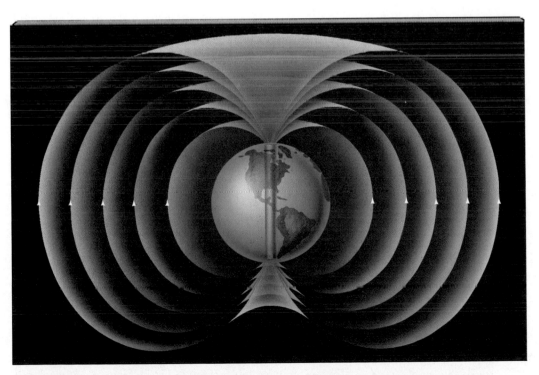

Figure 5.17 Earth's magnetic field resembles that of an enormous bar magnet situated inside our planet. The white arrowheads on the field lines indicate the direction in which a compass needle would point.

our planet. The *magnetic field lines*, which indicate the strength and direction of the field at any point in space, run from south to north, as indicated by the white arrowheads in Figure 5.17. The north and south *magnetic poles*, where the axis of the imaginary bar magnet intersects Earth's surface, are roughly aligned with Earth's spin axis.

Earth's magnetosphere contains two doughnut-shaped zones of high-energy charged particles, one located about 3000 km and the other 20,000 km above Earth's surface. These zones are named the **Van Allen belts**, after the American physicist whose instruments on board one of the first satellites first detected them. We call them "belts" because they are most pronounced near Earth's equator and because they completely surround the planet. Figure 5.18 shows how these invisible regions envelop Earth except near the North and South Poles.

The particles that make up the Van Allen belts originate in the solar wind. Traveling through space, neutral particles and electromagnetic radiation are unaffected by Earth's magnetism, but electrically charged particles are strongly influenced. As illustrated in Figure 5.19, a magnetic field exerts a force on a moving charged particle, causing the particle to spiral around the magnetic field lines. In this way, charged

particles—mainly electrons and protons—from the solar wind can become trapped by Earth's magnetism. Earth's magnetic field exerts electromagnetic control over these particles, herding them into the Van Allen belts. The outer belt contains mostly electrons; the much heavier protons accumulate in the inner belt.[3]

Particles from the Van Allen belts often escape from the magnetosphere near Earth's north and south magnetic poles, where the field lines intersect the atmosphere. Their collisions with air molecules create a spectacular light show called an **aurora** (Figure 5.20). This colorful display results when atmospheric molecules, excited upon collision with the charged particles, fall back to their ground states and emit visible light. ∞ (Sec. 2.6) Aurorae are most brilliant at high latitudes, especially inside the Arctic and Antarctic circles. In the north, the spectacle is called the *aurora borealis*, or *Northern Lights*. In the south, it is called the *aurora australis*, or *Southern Lights*.

---

[3]Just as Earth's magnetosphere influences the charged particles of the solar wind, the steady stream of incoming solar-wind particles also affects our magnetosphere. In reality, Earth's magnetosphere is not nearly as symmetrical as indicated in Figures 5.17–5.19. Instead, the sunward (daytime) side is squeezed toward Earth's surface, while the opposite side often has a long "tail" extending far (hundreds of thousands of kilometers) into space.

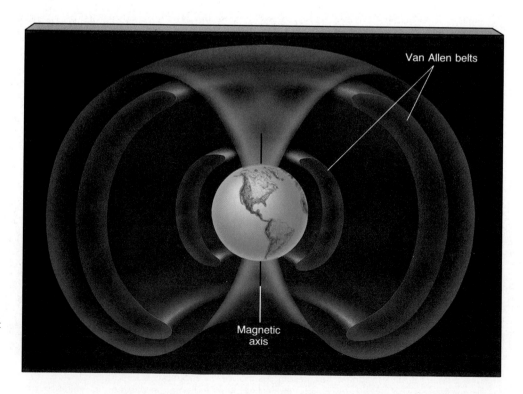

Figure 5.18 High above Earth's atmosphere, the magnetosphere (blue-green area) contains two doughnut-shaped regions (pink and purple areas) of magnetically trapped charged particles. These are the Van Allen belts.

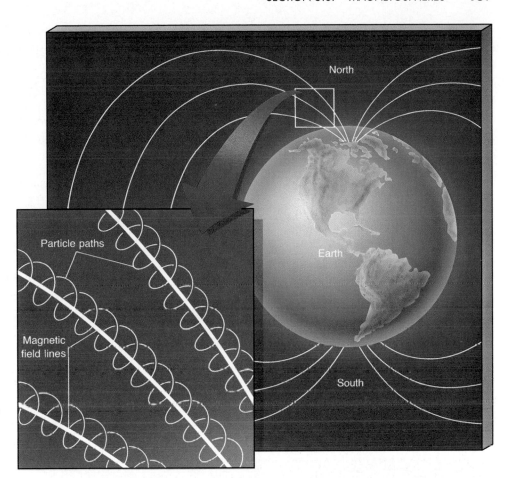

Figure 5.19 Charged particles in a magnetic field spiral around the field lines. Thus, the particles tend to become trapped by strong magnetic fields.

Figure 5.20 A colorful aurora results from the emission of light radiation after magnetospheric particles collide with atmospheric molecules. The aurora rapidly flashes across the sky, looking like huge wind-blown curtains glowing in the dark.

Earth's magnetosphere plays an important role in controlling many of the potentially destructive charged particles that venture near our planet. Without the magnetosphere, Earth's atmosphere—and perhaps the planet surface, too—would be bombarded by harmful particles, possibly damaging many forms of life. Some researchers have even suggested that had the magnetosphere not existed in the first place, life might never have arisen on planet Earth.

Earth's magnetic field is not an intrinsic part of our planet. Instead, it is thought to be continuously generated in Earth's core and exists only because the planet is rotating. As in the dynamos that run industrial machines, Earth's magnetism is produced by the spinning, electrically conducting, liquid metal core deep within our planet. Both rapid rotation *and* a conducting liquid core are needed for this mechanism to work.

Lunar Magnetism

No Earth-based observation or spacecraft measurement has ever detected any lunar magnetic field. Based on

our current understanding of how Earth's magnetic field is created, this is not surprising. As we have just seen, researchers believe that planetary magnetism requires a rapidly rotating liquid metal core. Because the Moon rotates slowly and because its core is probably neither molten nor particularly rich in metals, the absence of a lunar magnetic field is exactly what we expect.

5.7 Surface Activity

Continental Drift

7 Earth is geologically alive today. Its interior seethes, and its surface constantly changes. Many clear indicators of geological activity, in the form of earthquakes and volcanic eruptions, are scattered across our globe. Erosion has obliterated much of the evidence from ancient times, but modern exploration has documented the sites of recent activity.

Figure 5.21 is a map of the currently active areas of our planet. Nearly all these sites have experienced

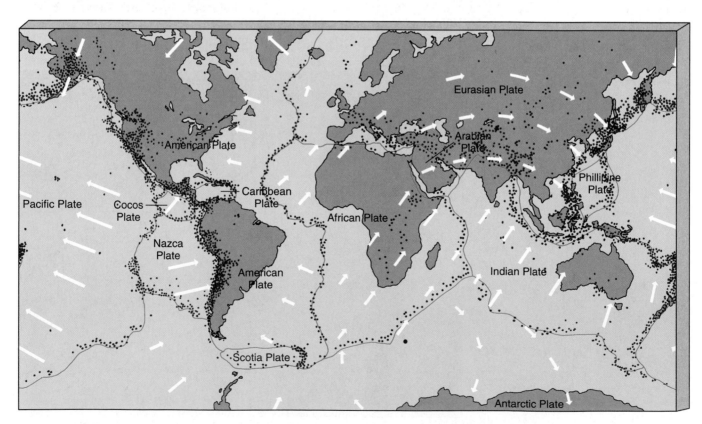

Figure 5.21 The red dots represent active sites where major volcanoes or earthquakes have occurred in the twentieth century. Taken together, the sites delineate vast plates, outlined in blue, that drift around on the surface of our planet. The white arrows show the general directions of plate motions.

surface activity during this century. The intriguing aspect of the figure is that the active sites are not spread evenly across our planet. Instead, they trace well-defined lines of activity, where crustal rocks shift (as in earthquakes) or mantle material upwells (as in volcanos). In the mid-1960s, it became clear that these lines are the outlines of gigantic *plates*, or slabs of Earth's surface. Most startling of all, these plates are slowly drifting around the surface of our planet. These plate motions have created the mountains, oceanic trenches, and many other large-scale features across the face of planet Earth. In fact, plate motions have shaped the continents, and the process is popularly known as *continental drift*. The technical term for the study of plate movement and the reasons for it is **plate tectonics**. Some plates are made mostly of continental landmasses, some are made of a continent plus a large part of an ocean floor, and some contain no continental land at all and are made solely of ocean floor. For the most part, the continents are just passengers riding on much larger plates.

The plates are not simply slowing to a stop after some initial and ancient movement. Rather, they continue to drift today, although at an extremely slow rate. Although typical speeds amount to only a few centimeters per year—about the same speed as your fingernails grow—this is well within the measuring capabilities of modern equipment. Laser ranging and other techniques now routinely track the relative motion of plates in many populated areas, such as California.

During the course of Earth history, the plates have had plenty of time to move large distances, even at their sluggish pace. For example, a drift rate of only 2 cm per year can cause two continents (for example, Europe and North America) to separate by 4000 km—the width of the Atlantic Ocean—in 200 million years. That may be a long time by human standards, but it represents only about 5 percent of Earth's age.

As the plates drift around, we might expect collisions to be routine. Indeed, plates do collide. However, unlike the case of two moving automobiles colliding and coming to a stop, colliding plates do not stop easily. They are driven by enormous forces. Instead of stopping they just keep crunching into one another. Figure 5.22 shows the result of a collision currently occurring between two continental landmasses. The folds of rocky crust, shown in the figure, create mountains—in this case, Mount Everest in the Himalayan

Figure 5.22 Mountain building results largely from plate collisions. Here, the folding of rock is visible (in the foreground) near Mount Everest, which is part of the Himalayan mountain range at the northern end of the Indian Plate.

range. This mountain system is still being formed today as the Indian Plate thrusts northward into the Eurasian Plate.

Not all plates collide head-on. As indicated by the arrows on Figure 5.21, many plates slide or shear past one another. A good example is the most famous active region in North America—the San Andreas Fault in California. Illustrated in Figure 5.23, this fault causes earthquake activity because the Pacific and North American plates are rubbing past one another. They are not moving in quite the same direction, nor at quite the same speed. Like parts in a poorly oiled machine, the motion of these two plates is neither steady nor smooth. The sudden, jerky movements that occur when they do move against each other are often strong enough to cause major earthquakes.

At still other locations, such as the seafloor under the Atlantic Ocean, the plates are moving apart. As they recede, new mantle material wells up between them, forming midocean ridges. Today, hot mantle material is rising through a crack all along the Mid-Atlantic Ridge, which extends, like a seam on an enormous baseball, all the way from the North Atlantic to the southern tip of South America. Radioactive dating indicates that material has been upwelling along the ridge more or less steadily for the past 200 million years. The Atlantic seafloor is slowly growing, as the North and South American plates move away from the Eurasian and African plates.

Figure 5.23 The San Andreas and associated faults in California. This fault system is the result of the North American and Pacific plates sliding past one another. The percentages are estimates of the probability of a major earthquake occurring at various locations along the faults. The inset at the top right shows a small part of the fault line separating the two plates.

What Drives the Plates?

What is responsible for the enormous forces that drag plates apart in some locations and ram them together in others? The answer (Figure 5.24) is *convection*—the same process we encountered earlier in our study of Earth's atmosphere. Each plate is made up of crust plus a small portion of upper mantle. Below the plates, at a depth of perhaps 50 km, the temperature is sufficiently high that the mantle at that depth is soft enough to flow, very slowly, although it is not molten.

This is a perfect setting for convection—warm matter underlying cool matter. The warm mantle rock rises, just as hot air rises in our atmosphere, and large circulation patterns become established. Riding atop these convection patterns are the plates. The circulation is extraordinarily sluggish. Semisolid rock takes millions of years to complete one convection cycle. Although the details are far from certain and remain controversial, many researchers believe that it is large-scale convection patterns near plate boundaries that cause the plates to move.

Plate Tectonics on the Moon

There is no evidence for tectonic motion of any kind on the Moon today—no obvious extensive fault lines, no significant seismic activity, no ongoing mountain building. Plate tectonics requires both a relatively thin outer rocky layer, which is easily fractured into conti-

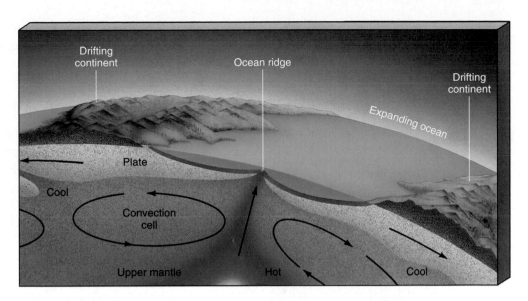

Figure 5.24 Plate drift is caused by convection patterns in the upper mantle that drag the plates across Earth's surface.

nent-sized pieces, and a soft, convective region under it, to make the pieces move. On the Moon, neither of these ingredients exists. The Moon's thick crust and solid upper mantle make it impossible for pieces of the surface to move relative to one another. There simply isn't enough energy left in the lunar interior for plate tectonics to work.

5.8 The Origin of the Moon

8 Both the similarities between the Moon and Earth and the differences conspire to confound many promising attempts to explain the Moon's existence. How the Moon formed remains uncertain, although several theories have been advanced.

The *sister theory*, or *coformation theory*, suggests that the Moon formed as a separate object near Earth and in much the same way as our own planet. The blob of material that eventually coalesced into Earth also gave rise to the Moon at about the same time. The two objects thus formed as a double-planet system. Although once favored by many astronomers, this idea suffers from a major flaw: the Moon differs from Earth in both density and composition, and these differences make it hard to understand how both bodies could have originated from the same preplanetary matter. It is also quite difficult to reconcile the double-planet formation scenario with the theory of solar system formation described in Section 4.3.

The *capture theory* maintains that the Moon formed far from Earth and was later captured by Earth.

In this way the density and composition of the two objects need not be similar, for the Moon presumably arose in a quite different region of the early solar system. The main objection to this theory is that the Moon's capture would be an extraordinarily difficult event, perhaps even an impossible one. Mathematical modeling indicates that Earth's gravitational pull probably could not have attracted the Moon in exactly the right way to capture it during a close encounter sometime in the past. Furthermore, while there are indeed significant composition differences between our world and its companion, there are also many similarities—particularly between the mantles of the two bodies—that make it unlikely that they formed entirely independently of one another.

A third, older theory—the *daughter*, or *fission*, *theory*—speculated that the Moon was originally part of Earth. The Pacific Ocean basin was often mentioned as the place from which protolunar matter may have been torn—the result of the rapid spin of the young, mostly molten Earth. Indeed, there are some chemical similarities between the matter in the Moon's outer mantle and that in Earth's Pacific basin. However, this theory offers no solution to the fundamental mystery of how Earth could possibly have been spinning so fast that it ejected an object as large as our Moon. Also, computer simulations indicate that the ejection of the Moon into a stable orbit simply would not have occurred. As a result, the daughter theory, in this form at least, is no longer taken seriously.

Today, many astronomers favor a hybrid of the capture and daughter theories. This idea—often called

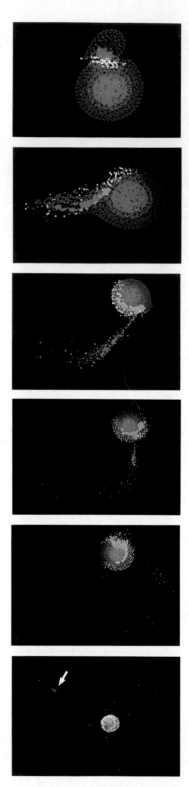

Figure 5.25 A simulated collision between Earth and an object the size of Mars. (The sequence proceeds from top to bottom and zooms out dramatically.) The arrow in the final frame shows the newly formed Moon. Red and blue colors represent rocky and metallic regions, respectively, and the direction of motion of the blue material in frames 2, 3, 4, and 5 is toward Earth. Note how most of the impactor's metallic core becomes part of Earth, leaving the Moon composed primarily of rocky material.

the *impact theory*—postulates a collision between a large, Mars-sized object and a youthful, molten Earth. As discussed in Section 4.3, such collisions were probably quite frequent in the early solar system. The collision presumed by the impact theory would have been more a glancing blow than a direct impact. Computer simulations of such a catastrophic event show that most of the bits and pieces of splattered Earth could have coalesced into a stable orbit, forming the Moon. Figure 5.25 shows one such simulation. If Earth had already formed an iron core by the time the collision occurred, the Moon could indeed have ended up with a composition similar to Earth's mantle. During the collision, any iron core in the impacting object would have been left behind, eventually to become part of Earth's core. Thus, both the Moon's overall similarity to Earth's mantle and its lack of a dense central core are explained.

5.9 Geological History of Earth and the Moon

8 Given all the data, can we construct a reasonably consistent history of Earth and the Moon after their formation? The answer seems to be yes. Many specifics are still debated, but a consensus now exists.

Sometime around 4.6 billion years ago, Earth and the Moon formed—Earth by accretion in the solar nebula, as described in Section 4.3, the Moon by a somewhat less certain chain of events, as just discussed. The approximate age of the oldest rocks discovered in the lunar highlands is 4.4 billion years, so we know that at least part of the lunar crust must already have solidified by that time. The oldest known Earth rocks are substantially younger—"only" 3.9 billion years old. This does not mean Earth formed later than the Moon, however; it means only that Earth's surface stayed molten a little longer or possibly that erosion on Earth has been more effective at hiding the details of our planet's distant past. At formation, the Moon was already depleted in heavy metals relative to Earth.

Earth was at least partially molten during most of its first billion years of existence. Denser matter sank toward the core while lighter material rose to the surface—in other words, Earth became differentiated. As mentioned in Section 5.4, the intense meteoritic bombardment that helped melt Earth at early times subsided 3.9 billion years ago. Radioactive heating in the interior continued even after Earth's surface cooled and solidified, but it too diminished with time.

As our planet cooled, it did so from the outside in because regions closest to the surface could most easily lose their excess heat to space. In this way, the surface developed a crust, and the differentiated interior attained the layered structure now implied by seismic studies. Today, radioactive heating continues throughout Earth, but there is probably not enough of it to melt any part of our planet. The high temperatures in the core are mainly the trapped remnant of a much hotter Earth that existed eons ago.

During the earliest phases of the Moon's existence—roughly the first half-billion years—the meteoritic bombardment was violent enough to heat, and keep molten, most of the surface layers, perhaps to a depth of 400 km in places. As on Earth, however, the intense heat derived from these collisions probably did not penetrate very deeply into the lunar interior because rock simply does not conduct heat well. Radioactivity probably heated the Moon a little but not sufficiently to transform it from a warm, semisolid object to a completely liquid one. The Moon is much less massive than Earth and did not contain enough radioactive elements to heat it much further. The Moon must have differentiated during this period; if it has a small iron core, that core also formed at this time.

About 3.9 billion years ago, when the heaviest bombardment ceased, the Moon was left with a solid crust dented with numerous large basins. The crust ultimately became the highlands, and the basins soon flooded with lava and became the maria. Between 3.9 and 3.2 billion years ago, lunar volcanism filled the basins with the basaltic material we see today. The age of the youngest maria—3.2 billion years—apparently indicates the time when this volcanic activity finally subsided.

Not all these great craters became flooded with lava, however. One of the youngest is the Orientale Basin (Figure 5.26), which formed about 3.9 billion years ago. It did not undergo much subsequent volcanism, and so we can recognize it as an impact crater rather than a mare. On the far side of the Moon, similar "unflooded" basins can be seen.

Because of Earth's gravitational pull, the lunar crust became thicker on the far side than on the near side. Therefore, lava from the interior had a shorter route through the crust to the surface on the Moon's Earth-facing side. As a result, relatively little volcanic activity occurred on the far side, and no large maria were created—the crust was simply too thick to allow that to occur.

Because, on average, their interior is closer to the surface, small objects cool more rapidly than large ones. Being so small, the Moon rapidly lost its internal heat to space. As a consequence, it cooled much faster than did Earth. As the Moon cooled, volcanic activity ended as the thickness of the solid surface layer increased. The crust is now far too thick for volcanism or plate tectonics to occur. With the exception of a few meters of surface erosion from eons of meteoritic bombardment, the lunar landscape has remained more or less structurally frozen for the past 3 billion years. The Moon is dead now, and it has been dead for a long time.

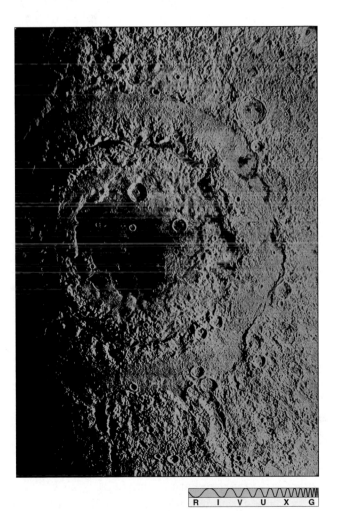

R I V U X G

Figure 5.26 A large lunar crater, called the Orientale Basin. The meteoroid that produced this crater upthrust much surrounding matter, which can be seen as concentric rings of cliffs called the Cordillera Mountains. The outermost ring is nearly 1000 km in diameter.

Chapter Review

Summary

The six main regions of Earth are (from inside to outside) a central metallic **core** (p. 139), which is surrounded by a thick rocky **mantle** (p. 139) and topped with a thin **crust** (p. 139). The liquid oceans on our planet's surface make up the **hydrosphere** (p. 139). Above the surface is the **atmosphere** (p. 139), which is composed primarily of nitrogen and oxygen. Higher still lies the **magnetosphere** (p. 139), where charged particles from the Sun are trapped by Earth's magnetic field.

The daily **tides** (p. 140) in Earth's oceans are caused by the gravitational effect of the Moon and the Sun; this gravitational effect raises **tidal bulges** (p. 140) in the oceans. The size of the tides depends on the orientations of the Sun and the Moon relative to Earth. A differential gravitational force is always called a **tidal force** (p. 141), even when no oceans or even planets are involved. The tidal interaction between Earth and the Moon is causing Earth's spin to slow and is responsible for the Moon's **synchronous orbit** (p. 138), in which the same side of the Moon always faces our planet.

Earth's atmosphere, composed primarily of nitrogen (78 percent), oxygen (21 percent), argon (0.9 percent), and carbon dioxide (0.03 percent), thins rapidly with altitude. **Convection** (p. 143) is the process by which heat is moved from one place to another by the upwelling or downwelling of a fluid, such as air or water. Convection occurs in the troposphere, the lowest region of Earth's atmosphere. It is the cause of surface winds and weather. Above the troposphere, in the stratosphere and mesosphere, the air is calm. At even higher altitudes, in the ionosphere, the atmosphere is kept ionized by high-energy radiation and particles from the Sun. Straddling the stratosphere and mesosphere is the **ozone layer** (p. 144), where incoming solar ultraviolet radiation is absorbed.

The **greenhouse effect** (p. 146) is the absorption and trapping by atmospheric gases (primarily carbon dioxide) of infrared radiation emitted by Earth's surface. By making it more difficult for Earth to radiate its energy back into space, the greenhouse effect makes our planet's surface some 40 K warmer than would otherwise be the case.

The Moon has no atmosphere because lunar gravity is too weak to retain any gases. The main surface features on the Moon are the dark **maria** (p. 147) and the lighter-colored **highlands** (p. 147). **Craters** (p. 148) of all sizes, caused by impacting meteoroids, are found everywhere on the lunar surface. The highlands are older than the maria and are much more heavily cratered. The rate at which craters are formed decreases rapidly with increasing crater size. Meteoritic impacts are the main source of erosion on the Moon's surface.

We study Earth's interior by observing how **seismic waves** (p. 152), produced by **earthquakes** (p. 152) just below Earth's surface, travel through the mantle. We study the upper mantle by analyzing the material brought to the surface when a **volcano** (p. 153) erupts. Seismic studies and mathematical modeling indicate that Earth's iron core consists of a solid **inner core** (p. 153) surrounded by a liquid **outer core** (p. 153). Earth's center is extremely hot—about the same temperature as the surface of the Sun. There is no volcanic activity on the Moon because all volcanism was stifled by the Moon's cooling mantle shortly after extensive lava flows formed the maria more than 3 billion years ago.

The density at Earth's center is much greater than the density of surface rocks. The process by which heavy material sinks to the center of a planet while lighter material rises to the surface is called **differentiation** (p. 154). The differentiation of Earth implies that our planet must have been at least partially molten in the past. This molten state was caused by a combination of the heat released during Earth's formation, the heat resulting from bombardment by material from interplanetary space, and the heat released by the decay of **radioactive** (p. 154) elements present in the material from which Earth formed.

Earth's magnetic field extends far beyond the surface of the planet. Charged particles from the solar wind are trapped by Earth's magnetic field lines to form the **Van Allen belts** (p. 156) that surround our planet. When particles from the Van Allen belts hit Earth's atmosphere, they heat and ionize the atoms there, causing the atoms to glow in an **aurora** (p. 156). Planetary magnetic fields are produced by the motion of rapidly rotating, electrically conducting fluid (such as molten iron) in a planet's core. The Moon rotates slowly, and lacks a conducting liquid core, which accounts for the absence of a lunar magnetic field.

Earth's surface is made up of enormous slabs, or plates. The slow movement of these plates across the surface is called continental drift, or **plate tectonics** (p. 159). Earthquakes, volcanism, and mountain building are associated with plate boundaries, where plates may collide, move apart, or rub against one another. The motion of the plates is thought to be driven by convection in Earth's mantle. On the Moon, the crust is too thick and the mantle too cool for plate tectonics to occur.

The most likely explanation for the formation of the Moon is that the newly formed Earth was struck by a Mars-sized object. The core of the impacting body remained behind as part of our planet, and debris splattered into space formed the moon.

Self-Test: True or False?

_____ 1. The average density of Earth is less than the density of water.

_____ 2. There is one high tide and one low tide per day at any given coastal location on Earth.

_____ 3. Because of the tides, Earth's rate of rotation is speeding up.

_____ 4. Because of the tides, the Moon is in a synchronous orbit around Earth.

_____ 5. Lunar maria are extensive lava-flow regions.

_____ 6. Large craters are formed frequently today on the surface of the Moon.

_____ 7. Except for the layer of air closest to Earth's surface, the ozone layer is the warmest part of the atmosphere.

_____ 8. Earth's atmosphere is composed primarily of oxygen.

_____ 9. Most of Earth's atmosphere lies within 30 km of the surface.

_____ 10. Water vapor and nitrogen are the primary greenhouse gases in Earth's atmosphere.

_____ 11. The Moon has no detectable atmosphere.

_____ 12. Earth's magnetic field is the result of our planet's large, permanently magnetized iron core.

_____ 13. The Moon has a weak magnetic field.

_____ 14. Motion of the crustal plates is driven by convection in Earth's upper mantle.

_____ 15. Samples of Earth's core are available from volcanos.

_____ 16. Some volcanic activity continues today on the surface of the Moon.

_____ 17. Like Earth, the Moon has a liquid metal core.

Self-Test: Fill in the Blank

1. Earth's radius is roughly _____ km.

2. The radius of the Moon is about _____ Earth's radius. (Give your answer as a simple fraction, not as a decimal.)

3. The most accurate method of determining the Earth–Moon distance is _____.

4. Of Earth's crust, mantle, outer core, and inner core, which layer is the thinnest?

5. Earth is unique among the planets in that it has _____ on its surface.

6. The tidal force is due to the _____ in the gravitational force from one side of Earth to the other.

7. The _____ on the Moon are dark, flat, roughly circular regions hundreds of kilometers in diameter.

8. Craters on the Moon are primarily caused by _____.

9. The lunar maria's dark, dense rock originally was part of the _____ of the Moon.

10. Earth's atmosphere is 78 percent _____ and 21 percent _____.

11. The troposphere is where the process called _____ occurs.

12. Sunlight is absorbed by Earth's surface and is reemitted in the form of _____ radiation.

13. A continued rise in the level of carbon dioxide in

Earth's atmosphere would _____ our planet's temperature.

14. When trapped electrons and protons from the magnetosphere eventually collide with the upper atmosphere, they produce an _____.

15. Observations of seismic waves imply that Earth's inner core is _____ and the outer core is _____.

16. For differentiation to have occurred, Earth's inte-rior must, at some time in the past, have been largely _____.

17. Because the Moon's average density is so much lower than Earth's average density, the Moon must contain less _____.

18. Continental drift, volcanism, earthquakes, faults, and mountain building can all be explained by the process known as _____.

Review and Discussion

1. Explain how the Moon produces tides in Earth's oceans.

2. What is a synchronous orbit? How did the Moon's orbit become synchronous?

3. The best place to aim a telescope or binoculars on the Moon is along the terminator line, the line between the Moon's light and dark hemispheres. Why? If you were standing on the lunar terminator, where would the Sun be in your sky? What time of day is it when you're standing on Earth's terminator line?

4. In what sense are the lunar maria "seas"?

5. What is the primary source of erosion on the Moon? Why is the average rate of lunar erosion so much less than on Earth?

6. Name two pieces of evidence indicating that the lunar highlands are older than the maria.

7. In contrast to Earth, the Moon undergoes extremes in temperature. Why?

8. The density of water in Earth's hydrosphere and the density of rocks in the crust are both lower than the average density of the planet as a whole. What does this fact tell us about Earth's interior?

9. How would our knowledge of Earth's interior change if our planet were geologically dead, as the Moon is?

10. Give two reasons geologists believe that part of Earth's core is liquid.

11. What clue does Earth's differentiation provide to our planet's history?

12. What is convection? What effect does it have on (a) Earth's atmosphere and (b) Earth's interior?

13. Use the concept of escape speed to explain why the Moon has no atmosphere.

14. What is the greenhouse effect? How does it influence Earth's surface temperature?

15. Is the greenhouse effect operating in Earth's atmosphere helpful or harmful? What are the consequences of an enhanced greenhouse effect?

16. Give a brief description of Earth's magnetosphere Why does the Moon have no magnetosphere?

17. What process has created the surface mountains, oceanic trenches, and other large-scale features on Earth's surface?

18. If the Moon had oceans like Earth's, what would the tidal effect be like there? How many high and low tides are there during a lunar day? How would the variations in tidal height compare to those on Earth?

19. Describe the theory of the Moon's origin currently favored by many astronomers

20. How is the varying thickness of the lunar crust related to the presence or absence of maria on the Moon?

Problems

1. Approximating Earth's atmosphere as a layer of gas that is 10 km thick and has a uniform density of 1.2 kg/m³, calculate the atmosphere's total mass. Compare this with the mass of Earth.

2. At 2 cm/yr, how long would it take a typical plate to traverse the present width of the Atlantic Ocean, about 6000 km?

3. Following an earthquake, how long would it take a seismic wave moving in a straight line with speed 5 km/s to reach the opposite side of Earth?

4. Using the rate given in the text for the formation of 10-km craters on the Moon, estimate how long it would take to cover the Moon with new craters of this size.

5. The *Hubble Space Telescope* has a resolution of about 0.05 arc second. What is the smallest object it can see on the surface of the Moon? Give your answer in meters.

Projects

1. Write or draw an explanation of spring tides and neap tides. Look in an almanac to find a month during which the Moon's *perigee* (the point in its orbit where the Moon is closest to Earth) and the day of new or full Moon more or less coincide. Then watch the newspapers and television weather reports for news of especially high spring tides. Do especially high tides actually occur on that day of the new or full Moon?

2. Consult an almanac to find out the date of the next fourth quarter Moon. Around 10:00 or 11:00 P.M. on that date, go to a country location where many stars can be seen sprinkled across the heavens. Around midnight, the Moon will rise. Can you see as many stars?

3. Go to a sporting goods store and get a tide table; many stores near the ocean provide them free. Choose a month and plot the height of one high and one low tide versus the day of the month. Now mark the dates when the primary phases of the Moon occur. How well does the phase of the Moon predict the tides?

4. Watch the Moon over a period of hours on a night when you can see one or more bright stars near it. Estimate how many Moon diameters the Moon moves per hour. Knowing the Moon is about 0.5° in diameter, how many degrees per hour does it move? What is your estimate of its orbital period?

The Terrestrial Planets

A Study in Contrasts

(Opposite page, background) The red planet, Mars, displays an array of fascinating surface features. Most prominent in this Viking-mission image is the Mariner Valley, a vast "canyon" extending for approximately 4000 km, roughly the width of the U. S.

(Inset A) A close-up of the Martian surface, rock-strewn and flat, seen through the eyes of the Viking 2 robot that soft-landed on the northern Utopian plains. The discarded canister is about 20 cm long. The scars in the "dirt" were made by the robot's shovel.

(Insert B) This 360° panorama was taken by the Pathfinder robot lander in 1997. The "Twin Peaks" on the horizon are 1–2 km away.

LEARNING GOALS

Studying this chapter will enable you to:

1 Explain how Mercury's rotation has been influenced by its orbit around the Sun.

2 Describe how the atmospheres of Venus and Mars differ from one another and from Earth's.

3 Compare the surface of Mercury with that of the Moon and the surfaces of Venus and Mars with that of Earth.

4 Describe how we know that Mars once had running water and a thick atmosphere.

5 Discuss the similarities and differences in the geological histories of the four terrestrial planets.

6 Explain why the atmospheres of Venus, Mars, and Earth are now so different from one another.

With Earth and the Moon as our guides, we now expand our field of view to study the other planets of the inner solar system. As we explore these worlds and seek to understand the similarities and differences among them, we begin our comparative study of the only planetary system we know.

Closest to the Sun lies Mercury, the smallest terrestrial planet. In many ways, it is kin to Earth's Moon, and much can be learned by comparing Mercury with our own satellite. Farther out lie Venus and Mars, both of which have properties that are more like Earth's, and we learn about these two terrestrial worlds by drawing parallels with Earth. It is possible that Venus, Earth, and Mars had many similarities when they formed. Yet Earth today is vibrant, teeming with life, while Venus is an uninhabitable inferno and Mars is a dry, dead world. What were the factors leading to these present conditions? In answering this question, we will discover that a planet's environment, as well as its composition, can play a critical role in determining its future.

6.1 Orbital and Physical Properties

Mercury, named for the messenger of the Roman gods, lies closest to the Sun and is visible above the horizon for at most two hours before the Sun rises or after it sets. Being farther from the Sun, Venus (named after the Roman goddess of love) is visible for a little longer—up to three hours, depending on the time of year. Sometimes known as the morning or evening "star" (because it is visible only at those times of day), Venus is the third brightest object in the entire sky. Only the Sun and the Moon are brighter. But Venus is most definitely not a star. Like all the planets, it shines by reflected sunlight. It is bright because almost all the sunlight reaching it is reflected from thick clouds high in its dense atmosphere. You can even see Venus in the daytime if you know where to look. The much fainter Mercury is visible to the naked eye only when the Sun's light is blotted out—just before dawn, just after sunset, and during a total solar eclipse.

Named for the bloody Roman god of war, orange-red Mars is also quite easy to spot in the night sky. Because of its less reflective surface, smaller size, and greater distance from the Sun, Mars does not appear as bright as Venus, as seen from Earth. However, at its brightest, Mars is still brighter than any star.

Table 6.1 expands Table 5.1 to cover physical properties of the remaining terrestrial planets; in addition, this expanded table shows two new properties: surface temperature and surface atmospheric pressure.

In seeking to understand the other terrestrial planets, astronomers are guided by our more detailed knowledge of Earth and the Moon. Mercury's high average density, for instance, tells us that this planet

Table 6.1 Some Properties of the Terrestrial Planets and Earth's Moon

	Mass (kg)	(Earth = 1)	Radius (km)	(Earth = 1)	Density (kg/m³)	Surface Gravity (Earth = 1)	Escape Speed (km/s)	Rotation Period (solar days)	Surface Temperature (K)	Surface Atmospheric Pressure (Earth = 1)
Mercury	3.3×10^{23}	0.055	2400	0.38	5400	0.38	4.3	59	100–700	—
Venus	4.9×10^{24}	6.82	6100	0.95	5300	0.90	10.4	-243^1	730	90
Earth	6.0×10^{24}	1.00	6400	1.00	5500	1.00	11.2	1.00	290	1.0
Mars	6.4×10^{23}	0.11	3400	0.53	3900	0.38	5.0	1.03	180–270	0.007
Moon	7.3×10^{22}	0.012	1700	0.27	3300	0.17	2.4	27.3	100–400	—

[1]The minus sign indicates retrograde rotation.

must have a substantial iron core, consistent with the condensation theory's account of its formation. ⚭ (Sec. 4.4) However, in many other respects Mercury is similar to Earth's Moon, leading us to use the Moon as a model for understanding Mercury's past. Venus and, to a lesser extent, Mars are similar to Earth, at least in terms of bulk properties, and so Earth provides the natural starting point for studies of those planets. For example, despite the lack of seismic data, we nevertheless assume that Venus has a metallic core and rocky mantle similar to those shown for Earth in Figure 5.1. Even the widely different atmospheres of these two worlds can be explained in familiar Earthly terms.

6.2 Rotation Rates Videos

Mercury's Curious Spin

1 Mercury is difficult to observe from Earth because of its closeness to the Sun. Even with a fairly large telescope, we can see Mercury only as a slightly pinkish, almost featureless disk. The largest ground-based telescopes can resolve features on the surface of Mercury about as well as we can perceive features on our Moon with our unaided eyes. Figure 6.1 is one of the few photographs of Mercury taken from Earth that show any indication of surface features. In the days before close-up images were obtainable from space, astronomers could only speculate about the faint, dark markings this photograph reveals.

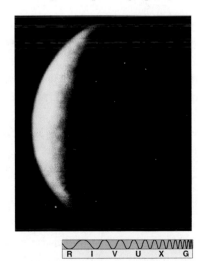

R   I   V   U   X   G

Figure 6.1 Photograph of Mercury taken from Earth with one of the largest ground-based optical telescopes. Only a few surface features are discernible.

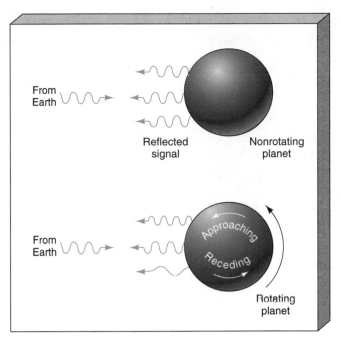

Figure 6.2 A radar beam reflected from a rotating planet yields information both about the planet's line-of-sight motion and about its rotation rate.

In principle, the ability to discern surface features, even indistinct ones, should allow us to measure Mercury's rotation rate simply by watching a particular region move around the planet. In the mid-nineteenth century, an Italian astronomer named Giovanni Schiaparelli did just that and concluded that Mercury always keeps one side facing the Sun, much as our Moon always presents only one face to Earth. The explanation suggested for this synchronous rotation was the same as for the Moon: the tidal bulge raised in Mercury by the Sun had modified the planet's rotation rate until the bulge always points directly at the Sun. ⚭ (Sec. 5.2) Although the surface features could not be seen clearly, the combination of Schiaparelli's observations and a plausible physical explanation was enough to convince most astronomers. The belief that Mercury rotated synchronously with its revolution about the Sun (once every 88 Earth days) persisted for almost a century.

In 1965 astronomers making radar observations of Mercury from the Arecibo radio telescope in Puerto Rico (Figure 3.19) discovered that this long-held view was in error. The technique they used is illustrated in Figure 6.2, which shows a radar signal reflecting from the surface of a hypothetical planet. Let's imagine, for the purpose of this discussion, that the pulse of outgoing radiation is of a single frequency.

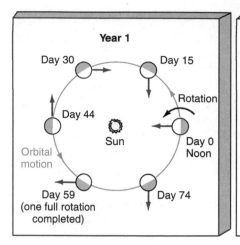
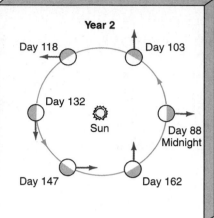
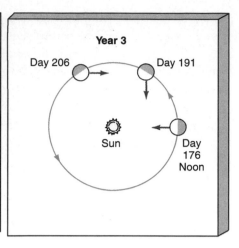

Figure 6.3 Mercury's orbital and rotational motions combine to produce a day that is two Mercury years long. The pink arrows represent an observer standing on the surface of the planet. At day 0 (center right in year 1 drawing), it is noon for our observer and the Sun is directly overhead. By the time Mercury has completed one full orbit around the Sun and moved from day 0 to day 88 (center right in year 2 drawing), it has rotated on its axis exactly 1.5 times, so that it is now midnight at the observer's location. After another complete orbit, it is noon once again on day 176 (center right in year 3 drawing).

The reflected signal may be modified in two important ways. First, if the planet has some overall line-of-sight velocity relative to Earth, the signal will be redshifted or blueshifted by the Doppler effect (see *More Precisely 2-2*). For simplicity, we take this line-of-sight velocity to be zero so that, on average, the frequency of the reflected signal is the same as that of the outgoing signal. Second, if the planet is rotating, the radiation reflected from the side moving toward us returns at a slightly higher frequency than does the radiation reflected from the receding side, simply as a consequence of the Doppler effect. (Think of the two hemispheres as being separate sources of radiation and moving at slightly different velocities, one toward us and one away from us.) What we see in the reflected signal is a spread of frequencies on either side of the original frequency. By measuring the extent of that spread, we can determine the planet's rotational speed.

In this way, astronomers found that the rotation period of Mercury is not 88 days, as had previously been believed, but 59 days, which is exactly two-thirds of Mercury's year. This state of affairs surely did not occur by chance. In fact, nineteenth-century astronomers were quite correct in thinking that Mercury's rotation was governed by the tidal effect of the Sun. However, the combination of the Sun's gravity and Mercury's eccentric orbit has caused the planet to set-

tle into a complex rotational state. Unable to come into a state of precisely synchronous rotation (because Mercury's orbital speed changes significantly from place to place in its orbit), Mercury did the next best thing. It presents the same face to the Sun not every time around, but every other time. Figure 6.3 illustrates the implications of this odd rotation for a hypothetical inhabitant of Mercury. The planet's solar day—the time from one noon to the next—is two Mercury years long!

The Sun also influences the tilt of Mercury's spin axis. Because of the Sun's tides, Mercury's rotation axis is almost exactly perpendicular to its orbit plane. Thus, the noontime Sun is always directly overhead for someone standing on the equator and always on the horizon for someone standing at either pole.

Venus and Mars

The same clouds whose reflectivity make Venus so easy to see in the night sky also make it impossible for us to discern any surface features, at least in visible light. Figure 6.4, one of the best photographs of Venus taken with an Earth-based telescope, shows an almost uniform yellow-white disk, with rare hints of clouds. Because of the cloud cover, astronomers did not know Venus's rotation period until the development of radar

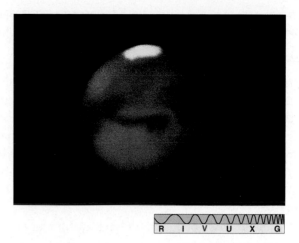

Figure 6.5 A deep-red (800-nm) image of Mars taken in 1991 at Pic du Midi, an exceptionally clear site in the French Alps. One of the planet's polar caps appears at the top, and numerous other surface markings are visible.

Figure 6.4 This photograph taken from Earth shows Venus with its creamy yellow mask of clouds.

techniques in the 1960s, when Doppler broadening of returning radar echoes indicated an unexpectedly sluggish 243-day rotation period. Furthermore, Venus's spin was found to be *retrograde*—that is, opposite that of Earth and most other solar system objects and in the direction opposite Venus's orbital motion. The planet's rotation axis is almost exactly perpendicular to its orbit plane, just as Mercury's is.

We have no "evolutionary" explanation for Venus's anomalous rotation. It is not the result of any known interaction with the Sun or with Earth or any other planet. As mentioned in Section 4.4, the best we

can do is to appeal to a random event that occurred during the final stages of solar system formation—a collision that just happened to leave Venus rotating in the manner we now observe.

In contrast to Mercury and Venus, surface markings are readily seen on Mars (Figure 6.5), allowing astronomers to track the planet's rotation. Mars rotates once on its axis every 24.6 hours—close to one Earth day. The planet's equator is inclined to the orbit plane at an angle of 25.2°, very similar to Earth's inclination of 23.5°. Thus, as Mars orbits the Sun, we find both daily and seasonal cycles, just as on Earth.

Figure 6.6 summarizes the rotations and revolutions of the four terrestrial planets.

Figure 6.6 The inner planets of the solar system—Mercury, Venus, Earth, and Mars—display widely different rotational properties. Although all orbit the Sun in the same direction and in nearly the same plane, Mercury's rotation is slow and prograde, Venus's is slow and retrograde, Earth's is fast and prograde, and Mars's is fast and prograde. Venus rotates clockwise as seen from above the plane of the ecliptic, but Mercury, Earth, and Mars all rotate counterclockwise.

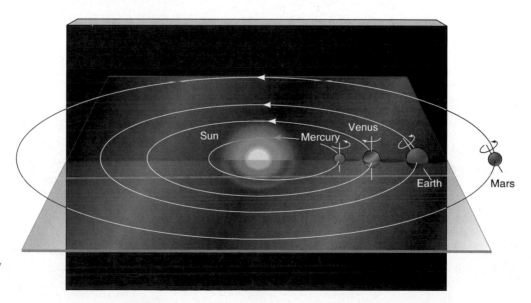

6.3 Atmospheres

Mercury

Astronomers have never observed, either from Earth or from space, any appreciable atmosphere on Mercury. Although the U.S. space probe *Mariner 10*[1] did find a trace of what was at first thought to be an atmosphere, the gas is now known to be temporarily trapped hydrogen and helium from the solar wind. Mercury holds this gas for just a few weeks before it leaks away into space.

The absence of an atmosphere on Mercury is easily explained. Mercury's closeness to the Sun means that its daytime surface temperature is as high as 700 K. The combination of high temperature and low mass (only 4.5 times the mass of the Moon, as Table 6.1 shows) means that any atmosphere Mercury might have once had escaped long ago (see *More Precisely 5-1*). Lacking an atmosphere, Mercury, like Earth's Moon, has no protection from the harsh environment of interplanetary space. Meteoroids, X rays, and ultraviolet radiation constantly rain down on its surface.

With no atmosphere to retain heat, Mercury's surface temperature falls to about 100 K during the planet's long night. Mercury's 600 K temperature range is the largest of any planet or moon in the solar system. Near the poles, where the Sun's light arrives almost parallel to the surface, the temperature remains low at all times. Recent Earth-based radar studies suggest that Mercury's polar temperature could be as low as 125 K and that the poles may be permanently covered with extensive thin sheets of water ice despite the planet's scorched equator.

Venus

2 In the 1930s, scientists used spectroscopy to measure the temperature of Venus's upper atmosphere and found it to be about 240 K, not much different from Earth's. Taking into account the cloud cover and Venus's nearness to the Sun, and assuming

that Venus had an atmosphere much like our own, researchers concluded that Venus might have an average surface temperature only a few degrees higher than Earth's. In the 1950s, however, when radio observations of the planet penetrated the cloud layer and gave the first indication of conditions near the surface, they revealed a temperature exceeding 600 K! Almost overnight, the popular conception of Venus changed from lush tropical jungle to arid, uninhabitable desert.

Since then, spacecraft data have revealed the full extent of the differences between the atmospheres of Venus and Earth. Venus's atmosphere is much more massive than our own, and it extends to a much greater height above the planet's surface. The surface pressure of Venus's atmosphere is about 90 times the pressure at sea level on Earth. This is equivalent to an (Earthly) underwater depth of about 1 km. (Unprotected humans cannot dive much below 100 m.) The surface temperature is a sizzling 730 K.

The dominant constituent (96.5 percent) of Venus's atmosphere is carbon dioxide. Almost all of the remaining 3.5 percent is nitrogen. Given Venus's similarity to Earth in mass, radius, and location in the solar system, it is commonly assumed that Venus must have started off looking something like Earth. However, there is no sign of the large amount of water vapor that would be present if a volume of water equivalent to Earth's oceans had once existed on Venus and later evaporated. If Venus started off with Earth-like composition, something happened to its water, for the planet is now an exceedingly dry place. Even the highly reflective clouds are composed not of water vapor, as on Earth, but of sulfuric acid droplets.

Venus's atmospheric patterns are much more evident when examined with equipment capable of detecting ultraviolet radiation. Some of Venus's upper-level clouds absorb this high-frequency radiation, thereby increasing the contrast. Figure 6.7 is an ultraviolet image taken in 1979 by the U.S. *Pioneer Venus* spacecraft from a distance of 200,000 km (compare the optical image shown in Figure 6.4). The large, fast-moving cloud patterns lie between 50 and 70 km above the surface. Upper-level winds reach speeds of 400 km/h relative to the planet. Below the clouds, extending down to an altitude of 30 km, is a layer of haze. Below 30 km, the air is clear.

---

[1]*Mariner 10* is one of the few spacecraft to have visited Mercury. In 1973 it was placed in an eccentric orbit around the Sun that brings it close to Mercury every 176 days (two Mercury years). The three close encounters it made with Mercury in 1974 and 1975 (after which it ran out of fuel) are the source of virtually all the detailed information we have on that planet.

Figure 6.7 Venus as photographed by the *Pioneer* spacecraft's cameras 200,000 km away from the planet. This image was made by capturing solar ultraviolet radiation reflected from the planet's clouds, which are composed mostly of sulfuric acid droplets, much like the corrosive acid in a car battery.

Mars

Well before the arrival of spacecraft, astronomers knew from Earth-based spectroscopy that the Martian atmosphere is quite thin and composed primarily of carbon dioxide. Spacecraft measurements confirmed these results, indicating that the atmospheric pressure is only about 1/150 the pressure of Earth's atmosphere at sea level. The Martian atmosphere is 95.3 percent carbon dioxide, 2.7 percent nitrogen, and 1.6 percent argon, plus small amounts of oxygen, carbon monoxide, and water vapor. While there is some superficial similarity in composition between the atmospheres of Mars and Venus, the two planets must have very different atmospheric histories: Mars's present-day "air" is more than 13,000 times thinner than that on Venus. Average surface temperatures on Mars are about 50 K cooler than on Earth.

6.4 The Surface of Mercury

Extension

3 Figure 6.8 shows a picture of Mercury taken when the *Mariner 10* spacecraft was 200,000 km from the planet, and Figure 6.9 shows a higher-resolution photograph of the planet from a distance of 20,000 km. The similarities to our Moon are striking. We see no sign of clouds, rivers, dust storms, or other aspects of weather. Much of the cratered surface bears a strong resemblance to the Moon's highlands. The crater walls are generally not as high as on the Moon, and the ejected material landed closer to the impact site, exactly as we would expect on the basis of Mercury's greater surface gravity (which is a little more than twice that of the Moon). Mercury, however, shows no extensive lava flow regions akin to the lunar maria.

As on the Moon, Mercury's craters are the result of meteoritic bombardment. The craters are not so densely packed as their lunar counterparts, however, and there are extensive **intercrater plains**. One likely explanation for Mercury's relative lack of craters is that

Figure 6.8 Mercury is imaged here as a mosaic of photographs taken by the *Mariner 10* spacecraft in the mid-1970s during its approach to the planet. At the time, the spacecraft was 200,000 km away from Mercury.

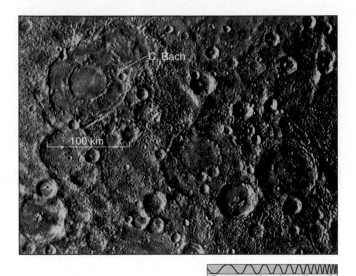

Figure 6.9 Another photograph of Mercury by *Mariner 10*, this time from a distance of about 20,000 km. The double-ringed crater at the upper left, named C. Bach and about 100 km across, exemplifies how many of the large craters on Mercury tend to form double, rather than single, rings. The reason is not yet understood.

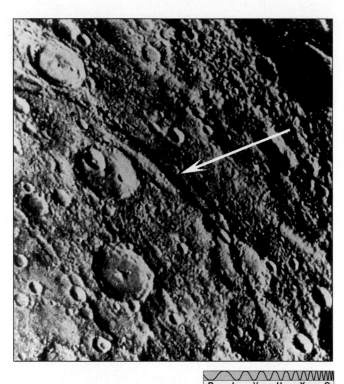

Figure 6.10 Discovery Scarp on Mercury's surface. This cliff appears to have formed when the planet's crust cooled and shrank early in its history. The shrinking caused the crust to split. Subsequently the crust on one side of the split moved upward relative to the crust on the other side, forming the scarp. Several hundred kilometers long and up to 3 km high in places, it runs diagonally across the center of the frame in this photograph.

the older impact craters have been filled in by volcanic activity, in much the same way as the Moon's maria filled in craters as they formed. However, Mercury's intercrater plains do not look much like mare material; they are much lighter in color and not as flat. The apparent absence of features associated with large-scale lava flows, along with the light color of the lava-flooded regions, suggests that Mercury's volcanic past was different from the Moon's. However, many of the details of how Mercury's landscape came to look the way it does remain unexplained.

Mercury has at least one type of surface feature not found on the Moon. Figure 6.10 shows a **scarp**, or cliff, that does not appear to be the result of volcanic or other familiar geological activity. The scarp cuts across several craters, which indicates that whatever produced it occurred after most of the meteoritic bombardment was over. Numerous scarps have been found in the *Mariner* images. Mercury shows no evidence for crustal motions, so the scarps could not have been formed by tectonic processes such as those responsible for fault lines on Earth. Instead, the scarps probably formed when the crust cooled, shrank, and split long ago; after the splitting, one side of the crack moved upward relative to the other side, forming the scarp

cliff face. If we can apply to Mercury the cratering age estimates we use for the Moon, the scarps appeared about 4 billion years ago.

Much of the discussion in Chapter 5 about the surface of Earth's Moon applies equally well to Mercury. ∞ (Sec. 5.4) Indeed, a lot of our understanding of Mercury's past is based on comparison with the Moon. Figure 6.11 shows the result of what was probably the last great event in the geological history of Mercury—an immense bull's-eye crater called the Caloris Basin, formed eons ago by the impact of a large asteroid. (Because of the orientation of the planet during *Mariner 10's* flybys, only half the basin is visible. The center is off the left-hand side of the photograph.) Compare this basin with the Orientale Basin on the Moon (Figure 5.26).

So large was the impact that created the Caloris Basin that it apparently sent strong seismic waves

R I V U X G

Figure 6.11 Mercury's most prominent geological feature—the Caloris Basin—measures about 1400 km across and is ringed by concentric mountain ranges (dashed lines) that reach more than 3 km high in places. This huge circular basin, only half of which shows in this *Mariner 10* photograph, is similar in size to the Moon's Mare Imbrium and spans more than half of Mercury's radius.

reverberating throughout the planet. On the opposite side of Mercury from Caloris there is a region of oddly rippled and wavy surface features known as **weird terrain**. Scientists believe that this terrain was produced when seismic waves from the Caloris impact traveled around the planet and converged on the point opposite the collision, causing large-scale disruption of the surface there.

6.5 The Surface of Venus

3 Although the clouds are thick and the surface totally shrouded, we are by no means ignorant of Venus's surface. Radar astronomers have bombarded the planet with radio signals, both from Earth and from a series of U.S. and Soviet spacecraft. Analysis of the radar echoes yields a map of the planet's surface; except for Figure 6.16, all the views of Venus in this section are "radargraphs" (not photographs) created in this

way. Most recently, the U.S. *Magellan* spacecraft provided very-high-resolution images of Venus's surface features.

Large-Scale Topography

Figure 6.12(a) shows a relatively low-resolution map of Venus made by *Pioneer Venus* in 1979. Surface elevation above the average radius of the planet's surface is indicated by color, with white representing the highest elevations, blue the lowest. (Note that the blue has nothing to do with oceans.) For comparison, Figure 6.12(b) shows a map of Earth to the same scale and at the same spatial resolution. Figure 6.13 is a 1995 mosaic of *Magellan* images of Venus.

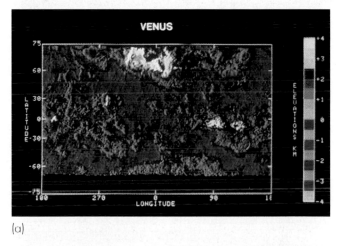

(a)

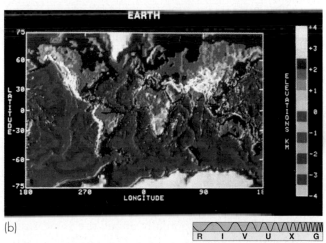

(b)

R I V U X G

Figure 6.12 (a) Radar map of the surface of Venus, based on *Pioneer Venus* data. Color represents elevation, with white the highest areas and blue the lowest. Some major surface features are indicated. (b) A similar map of Earth, at the same spatial resolution.

R I V U X G

Figure 6.13 A planetwide mosaic of Venus made from *Magellan* images. The largest "continent" on Venus, Aphrodite Terra, is the yellow "dragon-shaped" area across the center of the image.

Venus's surface appears to be mostly smooth, resembling rolling plains with modest highlands and lowlands. Only two continent-sized features, called Ishtar Terra and Aphrodite Terra, adorn the landscape, and these contain mountains comparable in height to those on Earth. The highest peaks rise some 14 km above the level of the deepest surface depressions. (The highest point on Earth, the summit of Mount Everest, lies about 20 km above the deepest section of the ocean floor.) The elevated "continents" occupy only 8 percent of Venus's total surface area. (Continents on Earth make up about 25 percent of our planet's surface.)

Ishtar Terra is an extensive uplifted plateau, 1500 km across at its widest point, in Venus's northern high latitudes. The projection of the *Pioneer* map in Figure 6.12 makes Ishtar Terra appear much larger than it really is (it is actually about the same size as Australia). The plateau is ringed by mountain ranges that include the highest peak (Maxwell Mons) on the planet.

The other continent-sized formation, Aphrodite Terra, is located on Venus's equator, south and east of Ishtar Terra. It is comparable in size to Africa. Before *Magellan's* arrival, some researchers had speculated that

Aphrodite Terra might have been the site of something equivalent to seafloor spreading on Earth, where two tectonic plates moved apart and molten rock rose to the surface in the gap between them, forming an extended ridge. (This is just what is happening today at Earth's Mid-Atlantic Ridge, which is clearly visible in Figure 6.12b.) However, the Magellan images now seem to rule out any tectonic activity on Venus, and the Aphrodite region shows no signs of spreading. The crust appears buckled and fractured, suggesting large compressive forces, and there seem to have been numerous periods when extensive lava flows occurred.

Volcanism and Cratering

Although erosion by the planet's atmosphere may play some part in obliterating surface features, the most important factor is volcanism, which appears to "resurface" the planet every few hundred million years. Many areas of Venus have volcanic features. Figure 6.14(a) shows a *Magellan* image of seven pancake-shaped lava domes, each about 25 km across. They probably formed when lava oozed out of the surface, formed the dome, then withdrew, leaving the crust to crack and subside. Lava domes such as these are found in several locations on Venus. Figure 6.14b shows a three-dimensional view of the domes.

The most common volcanoes on the planet are of the type known as **shield volcanoes**. Those on Earth are associated with lava welling up through a "hot spot" in the crust (like the Hawaiian Islands). They are built up over long periods of time by successive eruptions and lava flows. A characteristic of shield volcanoes is the formation of a *caldera*, or crater, at the summit when the underlying lava withdraws and the surface collapses. A large-shield volcano, called Gula Mons, is shown (in a computer-generated view) in Figure 6.14(c).

The largest volcanic structures on Venus are huge, roughly circular regions known as **coronae**. A large corona, called Aine, can be seen in Figure 6.15. Coronae are unique to Venus. They appear to have been caused by upwelling motions in the mantle that never developed into full-fledged convection as on Earth. Coronae generally have volcanoes both in and around them, and their rims usually show evidence of extensive lava flows into the plains below.

Two pieces of indirect evidence suggest that volcanism on Venus continues today. First, the level of

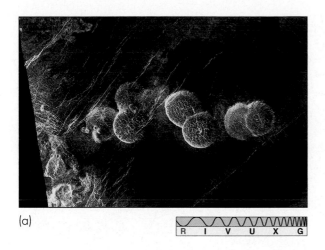

(a)

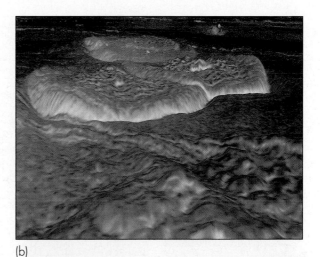

(b)

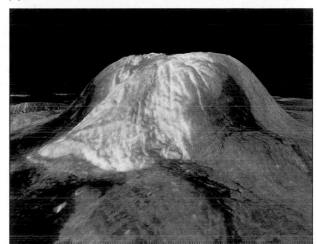

(c)

Figure 6.14 (a) A series of dome-shaped structures on Venus. They are the result of molten rock having bulged out of the ground and then retreated, leaving behind a thin, solid crust that subsequently cracked and subsided. *Magellan* found features like this in several locations on Venus. (b) A computer-generated three-dimensional representation of four of the domes. (c) A three-dimensional *Magellan* view of the large shield volcano known as Gula Mons. The volcanic caldera at the summit is about 100 km across. As in part (b), the vertical scale has been greatly exaggerated. The volcano is about 4 km high. Color in (b) and (c) is based on data returned by Soviet landers.

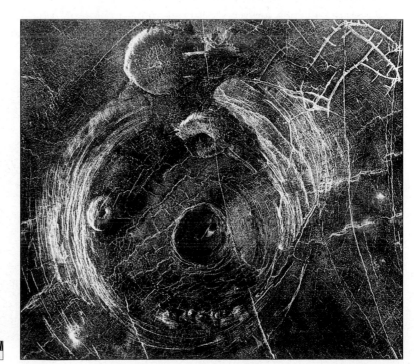

Figure 6.15 This corona, called Aine, lies in the plains south of Aphrodite Terra. It is about 300 km across. Notice the cracked and fissured surface surrounding it. Coronae are probably the result of upwelling mantle material's causing the surface to bulge outward.

Figure 6.16 One of the first true-color views of the surface of Venus, radioed back to Earth from the Russian *Venera 14* spacecraft, which made a soft landing in 1975. The amount of sunlight penetrating Venus's cloud cover is about the same as the amount reaching Earth's surface on a heavily overcast day.

R I V U X G

sulfur dioxide above Venus's clouds shows large and fairly frequent fluctuations. It is quite possible that these variations result from volcanic eruptions on the surface. If so, volcanism may be the primary cause of Venus's thick cloud cover. Second, orbiting spacecraft have observed bursts of radio energy from Aphrodite and other regions of the planet's surface. These bursts are similar to those produced by the lightning discharges that often occur in the plumes of erupting volcanoes on Earth, again suggesting ongoing activity. However, while these pieces of evidence are quite persuasive, they are still only circumstantial. No "smoking gun" (or erupting volcano) has yet been seen, so the case for active volcanism is not yet complete.

A few Soviet *Venera* spacecraft have landed on Venus's surface. Each survived for about an hour before being destroyed by the intense heat, its electronic cir-

cuitry melting in this planetary oven. Figure 6.16 shows one of the first photographs of the surface of Venus radioed back to Earth. The flat rocks visible in this image show little evidence of erosion and are apparently quite young, supporting the idea of ongoing surface activity. Later Soviet landers performed simple chemical analyses of the surface. Some of the samples studied were found to be predominantly basaltic, again implying a volcanic past. Others resembled terrestrial granite.

Not all the craters on Venus are volcanic in origin. Some were formed by meteoritic impact. The largest impact craters on Venus are generally circular, but those less than about 15 km in diameter can be quite asymmetric. Figure 6.17 shows a *Magellan* image of a relatively small impact crater, about 10 km across, in Venus's southern hemisphere. Geologists believe

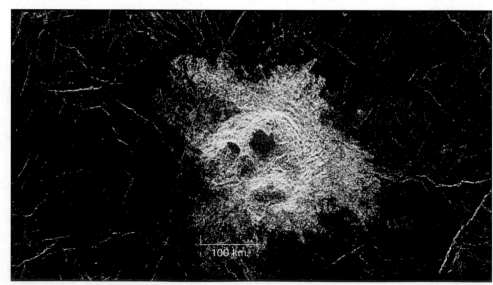

Figure 6.17 A *Magellan* image of an apparent multiple-impact crater in Venus's southern hemisphere. The irregular shape of the light-colored ejecta blanket seems to be the result of a meteorite that fragmented just prior to impact. The dark regions in the crater may be pools of solidified lava associated with the individual fragments.

100 km

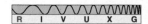

R I V U X G

that the light-colored region is an ejecta blanket—material thrown from the crater following the impact. Its irregular shape may be the result of a large meteorite that broke up just before impact, with the separate pieces hitting the surface near one another. Numerous impact craters, again identifiable by their ejecta blankets, can also be discerned in Figure 6.15.

6.6 The Surface of Mars *Video*

The View From Earth *Extension*

3 Earth-based observations of Mars at closest approach can distinguish surface features as small as 100 km across—about the same resolution as the unaided human eye can achieve when viewing the Moon. However, when Mars is closest to us and most easily observed, it is also full, so the angle of the Sun's rays does not permit us to see any topographical detail, such as craters or mountains. Even through a large telescope Mars appears only as a reddish disk with some light and dark patches and prominent polar caps (Figure 6.5). Figure 6.18 shows a *Hubble Space Telescope* image of Mars, along with a photograph taken by one of the two U.S. *Viking* spacecraft en route to the planet.

Mars's surface features undergo slow seasonal changes over the course of a Martian year—a consequence of Mars's axial tilt and somewhat eccentric orbit. The polar caps grow or shrink according to the seasons, almost disappearing during the Martian summer. The dark features also vary in size and shape. To fanciful observers around the start of the twentieth century, these changes suggested the annual growth of vegetation. It was but a small step from seeing polar ice caps and speculating about teeming vegetation to imagining a planet harboring intelligent life, perhaps not unlike us.

As with Venus, however, these speculations were not to be confirmed. The changing polar caps are mostly frozen carbon dioxide, not water ice as at Earth's North and South poles, and the dark regions are just highly cratered and eroded areas on the surface. During summer in the Martian southern hemisphere, planetwide dust storms sweep up the dry dust and carry it aloft, sometimes for months at a time, eventually depositing it elsewhere on the planet. Repeated covering and uncovering of the Martian landscape gives the impression from a distance of surface variability, but it is only the thin dust cover that changes.

Large-Scale Topography

A striking feature of the terrain of Mars is the marked difference between the northern and southern hemispheres (Figure 6.19). The northern hemisphere is made up largely of rolling volcanic plains somewhat like the lunar maria but much larger than any plains found on Earth or the Moon. They were apparently formed by eruptions involving enormous volumes of lava. The plains are strewn with blocks of volcanic rock, as well as with boulders blasted out of impact areas by infalling meteoroids. The southern hemisphere consists of heavily cratered highlands lying several kilometers above the level of the lowland north. Most of the dark regions visible from Earth are mountainous regions in the south.

Figure 6.18 (a) A *Hubble Space Telescope* image of Mars, taken while the planet was near closest approach to Earth in 1995. (b) A view of Mars taken from a *Viking* spacecraft during its approach in 1976. Some of the planet's surface features are visible at a level of detail completely unattainable from Earth.

(a)   R I V U X G   (b)

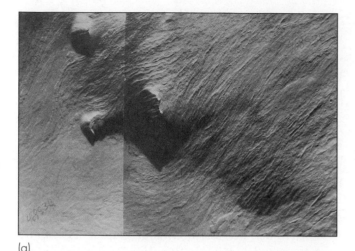

(a)

(b)

R I V U X G

Figure 6.19 (a) Mars's northern hemisphere consists of rolling volcanic plains (false color image.) (b) The southern Martian highlands are heavily cratered (true color). Both photographs show roughly the same scale, nearly 1000 km across.

roughly the size of North America, lies on the Martian equator and rises 10 km higher than the rest of the Martian surface. To the east and west of Tharsis lie wide depressions, hundreds of kilometers across and up to 3 km deep. Tharsis appears to be even less heavily cratered than the northern hemisphere, making it the youngest region on the planet. It is estimated to be only 2 to 3 billion years old. If we wished to extend the idea of "continents" from Earth and Venus to Mars, we would say that Tharsis is the only continent on the Martian surface. However, as on Venus, there is no sign of plate tectonics—the continent of Tharsis is not drifting as its earthly counterparts are.

Both U.S. *Viking* missions dispatched landers to the Martian surface. Figure 6.21 is the view from *Viking 1*, which touched down near the planet's equator just east of Tharsis. The photograph shows a windswept, gently rolling, desolate plain littered with rocks of all sizes, not unlike a high desert on Earth. This view may be quite typical of the low-latitude northern plains. The *Viking* landers performed numerous chemical

R I V U X G

Figure 6.20 The Tharsis region of Mars, 5000 km across, bulges out from the planet's equatorial region, rising to a height of about 10 km. The two large volcanoes on the left mark the approximate peak of the bulge. Dominating the center of the field of view is a vast "canyon" known as Valles Marineris—the Mariner Valley.

The northern plains are cratered much less than the southern highlands, suggesting that the northern surface is younger—perhaps 3 billion years old, compared with 4 billion in the south. In places, the boundary between the southern highlands and the northern plains is quite sharp. The surface level can drop by as much as 4 km in a distance of 100 km or so. Most scientists assume that the southern terrain is the original crust of the planet. How most of the northern hemisphere could have been lowered in elevation and flooded with lava remains a mystery.

Figure 6.20 is a large-scale view showing Mars's major geological feature, the *Tharsis bulge*. This region,

Figure 6.21 Panoramic view from the perspective of the *Viking 1* spacecraft now parked on the surface of Mars. The fine-grained soil and the rock-strewn terrain stretching toward the horizon are reddish. Containing substantial amounts of iron ore, the surface of Mars is literally rusting away. The sky is a pale pink, the result of airborne dust. (See also inset B of the chapter opener.)

analyses of the rock on Mars's surface. One important finding of these studies was the high iron content of the planet's surface. Chemical reactions between the iron-rich surface soil and trace amounts of oxygen in the atmosphere are responsible for the iron oxide ("rust") that gives Mars its characteristic red color.

Volcanism

Mars contains the largest known volcanoes in the solar system. Four particularly large ones are found on the Tharsis bulge, two of them visible in Figure 6.20. Biggest of all is Olympus Mons (Figure 6.22), which lies on the northwestern slope of Tharsis, just off the left edge of Figure 6.20. Olympus Mons is 700 km in diameter at its base—only slightly smaller than the state of Texas—and rises to a height of 25 km above the surrounding plains.

Like the volcanoes on Venus, those on Mars are not associated with plate motion but instead are shield volcanoes, sitting atop various hot spots in the Martian mantle. Spacecraft images of the Martian surface reveal many hundreds of volcanoes. Most of the largest

Figure 6.22 Olympus Mons, the largest volcano known on Mars or anywhere else in the solar system. Nearly three times taller than Mount Everest on Earth, this mountain measures about 700 km across the base, and its peak rises 25 km above the Martian surface. It seems currently inactive and may have been extinct for at least several hundred million years. (By comparison, the largest volcano on Earth, Hawaii's Mauna Loa, measures a mere 120 km across and peaks about 9 km above the Pacific Ocean floor.)

are associated with the Tharsis bulge, but many smaller ones are also found in the northern plains. It is not known whether any of them are still active. However, based on the extent of impact cratering on their slopes, some of them apparently erupted as recently as 100 million years ago.

The great height of Martian volcanoes is a direct consequence of the planet's low surface gravity. As a shield volcano forms and lava flows and spreads, the new mountain's height depends on its ability to support its own weight. The lower the surface gravity, the less the lave weights, and the higher the mountain can be. Mars has a surface gravity only 40 percent that of Earth, and its volcanoes rise roughly 2.5 times as high. (The surface gravity on Venus is almost equal to that on Earth, and Maxwell Mons on Venus and the Hawaiian shield volcanoes on Earth rise to roughly the same height, about 10 km, above their respective bases.)

The Martian "Grand Canyon"

Associated with the Tharsis bulge is a great "canyon" known as Valles Marineris (the Mariner Valley). Shown in its entirety cutting across the center of Figure 6.20, this enormous crack in the Martian surface is not really a canyon in the terrestrial sense, because running water played no part in its formation. Astronomers believe it was formed by the same crustal forces that pushed the Tharsis region upward, causing the surface to split and crack. Cratering studies suggest that the Valles Marineris is at least 2 billion years old. Similar (but smaller) cracks, formed in a similar way, have been found in the Aphrodite Terra region of Venus.

Valles Marineris runs for almost 4000 km along the Martian equator and extends about one-fifth of the way around the planet. At its widest, it is 120 km across, and it is as deep as 7 km in places. Like many other Martian surface features, it dwarfs Earthly competition. Earth's Grand Canyon in Arizona would easily fit into one of its side "tributary" cracks. It is so large that it can even be seen from Earth. We must emphasize, however, that this Martian feature was not carved by a river, nor is it a result of Martian plate tectonics. For some reason, the crustal forces that formed it never developed into full-fledged plate motion as on Earth.

Evidence for Water on Mars

4 Although the Valles Marineris was not formed by running water, photographic evidence reveals that liquid water once existed in great quantity on the surface of Mars. Two types of flow features are seen: **runoff channels** and **outflow channels**.

The runoff channels (Figure 6.23) are found in the southern highlands. They are extensive systems—sometimes hundreds of kilometers in total length—of interconnecting, twisting channels that merge into larger, wider channels. They bear a strong resemblance to river systems on Earth, and geologists believe that that is just what they are: the dried-up beds of long-gone rivers that once carried rainfall on Mars from the mountains down into the valleys. These runoff channels speak of a time 4 billion years ago (the age of the Martian highlands) when the atmosphere was thicker, the surface warmer, and liquid water widespread.

The outflow channels (Figure 6.24) are probably relics of catastrophic flooding on Mars long ago. They appear only in equatorial regions and generally do not form the extensive interconnected networks that characterize the runoff channels. Instead, the outflow channels are probably the paths taken by huge volumes of water draining from the southern highlands into the northern plains. Judging from the width and depth of the channels, the flow rates must have been truly enormous—perhaps as much as a hundred times greater than the 10^5 tons per second carried by the Amazon River, the largest river system on Earth. Flooding shaped the outflow channels about 3 billion years ago, about the same time as the northern volcanic plains formed.

Astronomers find no evidence for liquid water anywhere on Mars today, and the amount of water vapor in the Martian atmosphere is tiny. Yet the extent of the outflow channels indicates that a huge total volume of water existed on Mars in the past. Where did all that water go? One possible answer is that much of Mars's original water is now locked in a layer of **permafrost**, which is water ice lying just below the planet's surface. Figure 6.25 shows evidence for this permafrost layer in the form of a fairly typical Martian impact crater named Yuty. Unlike the lunar craters discussed in Chapter 5 (see Section 5-4), Yuty's ejecta blanket gives the distinct impression of a liquid that has splashed or flowed out of the crater. Most likely,

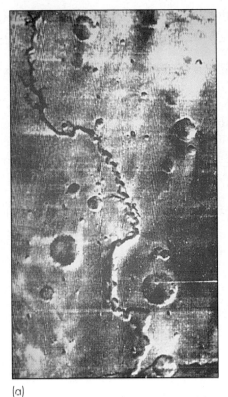

Figure 6.23 (a) This runoff channel on Mars is about 400 km long and 5 km wide. (b) The Red River running from Shreveport, Louisiana, to the Mississippi. Martian runoff channels and rivers on Earth differ mainly in that there is currently no liquid water in this, or any other, Martian channel.

(a)

(b)

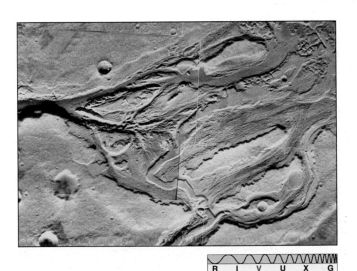

Figure 6.24 An outflow channel near the Martian equator bears witness to a catastrophic flood that occurred about 3 billion years ago.

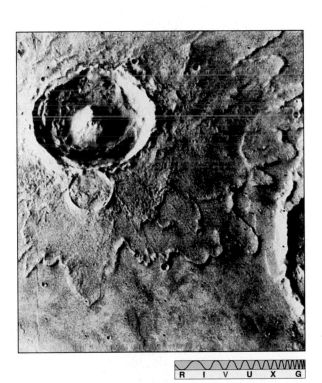

Figure 6.25 The ejecta from Mars's crater Yuty (18 km in diameter) evidently was once liquid. This type of crater is sometimes called a "splosh" crater.

the explosive impact heated and liquefied the permafrost, resulting in the fluid appearance of the ejecta.

Scientists believe that 4 billion years ago, as the Martian climate changed, the running water that formed the runoff channels began to freeze, forming the permafrost and drying out the river beds. Mars remained frozen for about a billion years, until volcanic (or some other) activity heated large regions of the surface, melting the permafrost and causing the flash floods that created the outflow channels. Subsequently, volcanic activity subsided, the water refroze, and Mars once again became a dry world.

6.7 Internal Structure and Geological History

Mercury

5 Mercury's magnetic field, discovered by *Mariner 10*, is about 100 times weaker than Earth's field. The discovery that Mercury has a magnetic field came as a surprise to planetary scientists, who, having detected no magnetic field in the Moon, expected Mercury to have none either. In Chapter 5, we saw how a combination of liquid metal core *and* rapid rotation is necessary for the production of a planetary magnetic field. ∞ (Sec. 5.6) Mercury certainly does not rotate rapidly, and it may also lack a liquid metal core. Yet a magnetic field undeniably surrounds it. Although weak, the field is strong enough to deflect the solar wind and create a small magnetosphere around the planet. Scientists have no clear understanding of its origin. Possibly it is simply a "fossil remnant" dating back to the distant past when the planet's core solidified.

Mercury's magnetic field and high average density (roughly 5400 kg/m^3) together imply that most of the planet's interior is dominated by a large, heavy, iron-rich core having a radius of perhaps 1800 km. Whether that core is solid or liquid is unknown. The ratio of core volume to total planet volume is greater for Mercury than for any other object in the solar system. Figure 6.26 illustrates the relative sizes and internal structures of Earth, the Moon, and Mercury.

Like the Moon, Mercury seems to have been geologically dead for roughly the past 4 billion years. Again as on the Moon, the lack of present-day geological activity on Mercury results from the mantle's being solid, preventing volcanism or tectonic motion.

Largely on the basis of studies of the Moon, scientists have pieced together the following outline of Mercury's early history.

When Mercury formed 4.6 billion years ago, it was already depleted in rocky material. This was largely a consequence of its location in the hot inner region of the early solar system, although collisions with asteroid-sized (and larger) bodies may also have stripped away some of its light mantle. During the next half-billion years, Mercury melted and differentiated, just as the other terrestrial worlds did. It suffered the same intense meteoritic bombardment as the Moon. Being more massive than the Moon, Mercury cooled more slowly, so its crust was thinner than the Moon's and volcanic activity more common. More craters were erased by lava, leading to the intercrater plains found by *Mariner 10*.

As the planet's large iron core formed and then cooled, the planet began to shrink, causing the surface to contract. This compression produced the scarps seen on Mercury's surface and may have prematurely terminated volcanic activity by squeezing shut the cracks and fissures on the surface. Thus, Mercury did not experience the subsequent extensive volcanic outflows that formed the lunar maria. Despite its larger mass and greater internal temperature, Mercury has probably been geologically inactive for even longer than the Moon.

Venus

Both U.S. and Soviet spacecraft have failed to detect any magnetosphere around Venus. Given that Venus's average density is similar to Earth's, it seems likely that Venus has an Earth-like overall composition and a partially molten iron-rich core. The lack of any detectable magnetic field, then, is almost surely the result of the planet's extremely slow rotation.

Because none of the *Venera* landers carried seismic equipment, no direct measurements of the planet's interior have ever been made, and theoretical models of the interior have very little hard data to constrain them. However, to many geologists the surface of Venus resembles that of the young Earth, at an age of perhaps a billion years. At that time, volcanic activity had already begun on Earth, but the crust was still relatively thin, and the convective processes in the mantle that drive plate tectonic motion were not yet established.

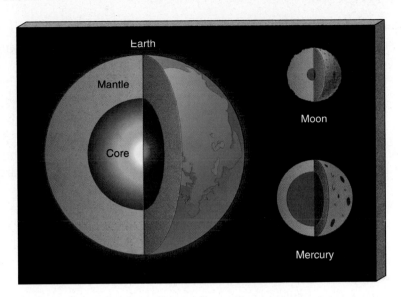

Figure 6.26 The internal structure of Earth, the Moon, and Mercury, drawn to the same scale. Note how large a fraction of Mercury's interior is core.

Why has Venus remained in that immature state and not developed plate tectonics as Earth did? That question remains to be answered. Some planetary geologists have speculated that the high surface temperature on Venus has inhibited evolution by slowing the planet's cooling rate. Possibly the high surface temperature has made the crust too soft for Earth-style plates to develop. Or perhaps the high temperature and soft crust led to more volcanism, tapping the energy that might otherwise have gone into convective motion. As further data are gathered, astronomers will eagerly compare the interior of Venus with that of Earth.

Mars

No Martian magnetic field has ever been detected. Because Mars is rotating rapidly, we must conclude that its core is nonmetallic, or nonliquid, or both.

Mars's small size means that any internal heat would have been able to escape more easily than in a larger planet like Earth or Venus. The evidence just noted for ancient surface activity, especially volcanism, suggests that at least parts of the Martian interior must have melted at some time in the past, but the lack of current activity and the absence of any significant magnetic field indicate that the melting was never as extensive as on Earth. Scientists now believe that Mars's core has a diameter of about 2500 km and is composed largely of iron sulfide (a compound about twice as dense as surface rock).

The history of Mars appears to be that of a planet where large-scale tectonic activity almost started but was stifled by the planet's rapidly cooling outer layers. On a larger, warmer planet, the large upwelling of material that formed the Tharsis bulge might have developed into full-fledged plate tectonic motion, but the Martian mantle became too rigid and the crust too thick for that to occur. Instead, the upwelling continued to fire volcanic activity, perhaps even up to the present day, but geologically much of the planet died 2 billion years ago.

6.8 Atmospheric Evolution on Earth, Venus, and Mars

6 Now that we have studied the structure and history of the three terrestrial worlds that still have atmospheres today— Venus, Earth, and Mars— we can return to the question of why their atmospheres are so different from one another. Let's begin our study with Earth. For purposes of comparison, it may help to bear in mind the chain of events that occurred on Earth as a baseline against which to gauge the histories of the other two planets.

The Development of Earth's Atmosphere

When Earth formed, any atmosphere it might have had—sometimes called the **primary atmosphere**—would have consisted of the gases most common in the early solar system: light gases such as hydrogen, helium, methane, ammonia, and water vapor, a far cry from the atmosphere we enjoy today. Almost all this light material, and especially any hydrogen or helium, escaped into space during the first half-billion or so years after Earth was formed. As discussed in Chapter 5, Earth's gravity is simply too weak to retain these light gases. ∞ (More Precisely 5-1)

Subsequently, Earth developed a **secondary atmosphere**, which was *outgassed* from the planet's interior as a result of volcanic activity. Volcanic gases are rich in water vapor, carbon dioxide, sulfur dioxide, and compounds containing nitrogen. As Earth's surface temperature fell and the water vapor condensed, oceans formed. Much of the carbon dioxide and sulfur dioxide became dissolved in the oceans or combined with surface rocks. Oxygen is such a reactive gas that any free oxygen that appeared at early times was removed as quickly as it formed. An atmosphere consisting largely of nitrogen slowly appeared.

The final major development in the story of Earth's atmosphere was the appearance of *life* in the oceans 3.5 billion years ago. Living organisms began to produce atmospheric oxygen and eventually the ozone layer formed, shielding the surface from the Sun's harmful radiation. Eventually, life spread to the land and flourished. The fact that oxygen is a major constituent (21 percent) of our present-day atmosphere is a direct consequence of the evolution of life on Earth.

The Runaway Greenhouse Effect on Venus

Given the distance of Venus from the Sun, the planet was not expected to be such a pressure cooker. Why is Venus so hot? And if, as we believe, Venus started off like Earth, why is its atmosphere now so different from Earth's?

The answer to the first question is fairly easy. Given the present composition of its atmosphere, Venus is hot because of the greenhouse effect. Recall from Chapter 5 that *greenhouse gases* in Earth's atmosphere, particularly water vapor and carbon dioxide, tend to warm our planet. ∞ (Sec. 5.3) By stopping the escape of much of the infrared radiation emitted by Earth's surface, these gases increase the planet's equilibrium temperature in much the same way as an

extra blanket keeps you warm on a cold night. Continuing the analogy a little further, the more blankets you place on your bed, the warmer you become. Similarly, the more greenhouse gases there are in a planet's atmosphere, the hotter the planet's surface will be.

Venus's dense atmosphere is made up almost entirely of a prime greenhouse gas, carbon dioxide (Figure 6.27). The thick carbon dioxide blanket absorbs about 99 percent of all the infrared radiation released from the surface of Venus and is the immediate cause of the planet's sweltering 730 K surface temperature.

The answer to the second question—why is Venus's atmosphere so different from Earth's?—is more complex. The initial stages of atmospheric development on Venus probably took place in more or less the same way as just described for our own planet: first a primary atmosphere of light gases that rapidly escaped into space, then an outgassed secondary atmosphere containing water, carbon dioxide, sulfur dioxide, and nitrogen-rich compounds.

On Earth, much of the secondary atmosphere became part of the planet surface, as carbon dioxide and sulfur dioxide dissolved in the oceans or combined with surface rocks. If all the dissolved or chemically combined carbon dioxide on Earth were released back into our present-day atmosphere, its new composition would be 98 percent carbon dioxide and 2 percent nitrogen, and it would have a pressure about 70 times its current value. In other words, apart from the presence of oxygen (which appeared on Earth only after the development of life) and water (whose absence on Venus will be explained shortly), Earth's atmosphere would look a lot like Venus's. The real difference between Earth and Venus, then, is that Venus's greenhouse gases never left the atmosphere the way they did on Earth.

When Venus's secondary atmosphere appeared, the temperature was higher than on Earth because Venus is closer to the Sun. The exact temperature is uncertain, however. If it was so high that water vapor could not become liquid, no oceans would have formed. Consequently, outgassed water vapor and carbon dioxide would have remained in the atmosphere, and the full greenhouse effect would have gone into operation immediately. If oceans did form and most of the greenhouse gases left the atmosphere to become

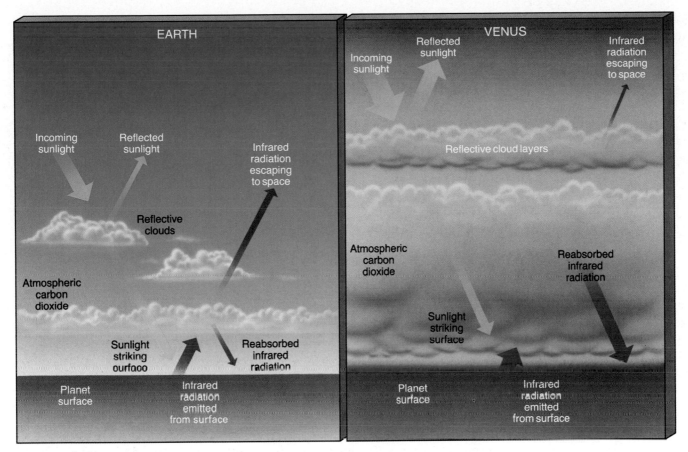

Figure 6.27 Because Venus's atmosphere is much deeper and denser than Earth's, a much smaller fraction of the infrared radiation leaving the planet's surface escapes into space. The result is a much stronger greenhouse effect than on Earth and a correspondingly hotter planet. The outgoing infrared radiation is not absorbed at a single point in the atmosphere. Instead, absorption occurs at all atmospheric levels. The arrows indicate only that absorption occurs, not that it occurs at any specific level.

dissolved in the water, the temperature must still have been sufficiently high to allow a process known as the **runaway greenhouse effect** to come into play.

To understand the runaway greenhouse effect, imagine that we took Earth from its present orbit and placed it in Venus's orbit. At its new distance from the Sun, the amount of sunlight hitting Earth's surface would be almost twice its present level, and so the planet would warm up. More water would evaporate from the oceans, leading to an increase in atmospheric water vapor. At the same time, the ability of both the oceans and surface rocks to hold carbon dioxide would diminish, allowing more carbon dioxide to enter the atmosphere. As a result, the greenhouse heating would increase, and the planet would warm still further, leading to a further increase in atmospheric greenhouse gases, and so on. Once started, the process would "run

away," eventually leading to the complete evaporation of the oceans, restoring all the original greenhouse gases to the atmosphere. Basically the same thing would have happened on Venus long ago, leading to the planetary inferno we see today.

The greenhouse effect on Venus was even more extreme in the past, when the atmosphere also contained water vapor. By intensifying the blanketing effect of the carbon dioxide, the water vapor helped the surface of Venus reach temperatures perhaps twice as hot as at present. At those high temperatures, the water vapor was able to rise high into the planet's upper atmosphere—so high that it was broken up by solar ultraviolet radiation into its components, hydrogen and oxygen. The light hydrogen rapidly escaped, the reactive oxygen quickly combined with other atmospheric gases, and all water on Venus was lost forever.

6-1 INTERLUDE ✦Extension✦

LIFE ON MARS?

Even before the *Viking* missions reached Mars in 1976, astronomers had abandoned hope of finding life on the planet. Scientists knew there were no large-scale canal systems, no surface water, almost no oxygen in the atmosphere, and no seasonal vegetation changes. The present lack of liquid water on Mars especially dims the chances for life there now. However, running water and possibly a dense atmosphere in the past may have created conditions suitable for the emergence of life long ago. In the hope that some form of microbial life might have survived to the present day, the *Viking* landers carried out experiments designed to detect biological activity. The accompanying pair of photographs show the robot arm of one of the landers digging a shallow trench.

(Before)

R I V U X G

(After)

All three *Viking* biological experiments assumed some basic similarity between hypothetical Martian bacteria and those found on Earth. A gas-exchange experiment offered a nutrient broth to any residents of a sample of Martian soil and looked for gases that would signal metabolic activity. A labeled-release experiment added compounds containing radioactive carbon to the soil, then waited for results signaling that Martian organisms had either eaten or inhaled this carbon. Finally, a pyrolitic-release experiment added radioactively tagged carbon dioxide to a sample of Martian soil and atmosphere, waited awhile, then removed the gas and tested the soil (by heating it) for signs that something had absorbed the tagged gas. In all cases, contamination by terrestrial bacteria was a major concern. Indeed, any release of Earth organisms would have invalidated these and all future such experiments on Martian soil. Both *Viking* landers were carefully sterilized prior to launch, and international agreement presently protects the Martian environment from contamination by future terrestrial probes. (How we will sterilize a mission with human astronauts is still unclear.)

Initially, all three experiments appeared to be giving positive signals! However, subsequent careful studies showed that all the results could be explained by inorganic (that is, nonliving) chemical reactions. Thus, we have no clear evidence for even microbial life on the Martian surface. The *Viking* robots detected peculiar reactions that mimic in some ways the basic chemistry of living organisms, but they did not detect life itself.

One criticism of the *Viking* experiments is that they searched only for life now living. Today, Mars seems locked in an ice age—the kind of numbing cold that would prohibit sustained life as we know it. If bacterial life did arise on an Earth-like early Mars, however, then we might be able to find its fossilized remains preserved on or near the Martian surface.

Surprisingly, one place to look for life on Mars is right here on Earth. Scientists think that some meteorites found on Earth's surface come from the Moon and from Mars. These meteorites were apparently blasted off these bodies long ago during an impact of some sort, thrown into space, and eventually trapped by Earth's gravity, ultimately to fall to the ground. The most fascinating of these rocks are surely those from the red planet—for one of them may harbor fossil evidence for past life on Mars!

The accompanying figure shows ALH84001, a blackened, 2-kg meteorite about 17 cm across, found in 1984 in Antarctica. Based on estimates of the cosmic-ray exposure it received before reaching Earth, the rock is thought to have been blasted off Mars 16 million years ago. Looking at this specimen through a microscope, scientists can see rounded orange-brown "globules" of carbonate minerals on the rock's shiny crust. Because carbonates form only in the presence of water, the presence of these globules suggests that carbon dioxide gas and liquid water existed near ground level at some point in Mars's history, a conclusion that planetary scientists had earlier drawn from studies of *Viking*'s orbital images of valleys apparently carved by water when the Martian climate was wetter and warmer.

In a widely viewed press conference in Washington, D.C., in the summer of 1996, a group of scientists argued, based on all the data accumulated from studies of ALH84001, that they had discovered fossilized evidence for life on Mars. The key pieces of evidence for primitive Martian life are as follows: (1) Bacteria on Earth can produce structures similar to the globules shown above. (2) The meteorite contains traces of *polycyclic aromatic hydrocarbons*—a tongue-twisting name for a class of complex organic molecules (usually abbreviated PAHs) that, while not directly involved in known biological cycles on Earth, occur among the decay products of plants and other organisms. (3) High-powered electron microscopes show that ALH84001 contains tiny, teardrop-shaped crystals of magnetite and iron sulfide embedded in places where the carbonate has dissolved. On Earth, bacteria are known to manufacture similar chemical crystals. (4) On very small scales, elongated and egg-shaped structures are seen within the carbon-

(continued)

ate globules. The researchers interpret these minute structures as fossils of primitive organisms.

The accompanying photomicrograph shows this fourth, and most controversial, piece of evidence—curved, rodlike structures that resemble bacteria on Earth. Scale is crucial here, however. The structures are only about 0.5 μm across, 30 times smaller than ancient bacterial cells found fossilized on Earth. Furthermore, several key tests have not yet been done, such as cutting through the suspected fossilized tubes to search for evidence of cell walls or semipermeable membranes, or of any internal cavities where body fluids would have resided. Nor has anyone yet found in ALH84001 any amino acids, the basic building blocks of life as we know it.

These results are very controversial. Many experts do not agree that life has been found on Mars—not even fossilized life. The skeptics maintain that all the evidence could be the result of chemical reactions not requiring any kind of biology. Carbonate compounds are common in all areas of chemistry, PAHs are found in many lifeless places (glacial ice, asteroid-belt meteorites, interstellar clouds, the exhaust fumes of your automobile), bacteria are not needed to produce crystals, and it remains unclear whether the tiny tubular structures shown below are animal, vegetable, or merely mineral. In addition, there is the huge problem of contamination—after all, ALH84001 was found on Earth and apparently sat in the Antarctic ice fields for 13,000 years before being picked up by meteorite hunters.

As things now stand, it's a matter of interpretation—at the frontiers of science, issues are not always as clear-cut as we might hope. Only additional analysis and new data will tell for sure if primitive Martian life existed long ago. Should the claim of life on Mars hold up against much skepticism in the scientific community, these findings will go down in history as one of the greatest scientific discoveries of all time. A fantastic result: we are—or at least were—not alone in the universe! Maybe.

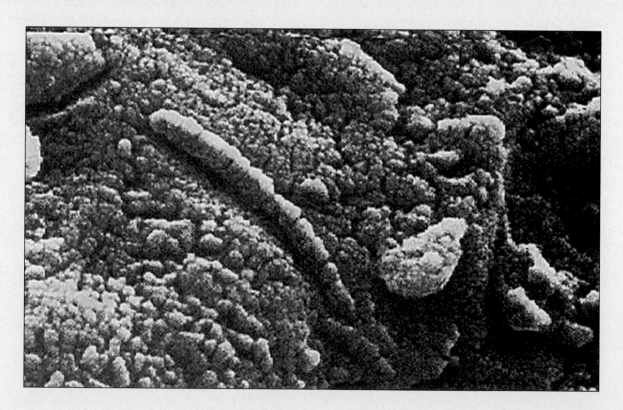

Evolution of the Martian Atmosphere

Presumably, Mars also had first a primary and then a secondary (outgassed) atmosphere early in its history. Around 4 billion years ago, Mars may have had a fairly dense atmosphere, perhaps even complete with blue skies and rain. Despite Mars's distance from the Sun, the greenhouse effect would have kept conditions fairly comfortable, and a surface temperature above 0°C seems quite possible.

Sometime during the next billion years, however, most of the Martian atmosphere disappeared. Possibly some of it was lost because of impacts with large bodies in the early solar system, and a large part may have leaked away into space because of the planet's weak gravity (as discussed in *More Precisely 5-1*). Most of the remainder probably became unstable in a kind of reverse runaway greenhouse effect. In this scenario, much of Mars's atmospheric carbon dioxide dissolved in the liquid water of the planet's rivers and lakes (and oceans, if any), ultimately to combine with Martian surface rocks. Calculations show that much of the Martian atmospheric carbon dioxide could have been depleted in this way in a relatively short period of time, perhaps as quickly as a few hundred million years. As the level of carbon dioxide declined and the greenhouse heating diminished, the planet cooled. The water froze out of the atmosphere, lowering still further the level of atmospheric greenhouse gases and accelerating the cooling.

The present level of water vapor in the Martian atmosphere is the maximum possible given the atmosphere's present density and temperature. Estimates of the total amount of water stored as permafrost or in the polar caps are quite uncertain, but it is likely that if all the water on Mars were to become liquid, it would cover the surface to a depth of several meters.

6.9 The Moons of Mars

The two Martian moons—named Phobos (Fear) and Deimos (Panic) for the horses that drew the Roman war god's chariot—were discovered in 1877 by American astronomer Asaph Hall. Both are small, irregularly shaped, and heavily cratered. The larger of the two, Phobos (Figure 6.28a), is about 28 km long and 20 km wide. Deimos (Figure 6.28b) is just 16 km long by 10 km wide.

Based on measurements of the moons' gravitational effects on orbiting spacecraft, astronomers have estimated their densities to be around 2000 kg/m³, far less than the density of Earth's Moon or any terrestrial

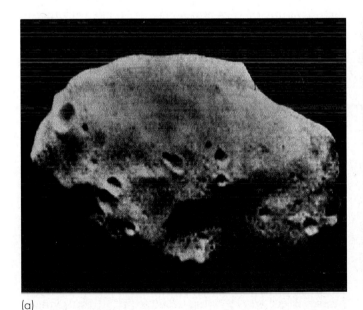

(a)

(b)

R I V U X G

Figure 6.28 In its trek around the Sun, Mars is accompanied by two tiny moons. Phobos and Deimos measure about 28 and 16 km across, respectively. Both are shaped like potatoes and are heavily cratered. (a) A *Mariner 9* photograph of Phobos, not much larger than Manhattan Island. (b) This photograph of the smaller moon Deimos was taken by a *Viking* orbiter. The field of view is only 2 km across, and most of the boulders shown are about the size of a house.

world, implying that the moons' compositions are quite *dissimilar* from that of Mars. Most likely, Phobos and Deimos did not form along with Mars, but instead are asteroids that were slowed and captured by the outer fringes of the early Martian atmosphere (which, as we have just seen, was probably much denser than the atmosphere today). It is even possible that the two moons are remnants of a single object that broke up during capture. Astronomers study them not to gain insight into Martian evolution but rather because they contain information about the very early solar system, before the major planets formed.

Chapter Review

Summary

In size and appearance, Mercury is similar to the Moon. Venus is comparable in mass and radius to Earth. Mars is smaller than Earth, but it still has many Earth-like characteristics.

Mercury's rotation rate is strongly influenced by the tidal effect of the Sun. Because of Mercury's eccentric orbit, the planet rotates not synchronously but exactly one and a half times for every one revolution around the Sun. Venus's rotation is slow and retrograde. Mars rotates at almost the same rate as Earth.

Mercury has no permanent atmosphere. The atmospheres of Venus and Mars are mainly carbon dioxide. Venus's atmosphere is extremely hot and 90 times denser than Earth's, while the density of the cool Martian atmosphere is only 0.7 percent that of Earth's.

Mercury's surface is heavily cratered, much like the lunar highlands. Mercury lacks lunarlike maria but has extensive **intercrater plains** (p. 175) and cliff-forming cracks, or **scarps** (p. 176), in its crust. The plains were caused by lava flows early in Mercury's history. The scarps were apparently formed when the planet's core cooled and shrank, causing the surface to crack. Mercury has a large impact crater called the Caloris Basin, the diameter of which is comparable to the radius of the planet. The impact forming this basin apparently sent violent shock waves around the planet, buckling the crust on the opposite side into a region of **weird terrain** (p. 177).

Because of a thick cloud cover, Venus's surface cannot be seen in visible light from Earth, but it has been thoroughly mapped by radar. Many lava domes and **shield volcanoes** (p. 178) have been found on the surface. Although no eruptions have ever been observed, the high level of sulfur dioxide above Venus and detected burst of radio energy lead astronomers to believe Venus is still volcanically active today. The planet's surface shows no sign of plate tectonics. Features called **coronae** (p. 178) are thought to have been caused by an upwelling of mantle material that never developed into full convective motion.

Mars's major surface feature is the Tharsis bulge. Associated with the bulge is the largest known volcano in the solar system and a huge crack, called the Valles Marineris, in the planet's surface. There is clear evidence that water once existed in great quantity on Mars. The **runoff channels** (p. 184) are the remains of ancient Martian rivers. The **outflow channels** (p. 184) are the paths taken by floods that cascaded from the southern highlands into the northern plains. Today, a large amount of that water may be locked up in the polar caps and in a layer of **permafrost** (p. 184) lying under the Martian surface.

Neither Venus nor Mars has any detectable magnetic field. Mercury's weak field seems to be a fossil remnant from long ago, dating to the time when the planet's iron core solidified.

Any **primary atmospheres** (p. 187) of light gases that the Earth, Venus, and Mars had at formation rapidly escaped. All the terrestrial planets probably developed **secondary atmospheres** (p. 188), outgassed from volcanoes, early in their lives. On Mercury, the atmosphere escaped, as on the Moon. On Earth, much of the outgassed material became absorbed in surface rocks or dissolved in the oceans. The development of life produced oxygen. On Venus, the **runaway greenhouse effect** (p. 189) has resulted in all the planet's carbon dioxide being left in the atmosphere, leading to the extreme conditions we observe today. On Mars, part of the secondary atmosphere escaped into space. Of the part that did not escape, most of the carbon dioxide is now locked up in surface rock, and most of the water vapor is now stored in the permafrost and the polar caps.

Mars has two small moons, which are probably asteroids captured after encountering the planet's atmosphere long ago.

Self-Test: True or False?

_____ 1. Mercury has no detectable atmosphere.

_____ 2. Mercury has nighttime low temperatures of 100 K, well below the freezing point of water.

_____ 3. Mercury's solar day is longer than its solar year.

_____ 4. Some volcanic activity continues today on the surface of Mercury.

_____ 5. Mercury's craters are more densely packed than craters on the Moon.

_____ 6. The average surface temperature of Venus is about 250 K.

_____ 7. Numerous large surface features on Venus can be seen from Earth-based observations made in the ultraviolet part of the spectrum.

_____ 8. The cloud cover of Venus is composed of sulfuric acid.

_____ 9. The surface of Venus is rougher than the surface of Earth, with higher highs and lower lows.

_____ 10. Lava flows are common on the surface of Venus.

_____ 11. There is strong circumstantial evidence that active volcanism continues on Venus.

_____ 12. Venus has a magnetic field similar to that of Earth.

_____ 13. Seasonal changes in the appearance of Mars are caused by vegetation on the surface.

_____ 14. Mars has the largest volcanoes in the solar system.

_____ 15. The northern hemisphere of Mars is much older than the southern hemisphere.

_____ 16. Olympus Mons is the largest impact crater on Mars.

_____ 17. There are many indications of plate tectonics on Mars.

_____ 18. Valles Marineris is similar in size to Earth's Grand Canyon.

_____ 19. The orange-red color of the surface of Mars is primarily due to rust (iron oxide) in its soil.

_____ 20. One of the two moons of Mars is larger than Earth's Moon.

Self-Test: Fill in the Blank

1. Mercury's iron core contains a _____ fraction of the planet's mass than does Earth's core.

2. Mercury's rate of rotation was first measured using _____.

3. Although Mercury's daytime temperatures are always very hot, it may still be possible for the planet to have sheets of water ice at its _____.

4. Mercury's _____ is about 1/100 that of Earth and was originally thought not to exist at all.

5. Because Venus has a mass and average _____ only slightly lower than Earth's, we expect its internal structure and evolution to be Earthlike.

6. Venus's rotation is unusual because it is _____.

7. The most abundant gas in the atmosphere of Venus is _____.

8. The runaway greenhouse effect on Venus was a result of the planet's being _____ to the Sun than is Earth.

9. The surface of Venus has been mapped using _____.

10. Huge, roughly circular regions on the surface of Venus are known as _____.

11. The surface of Venus appears to have been resurfaced by _____ every few hundred million years.

12. The main difficulties in using landers to study

Venus's surface are the planet's extremely high _____ and _____.

13. The southern hemisphere of Mars consists of heavily _____ highlands.

14. The Tharsis region of Mars is a large equatorial _____ .

15. Several large _____ are found on the Tharsis bulge.

16. The great height of Martian volcanoes is a direct result of the planet's low _____ .

17. The flowing-liquid appearance of the ejecta surrounding Martian impact craters is evidence of a layer of _____ just under the surface.

18. Runoff channels carried _____ from the southern mountains into the valleys.

19. Outflow channels are the result of catastrophic _____ .

20. Water flowed on the surface of Mars a few _____ years ago, but not now.

Review and Discussion

1. How fast does Mercury rotate? How and when was the planet's true rotation rate discovered?

2. What is a scarp? How are scarps thought to have formed? Why do scientists believe that scarps formed after most meteoritic bombardment ended?

3. In contrast to Earth, Mercury undergoes extremes in temperature. Why?

4. What do Mercury's magnetic field and large average density imply about the planet's interior?

5. How is Mercury's evolutionary history like that of the Moon? How is it different?

6. Mercury used to be called "the Moon of the Sun." Why do you think it had this name? What piece of scientific evidence proved the name inappropriate?

7. Why does Venus appear so bright to the naked eye?

8. Venus probably has a molten iron-rich core like Earth. Why doesn't it also have a magnetic field?

9. How did radio observations of Venus made in the 1950s change our conception of that planet?

10. What two features of Venus's atmosphere make the planet hostile to earthly life?

11. What are the main constituents of Venus's atmosphere? What are clouds in the upper atmosphere made of?

12. What is the runaway greenhouse effect, and how might it have altered the climate of Venus?

13. What is the evidence for active volcanoes on Venus?

14. If Venus had formed at Earth's distance from the Sun, what might its climate be like today? Explain.

15. Why is Mars red?

16. What is the evidence that water once flowed on Mars? Is there water on Mars today?

17. Why were Martian volcanoes able to become so large?

18. Given that Mars has an atmosphere and its composition is mostly carbon dioxide, why isn't there a significant greenhouse effect to warm its surface?

19. Do you think that sending humans to Mars in the near future is a reasonable goal? Why or why not?

20. Compare and contrast the evolution of the atmospheres of Mars, Venus, and Earth. Include a discussion of the importance of volcanoes and water.

Problems

1. How long does it take a radar signal to travel from Earth to Mercury and back?

2. Given that the *Hubble Space Telescope* orbits 600 km above Earth's surface with a period of 95 minutes, what was the orbit period of the *Magellan* spacecraft as it orbited 500 km above the surface of Venus?

3. Approximating Venus's atmosphere as a layer of gas that is 50 km thick and has a uniform density of 21 kg/m^3, calculate the total mass of this atmosphere. Compare your answer with the mass of Earth's atmosphere and with the mass of Venus.

4. Verify that the surface gravity on Mars is 40 percent that of Earth.

Projects

1. Try to spot Mercury in morning or evening twilight. (In the Northern Hemisphere, the best evening sightings of the planet take place in the spring, the best morning sightings in the fall.) What color is the planet? Do you think this is the planet's actual color, or does Mercury appear this way because we generally see it in a twilight sky?

2. Consult an almanac to determine the next time Venus will pass between Earth and the Sun. How many days before and after this event can you glimpse the planet with your naked eyes? Use the almanac to find out the next time Venus will pass on the far side of the Sun from Earth. How many days before and after this event can you glimpse the planet with the unaided eye?

3. Using a powerful pair of binoculars or a small telescope, examine Venus as it goes through its phases. Note each phase and its relative size. (You can compare its size to the field of view in a telescope; always use the same eyepiece for this.) Look at the planet every few days or once a week. Make a table of the shape of the phase, the size, and the relative brightness to the naked eye. After you have observed through a significant change in phase, can you see the correlations in these three properties first recognized by Galileo?

4. Track the motion of Mars relative to background stars for several months following its appearance in the predawn sky. (Consult an almanac to determine where Mars is in the sky this year.) You will see that the planet moves rapidly eastward, crossing many constellation boundaries, while the background stars shift continuously toward the west. Can you explain why Mars's orbit appears as it does?

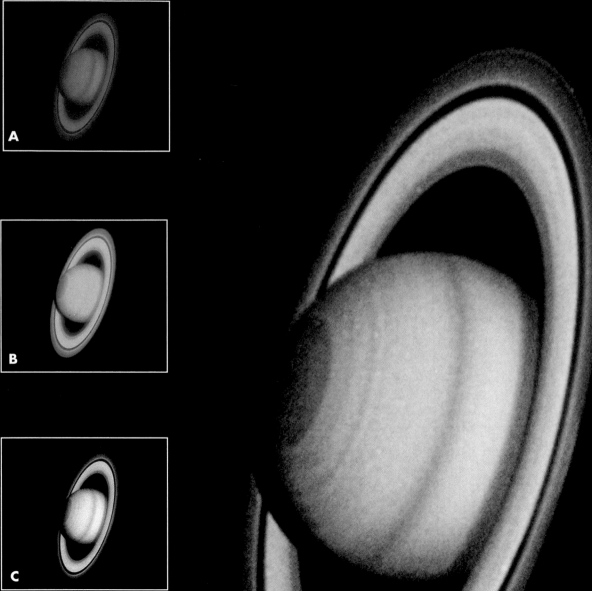

7 The Jovian Planets
Giants of the Solar System

(Opposite page, background) Through the cameras onboard the Hubble Space Telescope, the planet Saturn appears much as our naked eyes would see it if it were only twice as far as away as the Moon. Resolution here is 670 km, good enough to see clearly the band structure on the ball of the planet, as well as several gaps in the rings. True-color images like this one are made by combining separate images taken in red, green, and blue light, as shown in the insets.

(Inset A) Saturn as seen through a red filter, centered at a wavelength of 718 nm.

(Inset B) Saturn as seen through a green filter, centered at a wavelength of 547 nm.

(Inset C) Saturn as seen through a blue filter, centered at a wavelength of 439 nm.

LEARNING GOALS

Studying this chapter will enable you to:

1 Explain how both chance and calculation played major roles in the discovery of Uranus and Neptune.

2 Describe the similarities and the differences among the four jovian worlds.

3 Discuss some of the processes responsible for the properties of the jovian atmospheres.

4 Describe how the internal structure and composition of the jovian planets are inferred from external measurements.

5 Explain why three of the four jovian worlds radiate more energy into space than they receive from the Sun.

U.S. spacecraft visiting the outer planets over the past two decades have revealed these worlds in detail that was only dreamed of by centuries of Earth-bound astronomers. Beyond the orbit of Mars, the solar system is a very different place. In sharp contrast to the small, rocky, terrestrial bodies found near the Sun, the outer solar system presents us with a totally unfamiliar environment—huge gas balls, peculiar moons, planet-circling rings, and a wide variety of physical and chemical properties, many of which are only poorly understood even today. The jovian planets—Jupiter, Saturn, Uranus, and Neptune—differ from each other in many ways, but they have much in common, too. As with the terrestrial planets, we learn from their differences as well as from their similarities.

7.1 Observations of Jupiter and Saturn

The View from Earth *Animation*

Named after the most powerful god of the Roman pantheon, Jupiter is the third-brightest object in the night sky (after the Moon and Venus), making it very easy to locate and study from Earth. Ancient astronomers could not have known the planet's true size, but their name choice was very apt—Jupiter is by far the largest planet in the solar system.

Figure 7.1(a) is a ground-based photograph of Jupiter; Figure 7.1(b) is a *Hubble Space Telescope* image taken in December 1990. Notice in each image the alternating light and dark bands that cross the planet parallel to its equator, and also notice the large oval at the lower right of Figure 7.1(b). These atmospheric features, which we'll study in more detail later in this chapter, are quite unlike anything found on the inner planets. Also in contrast to the terrestrial worlds, Jupiter has many moons that vary greatly in size and other properties. The four largest are visible from Earth with a small telescope (or, for a few people, even with the naked eye). They are known as the *Galilean moons* after their discoverer, Galileo Galilei. All four can be seen in Figure 7.1(a).

Saturn, the next major body we encounter as we continue outward beyond Jupiter, was the most distant planet known to Greek astronomers. Named after the

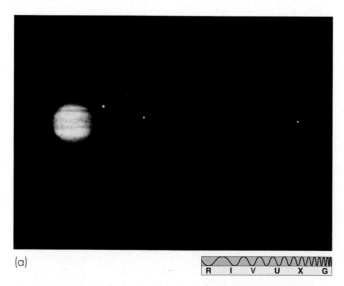

(a)

R I V U X G

(b)

R I V U X G

Figure 7.1 (a) Photograph of Jupiter made with an Earth-based telescope, showing the planet and its four Galilean moons. (b) A *Hubble Space Telescope* image of Jupiter, in true color. Features as small as a few hundred kilometers across are resolved.

(a)

R I V U X G

(b)

R I V U X G

Figure 7.2 (a) One of the best images of Saturn from an Earth-bound telescope. (b) Saturn as seen by the *Hubble Space Telescope*. Resolution is about 700 km.

father of Jupiter in Greco–Roman mythology, Saturn orbits at almost twice Jupiter's distance from the Sun. The planet's sidereal orbital period of 29.5 Earth years was the longest natural unit of time known to the ancient world. Saturn's greater distance from the Sun makes it considerably fainter (as seen from Earth) than either Jupiter or Mars.

Two images of Saturn—one from the ground and one from Earth orbit—are shown in Figure 7.2. The planet's banded atmosphere is somewhat similar to Jupiter's, but Saturn's atmospheric bands are much less distinct. Overall, the planet has a rather uniform butterscotch hue. Again like Jupiter and unlike the inner planets, Saturn has many moons orbiting it. Saturn's best-known feature, its spectacular *ring system*, is clearly visible in Figure 7.2. The moons and rings of the jovian planets are the subject of Chapter 8.

Spacecraft Exploration

Most of our detailed knowledge of the jovian planets has come from NASA spacecraft visiting those worlds. Of particular importance are the two *Voyager* probes and the recent *Galileo* mission.

The *Voyager* spacecraft left Earth in 1977, reaching Jupiter in March (*Voyager 1*) and July (*Voyager 2*) of 1979. Both craft subsequently used Jupiter's strong gravity to send them on to Saturn in a maneuver called a *gravity assist*. *Voyager 2* then took advantage of a rare planetary configuration that enabled it to use Saturn's gravity to propel it to Uranus and then on to Neptune in a spectacularly successful "grand tour" of the outer

planets. Each craft carried equipment to study planetary magnetic fields and magnetospheres, as well as radio, visible-light, and infrared sensors to analyze reflected and emitted radiation from the jovian planets and their moons. The data they returned revolutionized our knowledge of all the jovian worlds; much of the information presented in this and the next chapter came from *Voyager* sensors. The two *Voyager* craft are now headed out of the solar system, racing toward interstellar space.

Galileo was launched in 1989 and reached Jupiter in December 1995, after a circuitous route that took it three times through the inner solar system, receiving gravity assists from both Venus and Earth before finally reaching its target. The mission had two components: an atmospheric probe and an orbiter. The probe, slowed by a heat shield and a parachute, descended into Jupiter's atmosphere, making measurements and chemical analyses as it went. The orbiter executed a complex series of orbits through Jupiter's moon system, returning to some moons studied by *Voyager* and visiting others for the first time. The mission ended in December 1997.

7.2 The Discoveries of Uranus and Neptune

The British astronomer William Herschel discovered the planet Uranus in 1781. Because this was the first new planet discovered in recorded history, the event caused quite a stir. The story goes that

Herschel's first instinct was to name the new planet "Georgium Sidus" (Latin for "George's star") after his king, George III of England. The world was saved from a planet named George by the wise advice of another astronomer, Johann Bode, who suggested instead that the tradition of using names from ancient mythology be continued and that the planet be named after Uranus, the father of Saturn.

Uranus is just barely visible to the naked eye. It looks like a faint, undistinguished star—no wonder it went unnoticed by the ancients. Even through a large optical telescope, Uranus appears hardly more than a tiny, pale, slightly greenish disk. A good Earth-based image of Uranus is shown in Figure 7.3(a). Figure 7.3(b) is a close-up visible-light image taken by *Voyager 2* in 1986. Uranus's apparently featureless atmosphere contrasts sharply with the bands and spots visible in the atmospheres of the other jovian planets.

Once Uranus was discovered, astronomers set about charting its orbit. They quickly discovered a small discrepancy between the planet's predicted position and its observed position. Try as they might, they could not find an elliptical orbit that fit the planet's path to within the accuracy of their measurements. Half a century after Uranus's discovery, the discrepancy had grown to a quarter of an arc minute, far too big to be explained away as observational error. Uranus seemed to be violating Kepler's laws of planetary motion. ∞ (Sec. 1.6)

Astronomers realized that, although the Sun's strong gravitational pull determines Uranus's orbit nearly completely, the deviation meant that some unknown body was exerting a much weaker but still measurable gravitational force on Uranus. There had to be *another* planet in the solar system influencing Uranus's motion. In September 1845, after almost two years of work, an English mathematician named John Adams solved the problem of determining the new planet's mass and orbit. In June 1846 a French mathematician, Urbain Leverrier, independently came up with essentially the same answer. Later that year, a German astronomer named Johann Galle found the new planet within one or two degrees of the predicted position. The new planet was named Neptune, and Adams and Leverrier

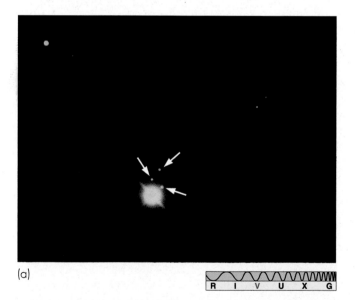

(a)

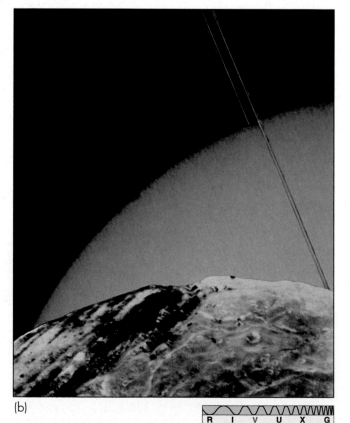

(b)

Figure 7.3 (a) Details are barely visible on this photograph of Uranus made with a large Earth-based telescope. (Arrows point to three of the planet's moons.) (b) A close-up view of Uranus sent back to Earth by *Voyager 2* while the spacecraft was whizzing past this gigantic planet at 10 times the speed of a rifle bullet. This montage shows Uranus's blue, featureless upper atmosphere, taken when *Voyager 2* was about 100,000 km away. The brown foreground is a *Voyager 2* image of Miranda, one of Uranus's moons. The greenish rings encircling the planet are an artist's conception.

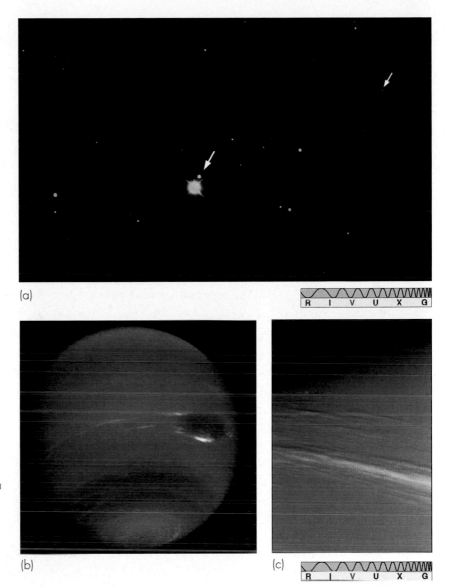

(a)

Figure 7.4   (a) Neptune and two of its moons (indicated by the arrows), imaged with a large Earth-based telescope. (b) Neptune as seen by *Voyager 2* from a distance of 1,000,000 km. (c) Another image of Neptune as seen by *Voyager 2*. The resolution in this image is about 10 km. The cloud streaks range in width from 50 to 200 km.

(b)

(c)

(but not Galle!) are now jointly credited with its discovery.

Neptune cannot be seen with the naked eye, but it is visible through binoculars or a small telescope. Through a large telescope, it appears as a small, bluish disk. Figure 7.4(a) shows a long exposure of Neptune and two of its moons. The planet is so distant from Earth that surface features are virtually impossible to discern. Even under the best observing conditions, only a few markings can be seen. These markings suggest multicolored cloud bands, with light bluish hues seeming to dominate. With *Voyager 2*'s arrival, much more detail emerged (Figure 7.4b and c). Superficially, at least, Neptune resembles a blue-tinted Jupiter, with atmospheric bands and spots clearly evident.

7.3 The Jovian Planets in Bulk

Overall Properties

Table 7.1 extends Table 6.1 to include the jovian planets. Figure 7.5 shows the four jovian planets to scale, along with Earth for comparison. Clearly, the jovian worlds dwarf their terrestrial counterparts. Their large sizes and relatively low average densities imply that these planets differ radically in both composition and structure from Earth and the other terrestrial worlds.

The two largest jovian planets are Jupiter and Saturn. Spectroscopic observations of their atmospheres and theoretical modeling of their interiors

Table 7.1 Planetary Properties

	(kg)	Mass (Earth = 1)	(km)	Radius (Earth = 1)	Density (kg/m³)	Escape Speed (km/s)	Rotation Period (solar days)	Axial Tilt (degrees)	Surface Temperature (K)	Surface Magnetic Field (Earth = 1)
Mercury	3.3×10^{23}	0.055	2,400	0.38	5400	4.3	59	0	100–700	0.01
Venus	4.9×10^{24}	1.00	6,100	0.95	5300	10	-243^1	179	730	0.0
Earth	6.0×10^{24}	1.00	6,400	1.00	5500	11	1.00	23	290	1.0
Mars	6.4×10^{23}	0.11	3,400	0.53	3900	5.0	1.03	24	180–270	0.0
Jupiter	1.9×10^{27}	318	71,000	11	1330	60	0.41	3	124	14
Saturn	5.7×10^{26}	95	60,000	9.5	710	36	0.43	27	97	0.7
Uranus	8.7×10^{25}	15	26,000	4.0	1240	21	-0.69^1	98	58	0.7
Neptune	1.0×10^{26}	17	25,000	3.9	1670	24	0.72	30	59	0.4

[1]The minus sign indicates retrograde rotation.

Figure 7.5 Jupiter, Saturn, Uranus, Neptune, and Earth, drawn to scale.

indicate that these gigantic worlds are composed primarily of hydrogen and helium. On Earth, these gases have very low densities (0.08 and 0.16 kg/m^3, respectively, at room temperature and atmospheric pressure). The large masses and strong gravitational fields of Jupiter and Saturn create enormous pressures in those planets' interiors, however, compressing hydrogen and helium to the much higher densities listed in Table 7.1. Because of Saturn's lower mass, the interior pressure in it is less than in Jupiter, leading to a lower average density.

Uranus and Neptune are similar to one another in overall properties and considerably smaller and less massive than either Jupiter or Saturn. Because of the lower masses of Uranus and Neptune, we would not expect the hydrogen and helium in their interiors to be compressed as much as in the two larger jovian worlds. Yet the average densities of Uranus and Neptune are actually *greater* than the density of Saturn, and similar to that of Jupiter. Astronomers interpret this to mean that both Uranus and Neptune have large, probably rocky, cores perhaps 10 times more massive than our entire planet. As we will see in Section 7.6, Jupiter and Saturn likely also have large, rocky cores, but in those two planets the cores represent a far smaller fraction of the total planet mass. (Note that the term *rocky* here refers to the chemical composition of the core, not to its physical state. At the high temperatures and pressures found deep in the jovian interiors, the core material bears little resemblance to rocks found on Earth's surface.)

The marked difference in composition between the terrestrial and the jovian planets is a consequence of the jovian planets' much stronger gravitational fields. The gravitational pull of a terrestrial planet is too weak to have retained any low-mass atmospheric molecules that may have been present when the planet formed. ∞ (*More Precisely 5-1*) In contrast, the jovian planets are massive enough to have retained even hydrogen. Very little of their original atmospheres have escaped since the birth of the solar system 4.6 billion years ago.

None of the jovian planets has a solid surface of any kind. Their gaseous atmospheres just become denser with depth (because of the pressure of the overlying layers), eventually becoming liquid in the interior. The surface we see from Earth is simply the top of the outermost layer of atmospheric clouds. Where relevant, the term *surface* in Table 7.1 and elsewhere in this chapter refers to this level in the atmosphere.

Rotation

You might think of measuring the rotation rate of a jovian planet by timing a surface feature as it moves around the planet, but there is a catch: as just mentioned, the jovian planets have no solid surface. With nothing to "tie them down," different parts of the atmosphere can and do move at different rates. This state of affairs—when the rotation rate is not constant from one location to another—is called **differential rotation**. Although differential rotation is not possible in solid objects like the terrestrial worlds, it is normal for fluid bodies such as the jovian planets.

In the case of Jupiter, the amount of differential rotation is small. Jupiter's equatorial regions rotate once every 9^h 50^m, while the higher latitudes take about 6 minutes longer to rotate once. On Saturn, the difference between equatorial and polar rotation rates is a little greater (26 minutes, again slower at the poles). On Uranus and Neptune, the differences are greater still—more than 2 hours in the case of Uranus—this time with the poles rotating more rapidly. These variations are the result of large-scale wind flows in the planets' atmospheres.

More meaningful measurements of overall planetary rotation rates are provided by measurements of magnetospheres. All four jovian worlds have strong magnetic fields and emit radiation at radio wavelengths. The strength of this radio emission varies with time and repeats itself periodically. Scientists assume that this period matches the rotation of the planet's deep interior, where (as on Earth) the magnetic field arises. The rotation periods listed in Table 7.1 were obtained in this way. There is no clear relationship between the interior and atmospheric rotation rates of the jovian planets. In Jupiter and Saturn, the interior rotation rate matches the polar rotation. In Uranus, the interior rotates more slowly than any part of the atmosphere; in Neptune, the reverse is true.

The observations used to measure a planet's rotation period also let us determine the orientation of its rotation axis. Jupiter's rotation axis is almost perpendicular to the orbit plane, the axial tilt being only 3°, quite small compared with Earth's tilt of 23°. The tilts of Saturn and Neptune (Table 7.1) are quite similar to those of Earth and Mars.

Each planet in our solar system seems to have some outstanding peculiarity. In the case of Uranus, it is rotation axis. Unlike the other major planets, which have their spin axes (very) roughly perpendicular to

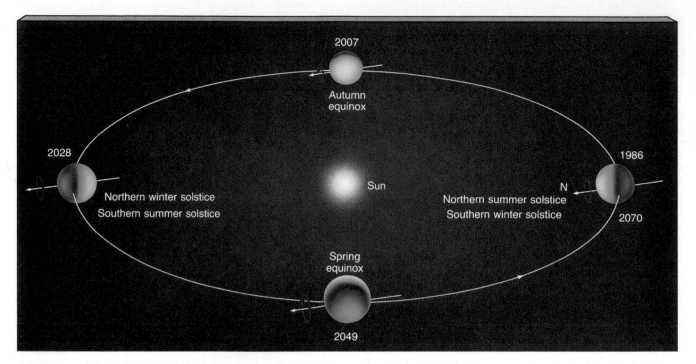

Figure 7.6 Because of Uranus's axial tilt of 98°, the planet experiences the most extreme seasons known in the solar system. The equatorial regions experience two "summers" (warm seasons, at the times of the two equinoxes) and two "winters" (cool seasons, at the times of the solstices) each year, and the poles are alternately plunged into darkness for 42 years at a time.

the ecliptic plane, Uranus's rotation axis lies almost within that plane. We might say that Uranus is "tipped over" onto its side, for its axial tilt is 98°. As a result, the "north" pole[1] of Uranus, at some point in the planet's orbit, points almost directly toward the Sun. Half a Uranian year later, the "south" pole faces the Sun, as illustrated in Figure 7.6. When *Voyager 2* encountered the planet in 1986, the north pole happened to be pointing nearly directly at the Sun, so it was midsummer in the northern hemisphere.

No one knows why Uranus is tilted in this way. Astronomers speculate that a catastrophic event, such as a grazing collision between the planet and another planet-sized body, might have altered the planet's spin axis. However, there is no direct evidence for such an occurrence and no theory to tell us how we should seek to confirm it.

7.4 Jupiter's Atmosphere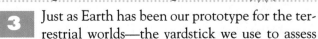

3 Just as Earth has been our prototype for the terrestrial worlds—the yardstick we use to assess

[1] We adopt here the convention that a planet's rotation is counterclockwise as seen from above the north pole (that is, planets rotate from west to east).

and compare the properties of the inner planets—Jupiter is our guide to the outer planets. We therefore begin our study of jovian atmospheres with Jupiter.

Overall Appearance and Composition

Visually, Jupiter is dominated by two atmospheric features (both clearly visible in Figure 7.1): a series of ever-changing atmospheric cloud bands arranged parallel to the equator and an oval atmospheric blob called the **Great Red Spot**. The cloud bands display many colors—pale yellows, light blues, deep browns, drab tans, and vivid reds, among others. The Red Spot is one of many features associated with Jupiter's weather. This egg-shaped region, the long diameter of which is approximately twice Earth's diameter, seems to be a hurricane that has persisted for hundreds of years.

The most abundant gas in Jupiter's atmosphere is molecular hydrogen (roughly 86 percent of all molecules), followed by helium (nearly 14 percent). Small amounts of atmospheric methane, ammonia, and water vapor are also found. Based on indirect studies of the planet's internal structure (such as measurements of the average density, as discussed in the previous section), scientists believe that hydrogen and helium make up the bulk of the planet as a whole.

None of the atmospheric gases just listed can, by itself, account for Jupiter's observed coloration. For example, frozen ammonia and water vapor would produce white clouds, not the many colors we see. Scientists believe that complex chemical processes occurring in Jupiter's turbulent atmosphere are responsible for these colors, although the details of the chemistry are not fully understood. Planetary scientists believe that the element sulfur plays an important role in influencing the cloud colors—particularly the reds, browns, and yellows—and that compounds containing the element phosphorus also contribute to the coloration.

The cloud chemistry is very sensitive to small changes in atmospheric conditions, such as pressure, temperature, and composition. Jupiter's atmosphere is in incessant, churning motion, causing these conditions to change from place to place and from hour to hour. The energy that powers the reactions comes in many forms: the planet's own interior heat, solar ultraviolet radiation, aurorae in the planet's magnetosphere, and lightning discharges within the clouds.

Atmospheric Bands

Astronomers describe the banded structure of Jupiter's atmosphere as consisting of a series of lighter-colored **zones** and darker **belts** crossing the planet. The zones and belts vary in both latitude and intensity during the year, but the general pattern is always present. These variations appear to be the result of convective motion in the planet's atmosphere. The zones lie above upward-moving convective currents in Jupiter's atmosphere. The belts are the downward part of the cycle, where material is generally sinking, as illustrated in Figure 7.7.

Because of the upwelling material below them, the zones are regions of high pressure; the belts, conversely, are low-pressure regions. Thus, the belts and zones are the planet's equivalents of the familiar high- and low-pressure systems that cause weather on Earth. A major difference is that Jupiter's rapid rotation causes these systems to wrap all the way around the planet, instead of forming localized circulating storms as on our own world. Because of the pressure difference, the zones lie slightly higher in the atmosphere than the belts. The temperature difference between the two (recall that temperature decreases with altitude) is the basic reason for their different colors.

Underlying the bands is an apparently very stable pattern of eastward and westward wind flow called Jupiter's **zonal flow** (Figure 7.8). The equatorial regions of Jupiter's atmosphere rotate faster than the planet as a whole, with an average flow speed of about

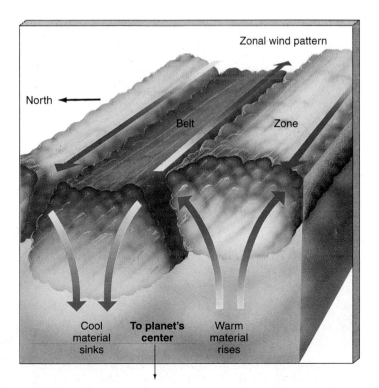

Figure 7.7 The colored bands in Jupiter's atmosphere are associated with vertical convective motion. Upwelling warm gas results in the lighter-colored zones; the darker belts lie atop lower-pressure regions, where cooler gas is sinking back down into the atmosphere. As on Earth, surface winds tend to blow in the direction from high-pressure regions to low-pressure regions. Jupiter's rotation channels these winds into an east–west flow pattern, as indicated by the three arrows drawn atop the belts and zones.

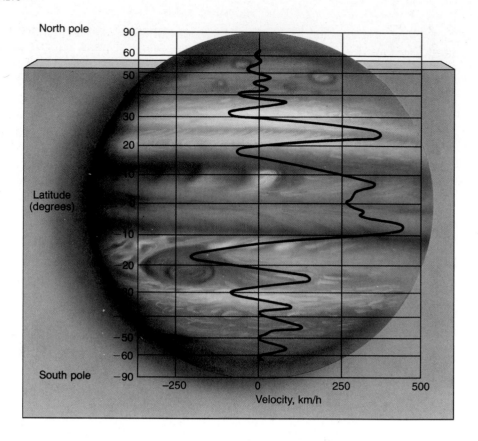

North pole

Latitude (degrees)

South pole

Velocity, km/h

Figure 7.8 The wind speed in Jupiter's atmosphere, measured relative to the planet's internal rotation rate (determined from studies of Jupiter's magnetic field). Positive velocity values mean the wind direction is toward the east, and negative velocity values mean the wind direction is toward the west. The alternations in wind direction are associated with the atmospheric band structure.

300 km/h in the easterly direction, somewhat similar to the jet stream on Earth. At higher latitudes, there are alternating regions of westward and eastward flow, roughly symmetric about the equator, with the flow speed generally diminishing toward the poles. Jupiter's belts and zones are closely related to the zonal flow pattern. Near the poles, where the zonal flow disappears, the band structure vanishes also.

Atmospheric Structure

Planetary scientists believe that Jupiter's clouds are arranged in several layers, with white ammonia clouds generally overlying the colored layers, whose composition we will discuss in a moment. Above the ammonia clouds lies a thin, faint layer of haze created by chemical reactions similar to those that cause smog on Earth. When we observe Jupiter's colors, we are looking down to many depths in the planet's atmosphere.

Figure 7.9 is a diagram of Jupiter's atmosphere, based on observations and computer models. Because the planet lacks a solid surface to use as a reference level for measuring altitude, scientists conventionally take the top of the troposphere (the turbulent region

containing the clouds we see) to lie at 0 km. With this level as the zero point, the troposphere's colored cloud layers all lie at negative altitudes in the diagram. As on other planets, weather on Jupiter is the result of convection in the troposphere.

The haze layer lies at the upper edge of Jupiter's troposphere, at an altitude of zero. The temperature at this level is about 110 K. Above the troposphere, as on Earth, the temperature rises as the atmosphere absorbs solar ultraviolet light. Below the haze layer, at an altitude of –30 km, lie white, wispy clouds of ammonia ice. At these cloud tops, the atmospheric temperature is 125–150 K; it increases rapidly with increasing depth.

A few tens of kilometers below the ammonia clouds, the temperature is a little warmer—above 200 K—and the clouds are made up mostly of droplets or crystals of ammonium hydrosulfide, produced by reactions between ammonia and hydrogen sulfide. At deeper levels, the ammonium hydrosulfide clouds give way to clouds of water ice or water vapor. The top of this lowest cloud layer, which is not seen in visible-light images of Jupiter, lies some 80 km below the top of the troposphere.

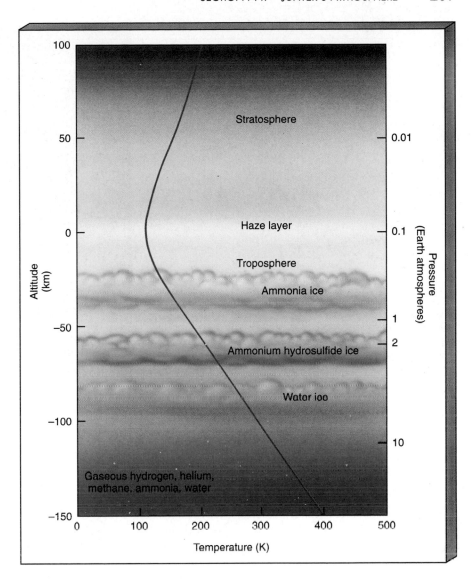

Figure 7.9 The vertical structure of Jupiter's atmosphere. Jupiter's clouds are arranged in three main layers, each with quite different colors and chemistry. The colors we see in photographs of the planet depend on the cloud cover. The white regions are the tops of the upper ammonia clouds. The yellows, reds, and browns are associated with the second cloud layer, which is composed of ammonium hydrosulfide ice. The lowest cloud layer is water ice and therefore bluish. However, the overlying layers are sufficiently thick that this level is not seen in visible light.

This description of Jupiter's atmosphere, based largely on *Voyager* data, was put to the test in December 1995 when the *Galileo* atmospheric probe arrived at the planet. The probe survived for about an hour before being crushed by atmospheric pressure at an altitude of –150 km (that is, right at the bottom of Figure 7.9). Initially, the data from the probe appeared to contradict many of the facts we have just reported! Preliminary analysis of the data indicated that Jupiter was much windier, hotter, and drier than expected, and severely depleted in helium. Furthermore, no clear evidence of the three-layered cloud structure depicted in Figure 7.9 was found. However, most of the discrepancies were the result of improperly calibrated instruments; *Galileo*'s revised findings on wind speed, temperature, and composition are in good agreement with

the picture presented above. In addition, the probe's entry location was in Jupiter's equatorial zone and, as luck would have it, coincided with an atypical hole almost devoid of upper-level clouds (Figure 7.10). The low water content may be normal for the hot, windy regions near Jupiter's equator.

Weather on Jupiter

In addition to the large-scale zonal flow, Jupiter has many smaller-scale weather patterns. The Great Red Spot is a prime example. The Spot, shown in Figure 7.11, was first reported by British scientist Robert Hooke in the mid-seventeenth century, and we are reasonably sure that it has existed continuously in one

Figure 7.10 The arrow on this *Hubble* image shows where the *Galileo* atmospheric probe plunged into Jupiter's cloud deck on December 7, 1995. The entry location was in Jupiter's equatorial zone and apparently almost devoid of upper-level clouds. Until its demise, the probe took numerous meteorological measurements, transmitting those signals to the overhead mother ship, which then relayed them to Earth.

form or another for more than 300 years. It may well be much older. *Voyager* observations showed the Spot to be a region of swirling, circulating winds, like a whirlpool or a terrestrial hurricane, a persistent and vast atmospheric storm. The size of the Spot varies, although its length averages about 25,000 km. It rotates around Jupiter at a rate similar to that of the planet's interior, suggesting that its roots lie far below the atmosphere.

The origin of the Spot's red color is unknown, as is its source of energy. It is generally supposed, however, that the Spot is somehow sustained by Jupiter's large-scale atmospheric motion and that its color is the result of chemical reactions in the planet's turbulent atmosphere. The gas flow around the Spot is counterclockwise, taking about 6 days to complete one circuit. Turbulent eddies form and drift away from the spot's edge; the center, however, remains tranquil in appearance, like the eye of a hurricane on Earth. The zonal motion north of the Spot is westward, while that to the south is eastward (Figure 7.11), supporting the idea that the Spot is confined and powered by the zonal flow. However, the details of how this occurs are still a matter of conjecture.

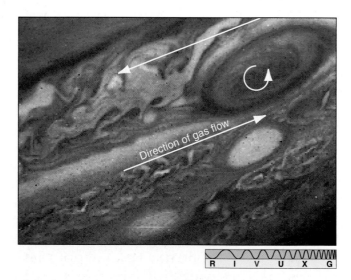

Figure 7.11 *Voyager 1* took this photograph of Jupiter's Great Red Spot from a distance of about 100,000 km. Resolution is about 100 km. Note the complex flow patterns to the left of both the Red Spot and the smaller white oval below it. For scale, Earth is about the size of the white oval. The arrows indicate direction of gas flow above, below, and inside the Great Red Spot.

The *Voyager* mission discovered many smaller spots that are also apparently circulating storm systems. Examples of these smaller systems are the **white ovals** seen in many images of Jupiter (see, for example, the one just south of the Great Red Spot in Figure 7.11). Their high cloud tops give these regions their white color. Like the Red Spot, white-oval storms rotate counterclockwise. The white oval shown in Figure 7.11 is known to be at least 40 years old.

Figure 7.12 shows a **brown oval**, a hole in the overlying clouds that allows us to look down into the lower atmosphere. For unknown reasons, brown ovals appear only in latitudes around 20° north. Although not as long-lived as the Red Spot, they can persist for many years.

Although we cannot explain their formation, we can at least offer a partial explanation for the longevity of storm systems on Jupiter. On Earth a large storm, such as a hurricane, forms over the ocean and may survive for many days, but it dies quickly once it encoun-

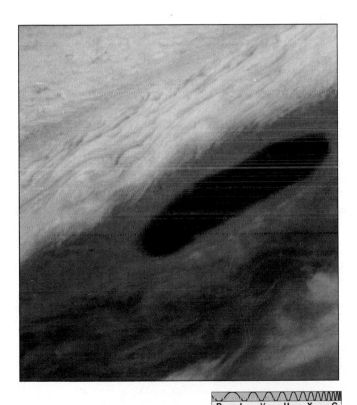

Figure 7.12 A brown oval in Jupiter's northern hemisphere is a break in the upper cloud layer, allowing us to see deeper in, to where the clouds are brown. The oval's long diameter is approximately equal to Earth's diameter.

ters land because the land disrupts the flow patterns that sustain the storm. Jupiter has no continents, so once a storm is established and has reached a size at which other storm systems cannot destroy it, apparently little affects it. The larger the system, the longer its lifetime.

7.5 The Atmospheres of the Outer Jovian Worlds

Video

Saturn's Atmospheric Composition

Saturn is not so colorful as Jupiter. Figure 7.13 shows yellowish and tan cloud bands that parallel the equator, but these regions display less atmospheric structure than do the bands on Jupiter. No obvious large spots or ovals adorn Saturn's cloud decks. Storms do exist, but the color changes that distinguish them on Jupiter are largely absent on Saturn.

Spectroscopic studies show that Saturn's atmosphere consists of molecular hydrogen (92.4 percent) and helium (7.4 percent), with traces of methane and ammonia. As on Jupiter, hydrogen and helium dominate; these most abundant elements never escaped from Saturn's atmosphere because of the planet's large mass and low temperature.

The fraction of helium on Saturn is less than on Jupiter or in the Sun. It is extremely unlikely that the processes that created the outer planets preferentially stripped Saturn of nearly half its helium or that the missing helium somehow escaped from the planet while the hydrogen remained behind. Instead, astronomers believe that at some time in Saturn's past the helium gas liquefied and then sank toward the center of the planet, reducing the abundance of that gas in the outer layers and leaving them relatively hydrogen-rich. The reason for this gas-to-liquid transformation is discussed in Section 7.7.

Figure 7.14 illustrates the vertical structure of Saturn's atmosphere. In many respects, it is similar to Jupiter's (Figure 7.9), except that Saturn's atmospheric temperature is a little lower because of its greater distance from the Sun and because its clouds are thicker than Jupiter's. Because Saturn, like Jupiter, lacks a solid surface, we again take the top of the troposphere

Figure 7.13 The banded structure of Saturn's atmosphere appears clearly in this image, taken as *Voyager 2* approached the planet. (Three of Saturn's moons appear at bottom; a fourth casts a black shadow on the cloud tops.) This image is approximately true color.

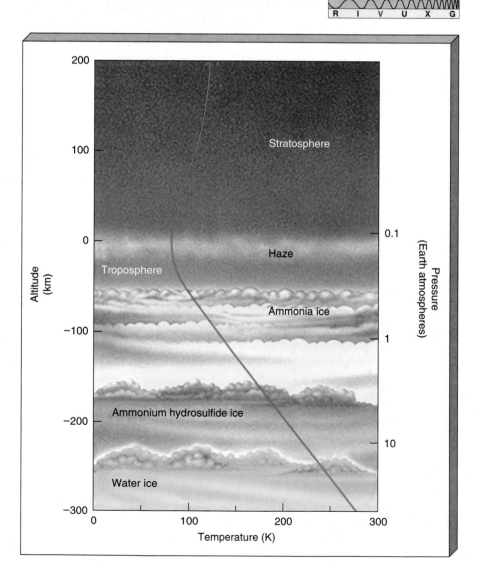

Figure 7.14 The vertical structure of Saturn's atmosphere. As with Jupiter, there are several cloud layers, but Saturn's weaker gravity results in thicker layers and a more uniform appearance.

R I V U X G

Figure 7.15 We see more structure in Saturn's cloud cover when computer processing and artificial color are used to enhance the image contrast, as in these *Voyager* images of the total planet and of a smaller, magnified piece of it.

as our reference level and set it to 0 km. The top of the visible clouds lies about 50 km below this level. As on Jupiter, the troposphere contains clouds arranged in three distinct layers, composed (in order of increasing depth) of ammonia ice, ammonium hydrosulfide ice, and water ice. Above them, at the very top of the troposphere, lies a layer of haze formed by the action of sunlight on Saturn's upper atmosphere.

The total thickness of the three cloud layers in Saturn's atmosphere is roughly 250 km—compared with a thickness of about 80 km on Jupiter—and each layer is thicker than its counterpart on Jupiter. The reason for this difference is Saturn's weaker gravity. At the haze level, Jupiter's gravitational field is nearly 2.5 times stronger than Saturn's, so Jupiter's atmosphere is pulled much more powerfully toward the center of the planet and the cloud layers are squeezed more closely together.

The colors of Saturn's cloud layers, as well as the planet's overall yellowish coloration, are likely due to the same basic cloud chemistry as on Jupiter. However, because Saturn's clouds are thicker, there are fewer

holes and gaps in the top layer, with the result that we rarely glimpse the more colorful levels below. Instead, we see only the topmost layer, which accounts for Saturn's less varied appearance.

Weather on Saturn

Computer-enhanced images of Saturn (Figure 7.15) clearly show the existence of bands, oval storm systems, and turbulent flow patterns, all looking very much like those on Jupiter. In addition, there is an apparently quite stable east–west zonal flow, although the wind speed is considerably greater than on Jupiter and shows fewer east–west alternations. The equatorial eastward jet stream, which moves at about 400 km/h on Jupiter, reaches a brisk 1500 km/h on Saturn, and it extends to much higher latitudes. Not until latitudes 40° north and south are the first westward flows found; latitude 40° N also marks the strongest bands on Saturn and the most obvious ovals and turbulent eddies. Astronomers do not fully understand the rea-

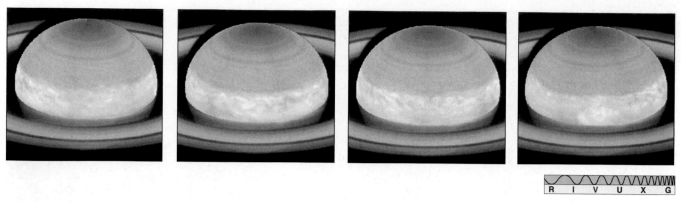

Figure 7.16 Circulating and evolving cloud systems on Saturn, imaged by the *Hubble Space Telescope*.

sons for the differences between Jupiter's and Saturn's flow patterns.

In September 1990, amateur astronomers detected a large white spot in Saturn's southern hemisphere, close to the equator. In November of that year, when the *Hubble Space Telescope* imaged the phenomenon in more detail, the spot had developed into a band of clouds completely encircling the planet's equator. Some of the *Hubble* images are shown in Figure 7.16. Astronomers believe that the white coloration arose from crystals of ammonia ice that formed when an upwelling plume of warm gas penetrated the cool upper cloud layers. Because the crystals were freshly formed, they had not yet been affected by the chemical reactions that color the planet's other clouds.

Storm systems like this are rare on Saturn. The last one visible from Earth appeared in 1933, but it was much smaller than the 1990 system and much shorter-lived, lasting for only a few weeks. The turbulent flow patterns seen in and around the 1990 spot have many similarities to the flow around Jupiter's Great Red Spot. Scientists hope that routine observations of such temporary atmospheric phenomena on the outer worlds will enable them to gain greater insight into the dynamics of planetary atmospheres.

Atmospheric Composition of Uranus and Neptune

Spectroscopic studies of sunlight reflected from Uranus's and Neptune's dense clouds indicate that the two planets' outer atmospheres are similar to those of Jupiter and Saturn. The most abundant gas is molecular hydrogen (84 percent), followed by helium (about

14 percent) and methane, which is more abundant on Neptune (about 3 percent) than on Uranus (2 percent). Helium is not depleted in the outer layers of either planet.

Ammonia, which plays such an important role on Jupiter and Saturn, is not present in any significant quantity in the atmosphere of either Uranus or Neptune. The reason for this is the low atmospheric temperatures of those planets. Ammonia gas freezes into ammonia ice crystals at about 70 K—cooler than the cloud-top temperatures of Jupiter and Saturn, but warmer than those of Uranus and Neptune—accounting for its absence on the two outermost jovian worlds.

The blue color of the outer jovian planets is the result of their relatively high percentages of methane and lower percentages of ammonia. Because methane absorbs long-wavelength red light quite efficiently, sunlight reflected from these planets' atmospheres is deficient in red and yellow photons and appears blue-green or blue. The greater the concentration of methane, the bluer the reflected light. Therefore Uranus, having less methane, looks blue-green, while Neptune, having more methane, looks distinctly blue.

Weather on Uranus and Neptune

Early observers of Uranus reported dusky atmospheric bands parallel to the equator. Some even claimed to have spotted bright spots reminiscent of the Great Red Spot on Jupiter. These ground-based observations were almost certainly erroneous, though, as *Voyager 2* detected only a few atmospheric features (Figure

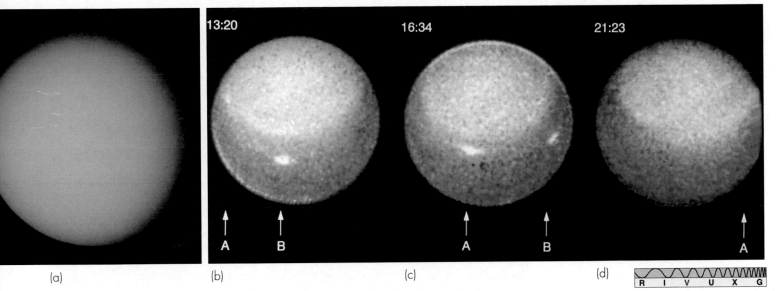

Figure 7.17 (a) This *Voyager* view of Uranus approximates the planet's true color but shows little else. (b), (c), and (d) Image-processed *Hubble Space Telescope* photographs made at roughly four-hour intervals, showing the motion of a pair of bright clouds (labeled A and B) in the planet's southern hemisphere. (The numbers at the top give the time of each photograph.)

7.17a), and even they became visible only after extensive computer enhancement.

Few clouds exist in Uranus's cold upper atmosphere; clouds of the kind found on Jupiter and Saturn are found only at lower, warmer atmospheric levels. The absence of high-level clouds means we must look deep into the planet's atmosphere to see any structure, and the bands and spots that characterize flow patterns on the other jovian worlds are largely washed out on Uranus by intervening stratospheric haze.

With computer-processed images, astronomers have learned that Uranus's atmospheric clouds and flow patterns move around the planet in the same direction as the planet's rotation, with wind speeds in the 200- to 500-km/h range (Figures 7.17b–d). Despite the planet's odd orientation—with the axis of rotation almost in the plane of the ecliptic and the north pole currently pointing almost directly towards the Sun, only the northern hemisphere presently receives any sunlight—Uranus's atmospheric bands are reminiscent of those found on Jupiter and Saturn.

Neptune's cloud and band structure is much more easily seen. Although it lies farther from the Sun, Neptune's upper atmosphere is slightly warmer than Uranus's. (The reasons for this will be discussed shortly.) The extra heat is probably responsible for the greater visibility of Neptune's atmospheric features; the heat makes the cloud layers less dense than they would be if they were cooler, with the result that they lie at higher levels in the atmosphere than on Uranus. The excess heat also causes Neptune's stratospheric haze to be thinner than the haze on Uranus. *Voyager 2* detected numerous white methane clouds (some of them visible in Figure 7.4b) lying about 50 km above the main cloud tops.

Neptune's equatorial winds blow from east to west with speeds of over 2000 km/h relative to the planet's interior. Why such a cold planet should have such rapid winds, and why they should be retrograde relative to the west-to-east rotation of the interior, is a mystery.

Neptune sports several storm systems similar in appearance to those seen on Jupiter (and assumed to be produced and sustained by the same basic processes). The largest such storm, the **Great Dark Spot** (Figure 7.18a), was discovered by *Voyager 2* in 1989. Comparable in size to Earth, the Spot was located near the planet's equator and exhibited many of the same general characteristics as the Great Red Spot on Jupiter. The flow around it was counterclockwise, as with the Red Spot, and there appeared to be turbulence where the winds associated with the Great Dark Spot interacted with the zonal flow to its north and south. The

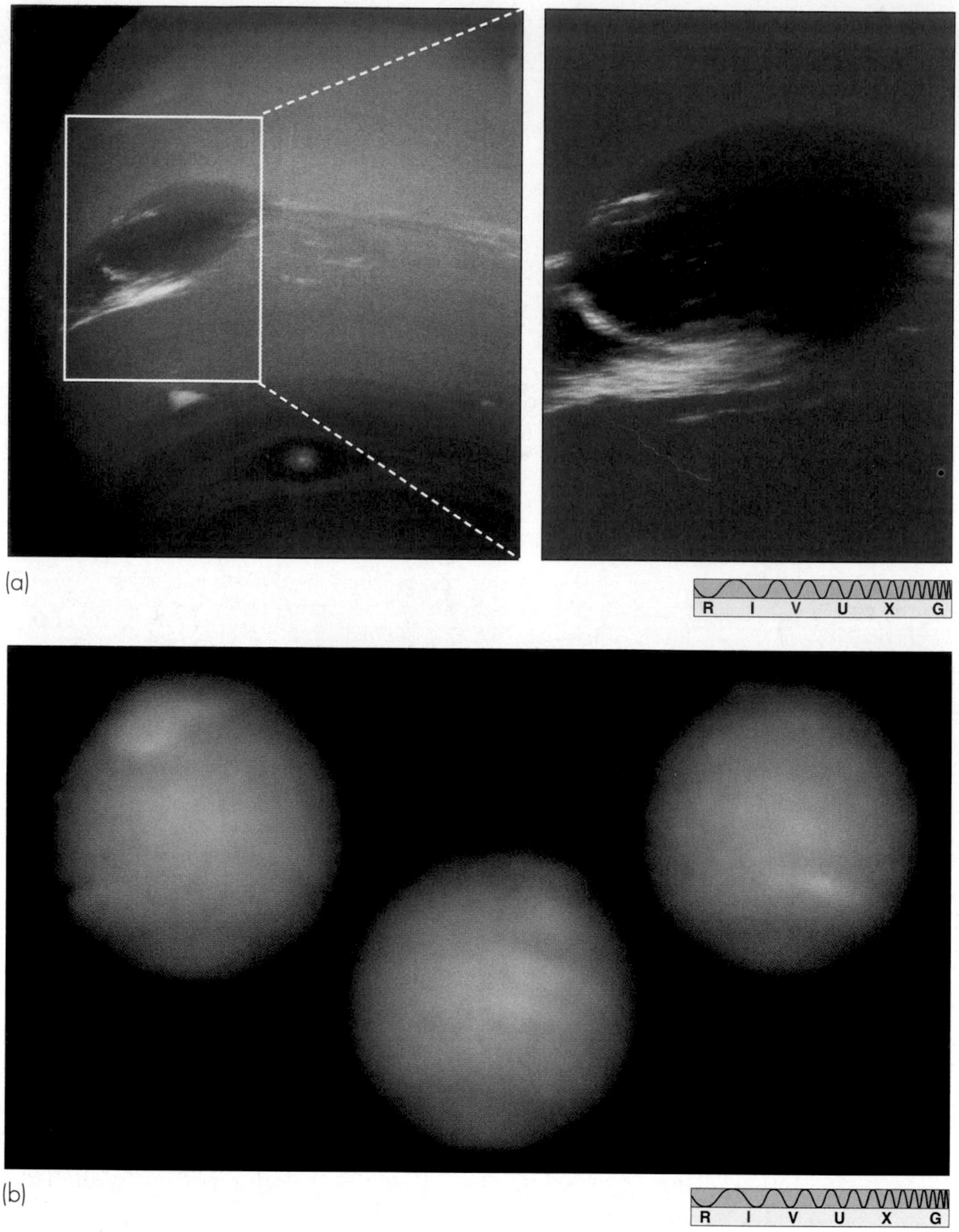

Figure 7.18 (a) Close-up views of the Great Dark Spot of Neptune, which astronomers believe to be a large storm system in the planet's atmosphere, possibly similar in structure to Jupiter's Great Red Spot. Resolution in the image on the right is about 50 km; the dark spot is roughly the size of Earth. (b) These three *Hubble Space Telescope* views of Neptune were taken about 10 days apart in 1994, when the planet was 4.5 billion km from Earth. The planet appears aqua because methane in its clouds has absorbed red light from the sunlight striking the planet, with the result that the reflected light we see is aqua. The cloud features (mostly methane ice crystals) are tinted pink here because they were imaged in the infrared, but they are white in visible light. Neptune apparently has a remarkably dynamic atmosphere that changes every few days. Notice that the Great Dark Spot has now disappeared.

flow around this and other spots may drive updrafts to high altitudes, where methane crystallizes out of the atmosphere to form the high-lying clouds. Astronomers did not have long to study the Dark Spot's properties, however. As shown in Figure 7.18(b), when the *Hubble Space Telescope* viewed Neptune in 1994, the Spot had disappeared.

7.6 Interior Structure

4 The visible cloud layers of the jovian planets are all less than a few hundred kilometers thick. How do we determine the conditions beneath

the clouds, in the unseen interiors of these distant worlds? We have no seismographic information on these regions, and no prospect of obtaining any, for the very good reason that there are no solid surfaces to experience a tremor! Nor do we happen to live on a more or less similar planet that we can use for comparison. Instead, we must use all available bulk data on each planet to construct a model of its interior that agrees with observations. Our statements about the interiors of the jovian worlds are really statements about the model that best fits the facts. However, because the interiors consist largely of hydrogen and helium—two simple gases whose physics we think we know—we are fairly confident that

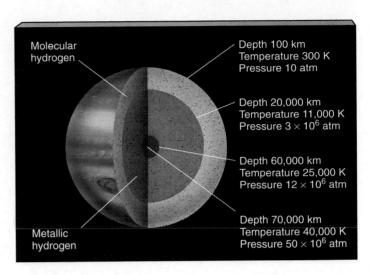

Figure 7.19 Jupiter's internal structure, as deduced from spacecraft measurements and theoretical modeling. The outer radius shown in the diagram represents the bottom of the cloud layers, 70,000 km from the planet's center and about 100 km below the visible cloud top. Pressure and temperature increase with depth, and the atmosphere gradually liquefies over the outermost few thousand kilometers. Below 20,000 km, the hydrogen behaves like a liquid metal. At the center of the planet lies a rocky core, somewhat terrestrial in composition but much larger than any of the inner planets. The temperature at the center is probably about 40,000 K, and the pressure there is thought to be 50 million Earth atmospheres.

the jovian internal structures are reasonably well understood.

As illustrated in Figure 7.19, both the temperature and the pressure of Jupiter's atmosphere increase with depth. At a depth of a few thousand kilometers, the gas makes a gradual transition into the liquid state. By a depth of about 20,000 km, the pressure is about 3 million times greater than atmospheric pressure on Earth. Under these conditions, the hot liquid hydrogen is compressed so much that it undergoes another transition, this time to a "metallic" state, with properties in many ways similar to a liquid metal. Of particular importance for Jupiter's magnetic field, this metallic hydrogen is an excellent conductor of electricity. The combination of rapid overall rotation and an extensive region of highly conductive fluid in its interior gives Jupiter by far the strongest planetary magnetic field in the solar system. ∞ (Sec. 5.6)

The existence of Jupiter's core is inferred from the planet's observed shape. The strong outward push arising from Jupiter's rotation causes a pronounced bulge at the planet's equator—in fact, Jupiter's equatorial radius exceeds its polar radius by nearly 7 percent. (The equatorial radius is listed in Table 7.1.) However, careful calculations indicate that the planet would be *more* flattened than it is if its core were composed of hydrogen and helium alone. Jupiter's observed degree of flattening implies a dense core having a mass perhaps 15 times the mass of Earth.

The precise composition of the core is unknown, but many planetary scientists think it consists of materials similar to those found in the terrestrial worlds: molten, perhaps semisolid, rock. Because of Jupiter's

high central pressure—approximately 50 million times that on Earth's surface, or 10 times that at Earth's center—the core must be compressed to quite high densities (perhaps twice the core density of Earth). Jupiter's core is probably not much more than 20,000 km in diameter. Its central temperature may be as high as 40,000 K.

Saturn has the same basic internal parts as Jupiter, but their relative proportions are different: Saturn's metallic hydrogen layer is thinner, and its central rocky core (again inferred from studies of the planet's polar flattening) is larger. Because of its lower mass, Saturn has a less extreme core temperature, density, and pressure than Jupiter. The central pressure is around a tenth of Jupiter's—not too different from the pressure at the center of Earth.

As we saw in Section 7.3, Uranus and Neptune have rocky cores slightly smaller that those found in Jupiter and Saturn—about the size of Earth and perhaps ten times Earth's mass. The pressure outside the cores of Uranus and Neptune is sufficiently low that hydrogen stays in its molecular form all the way into the planets' cores. Currently, astronomers theorize that, deep below the cloud layers, Uranus and Neptune may have high-density, "slushy" interiors containing thick layers of water clouds. It is also possible that much of the planets' ammonia could be dissolved in the water, simultaneously accounting for ammonia's absence at higher cloud levels and producing a thick, electrically conducting layer that could conceivably explain the planets' magnetic fields. At the present time, however, we don't know enough about the interiors to assess the correctness of this picture.

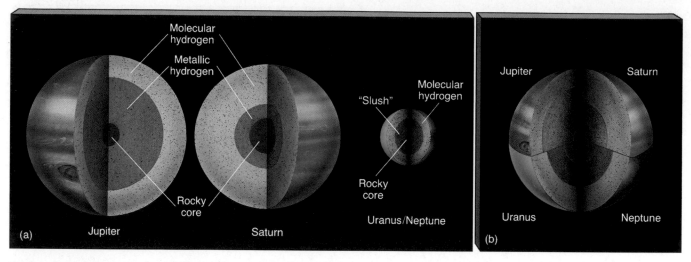

Figure 7.20 A comparison of the interior structures of the four jovian planets. (a) The planets drawn to scale. (b) The relative proportions of the various internal zones.

Our current state of knowledge is summarized in Figure 7.20, which compares the internal structures of the four jovian worlds.

7.7 Internal Heating

5 On the basis of Jupiter's distance from the Sun, astronomers expected to find the temperature of the cloud tops to be around 105 K. At that temperature, they reasoned, Jupiter would radiate back into space the same amount of energy as it receives from the Sun. Radio and infrared observations of the planet, however, revealed a black-body spectrum corresponding to a temperature of 125 K, implying (from Stefan's law) that Jupiter emits about twice as much energy as it receives from the Sun. ∞ (Sec. 2.4) Thus, unlike any of the terrestrial planets, Jupiter must have its own internal heat source.

Astronomers theorize that the source of Jupiter's excess energy is the slow escape of gravitational energy released during the planet's formation. As the planet took shape, some of its energy was converted to heat in the interior. That heat is still slowly leaking out through the planet's heavy atmospheric blanket, resulting in the excess emission we observe. Despite the huge amounts of energy involved—Jupiter emits about 4×10^{17} watts more than it receives from the Sun—the energy loss is slight compared with the planet's total energy. A simple calculation indicates that the average temperature of the interior of Jupiter falls by only about a millionth of a kelvin per year.

Infrared measurements indicate that Saturn too has an internal energy source. Saturn's cloud-top temperature is 97 K, substantially higher than the temperature at which Saturn would reradiate all the energy it receives from the Sun. In fact, Saturn radiates away almost *three times* more energy than it absorbs. However, the explanation behind Jupiter's excess energy—that the planet has a large reservoir of heat left over from its formation—doesn't work for Saturn. Saturn is smaller than Jupiter and so must have cooled more rapidly—rapidly enough that its original supply of energy should have been used up long ago. What then is happening inside Saturn to produce this extra heat?

The explanation for this strange state of affairs also explains the mystery of Saturn's atmospheric helium deficit. At the temperatures and high pressures found in Jupiter's interior, liquid helium dissolves in liquid hydrogen. In Saturn, where the internal temperature is lower, the helium doesn't dissolve so easily and tends to form droplets instead. The phenomenon is familiar to cooks who know that it is generally much easier to dissolve ingredients in hot liquids than in cold ones. Saturn probably started out with a fairly uniform mix of hydrogen and helium, but the helium tended to condense out of the surrounding hydrogen, much as water vapor condenses out of Earth's atmosphere to form a mist. The amount of helium condensation was greatest in the planet's cool outer layers, where the mist turned to rain about 2 billion years ago. A light shower of liquid helium has been falling through Saturn's interior ever since. This helium precipitation is responsible for depleting the outer layers of their helium content.

So we account for the unusually low abundance of helium in Saturn's atmosphere—much of it has rained down to lower levels—but what about the excess heating? The answer is simple: as the helium sinks toward the center, the planet's gravitational field compresses it and heats it up. The gravitational energy thus released is the source of Saturn's internal heat. In the distant future, the helium rain will stop and Saturn will cool until its outermost layers radiate only as much energy as they receive from the Sun. When that happens, the temperature at Saturn's cloud tops will be 74 K.

Uranus apparently has no internal energy source, radiating into space the same amount of energy as it receives from the Sun. It never experienced helium precipitation, and its initial supply of heat was lost to space long ago. What is surprising, though, given its other similarities to Uranus, is that Neptune *does* have an internal source of heat—the planet emits 2.7 times more energy than it receives from the Sun. The cause of this heating is still uncertain. Some scientists have suggested that Neptune's relatively high concentration of methane insulates the planet, helping to maintain its initially high internal temperature.

7.8 Jovian Magnetospheres

For decades, ground-based radio telescopes monitored radiation leaking from Jupiter's magnetosphere. It was only when spacecraft reconnoitered the planet in the mid-1970s, however, that astronomers realized the full extent of the planet's magnetic field. Jupiter is surrounded by a vast sea of energetic charged particles, mostly electrons and protons. This charge-filled region is somewhat similar to Earth's Van Allen belts but very much larger. ∞ (Sec. 5.6)

As in the case of Earth, the size and shape of the magnetosphere are determined by the interaction between the planetary magnetic field and the solar wind. Outside the field, solar wind particles flow freely away from the Sun, past the planet. Inside, their motions are governed by the Jupiter's magnetism.

Direct spacecraft measurements show Jupiter's magnetosphere to be almost 30 million km across (north to south), roughly a million times more volumi-

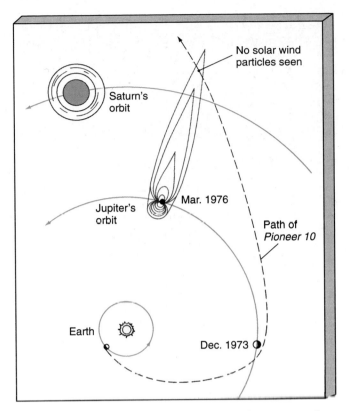

Figure 7.21   The *Pioneer 10* spacecraft (a forerunner to the *Voyager* missions) did not detect any solar-wind particles while moving far behind Jupiter in March 1976, indicating that the tail of Jupiter's magnetosphere extends beyond the orbit of Saturn.

nous than Earth's magnetosphere and far larger than the entire Sun. All four Galilean moons lie within it. On the sunward side, the boundary of Jupiter's magnetic influence on the solar wind lies about 3 million km from the planet. On the opposite side, however, the magnetosphere has a long "tail" extending away from the Sun at least as far as Saturn's orbit (at a distance of 4 A.U., or 600 *million* kilometers), as sketched in Figure 7.21.

Saturn's electrically conducting interior and rapid rotation produce a strong magnetic field and an extensive magnetosphere. Probably because of the considerably smaller mass of Saturn's metallic hydrogen zone, Saturn's basic magnetic field strength is only about 1/20 that of Jupiter, or about 1000 times greater than that of Earth. The magnetic field at Saturn's cloud tops is approximately the same as at Earth's surface.

Saturn's magnetosphere extends about 1 million km toward the Sun and is large enough to contain the planet's ring system and most of its moons. Its size varies with the strength of the solar wind, which tends

7-1 INTERLUDE

A COMETARY IMPACT

In July 1994, sky-watchers were treated to an exceedingly rare event—the collision of a comet (known as Shoemaker–Levy 9, after its discoverers) with the planet Jupiter. When discovered in March 1993, Shoemaker–Levy 9 appeared to have a curious squashed appearance. Higher-resolution images (shown below) revealed that the comet's flattened nucleus was made up of several large pieces, the largest no more than 1 km across. All the pieces were following the same orbit, but they were spread out along the comet's path like a string of pearls 1,000,000 km long.

What could have caused such an unusual object? Tracing the orbit backward in time, researchers calculated that, early in July 1992, the comet had approached to within about 100,000 km of Jupiter. They realized that the objects shown in the figure were the fragments produced when a previously normal comet was captured by Jupiter and torn apart by its strong gravitational field.

On its next approach to Jupiter, in July 1994, Shoemaker–Levy 9 struck the planet's upper atmosphere, plowing into it at a speed of more than 60 km/s and causing a series of enormous explosions, as illustrated in the figures below. Each impact created, for a period of a few minutes, a brilliant fireball hundreds of kilometers across and having a temperature of many thousands of kelvins. The energy released in each explosion was comparable to a billion terrestrial nuclear detonations. Every major telescope on Earth, the *Hubble Space Telescope*, *Galileo* (which was only 1.5 A.U. from the planet at the time), and even *Voyager 2* were watching. The effects on the planet's atmosphere and the vibrations produced throughout Jupiter's interior were observable for days after the impact.

One of the largest pieces of the comet, fragment G, produced a spectacular fireball on the southwestern part of Jupiter. This fireball is seen in the left-hand image at right, radiating strongly in the infrared (that is, giving off lots of heat). The colli-

S   G   D   → Toward Jupiter

R I V U X G

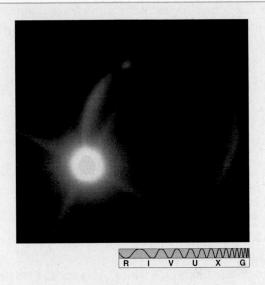

sions created several "black-eye" impact sites, each roughly the size of Earth, in Jupiter's southern hemisphere. One of the most prominent impact sites created by fragment G is clearly visible in the true-color visible-light image at right. The large, dark, crescent-shaped arc some 6000 km from the impact site is the result of material falling back onto Jupiter.

As best we can determine, none of the cometary fragments breached the jovian clouds. Only *Galileo* had a direct view of the impacts on the back side of Jupiter, and in every case the explosions seemed to occur high in the atmosphere, above the upper-most cloud layer. Most of the dark material seen in the images is probably pieces of the comet rather than parts of Jupiter. Spectral lines from silicon, magnesium, and iron were detected in the aftermath of the collisions, and the presence of these metals might explain the dark material observed near some of the impact sites. Water vapor was also detected spectroscopically, again apparently from the comet. The fallen material from the impacts spread slowly around Jupiter's bands and after five months reached completely around the planet. It will probably take years for all the cometary matter to settle into Jupiter's interior.

to push the sunward side of the magnetosphere closer to the planet. Like Jupiter's, Saturn's magnetosphere emits radio waves, but as luck would have it, they are reflected from Earth's ionosphere (they lie in the AM band) and were not detected until the *Voyager* craft approached the planet.

Voyager 2 found that both Uranus and Neptune have fairly strong magnetic fields—about 100 times stronger than Earth's field and 1/10 as strong as Saturn's. The magnetic fields at their cloud tops, about four Earth radii from their centers, are comparable in strength to Earth's surface field. Uranus and Neptune have substantial magnetospheres, populated largely by electrons and protons either captured from the solar wind or created from hydrogen gas escaping from the planets.

Chapter Review

Summary

Jupiter and Saturn were the outermost planets known to ancient astronomers. Uranus was discovered in the eighteenth century, by chance. Neptune was discovered after mathematical calculations of Uranus's slightly non-Keplerian orbit revealed the presence of an eighth planet.

Having no solid surface, the jovian planets display **differential rotation** (p. 205)—the rotation rate varies from place to place in the atmosphere. Radio waves emitted from the planets' magnetospheres provide a measure of their interior rotation rates. For unknown reasons, Uranus's spin axis lies nearly in the ecliptic plane, leading to extreme seasonal variations in solar heating on the planet as it orbits the Sun.

The cloud layers on all the jovian worlds are arranged into bands of light-colored **zones** (p. 207) and darker **belts** (p. 207) crossing the planets parallel to the equator. The bands are the result of (1) convection in the planets' interiors and (2) the planets' rapid rotation. Underlying the bands is a stable pattern of eastward or westward wind flow called the **zonal flow** (p. 207). The wind direction alternates as we move north or south away from the equator.

Jupiter's atmosphere consists of three main cloud layers. The colors we see are the result of chemical reactions at varying depths below the cloud tops. The main weather pattern on Jupiter is the **Great Red Spot**

(p. 206), a huge hurricane that apparently has been raging for at least three centuries. Other, smaller, atmospheric features—**white ovals** (p. 211) and **brown ovals** (p. 211)—are also observed. Similar systems are found on the other jovian planets, although they are less distinct on Saturn and Uranus. Short-lived storms on Saturn are occasionally seen. The **Great Dark Spot** (p. 215) on Neptune had many similarities to Jupiter's Red Spot.

The atmospheres of the jovian planets become hotter and denser with depth, eventually becoming liquid. In Jupiter and Saturn, the interior pressures are so high that the hydrogen near the center exists in a liquid-metallic form rather than as molecular hydrogen. In Uranus and Neptune, this change from molecular hydrogen to metallic hydrogen apparently does not occur. All four planets have large cores, of rocky composition.

Three jovian planets radiate more energy than they receive from the Sun. On Jupiter and Neptune, the source of this energy is most likely heat left over from the planet's formation. On Saturn, the heating is the result of helium precipitation in the interior, where helium liquefies and forms droplets, which then fall toward the center of the planet. This process is also responsible for reducing the amount of helium observed in Saturn's surface layers.

Self-Test: True or False?

_____ **1.** Uranus was discovered by Galileo.

_____ **2.** After the discovery of Uranus, astronomers started looking for other planets and quickly discovered Neptune.

_____ **3.** Jupiter has more than 300 times the mass of Earth and twice the mass of all the other planets combined.

_____ **4.** Neptune has a smaller radius than Uranus

_____ 5. There is no evidence to suggest that either Jupiter or Saturn has a rocky core.

_____ 6. There is no evidence to suggest that either Uranus or Neptune has a rocky core.

_____ 7. Jupiter's solid surface lies just below the cloud layers visible from Earth.

_____ 8. The thickness of Jupiter's cloud layer is less than 1 percent of the planet's radius.

_____ 9. Jupiter has only one large storm system.

_____ 10. In general, a storm system in Jupiter's atmosphere is much longer-lived than a storm system in Earth's atmosphere.

_____ 11. The element helium plays an important role in producing the colors in jovian atmospheres.

_____ 12. Unlike that of Jupiter, the rotation of Saturn is slow and shows little differential rotation.

and therefore a smaller mass and lower density.

_____ 13. The rotation rates of Uranus and Neptune are almost identical.

_____ 14. The atmosphere of Saturn contains only half the helium of Jupiter's atmosphere, but overall Saturn probably has a normal helium abundance.

_____ 15. Like the other jovian worlds, Uranus has prominent belts and zones in its atmosphere.

_____ 16. The strength of Jupiter's magnetic field is similar to the strength of Earth's field.

_____ 17. Saturn's magnetic field is much weaker than Jupiter's and about as strong as Earth's.

_____ 18. Both Uranus and Neptune have layers of metallic hydrogen surrounding their central cores.

_____ 19. The fraction of Uranus and Neptune that is core is larger than the fraction of Jupiter or Saturn that is core.

Self-Test: Fill in the Blank

1. The _____ of Jupiter indicates that the planet's overall composition differs greatly from that of the terrestrial planets.

2. The planet _____ is blue-green and virtually featureless.

3. The planet _____ is dark blue, with white clouds and visible storm systems.

4. The overall coloration of Uranus and Neptune is caused by the gas _____.

5. The main constituents of Jupiter and Saturn are _____ and _____.

6. Jupiter's clouds consist of a series of light-colored _____ and darker _____.

7. Jupiter's Great Red Spot has similarities to _____ on Earth.

8. The long diameter of the Great Red Spot is about _____ times that of Earth.

9. Features in the atmospheres of Saturn and Jupiter are mostly hidden by an upper layer of frozen _____ clouds.

10. Saturn's cloud layers are thicker than those of Jupiter because of Saturn's weaker _____.

11. The Great Dark Spot was a storm system on _____.

12. Jupiter's rapid _____ produces a significant equatorial bulge.

13. Uranus's rotation axis is almost _____ to the ecliptic plane.

14. Although often referred to as a gaseous planet, Jupiter is mostly _____ in its interior.

15. Jupiter emits about _____ times more radiation than it receives from the Sun.

16. Jupiter's magnetic field is generated by its rapid rotation and by the element _____, which becomes metallic in the interior of the planet.

17. Saturn's excess energy emission is caused by _____.

Review and Discussion

1. How was Uranus discovered?

2. Why did astronomers suspect an eighth planet beyond Uranus?

3. Why do you think the discovery of Uranus in 1781 was so surprising? Might there be similar surprises in store for today's astronomers?

4. Why have the jovian planets retained most of their original atmospheres?

5. What is differential rotation, and what evidence do we have that it occurs on Jupiter?

6. Why does Saturn have a less varied appearance than Jupiter?

7. Compare the thicknesses of Saturn's various layers (clouds, molecular hydrogen, metallic hydrogen, and core) to the equivalent layers in Jupiter.

8. What is the Great Red Spot? What is known about the source of its energy?

9. Describe the cause of the colors in the jovian planets' atmospheres.

10. Briefly describe the weather on Jupiter.

11. What is Jupiter thought to be like beneath its clouds? Why do we think this?

12. Compare and contrast the atmospheres of the jovian planets, describing how the differences affect the appearance of each.

13. Explain the theoretical model that accounts for Jupiter's internal heat source.

14. Describe the theoretical model that accounts for Saturn's missing helium and the surprising amount of heat the planet radiates.

15. What is unusual about the rotation of Uranus?

16. What is responsible for Jupiter's enormous magnetic field?

17. How are the interiors of Uranus and Neptune thought to differ from those of Jupiter and Saturn?

18. Give a brief description of how much information you think we might have about the outer solar system without the *Voyager* and *Galileo* missions.

Problems

1. Verify the statement made in the text that the force of gravity at Jupiter's cloud tops is nearly 2.5 times that at Saturn's cloud tops.

2. How long does it take for Saturn's equatorial flow, moving at 1500 km/h, to encircle the planet?

3. What is the round-trip travel time of light from Earth to Jupiter (at the minimum Earth–Jupiter distance of 4.2 A.U.)? How far would a spacecraft orbiting Jupiter at a speed of 20 km/s travel in that time?

Projects

1. Look in an almanac to find out where Jupiter is now. What constellation is it in? Do any stars in the sky look as bright as Jupiter?

2. Saturn moves more slowly among the stars than any other visible planet. It crosses a constellation boundary only once every few years. Look in an almanac to see where the planet is now, and find Saturn in the sky. What constellation is it in? Can you explain why it moves so slowly, in contrast to Mars or Jupiter? How many stars in the night sky are brighter than Saturn at its brightest? What do you notice about the difference in appearance between Saturn and a star of about equal brightness?

3. The major astronomy magazines *Sky & Telescope* and *Astronomy* generally print, in their January issues, charts showing the whereabouts of the planets. Consult the charts in one of these magazines, and locate Neptune and Uranus in the sky. Uranus may be visible to the naked eye, but binoculars make the search much easier. (Hint: Uranus shines more steadily than do the background stars.) With the eye alone, can you detect a color to Uranus? Does color become more intense through binoculars?

4. The search for Neptune requires a much more determined effort! A telescope is best, but high-powered binoculars mounted on a steady support also reveal the planet. If you have an opportunity to see Neptune and Uranus through a telescope (they are close together on the sky for the rest of this century), contrast their colors. Which planet appears bluer? Through a telescope, can you see that Uranus appears as a disk rather than a point? Does Neptune also look like a disk, or does it look more like a pinpoint of light?

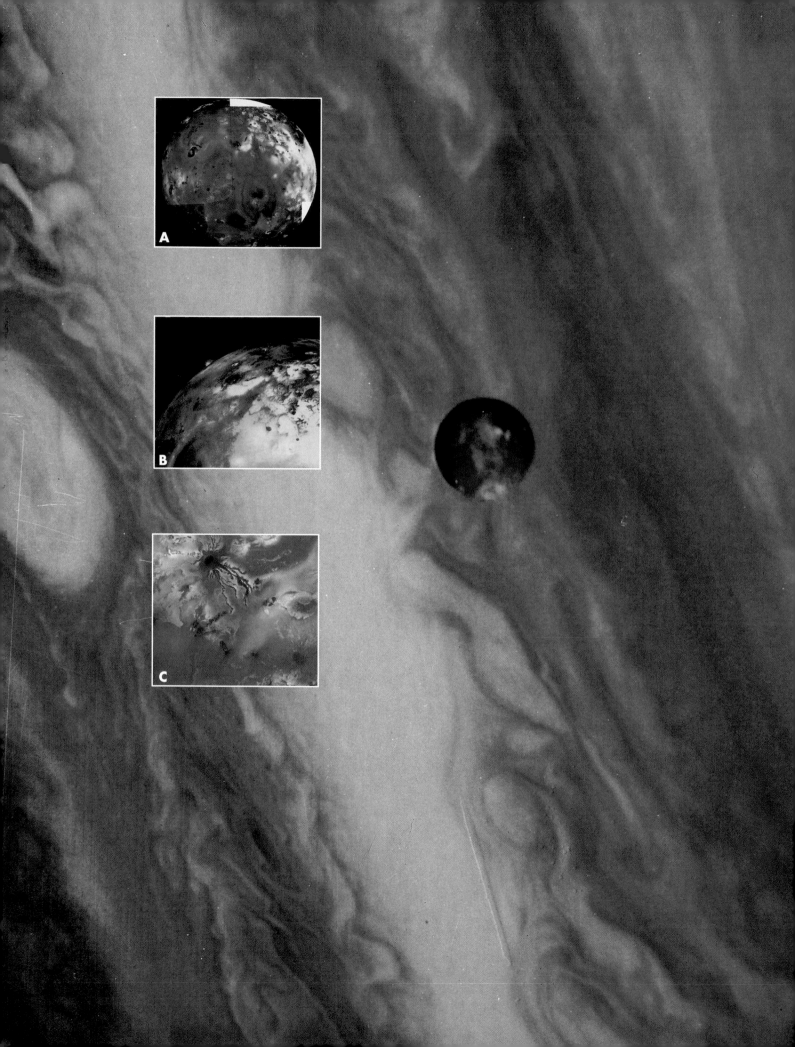

8 Moons, Rings, and Pluto

Small Worlds Among Giants

(Opposite page, background) In this close-up image of Jupiter, taken by the Voyager 2 spacecraft, the moon Io is clearly visible. One of the most peculiar moons in the solar system, and certainly the most active, Io's multi-hued surface is caused by various chemical compounds deposited by currently active volcanos.

(Inset A) Photo of Io taken by Voyager 1 with a remarkable resolution of 7 km. The red color results largely from sulfur compounds, the white areas probably covered with sulfur dioxide frost.

(Inset B) At top left, a plume from an erupting volcano, known as Prometheus, is clearly seen at the limb of Io against the blackness of space.

(Inset C) Many lava flows can be seen on Io, like this one stretching for several hundred km.

LEARNING GOALS

Studying this chapter will enable you to:

1 Describe how the Galilean moons form a miniature solar system around Jupiter and exhibit a wide range of properties.

2 Understand that Titan, Saturn's largest moon, has an atmosphere in which complex chemical reactions take place.

3 Explain what the medium-sized jovian moons reveal about the outer solar system.

4 Describe the nature and detailed structure of Saturn's rings.

5 Discuss the differences between the rings of Saturn and those of the other jovian planets.

6 Explain how the Pluto–Charon system is fundamentally different from all other planet–moon systems.

All four jovian planets have extensive systems of moons and rings that exhibit fascinating variety and complexity. The six largest jovian moons have many planetary features, providing us with insight into the terrestrial worlds. Two of these moons, Jupiter's Europa and Saturn's Titan, are prime candidates for study in the search for life elsewhere in the solar system. Like the moons, the rings of the jovian planets differ greatly from one to the next. The rings of Saturn—the best-known ring system among the jovian planets—are one of the most spectacular displays in the sky. Pluto is a planet, not a moon, but it is very much smaller than even the terrestrial worlds and generally seems much more moonlike than planetlike. For that reason, we study it here along with the moons and rings of its giant neighbors.

8.1 Small Bodies in the Outer Solar System

All the jovian planets have extensive moon systems consisting of a few fairly large moons, long known from Earth, plus many smaller moonlets discovered by the *Voyager* spacecraft. Jupiter has at least 16 moons, Saturn 18, Uranus 15, and Neptune 8. The reflected light from the known moons suggests that most are covered with snow and ice—in fact, many of them are probably made almost entirely of water ice. They fall into three groups: *large bodies*, comparable in size to Earth's Moon, having radii greater than 1000 km (Table 8.1); *medium-sized bodies*, with diameters ranging from 400 to 1500 km; and *small bodies*, which are irregularly shaped chunks of ice less than 300 km in diameter, most only a few tens of kilometers across.

All four jovian worlds also have planetary ring systems girdling their equators. The properties and appearance of each planet's rings are intimately connected with the planet's small and medium-sized moons, with many of the inner jovian moons orbiting close to (or even within) the parent planet's rings. Saturn's rings are by far the best known and best observed, and we will study them in greatest detail. However, the rings of the other jovian planets also have much to tell us about the environment in the outer solar system.

Pluto is defined as a planet because it orbits the Sun, so why study it here? The answer is that Pluto does not fit into the terrestrial–jovian classification. In fact, it has far more similarities to the icy moons of the outer planets than to any planet. Pluto is most similar in mass, radius, and composition to Triton, the largest

Table 8.1 The Large Moons of the Solar System

Name	Parent Planet	Distance from Parent Planet		Orbit Period (Earth days)	Radius (km)	Mass (Earth's Moon = 1)	Density (kg/m^3)
		(km)	(planet radii)				
Moon	Earth	384,000	60.2	27.3	1740	1.00	3300
Io	Jupiter	422,000	5.91	1.77	1820	1.22	3600
Europa	Jupiter	671,000	9.40	3.55	1570	0.65	3000
Ganymede	Jupiter	1,070,000	15.0	7.16	2630	2.02	1900
Callisto	Jupiter	1,880,000	26.4	16.7	2400	1.47	1900
Titan	Saturn	1,220,000	20.3	16.0	2580	1.83	1900
Triton	Neptune	354,000	14.3	−5.88[1]	1380	0.29	2100

[1]The minus sign indicates a retrograde orbit.

R I V U X G

Figure 8.1 *Voyager 1* spacecraft images of the four Galilean moons of Jupiter. Shown here to scale as they would appear from a distance of about 1 million km are (clockwise from upper left) Io, Europa, Callisto, and Ganymede.

moon of Neptune, Pluto's nearest planetary neighbor. Is Pluto an escaped moon, or would *any* planet that formed in the outermost solar system look like Pluto? These are some of the questions we ask as we round out our tour of the solar system with a study of its outermost known member.

8.2 The Galilean Moons of Jupiter

1 Jupiter's four largest moons—called the Galilean moons, as mentioned in Section 7.1—travel in nearly circular, prograde orbits about Jupiter and move in its equatorial plane. Moving outward from the planet, they are named Io, Europa, Ganymede, and Callisto, after the mythical attendants of the Roman god Jupiter. The two *Voyager* spacecraft and, more recently, the *Galileo* mission have obtained some remarkable images of these bodies, allowing us to dis-

cern fine surface detail on each and greatly expanding our knowledge of these small, distant worlds. Figure 8.1 compares the appearances and sizes of the four Galilean satellites, and Figure 8.2 shows two of them alongside Jupiter.

The Galilean moons range in size from slightly smaller than Earth's Moon (Europa) to slightly larger than Mercury (Ganymede). The moons' densities decrease with increasing distance from Jupiter in a manner reminiscent of the falloff in density of the terrestrial planets with increasing distance from the Sun. Recent measurements made by the *Galileo* spacecraft indicate that Io has a large metallic core, most likely composed of iron and making up nearly half the moon's diameter. Io's outer layers, and essentially all of Europa, are thought to be rocky, possibly similar to the crusts of the terrestrial planets. Ganymede and Callisto are of more lightweight overall composition. Low-density materials, such as water ice, may account for as much as half their total mass, surrounding rocky cores.

Many astronomers think that the formation of Jupiter and its Galilean satellites may have mimicked

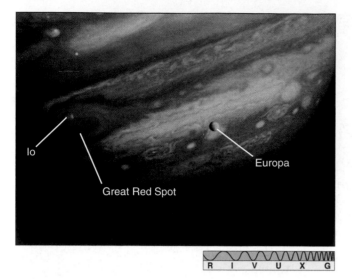

Figure 8.2 *Voyager 1* took this photograph of Jupiter with ruddy Io on the left and pearl-like Europa toward the right. Note the scale of objects here: Io and Europa are each comparable in size to our Moon; the Great Red Spot is roughly twice as large as Earth.

on a small scale the formation of the Sun and the inner planets. For that reason, studies of the Galilean moon system can provide us with valuable insight into the processes that created our own world.

Io

Io is the most geologically active object in the entire solar system. Its mass and radius are fairly similar to those of Earth's Moon, but there the resemblance ends. Shown in Figure 8.3, Io's surface is a collage of oranges, yellows, and blackish browns—resembling a giant pizza in the minds of some startled *Voyager* scientists. As the spacecraft glided past Io, an outstanding discovery was made: Io has active volcanoes! *Voyager 1* photographed eight erupting volcanoes; six were still erupting when *Voyager 2* passed by four months later. The inset in Figure 8.3 shows one volcano ejecting matter to an altitude of over 200 km. The gases are spewed forth at speeds up to 2 km/s, quite unlike the (relatively) sluggish lava that emanates from Earth's insides.

The orange color immediately surrounding the volcanoes most likely results from sulfur compounds in the ejected material. In contrast to the other Galilean moons, Io's surface is neither cratered nor streaked. Rather, it is exceptionally smooth, apparently the result of molten matter constantly filling in any dents and cracks. From this observation we conclude that

this remarkable moon has the youngest surface of any known object in the solar system. Io has a thin, temporary atmosphere made up mainly of sulfur dioxide, presumably the result of gases ejected by volcanic activity.

What causes Io's volcanism? The moon is far too small to have geological activity like that on Earth. Io would be long-dead, as is our own Moon, if its internal heat were the only energy available. In fact, the source of Io's energy is external—namely, Jupiter's gravity. Io orbits very close to Jupiter, at just under six Jupiter radii from the center (Table 8.1). As a result, Jupiter's huge gravitational field produces strong tidal forces on Io; ultimately, these forces are responsible for the moon's spectacular activity.

If Io were Jupiter's only satellite, it would long ago have come into a state of synchronous rotation with the planet, just as our own Moon did. ∞ (Sec. 5.2) In that case, Io would move in a perfectly circular

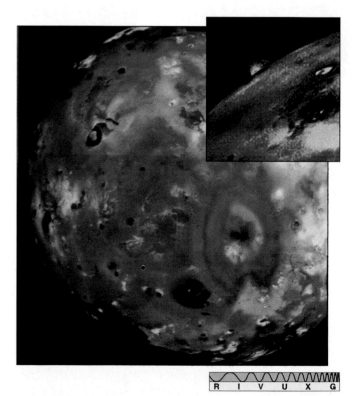

Figure 8.3 Jupiter's innermost moon, Io, is quite different from the other three Galilean satellites. Its surface is kept smooth and brightly colored by constant volcanism. The resolution of this image is about 7 km. The dark, circular features are volcanoes. The inset shows one of Io's volcanoes erupting as *Voyager 1* flew past this fascinating moon. Surface features here are resolved to a few kilometers. The volcano's umbrellalike profile shows clearly against the darkness of space. The plume measures about 100 km high and 300 km across.

orbit with one face permanently turned toward Jupiter. The tidal bulge would be stationary with respect to the moon, and there would be no internal stresses and no volcanism. But Io is not alone. As it orbits, it is constantly tugged by the gravity of its nearest large neighbor, Europa. These tugs are small and not enough to cause any great tidal effect in and of themselves, but they are sufficient to make Io's orbit slightly noncircular, preventing the moon from settling into a precisely synchronous state. As a result, seen from Jupiter, Io wobbles slightly from side to side as it moves.

The conflicting forces acting on Io result in enormous tidal stresses that continually flex and squeeze its interior. Just as repeated back-and-forth bending of a wire can produce heat through friction, Io is constantly energized by this ever-changing distortion. The large amount of heat generated within Io causes huge jets of gas and molten rock to squirt out of the surface. It is likely that much of Io's interior is soft or molten, with only a relatively thin solid crust overlying it. Io's volcanism is a primary source of charged particles in the inner regions of Jupiter's magnetosphere.

Europa

Europa (Figure 8.4) is a world very different from Io. Europa has relatively few craters on its surface, again suggesting geologic youth—recent activity has erased the scars of ancient meteoric impacts. Europa's surface displays a vast network of lines criss-crossing bright, clear fields of water ice. Many of these linear bands extend halfway around the satellite.

Some planetary scientists have suggested that Europa's fractured surface is the result of a form of tectonic activity, one involving ice rather than rock. Others theorize that Europa is covered completely by an ocean of liquid water whose surface layers are frozen. They attribute the cracks to the tidal influence of Jupiter and the gravitational pulls of the other Galilean satellites, although these forces are weaker than those powering Io's volcanic activity.

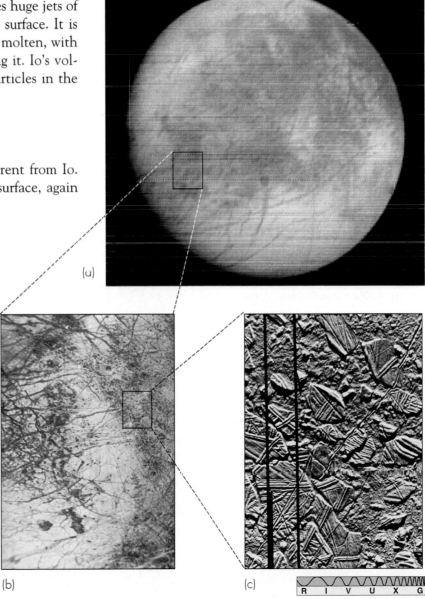

Figure 8.4 (a) A *Voyager 1* mosaic of Europa. Resolution is about 5 km. (b) Europa's icy surface is only lightly cratered, indicating that some ongoing process must be obliterating impact craters soon after they form. The origin of the cracks criss-crossing the surface is uncertain. (c) At 20-m resolution—the width of a typical house—this image from the *Galileo* spacecraft shows a smooth yet tangled surface resembling the huge ice flows that reside in Earth's polar regions.

(a)

(b)

(c)

R I V U X G

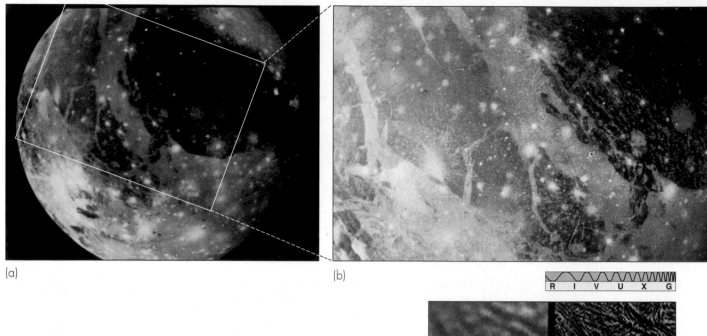

(a)

(b)

R   I   V   U   X   G

(c)

Figure 8.5 (a) and (b) *Voyager 2* images of Ganymede. The dark regions are the oldest parts of the moon's surface and probably represent its original icy crust. The largest dark region visible here is called Galileo Regio. It spans some 320 km. The lighter, younger regions are the result of flooding and freezing that occurred within a billion years or so of Ganymede's formation. The light-colored spots are recent impact craters. (c) Grooved terrain on Ganymede may have been caused by a process similar to plate tectonics on Earth. The image on the left is what was seen by *Voyager* in 1979; that on the right is what was seen by *Galileo* in 1996. The 55-by-35-km area shown here reveals a multitude of ever-smaller ridges, valleys, and craters, right down to the resolution limit of *Galileo*'s camera (about 100 m, the length of a football field).

Recent *Galileo* images seem to support the idea that Europa has a liquid ocean beneath its icy crust. Figure 8.4(b) is a high-resolution image of this weird moon, showing ice-filled surface cracks similar to those seen in Earth's polar ice floes. Figure 8.4(c) shows what appear to be "icebergs"—flat chunks of ice that have been broken apart and reassembled, perhaps by the action of water below. Mission scientists speculate that Europa's ice may be several kilometers thick and that there may be a 100-km-deep ocean below it.

If Europa does have an ocean of liquid water below its surface ice, it opens up many interesting avenues of speculation about the possible development of life there. In the rest of the solar system, only Earth has liquid water on or near its surface, and most scientists agree that water played a key role in the appearance of life here (see Chapter 18). However, bear in mind that the existence of water does not *necessarily* imply the

emergence of life. Europa is a very hostile environment compared with Earth. The surface temperature on Europa is just 130 K, and the atmospheric pressure is only a billionth the pressure on our planet.

Ganymede and Callisto

Ganymede, shown in Figure 8.5, is the largest moon in the solar system, exceeding not only Earth's Moon but also the planets Mercury and Pluto in size. It has many impact craters on its surface, along with patterns of dark and light markings reminiscent of the highlands and maria on Earth's Moon. In fact, Ganymede's history has many parallels with that of the Moon (with water ice replacing lunar rock).

As with the inner planets, we can estimate ages on Ganymede by counting craters. The darker regions are the oldest parts of Ganymede's surface. They are

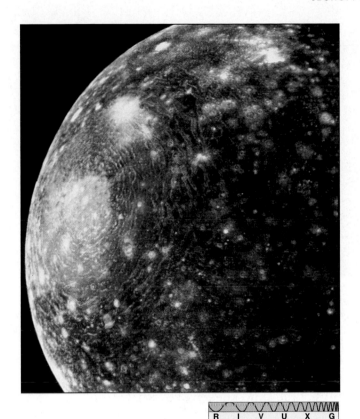

R   I   V   U   X   G

Figure 8.6 Callisto is similar to Ganymede in composition but is more heavily cratered. The large series of concentric ridges visible at left is known as Valhalla. Extending nearly 1500 km from the basin center, the ridges formed when "ripples" from a large meteoritic impact froze before they could disperse completely. The resolution here is around 10 km.

the original icy surface, just as the ancient highlands on our own Moon are its original crust. Ganymede's surface has darkened with age as micrometeorite dust has slowly covered it. Being much less heavily cratered, the light-colored parts of Ganymede must be younger. They probably formed in a manner similar to the maria on the Moon. Intense meteoritic bombardment caused liquid water—Ganymede's counterpart to our own Moon's molten lava—to upwell from the interior, flooding the impacting regions before solidifying.

Not all of Ganymede's surface features follow the lunar analogy. Ganymede has a system of grooves and ridges (Figures 8.5b, c) that may have resulted from crustal tectonic motion, much as Earth's surface undergoes mountain building and faulting at plate boundaries. Ganymede's large size indicates that its original radioactivity probably helped to heat and differentiate its partly rocky interior, after which the moon cooled

and the crust cracked. Ganymede seems to have had some early plate tectonic activity, but the process apparently stopped about 3 billion years ago when the cooling crust became too thick. The *Galileo* data suggest that the surface of Ganymede may be older than was previously thought. With the improved resolution of that spacecraft's images, some regions believed to have been smooth, and hence young, are now seen to be heavily splintered by fractures and thus probably very old.

Callisto, shown in Figure 8.6, is in many ways similar in appearance to Ganymede, although the latter has fewer craters and more fault lines. Callisto's most obvious feature is a huge series of concentric ridges surrounding two large basins, one clearly visible in Figure 8.6. The ridges resemble the ripples made as a stone hits water, but on Callisto they probably resulted from a cataclysmic impact with a meteorite. The upthrust ice was partially melted, but it resolidified quickly, before the ripples had a chance to subside. Today, both the ridges and the rest of the crust are frigid ice and show no obvious signs of geological activity. Apparently, Callisto froze before plate tectonic or other activity could start. The density of impact craters on the large impact basins indicates that they formed long ago, perhaps 4 billion years in the past.

8.3 The Large Moons of Saturn and Neptune

Titan

2 The largest and most intriguing of Saturn's moons is Titan (Figure 8.7). Scientists believe that Titan's internal composition and structure must be similar to Ganymede's and Callisto's because these three moons have similar masses and radii and hence similar average densities. Thus, Titan probably contains a rocky core surrounded by a thick mantle of water ice. The nature of its surface is unknown.

From Earth, Titan is visible only as a barely resolved reddish disk. Long before the *Voyager* missions, astronomers knew from spectroscopic observations that the moon's reddish coloration is caused by something quite special—an atmosphere. So anxious were mission planners to obtain a closer look that they

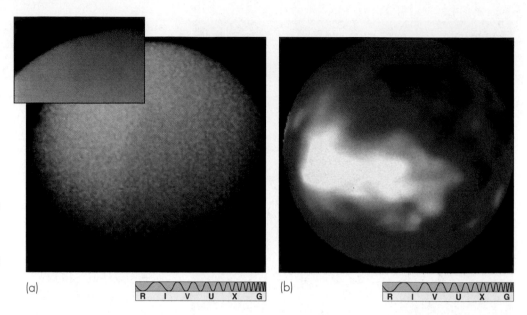

Figure 8.7 (a) Titan, larger than Mercury and roughly half the size of Earth, was photographed from only 4000 km away as the *Voyager 1* spacecraft passed by in 1980. All we can see here is Titan's upper cloud deck. The inset shows a contrast-enhanced image of the haze layer (falsely colored blue). (b) An infrared image from the *Hubble Space Telescope* shows the same surface features, the bright area being nearly 400 km across, about the size of Australia.

(a) R I V U X G   (b) R I V U X G

programmed *Voyager 1* to pass very close to Titan, even though that meant the spacecraft could not then use Saturn's gravity to continue on to Uranus and Neptune.

Unfortunately, despite *Voyager 1*'s close pass, the moon's surface remains a mystery. A thick, uniform haze layer, similar to the smog found over many cities on Earth, envelops the moon and completely obscured the spacecraft's view. Titan's atmosphere is thicker and denser even than Earth's, and it is certainly more substantial than that of any other moon. Prior to *Voyager 1*'s arrival in 1980, only methane and a few other simple hydrocarbons had been conclusively detected on Titan. Radio and infrared observations from the spacecraft showed that the atmosphere is made up mostly of nitrogen (roughly 90 percent) and possibly argon (up to 10 percent), with a few percent of methane and trace amounts of other gases.

Why does Titan have such a thick atmosphere when Ganymede and Callisto do not? The answer lies largely in Titan's temperature, which is substantially lower than that of Jupiter's moons because of Saturn's greater distance from the Sun. At this lower temperature, a moon is more likely to retain an atmosphere. Also, gases may have become trapped in the surface ice when the moon formed. As Titan's internal radioactivity warmed the moon, the ice released the trapped gases, forming a thick methane–ammonia atmosphere. Sunlight split the ammonia into hydrogen, which escaped into space, and nitrogen, which remained in

the atmosphere. The methane, which was less easily broken apart, survived intact. Together with argon outgassed from Titan's interior, these gases form the basis of the atmosphere we see today.

Triton

Neptune's large moon, Triton, is the smallest of the six large moons in the solar system, with about half the mass of the next smallest, Europa. Lying 4.5 billion km from the Sun, and possessing an icy surface that reflects much of the solar radiation reaching it, Triton has a temperature of just 37 K. *Voyager 2* found that this moon has an extremely thin nitrogen atmosphere and a solid, frozen surface probably composed primarily of water ice.

A *Voyager 2* mosaic of Triton's south polar region is shown in Figure 8.8. The moon's low temperatures produce a layer of nitrogen frost that forms, evaporates, and reforms seasonally over the polar caps. The frost is visible as the pinkish region at the bottom right of Figure 8.8. Overall, there is a marked lack of cratering on Triton, presumably indicating that surface activity has obliterated the evidence of most impacts. There are many other signs of an active past. Triton's face is scarred by large fissures similar to those seen on Ganymede, and the moon's odd cantaloupe-like terrain may indicate repeated faulting and deformation over the moon's lifetime. In addition, Triton has

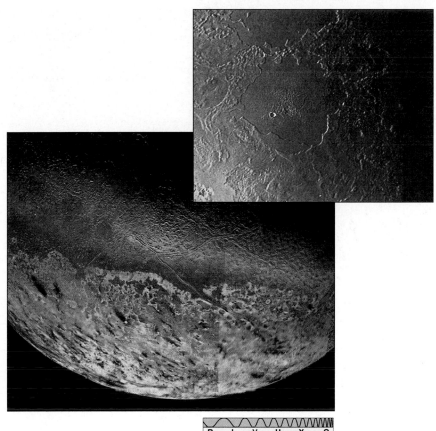

Figure 8.8 The south polar region of Triton, showing a variety of terrains ranging from deep ridges and gashes to what appear to be frozen water lakes, all indicative of past surface activity. The pinkish region at the lower right is nitrogen frost, forming the moon's polar cap. Resolution is about 4 km. Scientists think that the roughly circular lakelike feature shown in the inset may have been caused by the eruption of an ice volcano. The water "lava" has since solidified, leaving a smooth surface. The lake is about 200 km in diameter. The absence of craters indicates that the eruption was a relatively recent event. Details in this image are resolved to a remarkable 1 km.

R I V U X G

numerous frozen lakes of water ice, which are believed to be volcanic in origin.

Triton's surface activity is not just a thing of the past. As *Voyager 2* passed the moon, its cameras detected great jets of nitrogen gas erupting several kilometers into the sky. These geysers may result when liquid nitrogen below Triton's surface is heated and vaporized by some internal energy source, or perhaps even by the Sun's feeble light. Scientists conjecture that nitrogen geysers may be very common on Triton and are perhaps responsible for much of the moon's thin atmosphere.

Triton is unique among the large moons of the solar system in that its orbit around Neptune is retrograde. Also, with an orbital inclination of about 20°, Triton is the only large jovian moon not to orbit in its parent planet's equatorial plane. The event or events that placed Triton on a tilted, retrograde orbit are the subject of considerable speculation. Triton's peculiar orbit and surface features suggest to some astronomers that the moon did not form as part of the Neptune sys-

tem but instead was captured by the planet, perhaps not too long ago. Other astronomers, basing their views on Triton's chemical composition, maintain that it formed as a normal moon but was later kicked into its present orbit by some catastrophic event, such as an interaction with another, similar-sized body. The surface deformations on Triton certainly suggest fairly violent and relatively recent events in the moon's past.

8.4 The Medium-Sized Jovian Moons

Saturn

3 Table 8.2 lists the 12 medium-sized satellites of the solar system. The six that orbit Saturn are shown in Figure 8.9. In order of increasing distance from the planet, they are called Mimas, Enceladus, Tethys, Dione, Rhea, and Iapetus. All move on nearly circular paths and are tidally locked by Saturn's gravity

Table 8.2 The Medium-Sized Moons of the Solar System

Name	Parent Planet	Distance from Parent Planet (km)	Distance from Parent Planet (planet radii)	Orbit Period (Earth days)	Radius (km)	Mass (Moon = 1)	Density (kg/m³)
Mimas	Saturn	186,000	3.10	0.94	200	0.00054	1200
Enceladus	Saturn	238,000	3.97	1.37	250	0.0011	1200
Tethys	Saturn	295,000	4.92	1.89	530	0.010	1300
Dione	Saturn	377,000	6.28	2.74	560	0.015	1400
Rhea	Saturn	527,000	8.78	4.52	770	0.0034	1300
Iapetus	Saturn	3,560,000	59.3	79.3	720	0.026	1200
Miranda	Uranus	130,000	5.09	1.41	240	0.0011	1300
Ariel	Uranus	191,000	7.48	2.52	580	0.018	1600
Umbriel	Uranus	266,000	10.4	4.14	590	0.018	1400
Titania	Uranus	436,000	17.1	8.71	800	0.048	1600
Oberon	Uranus	583,000	22.8	13.5	770	0.040	1500
Proteus	Neptune	118,000	4.76	1.12	210	—	—

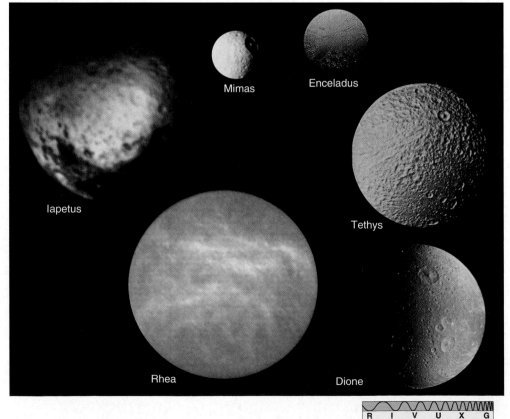

Figure 8.9 Saturn's six medium-sized satellites, drawn to scale. All are heavily cratered. Iapetus shows a clear contrast between its light-colored (icy) trailing surface, at the top and center in this image, and its dark leading hemisphere, at the bottom. The light-colored wisps covering essentially the whole central portion of this image of Rhea are thought to be water that was released from the moon's interior during some long-ago period of activity and then froze on the surface. Dione and Tethys show evidence of ancient geological activity of some sort. Enceladus appears to still be volcanically active, but the cause of the volcanism is unexplained. Mimas's main surface feature is the large crater Herschel, plainly visible in this *Voyager* image; the impact that caused this crater must have come very close to shattering the moon.

R I V U X G

into synchronous rotation (with the result that the same side of each satellite always faces the planet). They therefore have permanently *leading* and *trailing* faces as they move in their orbits, a fact that is important in understanding their often asymmetrical surface markings. Based on the densities of Saturn's medium-sized moons, scientists believe that, like Titan, all are composed largely of rock and water ice.

The outermost medium-sized moon is Iapetus. Its dark, leading face reflects only about 3 percent of the sunlight reaching it, while the icy trailing side reflects around 50 percent. Researchers are divided on the origin of this dark coloration. Spectroscopic studies suggest to some scientists that the dark material originates on or within the moon. Other scientists, however, maintain that it comes from outside and that Iapetus is simply sweeping up external dust as it orbits. For now, the answer remains unknown.

Moving inward, we come next to Rhea, the largest of Saturn's six medium-sized satellites. Its icy surface is very reflective and heavily cratered. The crater density is similar to that in our Moon's highlands, indicating that Rhea's surface is old, and there is no evidence of extensive geological activity. Rhea's only real riddle is the light-colored cloud streaks covering almost the entire face of the image shown in Figure 8.9. Called *wispy terrain*, this prominent feature appears only on Rhea's trailing side; the leading face shows no such markings and instead features only craters. Astronomers believe that the "wisps" were caused by some event in the distant past during which water was released from the interior and condensed on the surface. Any similar markings on the leading side have presumably been long since obliterated by cratering, which should be much more frequent on the satellite's forward-facing surface.

Continuing inward from Rhea, we come to Dione and Tethys. Like Rhea, they have reflective surfaces that are heavily cratered. However, each shows signs of surface activity, too. Dione's trailing face has prominent bright streaks, which are probably similar to Rhea's wispy terrain. Dione also has "maria" of sorts, where flooding appears to have obliterated the older craters. The cracks on Tethys may have been caused by cooling and shrinking of the surface layers or, more probably, by meteoritic bombardment.

Enceladus is so bright and shiny—reflecting virtually 100 percent of the sunlight falling on it—that astronomers believe its surface must be completely coated with fine crystals of pure ice, possibly the icy "ash" of water volcanoes. Although no volcanoes have actually been observed, there seems to be strong circumstantial evidence for volcanism on the satellite. Much of the moon's surface is devoid of impact craters, which seem to have been erased by what look like lava flows, except that the "lava" is water, temporarily liquefied during recent internal upheavals and now frozen again. It is not known why there should be such activity on so small a moon.

The innermost medium-sized moon, Mimas, orbits close to Saturn's rings and causes observable effects in their structure. Possibly because of its proximity to the rings, Mimas is heavily cratered. Its chief surface feature is the enormous Herschel crater shown in the Figure 8.9 image of Mimas. It is possible that the debris produced by impacts such as the one that formed Herschel is responsible for creating or maintaining the spectacular rings we see.

Uranus

The five medium-sized moons of Uranus, shown in Figure 8.10, are similar in many respects to the six moons of Saturn just discussed. Their densities suggest composition of ice and rock, and all revolve in the planet's equatorial plane, almost perpendicular to the ecliptic, in circular, tidally locked orbits. Because the satellites share Uranus's odd orientation, they also share its extreme seasons. ∞ (Sec. 7.3)

The two outermost moons, Titania and Oberon, are heavily cratered. Oberon shows little indication of any geological activity; Titania may have experienced some activity in the distant past. The overall appearance of these moons is similar to Saturn's moon Rhea, except that they lack Rhea's wispy streaks. Umbriel, the darkest of Uranus's satellites, also displays little or no evidence of past surface activity. Its only mark of distinction is a bright spot about 30 km across, of unknown origin, on its sunward side. Ariel is similar in size to Umbriel and appears to have experienced some activity in the past. It shows signs of resurfacing in places and exhibits surface cracks a little like those seen on Saturn's moon Tethys. These cracks may be

Figure 8.10 The five medium-sized moons of Uranus, to scale. In order of increasing distance from the planet, the moons are Miranda, Ariel, Umbriel, Titania, and Oberon. Note the variations in brightness among them. Earth's moon is also shown for comparison.

due to meteoritic impact, as on Tethys. Alternatively, Ariel's activity might have occurred as internal forces and external tidal stresses (due to Uranus) distorted the moon and cracked its surface.

Strangest of all Uranus's icy moons is Miranda, shown in Figure 8.11. Before the *Voyager 2* encounter, astronomers thought that Miranda would most resemble Mimas, the moon of Saturn whose size and location it most closely approximates. However, instead of being a relatively uninteresting, cratered, geologically inactive world, Miranda displays a wide range of surface terrains, including ridges, valleys, large oval faults,

and many other geological features. A close-up view of part of this strange landscape is presented in the inset in Figure 8.11. To explain why Miranda seems to combine so many types of surface feature, some researchers have hypothesized that this baffling object has been catastrophically disrupted several times (from within or without), with the pieces falling back together in a chaotic, jumbled way. Certainly, the frequency of large craters on the moons of the outer planets suggests that destructive impacts may once have been quite common. It will be a long time, though, before we can obtain more detailed information to test this theory.

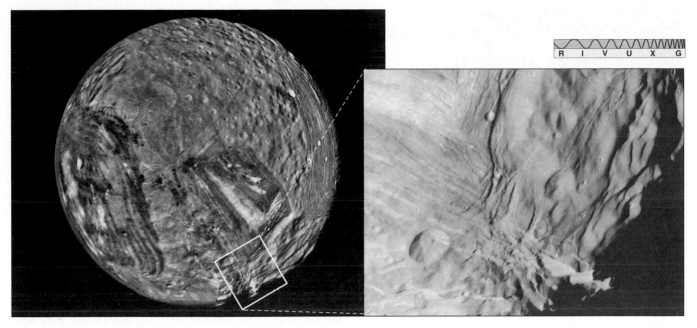

Figure 8.11 Miranda, the asteroid-sized, innermost moon of Uranus, photographed by *Voyager 2*. The fractured surface suggests a violent past, but the cause of the grooves and cracks is presently unknown. The resolution in the inset is about 2 km.

8.5 Saturn's Spectacular Rings

The View from Earth

4 Astronomers now know that all four jovian planets have rings, but Saturn's are by far the brightest, the most extensive, and the most beautiful. Galileo saw them first in 1610, but he did not recognize what he saw as a planet with a ring. At the resolution

of his small telescope, the rings looked like bumps on the planet or perhaps like a triple planet of some sort. In 1659, the Dutch astronomer Christian Huygens realized what the bumps are—a thin, flat ring completely encircling the planet.

Because Saturn's rings lie in the planet's equatorial plane, their appearance (as seen from Earth) changes seasonally, as shown in Figure 8.12. As Saturn orbits the Sun, the angle at which the rings are illumi-

Figure 8.12 Over time, Saturn's rings change their appearance to terrestrial observers as the tilted ring plane orbits the Sun. These images are separated by a few months, having been taken in true color by the *Hubble Space Telescope* in the summer and fall of 1995. (a) The ring plane is precisely edge-on as viewed from Earth; several moons are visible, including Titan's shadow (black dot). (b) The rings have begun to tilt.

Table 8.3 The Rings of Saturn

Ring	Inner Radius (km)	Inner Radius (planet radii)	Outer Radius (km)	Outer Radius (planet radii)	Width (km)
D	60,000	1.00	74,000	1.23	14,000
C	74,000	1.23	92,000	1.53	18,000
B	92,000	1.53	117,600	1.96	25,600
A	122,200	2.04	136,800	2.28	14,600
F	140,500	2.34	140,600	2.34	100
E	210,000	3.50	300,000	5.00	90,000

nated varies. When the planet's north or south pole is tipped toward the Sun, which means it is either summer or winter on Saturn, the highly reflective rings are at their brightest. In Saturn's spring and fall, the rings are close to being edge-on, both to the Sun and to observers on Earth, and so they seem to disappear. One important deduction we can make from this simple observation is that the rings are very thin. In fact, we now know that their thickness is less than a few hundred meters, even though they are more than 200,000 km in diameter.

The main ring features are marked on Figure 8.13. The dark gap about two-thirds of the way out from the inner edge is named the **Cassini Division**. Outside the Cassini Division lies the **A ring**. The inner ring, between the Cassini Division and the planet, is in reality two rings, the **B** and **C rings**. A smaller gap, known as the **Encke Division**, is found in the outer part of the A ring. Its width is 270 km. No finer ring details are visible from Earth. Of the three main rings, the B ring is brightest, followed by the somewhat fainter A ring, and then by the almost translucent C ring. The additional rings listed in Table 8.3 are not visible in Figure 8.13.

Ring Properties

A question that perplexed scientists and mathematicians for centuries was, "What are Saturn's rings?" By the middle of the nineteenth century, careful dynamical and thermodynamic arguments had conclusively proved that the rings could not be made of solid sheets, nor could they be liquid or gas! What is left? In 1857, Scottish physicist James Clerk Maxwell suggested that the rings are composed of a great number of small solid *particles*, all independently orbiting Saturn, like so many tiny moons. That inspired speculation was verified in 1895, when astronomer James Keeler measured the Doppler shift of sunlight reflected from the rings and showed that the speeds thus determined were exactly those expected for separate particles moving in circular orbits. The orbital speed decreases with increasing distance from the planet, in accordance with Kepler's laws and Newton's law of gravity.

C ring
B ring
Cassini division
A ring
Encke division

R I V U X G

Figure 8.13 Saturn as seen by the *Hubble Space Telescope*. The main ring features are marked.

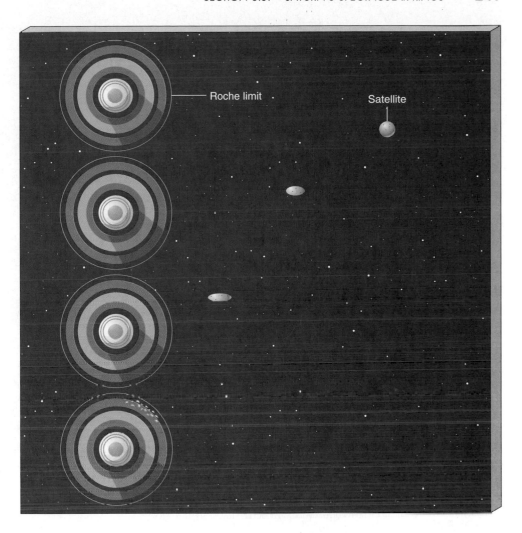

Figure 8.14   The tidal field of a planet first distorts and then destroys a moon that strays too close.

What sort of particles make up the rings? Their highly reflective nature suggests that they are made of ice, and infrared observations in the 1970s confirmed that water ice is indeed a prime ring constituent. Radar observations and later *Voyager* studies of scattered sunlight showed that the sizes of the particles range from fractions of a millimeter to tens of meters in diameter, with most particles being about the size (and composition) of a large snowball. The rings are truly thin—perhaps only a few tens of meters thick in places. Stars can occasionally be seen through them, like the beams from automobile headlights penetrating an open-weave curtain.

The Roche Limit

But why a ring of particles at all? To answer this question, consider the fate of a small moon orbiting close to a massive planet such as Saturn. The moon is held together by internal forces—either its own gravity or, for a small body, its own tensile strength (its resistance to stretching). As we bring our hypothetical moon closer to the planet, the tidal force exerted on it by the planet increases. Recall from Chapter 5 that the effect of a tidal force is to stretch the moon along the direction to the planet—that is, to create a tidal bulge. ∞ (Sec. 5.2) Recall also that the tidal force increases rapidly as distance from the planet decreases. As the moon is brought closer to the planet, it reaches a point where the tidal force becomes greater than the internal forces holding it together. At that point, the moon is torn apart by the planet's gravity, as shown in Figure 8.14. From that point on, the pieces of the satellite have their own individual orbits around that planet, eventually spreading all the way around it in the form of a ring.

For any given planet and any given moon, this critical distance inside of which the moon is destroyed

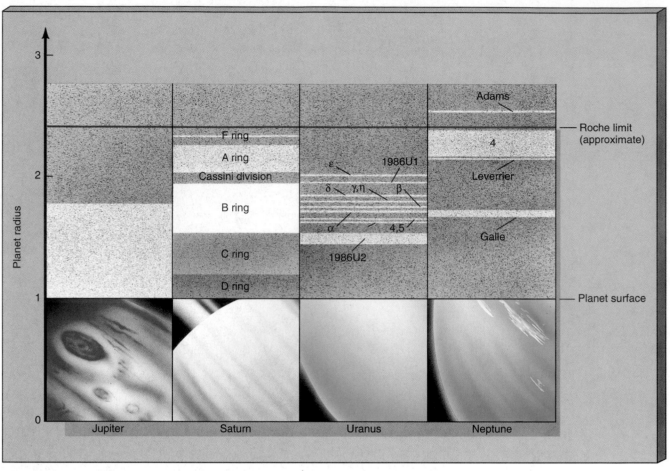

Figure 8.15 The rings of Jupiter, Saturn, Uranus, and Neptune. All distances are expressed in planetary radii. The red line represents the Roche limit. All the rings lie within (or else very close to) the Roche limit of the parent planet.

is known as the **Roche limit**, after the nineteenth-century French mathematician Edouard Roche, who first calculated it. If our hypothetical moon is held together by its own gravity and if its average density is similar to that of the parent planet (both reasonably good assumptions for Saturn's larger moons), then the Roche limit is 2.4 times the radius of the planet. Thus, for Saturn, the radius of which is about 60,000 km, no moon can survive within 144,000 km of the planet's center, which is about 7000 km beyond the outer edge of the A ring. The rings of Saturn occupy the region inside Saturn's Roche limit. These considerations apply equally well to the other jovian worlds. As indicated in Figure 8.15, all four ring systems in our solar system are found within (or very close to) the Roche limit of their parent planet.

Note that the calculation of the Roche limit applies only to moons massive enough for their own gravity to be the dominant binding force. Sufficiently small moons can survive within the Roche limit because they are mainly held together not by gravity but rather by interatomic (electromagnetic) forces.

The View from Voyager

The 1979 *Voyager* flybys of Saturn changed forever our view of this spectacular region in our cosmic backyard, revealing the rings to be vastly more complex than astronomers had previously imagined. As the *Voyager* probes approached Saturn, it was found that the main rings are actually composed of tens of thousands of narrow **ringlets** (Figure 8.16). Furthermore, this fine structure is not fixed but instead varies both with time and with position in the rings. Although the process is not fully understood, the mutual gravitational attraction of the myriad ring particles (as well as the effects of

Figure 8.16 *Voyager 2* took this false-color close-up of the ring structure just before flying through the tenuous outer rings of Saturn. Earth is superposed, to scale, for a size comparison.

R   I   V   U   X   G

Saturn's inner moons) enables waves of matter to form and move in the plane of the rings, like ripples on the surface of a pond.

Although the ringlets are probably the result of waves in the rings, the gaps are not. The narrower gaps—about 20 of them—are most likely swept clean by the action of small moonlets embedded in them. Because these moonlets are larger (perhaps 10 or 20 km in diameter) than the largest ring particles, they simply scoop up ring material as they go. Despite many careful searches of the *Voyager* images, only one of these moonlets has so far been found. Astronomers have found indirect evidence for them, in the form of "wakes" that they leave in the rings, but no other direct

sightings have occurred. Nevertheless, moonlets are still regarded as the best explanation for the small gaps.

The largest gap—the Cassini Division—has a different origin. It owes its existence to the gravitational influence of Saturn's innermost medium-sized moon, Mimas. Over time, particles orbiting within the Division have been deflected by Mimas's gravity into different orbits, causing them to collide with other ring particles, eventually finding their way into new circular orbits at other radii. The net result is that the number of ring particles in the Cassini Division is greatly reduced.

Outside the A ring lies perhaps the strangest ring of all: the faint, narrow **F ring** (Figure 8.17). It was dis-

Figure 8.17 (a) Saturn's narrow F ring appears to contain kinks and braids, making it unlike any of Saturn's other rings. Its thinness, and possibly its other peculiarities too, can be explained by the effects of two shepherd satellites that orbit the ring, one a few hundred kilometers inside the ring and the other a few hundred kilometers outside it. (b) One of the shepherding satellites, roughly 100 km in diameter, can be seen at the right edge of the frame.

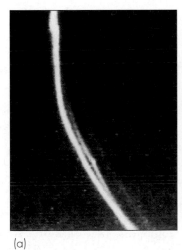

(a)

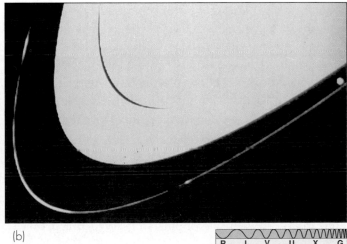

(b)

R   I   V   U   X   G

covered by *Pioneer* in 1979, but its full complexity became evident only when *Voyager* took a closer look. Its oddest feature is that it looks for all the world as though it is made up of several strands braided together! This remarkable discovery sent dynamicists scrambling in search of an explanation. It now seems as though the ring's intricate structure, as well as its thinness, arises from the influence of two small moons, known as **shepherd satellites**, that orbit about 1000 km on either side of it. Their gravitational effect gently guides any particle straying too far out of the F ring back into the fold. However, the details of how the shepherd moons produce braids and why the moons are there at all, in such similar orbits, remain unclear.

   Voyager found a faint series of rings, now known collectively as the **D ring**, inside the inner edge of the C ring, stretching down almost to Saturn's cloud tops. The D ring contains relatively few particles and is so dark that it is completely invisible from Earth. Another faint ring, also a *Voyager* discovery, lies well outside the main ring structure. Known as the **E ring**, it appears to be associated with the moon Enceladus.

8.6 The Rings of Jupiter, Uranus, and Neptune

5 One of the many remarkable findings of the 1979 *Voyager* missions was the discovery of a faint ring of matter encircling Jupiter in the plane of the planet's equator (Figure 8.18). This ring lies roughly 50,000 km above the top cloud layer, inside the orbit of the innermost moon. A thin sheet of material may extend all the way down to Jupiter's cloud tops, but most of the ring is confined within a region only a few thousand kilometers across. The outer edge of the ring is quite sharply defined. In the direction perpendicular to the equatorial plane, the ring is only a few tens of kilometers thick. The small, dark particles that make up the ring may have been chipped off by meteorite impacts on two small moons—discovered by *Voyager*—that lie very close to the ring.

   The ring system surrounding Uranus was discovered in 1977, when astronomers observed the planet

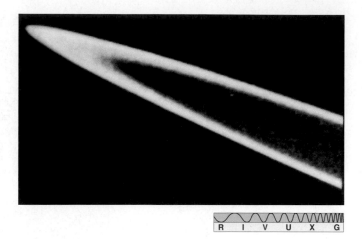

Figure 8.18 Jupiter's faint ring as photographed by *Voyager* 2. Made up of dark fragments of rock and dust possibly chipped off the innermost moons by meteorites, the ring was unknown before the two *Voyager* spacecraft arrived at the planet.

R I V U X G

passing in front of a bright star, momentarily dimming the star's light. Such a **stellar occultation** (Figure 8.19) happens a few times per decade and allows astronomers to measure planctary structures that are too small and faint to be detected directly. The 1977 observation was aimed at studying Uranus's atmosphere by watching how it absorbed starlight. However, 40 minutes before and after the planet occulted (hid) the star, the flickering starlight revealed the presence of a set of rings. The discovery was particularly exciting because, at the time, only Saturn was known to have a ring system.

   The ground-based observations revealed 9 thin rings encircling Uranus. Two more were discovered by *Voyager* 2. The locations of all 11 rings are indicated in Figure 8.15. The 9 rings known from Earth-based observation are shown in Figure 8.20, an image which makes it readily apparent that Uranus's rings are quite different from those of Saturn. While Saturn's rings are bright and wide with relatively narrow, dark gaps between them, those of Uranus are dark, narrow, and widely spaced. All but two are less than 10 km wide, and the spacing between them ranges from a few hundred to about a thousand kilometers. However, like Saturn's rings, the rings of Uranus are all less than a few tens of meters thick (that is, measured in the direction perpendicular to the ring plane).

   Like the F ring of Saturn, Uranus's narrow rings require shepherd satellites to keep them from diffusing

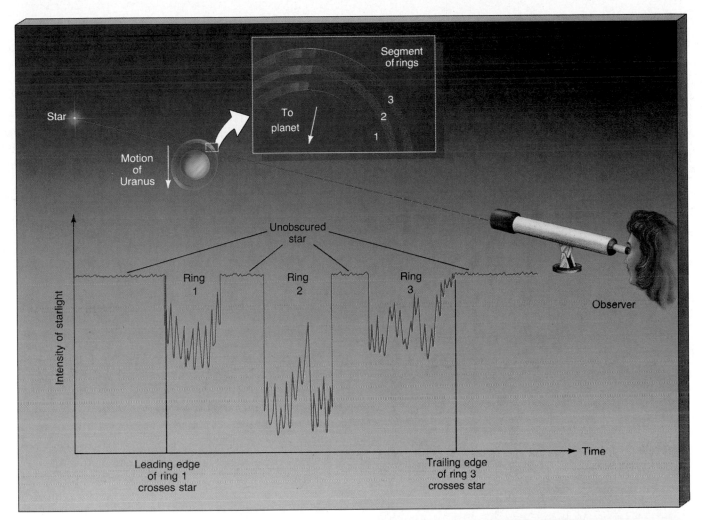

Figure 8.19   Occultation of starlight allows astronomers to detect fine detail on a distant body passing in front of a star. The rings of Uranus were discovered using this technique.

away. In fact, the theory of shepherd satellites was first worked out to explain the rings of Uranus before *Voyager 2*'s Saturn encounter. During its 1986 flyby, *Voyager 2* detected the shepherds of the Epsilon ring (Figure 8.21). Many other, undetected shepherd satellites must also exist.

As shown in Figure 8.22, Neptune is surrounded by four dark rings. Three are quite narrow, like the rings of Uranus, and one is quite broad and diffuse, more like Jupiter's ring. The outermost ring is noticeably clumped in places. From Earth we see not a complete ring, but only partial arcs; the unseen parts of the ring are simply too thin (unclumped) to be detected. The connection between the rings and the planet's small inner satellites has not yet been firmly estab-

R   I   V   U   X   G

Figure 8.20   The main rings of Uranus, as imaged by *Voyager 2*. All nine of the rings known before *Voyager 2*'s arrival can be seen in this photo. Resolution is about 10 km, which is just about the width of most of these rings. The two rings discovered by *Voyager 2* are too faint to be seen here.

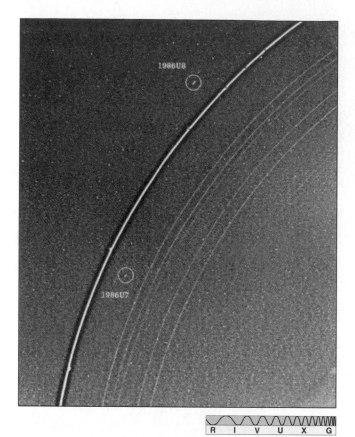

Figure 8.21 These two small moons, named Cordelia and Ophelia, were discovered by *Voyager 2* in 1986. They shepherd Uranus's Epsilon ring, keeping it from diffusing away.

lished, although many astronomers now believe the clumping is caused by shepherd satellites.

8.7 The Formation of Planetary Rings

The dynamic behavior observed in the rings of Saturn and those of the other planets implies to many researchers that the rings are quite young—perhaps no more than 50 million years old, or 100 times younger than the solar system. If the rings are indeed so young, then either they are replenished from time to time, perhaps by fragments of moons chipped off by meteoritic impact, or they are the result of a relatively recent, possibly catastrophic event in the planet's system—a moon torn apart by tidal forces, or destroyed in an impact with a large comet or even with another moon. Astronomers normally prefer not to invoke catastrophic events to explain observed phenomena, but the more we learn of the universe, the more we realize that catastrophe probably plays an important role in its development.

Astronomers estimate that the total mass of Saturn's rings is enough to make a satellite about

Figure 8.22 Neptune's faint rings. In this long-exposure image, the planet (center) is heavily overexposed and has been artificially blotted out to make the rings easier to see. One of the two fainter rings lies between the inner bright ring and the planet. The other lies between the two bright rings.

250 km in diameter. If such a satellite strayed inside Saturn's Roche limit or was destroyed (perhaps by a collision) near that radius, a ring system could have resulted. While we do not know if Saturn's rings originated in this way, this is the likely fate of Neptune's large moon Triton. Because of its retrograde orbit, the tidal bulge that Triton raises on Neptune tends to make the moon spiral *toward* the planet rather than away from it (as our Moon moves away from Earth). ∞ (Sec. 5.2) Triton is thus doomed to be torn apart by Neptune's tidal gravitational field, probably in no more than 100 million years or so, the time required for the moon's inward spiral to bring it within Neptune's Roche limit. By that time, it is quite possible that Saturn's ring system will have disappeared, and so Neptune will then be the planet in the solar system with spectacular rings!

Is there some standard way in which rings form around a planet? And is there a standard manner in which ring systems evolve? Or do the processes of ring formation and evolution depend entirely on the particular planet in question? If, as now appears to be the case, ring systems are relatively short-lived, their formation must be a fairly common event. Otherwise, we would not expect to find rings around all four jovian planets at once. However, there are also many indications that the individual planetary environment plays an important role in determining a ring system's appearance and longevity. Although many aspects of ring formation and evolution are now understood, no comprehensive theory yet exists.

8.8 Pluto and Its Moon

The Discovery of Pluto

6 In Chapter 7 ∞ (Sec. 7.2) we saw how irregularities in Uranus's orbit led to the discovery of Neptune in the mid-nineteenth century. By the end of that century, observations of the orbits of Uranus and Neptune suggested that Neptune's influence was not sufficient to account for all of the observed irregularities in Uranus's motion. Furthermore, it seemed that Neptune itself was being affected by some other unknown body, and astronomers hoped to pinpoint the location of this new planet using tech-

niques similar to those that had guided them to the discovery of Neptune.

One of the most ardent searchers was Percival Lowell, a capable and persistent observer and one of the best-known astronomers of his day. Basing his investigation primarily on the motion of Uranus (Neptune's orbit was still relatively poorly determined at the time), Lowell set about calculating where the supposed ninth planet should be. He searched for it, without success, during the decade preceding his death in 1916. In 1930, the American astronomer Clyde Tombaugh, working with improved equipment and photographic techniques at the Lowell Observatory, finally succeeded in finding Lowell's ninth planet, only 6° away from Lowell's predicted position. The new planet was named Pluto for the Roman god of the dead who presides over eternal darkness (and also because its first two letters and its astrological symbol ♇ are Lowell's initials). Its discovery was announced on March 13, 1930, Lowell's birthday.

Unfortunately, it now appears that the supposed irregularities in Neptune's orbit and the extra irregularities in the motion of Uranus do not really exist. Furthermore, the mass of Pluto, measured accurately only in the 1980s, is far too small to have caused them anyway. In the end, the discovery of Pluto owed much more to dumb luck than to elegant mathematics!

At nearly 40 A.U. from the Sun, Pluto is often hard to distinguish from the background stars. As Figure 8.23 shows, the planet appears considerably fainter than many stars. Like Neptune, it is never visible to the naked eye. Unlike the other outer planets, there is no present or proposed space mission that will suddenly and radically improve our knowledge of this distant world. Pluto is the only planet in the solar system not studied at close range by unmanned spacecraft, and there is little prospect of such a visit in the foreseeable future. Every new discovery about this remote world is the result of painstaking observations either from Earth or from Earth-orbiting instruments.

Figure 8.24 is a composite map of Pluto's surface obtained by combining many *Hubble Space Telescope* images of the planet. It shows surface detail at about the same level as can be seen on Mars with a small telescope. Other than the bright polar caps, none of surface features has been conclusively identified. Some of the dark regions may be craters or impact basins, as on Earth's Moon.

Figure 8.23 These two photographs, taken one night apart, show the motion of Pluto (arrow) projected against a field of much more distant stars. Most of Pluto's apparent motion in these two frames is due to Earth's orbital motion.

Charon

Pluto is so far away that little is known of its physical nature. Until the late 1970s, studies of variations in its reflected light suggested a rotation period of nearly a week, but measurements of its mass and diameter were very uncertain. Estimates of Pluto's mass ranged as high as a few tenths the mass of Earth.

All this changed in July 1978, when astronomers at the U.S. Naval Observatory discovered that Pluto has a satellite. It is now named Charon, after the mythical boatman who ferried the dead across the river Styx into Hades, Pluto's domain. The discovery photograph of Charon is shown in Figure 8.25(a). Charon is the large white bump at the top right of the image. In 1990, the *Hubble Space Telescope* imaged the Pluto–Charon system (Figure 8.25b), the improved resolution of that instrument clearly separating the two bodies. Based on studies of Charon's orbit, astronomers have measured the mass of Pluto to great accuracy. It is 0.0025 Earth masses (1.5×10^{22} kg), a value that is more like the mass of a moon than of a planet.

Before Charon was discovered, Pluto's radius was also poorly known. The planet's angular size as seen from Earth is much less than 1″, so its true diameter is blurred by the effects of Earth's turbulent atmosphere.

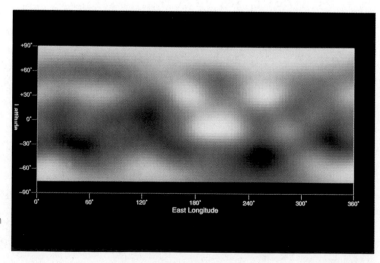

Figure 8.24 A surface map of Pluto—not a photograph but rather a modeled view created by carefully combining 24 *Hubble* images.

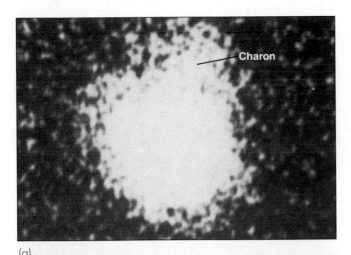

Figure 8.25 (a) The discovery photograph of Pluto's moon Charon. The moon is the small bump on the top right portion of the image. (b) The Pluto–Charon system, as seen by the *Hubble Space Telescope*. The angular separation of the planet and its moon is about 0.9 arc second.

Charon's discovery gave astronomers new insight into the system. By pure chance, Charon's orbit over the six-year period from 1985 to 1991 produced for Earth viewers a series of eclipses as Pluto and Charon repeatedly passed in front of one other, as seen from our vantage point. With more good fortune, these eclipses took place while Pluto was closest to the Sun, making for the best possible Earth-based observations.

Basing their calculations on the variations in light as Pluto and Charon periodically hid each other, astronomers computed the masses and radii of both

bodies and determined their orbit plane. By studying sunlight reflected from Pluto's surface, astronomers also found that the two objects are tidally locked as they orbit each other. Pluto's diameter is 2300 km, about one-fifth that of Earth. Charon is about 1300 km across and orbits 19,700 km from Pluto. It has an orbital period of 6.4 days, the same as the rotation period of each body. If planet and moon have the same composition (a reasonable assumption), Charon's mass must be about one-sixth that of Pluto, giving the Pluto–Charon system by far the largest satellite-to-planet mass ratio in the solar system. As shown in Figure 8.26, Charon's orbit and the spin axes of both planet and moon are inclined at an angle of 118° to the plane of the ecliptic. Thus, Pluto is the third planet in the solar system found to have retrograde rotation (Venus and Uranus being the other two).

Pluto's Origin

Pluto's average density of 2000 kg/m^3 is too low for a terrestrial planet but far too high for a jovian one. Instead, the mass, radius, and density of Pluto are just what we expect for one of the icy moons of a jovian planet. In fact, Pluto is quite similar in mass and radius to Neptune's large moon, Triton. This similarity has fueled much speculation, both about Pluto's origin and about the strange state of Neptune's moon system.

Because Pluto is neither terrestrial nor jovian in its makeup, and because of its similarity to the icy moons of the outer planets, some researchers suspect that it is not a true planet at all. Instead, it may be an escaped planetary moon or a large icy chunk of debris left over from the formation of the solar system. This idea is bolstered by Pluto's eccentric, inclined orbit, which is quite unlike the orbits of the other known planets. Given Triton's unusual retrograde path around Neptune and the fact that Pluto's eccentric orbit overlaps that of Neptune, it once seemed conceivable that Pluto was once a moon of Neptune and that Pluto and Triton were involved in a relatively recent encounter that resulted in the ejection of Pluto and the reversal of Triton's motion. Since 1978, however, the explanation of Pluto's origin has been greatly complicated by the presence of Charon. It was much easier to suppose that Pluto was an escaped moon before we learned that it has a moon of its own.

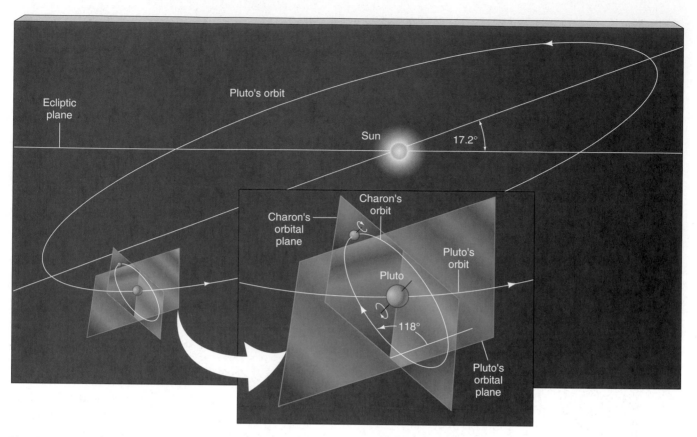

Figure 8.26 Charon's path around Pluto is circular, synchronous, and inclined at 118° to the orbit plane of the Pluto–Charon system about the Sun. The Pluto–Charon orbit plane is itself inclined at 17.2° to the plane of the ecliptic.

Pluto may be just what it seems—a planet that formed in its current orbit, possibly even with its own moon right from the outset. Because we know so little about the environment in the outer solar system, we cannot rule out the possibility that planets beyond Neptune simply look like Pluto. There is good evidence for large chunks of ice circulating in interplanetary space beyond the orbits of the giant planets (in the Kuiper Belt, and many researchers believe that there may have been thousands of Pluto-sized objects initially present in the outer solar system. ∞ (Sec. 4.2)

The capture of a few of these objects by the jovian planets would explain the strange moons of the outer worlds, especially Triton. And if there were enough moon-sized chunks originally orbiting beyond Neptune, it is quite plausible that Pluto could have captured Charon following a collision (or near-miss) between the two. At present, our scant knowledge of the compositions of the two bodies does not allow us to confirm or disprove either the "co-formation" or the "capture" theory of how the Pluto–Charon system originated.

Chapter Review

Summary

There are six large moons in the outer solar system, each comparable to or larger than Earth's Moon in size and mass. In addition, there are 12 medium-sized moons and many smaller ones.

The four Galilean moons of Jupiter have densities that decrease with increasing distance from the planet. The innermost, Io, has active volcanoes powered by the constant flexing of the moon by Jupiter's tidal forces. Europa has a cracked, icy surface that may possibly conceal an ocean of liquid water. Ganymede and Callisto have ancient, heavily cratered surfaces. Ganymede shows some evidence of past geological activity, but Callisto shows none.

Saturn's large moon Titan has a thick atmosphere that obscures the moon's surface and may be the site of complex cloud and surface chemistry. Neptune's moon Triton has a fractured surface of water ice and a thin atmosphere of nitrogen, probably produced by nitrogen "geysers" on its surface. Triton is the only large moon in the solar system to have a retrograde orbit around its parent planet. This orbit is unstable and will eventually cause Triton to be torn apart by Neptune's gravity.

The medium-sized moons of Saturn and Uranus are made up predominantly of rock and water ice. Many of them are heavily cratered and in some cases must have come close to being destroyed by the meteoritic impacts whose craters we now see.

From Earth, the main visible features of Saturn's rings are the **A, B,** and **C rings** and the **Cassini** and **Encke Divisions** (p. 240). The Cassini Division is a dark region between the A and B rings. The Encke Division lies near the outer edge of the A ring. The rings are made up of trillions of individual particles, ranging in size from dust grains to boulders. Their total mass is comparable to that of a small moon. Both divisions are dark because they are almost empty of ring particles.

The **Roche limit** (p. 242) of a planet is the distance within which the planet's tidal field overwhelms the internal gravity of an orbiting moon, tearing the moon apart and forming a ring. All the rings of the four jovian planets lie inside (or very close to) their parent planets' Roche limits.

When the *Pioneer* and *Voyager* probes reached Saturn, they found that the rings are made up of tens of thousands of narrow **ringlets** (p. 242). Interactions between the ring particles and the planet's inner moons are responsible for much of the fine structure observed.

Saturn's narrow **F ring** (p. 243), discovered by the *Pioneer* probe, lies just outside the A ring. It has a kinked, braided structure, apparently caused by two small **shepherd satellites** (p. 244) that orbit close to it and prevent it from breaking up. The *Voyager* probes also discovered the faint **D ring** (p. 244), lying between the C ring and Saturn's cloud layer, and the **E ring** (p. 244), apparently associated with the moon Enceladus.

Jupiter has a faint, dark ring extending down to the planet's cloud tops. Uranus has a series of dark, narrow rings first detected from Earth by **stellar occultation** (p. 244), which occurs when a body passing between Earth and some distant star obscures the light received from that star. Shepherd satellites keep Uranus's rings from breaking apart. Neptune has three narrow rings like Uranus's and one broad ring, like Jupiter's.

Many scientists believe that planetary rings have a lifetime of only a few tens of millions of years. The fact that we see rings around all four jovian planets means that the rings must constantly be re-formed or replenished, perhaps by material chipped off planetary moons by meteoritic impact or perhaps by the tidal destruction of entire moons.

Pluto was discovered in 1930 after a laborious search for a planet that was supposedly affecting Uranus's orbital motion. We now know that Pluto is far too small to have had any detectable influence on Uranus's path. Pluto has a moon, Charon, the mass of which is about one-sixth that of Pluto. Studies of Charon's orbit around Pluto have allowed the masses and radii of both bodies to be accurately determined. Pluto's properties are far more moonlike than planetlike. The origin of the Pluto–Charon system is unknown.

Self-Test: True or False?

_____ **1.** The densities of the Galilean moons increase with increasing distance from Jupiter.

_____ **2.** Io has a noticeable lack of impact craters on its surface.

_____ **3.** The surface of Europa is completely covered by water ice.

_____ **4.** Ganymede shows evidence of ancient plate tectonics.

_____ **5.** Most of Saturn's medium-sized moons rotate synchronously with their orbits around Saturn.

_____ 6. Saturn is unique among the planets in having a ring system.

_____ 7. A typical particle in Saturn's rings is more than 100 m in diameter.

_____ 8. Although Saturn's ring system is tens of thousands of kilometers wide, it is only a few tens of meters thick.

_____ 9. Saturn has small and medium-sized moons but no large ones.

_____ 10. Water ice predominates in Saturn's moons.

_____ 11. Titan's atmosphere is denser than Earth's.

_____ 12. Titan's surface is obscured by thick cloud layers.

_____ 13. The orbits of Uranus's moons share the 98° tilt of the planet's rotation axis.

_____ 14. Uranus has no large moons.

_____ 15. The surfaces of the largest Uranian moons are darker than similar moons of Saturn.

_____ 16. Triton's surface has a marked lack of cratering, indicating significant amounts of surface activity.

_____ 17. Pluto's moon, Charon, is small compared with Pluto.

_____ 18. Pluto is larger than Earth's Moon.

Self-Test: Fill in the Blank

1. The Galilean moon _____ is larger than Mercury.

2. In contrast to the inner two Galilean moons, the outer two have compositions that include significant amounts of _____.

3. Io is the only moon in the solar system having active _____.

4. As viewed from Earth, Saturn's ring system is conventionally divided into _____ broad rings.

5. The Cassini Division lies between the _____ and _____ rings.

6. Saturn's rings exist because they lie within Saturn's _____.

7. Two small moons, known as _____ satellites, are responsible for the unusually complex form of the F ring.

8. The composition of Titan's atmosphere is 90 percent _____.

9. There may be large amounts of liquid _____ under the frozen surface of Europa.

10. The Uranian moon that shows the greatest amount of geological activity and disruption over time is _____.

11. Triton's orbit is unusual because it is _____.

12. Neptune's moon Triton has a thin atmosphere made of _____.

13. Nitrogen geysers are found on _____.

14. Overall, Pluto is most similar to which object in the solar system? _____.

Review and Discussion

1. In what sense are Jupiter and its moons like a miniature solar system?

2. How does the density of the Galilean moons vary with increasing distance from Jupiter? Is there a trend to this variation? If there is such a trend, explain the reason for it.

3. What is the source of Io's volcanic activity?

4. How does the amount of cratering vary among the Galilean moons? Does it depend on their location? If so, how?

5. Why is there speculation that the Galilean moon Europa might be an abode for life?

6. Seen from Earth, Saturn's rings sometimes appear broad and brilliant but at other times seem to disappear. Why?

7. What is the Roche limit, and what is its relevance to planetary rings?

8. Why do many astronomers think Saturn's rings formed quite recently?

9. When _Voyager 1_ passed Saturn in 1980, why

didn't it see the surface of Saturn's largest moon, Titan?

10. What effect does Mimas have on the rings of Saturn?

11. Compare and contrast Titan with Jupiter's Galilean moons.

12. Do you think it would be worthwhile to send a spacecraft to explore Titan? Why or why not?

13. Describe the behavior of shepherd satellites.

14. What is unique about Miranda? Give a possible explanation for this uniqueness.

15. How does Neptune's moon system differ from those of the other jovian worlds? What do these differences suggest about the origin of Neptune's moon system?

16. What is the predicted fate of Triton?

17. How do the rings of Neptune differ from those of Uranus and Saturn?

18. How was Pluto discovered?

19. In what respect is Pluto more like a moon than a jovian or terrestrial planet?

20. How were the masses and radii of Pluto and Charon determined?

Problems

1. Io orbits Jupiter once every 42 hours at a distance of six planetary radii. If Jupiter completes one rotation every 10 hours and a satellite is to orbit the planet synchronously, at what distance from the planet must that satellite be located? (Use Kepler's third law.) ∞ (Sec. 1.7)

2. What is the orbital speed, in kilometers per second, of ring particles in the inner part of the B ring? Compare with the speed of a satellite in low Earth orbit. Why are these speeds so different from each other?

3. Show that Titan's surface gravity is about one-seventh that of Earth.

4. How close is Charon to Pluto's Roche limit?

Projects

1. Use binoculars to look at Jupiter. Be sure to hold them steady (try propping your arms up on the hood of a car or sitting down and bracing them against your knees). Can you see any of Jupiter's four largest moons? If you come back the following evening, the moons' relative positions will have changed. Have some changed more than others? Before observing, look up the positions of the Galilean moons in a current magazine such as *Astronomy* or *Sky & Telescope*. Identify each moon. Watch Io over a period of at least an hour. Can you see its motion? Do the same for Europa.

2. Binoculars may not reveal the rings of Saturn, but most small telescopes will. Use a telescope to look at Saturn. Does Saturn appear to be spherical, or is it slightly flattened at its poles? Examine the rings. Are they tilted toward Earth, or do you seem to be looking at them edge-on? Can you see a dark line in the rings? This is the Cassini Division. It once was thought to be a gap in the rings, but the *Voyager* spacecraft discovered that it is filled with tiny ringlets. Can you see the shadow of the rings on the surface of Saturn?

3. While looking at Saturn through a telescope, can you see any of its moons? The moons line up with the rings; Titan is often the farthest out, and always the brightest. How many moons can you see? Use an almanac to identify each one you find.

4. Pluto does not look like a circular disk, even through a powerful telescope. It always looks like a star. To see Pluto at all, you must have at least an 8-inch telescope. Then you must locate the field of stars in which it currently resides (check the January issue of *Astronomy* for this year's chart, or *Sky & Telescope* for the months when the planets are visible). Draw a picture of all the stars you see in that field of view. Come back a few nights later, and draw a picture of the field again. The "star" that has moved is Pluto!

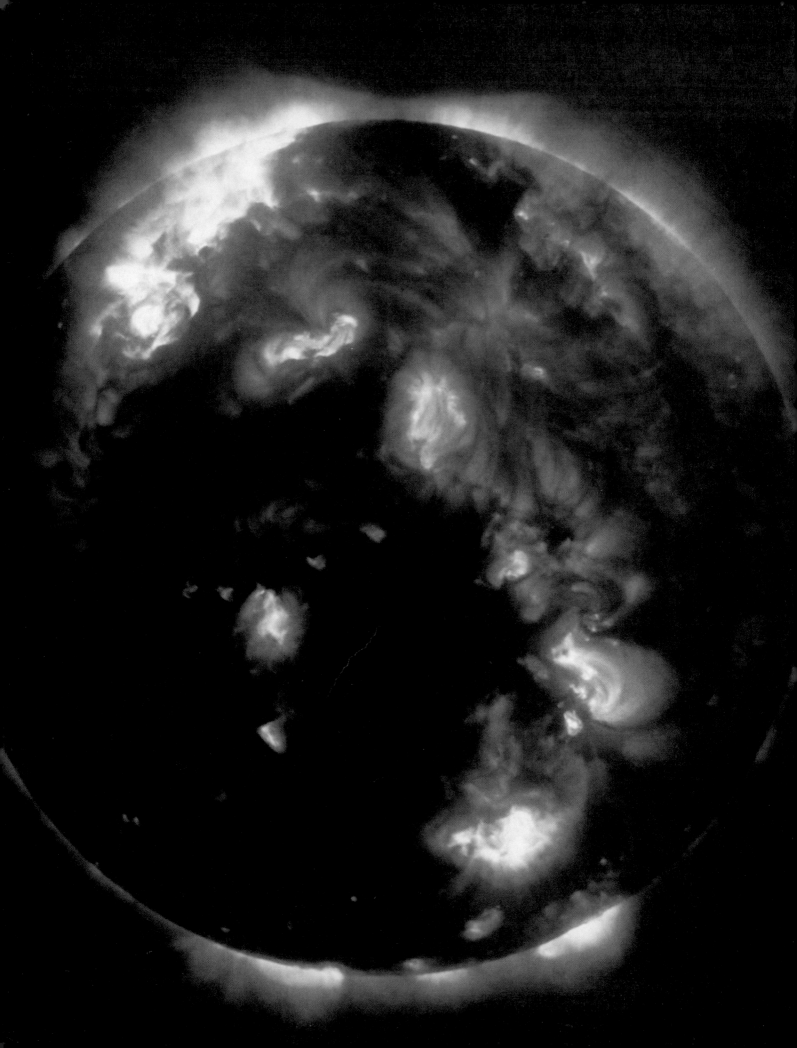

The Sun

Our Parent Star

(Opposite page) This spectacular image of the Sun was made by capturing X rays emitted by our star's most active regions. It was taken by a camera on a rocket lofted shortly before the total solar eclipse of July 1991. (Note the shadow of the Moon approaching from the west, at top.) The brightest regions have temperatures of about 3 million kelvins.

LEARNING GOALS

Studying this chapter will enable you to:

1 Summarize the overall properties of the Sun.

2 Explain how energy travels from the solar core, through the interior, and out into space.

3 Name the Sun's outer layers and describe what those layers tell us about the Sun's surface composition and temperature.

4 Discuss the nature of the Sun's magnetic field and its relationship to the various types of solar activity.

5 Outline the process by which energy is produced in the Sun's interior.

6 Explain how observations of the Sun's core challenge our present understanding of solar physics.

Living in the solar system, we have the chance to study, at close range, perhaps the most common type of cosmic object—a star. Our Sun is a star, and a fairly average star at that, but it has one unique feature: it is very close to us—300,000 times closer than our next nearest neighbor, Alpha Centauri. Whereas Alpha Centauri is 4.3 light years distant, the Sun is only 8 light minutes away from us. Consequently, we know far more about the Sun than about any of the other distant points of light in the universe. A sizable fraction of all our astronomical knowledge is based on modern studies of the Sun. Just as we studied our parent planet, Earth, to set the stage for our exploration of the solar system, we now study our parent star, the Sun, as the next step in our exploration of the universe.

9.1 The Sun in Bulk

1 The Sun is the sole source of light and heat for the maintenance of life on Earth. It is a **star**, a glowing ball of gas held together by its own gravity and powered by nuclear fusion at its center. In its physical and chemical properties, the Sun is very similar to most other stars, regardless of when and where they formed. Indeed, our Sun appears to lie right in the middle of the observed ranges of stellar mass, radius, brightness, and composition. Far from detracting from our interest in the Sun, this very mediocrity is one of the main reasons we study it—we can apply our knowledge of solar phenomena to so many other stars in the universe.

Physical Properties and Overall Structure

Table 9.1 lists some basic solar data. The Sun's radius is determined by measuring its angular size (0.5°) and then employing elementary geometry, ∞ (Sec. P.3) its mass by application of Newton's laws of motion and gravity to the observed orbits of the planets. ∞ (Sec. 1.7) The solar rotation period is found by timing sunspots ∞ (Sec. 1.5) and other surface features as they cross the Sun. Surface temperature is measured by applying the radiation laws ∞ (Sec. 2.4) to the observed solar spectrum. We will return to the final quantity listed in Table 9.1—the Sun's *luminosity*—in a moment. Having a radius of more than 100 Earth radii, a mass of more than 300,000 Earth masses, and a surface temperature well above the melting point of any known material, the Sun is clearly a body very different from any other we have encountered so far.

The Sun has a surface of sorts—not a solid surface (the Sun contains no solid material), but rather that part of the brilliant gas ball we perceive with our eyes or view through a heavily filtered telescope. This "surface"—the part of the Sun that emits the radiation we see—is called the **photosphere**. Its radius (listed as the radius of the Sun in Table 9.1) is about 700,000 km. However, its thickness is probably no more than 500 km—less than 0.1 percent of the radius—which is why we perceive the Sun as having a well-defined, sharp edge (Figure 9.1).

The main regions of the Sun are illustrated in Figure 9.2. Just above the photosphere is the Sun's lower atmosphere, called the **chromosphere**, about 1500 km thick. From 1500 km to 10,000 km above the top of the photosphere lies a region called the **transition zone**, where the temperature rises dramatically. Above 10,000 km, and stretching far beyond, is a tenuous (thin), hot upper atmosphere, the solar **corona**. At still greater distances, the corona turns into the solar wind, ∞ (Sec. 4.2) which flows away from the Sun and permeates the entire solar system.

Extending down some 200,000 km below the photosphere is the **convection zone**, a region where the material of the Sun is in constant convective motion. Below the convection zone lies the **radiation zone**, where solar energy is transported toward the sur-

Table 9.1 Some Solar Properties	
Radius	6.96×10^8 m
Mass	1.99×10^{30} kg
Average density	1410 kg/m³
Rotation period	24.9 days (equator); 29.8 days (poles)
Surface temperature	5780 K
Luminosity	3.86×10^{26} W

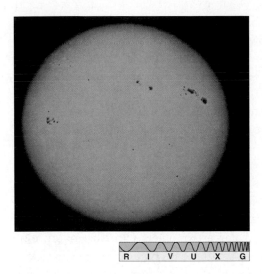

Figure 9.1 This photograph of the Sun, taken through a heavily filtered telescope, shows a sharp edge, although the Sun, like all other stars, is made of a gradually thinning gas. The edge appears sharp only because the solar photosphere is so thin.

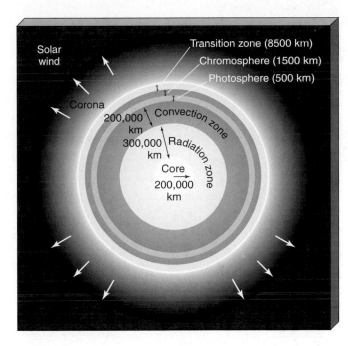

Figure 9.2 The main regions of the Sun, not drawn to scale, with some physical dimensions labeled.

face by radiation rather than by convection. The term *solar interior* is often used to mean both the radiation and convection zones. The central **core**, roughly 200,000 km in radius, is the site of powerful nuclear reactions that generate the Sun's enormous energy output.

Luminosity

All but one of the properties listed in Table 9.1 are familiar from our studies of the planets. However, the Sun has an additional property, perhaps the most important of all from the point of view of life on Earth—it radiates a great deal of energy into space, uniformly (we assume) in all directions.

By holding a light-sensitive device—a solar cell, perhaps—perpendicular to the Sun's rays, we can measure how much solar energy is received per square meter of surface area every second. Imagine our detector as having a surface area of 1 m² and as being placed at the top of Earth's atmosphere. The amount of solar energy reaching this surface each second is a quantity known as the **solar constant**. Its value is approximately 1400 watts per square meter (W/m²)[1]. Most of

this solar energy reaches Earth's surface. Thus, for example, a sunbather's body having a total surface area of about 0.5 m² receives solar energy at a rate of nearly 700 watts, roughly equivalent to the output of a typical electric room heater or about ten 75-W light bulbs.

Let us now ask about the *total* amount of energy radiated in all directions from the Sun, not just the small fraction intercepted by our detector or by Earth. Imagine a three-dimensional sphere that is centered on the Sun and is just large enough that its surface intersects Earth's center (Figure 9.3). The sphere's radius is 1 A.U., and its surface area is therefore $4\pi \times (1 \text{ A.U.})^2$, or approximately 2.8×10^{23} m². Multiplying the rate at which solar energy falls on each square meter of the sphere (that is, the solar constant) by the total surface area of our imaginary sphere, we can determine the total rate at which energy leaves the Sun's surface. This quantity is known as the **luminosity** of the Sun. It turns out to be just under 4×10^{26} W.

The Sun is an enormously powerful source of energy. *Every second*, it produces an amount of energy equivalent to the detonation of about 100 billion 1-megaton nuclear bombs. Put another way, the solar luminosity is equivalent to 4 trillion trillion 100-W light bulbs shining simultaneously—about 10^{19} dollars'

[1]The SI unit of energy is the *joule* (J). Probably more familiar to most readers is the closely related unit *watt* (W), which measures *power*, defined as the rate at which energy is emitted or expended by an object. One watt is equal to one joule per second; for instance, a light bulb with a power rating of 100 W radiates 100 J of energy each second.

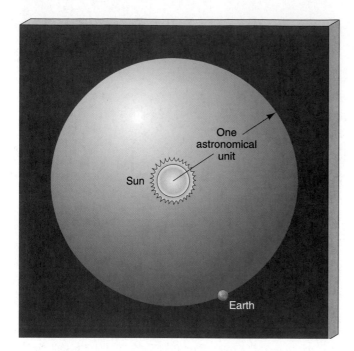

Figure 9.3 We can draw an imaginary sphere around the Sun so that the sphere's surface passes through Earth's center. The radius of this imaginary sphere is 1 A.U. By multiplying the sphere's surface area by the solar constant, we can measure the Sun's luminosity, the amount of energy it emits each second.

worth of energy radiated every second (at 1997 U.S. rates)!

9.2 The Solar Interior

Modeling the Structure of the Sun

Lacking any direct measurements of the Sun's interior, astronomers must use indirect means to probe the inner workings of our parent star. To accomplish this, they construct mathematical models of the Sun, combining all available observations with theoretical insight into solar physics. The result is the **Standard Solar Model**, which has gained widespread acceptance among astronomers.

To test and refine the Standard Solar Model, researchers are eager to obtain information about the solar interior. An important technique for probing the Sun beneath the photosphere emerged in the 1960s, when it was discovered that the surface of the Sun vibrates like a complex set of bells. These vibrations, illustrated in Figure 9.4, are the result of internal pressure ("sound") waves that reflect off the photosphere and repeatedly cross the solar interior. Because these waves can penetrate deep inside the Sun, analysis of

their surface patterns allows scientists to study conditions far below the Sun's surface. This process is similar to the way in which seismologists study Earth's interior by observing seismic waves produced by earthquakes. ∞ (Sec. 5.5) For this reason, study of solar surface patterns is usually called **helioseismology**, even though solar pressure waves have nothing whatever to do with solar seismic activity (which doesn't exist).

The most extensive study of solar oscillations (vibrations) is the ongoing GONG (short for Global Oscillations Network Group) project. By making continuous observations of the Sun from many clear sites around Earth, solar astronomers can obtain uninterrupted high-quality solar data spanning many days and even weeks—almost as though Earth were not rotating and the Sun never set. Analysis of these data provides important additional information about the temperature, density, rotation, and convective state of the solar interior, allowing detailed comparisons between theory and reality to be made.

Figure 9.5 shows the solar density and temperature according to the Standard Solar Model, plotted as functions of distance from the Sun's center. Notice how the density drops sharply at first, then more slowly

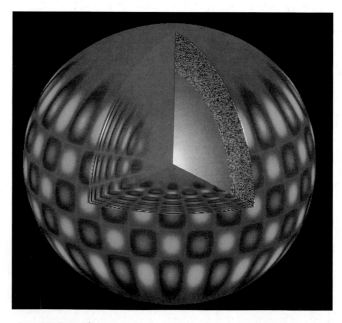

Figure 9.4 The Sun vibrates in a very complex way as sound waves of many frequencies move through its interior. By observing the motion of the solar surface, scientists can determine the wavelengths and frequencies of the individual waves and thus deduce information about the solar interior not obtainable by other means. The alternating patches of color in this rendering represent gas moving inward (red) and outward (blue).

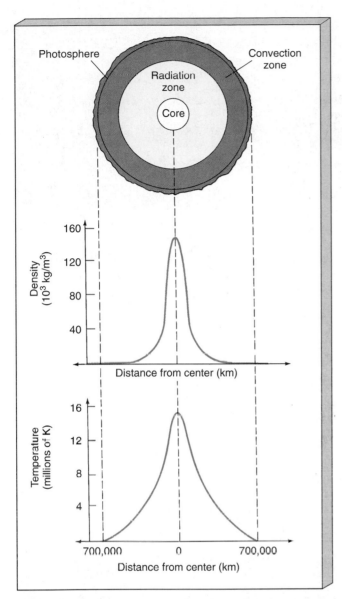

Figure 9.5 Theoretically modeled profiles of density and temperature in the Sun's interior.

best vacuum that physicists can create in laboratories on Earth.

As shown in Figure 9.5, the solar temperature also decreases with increasing radius but not as rapidly as the density. Computer models indicate a central temperature of about 15,000,000 K, consistent with the minimum 10,000,000 K needed to initiate the nuclear reactions known to power all stars. The temperature decreases steadily, reaching the observed value of 5800 K at the photosphere.

Energy Transport

2 The very hot solar interior ensures violent and frequent collisions among gas particles. Particles move in all directions with high velocities, bumping into one another unceasingly. In and near the core, the extremely high temperatures guarantee that the gas is completely ionized. Recall from Chapter 2 that under less extreme conditions, atoms absorb photons that can boost the atoms' electrons to more excited states. ∞ (Sec. 2.6) Because there are no electrons left on the completely ionized atoms to capture the photons, however, the deep solar interior is quite transparent to radiation. Only occasionally does a photon scatter off a free electron. As a result, the energy produced by nuclear reactions in the core travels outward toward the surface in the form of radiation with relative ease.

As we move outward from the core, the temperature falls, atoms collide less frequently and less violently, and more electrons manage to remain bound to their parent nuclei. With more and more atoms retaining electrons that can absorb the outgoing radiation, the gas in the interior changes from being relatively transparent to being almost totally opaque. By the outer edge of the radiation zone, *all* of the photons produced in the Sun's core have been absorbed. Not one of them reaches the surface. But what happens to the energy they carry?

The photons' energy must travel beyond the Sun's interior. If it did not, the Sun would have exploded long ago. That we see sunlight—visible energy—proves that some energy escapes. That energy reaches the solar surface by *convection*—the same basic physical process we saw in our study of Earth's atmosphere, although it operates in a very different environment in the Sun. ∞ (Sec. 5.3) Convection can

near the photosphere. The variation in density is large, ranging from a core value of about 150,000 kg/m^3, 20 times the density of iron, to an intermediate value (at 350,000 km) of about 1000 kg/m^3, the density of water, to an extremely small photospheric value of 2 × 10^{-4} kg/m^3, about 10,000 times less dense than air at Earth's surface. The *average* density of the Sun (Table 9.1) is 1400 kg/m^3. Because the density is so high in the core, we calculate that roughly 90 percent of the Sun's mass is contained within the inner half of its radius. The solar density continues to decrease out beyond the photosphere, reaching values as low as 10^{-23} kg/m^3 in the far corona—about as thin as the

Figure 9.6 Transport of energy in the Sun's convection zone. We can visualize this region as a boiling, seething sea of gas. Each convective cell at the top of the convection zone is about 1000 km across. The cells are arranged in tiers, with cells of progressively smaller size at increasing distance from the center. (This is a highly simplified diagram; there are many different cell sizes, and they are not so neatly arranged.)

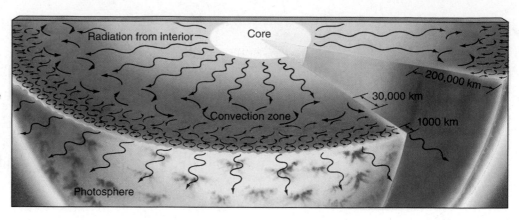

occur whenever cooler material overlies warmer material, and this is just what happens in the Sun's convection zone. Hot solar gas physically moves outward, while cooler gas above it sinks, creating a characteristic pattern of convection cells. All through the convection zone, energy is transported to the surface by physical motion of the solar gas. (Remember that there is no physical movement of material when radiation is the energy-transport mechanism; convection and radiation are *fundamentally different* ways in which energy can be transported from one place to another.) Figure 9.6 is a diagram of solar convection cells, where columns of hot gas rise, cool, and descend.

In reality, the convection zone is much more complex than we have just described. There is a hierarchy of convection cells, organized in tiers at different depths. The deepest tier, about 200,000 km below the photosphere, is thought to contain cells tens of thousands of kilometers in diameter. Energy is carried upward through a series of progressively smaller cells, stacked one upon another until, at a depth of about 1000 km below the photosphere, the individual cells are about 1000 km across. The top of this uppermost tier of convection is the visible solar surface, where astronomers can directly observe the cell sizes. Information about convection at deeper levels is inferred mostly from computer models of the solar interior.

At some distance from the core, the solar gas becomes too thin to sustain further upwelling by convection. This distance roughly coincides with the photospheric surface we see. Convection does not proceed into the solar atmosphere. There is simply not enough gas there. The density is so low that the gas becomes transparent and radiation once again becomes the mechanism of energy transport. Photons reaching the

photosphere escape more or less freely into space, and the photosphere emits thermal radiation like any other hot object.

Granulation

Figure 9.7 is a high-resolution photograph of the solar surface taken from the *Skylab* space station. The visible surface is highly mottled with regions of bright and dark gas known as *granules*. This **granulation** of the solar surface is a direct reflection of motion in the convection zone. Each bright granule measures about 1000 km across—comparable in size to a continent on Earth—and has a lifetime of between 5 and 10 min-

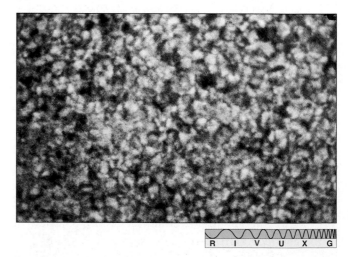

Figure 9.7 *Skylab* photograph of the granulated solar photosphere. Typical solar granules are comparable in size to Earth's continents. The bright portions of the image are regions where hot material is upwelling from below, as shown in Figure 9.6. The dark regions correspond to cooler gas that is sinking back down into the interior.

utes. Together, several million granules constitute the top layer of the convection zone, immediately below the photosphere.

Each granule forms the topmost part of a solar convection cell. Spectroscopic observation of the photosphere within and around the bright regions shows direct evidence for the upward motion of gas as it "boils" up from within, proving that convection really does occur at or just below the photosphere. Spectral lines detected from the bright granules appear slightly bluer than normal, indicating Doppler-shifted matter coming toward us with a speed of about 1 km/s. ∞ (*More Precisely 2-2*) Conversely, spectroscopes focused on the dark granules show the same spectral lines to be redshifted, indicating matter moving away from us.

The brightness variations of the granules result from differences in temperature. The upwelling gas is hotter and therefore, by Stefan's law, emits more radiation than the cooler downwelling gas. ∞ (Sec. 2.4) The adjacent bright and dark gases appear to contrast considerably, but in reality their temperature difference is less than about 500 K.

Careful measurements also reveal a much larger-scale flow beneath the solar surface. **Supergranulation** is a flow pattern quite similar to granulation except that supergranulation cells measure some 30,000 km across. As with granulation, material upwells at the center of the cells, flows across the surface, then sinks down again at the edges. Scientists believe that supergranules are the imprint on the photosphere of the deepest tier of large convective cells depicted in Figure 9.6.

9.3 The Solar Atmosphere

Composition

3 Astronomers can glean an enormous amount of information about the Sun from analysis of absorption lines ∞ (Sec. 2.5) that arise in the photosphere and chromosphere. Figure 9.8 (see also Figure 2.15) is a detailed visible spectrum of the Sun, spanning the range of wavelengths from 360 to 690 nm. Notice the intricate, dark Fraunhofer absorption lines superposed on the background continuous spectrum.

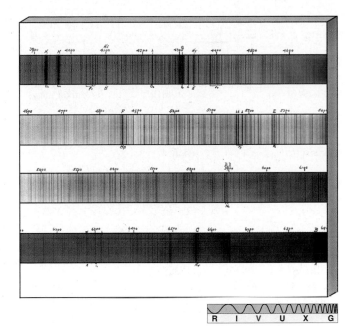

Figure 9.8 A spectrum of our Sun in the visible domain shows thousands of dark absorption lines, indicating the presence, in the lower solar atmosphere, of 67 elements in various stages of excitation and ionization.

Tens of thousands of spectral lines have been observed and cataloged in the solar spectrum, although there are not nearly this many elements in the Sun. The reason for the large number of lines is that most elements are present in many different states of excitation and ionization. ∞ (Sec. 2.6) They therefore absorb photons having many different energies, even in the relatively narrowly defined visible range, giving rise to the multiplicity of lines we see. The more complex the element, the more lines it can produce. For example, hundreds of lines are attributed to just the element iron. In all, 67 elements have been identified in the solar spectrum. More elements probably exist there, but they are present in such small quantities that our instruments are not sensitive enough to detect them.

Table 9.2 lists the 10 most common elements in the Sun, as determined from spectroscopic observations. Notice that hydrogen is by far the most abundant element, followed by helium. This distribution is just what we saw on the jovian planets, and it is what we will find for the universe as a whole. Strictly speaking, analysis of spectral lines allows us to draw conclusions only about the part of the Sun where the lines form—the photosphere and chromosphere. However,

Table 9.2 The Composition of the Sun

Element	Abundance (percentage of total number of atoms)	Abundance (percentage of total mass)
Hydrogen	91.2	71.0
Helium	8.7	27.1
Oxygen	0.078	0.97
Carbon	0.043	0.40
Nitrogen	0.0088	0.096
Silicon	0.0045	0.099
Magnesium	0.0038	0.076
Neon	0.0035	0.058
Iron	0.0030	0.14
Sulfur	0.0015	0.040

most astronomers believe that, with the exception of the solar core (where nuclear reactions are steadily changing the composition—see Section 9.5), the data in Table 9.2 are representative of the entire Sun.

The Chromosphere

The lower part of the Sun's atmosphere, the chromosphere, emits very little light of its own and cannot be observed visually under normal conditions. The photosphere is just too bright, dominating the chromosphere's radiation. The relative dimness of the chromosphere results from its low density—large numbers of photons simply cannot be emitted by a tenuous gas containing very few atoms per unit volume. Still, although it is not normally seen, astronomers have long been aware of the chromosphere's existence. Figure 9.9 shows the Sun during an eclipse in which the photosphere is obscured by the Moon but the chromosphere is not. The chromosphere's characteristic reddish hue is plainly visible. This coloration is due to the red emission line of hydrogen, which dominates the chromospheric spectrum. ∞ (Sec. 2.6)

The chromosphere is far from tranquil. Every few minutes, small solar storms erupt there, expelling jets of hot matter known as *spicules* into the Sun's upper atmosphere (Figure 9.10). These long, thin spikes of matter leave the Sun's surface at typical speeds of about 100 km/s, reaching several thousand kilometers above the photosphere. Spicules are not spread evenly across the solar surface. Instead, they cover only about 1 percent of the total area, tending to accumulate around the edges of supergranules. The Sun's magnetic field is also known to be somewhat stronger than average in those regions. Scientists speculate that the downwelling material there tends to strengthen the solar magnetic field and that spicules are the result of magnetic disturbances in the Sun's churning outer layers.

The Transition Zone and Corona

During the brief moments of a solar eclipse, if the Moon's angular size is large enough that both the photosphere and the chromosphere are blocked, the ghostly solar corona can be seen, as in Figure 9.11. With the photospheric light removed, the pattern of spectral lines changes dramatically. The intensities of the usual lines change, the spectrum shifts from absorption to emission, and an entirely new set of spectral lines suddenly appears.

These new lines arise because atoms in the corona have lost several more electrons than atoms in

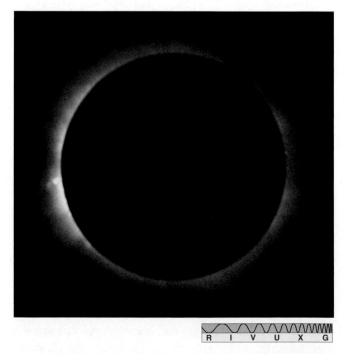

R I V U X G

Figure 9.9 This photograph of a total solar eclipse shows the solar chromosphere, a few thousand kilometers above the Sun's surface.

Figure 9.10 Solar spicules, which are short-lived, narrow jets of gas that typically last mere minutes, can be seen sprouting up from the chromosphere in this image of the Sun. The spicules are the thin, dark, spikelike regions. They appear dark against the face of the Sun because they are cooler than the photosphere.

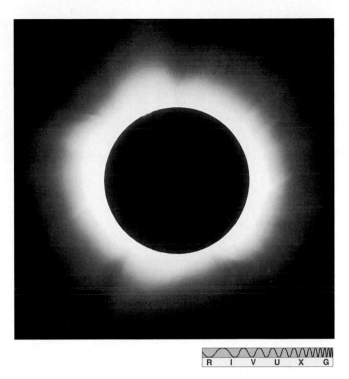

Figure 9.11 When both the photosphere and the chromosphere are obscured by the Moon during a solar eclipse, the faint solar corona becomes visible. This photograph shows clearly the emission of radiation from the corona.

the photosphere—that is, the coronal atoms are much more highly ionized. Therefore, their internal electronic structure, and hence their spectra, are quite different from the structure and spectra of atoms in the photosphere. For example, astronomers have identified coronal lines corresponding to iron atoms having as many as 13 of their normal 26 electrons missing. (In the photosphere, most iron atoms have lost only 1 or 2 electrons.) The cause of this extensive electron stripping is the high coronal temperature. The degree of ionization inferred from spectra observed during solar eclipses tell us that the gas temperature of the upper chromosphere exceeds that of the photosphere. The temperature of the solar corona, where even more ionization is seen, is higher still.

Figure 9.12 shows how the temperature of the Sun's atmosphere varies with altitude. The temperature decreases to a minimum of about 4500 K some 500 km above the photosphere, after which it rises steadily. About 1500 km above the photosphere, in the transition zone, the gas temperature begins to rise rapidly,

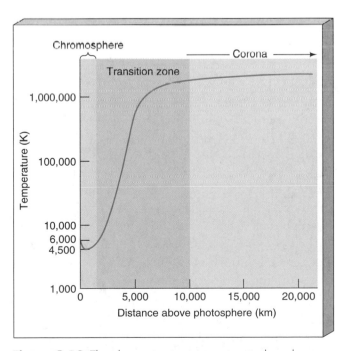

Figure 9.12 The change in gas temperature in the solar atmosphere is dramatic. The temperature reaches a minimum of 4500 K in the chromosphere and then rises sharply in the transition zone, finally leveling off at more than 1,000,000 K in the corona.

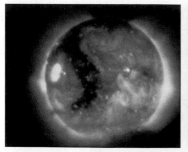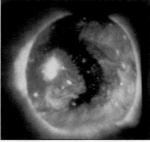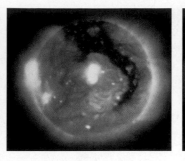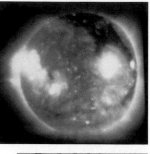

Figure 9.13 Images of X-ray emission from the Sun observed by the *Skylab* space station. These frames were taken at one-day intervals. Note the dark, boot-shaped coronal hole traveling from left to right, where the X-ray observations outline in dramatic detail the abnormally thin regions through which the high-speed solar wind streams forth.

R I V U X G

reaching more than 1,000,000 K at an altitude of 10,000 km. Thereafter, in the corona, the temperature remains roughly constant.

The cause of the rapid temperature rise in the transition zone is not fully understood. The temperature profile there runs contrary to intuition—moving away from a heat source, we would normally expect the temperature to diminish, but this is not the case for Sun's upper atmosphere. The transition zone and corona must have another energy source. Astronomers now believe that magnetic disturbances in the photosphere—a little like spicules, but on a much larger scale—are ultimately responsible for heating the transition zone and corona.

The Solar Wind

Electromagnetic radiation and fast-moving particles—mostly protons and electrons—escape from the Sun all the time. The radiation moves away from the photosphere at the speed of light, taking 8 minutes to reach Earth. The particles travel more slowly, although at the still-considerable speed of about 500 km/s, reaching Earth in a few days. This constant stream of escaping solar particles is the solar wind.

The solar wind results from the high temperature of the corona. About 10,000,000 km above the photosphere, the coronal gas is hot enough to escape the Sun's gravity, and it begins to flow outward into space. At the same time, the solar atmosphere is continuously replenished from below. If that were not the case, the corona would disappear in about a day. The Sun is, in effect, "evaporating"—constantly shedding mass through the solar wind. The wind is an extremely thin medium, however. Although it carries away about

a million tons of solar matter each second, less than 0.1 percent of the Sun has been lost since the solar system formed 4.6 billion years ago.

The Sun in X Rays

What sort of radiation is emitted by a gas having a temperature of 1,000,000 K? Unlike the 5800 K photosphere, which emits most strongly in the visible part of the electromagnetic spectrum, the hotter coronal gas radiates at much higher frequencies—primarily in X rays. ∞ (Sec. 2.4) For this reason, X-ray telescopes have become important tools in studying the solar corona. Figure 9.13 shows X-ray images of the Sun. The full corona extends well beyond the regions shown, but the density of coronal particles emitting the radiation diminishes rapidly with distance from the Sun. The intensity of X-ray radiation farther out is too dim to be seen here.

In the mid-1970s, instruments aboard NASA's *Skylab* space station revealed that the solar wind escapes mostly through solar "windows" called **coronal holes**. The dark area moving from left to right in Figure 9.13 represents a coronal hole. Not really holes, such structures are simply deficient in matter—vast regions of the Sun's atmosphere where the density is about 10 times lower than it is in the rest of the corona. In coronal holes, the solar magnetic field lines extend from the photosphere far out into interplanetary space. Charged particles tend to follow the field lines, so they can escape. In other parts of the corona, the solar magnetic field lines stay close to the Sun. Because the lines stay close to the Sun, keeping charged particles near the surface and inhibiting the outward flow of the solar wind, the density remains (relatively) high.

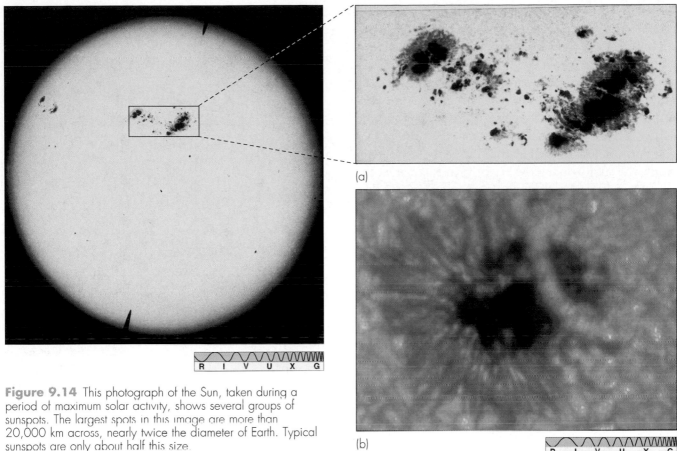

(a)

(b)

R I V U X G

Figure 9.14 This photograph of the Sun, taken during a period of maximum solar activity, shows several groups of sunspots. The largest spots in this image are more than 20,000 km across, nearly twice the diameter of Earth. Typical sunspots are only about half this size.

Figure 9.15 (a) The largest pair of sunspots in Figure 9.14. Each spot consists of a cool, dark inner region called the umbra surrounded by a warmer, less dark region called the penumbra. The spots appear dark because they are slightly cooler than the surrounding photosphere. (b) A high-resolution, true-color image of a single sunspot shows details of its structure as well as much surface granularity surrounding it. This spot is about the size of Earth.

The largest coronal holes can be hundreds of thousands of kilometers across. Structures of this size are seen only a few times each decade. Smaller holes—perhaps only a few tens of thousand kilometers in size—are much more common, appearing every few hours.

9.4 The Active Sun

4 Most of the Sun's luminosity results from continuous emission from the photosphere. This radiation arises from what we call the **quiet Sun**, the underlying predictable star that blazes forth day after day. This steady behavior contrasts with the sporadic, unpredictable radiation of the **active Sun**, a much more irregular component of our star's energy output characterized by explosive, unpredictable behavior. This aspect of solar radiation contributes little to the Sun's total luminosity and has little effect on the evolution of the Sun as a star, but it does affect us directly here on Earth. The size and duration of coronal holes are strongly influenced by the level of solar activity, as is the strength of the solar wind.

Sunspots

Figure 9.14 is an optical photograph of the entire Sun, showing numerous dark blemishes on the surface. First studied in detail by Galileo, ∞ (Sec. 1.5) these "spots" provided one of the first clues that the Sun is not a perfect, unvarying creation but rather a place of constant change. These dark areas are called **sunspots**. They typically measure about 10,000 km across, about the size of Earth. At any given instant, the Sun may have hundreds of sunspots, or it may have none at all.

Studies of sunspots show an *umbra*, or dark center, surrounded by a grayish *penumbra*. The close-up view of a pair of sunspots in Figure 9.15(a) shows both of these dark areas against the bright background of the undisturbed photosphere. This gradation in darkness indicates a gradual change in photospheric temperature. In other words, sunspots are simply relatively *cooler* regions of the photospheric gas. The temperature of the umbra is about 4500 K, that of the penumbra 5500 K. The spots, then, are certainly composed of hot gases. They seem dark only because they appear against an even brighter background (the 5800 K photosphere). If we could magically remove a sunspot from the Sun (or just block out the rest of the Sun's emission), the spot would glow brightly, just like any other hot object having a temperature of roughly 5000 K.

Sunspots are not steady. Most change their size and shape, and all come and go. Individual spots may last anywhere from 1 to 100 days. A large group of spots typically lasts 50 days.

Solar Magnetism

What causes a sunspot? Why is it cooler than the surrounding photosphere? The answers to these questions involve the Sun's magnetism. As we saw in Chapter 2, analysis of spectral lines can yield information about the magnetic field where the lines originate. ∞ (Sec. 2.7) The magnetic field in a typical sunspot is about 1000 times greater than the field in the surrounding, undisturbed photosphere (which is itself several times stronger than Earth's magnetic field). Scientists believe that sunspots are cooler than their surroundings because these abnormally strong fields tend to block (or redirect) the convective flow of hot gas, which is normally toward the surface of the Sun.

Another indicator of the magnetic nature of sunspots is their grouping. They almost always come in pairs, and the magnetic fields observed in the two members of any pair are always opposite one another; the members of the pair are said to have opposite *magnetic polarities*. As illustrated in Figure 9.16, magnetic field lines emerge from the interior through one member of a sunspot pair, loop through the solar atmosphere, then reenter the photosphere through the other member of the pair. What's more, *all* the sunspot pairs in the same solar hemisphere (north or south) at any instant have the same magnetic configuration. If the magnetic field lines are directed into the Sun in one leading spot (measured in the direction of the Sun's rotation), they are inwardly directed in all leading spots in that hemisphere. In the other hemisphere at the same time, all sunspot pairs have the *opposite*

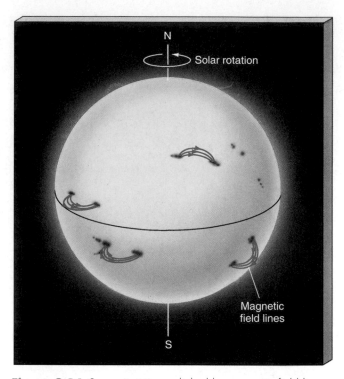

Figure 9.16 Sunspot pairs are linked by magnetic field lines. The Sun's magnetic field lines emerge from the surface through one member of a pair and reenter the Sun through the other member. The leading members of all sunspot pairs in the solar northern hemisphere have the same polarity. If the magnetic field lines are directed into the Sun in one leading spot, they are inwardly directed in all other leading spots in that hemisphere. The same is true in the southern hemisphere, except that the polarities are always opposite those in the north.

polarity. Despite the irregular shape of sunspots, these correlations suggest a high degree of order in the solar magnetic field.

Observations of sunspots indicate that the Sun does not rotate as a solid body. Instead, it spins *differentially*—faster at the equator and more slowly at the poles, like Jupiter and Saturn. The solar photosphere rotates once every 25 days at the equator, but only once in 30 days at the poles. The combination of differential rotation and convection radically affects the character of the Sun's magnetism.

As illustrated in Figure 9.17, the Sun's differential rotation distorts the solar magnetic field, "wrapping" it around the solar equator and eventually causing any originally north–south magnetic field to reorient itself in an east–west direction. Convection then causes the magnetized gas to upwell toward the surface, twisting and tangling the magnetic field pattern. In some places, the field lines becomes kinked like a knot in a garden hose, causing the field strength

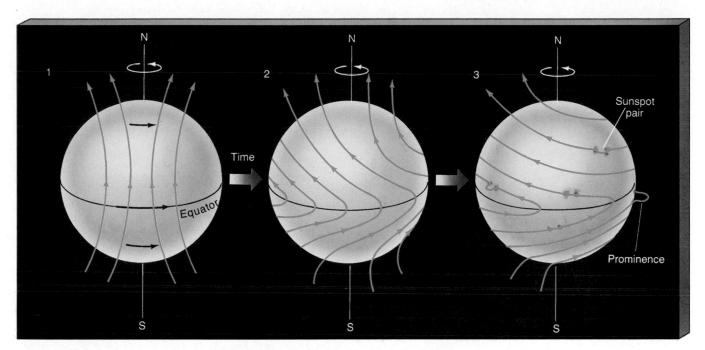

Figure 9.17 The Sun's differential rotation wraps and distorts the solar magnetic field. Occasionally, the field lines burst out of the surface and loop through the lower atmosphere, thereby creating a sunspot pair. The underlying pattern of the solar field lines explains the observed pattern of sunspot polarities. (If the loop happens to occur near the edge of the Sun and is seen against the blackness of space, we see a phenomenon called a prominence, described shortly.)

to increase. Occasionally, the field becomes so strong that it overwhelms the Sun's gravity, and a "tube" of field lines bursts out of the surface and loops through the lower atmosphere, forming a sunspot pair. The general east–west organization of the underlying solar field accounts for the observed polarities of the sunspot pairs in each hemisphere.

The Solar Cycle

Not only do sunspots come and go with time, but their numbers and distribution across the face of the Sun also change in a fairly regular fashion. Centuries of observations have established a clear **sunspot cycle**. Figure 9.18(a) shows the number of sunspots observed each year during the twentieth century. The average number of spots reaches a maximum every 11 or so years and then falls off almost to zero before the cycle begins afresh. The latitudes at which sunspots appear vary as the sunspot cycle progresses. Individual sunspots do not move up or down in latitude, but new spots appear closer to the equator as older ones at higher latitudes fade away over the course of the 11-

year cycle. Figure 9.18(b) is a plot of observed sunspot latitudes as a function of time.

In fact, the 11-year sunspot cycle is only half of a 22-year **solar cycle**. For the first 11 years of the solar cycle, the leading spots of all sunspot pairs in the same solar hemisphere have one polarity, and spots in the other hemisphere have the opposite polarity (Figure 9.16). These polarities then reverse their signs for the next 11 years.

Astronomers believe that the Sun's magnetic field is both generated and amplified by the constant stretching, twisting, and folding of magnetic field lines that results from the combined effects of differential rotation and convection. The theory is essentially the same "dynamo" theory that accounts for the magnetic fields of Earth and the jovian planets, except that the solar dynamo operates much faster and on a much larger scale. One important prediction of this theory is that the Sun's magnetic field should rise to a maximum, then fall to zero and reverse itself in a more or less periodic way, just as observed. Solar surface activity, such as the sunspot cycle, simply follows the changes in the underlying magnetic field.

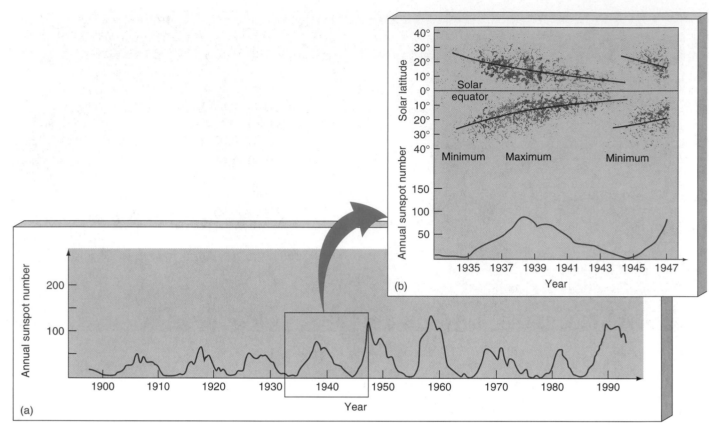

Figure 9.18 (a) Annual number of sunspots throughout the twentieth century, showing 5-year averages of annual data to make long-term trends more evident. The (roughly) 11-year solar cycle is clearly visible. At the time of minimum solar activity, hardly any sunspots are seen. About 4 years later, at the time of maximum solar activity, as many as 200 spots are observed per year. (b) Sunspots cluster at high latitudes when solar activity is at a minimum. They appear at lower and lower latitudes as the number of sunspots peaks. As the next minimum is approached, the spots get closer and closer to the equator. By the time the minimum is reached, there are no spots at the equator, we see a new flareup of spots at high latitudes, and the cycle repeats. The most recent sunspot maximum occurred in 1990.

Figure 9.19 plots all sunspot data recorded since the invention of the telescope. As can be seen, the 11-year "periodicity" of the solar sunspot cycle is far from regular. Not only does the period vary from 7 to 15 years, but the sunspot cycle has disappeared entirely in the relatively recent past. The lengthy period of solar inactivity that extended from 1645 to 1715 is called the *Maunder minimum*, after the British astronomer who drew attention to these historical records. The corona was apparently also less prominent during total solar eclipses occurring during that period, and Earth aurorae were sparse throughout the late seventeenth century. Lacking a complete understanding of the solar cycle, we cannot easily explain how it could shut down entirely. Most astronomers suspect changes in the Sun's convection zone and/or its rotation pattern, but the specific cause of the Sun's century-long variations remains a mystery.

Active Regions

Sunspots are relatively gentle aspects of solar activity. However, the photosphere surrounding them occasionally erupts violently, spewing forth into the corona large quantities of energetic particles. The sites of these explosive events are known simply as **active regions**. Most pairs or groups of sunspots have active regions associated with them. Like all other aspects of solar activity, these phenomena tend to follow the solar cycle and are most frequent and violent around the time of sunspot maximum.

Figure 9.20 shows two solar **prominences**—loops or sheets of glowing gas ejected from an active region on the solar surface. Prominences move through the inner parts of the corona under the influence of the Sun's magnetic field. Magnetic instabilities in the strong fields found in and near sunspot groups may

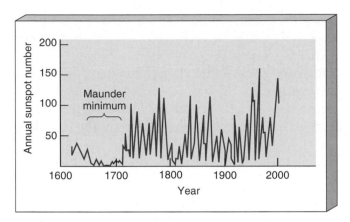

Figure 9.19 Number of sunspots occurring each year. Note the approximate 11-year "periodicity" and the absence of spots during the late seventeenth century.

cause the prominences, although the details are still not completely understood. A typical solar prominence measures some 100,000 km in extent, nearly 10 times the diameter of planet Earth. Some may persist for days or even weeks. Prominences as large as that shown in Figure 9.20(b) (which traversed almost half a million kilometers of the solar surface) are less common and usually appear only at times of greatest solar activity.

Flares are another type of solar activity observed near active regions (Figure 9.21). Also the result of

magnetic instabilities, flares are even more violent (and even less well understood) than prominences. They flash across a region of the Sun in minutes, releasing enormous amounts of energy as they go. Satellite observations have demonstrated that X-ray and ultraviolet emissions are especially intense in the extremely compact hearts of flares, where temperatures can reach 100,000,000 K. So energetic are these cataclysmic explosions that some researchers have likened flares to bombs exploding in the lower regions of the Sun's atmosphere.

A major flare can release as much energy as the largest prominences, but it does so in a matter of minutes or hours rather than days or weeks. Unlike the gas that makes up the characteristic loop of a prominence, the particles produced by a flare are so energetic that the Sun's magnetic field is unable to hold them and shepherd them back to the surface. Instead, the particles are simply blasted into space by the violence of the explosion.

The Changing Solar Corona

The solar corona also varies in step with the sunspot cycle. The photograph of the corona in Figure 9.11

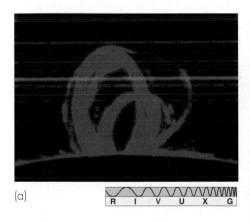

(a)

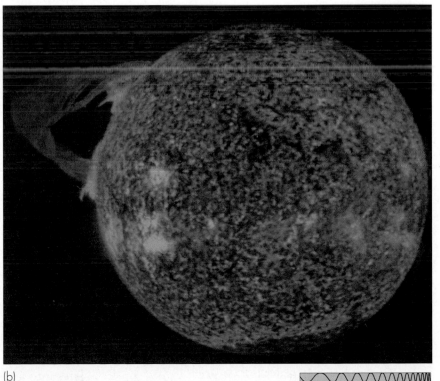

(b)

Figure 9.20 (a) The looplike structure of this prominence clearly reveals the magnetic field lines connecting the two members of a sunspot pair. (b) This image of a particularly large solar prominence was observed by ultraviolet detectors aboard the *Skylab* space station in 1979.

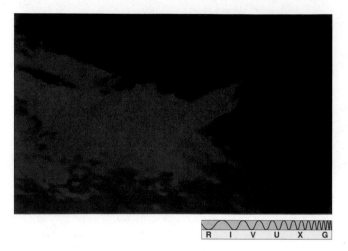

R I V U X G

Figure 9.21 Much more violent than a prominence, a solar flare is an explosion on the Sun's surface that sweeps across an active region in a matter of minutes, accelerating solar material to high speeds and blasting it into space.

shows the quiet Sun at sunspot minimum. The corona is fairly regular in appearance and surrounds the Sun more or less uniformly. Compare this image with Figure 9.22, which was taken in 1991, close to the most recent peak of the sunspot cycle. The active corona here is much more irregular than the one in Figure 9.11 and extends farther from the solar surface. The streamers of coronal material pointing away from the Sun are characteristic of this phase.

Astronomers think that the corona is heated primarily by solar surface activity, particularly promi-

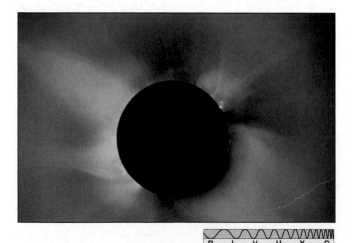

R I V U X G

Figure 9.22 Photograph of the solar corona during the eclipse that occurred in July 1991, near the peak of the sunspot cycle. At these times, the corona is much less regular and much more extended than at sunspot minimum (compare Figure 9.11). Astronomers believe that coronal heating is caused by surface activity on the Sun. The changing shape and size of the corona are the direct result of variations in prominence and flare activity over the course of the solar cycle.

nences and flares, which can inject large amounts of energy into the corona, greatly distorting its shape. Extensive disturbances often move through the corona above an active site in the photosphere, distributing the energy throughout the coronal gas. Given this connection, it is hardly surprising that both the appearance of the corona and the strength of the solar wind are closely correlated with the solar cycle.

9.5 The Heart of the Sun
Nuclear Fusion *Extension*

5 Compared with familiar terrestrial events— storms, tornadoes, even violent volcanic eruptions and earthquakes—the spots and flares of the active Sun are enormously energetic. The much greater *steady* emission of energy from the quiet Sun simply staggers the imagination. The Sun is somehow able to produce huge amounts of energy, and, according to Earth's fossil record, it has been doing so for at least the last several billion years. What forces are at work in the Sun's core to produce such energy? By what process does the Sun shine, day after day, year after year, eon after eon? Answers to these questions are central to all of astronomy. Without them, we can understand neither the physical existence of stars and galaxies in the universe nor the biological existence of life on Earth.

Only one known energy-generation mechanism can conceivably account for the Sun's enormous energy output. That process is **nuclear fusion**—the combining of light nuclei into heavier ones. We can represent a typical fusion reaction symbolically as

$$\text{nucleus } 1 + \text{nucleus } 2 \rightarrow \text{nucleus } 3 + \text{energy}.$$

For powering the Sun, the most important piece of this equation is the energy produced. The key point here is that, during a fusion reaction, the total amount of mass *decreases*—the mass of nucleus 3 is less than the combined masses of nuclei 1 and 2. The lost mass is converted to energy in accordance with Einstein's famous equation $E = mc^2$. In words, this equation says that, to determine the amount of energy corresponding to a given mass, simply multiply the mass by the square of the speed of light (c in the equation). The speed of light is so large that even a small amount of mass translates into an enormous amount of energy.

The production of energy by a nuclear fusion reaction is governed by the **law of conservation of mass and energy**, which states that the sum of mass and energy (properly converted to the same units, using Einstein's equation) must always remain constant in any physical process. There are no known exceptions. In the case of fusion reactions in the solar core, the energy is produced primarily in the form of electromagnetic radiation. The light we see coming from the Sun means that the Sun's mass must be slowly but steadily decreasing with time.

The Proton–Proton Chain

The most common element in the Sun is hydrogen, and it is the fusion of two hydrogen nuclei—that is, two protons—that begins the energy-generation process in the Sun's core. Recall from Chapter 2 that protons, having the same (positive) electrical charge, repel one another, and this repulsion increases rapidly as the distance between the protons decreases. ∞ (Sec. 2.2) As a result, to get close enough together to fuse, two protons must be slammed together at speeds in excess of a few hundred kilometers per second. For this to occur, the temperature of the solar gas must be at least 10^7 K. Such conditions are found in the core of the Sun as well as in the cores of all other stars. The proton–proton fusion reaction creates a nucleus of *deuterium*, or "heavy" hydrogen, consisting of a proton and a neutron. In addition to the energy released, two by-products are produced—a *positron* and a *neutrino*.

The **positron** is a positively charged electron— that is, its properties are identical to those of a normal, negatively charged electron except that it has a positive charge. The newly created positrons do not survive long in the Sun's core. They find themselves in the midst of a sea of electrons, with which they interact immediately and violently. Positrons and electrons annihilate one another, producing energy in the form of gamma rays.

The other product of the reaction, the **neutrino**, is chargeless and virtually massless. (The name derives from the Italian for "little neutral one.") Neutrinos move at (or nearly at) the speed of light and interact with hardly anything. They can penetrate, without stopping, several light years of lead. Yet despite their elusiveness, they can be detected with carefully constructed instruments. Below we discuss some rudimentary neutrino "telescopes" and the important contribution they have made to solar astronomy.

The deuterium nucleus produced by proton–proton fusion soon combines with another proton, resulting in a form of helium called helium-3 (which consists of two protons and a single neutron) and the release of more energy. The final step in the chain of events powering the Sun is the collision of two helium-3 nuclei to create a nucleus of "normal" helium (helium-4), which contains two protons and two neutrons. Two additional protons are produced—fuel for future nuclear reactions—and still more energy is released.

The net effect of all these steps is this: four hydrogen nuclei (protons) combine to create one helium-4 nucleus, some gamma-ray radiation, and two neutrinos. The entire sequence is called the **proton–proton chain**:

$$4 \text{ protons} \rightarrow \text{helium-4} + 2 \text{ neutrinos} + \text{energy}.$$

The process is illustrated in Figure 9.23. Gargantuan quantities of protons are fused in this way within the core of the Sun each and every second. Although the energy is produced in the form of gamma rays, as it passes through the cooler layers of the solar interior, the radiation's black-body spectrum steadily shifts toward lower and lower temperatures. The energy eventually escapes from the photosphere, mainly in the form of infrared and visible radiation. The neutrinos escape unhindered into space at the speed of light. The helium stays put in the core.

Energy Generated by the Proton–Proton Chain

Careful laboratory experiments have determined the masses of all the particles involved in the conversion of four protons to a helium-4 nucleus: the total mass of the protons is 6.6943×10^{-27} kg, the mass of the helium-4 nucleus is 6.6466×10^{-27} kg, and the neutrinos are virtually massless. The difference between the total mass of the protons and the mass of the helium nucleus, 0.0477×10^{-27} kg, is not great, but it is easily measurable.

Multiplying the vanished mass by the square of the speed of light yields 4.3×10^{-12} J. This is the energy produced in the form of radiation when 6.7×10^{-27} kg (the rounded-off mass of the four protons) of

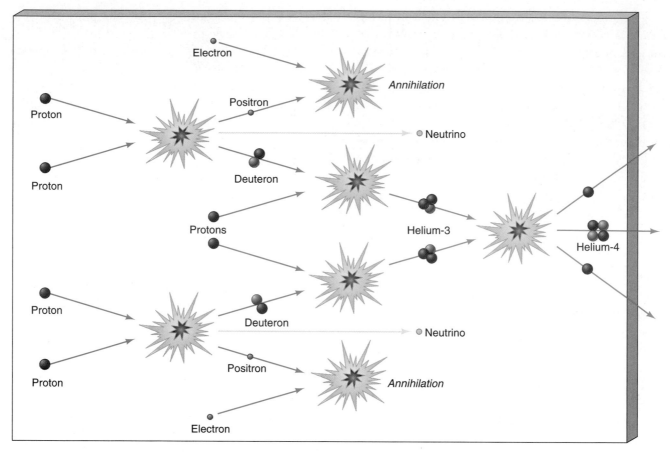

Figure 9.23 The proton–proton chain. A total of six protons (and two electrons) are converted to two protons, one helium-4 nucleus, and two neutrinos. The two leftover protons are available as fuel for new proton–proton reactions, so the net effect is that four protons are fused to form one helium-4 nucleus. Energy, in the form of gamma rays, is produced in each reaction. (A deuteron is a deuterium nucleus.)

hydrogen fuses to helium. It follows that fusion of 1 kg of hydrogen generates $4.3 \times 10^{-12}/6.7 \times 10^{-27} = 6.4 \times 10^{14}$ J. To fuel the Sun's present energy output, hydrogen must be fused to helium in the core at a rate of 600 million tons per second—a lot of mass, but only a tiny fraction of the total amount available. The Sun will be able to sustain this rate of burning for about another 5 billion years. This same basic process is responsible for the light emitted by almost all the stars we see.

Observations of Solar Neutrinos

6 Because the gamma-ray energy created in the proton–proton chain is transformed into visible and infrared radiation by the time it emerges from the Sun, astronomers have no direct electromagnetic evidence of the core nuclear reactions. Instead, the *neutrinos* created in the proton–proton chain are our best bet for learning about conditions in the solar core. They travel cleanly out of the Sun, interacting with virtually nothing, and escape into space a few seconds after being created.

Of course, the fact that they can pass through the entire Sun without interacting also makes neutrinos difficult to detect on Earth! Nevertheless, they do interact a little more strongly with some elements— chlorine and gallium, for example—than with others, and this knowledge can be used in the construction of neutrino-detection devices (Figure 9.24). Over the past three decades, four experiments have been designed to detect solar neutrinos reaching Earth's sur-

Figure 9.24 This swimming-pool-sized detector was a "neutrino telescope" of sorts, buried underground in a South Dakota gold mine. It contained 400,000 liters of dry-cleaning fluid, a chemical containing chlorine. About once every two days, one of the chlorine atoms in the tank was converted to argon by interaction with a neutrino from the Sun. This particular experiment, which operated from 1968 until 1993, detected fewer than half the number of neutrinos predicted by theory.

face. The four are of widely different design and are sensitive to neutrinos of very different energies. The four disagree somewhat on the details of their findings, but they seem to agree on one important point: the number of solar neutrinos reaching Earth is substantially *less* (by 30 to 50 percent) than the prediction of the Standard Solar Model. This discrepancy is known as the **solar neutrino problem**.

Most scientists are confident that the experimental results are to be trusted, so how do we explain this clear disagreement between theory and observation? There are really only two possibilities. Either neutrinos are not produced in the solar core as frequently as we think, or not all of them theoretically predicted to make it to Earth do in fact make it that far. Let us consider these alternatives in turn.

If the temperature in the solar core were lower, then the number of neutrinos predicted by theory would also be lower. If the center of the Sun were about 10 percent cooler than in the Standard Solar Model—about 13,500,000 K, rather than the 15,000,000 K noted in Section 9.2—helium-4 would still be produced, but its production would be accompanied by fewer neutrinos detectable by the experiments mentioned above. However, lowering the core temperature would also lower the Sun's luminosity, and most theorists agree that the numerical models could not be in error by as much as 1,500,000 K while

remaining consistent with all other solar observations. In addition, observations by the GONG group (Section 9.2) seem to rule out a central temperature much below 15,000,000 K. Most astronomers regard it as very unlikely that the resolution of the solar neutrino problem will be found in the nuclear physics of the Sun's interior.

Instead, the properties of the neutrinos themselves may provide the answer. If neutrinos do have a minute amount of mass (and this has yet to be proved), it may be possible for them to change their properties, even to transform themselves into other particles, during their 8-minute flight from the solar core to Earth, through a process known as **neutrino oscillation**. In this picture, neutrinos are produced in the Sun at the rate required by the Standard Solar Model, but some of them turn into something else (in particle-physics jargon, they are said to "oscillate" into other particles) on their way to Earth and so go undetected. Proposed experiments near neutrino-producing nuclear reactors on Earth may be able to test this idea within the next few years.

Where are the missing neutrinos? Is the proton–proton chain operating as we think? Do we *really* know what processes are at work deep in the hearts of stars? For now, the mystery of the solar neutrinos remains unsolved, although most physicists favor the neutrino-oscillation explanation. Virtually all researchers concur—or at least hope—that the correct interpretation of the solar neutrino problem will not tear apart the theoretical fabric of the proton–proton chain. Most believe that the description of solar fusion we have presented in this chapter is basically right and that our understanding of neutrino physics just needs to be fine-tuned. However, should drastic measures be needed to solve the solar puzzle, we may yet have to return to the drawing board to answer one of the most fundamental scientific questions of all: How does a star shine?

Chapter Review

Summary

Our Sun is a **star** (p. 256), a glowing ball of gas held together by its own gravity and powered by nuclear fusion at its center. The solar **photosphere** (p. 256) is the region at the Sun's surface from which essentially all of the visible light is emitted. Above the photosphere lies the **chromosphere** (p. 256), which is the Sun's lower atmosphere. In the **transition zone** (p. 256) above the chromosphere, the temperature increases from a few thousand to around a million kelvins. Above the transition zone is the Sun's thin, hot upper atmosphere, the solar **corona** (p. 256). The main interior regions of the Sun are the **core** (p. 257), where nuclear reactions generate energy; the **radiation zone** (p. 256), where the energy travels outward in the form of electromagnetic radiation; and the **convection zone** (p. 256), where the Sun's matter is in constant convective motion.

The amount of solar energy reaching a 1-m^2 area at the top of Earth's atmosphere each second is a quantity known as the **solar constant** (p. 257). The Sun's **luminosity** (p. 257) is the total amount of energy radiated from the solar surface per second. It is determined by multiplying the solar constant by the area of an imaginary sphere of radius 1 A.U.

Much of our knowledge of the solar interior comes from mathematical models. The model that best fits the observed properties of the Sun is the **Standard Solar Model** (p. 258). **Helioseismology** (p. 258)—the study of vibrations of the solar surface caused by pressure waves in the interior—provides further insight into the Sun's structure.

The effect of the solar convection zone can be seen on the surface in the form of **granulation** (p. 260) of the photosphere. As hotter (and therefore brighter) gas rises and cooler (dimmer) gas sinks, a characteristic mottled appearance results. Lower levels in the convection zone also leave their mark on the photosphere in the form of larger transient patterns called **supergranulation** (p. 261).

At a distance of about 12 solar radii, the gas in the corona is hot enough to escape the Sun's gravity, and the corona begins to flow outward as the solar wind. Most of the solar wind flows from low-density regions of the corona called **coronal holes** (p. 264).

The steady component of the Sun's energy production is known as the **quiet Sun** (p. 265). Superimposed on that is the much more erratic emission of the **active Sun** (p. 265). Solar activity is generally associated with disturbances in the Sun's magnetic field.

Sunspots (p. 265) are Earth-sized regions on the solar surface that are a little cooler than the surrounding photosphere. They are regions of intense magnetism. Both the numbers and locations of sunspots vary in a roughly 11-year **sunspot cycle** (p. 267). The sunspot cycle is quite irregular, its length varying from 7 to 15 years, and there have been times in the past when no sunspots were seen for long periods. The overall direction of the solar magnetic field reverses from one sunspot cycle to the next. The 22-year cycle that results when the direction of the field is taken into account is called the **solar cycle** (p. 267).

Solar activity tends to be concentrated in **active regions** (p. 268) associated with sunspot groups. **Prominences** (p. 268) are loop- or sheetlike structures produced when hot gas ejected by activity on the solar surface interacts with the Sun's magnetic field. **Flares** (p. 269) are violent surface explosions that blast particles and radiation into interplanetary space.

The Sun generates energy by converting hydrogen to helium in its core by the process of **nuclear fusion** (p. 270). When four protons are converted to a helium nucleus in the **proton–proton chain** (p. 271), some mass is lost. The **law of conservation of mass and energy** (p. 271) requires that this mass appear as energy, eventually resulting in the light we see.

Some particles produced during nuclear fusion in the solar core are **positrons** (p. 271), "positively charged electrons" that quickly react with normal electrons in the Sun's core, being destroyed in the reaction and generating gamma rays; deuterium nuclei, each nucleus consisting of a proton and a neutron; and **neutrinos** (p. 271), massless (or perhaps only nearly massless) particles that escape from the Sun.

Despite their elusiveness, it is possible to detect a small fraction of the neutrinos streaming from the Sun. The observations lead to the **solar neutrino problem** (p. 273)—substantially fewer neutrinos are observed than are predicted by theory. The resolution to this problem is unclear. A leading explanation is that **neutrino oscillations** (p. 273) convert some neutrinos to other (undetected) particles en route from the Sun to Earth.

Self-Test: True or False?

_____ 1. The Sun is a typical star.

_____ 2. The average density of the Sun is significantly greater than the density of Earth.

_____ 3. The Sun's diameter is about 10 times that of Earth.

_____ 4. The Sun's mass is comparable to the mass of Jupiter.

_____ 5. The Sun's differential rotation indicates that it is not solid.

_____ 6. In the solar convection zone, the gas is fully ionized.

_____ 7. Helioseismology involves the study of "sunquakes" occurring in the photosphere.

_____ 8. Convection involves cool gas rising to the solar surface and hot gas sinking into the interior.

_____ 9. Absorption lines in the solar spectrum are produced mainly in the corona.

_____ 10. There are as many absorption lines in the solar spectrum as there are elements present in the Sun.

_____ 11. The faintness of the chromosphere is a direct result of its low temperature.

_____ 12. The temperature of the solar corona decreases with increasing radius.

_____ 13. Sunspots are regions of intense magnetic fields.

_____ 14. Coronal holes are low-density regions of the upper solar atmosphere.

_____ 15. Prominences are large flames erupting from the burning surface of the Sun.

_____ 16. Neutrinos are neutrons traveling close to the speed of light.

_____ 17. Nuclear fusion releases energy because the total mass of the nuclei involved increases.

_____ 18. The nuclear reactions that power the Sun create energy in the form of gamma rays.

Self-Test: Fill in the Blank

1. The part of the Sun we see is called the _____.

2. Traveling outward from the photosphere, the three main regions of the solar atmosphere are the ____, the ____, and the _____.

3. Below the solar surface, in order of increasing depth, lie the _____ zone, the _____ zone, and the _____.

4. The _____ seen on the surface of the Sun is evidence of convective cells.

5. The Sun appears to have a well-defined edge because the thickness of the _____ is only 0.1 percent of the solar radius.

6. _____ is the most abundant element in the Sun.

7. _____ is the second most abundant element in the Sun.

8. The two most abundant elements in the Sun make up about _____ percent of its composition.

9. The gas in the corona is highly _____.

10. The _____ is formed as the hot outer regions of the corona expand into interplanetary space.

11. Sunspots appear dark because they are _____ than the surrounding gas of the photosphere.

12. The sunspot cycle is _____ years long; the solar cycle is _____ as long.

13. A _____ is a violent explosive event in a solar active region.

14. The entire solar luminosity is produced in the _____ (give the region) of the Sun.

15. The net result of the proton–proton chain is that _____ protons are fused into a nucleus of _____, two _____ are emitted, and energy is released in the form of _____.

16. The solar neutrino problem is the fact that astronomers observe too _____ neutrinos coming from the Sun.

Review and Discussion

1. Name and briefly describe the main regions of the Sun.

2. How massive is the Sun, compared with Earth?

3. How hot is the solar surface? The solar core?

4. How do scientists construct models of the Sun?

5. Describe how energy generated at the center of the Sun reaches Earth.

6. Why does the Sun appear to have a sharp edge?

7. What evidence do we have for solar convection?

8. What is the solar wind?

9. What is the cause of sunspots, flares, and prominences?

10. What is the Maunder minimum?

11. What fuels the Sun's enormous energy output?

12. What is the law of conservation of mass and energy? How is it relevant to nuclear fusion in the Sun?

13. What are the ingredients and the end result of the proton–proton chain? Why is energy released in the process?

14. Why are scientists trying so hard to detect solar neutrinos?

15. Describe the two possible solutions to the solar neutrino problem.

16. What would we observe on Earth if the Sun's internal energy source suddenly shut off? Would the Sun darken instantaneously? If not, how long do you think it might take—minutes, days, years, millions of years—for the Sun's light to begin to fade? Repeat the question for solar neutrinos.

Problems

1. Given that the solar spectrum corresponds to a temperature of 5800 K and peaks at a wavelength of 500 nm, use Wien's law (peak wavelength $\propto$ $1/T$, where T is the temperature in kelvins; see Section 7.4) to determine the wavelength corresponding to the peak of the black-body curve (a) in the core of the Sun, where the temperature is 10^7 K, (b) in the solar convection zone (10^5 K), and (c) just below the solar photosphere (10^4 K). What form (visible, infrared, X-ray, etc.) does the radiation take in each case?

2. If convected solar material moves at 1 km/s, how long does it take to flow across the 1000-km expanse of a typical granule? Compare this length of time with the roughly 10-minute lifetimes observed for most solar granules.

3. Use Stefan's law (flux $\propto T^4$, where T is the temperature in kelvins; see Section 9.4) to calculate how much less energy is emitted per unit area of a 4500-K sunspot than is emitted per unit area of the surrounding 5800-K photosphere. Give your answer in the form of a fraction.

4. How long does it take the Sun to convert 1 Earth mass of hydrogen to helium?

Projects

The projects given here all require a special solar filter. Such filters are easily purchased from various sources. **NEVER LOOK DIRECTLY AT THE SUN WITHOUT A FILTER!**

1. An appropriately filtered telescope will easily show you sunspots, although the numbers will vary considerably during the sunspot cycle. Count the number of sunspots you see on the Sun's surface. Notice that sunspots often come in pairs or groups. Come back and look again a few days later and you'll see that (a) the Sun's rotation has caused the spots to move and (b) the appearance of the spots has changed. If you see a sufficiently large sunspot (or, more likely, sunspot group), continue to watch it as the Sun rotates. It will be out of view for about two weeks. Can you determine the rotation period of the Sun from these observations?

2. Solar granulation is not too hard to see. The atmosphere of Earth is most stable in the morning hours. Observe the Sun on a cool morning, one or two hours after it has risen, using a properly filtered telescope for safety. Use high magnification and look initially at the middle of the Sun's disk. Can you see changes in the granulation pattern? They are there but are not always easy to see.

3. View some solar prominences and flares through a properly filtered telescope. Hydrogen-alpha filters, which transmit only a narrow band of radiation around the characteristic red spectral line of hydrogen, are commercially available for small telescopes. Using such a filter, you can often see prominences and flares even during times of sunspot minimum. Because you are actually viewing the chromosphere rather than the photosphere, the Sun looks quite different and can be very impressive in appearance.

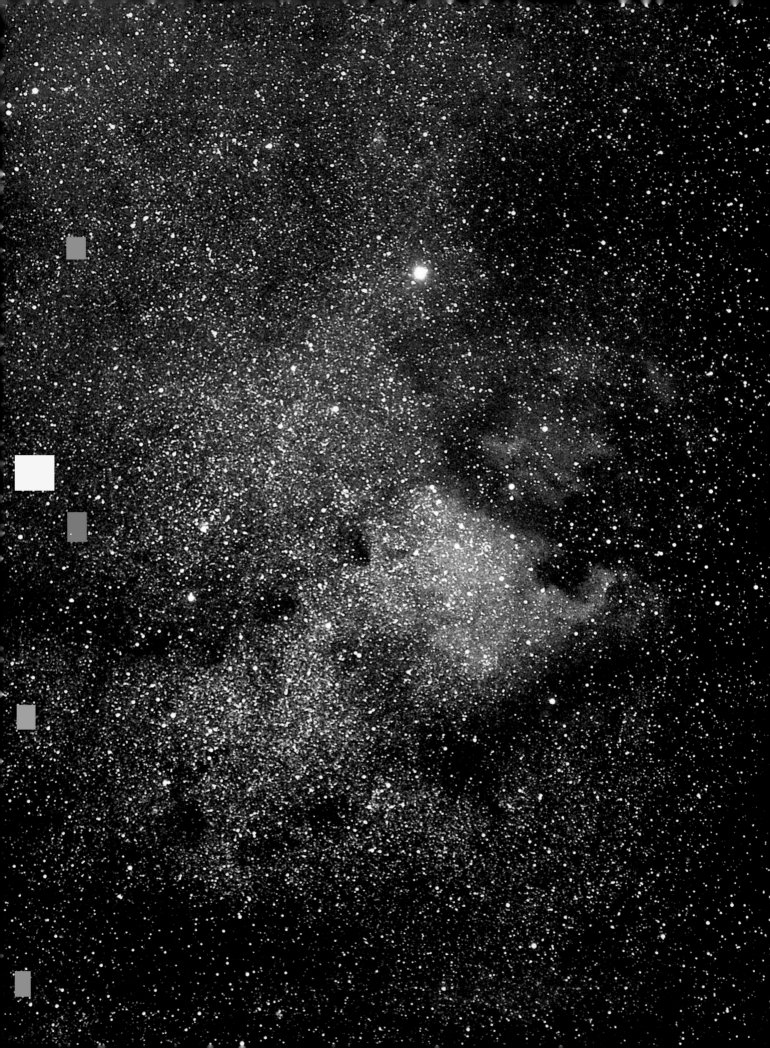

Measuring the Stars

Giants, Dwarfs, and the Main Sequence *www*

LEARNING GOALS

Studying this chapter will enable you to:

1 Explain how stellar distances are determined.

2 Discuss stellar motion and how this motion is measured from Earth.

3 Explain how physical laws are used to estimate stellar sizes.

4 Distinguish between luminosity and apparent brightness, and explain how stellar luminosity is determined.

5 Explain the usefulness of classifying stars according to their colors, surface temperatures, and spectral characteristics.

6 Describe how an H–R diagram is constructed and used to identify stellar properties.

7 Explain how stellar masses are measured and how mass is related to other stellar properties.

8 Distinguish between open and globular star clusters, and explain why the study of star clusters is important to astronomers.

We have studied Earth, the Moon, the solar system, and the Sun. As we continue our inventory of the contents of the universe, we now move away from our local environment and out into the depths of space. Our primary goal in this chapter is to comprehend the nature of stars, both the familiar members of the constellations we see each night and the myriad more distant stars too faint to be detected by our unaided eyes. Rather than studying the individual peculiarities of these faraway bodies, however, we instead concentrate on determining the physical and chemical properties they share. There is order in the legions of stars scattered across the sky. Just as we compared planets in our study of the solar system, we now compare and catalog the stars in order to further our understanding of the galaxy and the universe we inhabit.

10.1 The Distances to the Stars

Stellar Parallax

1 Recall from the Prologue how we can use *parallax* to measure distances to terrestrial and solar system objects. Parallax is an object's apparent shift relative to some distant background as the observer's point of view changes. ∞ (Sec. P.3) To measure parallax, we must observe the object from both ends of some baseline and measure the angle through which the line of sight to the object shifts. In astronomical contexts, we usually determine parallax by comparing photographs made from the two ends of the baseline.

As the distance to the object increases, or the length of the baseline shrinks, the parallax becomes smaller and therefore harder to measure. Accordingly, a long baseline is essential for measuring the distance to a very remote object. The stars are so far away from us that even Earth's diameter is too short a baseline for determining their distances. However, by comparing observations made of a star at different times of the year, as shown in Figure 10.1, we can effectively extend the baseline to the diameter of Earth's orbit around the Sun, 2 A.U. Only with this enormously longer baseline do some stellar parallaxes become measurable.

As indicated in Figure 10.1, a star's parallactic angle—or, more commonly, just its "parallax"—is conventionally defined to be *half* its apparent shift relative to the background as we move from one side of Earth's orbit to the other.

The parallaxes of even the closest stars are very small, so astronomers generally find it convenient to measure parallax in arc seconds rather than in degrees. If we ask at what distance from the Sun a star must lie in order for its observed parallax to be exactly 1″, we get an answer of 206,265 A.U., or 3.1×10^{16} m. Astronomers call this distance 1 **parsec** (1 pc), from "*p*arallax in arc *sec*onds." Because parallax decreases as distance increases, we can relate a star's parallax to its distance from the Sun by the following simple formula

$$\text{distance (in parsecs)} = \frac{1}{\text{parallax (in arc seconds)}}.$$

Thus, a star with a measured parallax of 1″ lies at a distance of 1 pc from the Sun. The parsec is defined so as to make the conversion between parallactic angle and distance easy. An object with a parallax of 0.5″ lies at a distance of 1/0.5 = 2 pc, an object with a parallax of 0.1″ lies at 1/0.1 = 10 pc, and so on. One parsec is approximately equal to 3.3 light years.

Our Nearest Neighbors

The closest star to the Sun is called Proxima Centauri. This star is a member of a triple-star system (three stars orbiting one another, bound together by gravity) known as the Alpha Centauri complex. Proxima Centauri displays the largest known stellar parallax, 0.76″, which means that it is about 1/0.76 = 1.3 pc away—about 270,000 A.U., or 4.3 light years. That's the *nearest* star to Earth—at almost 300,000 times the Earth–Sun distance! This is a fairly typical interstellar distance in the Milky Way Galaxy.

Vast distances can sometimes be grasped by means of analogies. Imagine Earth as a grain of sand orbiting a golfball-sized Sun at a distance of 1 m. The nearest star, also a golfball-sized object, is then 270 *kilometers* away. Except for the other planets in our solar system, themselves ranging in size from grains of sand to small marbles and all lying within 50 m of

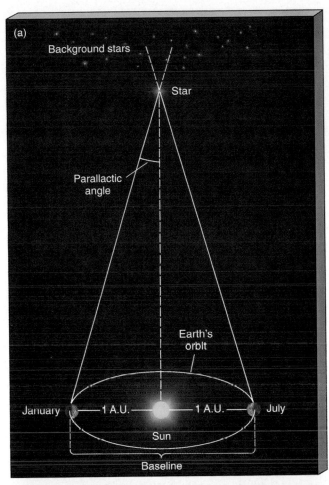

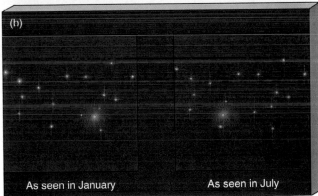

Figure 10.1 (a) The geometry of stellar parallax. For observations made 6 months apart, the baseline is twice the Earth–Sun distance, or 2 A.U. (b) The parallactic shift is usually measured photographically. Images of the same region of the sky made at different times of the year are used to determine a star's apparent movement relative to background stars.

the "Sun," nothing else of consequence exists in the 270 km separating the Sun and the other star.

The next nearest neighbor to the Sun beyond the Alpha Centauri system is called Barnard's Star. Its

parallax is 0.55", so it lies at a distance of 1.8 pc, or 6.0 light years—370 km in our model. All told, fewer than 100 stars lie within 5 pc (1000 km in our model) of the Sun. Such is the void of interstellar space.

As we mentioned in Chapter 3, ground-based images of stars are generally smeared out into a disk of radius 1" or so by turbulence in Earth's atmosphere. ∞ (Sec. 3.3) However, astronomers have special equipment that can routinely measure stellar parallaxes of 0.03" or less, corresponding to stars lying within about 30 pc (100 ly) of Earth. Several thousand stars lie within this range. Most are much dimmer than the Sun and invisible to the naked eye. Recently released data from the European *Hipparcos* satellite have extended the range of accurately measured parallaxes to well over 100 pc. Even so, the vast majority of stars in our Galaxy are far more distant than this.

10.2 Stellar Motion

2 In addition to the apparent motion caused by parallax, stars have real motion, too. In other words, stars travel through space. This stellar motion has two components. The *transverse component* measures a star's motion perpendicular to our line of sight—in other words, its motion across the sky. The *radial component* measures a star's movement along our line of sight—toward us or away from us.

The annual movement of a star across the sky, as seen from Earth (and corrected for parallax), is called **proper motion**. Proper motion and transverse motion of a star describe the same component of a star's velocity. The main difference is that transverse motion is measured in linear units, such as kilometers per second, and proper motion, like parallax, is measured in terms of angular displacement. The angles involved are typically very small, and proper motion is usually expressed in arc seconds per year. Stars' transverse velocities can be quite large—tens or even hundreds of kilometers per second. However, because of their great distances from the Sun, it usually takes many years for us to discern their movement across the sky.

Figure 10.2 shows two photographs of the sky around Barnard's Star, made on the same day of the year, but 22 years apart. If the two images were superimposed, the images of all the other stars in the field of view would coincide, but those of Barnard's Star would not—Barnard's Star moved during this time interval.

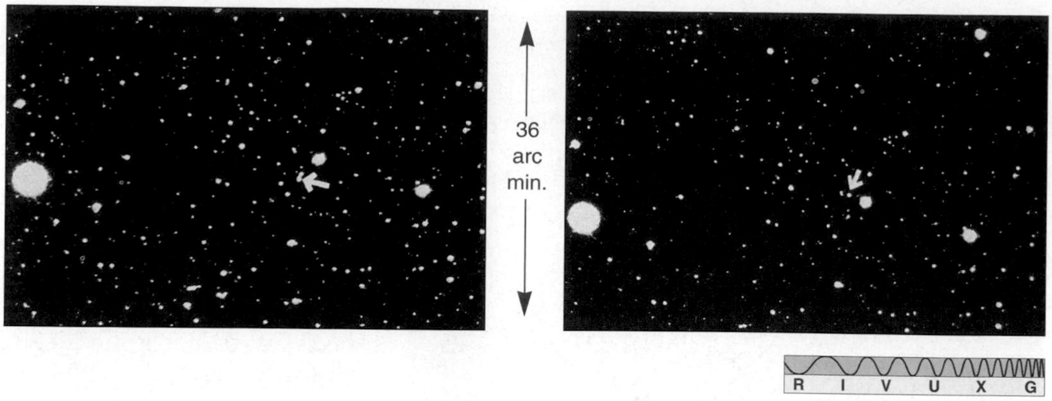

Figure 10.2
Comparison of two photographs taken 22 years apart shows evidence of real space motion for Barnard's Star (denoted by an arrow).

36 arc min.

R I V U X G

Because Earth was at the same point in its orbit when the photographs were taken, the displacement cannot be due to parallax caused by Earth's motion around the Sun. We conclude that the observed displacement indicates real space motion of Barnard's Star relative to the Sun. Careful measurements show that Barnard's Star moved 227″ during the 22-year interval. The proper motion of Barnard's Star is therefore 227″/22 years, or 10.3″/yr. This is the largest known proper motion of any star. Only a few hundred stars have proper motions greater than 1″/yr.

A star's transverse velocity is easily calculated once its proper motion and its distance are known. At the distance of Barnard's Star (1.8 pc), an angle of 10.3″ corresponds to a physical displacement of 0.00009 pc, or about 2.8 billion km. Barnard's Star takes a year to travel this distance, so its transverse velocity is 2.8 billion km/(3.2 × 10⁷ s/yr), or 88 km/s.

As another example, consider the three-dimensional motion of our nearest neighbor, the Alpha Centauri system, sketched in Figure 10.3 in relation to our own solar system. Alpha Centauri's proper motion is about 3.5″/yr. At that system's distance of 1.3 pc, this implies a transverse velocity of 22 km/s. The other component of Alpha Centauri's motion—the radial velocity—can be determined using the Doppler effect, as discussed in Chapter 2. ∞ *(More Precisely 2-2)* Spectral lines from Alpha Centauri are slightly blueshifted, allowing astronomers to measure the star system's radial velocity (relative to the Sun) as 20 km/s toward us.

What is the true space motion of Alpha Centauri? Given that its radial motion is toward us, will this alien system collide with our own some time in the future? The answer is no—Alpha Centauri's

transverse velocity will steer it well clear of the Sun. We can combine the transverse and radial velocities according to the Pythagorean theorem. The total velocity is $\sqrt{22^2 + 20^2}$, or about 30 km/s, in the direction shown by the horizontal red arrow in Figure 10.3. As that figure indicates, Alpha Centauri will get no closer to us than about 1 pc, and that won't happen until 280 centuries from now.

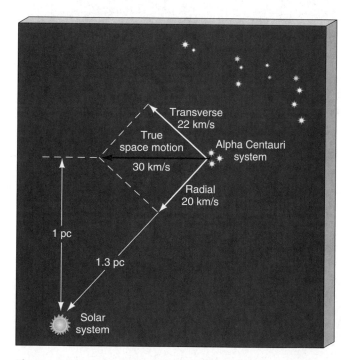

Figure 10.3 The motion of the Alpha Centauri star system drawn relative to our solar system. The transverse component of the velocity has been determined by observing the system's proper motion. The radial component is measured using the Doppler shift of lines in Alpha Centauri's spectrum. The true space velocity, indicated by the red arrow, results from the combination of the two.

10.3 Stellar Sizes *Extensions*

Direct and Indirect Measurements

3 Most stars are unresolvable points of light in the sky, even when viewed through the largest telescopes. Still, a few are big enough, bright enough, and close enough to allow us to measure their sizes *directly*. In an optical technique known as *speckle interferometry*, many short-exposure images of a star, each too brief for Earth's turbulent atmosphere to smear it out into a seeing disk, are combined to make a high-resolution map of the star's surface. In some cases, the results are detailed enough to allow a few surface features to be distinguished (Figure 10.4). As adaptive optics techniques continue to improve, it is becoming possible to perform this process in real time, allowing very-high-resolution stellar images to be made. ∞ (Sec. 3.3)

By measuring a star's angular size and knowing its distance from the Sun, astronomers can determine its radius by simple geometry. Optical astronomers have directly measured the sizes of a few dozen stars in this way. However, most stars are too distant or too small for this to be possible. Their sizes must be inferred by

more indirect means, using Stefan's law, which we studied in Chapter 2. ∞ (Sec. 2.4)

According to Stefan's law, the rate at which a star emits energy into space—the star's *luminosity*—is proportional to the fourth power of the star's *surface temperature*. ∞ (Sec. 9.1) However, the luminosity must also depend on the star's *surface area*, because large bodies radiate more energy than do small bodies having the same surface temperature. Because the surface area of a star is proportional to the square of its radius, we can combine these proportionalities to say

$$\text{luminosity} \propto \text{radius}^2 \times \text{temperature}^4,$$

where $\propto$ is the symbol representing proportionality. This **radius–luminosity–temperature relationship** is important because it demonstrates that knowledge of a star's luminosity and temperature can yield an estimate of its radius. In other words, it gives us an *indirect* way of determining stellar sizes.

Giants and Dwarfs

Let's consider some examples to clarify these ideas. The star Mira has a surface temperature of about 3000 K and a luminosity of 1.6×10^{29} W. Thus, its surface temperature is roughly half, and its luminosity about 400 times, the corresponding quantities for our Sun. The radius–luminosity–temperature relationship, rearranged into the more convenient form

$$\text{radius} \propto \frac{\sqrt{\text{luminosity}}}{\text{temperature}^2}$$

therefore implies that the star's radius is $\sqrt{400}/0.5^2 = 80$ times that of our Sun. A star as large as Mira is known as a *giant*. More precisely, **giants** are stars having radii between 10 and 100 times that of the Sun. Even larger stars, ranging up to 1000 solar radii in size, are known as **supergiants**. Since the color of any 3000 K object is red, Mira is a **red giant**.

Now consider Sirius B—a faint companion to Sirius A, the brightest star in the night sky. Sirius B's surface temperature is roughly 24,000 K, four times that of the Sun. Its luminosity is 10^{25} W, about 0.04 times the solar value. Substituting these quantities into our radius–luminosity–temperature relationship, we obtain a radius of $\sqrt{0.04}/4^2 = 0.01$ times the solar radius—roughly the size of Earth. Sirius B is much hot-

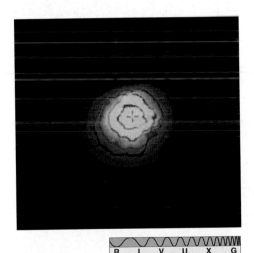

Figure 10.4 The giant star Betelgeuse (shown here in false color) is close enough for us to directly resolve its size, along with some surface features thought to be storms similar to those that occur on the Sun. Betelgeuse is such a huge star (300 times the size of the Sun) that its photosphere spans roughly the size of Mars's orbit. Most of the surface features discernible here are larger than the entire Sun.

R I V U X G

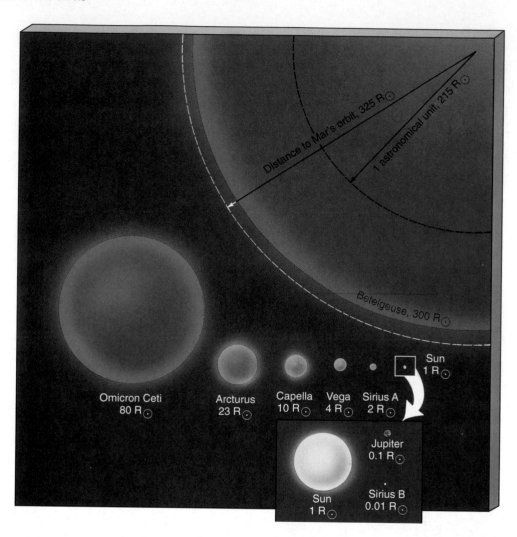

Figure 10.5 *Star sizes vary greatly. Shown here are the estimated sizes of several well-known stars, including a few of those discussed in this chapter. The symbol "R⊙" means "solar radius."*

ter but smaller and dimmer (less luminous) than our Sun. Such a star is known as a *dwarf*. In astronomical parlance, the term **dwarf** refers to any star of radius comparable to or smaller than the radius of the Sun (including the Sun itself). Because any 24,000 K object glows bluish-white, Sirius B is an example of a **white dwarf**.

The radii of the vast majority of stars (mostly measured using the radius–luminosity–temperature relationship) range from less than 0.01 to over 100 times the radius of the Sun. Figure 10.5 illustrates the estimated sizes of a few well-known stars.

10.4 Luminosity and Apparent Brightness

4 Luminosity is an *intrinsic* property of a star—it does not depend in any way on the location or motion of the observer. It is sometimes referred to as

the star's **absolute brightness**. However, when we look at a star, we see not its luminosity, but rather its **apparent brightness**—the amount of energy striking unit area of some light-sensitive surface or device (such as a CCD chip or a human eye) per unit time. In this section, we discuss how these important quantities are related to one another.

Another Inverse-Square Law

Figure 10.6 shows light leaving a star and traveling through space. Moving outward, the radiation passes through imaginary spheres of increasing radius surrounding the source. The amount of radiation leaving the star per unit time—the star's luminosity—is constant, so the farther the light travels from the source, the less energy passes through each unit of area. Think of the energy as being spread out over an ever-larger area, and therefore spread more thinly, or "diluted," as

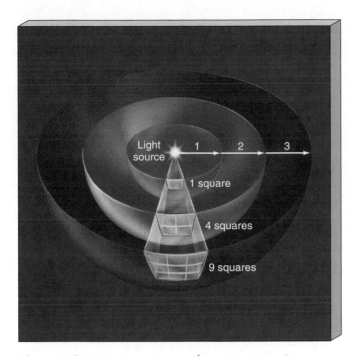

Figure 10.6 As it moves away from a source such as a star, radiation is steadily diluted while spreading over progressively larger surface areas (depicted here as sections of spherical shells). Thus, the amount of radiation received by a detector (the source's apparent brightness) varies inversely as the square of its distance from the source.

it expands into space. Because the area of a sphere grows as the square of the radius, the energy per unit area—the star's apparent brightness—is inversely proportional to the square of the distance from the star. Doubling the distance from a star makes it appear 2^2, or 4, times dimmer. Tripling the distance reduces the apparent brightness by a factor of 3^2, or 9, and so on.

Of course, the star's luminosity also affects its apparent brightness. Doubling the luminosity doubles the energy crossing any spherical shell surrounding the star and hence doubles the apparent brightness. We can therefore say that the apparent brightness of a star is directly proportional to the star's luminosity and inversely proportional to the square of its distance:

$$\text{apparent brightness} \propto \frac{\text{luminosity}}{\text{distance}^2}.$$

Thus, two identical stars can have the same apparent brightness if (and only if) they lie at the same distance from Earth. However, as illustrated in Figure 10.7, two nonidentical stars can also have the same apparent brightness if the more luminous one lies farther away. A bright star (that is, one having large apparent bright-

ness) is a powerful emitter of radiation (high luminosity), is near Earth, or both. A faint star (small apparent brightness) is a weak emitter (low luminosity), is far from Earth, or both.

Determining a star's luminosity is a twofold task. First, the astronomer must determine the star's apparent brightness by measuring the amount of energy detected through a telescope in a given amount of time. Second, the star's distance must be measured—by parallax for nearby stars and by other means (to be discussed later) for more distant stars. The luminosity can then be found using the inverse-square law. This is basically the same reasoning we used earlier in our discussion of how astronomers measure the solar luminosity. ∞ (Sec. 9.1)

The Magnitude Scale

Instead of measuring apparent brightness in SI units (for example, watts per square meter, the unit in which we expressed the solar constant in Chapter 9), optical astronomers find it more convenient to work in terms of a construct called the **magnitude scale**. The scale dates back to the second century B.C., when the Greek astronomer Hipparchus ranked the naked-eye stars into six groups. The brightest stars were categorized as

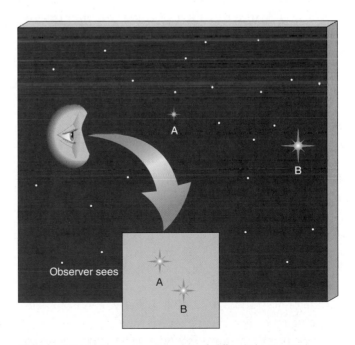

Figure 10.7 Two stars A and B of different luminosities can appear equally bright to an observer on Earth if the brighter star B is more distant than the fainter star A.

first magnitude. The next brightest stars were labeled second magnitude, and so on, down to the faintest stars visible to the naked eye, which were classified as sixth magnitude. The range 1 (brightest) through 6 (faintest) spanned all the stars known to the ancients. Notice that a *large* magnitude means a *faint* star.

When astronomers began using telescopes with sophisticated detectors to measure the light received from stars, they quickly discovered two important facts about the magnitude scale. First, the 1–6 magnitude range defined by Hipparchus spans about a factor of 100 in apparent brightness—a first-magnitude star is approximately 100 times brighter than a sixth-magnitude star. Second, the physiological characteristics of the human eye are such that each magnitude change of 1 corresponds to a factor of about 2.5 in apparent brightness. In other words, to the human eye a first-magnitude star is roughly 2.5 times brighter than a second-magnitude star, which is roughly 2.5 times brighter than a third-magnitude star, and so on. (By combining factors of 2.5, we confirm that a first-magnitude star is indeed $(2.5)^5 \approx 100$ times brighter than a sixth-magnitude star.)

Modern astronomers have modified and extended the magnitude scale in a number of ways. First, we now *define* a change of 5 in the magnitude of an object (going from magnitude 1 to magnitude 6, say, or from magnitude 7 to magnitude 2) to correspond to *exactly* a factor of 100 in apparent brightness. Second, because we are really talking about apparent (rather than absolute) brightnesses, the numbers in Hipparchus's ranking system are called **apparent magnitudes.** Third, the scale is no longer limited to whole numbers—a star of apparent magnitude 4.5 is intermediate in apparent brightness between a star of apparent magnitude 4 and one of apparent magnitude 5. Finally, magnitudes outside the range 1–6 are allowed—very bright objects can have apparent magnitudes much less than 1, and very faint objects can have apparent magnitudes far greater than 6.

Figure 10.8 illustrates the apparent magnitudes of some astronomical objects, ranging from the Sun, at –26.8, to the faintest object detectable by the *Hubble* or *Keck* telescopes, an object having an apparent magnitude of 30—about as faint as a firefly seen from a distance equal to Earth's diameter.

Apparent magnitude measures a star's apparent brightness when seen at the star's actual distance from the Sun. To compare intrinsic, or absolute, properties

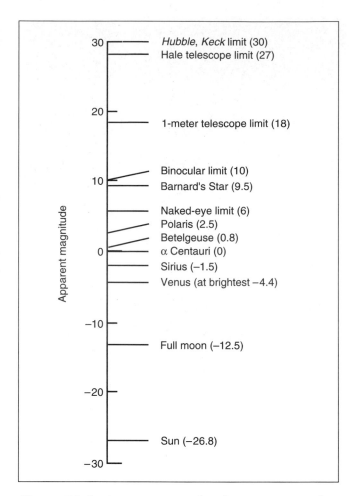

Figure 10.8 Apparent magnitudes of some astronomical objects. The original magnitude scale was defined so that the brightest stars in the night sky had magnitude 1 while the faintest stars visible to the naked eye had magnitude 6. It has since been extended to cover much brighter and much fainter objects. A increase of 1 in apparent magnitude corresponds to a decrease in apparent brightness of a factor of approximately 2.5.

of stars, however, astronomers imagine looking at all stars from a standard distance of 10 pc. There is no particular reason to use 10 pc—it is simply convenient. A star's **absolute magnitude** is its apparent magnitude when it is placed at a distance of 10 pc from the observer. Because distance is fixed in this definition, absolute magnitude is a measure of a star's absolute brightness, or luminosity.

When a star farther than 10 pc away from us is moved to a point 10 pc away, its apparent brightness increases and hence its apparent magnitude decreases. Stars more than 10 pc from Earth therefore have apparent magnitudes that are greater than their absolute magnitudes. For example, if a star at a distance of 100 pc were moved to the standard 10-pc distance, its distance would decrease by a factor of 10, so (by the inverse-square law) its apparent brightness would increase by a factor of 100. Its apparent magnitude (by definition) would therefore decrease by 5. In other words the star's absolute magnitude exceeds its apparent magnitude by 5.

For stars closer than 10 pc, the reverse is true. An extreme example is our Sun. Because of its proximity to Earth, it appears very bright and thus has a large negative apparent magnitude (Figure 10.8). However, the Sun's absolute magnitude is 4.8. If the Sun were moved to a distance of 10 pc from Earth, it would be only slightly brighter than the faintest stars visible in the night sky.

The numerical difference between a star's absolute and apparent magnitudes is a measure of the distance to the star. Knowledge of a star's apparent magnitude and of its distance allows us to compute its absolute magnitude. Alternatively, knowledge of a star's apparent and absolute magnitudes allows us to determine its distance.

10.5 Temperature and Color

Looking at the night sky, you can tell at a glance which stars are hot and which are cool. In Figure 10.9, which shows the constellation Orion as it appears through a small telescope, the colors of the cool red star Betelgeuse and the hot blue star Rigel are clearly evident. However, to obtain these stars' temperatures (3000 K for Betelgeuse and 15,000 K for Rigel), more detailed observations are required.

Astronomers can determine a star's surface temperature by measuring its apparent brightness (radiation intensity) at several frequencies, then matching the observations to the appropriate black-body curve. ∞ (Sec. 2.4) In the case of the Sun, the theoretical curve that best fits the emission describes a 5800-K emitter. The same technique works for any other star, regardless of its distance from Earth.

Because the basic shape of the black-body curve is so well understood, astronomers can estimate a star's temperature using as few as *two* measurements at selected wavelengths. This is accomplished through the use of telescope filters that block out all radiation except that within specific wavelength ranges. For example, a B (blue) filter rejects all radiation except for a certain range of violet to blue light. Similarly, a V

Figure 10.9 The constellation Orion as it appears through a small telescope or binoculars. The different colors of the member stars are easily distinguished. The bright red star at the upper left is Betelgeuse (α); the bright blue-white star at the lower right is Rigel (β). (Compare with Figure P.6.) The scale of this photograph is about 3° across.

(visual) filter passes only radiation in the green–yellow range (the part of the spectrum to which human eyes happen to be particularly sensitive). Figure 10.10 shows how these filters admit different amounts of light for objects of different temperatures.

In curve (a) of Figure 10.10, corresponding to a very hot 30,000-K emitter, considerably more radiation is received through the B filter than through the V filter. In curve (b), the temperature is 10,000 K, and the B and V intensities are about the same. In the cool 3000-K curve (c), far more energy is received in the V range than in the B range. In each case, it is possible to reconstruct the entire black-body curve on the basis of only those two measurements because no other black-body curve can be drawn through both measured points. To the extent that a star's spectrum is well approximated as a black-body, measurements of the B and V intensities are enough to specify the star's black-body curve and thus yield its surface temperature.

Table 10.1 lists surface temperatures and colors for a few well-known stars.

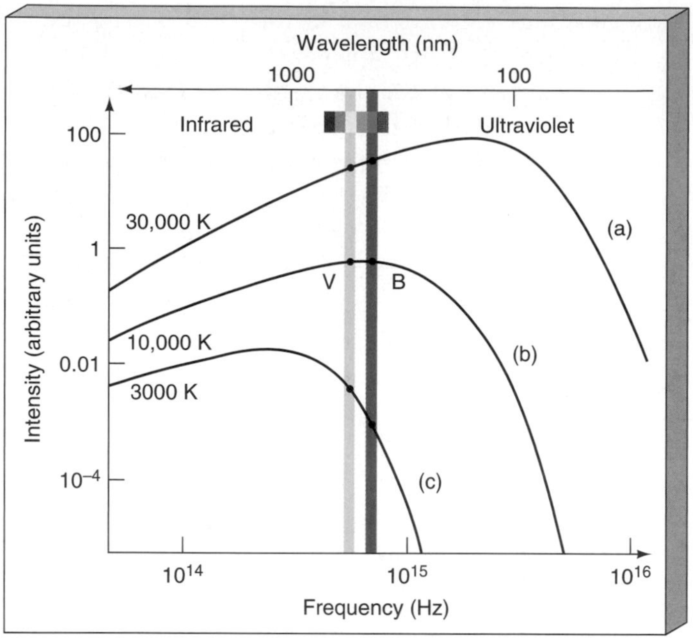

Figure 10.10 Black-body curves for three temperatures, along with the locations of the B (blue) and V (visual) filters. Star (a) is very hot—30,000 K—so its B intensity is considerably greater than its V intensity. Star (b) has roughly equal B and V readings and so appears white. Its temperature is about 10,000 K. Star (c) is red. Its V intensity greatly exceeds the B value, and its temperature is 3000 K.

Table 10.1 Stellar Colors and Temperatures

Surface Temperature (K)	Color	Familiar Examples
30,000	electric blue	
20,000	blue	Rigel
10,000	white	Vega, Sirius
7,000	yellow-white	Canopus
6,000	yellow	Sun, Alpha Centauri
4,000	orange	Arcturus, Aldebaran
3,000	red	Betelgeuse, Barnard's Star

10.6 The Classification of Stars

5 Astronomers can use color and temperature to classify stars reasonably well, but they often use a more detailed classification scheme. This scheme incorporates additional knowledge of stellar physics obtained through spectroscopy, the study of spectral-line radiation.

Detailed Spectra

Figure 10.11 compares the spectra of seven stars, arranged in order of decreasing surface temperature. All the spectra extend from 400 to 700 nm (4000–7000 Å), and each shows a series of dark absorption lines superimposed on a background of continuous color, like those of the Sun. ∞ (Sec. 9.3) However, the line patterns differ greatly from one star to another. Some stars display strong lines in the long-wavelength part of the spectrum, others have their strongest lines at short wavelengths. Still others show strong absorption lines spread across the whole visible spectrum. What do these differences tell us?

Although spectral lines of many elements are present at widely varying strengths, the differences among the spectra in Figure 10.11 are not due to differences in chemical composition—all seven stars are in fact more or less solar in their makeup. Instead, the differences are due almost entirely to the stars' *temperatures*. The spectrum at the top of Figure 10.11 is exactly what we expect from a star having solar composition and a surface temperature of about 30,000 K, the second from a 20,000-K star, and so on, down to the 3000-K star at the bottom.

The spectra of stars with surface temperatures exceeding 25,000 K usually show *strong* absorption lines of singly ionized helium and multiply ionized

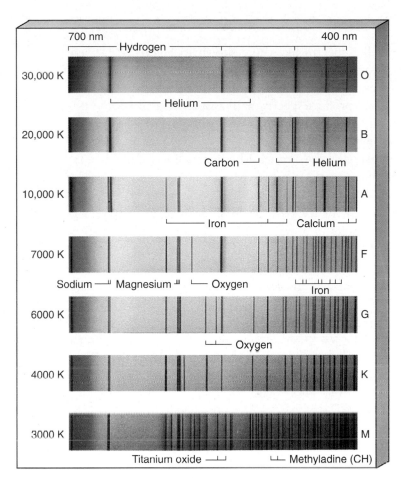

Figure 10.11 Comparison of spectra observed for seven stars having a range of surface temperatures. The spectra of the hottest stars, at the top, contain lines for helium and (although not indicated in the O spectrum shown here) lines for multiply-ionized heavy elements. In the coolest stars, at the bottom, there are no lines for helium, but lines of neutral atoms and molecules are plentiful. At intermediate temperatures, hydrogen lines are strongest. All seven stars have about the same chemical composition.

heavier elements, such as oxygen, nitrogen, and silicon (these latter lines are not shown in the spectra of Figure 10.11). These strong lines are not seen in the spectra of cooler stars because only very hot stars can excite and ionize these tightly bound atoms. In contrast, the hydrogen absorption lines in the spectra of very hot stars are relatively *weak*. The reason is not a lack of hydrogen—it is by far the most abundant element in all stars. At these high temperatures, however, much of the hydrogen is ionized, so there are few intact hydrogen atoms to produce strong spectral lines.

In cooler stars, with surface temperatures of around 10,000 K, hydrogen is responsible for the strongest absorption lines. This temperature is just right for electrons to move frequently between hydrogen's second and third orbitals, producing the characteristic red line at 656.3 nm (see Section 2.6, particularly Figure 2.20). Lines of tightly bound atoms—elements such as helium, oxygen, and nitrogen, which need lots of energy for excitation or ionization—are

rarely observed in the spectra of 10,000-K stars, whereas lines from more loosely bound atoms—such as calcium and titanium—are fairly common.

Even cooler stars, with surface temperatures of only a few thousand kelvins, show extremely weak hydrogen lines. As with the hottest stars, cooler stars do not have much electron traffic between the second and higher orbitals in hydrogen atoms. But unlike the hottest stars, in which most hydrogen atoms are highly excited or ionized, the hydrogen lines in cool stars are weak because most of the electrons are in the ground state. The most intense lines are due to weakly excited heavy atoms; no lines from ionized elements are seen. In fact, the average energy of the photons leaving the surface of the coolest stars is less than that needed to destroy some molecules, and many of the observed absorption lines are produced by molecules rather than by atoms (note the lines of the molecules titanium oxide and methyladine at the bottom of Figure 10.11).

Spectral Classification

Stellar spectra like those shown in Figure 10.11 were obtained for numerous stars well before the start of the twentieth century. Lacking full understanding of how atoms produce spectra, early workers classified stars primarily according to their hydrogen-line intensities. They adopted an A, B, C, D, . . . scheme in which A stars, with the strongest hydrogen lines, were thought to have more hydrogen than did B stars, and so on. The classification extended as far as the letter P.

In the 1920s, scientists began to understand the intricacies of atomic structure and the causes of spectral lines. Astronomers quickly realized that stars could be more meaningfully classified according to surface temperature. Instead of adopting an entirely new scheme, however, they chose to shuffle the existing alphabetical categories—those based on the strength of hydrogen lines—into a new sequence based on temperature. In the modern scheme, in order of decreasing temperature, the letters now run O, B, A, F, G, K, M. (The other letter classes have been dropped.) These stellar designations are called **spectral classes** (or *spectral types*). Use the mnemonic "**O**h, **Be A F**ine **G**uy, **K**iss **Me**" to remember them in the correct order.

Astronomers further subdivide each lettered spectral class into 10 subdivisions, denoted by the numbers 0–9. By convention, the lower the number, the hotter the star. Thus, for example, our Sun is classified as a G2 star (a little cooler than G1 and a little hotter than G3), Vega is A0, Barnard's Star is M5, Betelgeuse is M2, and so on. Table 10.2 lists the main properties of each stellar spectral class for the stars presented in Table 10.1.

10.7 The Hertzsprung–Russell Diagram

6 Astronomers use luminosity and surface temperature to classify stars in much the same way that height and weight might serve to classify the bulk properties of human beings. We know that people's height and weight are well correlated—tall persons tend to weigh more than short persons. We might nat-

Table 10.2 Stellar Spectral Classes

Spectral Class	Surface Temperature (K)	Prominent Absorption Lines	Familiar Examples
O	30,000	Ionized helium strong; multiply ionized heavy elements; hydrogen faint	
B	20,000	Neutral helium moderate; singly ionized heavy elements; hydrogen moderate	Rigel (B8)
A	10,000	Neutral helium very faint; singly ionized heavy elements; hydrogen strong	Vega (A0), Sirius (A1)
F	7,000	Singly ionized heavy elements; neutral metals; hydrogen moderate	Canopus (F0)
G	6,000	Singly ionized heavy elements; neutral metals; hydrogen relatively faint	Sun (G2), Alpha Centauri (G2)
K	4,000	Singly ionized heavy elements; neutral metals strong; hydrogen faint	Arcturus (K2), Adebaran (K5)
M	3,000	Neutral atoms strong; molecules moderate; hydrogen very faint	Betelgeuse (M2) Barnard's Star (M5)

the horizontal scale so that temperature increases conventionally to the right would play havoc with historical precedent.

The Main Sequence

The few stars plotted in Figure 10.12 give little indication of any particular connection between stellar properties. However, as Hertzsprung and Russell plotted more and more stellar temperatures and luminosities, they found that a relationship does in fact exist. Stars are *not* uniformly scattered across the H–R diagram. Instead, most are confined to a fairly well-defined band stretching diagonally from top left (high-temperature, high-luminosity) to bottom right (low-temperature, low-luminosity). In other words, cool stars tend to be faint (less luminous) and hot stars tend to be bright (more luminous). This band of stars spanning the H–R diagram is known as the **main sequence**. Figure 10.13 is an H–R diagram for stars lying within 5 pc of the Sun. Note that the vast majority of stars in the solar neighborhood lie on the main sequence.

The surface temperatures of main-sequence stars range from about 3000 K (spectral class M) to more than 30,000 K (spectral class O). This temperature range is relatively small—only a factor of 10. In contrast, the observed range in luminosities is very large, covering eight orders of magnitude (that is, a factor of 100 million), from 10^{-4} to 10^4 times the luminosity of the Sun.

Astronomers can use the radius–luminosity–temperature relationship (Section 10.3) to estimate the radii of main-sequence stars from the stars' temperatures and luminosities. They find that, in order to account for the observed range in luminosities, stellar radii must also vary along the main sequence. The faint, red M-type stars in the bottom right of the H–R diagram are only about 1/10 the size of the Sun, whereas the bright, blue O-type stars in the upper left are about 10 times larger than the Sun.

The dashed lines in Figure 10.13 represent constant stellar radius, meaning that any star lying on a given line has the same radius, regardless of its luminosity or temperature. Along a constant-radius line, the radius–luminosity–temperature relationship implies

$$\text{luminosity} \propto \text{temperature}^4.$$

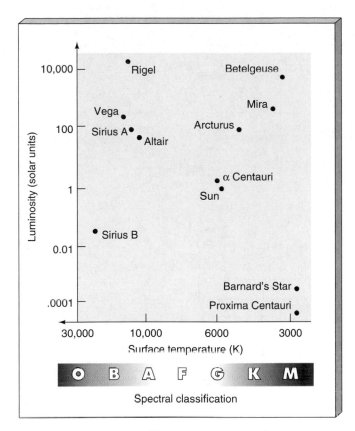

Figure 10.12 A plot of luminosity against surface temperature (or spectral class), known as an H–R diagram, is a useful way to compare stars. Plotted here are the data for some stars mentioned earlier in the text. The Sun, of course, has a luminosity of 1 solar unit. Its temperature, read off the bottom scale, is 5800 K—a G-type star. Similarly, the B-type star Rigel, at top left, has a temperature of about 15,000 K and a luminosity more than 10,000 times that of the Sun. The M-type star Proxima Centauri, at bottom right, has a temperature of less than 3000 K and a luminosity less than 1/10,000 that of the Sun.

urally wonder if the two basic stellar properties are also related in some way.

Figure 10.12 plots luminosity versus temperature for a few well-known stars. A figure of this sort is called a *Hertzsprung–Russell diagram*, or **H–R diagram**, after Danish astronomer Ejnar Hertzsprung and U.S. astronomer Henry Norris Russell, who independently pioneered the use of such plots in the second decade of the twentieth century. The vertical luminosity scale, expressed in units of solar luminosity (3.9×10^{26} W), extends over a large range, from 10^{-4} to 10^4. The Sun appears right in the middle of the luminosity range, at a luminosity of 1. Surface temperature is plotted on the horizontal axis, although in the unconventional sense of temperature increasing to the *left*, so that the spectral sequence O, B, A, . . . reads left to right. To change

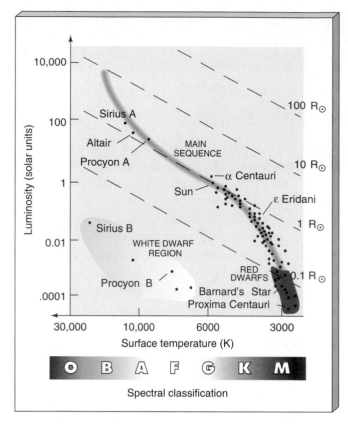

Figure 10.13 Most stars have properties within the shaded region of the H–R diagram known as the main sequence. The points plotted here are for stars lying within about 5 pc of the Sun. Each dashed diagonal line corresponds to a constant stellar radius, allowing us to indicate stellar size on the same diagram as stellar luminosity and temperature. (Here and below, the symbol ⊙ stands for the Sun.)

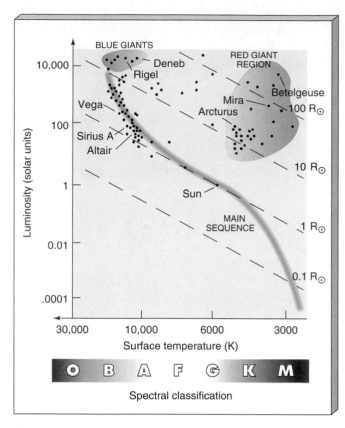

Figure 10.14 An H–R diagram for the 100 brightest stars in the sky. Such a plot is biased in favor of the most luminous stars—which appear toward the upper left—because we can see them more easily than we can the faintest stars. (Compare with Figure 10.13, which shows only the closest stars.)

By including such lines on our H–R diagrams, we can indicate stellar temperatures, luminosities, and radii on a single plot.

We see a very clear trend as we traverse the main sequence from top to bottom. At one end, the stars are large, hot, and very luminous. Because of their size and color, they are referred to as **blue giants**. The very largest are called **blue supergiants**. At the other end, stars are small, cool, and faint. They are known as **red dwarfs**. Our Sun lies right in the middle.

Figure 10.14 shows an H–R diagram for a different group of stars—the 100 stars of known distance having the greatest apparent brightness, as seen from Earth. Notice that here there are many more stars at the upper end of the main sequence than at the lower end. The reason for this excess of blue giants is simple—we can see very luminous stars a long way off. The stars shown in Figure 10.14 are scattered through a much greater volume of space than those in Figure 10.13, but the Figure 10.14 sample is heavily biased

toward the most luminous objects. In fact, of the 20 brightest stars in the sky, only 6 lie within 10 pc of us; the rest are visible, despite their great distances, because of their high luminosities.

If very luminous blue giants are overrepresented in Figure 10.14, low-luminosity red dwarfs are surely underrepresented. In fact, no dwarfs appear on this diagram. This absence is not surprising, because low-luminosity stars are difficult to observe from Earth. In the 1970s, astronomers began to realize that they had greatly underestimated the number of red dwarfs in our galaxy. As hinted at by the H–R diagram in Figure 10.13, which shows an unbiased sample of stars in the solar neighborhood, red dwarfs are actually the most common type of star in the sky. They probably account for upward of 80 percent of all stars in the universe.

White Dwarfs and Red Giants

Some of the points plotted in Figures 10.12–10.14 clearly do not lie on the main sequence. One such point in Figure 10.12 represents Sirius B, the white dwarf we met in Section 10.3. Its surface temperature (24,000 K) is about four times that of the Sun, but its luminosity is only 0.04 times the solar value. A few more such faint bluish-white stars can be seen in the bottom left-hand corner of Figure 10.13, in the **white dwarf region**.

Also shown in Figure 10.12 is Mira, with a surface temperature (3000 K) about half that of the Sun, but with luminosity some 400 times greater than the Sun's. Another point represents Betelgeuse, the ninth brightest star in the sky, a little cooler than Mira but more than 30 times more luminous. The upper right-hand corner of the H–R diagram (marked on Figure 10.14), where these stars are found, is called the **red giant region**. No red giants are found within 5 pc of the Sun (Figure 10.13), but many of the brightest stars seen in the sky are in fact red giants (Figure 10.14). Red giants are relatively rare, but they are so bright that they are visible to very great distances.

Although dwarfs and giants give some feeling for the extreme properties of stars, most stars have properties much more like our Sun's and lie on the main sequence. About 90 percent of all stars in our solar neighborhood, and presumably a similar percentage elsewhere in the universe, are main-sequence stars. About 9 percent of stars are white dwarfs, and 1 percent are red giants.

10.8 Extending the Cosmic Distance Scale

Spectroscopic Parallax

In Chapter 1 we introduced the first "rung" on a "ladder" of distance-measurement techniques that will ultimately carry us to the edge of the observable universe. That rung is radar ranging on the inner planets. ∞ (Sec. 1.6) It establishes the scale of the solar system and defines the astronomical unit. Earlier in this chapter, we discussed a second rung on the cosmic distance ladder—stellar parallax—which is based on the first, since Earth's orbit is the baseline. Now, having used the first two rungs to determine the distances and other physical properties of many nearby stars, we use that knowledge to construct a third rung on the ladder: **spectroscopic parallax**.[1] As illustrated schematically in

[1]This name is very misleading, as the method has nothing in common with stellar (geometric) parallax other than its use as a means of determining stellar distances.

Figure 10.15, this third rung expands our cosmic field of view still deeper into space.

We have already discussed the connections between absolute brightness (luminosity), apparent brightness, and distance. Knowledge of a star's apparent brightness and its distance allows us to determine its luminosity using the inverse-square law. But we can also turn the problem around. If we somehow knew a star's luminosity and then measured its apparent brightness, the inverse-square law would tell us its distance from the Sun.

Consider another analogy. Most of us have a rough idea of the approximate intrinsic brightness (that is, luminosity) of a typical traffic signal. Suppose we are driving down an unfamiliar street and see a red traffic light in the distance. Our knowledge of the light's luminosity enables us immediately to make a mental estimate of its distance. A light that appears relatively dim (low apparent brightness) must be quite distant (assuming it's not just dirty); a bright one must be relatively close. *A measurement of the apparent brightness of a light source, combined with some knowledge of its luminosity, can yield an estimate of its distance.* For stars, the trick is to find an independent measure of luminosity without knowing the distance. The H–R diagram can provide just that.

The main sequence represents a fairly close correlation between temperature and luminosity for most stars, with the exception of a few giants and dwarfs. Thus, the main sequence tells us about the *average*

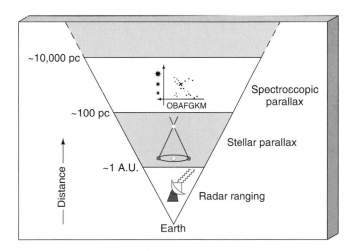

Figure 10.15 Knowledge of a star's luminosity and apparent brightness can yield an estimate of its distance. Astronomers use this third rung on our distance ladder, called spectroscopic parallax, to measure distances as far out as individual stars can be clearly discerned—several thousand parsecs.

properties of most stars. Let's imagine for a moment that the main sequence is a *line*, rather than a somewhat fuzzy band, in the H–R diagram, and that *all* stars lie on the main sequence. From a star's spectrum, we can determine its surface temperature. If the star lies on the main sequence, then there is only one possible luminosity corresponding to that temperature. We can read the star's luminosity directly off a graph such as Figure 10.13, and then determine its distance by measuring its apparent brightness and using the inverse-square law. The existence of the main sequence allows us to make a connection between an easily measured quantity (temperature) and a star's luminosity, which would otherwise be unknown. The term *spectroscopic parallax* refers to this whole process of using stellar spectra to infer distances.

Spectroscopic parallax can be used to determine stellar distances out to several thousand parsecs. Beyond that, spectra and colors of individual stars are difficult to obtain. The "standard" main sequence is obtained from H–R diagrams of stars whose distances can be measured by stellar (geometric) parallax, so the method of spectroscopic parallax is calibrated using nearby stars. Note that, in using this method, we are assuming (without proof) that distant stars are basically similar to nearby stars—in particular, that *distant stars fall on the same main sequence as nearby stars*. Only by making this assumption can we use spectroscopic parallax to expand the boundaries of our distance-measurement techniques.

Of course, the main sequence is not really a line in the H–R diagram: it has some thickness. For example, the luminosity of a main-sequence G2 star can be anywhere from about 0.5 to 1.5 times the luminosity of the Sun. As a result, there is an uncertainty in the luminosity used in the method of spectroscopic parallax, and hence an uncertainty in the distance. Distances obtained by spectroscopic parallax are probably accurate to no better than 25 percent. Although this may not seem very accurate—a cross-country traveler in the United States would hardly be impressed to be told that the best estimate of the distance between Los Angeles and New York is somewhere between 3000 and 5000 km—it illustrates the point that, in astronomy, even something as simple as the distance to another star can be very difficult to measure. Still, an estimate with an uncertainty of ±25 percent is far better than no estimate at all.

Luminosity Class

If a star happens to be a red giant or a white dwarf, its distance determined by spectroscopic parallax will be incorrect. We could argue that, because roughly 90 percent of all stars lie on the main sequence, the assumption that a star is a main-sequence star is valid 9 out of 10 times. In fact, astronomers can do much better than that. We mentioned in Chapter 2 that detailed analysis of the *widths* of spectral lines can provide information on the pressure, and hence the density, of the gas where the line formed. ∞ (Sec. 2.7) The atmosphere of a red giant is much less dense than that of a main-sequence star, and this in turn is much less dense than the atmosphere of a white dwarf. By studying a star's spectral lines, astronomers can usually tell with a high degree of confidence whether or not it lies on the main sequence.

Over the years, astronomers have developed a system for classifying stars according to the widths of their spectral lines. Because line width is particularly sensitive to density in the stellar photosphere, and because this density in turn is well correlated with luminosity, this stellar property has come to be known as **luminosity class**. This classification scheme provides a means for astronomers to distinguish supergiants from giants, giants from main-sequence stars, and main-sequence stars from white dwarfs.

The standard stellar luminosity classes are given in Table 10.3. Their locations on the H–R diagram are indicated in Figure 10.16. Now we have a way of specifying a star's location in the diagram in terms of properties that are measurable by purely spectroscopic

Table 10.3 Stellar Luminosity Classes	
Class	**Description**
Ia	Bright supergiants
Ib	Supergiants
II	Bright giants
III	Giants
IV	Subgiants
V	Main-sequence stars/dwarfs

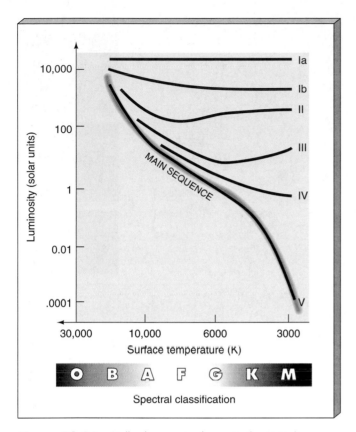

Figure 10.16 Stellar luminosity classes in the H–R diagram. A star's location in the diagram could be specified by its spectral type and luminosity class instead of by its temperature and luminosity.

a measure of the star's physical state. Knowledge of luminosity classes allows us to use spectroscopic parallax with some confidence that we are not accidentally counting a red giant or a white dwarf as a main-sequence star and making a huge error in our distance estimate as a result.

10.9 Stellar Mass

7 What ultimately determines a star's position on the main sequence? The answer is its *mass* and its *composition*. These are fundamental properties of any star, set once and for all at the time of its birth. Together, they uniquely determine the star's internal structure, its external appearance, and even (as we will see in Chapter 12) its future evolution. The ability to measure these two key stellar properties is therefore of the utmost importance if we are to understand how stars work. We have already seen how spectroscopy is used to determine a star's composition. Now let's turn to the problem of finding a star's mass.

As with all other objects, we measure a star's mass by observing its gravitational influence on some nearby body—another star, perhaps, or a planet. If we know the distance between the two bodies, then we can use Newton's laws to calculate their masses.

Binary Stars

Most stars are members of multiple-star systems—groups of two or more stars in orbit around one another. The majority of stars are found in **binary-star**

means; spectral type and luminosity class locate a star just as surely as do temperature and luminosity. The full specification of a star's spectral properties includes its luminosity class. For example, the Sun, on the main sequence, is of class G2V, Vega is A0V, the red dwarf Barnard's Star is M5V, the red supergiant Betelgeuse is M2Ia, and so on.

Consider, for example, a K7-type star (Table 10.4) with a surface temperature of approximately 4000 K. If the widths of the star's spectral lines tell us that it lies on the main sequence (that is, it is a K7V star), then its luminosity is about 0.1 times the solar value. If its spectral lines are observed to be narrower than lines normally found in main-sequence stars, the star may be recognized as a giant, with a luminosity of 20 solar luminosities (Figure 10.16). If the lines are very narrow, the star might instead be classified as a K7Ib supergiant, brighter by a further factor of 150, at 3000 solar luminosities. In this way, the observed width of the star's spectral lines translates directly into

Table 10.4 Variation in Stellar Properties Within a Spectral Class

Surface Temperature (K)	Luminosity (solar luminosities)	Radius (solar radii)	Object
4000	0.1	0.7	K7V (main-sequence star)
4000	20	10	K7III (giant)
4000	3000	100	K7Ib (supergiant)

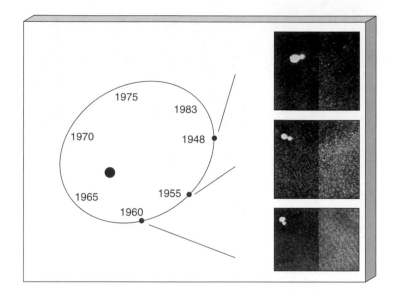

Figure 10.17 The periods and separations of binary stars can be observed directly if each star is clearly seen.

systems, which consist of two stars in orbit about their common center of mass, held together by their mutual gravitational attraction. (The Sun is not part of a multiple-star system; if it has anything at all uncommon about it, it is this lack of stellar companions.)

Astronomers classify binary-star systems (or *binaries*) according to their appearance from Earth and the ease with which they can be observed. **Visual binaries** (Figure 10.17) have widely separated members bright enough to be observed and monitored separately. Others, known as **spectroscopic binaries,** are too distant from us to be resolved into separate stars, but they can be indirectly perceived by monitoring the back-and-forth Doppler shifts of their spectral lines as the stars orbit one another. Recall that motion toward an observer blueshifts the lines, and motion away from the observer redshifts them. ∞ *(More Precisely 2-2)*

In a *double-line* spectroscopic binary, two distinct sets of spectral lines—one for each component star—shift back and forth as the stars move. Because we see particular lines alternately approaching and receding, we know that the objects emitting the lines are in orbit. In the more common *single-line* systems (Figure 10.18), one star is too faint for its spectrum to be distinguished, so we see only one set of lines shifting back and forth. This shifting means that the detected star must be in orbit around another star, even though the companion cannot be observed directly.

In the much less common **eclipsing binaries,** the orbital plane of the pair of stars is almost edge-on to our line of sight. In this situation, we observe a periodic decrease of starlight intensity as one member of the binary passes in front of the other (Figure 10.19). By studying the variation of the light from an eclipsing

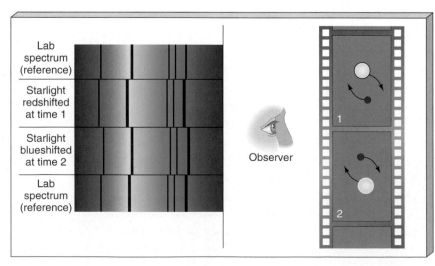

Figure 10.18 Binary properties can be determined indirectly by measuring the periodic Doppler shift of one star relative to the other as they move in their orbits. The diagram shows a single-line system, in which only one spectrum (from the brighter component) is visible. (The observer is situated to the left of the diagram. Reference laboratory spectra are shown at top and bottom.)

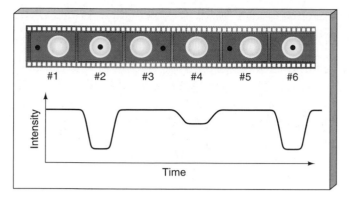

Figure 10.19 If the two stars in a binary happen to eclipse one another, additional information on their radii and masses can be obtained by observing the periodic decrease in starlight intensity as one star passes in front of the other.

binary system—called the binary's **light curve**—we can derive detailed information not only about the stars' orbits and masses but also about their radii.

Note that these categories are not mutually exclusive. For example, an eclipsing binary may also be (and in fact often is) a spectroscopic binary system.

By observing the orbits of the stars in a binary, the back-and-forth motion of the stellar spectral lines, or the dips in the binary's light curve—whatever information is available—we can measure the binary's orbital period. Observed binary periods span a broad range—from hours to centuries. Doppler-shift measurements give us information on the orbital velocities of the member stars. In addition, if the distance to a visual binary is known, the size (semi-major axis) of its orbit can be determined by simple geometry.

Knowledge of a binary's period and orbit size is all we need to determine the combined mass of the component stars, using the modified form of Kepler's third law discussed in Chapter 1. ∞ (Sec. 1.7) Additional observations are needed to determine the individual masses of the components. For example, in any binary system, each star orbits the system's center of mass. Measuring the distance from each star to the center of mass of a visual binary yields the ratio of the stellar masses. Knowing both the sum of the masses and their ratio, we can then calculate the mass of each star.

The individual component masses of a single-line spectroscopic binary system generally cannot be determined (unless the binary happens also to be an eclipsing system, in which case the extra information allows the mass ratio to be inferred). If a binary is too distant, or if only one component is visible, we cannot deter-

mine the component masses, only their sum. Nevertheless, for many nearby systems, individual masses can be obtained.

Virtually all we know about stellar masses is based on observations of binary systems. Consider, for example, the nearby binary system comprising the bright star Sirius A and its faint companion Sirius B. Observations of their orbit show that the sum of their masses is three times the mass of the Sun. Further observations show Sirius A to have roughly twice the mass of its companion. It follows that the mass of Sirius A is 2 solar masses and that of Sirius B is 1 solar mass.

Mass and Other Stellar Properties

Figure 10.20 is an H–R diagram showing how stellar mass varies along the main sequence. There is a clear progression from low-mass red dwarfs to high-mass blue giants. With few exceptions, main-sequence stars range in mass from about 0.1 to 20 times the mass of

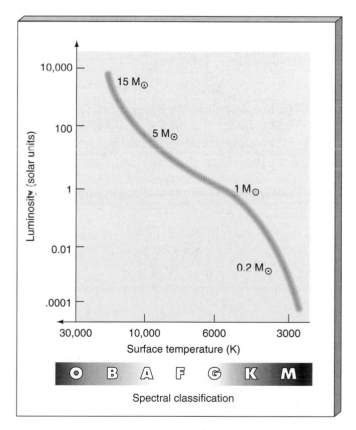

Figure 10.20 More than any other stellar property, mass determines a star's position on the main sequence. Low-mass stars are cool and faint; they lie at the bottom of the main sequence. Very massive stars are hot and bright; they lie at the top of the main sequence.

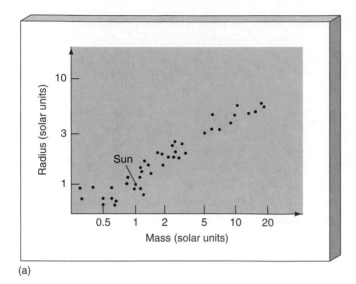

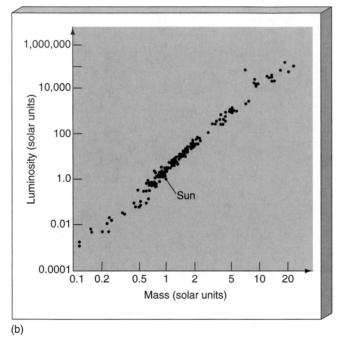

Figure 10.21 (a) Dependence of stellar radius on mass for main-sequence stars. The radius increases roughly in proportion to the mass over much of the range. (b) Dependence of luminosity on mass. The luminosity increases roughly as the cube of the mass.

the Sun. The hot O- and B-type stars are generally about 10 to 20 times more massive than our Sun. The coolest K- and M-type stars contain only a few tenths of a solar mass. Because all other stellar properties are determined by a star's mass, we can say that *the mass of a star at the time of formation determines its location on the main sequence.*

Figure 10.21 illustrates how the radii and luminosities of main-sequence stars depend on mass. The two plots, called the *mass–radius* and *mass–luminosity*

relationships, are based on observations of binary-star systems. Along the main sequence, both radius and luminosity increase with mass. As a rule of thumb, radius rises in direct proportion to mass, whereas luminosity increases much faster—more like the *cube* of the mass. For example, a 2-solar-mass main-sequence star has a radius roughly twice that of the Sun and a luminosity of 8 (2^3) solar luminosities; a 0.2-solar-mass main-sequence star has a radius of about 0.2 solar radii and a luminosity of 0.008 (0.2^3) solar luminosities.

Stellar Lifetimes

The rapid rate of nuclear burning deep inside a star releases vast amounts of energy per unit time. How long can the fire continue to burn? We can estimate a star's lifetime by dividing the amount of fuel available (the mass of the star) by the rate at which the fuel is being consumed (the star's luminosity), so we can say

$$\text{stellar lifetime} \propto \frac{\text{stellar mass}}{\text{stellar luminosity}}.$$

For example, O and B stars have masses 10 to 20 times that of the Sun and luminosities thousands of times higher than the solar luminosity. Accordingly, these massive stars can survive only for short times. Their nuclear reactions proceed so rapidly that their fuel is quickly depleted despite their large masses. Using the mass–luminosity relationship, we see that the lifetime of a 20-solar-mass O star is roughly $20/20^3 = 1/400$ of the (10-billion-year) lifetime of the Sun, or about 25 million years. We can be sure that all the O and B stars we now observe are quite young—less than a few tens of millions of years old. Massive stars older than that have already exhausted their fuel and no longer emit large amounts of energy. They have, in effect, died.

At the opposite end of the main sequence, the cooler K and M stars have less mass than our Sun. With their low core densities and temperatures, their proton–proton reactions churn away rather sluggishly, much more slowly than those in the Sun's core. The small energy release per unit time leads to low luminosities for these stars, so they have very long lifetimes. Many of the K- and M-type stars now visible in the sky will shine on for at least another trillion years.

Table 10.5 compares some key properties of several well-known main-sequence stars, arranged in order of decreasing mass. Notice that the central tempera-

Table 10.5 Key Properties of Some Well-Known Stars

Star	Spectral/ Luminosity Class	Mass (solar masses)	Central Temperature (10^6 K)	Luminosity (solar luminosities)	Estimated Lifetime (10^6 years)
Rigel	B8Ia	10	30	44,000	20
Sirius	A1V	2.3	20	23	1,000
Alpha Centauri	G2V	1.1	17	1.4	7,000
Sun	G2V	1.0	15	1.0	10,000
Proxima Centauri	M5V	0.1	0.6	0.00006	>1,000,000

ture differs very little from one star to another and that the spread in stellar luminosities and lifetimes is very large.

10.10 Star Clusters

8 When trying to obtain an H–R diagram for a distant region of the Galaxy, astronomers face a problem. In order to plot the diagram, we must know luminosities; and in order to know luminosities, we must know distances. Thus it would seem impossible to construct H–R diagrams for stars more distant than 100 pc or so, the maximum distance measurable by stellar parallax. (We can't use the method of spectroscopic parallax because that method *assumes* the properties of the main sequence.)

In some circumstances, however, it is possible to plot an H–R diagram for very distant stars even though their distances are not known. If we observe a group of stars that all lie at the same distance from us, then comparing *apparent* brightnesses is equivalent to comparing *absolute* brightnesses. Why? Because as the radiation travels toward Earth, the brightness of every star in the group is diminished by the same amount, according to the inverse-square law. By measuring and plotting apparent brightnesses, we can create a "relative" H–R diagram for the group that looks (apart from the numbers on the vertical axis) exactly the same as an H–R diagram based on luminosities. Such an easily recognizable group of distant stars is called a **star cluster**.

Star clusters can contain anywhere from a few dozen to a million stars in a region a few parsecs across. Astronomers believe that all the stars in a given clus-

ter formed at the same time out of the same cloud of interstellar gas, and under the same environmental conditions. Thus, when we look at a star cluster, we are looking at a group of stars that all have the same age, are similar in chemical composition, and lie in the same region of space, at essentially the same distance from Earth. Unlike the stars plotted in Figures 10.12–10.14, which differ in mass, age, and (to a lesser extent) composition, the only factor distinguishing one star in a cluster from another in the same cluster is mass.

Clusters are therefore near-ideal "laboratories" for stellar studies—not in the sense that astronomers can perform experiments on the stars in them, but because the properties of the stars are very tightly constrained. Hence theoretical models of star formation and evolution can be compared with reality without the major complications introduced by broad spreads in age, chemical composition, and place of origin. Clusters are of central importance to astronomers who wish to understand how stars evolve in time.

Open Clusters

Figure 10.22(a) shows a small star cluster called the Pleiades, or Seven Sisters, a well-known naked-eye object in the constellation Taurus. Individual stellar colors provide an estimate of the surface temperature of each star in the cluster. Luminosities follow directly from measurements of apparent brightness and the cluster's distance (which in this case is known to be about 120 pc). Figure 10.22(b) shows the cluster H–R diagram obtained from these data. This type of loose, irregular cluster, found mainly in the strip across the sky known as the Milky Way, is called an **open cluster**.

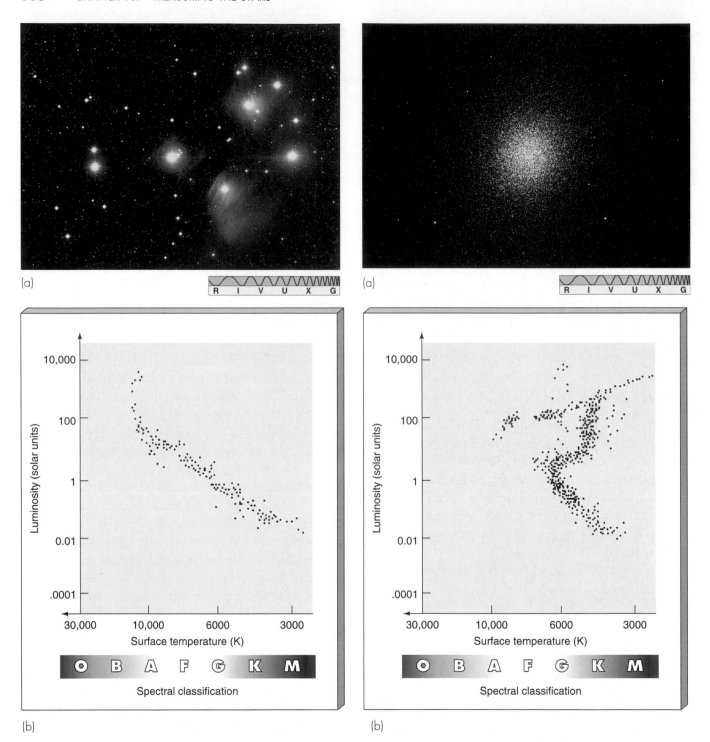

(a)

(b)

Figure 10.22 (a) The Pleiades cluster (also known as the Seven Sisters, or M45) lies about 120 pc from the Sun. (b) An H–R diagram for the stars of this well-known open cluster.

Figure 10.23 (a) The globular cluster Omega Centauri is approximately 5000 pc from Earth and some 40 pc in diameter. (b) An H–R diagram for some of its stars.

Open clusters typically contain from a few tens to a few thousands of stars and are a few parsecs across.

Figure 10.22(b) shows that the Pleiades cluster contains stars falling in all parts of the main sequence. The blue stars must be relatively young, for, as we have

seen, they burn their fuel rapidly. If all the stars in the cluster formed at the same time, then the red stars must be young too. Thus, even though we have no direct evidence of the cluster's birth, we can estimate its age as less than 20 million years, the lifetime of an O star.

Globular Clusters

A second type of stellar cluster, of which a representative is shown in Figure 10.23(a), is called a **globular cluster**. Globular clusters are much more tightly knit than open clusters. All globular clusters are roughly spherical (which accounts for their name) and contain hundreds of thousands, and sometimes millions, of stars spread out over about 50 pc. As with open clusters, the entire assemblage is held together by gravity.

Figure 10.23(b) shows an H–R diagram for this cluster, which is called Omega Centauri. Notice its many differences from Figure 10.22(b)—globular clusters are a very different stellar environment from open clusters like the Pleiades. The distance to this globular cluster has been determined by a variation on the method of spectroscopic parallax, applied to the entire cluster rather than to individual stars. By calculating the distance at which the apparent brightnesses of the cluster's stars taken as a whole best match a standard main sequence and theoretical models of stellar evolution, the cluster is found to lie about 5000 pc from Earth.

The most outstanding feature of globular clusters is their lack of O and B stars. This deficit is clear in Figure 10.23(b). Low-mass red stars and intermediate-mass yellow stars abound, but high-mass white or blue stars are very rare. In fact, globular clusters contain no main-sequence stars with masses greater than about 0.8 times the mass of the Sun. (The A stars in this plot are stars at a much later stage in their evolution that happen to be passing through the location of the upper main sequence.) Apparently, globular clusters formed long ago; the more massive O through F stars have long since exhausted their nuclear fuel and disappeared from the main sequence (in fact, becoming the red giants and other luminous stars above the main sequence, as we will discuss in Chapter 12).

On the basis of these and other observations, astronomers estimate that all globular clusters are at least 10 billion years old. They contain the oldest known stars in our Galaxy. As such, globular clusters are considered to be remnants of the earliest stages of our Galaxy's existence.

We will never be able to watch a single star move through all its evolutionary phases. The lifetimes of humans—even of human civilizations—are far too short compared with the lifetimes of even the shortest-lived O and B stars. Instead, we must observe stars as they presently exist, through snapshots taken at specific moments in their life cycles. The H–R diagram is just such a snapshot. By studying stars of different ages or, even better, by studying stars in clusters, where the ages are known to be the same, we can patch together an understanding of a star's "life story" without having to follow a few individuals from birth to death. Such evolutionary studies will be the subject of the next two chapters.

Chapter Review

Summary

The distances to the nearest stars can be measured by stellar parallax. A star with a parallax of one arc second is 1 **parsec** (p. 280)—about 3.3 light years—away from the Sun. Stars have real motion through space as well as apparent motion. A star's **proper motion** (p. 281), which is its true motion across the sky, is a measure of the star's velocity perpendicular to our line of sight. The star's radial velocity—along our line of sight—is measured by the Doppler shift of its spectral lines.

Only a few stars are large enough and close enough to Earth that their radii can be measured directly. The sizes of most stars are estimated indirectly through the **radius–luminosity–temperature relation**ship (p. 283). Stars are categorized as **dwarfs** (p. 284) comparable in size to or smaller than the Sun, **giants** (p. 283) up to 100 times larger than the Sun, and **supergiants** (p. 283) more than 100 times larger than the Sun. In addition to "normal" stars such as the Sun, two other important classes of star are **red giants** (p. 283), which are large, cool, and luminous, and **white dwarfs** (p. 284), which are small, hot, and faint.

The **absolute brightness** (p. 284) of a star is equivalent to its luminosity. The **apparent brightness** (p. 284) of a star is the rate at which energy from the star reaches a detector. Apparent brightness falls off as the inverse square of the distance. Optical astronomers use the **magnitude scale** (p. 285) to express and compare stellar brightnesses. The greater the magnitude, the fainter the star; a difference of 5 magnitudes corresponds to a factor of 100 in brightness. **Apparent magnitude** (p. 286) is a measure of apparent brightness. The **absolute magnitude** (p. 286) of a star is the apparent magnitude it would have if placed at a standard distance of 10 pc from the viewer. It is a measure of the star's luminosity.

Astronomers often measure the temperatures of stars by measuring their brightnesses through two or more optical filters, then fitting a black-body curve to the results. Astronomers also classify stars according to the absorption lines in their spectra. The lines seen in the spectrum of a given star depend mainly on the star's temperature, and spectroscopic observations of stars provide an accurate means of determining both stellar temperatures and stellar composition. The standard stellar **spectral classes** (p. 290), in order of decreasing temperature, are O B A F G K M.

A plot of stellar luminosity versus stellar spectral class (or temperature) is called an **H–R diagram** (p. 291). About 90 percent of all stars plotted on an H–R diagram lie on the **main sequence** (p. 291), which stretches from hot, bright **blue supergiants** (p. 292) and **blue giants** (p. 292), through intermediate stars such as the Sun, to cool, faint **red dwarfs** (p. 292). Most main-sequence stars are red dwarfs; blue giants are quite rare. About 9 percent of stars lie in the **white dwarf region** (p. 293); the remaining 1 percent lie in the **red giant region** (p. 293).

By careful spectroscopic observations, astronomers can determine a star's **luminosity class** (p. 294), allowing them to distinguish main-sequence stars from red giants or white dwarfs of the same spectral type (or color). Once a star is known to be on the main sequence, measurement of its spectral type allows its luminosity to be estimated and its distance to be measured. This method of distance determination, which is valid for stars up to several thousand parsecs from Earth, is called **spectroscopic parallax** (p. 293).

Most stars are not isolated in space but instead orbit other stars in **binary-star systems** (p. 295). In a **visual binary** (p. 296), both stars can be seen and their orbits charted. In a **spectroscopic binary** (p. 296), the stars cannot be resolved, but their orbital motion can be detected spectroscopically. In an **eclipsing binary** (p. 296), the orbits are oriented in such a way that one star periodically passes in front of the other as seen from Earth and dims the light we receive. The binary's **light curve** (p. 297) is a plot of its apparent brightness as a function of time.

Studies of binary stars often allow stellar masses to be measured. The mass of a star determines its size, temperature, and brightness. Fairly well-defined mass–radius and mass–luminosity relationships exist for main-sequence stars. Hot blue giants are much more massive than the Sun; cool red dwarfs are much less massive.

The lifetime of a star can be estimated by dividing its mass by its luminosity. High-mass stars burn their fuel rapidly and have much shorter lifetimes than the Sun. Low-mass stars consume their fuel slowly and may remain on the main sequence for trillions of years.

Many stars are found in compact groups known as **star clusters** (p. 299). **Open clusters** (p. 299), which are loose, irregular clusters that typically contain from a few tens to a few thousands of stars, are found mostly in the strip of sky known as the Milky Way. They typically contain many bright blue stars, indicating that they formed relatively recently. **Globular clusters** (p. 301) are much more tightly knit than open clusters, are spherical, and may contain millions of stars. They include no main-sequence stars more massive than the Sun, indicating that they formed long ago. Globular clusters are believed to date back to the formation of our Galaxy.

Self-Test: True or False?

_____ **1.** One parsec is a little more than 200,000 A.U.

_____ **2.** There are no stars within 1 pc of the Sun.

_____ **3.** Parallax can be used to measure stellar distances out to about 10,000 pc.

_____ **4.** Most stars have radii between 0.1 and 10 times the radius of the Sun.

_____ **5.** Star A appears brighter than star B, as seen from Earth. Therefore, star A must be closer to Earth than star B.

_____ **6.** Star A and star B have the same luminosity, but star B is twice as distant as star A. Therefore, star A appears four times brighter than star B.

_____ 7. A star of apparent magnitude 5 looks brighter than one of apparent magnitude 2.

_____ 8. Differences among stellar spectra are mainly due to differences in star composition.

_____ 9. Cool stars have very strong hydrogen lines in their spectra.

_____ 10. A G9 star is cooler than a G5 star.

_____ 11. Red dwarfs lie in the lower left part of the H–R diagram.

_____ 12. The brightest stars visible in the night sky are all found in the upper part of the H–R diagram.

_____ 13. In a spectroscopic binary, the orbital motion of the component stars appears as variations in their radial velocities.

_____ 14. It is impossible to have a one-billion-year-old O or B main-sequence star.

_____ 15. Globular clusters generally contain between a few hundred and a few thousand stars.

Self-Test: Fill in the Blank

1. Parallax measurements of the distances to the stars nearest Earth use _____ as a baseline.

2. The radial velocity of a star is determined by observing its _____ and using the _____ effect.

3. To determine the transverse velocity of a star, both its _____ and its _____ must be known.

4. The radius of a star can be indirectly determined if the _____ and _____ of the star are known.

5. The smallest stars normally plotted on the H–R diagram are _____.

6. Observations of stars through B and V filters are used to determine stellar _____.

7. The hottest stars show little evidence of hydrogen in their spectra because hydrogen is mostly _____ at these temperatures.

8. The coolest stars show little evidence of hydrogen in their spectra because hydrogen is mostly _____ at these temperatures.

9. The Sun has a spectral type of _____.

10. The H–R diagram is a plot of _____ on the horizontal scale versus _____ on the vertical scale.

11. The band of stars extending from the top left of the H–R diagram to its bottom right is known as the _____.

12. The large, cool stars found at the upper right of the H–R diagram are _____.

13. The small, hot stars found at the lower left of the H–R diagram are _____.

14. _____-star systems are important for providing measurements of stellar masses.

15. Going from spectral type O to M along the main sequence, stellar masses _____.

16. A group of many stars bound together by gravity is called a _____.

Review and Discussion

1. How is stellar parallax used to measure the distances to stars?

2. What is a parsec? Compare it to the astronomical unit.

3. Explain two ways in which a star's real space motion translates into motion observable from Earth.

4. Describe some characteristics of red giants and white dwarfs.

5. What is the difference between absolute and apparent brightness?

6. How do astronomers measure star temperatures?

7. Briefly describe how stars are classified according to their spectral characteristics.

8. What information is needed to plot a star on the Hertzsprung-Russell diagram?

9. What is the main sequence? What basic property of a star determines where it lies on the main sequence?

10. How are distances determined using spectroscopic parallax?

11. Which stars are most common in the Milky Way Galaxy? Why don't we see many of them in H–R diagrams?

12. How can stellar masses be determined by observing binary star systems?

13. If a high-mass star starts off with much more fuel than a low-mass star, why doesn't the high-mass star live longer?

14. Compare and contrast the properties of open star clusters and globular star clusters.

15. In general, is it possible to determine the age of an individual star simply by noting its position on an H–R diagram?

Problems

1. How far away is the star Spica, whose parallax is 0.013″?

2. A certain star has a temperature twice that of the Sun and a luminosity 64 times that of the Sun. What is its radius, in units of the solar radius?

3. Two stars—A and B, of luminosities 0.5 and 4.5 times the luminosity of the Sun, respectively—are observed to have the same apparent brightness. Which one is more distant, and how much farther away is it than the other?

4. Two stars—A and B, of absolute magnitudes 3 and 8, respectively—are observed to have the same apparent magnitude. Which one is more distant, and how much farther away is it than the other?

5. Given that the Sun's lifetime is about 10 billion years, estimate the life expectancy of (a) a 0.2-solar-mass, 0.01-solar-luminosity red dwarf, (b) a 3-solar-mass, 30-solar-luminosity star, (c) a 10-solar-mass, 1000-solar-luminosity blue giant.

Projects

1. Every winter, you can find an astronomy lesson in the evening sky. The Winter Circle is an *asterism*—or pattern of stars—made up of six bright stars in five constellations: Sirius, Rigel, Betelgeuse, Aldebaran, Capella, and Procyon. These stars span nearly the entire range of colors (and therefore temperatures) possible for normal stars. Rigel is a B star, Sirius an A, Procyon an F, Capella a G, Aldebaran a K, and Betelgeuse an M. The color differences of these stars are easy to see. Why do you suppose there is no O star in the Winter Circle?

2. Summer is a good time to search with binoculars for open star clusters. Open clusters are generally found in the hazy band of the Milky Way arcing across your night sky. If you are far from city lights and looking at an appropriate time of night and year—you can simply sweep with your binoculars along the Milky Way. Numerous clumps of stars will pop into view. Many will turn out to be open clusters.

3. Globular clusters are harder to find. They are intrinsically larger, but they are also much farther away and therefore appear smaller in the sky. The most famous globular cluster visible from the Northern Hemisphere is M13 in the constellation Hercules, visible on spring and summer evenings. This cluster contains half a million or so of the Galaxy's most ancient stars. Through binoculars, it may be glimpsed as a little ball of light, located in the constellation Hercules, about one-third of the way from the star Eta to the star Zeta. Telescopes reveal this cluster as a magnificent, symmetrical grouping of stars.

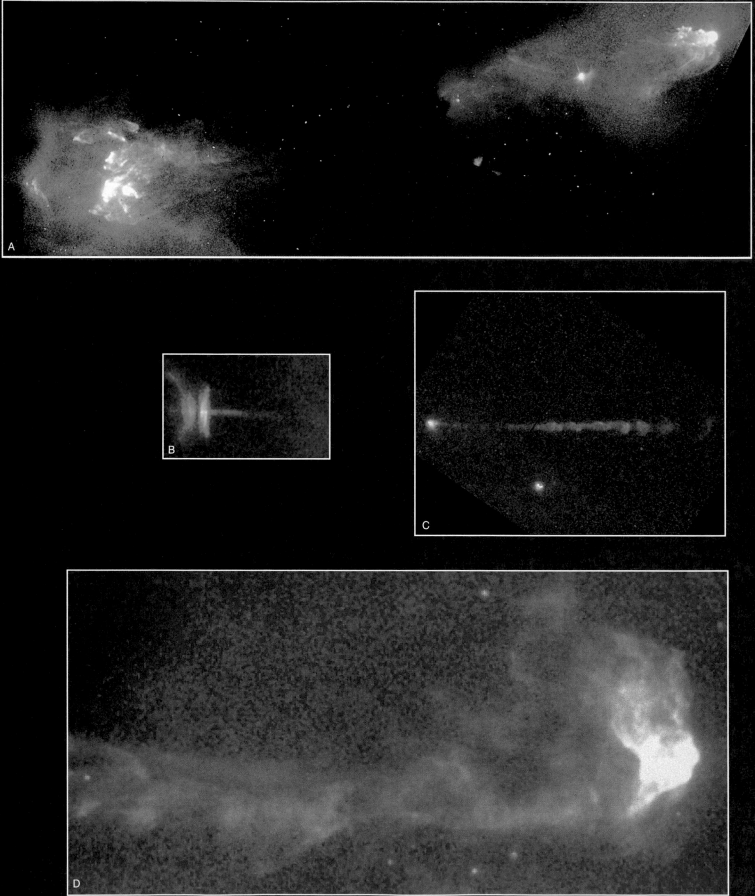

11

The Interstellar Medium

Birthplace of Stars

(Photo A) Disks and jets pervade the universe on many scales. This image shows a small region near the Orion Nebula known as HH1/HH2, whose twin jets have blasted outward for several trillion km (nearly half a light year) before colliding with interstellar matter.

(Photo B) This image of HH30, spanning approximately 250 billion km, or about 0.01 pc, shows a thin jet (in red) emanating from a circumstellar disk (at left in grey) encircling a nascent star.

(Photo C) One of the HH34's jets is longer, reaching some 600 billion km.

(Photo D) HH47 is more than a trillion km in length, or nearly 0.1 pc, its jets creating bow shocks in the process.

LEARNING GOALS

Studying this chapter will enable you to:

1 Summarize the composition and physical properties of the interstellar medium.

2 Describe the characteristics of emission nebulae, and explain their significance in the life cycle of stars.

3 Discuss the nature of dark interstellar clouds.

4 Specify the radio techniques used to probe the nature of interstellar matter.

5 Summarize the sequence of events leading to the formation of a star like our Sun.

6 Explain how the formation of a star is affected by its mass.

7 Describe some of the observational evidence supporting the modern theory of star formation.

Stars and planets are not the only inhabitants of our Galaxy. The space around us harbors invisible matter throughout the dark voids between the stars. The density of this interstellar matter is extremely low—approximately a trillion trillion times less dense than matter in either stars or planets, far more tenuous than the best vacuum attainable on Earth. Only because the volume of interstellar space is so vast does the mass of interstellar matter amount to anything at all. So why bother to study this near-perfect vacuum? We do so for three important reasons. First, there is as much mass in the "voids" between the stars as there is in the stars themselves. Second, interstellar space is the region out of which new stars are born. Third, it is the region into which some old stars explode at death. It is one of the most significant crossroads through which matter passes in the history of our universe.

11.1 Interstellar Matter

1 Figure 11.1 shows a much greater expanse of universal "real estate" than anything we have studied thus far. The bright regions are congregations of innumerable stars, merging together into a continuous blur at the resolution of our telescope. However, the dark areas are *not* simply "holes" in the stellar distribution. They are regions of space where *interstellar matter* obscures the light from stars beyond.

From Figure 11.1, it is evident that the dark interstellar matter is distributed very unevenly. In some directions, it is largely absent, allowing astronomers to study objects literally billions of parsecs from the Sun.

In other directions, there are small amounts of interstellar matter, so the obscuration is moderate, preventing us from seeing objects more than a few thousand parsecs away, but allowing us to study nearby stars. Still other regions are so heavily obscured by interstellar matter that starlight from even relatively nearby stars is completely absorbed before reaching Earth.

Gas and Dust

The matter between the stars is collectively termed the **interstellar medium**. It is made up of two components—gas and dust—intermixed throughout all of space.

Figure 11.1 A wide-angle photograph of a great swath of space, showing regions of brightness (vast fields of stars) as well as regions of darkness (where interstellar matter obscures the light from more distant stars). The field of view is roughly 30° across.

R   I   V   U   X   G

Interstellar gas is made up mainly of individual atoms and small molecules. Regions containing such small particles are transparent to nearly all types of electromagnetic radiation. Apart from numerous narrow atomic and molecular absorption lines, gas alone does not block radiation to any great extent. Interstellar dust is more complex. It consists of clumps of atoms and molecules—not unlike chalk dust and the microscopic particles that make up smoke, soot, or fog. Light from distant stars cannot penetrate the densest accumulations of interstellar dust any more than a car's headlights can illuminate the road ahead in a thick fog. The typical diameter of an interstellar dust particle— or **dust grain**—is about 10^{-7} m, comparable in size to the wavelength of visible light.

As a rule of thumb, a beam of light can be absorbed or scattered only by particles having diameters comparable to or larger than the wavelength of the radiation involved, and the amount of obscuration (absorption or scattering) produced by particles of a given size increases with decreasing wavelength. Consequently, dusty regions of interstellar space are transparent to long-wavelength radio and infrared radiation but opaque to shorter-wavelength optical, ultraviolet, and X-ray radiation.

Because the opacity of the interstellar medium increases with decreasing wavelength, light from distant stars is preferentially robbed of its higher-frequency ("blue") components. Hence, in addition to being generally diminished in overall brightness, stars also tend to appear redder than they really are (Figure 11.2). This effect, known as **reddening**, is essentially similar to the process that produces spectacular red sunsets here on Earth.

As indicated in Figure 11.2, interstellar dust can change a star's apparent brightness and color, but absorption lines in the original stellar spectrum are still recognizable in the radiation reaching Earth, so the star's spectral class can be determined. Astronomers can use this fact to study the interstellar medium. From a main-sequence star's spectral class astronomers learn its true luminosity and color. ∞ (Sec. 10.7) They can then measure the degree to which the starlight has been diminished and reddened en route to Earth. This information allows them to estimate both the numbers and the sizes of interstellar dust particles along the line of sight to the star. By repeating these measurements for stars in many different directions and at many different distances from Earth, astronomers have built up a picture of the distribution and

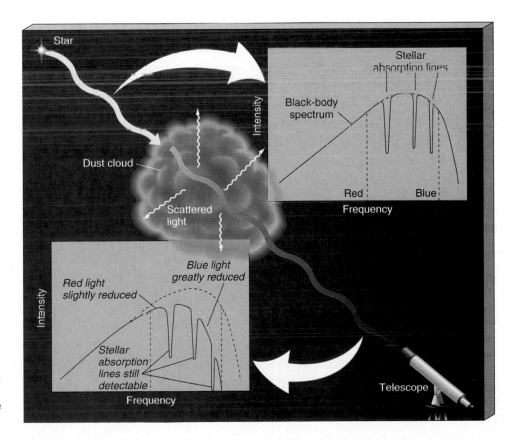

Figure 11.2 Starlight passing through a dusty region of space is both dimmed and reddened, but spectral lines are still recognizable in the light that reaches Earth.

properties of the interstellar medium in the solar neighborhood.

Temperature and Density

The temperature of the interstellar medium ranges from a few kelvins to a few thousand kelvins, depending on proximity to a star or some other source of radiation. Generally, we can take 100 K as an average temperature of a typical dark region of interstellar space. Compare this with 273 K, at which water freezes, and 0 K, at which atomic and molecular motions all but cease. Interstellar space is very cold.

Gas and dust are found everywhere in interstellar space—no part of our Galaxy is truly devoid of matter. However, the density of the interstellar medium is extremely low. The gas averages roughly 10^6 atoms per cubic meter—just 1 atom per cubic centimeter. Interstellar dust is even rarer: on average, there is only 1 dust particle for every trillion or so atoms. The space between the stars is populated with matter so thin that harvesting all the gas and dust in an interstellar region the size of Earth would yield barely enough matter to make a pair of dice.

How can such fantastically sparse matter diminish light so effectively? The key is size—interstellar space is vast. The typical distance between stars (1 pc or so in the vicinity of the Sun) is much, much greater than the size of a typical star (around 10^{-7} pc). Stellar and planetary sizes pale in comparison to the vastness of interstellar space. Thus, matter can accumulate, regardless of how thinly spread. For example, an imaginary cylinder 1 m^2 in cross section and extending from Earth to Alpha Centauri would contain more than 10 billion dust particles. Over huge distances, dust particles accumulate slowly but surely, to the point where they block visible light and other short-wavelength radiation. Even though the density of interstellar matter is extremely low, space in the vicinity of the Sun contains about as much mass in the form of interstellar gas and dust as exists there in the form of stars.

Composition

The composition of interstellar gas is reasonably well understood from spectroscopic studies of interstellar absorption lines formed when light from a distant star interacts with gas along the line of sight. ∞ (Sec. 2.5) The gas absorbs some of the stellar radiation in a manner that depends on its own temperature, density, and elemental abundance. The absorption lines thus produced contain information about dark interstellar matter, just as stellar absorption lines reveal the properties of stars. Because the interstellar absorption lines are produced by cold, low-density gas, astronomers can easily distinguish them from the much broader absorption lines formed in the star's hot lower atmosphere. ∞ (Sec. 2.7)

In most cases, the elemental abundances detected in interstellar gas mirror those found in other astronomical objects, such as the Sun, the stars, and the jovian planets. Most of the gas—about 90 percent—is atomic or molecular hydrogen; some 9 percent is helium, and the remaining 1 percent consists of heavier elements. The abundances of several of the heavy elements, such as carbon, oxygen, silicon, magnesium, and iron, are much lower in interstellar gas than in our solar system or in stars. The most likely explanation for this finding is that substantial quantities of these elements have been used to form the interstellar dust.

In contrast to interstellar gas, the composition of interstellar dust is currently not well known. We have some infrared evidence for silicates, carbon, and iron—the same elements that are underabundant in the gas—lending support to the theory that interstellar dust forms out of interstellar gas. The dust probably also contains some "dirty ice," a frozen mixture of water ice contaminated with trace amounts of ammonia, methane, and other compounds. This composition is quite similar to that of cometary nuclei in our own solar system. ∞ (Sec. 4.2)

11.2 Interstellar Clouds

Emission Nebulae

2 Figure 11.3 is a mosaic of photographs showing a region of the sky even larger than that shown in Figure 11.1. The patchy obscuration of background stars by foreground dust is plainly evident. From our vantage point on Earth, this assemblage of stellar and interstellar matter stretches all the way across the sky. On a clear night, it is visible to the naked eye as the Milky Way. In Chapter 14 we will come to recognize this band as the flattened disk, or *plane*, of our own Galaxy.

Figure 11.4 shows a 12°-wide swath of the Milky Way in the general direction of the constellation

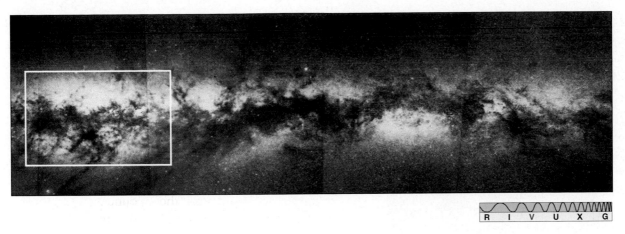

R I V U X G

Figure 11.3 A mosaic of the Milky Way. Photographed almost from horizon to horizon, and thus extending over nearly 180°, this band contains high concentrations of stars as well as interstellar gas and dust. The field of view is several times wider than that of Figure 11.1, whose outline is superimposed on this image.

Sagittarius, as photographed from Earth. The field of view is mottled with stars and interstellar matter. In addition, several large fuzzy patches of light are clearly visible. These fuzzy objects, labeled M8, M16, M17, and M20, correspond to the 8th, 16th, 17th, and 20th objects in a catalog compiled by Charles Messier, an eighteenth-century French astronomer. Today they are known as **emission nebulae**—glowing clouds of hot interstellar matter.

Recall from Chapter 4 ∞ (Sec. 4.3) that a *nebula* is a cloud of interstellar gas and dust. Historically, astronomers used the term to refer to any "fuzzy" patch (bright or dark) on the sky—any region of space clearly distinguishable through a telescope but not sharply defined like a star or a planet. We now know that many (though not all) nebulae are clouds of interstellar dust and gas. If they happen to obscure stars lying behind them, we see a dark patch. If something within the cloud—a group of hot young stars, for example—causes it to glow, the nebula appears bright instead.

We can gain a better appreciation of these nebulae by examining progressively smaller fields of view. Figure 11.5 is an enlargement of the region near the bottom of Figure 11.4, showing M20 at the top and M8 at the bottom. Figure 11.6 is an enlargement of the top of Figure 11.5, presenting a close-up of M20 and its immediate environment. The total area displayed measures some 10 pc across. Emission nebulae are among the most spectacular objects in the universe, yet they appear only as small, undistinguished patches of light when viewed in the larger context of the

Milky Way, as in Figure 11.4. Perspective is crucial in astronomy.

These nebulae are regions of glowing, ionized gas. At or near the center of each is at least one newly formed hot O or B star producing copious amounts of ultraviolet light. As ultraviolet photons travel outward from the star, they ionize the surrounding gas. As electrons recombine with nuclei, they emit visible radiation, causing the gas to glow. Figure 11.7 shows enlargements of two of the other nebulae visible in Figure 11.4. Notice the predominant red coloration of the emitted radiation—the result of hydrogen atoms emitting light in the red part of the visible spectrum—and the hot bright stars embedded within the glowing nebular gas. ∞ (Sec. 2.6) The interaction between stars and gas is particularly striking in Figure 11.7(b). The three dark "fingers" visible in this spectacular *Hubble Space Telescope* image are part of the interstellar cloud from which the stars formed. The rest of the cloud in the vicinity of the new stars has already been dispersed by their radiation; the fuzz around the edges of the fingers, especially at top right and center, is the result of this process.

Woven through the glowing nebular gas, and plainly visible in Figures 11.5–11.7, are dark lanes of obscuring dust. Recent studies have demonstrated that these **dust lanes** are part of the nebulae and are not just unrelated dust clouds that happen to lie along our line of sight. This relationship is again most evident in Figure 11.7(b), where the long fingers of gas and dust are simultaneously silhouetted against background nebular emission and illuminated by foreground nebular stars.

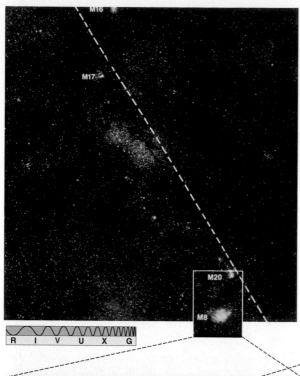

R I V U X G

Figure 11.4 A photograph of a small portion (about 12° across) of the region of the sky shown in Figure 11.1, displaying higher-resolution evidence for stars, gas, and dust as well as several distinct fuzzy patches of light, known as emission nebulae. The center line of the Milky Way is marked with a dashed white line.

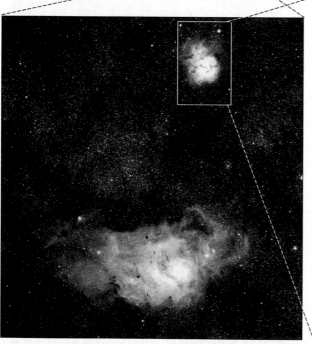

R I V U X G

Figure 11.5 An enlargement of the bottom of Figure 11.4, showing M20 (top) and M8 (bottom) more clearly.

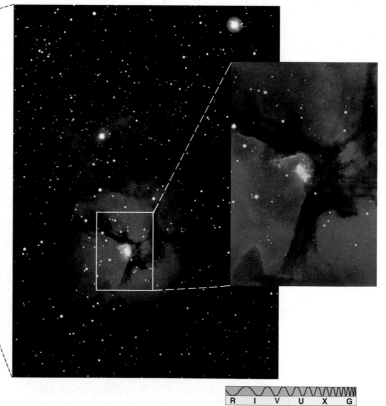

R I V U X G

Figure 11.6 Further enlargements of the top of Figure 11.5, showing only M20 and its interstellar environment. The nebula (red) is about 4 pc in diameter. It is often called the Trifid Nebula because of the three dust lanes that trisect its midsection (insert). The bluish region immediately above M20 is unrelated to the red emission nebula and is caused by starlight reflected from intervening dust particles. It is called a *reflection nebula*.

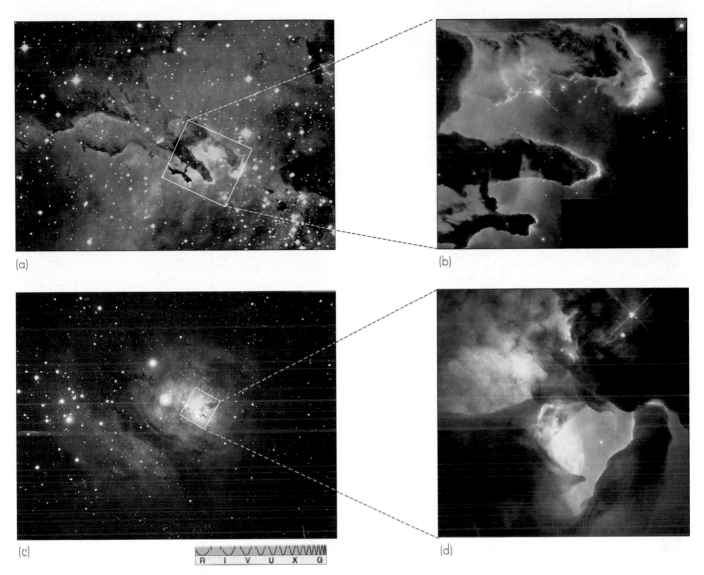

(a)

(b)

(c)

R I V U X G

(d)

Figure 11.7 Enlargements of selected portions of Figure 11.4. (a) M16, the Eagle Nebula. (b) A recent *Hubble* image of huge columns of cold gas and dust inside M16; delicate sculptures created by the action of stellar ultraviolet radiation on the original cloud. (c) M8, the Lagoon Nebula. (d) A high-resolution view of the core of M8, a region known as the Hourglass. Notice the irregular shape of the emitting regions, the characteristic red color of the light, the bright stars within the gas, and the patches of obscuring dust.

Nebular spectra tell us a great deal about interstellar gas. Because at least one hot star resides near the center of any emission nebula, you might think that the combined spectrum of star and nebula would be hopelessly confused. In fact, it is not. As with cloud absorption lines, we can easily distinguish nebular spectra from stellar spectra because the physical conditions in stars and nebulae differ so greatly. In particular, emission nebulae are made of hot, thin gas that, as we saw in Chapter 2, yields detectable *emission* lines. ∞ (Sec. 2.5) When our spectroscope is trained on

the star at the center of the nebula, we see a familiar stellar spectrum consisting of a black-body–like continuous spectrum and absorption lines, on which are superimposed emission lines from the nebular gas. When no star appears in the field of view, only the emission lines are seen.

Figure 11.8 is a typical nebular emission spectrum spanning part of the visible and near-ultraviolet wavelength interval. Numerous emission lines can be seen, and information on the nebula can be extracted from all of them. The results of analyses of many nebular

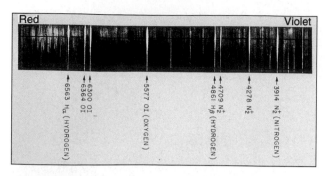

Figure 11.8 The emission spectrum of a typical emission nebula, showing light intensity as a function of frequency over the visible portion of the electromagnetic spectrum.

spectra show abundances close to those found in the Sun and other stars and elsewhere in the interstellar medium.

Unlike stars, nebulae are large enough for their sizes to be measurable by simple geometry. Coupling this size information with estimates of the amount of matter along our line of sight (as revealed by a nebula's total light emission), we can find the nebula's density. Generally, emission nebulae have only a few hundred particles, mostly protons and electrons, in each cubic centimeter—a density some 10^{22} times lower than that of a typical planet. Spectral-line widths imply that the gas atoms and ions have temperatures around 8000 K. Table 11.1 lists some vital statistics for the nebulae shown in Figure 11.4.

Dark Dust Clouds

3 Emission nebulae are only one small component of interstellar space. Most of space—in fact, more than 99 percent of it—is devoid of such

regions and contains no stars. It is simply dark. Look again at Figures 11.3 and 11.4, or just ponder the evening sky. The dark regions are by far the most representative of interstellar space.

Within these dark voids between the nebulae and the stars lurks another distinct type of astronomical object, the **dark dust cloud**. These clouds are cooler than their surroundings (with temperatures as low as a few tens of kelvins), and thousands or even millions of times denser. In some regions, densities exceeding 10^9 atoms/m^3 (1000 atoms/cm^3) are found—about as tenuous as the best laboratory vacuum on Earth. These clouds bear little resemblance to terrestrial clouds, however. Most of them are bigger than our solar system, and some are many parsecs across. (Yet even so, they make up no more than a few percent of the entire volume of interstellar space.) Despite their name, these clouds are made up primarily of gas, just like the rest of the interstellar medium. However, their absorption of starlight is due almost entirely to the dust they contain.

Figure 11.9(a) is an optical photograph of a typical dark dust cloud. Pockets of intense blackness mark regions where the dust and gas are especially concentrated and the light from background stars is completely obscured. Note the long "streamers" of (relatively) dense dust and gas. This cloud clearly is not spherical. Indeed, most dark dust clouds are very irregularly shaped. The bright patches within the dark region are emission nebulae in the foreground. Some of them are part of the cloud, where newly formed stars near the surface have created a "hot spot" in the cold, dark gas. Others have no connection to the cloud and just happen to lie along our line of sight. Like all dark dust clouds, this cloud is too cold to emit any visible

Table 11.1 Nebular Properties					
Object	Approx. Distance (pc)	Average Diameter (pc)	Density (10^6 particles/m^3)	Mass (solar masses)	Temperature (K)
M8	1200	14	80	2600	7500
M16	1800	8	90	600	8000
M17	1500	7	120	500	8700
M20	900	4	100	150	8200

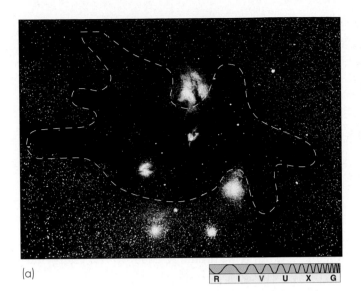

(a)

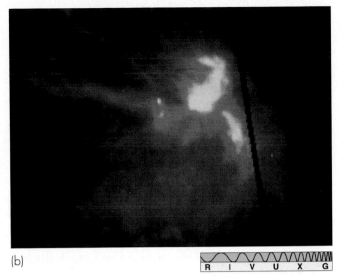

(b)

Figure 11.9   (a) A typical dark dust cloud, known as Rho Ophiuchi, is "visible" only because it blocks the light coming from stars lying behind it. The approximate outline of the cloud is indicated by the dashed line. (The scale of the image is 3 pc on a side.) (b) An infrared map of the same region, to roughly the same scale. The infrared emission, and therefore the dust that produces it, display a structure similar to the cloud's visual image. The very bright source of infrared radiation near the top of the cloud comes from a hot emission nebula, which can also be seen in the optical image. (The black diagonal streak at right is an instrumental effect.)

light. However, it does radiate strongly at longer wavelengths. Figure 11.9(b) shows an infrared image of the same region.

Figure 11.10 shows another striking example of a dark dust cloud—the Horsehead Nebula in Orion. This curiously shaped finger of gas and dust projects out from the much larger dark cloud in the bottom half of the image and stands out clearly against the red glow of a background emission nebula.

11.3 Probing the Interstellar Medium

4 A basic problem with the optical spectroscopic technique described in Section 11.1 is that it allows us to examine the interstellar medium only along the line of sight to a distant star. To form an absorption line, there has to be a background source of

Figure 11.10   The Horsehead Nebula in Orion is a striking example of a dark dust cloud, silhouetted against the bright background of an emission nebula. The "neck" of the horse is about 0.25 pc across. The dark region lies roughly 1500 pc from Earth.

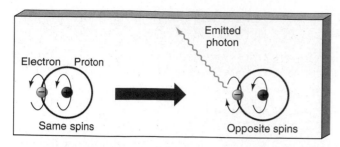

Figure 11.11 A ground-level hydrogen atom changing from a higher-energy state (electron and proton spinning in the same direction) to a lower-energy state (spinning in opposite directions). The emitted photon carries away an energy equal to the energy difference between the two spin states.

radiation to absorb. The need to see stars through interstellar matter also restricts this approach to relatively local regions, within a few thousand parsecs of the Sun. Beyond that distance, stars are completely obscured, and optical observations are impossible. As we have seen, infrared observations provide a means of viewing the emission from some clouds, they do not completely solve the problem, as only the denser, dustier clouds emit enough infrared radiation for astronomers to study them in that part of the spectrum.

To probe interstellar space more thoroughly, we need a more general, more versatile observational method—one that does not rely on conveniently located stars. In short, we need a way to detect cold interstellar matter anywhere in space through its *own* radiation. This may sound impossible, but such observational techniques do in fact exist. They rely on low-energy *radio* emissions produced by the interstellar gas itself.

21-Centimeter Radiation

Recall that a hydrogen atom has one electron orbiting a single-proton nucleus. ∞ (Sec. 2.6) Besides its orbital motion around the central proton, the electron also has some rotational motion—that is, *spin*—about its own axis. The proton also spins. This model parallels a planetary system, in which, in addition to the orbital motion of a planet about a central star, both the planet (electron) and the star (proton) rotate about their axes.

The laws of physics dictate that there are just two possible spin configurations for a hydrogen atom in its ground state. As illustrated in Figure 11.11, the elec-

tron and proton can either rotate in the *same* direction, with their spin axes parallel, or they can rotate in *opposite* directions, with their axes parallel, but oppositely oriented. The former configuration has slightly higher energy than the latter.

All matter in the universe tends to achieve its lowest possible energy state, and interstellar gas is no exception. A slightly excited hydrogen atom having the electron and proton spinning in the same direction eventually drops down to the less energetic, opposite-spin state as the electron suddenly and spontaneously reverses its spin. As with any such change, the transition from a high-energy state to a low-energy state releases a photon with energy equal to the energy difference between the two states.

Because the energy difference between the two states is very small, the energy of the emitted photon is very low. Consequently, the wavelength of the radiation is long—about 21 cm, roughly the width of this book. That wavelength lies in the radio portion of the electromagnetic spectrum. Researchers refer to the spectral line that results from this hydrogen-spin-flip process as the **21-centimeter line**. It provides a vital probe into any region of the universe containing atomic hydrogen gas. Needing no visible starlight to help calibrate their signals, radio astronomers can observe any interstellar region that contains enough hydrogen to produce a detectable signal. Even the low-density regions between dark dust clouds can be studied.

If all atoms eventually fall into their lowest-energy configuration, why isn't all the hydrogen in the universe in the lower-energy state by now? Why do we see 21-cm radiation today? The answer is that the energy difference between the two states shown in Figure 11.11 is comparable to the energy of a typical atom at a temperature of 100 K or so. As a result, atomic collisions in the interstellar medium are energetic enough to boost the electron up into the higher-energy configuration and so maintain comparable numbers of hydrogen atoms in either state. At any instant, any sample of interstellar hydrogen contains many atoms in the upper level, so 21-cm radiation is always emitted.

Of great importance is the fact that the wavelength of 21-cm radiation is much larger than the typical size of interstellar dust particles. Accordingly, this radio radiation reaches Earth completely unscattered by interstellar debris. The opportunity to observe interstellar space well beyond a few thousand parsecs, and in directions lacking background stars, makes 21-cm observations among the most important and useful in all of astronomy.

Molecular Spectral Lines

In certain interstellar regions of cold (typically 20 K), neutral gas, densities can reach as high as 10^{12} particles/m^3. Until the late 1970s, astronomers regarded these regions simply as abnormally dense interstellar clouds, but it is now recognized that they belong to an entirely new class of interstellar matter. The gas particles in these regions are not atoms but molecules. Because of the predominance of molecules in these dense interstellar regions, they are known as **molecular clouds**. Only within recent years have astronomers begun to appreciate the vastness of these clouds. They literally dwarf even the largest emission nebulae, which were previously thought to be the most massive residents of interstellar space.

As noted in Chapter 2, molecules emit radiation as they change from one rotational state to another. ∞ (Sec. 2.6) The energy differences between rotational states are generally very small, so the emitted radiation is usually in the radio range. We are fortunate that molecules emit radio waves because they are invariably found in the densest and dustiest parts of interstellar space, from which only low-frequency radio radiation can escape.

Why are molecules found only in the densest and darkest of the interstellar clouds? One possible reason is that the dust protects the fragile molecules from the harsh interstellar environment—the same absorption that prevents high-frequency radiation from getting out to our detectors also prevents it from getting in to destroy the molecules. Another possibility is that the dust acts as a catalyst that helps form the molecules. The dust grains provide both a place where atoms can stick and react and a means of dissipating any heat associated with the reaction, which might otherwise destroy the newly formed molecules. Probably the dust plays both roles. The close association between dust grains and molecules in molecular clouds argues strongly in favor of this picture, but the details are still being debated.

In mapping molecular clouds, radio astronomers are faced with a problem. Molecular hydrogen (H_2) is by far the most common constituent of these clouds, but unfortunately, despite its abundance, this molecule does not emit or absorb radio radiation. It emits only short-wavelength ultraviolet radiation, so it cannot easily be used as a probe of cloud structure. Nor are 21-cm observations helpful—they are sensitive only to *atomic* hydrogen, not to the *molecular* form of the gas.

With hydrogen effectively ruled out as a probe of molecular clouds, astronomers must use observations of other molecules to study the dark interiors of these dusty regions. Molecules such as carbon monoxide (CO), hydrogen cyanide (HCN), ammonia (NH_3), water (H_2O), formaldehyde (H_2CO), and about 60 others, some of them quite complex, are now known to exist in interstellar space. These molecules are found only in very small quantities—they are generally 1 million to 1 billion times less abundant than H_2—but they are important as *tracers* of a cloud's structure and physical properties. They are produced by chemical reactions within molecular clouds. When we observe them, we know that the regions under study must also contain high densities of molecular hydrogen, dust, and other important constituents.

For example, Figure 11.12 shows a contour map of the distribution of formaldehyde molecules in the immediate vicinity of the M20 nebula. It was made by observing radio spectral lines of formaldehyde at various locations and then drawing contours connecting regions of similar abundance. Notice that the amount of formaldehyde (and, we assume, the amount of hydrogen) peaks well away from the visible nebula.

Radio maps of interstellar gas and infrared maps of interstellar dust reveal that molecular clouds do not exist as distinct and separate objects in space. Rather, they make up huge **molecular cloud complexes**, some spanning as much as 50 pc across and containing enough gas to make millions of stars like our Sun. About 1000 such complexes are known in our Galaxy.

The very existence of molecules has forced astronomers to rethink and reobserve interstellar space. In doing so, they have begun to realize that this active and interesting domain is far from the void suspected by theorists not so long ago. Regions of space recently thought to contain nothing more than galactic "garbage"—the cool, tenuous darkness among the

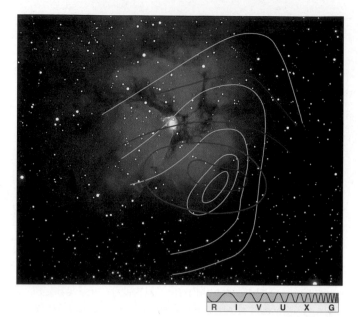

R I V U X G

Figure 11.12 Contour map of the amount of formaldehyde near the M20 nebula, demonstrating how formaldehyde is especially abundant in the darkest interstellar regions. Other kinds of molecules have been found to be similarly distributed. The contour values increase from the outside to the inside, so the maximum density of formaldehyde lies just to the bottom right of the visible nebula. The green and red contours outline the intensity of formaldehyde lines at different frequencies.

stars—now play a critical role in our understanding of stars and the interstellar medium from which they are born.

11.4 The Formation of Stars Like the Sun

5 Let us now turn our attention to the relationship between the interstellar medium and the stars in our Galaxy. How do stars form? What factors determine their masses, luminosities, and spatial distribution? In short, what basic processes are responsible for the appearance of our night sky?

Gravity and Heat

Simply stated, star formation begins when part of the interstellar medium—one of the cold, dark clouds discussed in Section 11.2—starts to collapse under its own weight. The cloud heats up as it shrinks, and eventually its center becomes hot enough for nuclear burning to begin. At that point, the contraction stops and a star is born. But what determines which interstellar clouds start to contract? For that matter, since all clouds exert a gravitational pull, why didn't they all collapse long ago? To begin to answer these questions, let us consider a small portion of a large interstellar cloud. Concentrate, first, on just a few atoms, as shown in Figure 11.13.

Even though the cloud's temperature is very low, each atom has some random motion (that is, the cloud still contains some heat. ∞ (More Precisely 2-1) Each atom is also influenced by the gravitational attraction of all its neighbors. The gravitational force is not large, however, because the mass of each atom is so small. Even when a few atoms accidentally cluster for an instant, as shown in Figure 11.13(b), their combined gravity is insufficient to bind them into a lasting, distinct clump of matter. This accidental cluster disperses as quickly as it formed. The effect of heat—the random motion of the atoms—is much stronger than the effect of gravity.

Now concentrate on a larger group of atoms—50, 100, 1000, even a million, each gravitationally pulling on all the others. The force of gravity is now stronger than before. Would this many atoms exert a combined gravitational attraction strong enough to prevent the clump from dispersing? The answer—at least under the conditions found in interstellar space—is still no. The gravitational attraction of this mass of atoms is still far too weak to overcome the effect of heat.

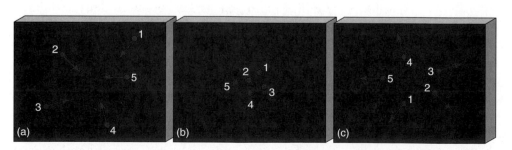

Figure 11.13 The motions of a few atoms within an interstellar cloud are influenced by gravity so slightly that the atoms' paths are hardly changed (a) before, (b) during, and (c) after an accidental, random encounter.

Table 11.2 Prestellar Evolution of a Solar-Type Star

Stage	Approximate Time to Next Stage (yr)	Central Temperature (K)	Surface Temperature (K)	Central Density (particles/m³)	Diameter[1] (km)	Object
1	2×10^6	10	10	10^9	10^{14}	Interstellar cloud
2	3×10^4	100	10	10^{12}	10^{12}	Cloud fragment
3	10^5	10,000	100	10^{18}	10^{10}	Cloud fragment/protostar
4	10^6	1,000,000	3000	10^{24}	10^8	Protostar
5	10^7	5,000,000	4000	10^{28}	10^7	Protostar
6	3×10^7	10,000,000	4500	10^{31}	2×10^6	Star
7	10^{10}	15,000,000	6000	10^{32}	1.5×10^6	Main-sequence star

[1]For comparison, recall that the diameter of the Sun is 1.4×10^6 km and that of the solar system roughly 1.5×10^{10} km.

How many atoms must be accumulated in order for their collective pull of gravity to prevent them from dispersing back into interstellar space? The answer, even for a typical cool (100 K) cloud, is a truly huge number. Nearly 10^{57} atoms are required—much larger than the 10^{25} grains of sand on all the beaches of the world, even larger than the 10^{51} elementary particles that constitute all the atomic nuclei in our entire planet. There is simply nothing on Earth comparable to a star.

An interstellar cloud is maintained in equilibrium by a balance between two basic opposing influences: gravity (which is always directed inward) and heat (in the form of outwardly directed pressure). Star formation begins when gravity begins to dominate over heat, causing the cloud to lose its equilibrium and start to contract. Only when the cloud has undergone radical changes in its internal structure is equilibrium finally restored.

Table 11.2 lists seven evolutionary stages that an interstellar cloud goes through prior to becoming a main-sequence star like our Sun. These stages are characterized by different central temperatures, surface temperatures, central densities, and radii of the prestellar object. They trace its progress from a quiescent interstellar cloud to a genuine star. The numbers given in Table 11.2 and the following discussion are valid *only* for stars of approximately the same mass as the

Sun. In the next section we will relax this restriction and consider the formation of stars of other masses.

Stage 1—An Interstellar Cloud

The first stage in the star-formation process is a dense interstellar cloud—the core of a dark dust cloud or perhaps a molecular cloud. These clouds are truly vast, sometimes spanning tens of parsecs (10^{14}–10^{15} km) across. Typical temperatures are about 10 K throughout, with a density of perhaps 10^9 particles/m³. Stage 1 clouds contain thousands of times the mass of the Sun, mainly in the form of cold atomic and molecular gas. (The dust they contain is important for cooling the cloud as it contracts and also plays a crucial role in planet formation, but it constitutes a negligible fraction of the total mass.) ∞ (Sec. 4.3)

The initial collapse occurs when a sufficiently massive pocket of gas becomes gravitationally unstable. Perhaps it is squeezed by some external event, such as the pressure wave produced when a nearby O or B star forms and ionizes its surroundings, or perhaps it simply cools below the temperature at which its internal pressure is no longer sufficient to support it against its own gravity. Whatever the cause, theory suggests that once the collapse begins, fragmentation into smaller and smaller clumps of matter naturally follows. As illus-

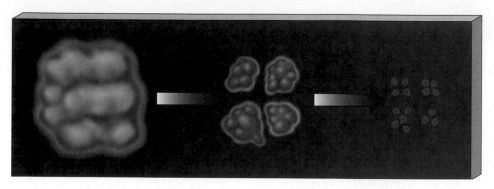

Figure 11.14 As an interstellar cloud contracts, gravitational instabilities cause it to fragment into smaller pieces. The pieces also contract and fragment, eventually forming many tens or hundreds of individual stars.

trated in Figure 11.14, a typical cloud can break up into tens, hundreds, even thousands, of fragments, each imitating the shrinking behavior of the parent cloud and contracting ever faster. The whole process, from a single quiescent cloud to many collapsing fragments, takes a few million years.

In this way, depending on the precise conditions under which fragmentation takes place, an interstellar cloud can produce either a few dozen stars, each much larger than our Sun, or a whole cluster of hundreds of stars, mostly comparable to or smaller than our Sun. There is little evidence for stars born in isolation, one star from one cloud. Most stars—perhaps all stars—appear to originate as members of multiple systems or star clusters. The Sun, which is now found alone and isolated in space, probably escaped from the multiple-star system where it formed after an encounter either with another star or with some much larger object (such as a molecular cloud).

The process of continued fragmentation is eventually stopped by the increasing density within the shrinking cloud. As fragments continue to contract, they eventually become so dense that radiation cannot get out easily. The trapped radiation causes the temperature to rise, the pressure to increase, and the fragmentation to stop. However, the contraction continues.

Stages 2 and 3—A Contracting Cloud Fragment

As it enters stage 2, a fragment destined to form a star like the Sun—the end product of the process sketched in Figure 11.14—contains between 1 and 2 solar masses of material. Estimated to span a few hundredths of a parsec across, this fuzzy, gaseous blob is still about 100 times the size of our solar system. Its central density is about 10^{12} particles/m^3.

Even though the fragment has shrunk substantially, its average temperature is not much different from that of its parent cloud. The reason is that the gas constantly radiates large amounts of energy into space. The material of the fragment is so thin that photons produced anywhere within it easily escape without being reabsorbed, so virtually all the energy released in the contraction is radiated away and does not cause any significant increase in temperature. Only at the center, where the radiation must traverse the greatest amount of material in order to escape, is there any appreciable temperature increase. The gas there might be as warm as 100 K by this stage. For the most part, however, the fragment stays cold as it shrinks.

Several tens of thousands of years after it first began contracting, a stage 2 fragment has shrunk by the start of stage 3 to a gaseous sphere with a diameter roughly the size of our solar system (still 10,000 times the size of our Sun). The inner regions have become opaque to their own radiation and so have started to heat up considerably, as noted in Table 11.2. The central temperature has reached about 10,000 K—hotter than the hottest steel furnace on Earth. However, the temperature near the edge of the fragment has not increased much. The gas there is still able to radiate its energy into space and so remains cool. The central density by this time is approximately 10^{18} particles/m^3 (still only 10^{-9} kg/m^3 or so).

For the first time, our fragment is beginning to resemble a star. The dense, opaque region at the center is called a **protostar**—an embryonic object perched at the dawn of star birth. Its mass increases as more and more material rains down on it from outside, although its radius continues to shrink because its pressure is still unable to overcome the relentless pull of gravity. By the end of stage 3, we can distinguish a "surface" on the protostar—its *photosphere*. Inside the photosphere, the protostellar material is opaque to the radiation it emits.[1] From here on, the surface temperatures listed in Table 11.2 refer to the photosphere and not to the edge of the collapsing fragment, whose temperature remains low.

Stages 4 and 5—Protostellar Evolution

As the protostar evolves, it shrinks, its density increases, and its temperature rises, both in the core and at the photosphere. Some 100,000 years after the fragment formed, it reaches stage 4, where its center seethes at about 1,000,000 K. The electrons and protons ripped from atoms whiz around at hundreds of kilometers per second, but the temperature is still short of the 10^7 K needed to ignite the proton–proton nuclear reactions that fuse hydrogen into helium. Still much larger than the Sun, our gassy heap is now about the size of Mercury's orbit. Heated by the material falling on it from above, its surface temperature has risen to a few thousand kelvins.

Knowing the protostar's radius and surface temperature, and using the radius–luminosity–temperature relationship, ∞ (Sec. 10.3) we can calculate its luminosity. This turns out to be around 1000 times the luminosity of the Sun. Even though the protostar has a surface temperature only about half that of the Sun, its radius is roughly a hundred times larger, making its total luminosity very large indeed—in fact, much greater than the luminosity of most main-sequence stars. Because nuclear reactions have not yet begun in the protostar's core, this luminosity is due entirely to the release of gravitational energy as the protostar continues to shrink and material from the surrounding fragment (which we called the solar nebula back in Chapter 4) rains down on its surface. ∞ (Sec. 4.3)

By the time stage 4 is reached, our protostar's physical properties can be plotted on a Hertzsprung–Russell (H–R) diagram. ∞ (Sec. 10.7) At each

phase of a star's evolution, its surface temperature and luminosity can be represented by a single point on the diagram. The motion of that point around the diagram as the star evolves is known as the star's **evolutionary track**. It is a graphical representation of a star's life. The red track on Figure 11.15 depicts the approximate path followed by our interstellar cloud fragment since it became a protostar at the end of stage 3 (which itself lies off the right-hand edge of the figure). Figure 11.16 is an artist's sketch of an interstellar gas cloud proceeding along the evolutionary path outlined so far.

Our protostar is still not in equilibrium. Even though its temperature is now so high that outward-directed pressure has become a powerful countervailing influence against gravity's inward pull, the balance is not yet perfect. The protostar's internal heat gradually diffuses out from the hot center to the cooler surface, where it is radiated away into space. As a result, the contraction slows but does not stop completely. From our perspective on Earth, this is quite fortunate: if the

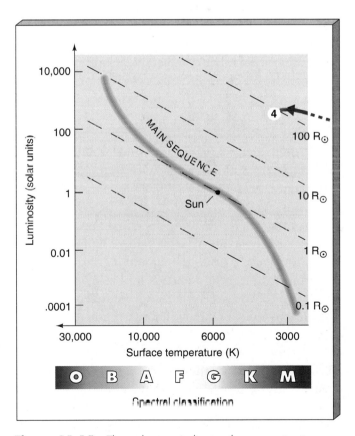

Figure 11.15 The red arrow indicates the approximate evolutionary track followed by an interstellar cloud fragment prior to becoming a stage-4 protostar. (The circled numbers on this and subsequent H–R plots refer to the prestellar evolutionary stages listed in Table 11.2 and described in the text.) Recall from Chapter 10 that "$R_\odot$" means the radius of the Sun.

---

[1]Note that this is the same definition of *surface* we used for the Sun in Chapter 9. ∞ (Sec. 9.1)

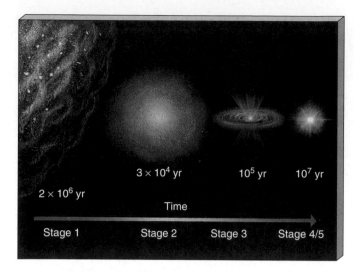

Figure 11.16 Artist's conception of the changes in an interstellar cloud during the early evolutionary stages outlined in Table 11.2. (Not drawn to scale.) The duration of each stage, in years, is indicated.

heated gas were somehow able to counteract gravity completely before the star reached the temperature and density needed to start nuclear burning in its core, the protostar would simply radiate away its heat and never become a true star. The night sky would be abundant in faint protostars but completely lacking in the genuine article. Of course, there would be no Sun either, so it is unlikely that we, or any other intelligent life form, would exist to appreciate these astronomical subtleties.

After stage 4, the protostar moves downward on the H–R diagram (toward lower luminosity) and slightly to the left (toward higher temperature), as shown in Figure 11.17. By stage 5, the protostar has shrunk to about 10 times the size of the Sun, its surface temperature is about 4000 K, and its luminosity has fallen to about 10 times the luminosity of the Sun. At this point, the central temperature has reached about 5,000,000 K. The gas is completely ionized by now, but the protons still do not have enough thermal energy for nuclear fusion to begin.

Protostars often exhibit violent surface activity during this phase of their evolution, resulting in extremely strong protostellar winds, much denser than that of our Sun. As mentioned in Section 4.3, this portion of the evolutionary track is often called the **T Tauri star**, after T Tauri, the first "star" (actually protostar) to be observed in this stage of prestellar development.

Events proceed more slowly as the protostar approaches the main sequence. The initial contraction and fragmentation of the interstellar cloud occurred quite rapidly, but by stage 5, as the protostar nears the status of a full-fledged star, its evolution slows. The cause of this slowdown is heat—even gravity must struggle to compress a hot object. The contraction is governed largely by the rate at which the protostar's internal energy can be radiated away into space. The greater this radiation of internal energy—that is, the more energy that moves through the star to escape from its surface, the faster the contraction occurs. As the luminosity decreases, so too does the contraction rate.

Stages 6 and 7—A Newborn Star

Some 10 million years after its first appearance, the protostar finally becomes a true star. By stage 6, when our roughly 1-solar-mass object has shrunk to a radius of about 1,000,000 km, the contraction has raised the

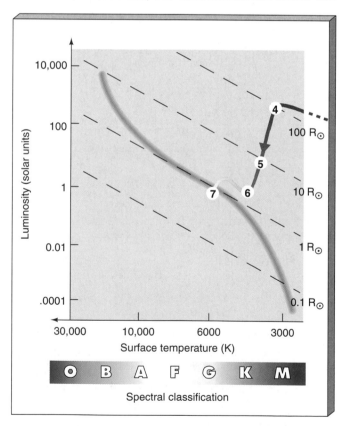

Figure 11.17 The changes in a protostar's observed properties are shown by the path of decreasing luminosity, from stage 4 to stage 6. At stage 7, the newborn star has arrived on the main sequence.

central temperature to 10,000,000 K, enough to ignite nuclear burning. Protons begin fusing into helium nuclei in the core, and a star is born. As shown in Figure 11.17, the star's surface temperature at this point is about 4500 K, still a little cooler than the Sun. Even though the radius of the newly formed star is slightly larger than that of the Sun, its lower temperature means that its luminosity is slightly less than (actually, about two-thirds of) the present solar value.

Over the next 30 million years or so, the stage 6 star contracts a little more. In making this slight adjustment, the central density rises to about 10^{37} particles/m^3 (more conveniently expressed as 10^5 kg/m^3), the central temperature increases to 15,000,000 K, and the surface temperature reaches 6000 K. By stage 7, the star finally reaches the main sequence just about where our Sun now resides. Pressure and gravity are finally balanced, and the rate at which nuclear energy is generated in the core exactly matches the rate at which energy is radiated from the surface.

The evolutionary events just described occur over the course of 40–50 million years. Although this is a long time by human standards, it is still less than 1 percent of the Sun's lifetime on the main sequence. Once an object begins fusing hydrogen and establishes a "gravity-in/pressure-out" equilibrium, it burns steadily for a very long time. The star's location on the H–R diagram will remain virtually unchanged for the next 10 billion years.

11.5 Stars of Other Masses

The Zero-Age Main Sequence

6 The numerical values and evolutionary track just described are valid only for 1-solar-mass stars. The temperatures, densities, and radii of prestellar objects of other masses exhibit similar trends, but the details differ, in some cases quite considerably. Perhaps not surprisingly, the most massive fragments formed within interstellar clouds tend to produce the most massive protostars and eventually the most massive stars. Similarly, low-mass fragments give rise to low-mass stars.

Figure 11.18 compares the theoretical premain-sequence evolutionary track taken by our Sun with the corresponding tracks of a 0.3-solar-mass star and a

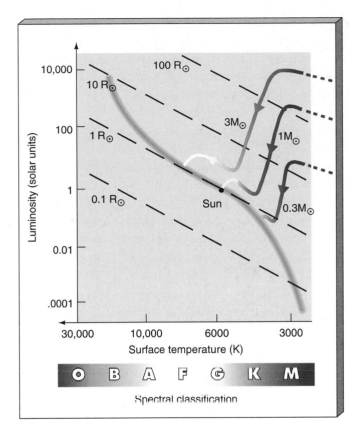

Figure 11.18 Prestellar evolutionary paths for stars more massive and less massive than our Sun.

3-solar-mass star. All three tracks traverse the H–R diagram in the same general manner, but cloud fragments that eventually form stars more massive than the Sun approach the main sequence along a higher track on the diagram, while those destined to form less massive stars take a lower track. The *time* required for an interstellar cloud to become a main-sequence star also depends strongly on its mass. The most massive fragments contract into O stars in a mere million years, roughly 1/50 the time taken by the Sun. The opposite is the case for prestellar objects having masses much less than our Sun. A typical M star, for example, requires nearly a billion years to form.

Whatever the mass, the endpoint of the prestellar evolutionary track is the main sequence. A star is considered to have reached on the main sequence when hydrogen burning begins in its core and the star's properties settle down to stable values. The main-sequence band predicted by theory is usually called the **zero-age main sequence** (ZAMS). It agrees quite well with main sequences observed for stars in the vicinity of the Sun and those observed for stars in more distant star clusters.

It is important to realize that *the main sequence is not an evolutionary track—stars do not evolve along it.* Rather, it is a "waystation" on the H–R diagram where stars stop and spend most of their lives—low-mass stars at the bottom, high-mass stars at the top. Once on the main sequence, a star stays in essentially the same location in the H–R diagram during its whole time as a stage 7 object. (In other words, a star that arrives on the main sequence as, say, a G star can never "work its way up" to become a B or an O main sequence star, or move down to become an M-type red dwarf. As we will see in Chapter 12, the next stage of stellar evolution occurs when a star leaves the main sequence. Any star leaving the main sequence and entering this next stage has pretty much the same surface temperature and luminosity it had when it arrived on the main sequence millions (or billions) of years earlier.

Failed Stars

Some cloud fragments are too small ever to become stars. Jupiter is a good example. It contracted under the influence of gravity, and the resultant heat is still detectable, but the planet does not have enough mass for gravity to crush its matter to the point of nuclear ignition. Pressure and gravity came into equilibrium before the central temperature became hot enough to fuse hydrogen. Jupiter never evolved beyond the protostar stage. If it had continued to accumulate gas from the solar nebula, it might eventually have become a star (to the detriment of life on Earth), but virtually all the matter present during the formative stages of our solar system is now gone, swept away by the solar wind.

Rather than turning into stars, low-mass fragments will continue to cool, eventually becoming compact, dark "clinkers"—cold fragments of unburned matter—in interstellar space. On the basis of theoretical modeling, astronomers believe that the minimum mass of gas needed to generate core temperatures high enough to begin nuclear fusion is about 0.08 solar masses.

Vast numbers of Jupiter-like objects may well be scattered throughout the universe—fragments frozen in time somewhere in the cloud contraction phase. Small, faint, and cool (and growing ever colder), they are known collectively as **brown dwarfs**. ∞ (*Interlude 4-2*) Our technology currently has great difficulty in detecting them, be they planets associated with stars or interstellar cloud fragments far from any star. We can telescopically detect stars and spectroscopically infer atoms and molecules, but, while recent advances in observational hardware and image-processing techniques have now identified several likely brown dwarf candidates, astronomical objects of intermediate size outside our solar system remain very hard to see. Interstellar space could contain many cold, dark Jupiter-sized objects without our knowing it. Conceivably, they might even account for more mass than we observe in the form of stars and interstellar gas combined.

11.6 Observations of Star Formation

7 How can we verify the theoretical predictions just outlined? The age of our entire civilization is far shorter than the time needed for a cloud to contract and form a star. We can never observe individual objects proceed through the full panorama of star birth. We can, however, observe many different objects—interstellar clouds, protostars, young stars approaching the main sequence—as they appear today at different stages of their evolutionary cycles. Here we cite just one example of each evolutionary stage. Each observation is like part of a jigsaw puzzle. When properly oriented relative to all the others, the pieces can be used to build up a picture of the full life cycle of a star.

Prestellar objects at stages 1 and 2 are not yet hot enough to emit much infrared radiation, and certainly no optical radiation arises from their dark, cool interiors. The best way to study these early stages of cloud contraction and fragmentation is to observe the radio emission from interstellar molecules within the clouds. Consider again M20, the splendid emission nebula studied earlier, shown again in Figure 11.19. The huge, dark molecular cloud surrounding the visible nebula is the stage 1 cloud, with a density of some 10^8 particles/m^3 and a temperature of around 20 K. The hot O star ultimately responsible for the nebula's optical emission probably lies at or near stage 7. How-

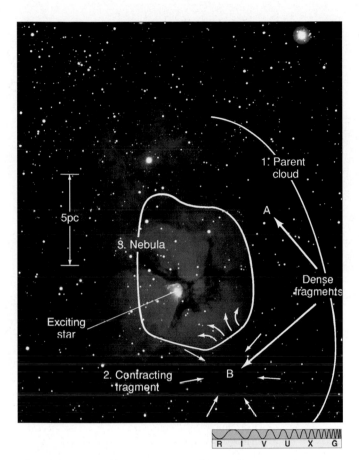

R I V U X G

Figure 11.19 The M20 region shows observational evidence for three broad phases in the birth of a star: (1) the parent cloud (stage 1 of Table 11.2), (2) a contracting fragment (between stages 1 and 2), and (3) the emission nebula (M20 itself) resulting from the formation of one or more massive stars (stage 6 or 7).

ever, the dark regions just outside the nebula also provide evidence for the early contraction phase of star formation.

The totally obscured regions labeled A and B in Figure 11.19 are warm, dense cloud fragments with temperatures of around 100 K and densities exceeding 10^9 particles/m^3. Doppler shifts of the radio lines observed in the vicinity of region B, which corresponds to the peak in radio intensity shown in Figure 11.12, indicate that it is contracting. Less than a light year across, this region has a total mass over 1000 times the mass of the Sun—considerably more than the mass of M20 itself. It lies somewhere between stages 1 and 2.

Figure 11.20 shows another star-forming region, the Orion complex. Lit from within by several O stars, the bright Orion Nebula is partly surrounded by a vast molecular cloud that extends well beyond the roughly

5×10 parsec region bounded by the photograph in Figure 11.20(b). The Orion molecular cloud harbors several smaller sites of intense radio emission from molecules deep within its core. Their extent, shown in Figures 11.20(c) and 11.20(d), measures about 10^{10} km, about the diameter of our solar system. The gas density of these smaller regions is about 10^{15} particles/m^3, much higher than the density of the surrounding cloud. Although the temperatures of these regions cannot be estimated reliably, many researchers regard the regions as objects well on their way to stage 3. We cannot determine if these regions will eventually form stars like the Sun, but it does seem likely that these intensely emitting regions are on the threshold of becoming protostars.

In the hunt for and study of objects at more advanced stages of star formation, radio techniques become less useful because stages 4, 5, and 6 have higher and higher temperatures. By Wien's law, their emission shifts toward shorter wavelengths, and so these objects shine most strongly in the infrared. ∞ (Sec. 2.4) One particularly bright infrared emitter, known as the Becklin–Neugebauer object (see Figure 11.23c), was detected in the core of the Orion molecular cloud in the 1970s. Its luminosity is around 1000 times the luminosity of the Sun. Most astronomers agree that this warm, dense blob is a high-mass protostar, probably around stage 4.

Until the *Infrared Astronomy Satellite* was launched in the early 1980s, astronomers were aware only of very massive stars forming in clouds far away. *IRAS* showed that stars are forming much closer to home, and some of these protostars have masses comparable to that of our Sun. Figure 11.21 shows a premier example of a solar-mass protostar—Barnard 5. Its infrared heat signature is that expected of a stage 5 object.

The energy sources for some infrared objects seem to be luminous hot stars that are hidden from optical view by surrounding dark clouds. Their radiation is mostly absorbed by a "cocoon" of dust, then reemitted by the dust as infrared radiation. Two considerations support the idea that the hot stars responsible for heating the clouds have only recently ignited: (1) dust cocoons are predicted to disperse quite rapidly once their central stars form, and (2) these objects are invariably found in the dense cores of molecular clouds. The central stars probably lie near stage 6.

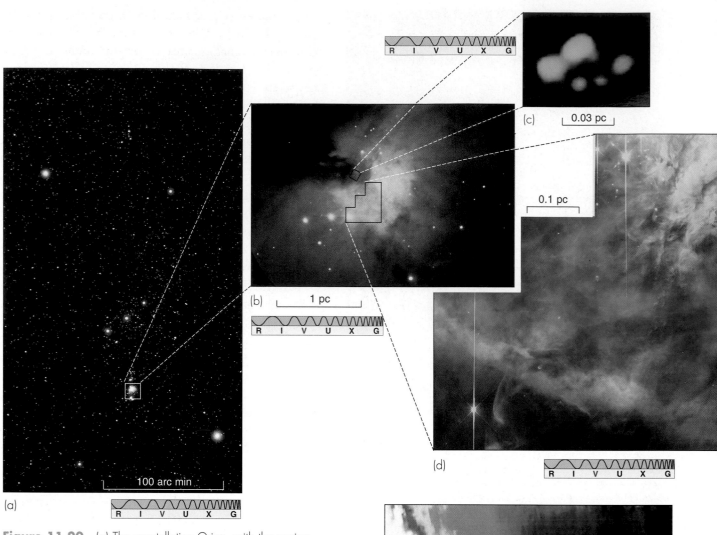

Figure 11.20 (a) The constellation Orion, with the region around its famous emission nebula marked by a rectangle. The Orion Nebula is the middle "star" of Orion's sword. (b) Enlargement of the framed region of part (a), suggesting how the nebula is partly surrounded by a vast molecular cloud. Various parts of this cloud are probably fragmenting and contracting, with even smaller sites forming protostars. (c) and (d) Some of the evidence for those protostars: (c) false-color radio image of some intensely emitting molecular sites; (d) real-color visible image of embedded nebular "knots" thought to harbor protostars.

Figure 11.21 An infrared image of the nearby region containing the source Barnard 5 (indicated by the arrow). On the basis of its temperature and luminosity, Barnard 5 appears to be a protostar around stage 5 on the H–R diagram.

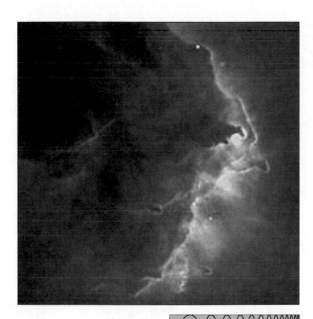

| R | I | V | U | X | G |

Figure 11.22 This enlargement of Figure 11.7(b) shows evidence of "evaporating gaseous globules" (EGGs), star-forming fragments dense enough to survive the onslaught of ultraviolet radiation from newly formed hot, young stars in their vicinity. The EGGs lie at the tips of narrow columns of gas projecting from the tip of the largest finger of gas in Figure 11.7(b); the columns survive because they lie in the EGGs' shadows and so are shielded from the radiation dispersing the rest of the cloud.

11.7 Star Clusters

7 We have seen how a portion of an interstellar cloud can become unstable, collapse, and fragment into stars. Let us take a moment to ask what happens next—not to the newborn stars themselves (that is the subject of the next two chapters), but to the galactic environment in which they formed.

The end result of cloud collapse is a group of stars, all formed from the same parent cloud and lying in the same region of space—in other words, a *star cluster.* ∞ (Sec. 10.10) As a by-product of cluster formation, a certain amount of unused gas and dust remains. How many stars form, and of what type? How much gas is left over? What does the collapsed cloud look like once star formation has run its course? At present, the answers to these questions are still very sketchy; they await a more thorough understanding of the star-formation process.

In general, the more massive the collapsing region, the more stars are likely to form there. In addition, we know from H–R diagrams that low-mass stars are much more common than high-mass ones. ∞ (Sec. 10.7) However, the precise number of stars of any given mass or spectral type depends in a complex (and poorly understood) way on conditions within the parent cloud. The same is true of the *efficiency* of star formation—that is, the fraction of the total mass that actually finds its way into stars—which determines the amount of leftover material. However, if, as is usually the case, one or more O or B stars form, their intense radiation and winds will cause the surrounding gas to disperse, leaving behind a young star cluster.

Figure 11.22 is a contrast-enhanced enlargement of Figure 11.7(b), showing several narrow columns of gas protruding from the top of the largest (topmost) pillar in the earlier figure. The narrow columns remain behind as the rest of the cloud evaporates because their dense tips, which are thought to contain solar system–sized stage 3 fragments, shield the rest of the column from ultraviolet radiation. These observations clearly illustrate the important role played by environment in the star-formation process. They show how the first massive stars to form tend to prevent the formation of additional high-mass stars by disrupting the environment in which stars are growing. This is one reason why low-mass stars are so much more common than high-mass stars; it also helps explain the existence of brown dwarfs, by providing a natural way in which star formation can stop before nuclear fusion begins in the growing stellar core.

Until recently, the existence of star clusters within emission nebulae was largely conjecture. The stars cannot be seen optically because they are obscured by dust. Infrared observations have now clearly demonstrated that stars really are found within star-forming regions! Figure 11.23 compares optical and infrared views of the central regions of the Orion Nebula. The optical image in Figure 11.23(a) shows the Trapezium, the group of four bright stars responsible for ionizing the nebula. However, the false-color infrared image in Figure 11.23(c) reveals an extensive cluster of stars within and behind the visible nebula. The Becklin–Neugebauer object (Section 11.6) can be seen as the central yellow spot within this region; it is thought to be a dust-shrouded B star just beginning to form its own emission nebula. These remarkable images show many stages of star formation.

For every O or B giant, tens or even hundreds of G, K, and M dwarfs may form. Thus, even a modest emission nebula can give rise to a fairly extensive col-

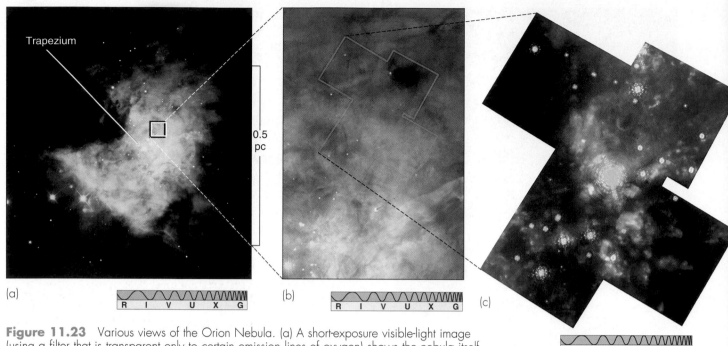

Figure 11.23 Various views of the Orion Nebula. (a) A short-exposure visible-light image (using a filter that is transparent only to certain emission lines of oxygen) shows the nebula itself and four bright O stars known as the Trapezium. (b) A magnified view of a smaller part of the nebula shows much irregular gas and dust, but few obvious stars which are hidden in the dust. (c) This short-exposure infrared image, acquired by the *Hubble* telescope in 1997, shows several faint red stars emerging from the nebular gas; the brightest star is known as the Becklin–Neugebauer object.

lection of stars. A typical open star cluster, like that shown in Figure 11.24, may measure 10 pc across and contain 1000 or more stars. Less massive, but more extended, clusters are usually known as **associations**. These typically contain no more than 100 stars but may span many tens of parsecs. Associations tend to be rich in very young stars. It is quite likely that the main difference between associations and open clusters is simply the efficiency with which stars formed from the parent cloud.

Eventually, star clusters dissolve into individual stars. Encounters between stars tend to eject the lightest stars from the cluster and the tidal gravitational field of the Milky Way Galaxy strips away stars that stray too far from the cluster center. Occasional distant encounters with giant molecular clouds also tend to remove cluster stars; a near miss may even disrupt a cluster entirely. As a result of all these influences, most clusters break up in less than a few hundred million years, although the actual lifetime depends on the cluster's mass. Loosely bound associations may survive for only a few tens of millions of years, whereas some very massive open clusters are known from their H–R diagrams to be billions of years old. In a sense, only when

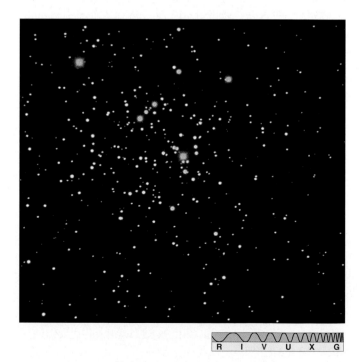

Figure 11.24 The Jewel Box cluster is a relatively young open cluster in the southern part of the Northern Hemisphere sky. Many bright stars appear in this image, but the cluster contains many more low-mass, less luminous stars. Because some red giants appear among the cluster's blue main-sequence stars, we can estimate the age of the cluster to be about 10 million years.

a star's parent cluster has completely dissolved is star formation really complete.

Take another look at the night sky. Ponder all that cosmic activity while gazing upward some clear, dark evening. After studying this chapter, you may find you have to modify your view of the night sky. Even the seemingly quiet darkness is dominated by continual change.

Chapter Review

Summary

The **interstellar medium** (p. 308) occupies the space between stars. It is made up of cold (less than 100 K) gas, mostly atomic or molecular hydrogen and helium, and **dust grains** (p. 309). Interstellar dust is very effective at blocking our view of distant stars, even though the density of the interstellar medium is very low. The spatial distribution of interstellar matter is very patchy. The dust preferentially absorbs short-wavelength radiation, leading to a distinct **reddening** (p. 309) of light passing through interstellar clouds.

Emission nebulae (p. 311) are extended clouds of hot, glowing interstellar matter. Associated with star formation, emission nebulae are caused by hot O and B stars heating and ionizing their surroundings. Studies of the emission lines produced by excited nebular atoms allow astronomers to measure the nebula's properties. Nebulae are often crossed by dark **dust lanes** (p. 311)—part of the larger cloud from which they formed.

Dark dust clouds (p. 314) are cold, irregularly shaped regions in the interstellar medium that diminish or completely obscure the light from background stars. Astronomers learn about these clouds by studying the absorption lines they produce in starlight that passes through them. Another way to observe cold, dark regions of interstellar space is through **21-centimeter radiation** (p. 316), which is produced whenever the electron in a hydrogen atom reverses its spin, changing its energy very slightly in the process. This radio radiation is important because it is emitted by all cool atomic hydrogen gas, even if the gas is undetectable by other means. In addition, 21-centimeter radiation is not appreciably absorbed by the interstellar medium, so radio astronomers making observations at this wavelength can "see" to great distances.

The interstellar medium also contains many cold, dark **molecular clouds** (p. 317), which are observed mainly through the radio radiation emitted by the molecules they contain. Dust within these clouds probably both protects the molecules and acts as a catalyst to help them form. As with other interstellar clouds, hydrogen is by far the most common constituent in molecular clouds, but molecular hydrogen is very hard to observe. Astronomers usually study these clouds through observations of other "tracer" molecules that are less common but much easier to detect.

Astronomers believe that molecular clouds are likely sites of future star formation. Often, several molecular clouds are found close to one another, forming an enormous **molecular cloud complex** (p. 317) millions of times more massive than the Sun.

Stars form when an interstellar cloud collapses under its own gravity and breaks up into smaller pieces. The evolution of the contracting cloud—the changes in its temperature and luminosity—can be conveniently represented as an **evolutionary track** (p. 321) on the Hertzsprung–Russell diagram. A cold interstellar cloud containing a few thousand solar masses of gas can fragment into tens or hundreds of smaller clumps of matter, from which stars eventually form.

As a collapsing prestellar fragment heats up and becomes denser, it eventually becomes a **protostar** (p. 320)—a warm, very luminous object that emits radiation mainly in the infrared portion of the electromagnetic spectrum. At this stage of its evolution, the protostar is also known as a **T Tauri star** (p. 322), after the first object of this type discovered.

Eventually, a protostar's central temperature becomes high enough for hydrogen fusion to begin, and the protostar becomes a star. For a star like the Sun, the whole formation process takes about 50 million years. More massive stars pass through similar stages, but much more rapidly. Stars less massive than the Sun take much longer to form. The **zero-age main sequence** (p. 323) is the main-sequence band predicted by stellar evolutionary theory. It agrees quite well with observed main sequences.

Mass is the key property for determining a star's characteristics and life span. The most massive stars have the shortest formation times and the shortest main-sequence lifetimes. At the other extreme, some low-mass fragments never reach the point of nuclear ignition. These objects not massive enough to fuse hydrogen to helium are called **brown dwarfs** (p. 324). The universe may be populated with a vast number of them.

Many of the objects predicted by the theory of star formation have been observed in real astronomical objects. The dark interstellar regions near emission nebulae often provide evidence for cloud frag0mentation and protostars. Radio telescopes are used for studying the early phases of cloud contraction and frag-

mentation; infrared observations allow us to see later stages of the process. Many well-known emission nebulae, lit by several O stars, are partially engulfed by molecular clouds, parts of which are probably fragmenting and contracting, with smaller sites forming protostars.

A single collapsing and fragmenting cloud can give rise to hundreds or thousands of stars—a star cluster. Infrared observations have revealed young star clusters in several emission nebulae. Loosely bound groups containing smaller numbers of newborn stars are called stellar **associations** (p. 328). Eventually, star clusters and associations break up into individual stars, although the process may take billions of years to complete.

Self-Test: True or False?

_____ 1. Interstellar matter is evenly distributed throughout the Milky Way Galaxy.

_____ 2. In the vicinity of the Sun, there is about as much mass in the form of interstellar matter as in the form of stars.

_____ 3. There is a lack of heavy elements in interstellar gas because they go into making interstellar dust.

_____ 4. Emission nebulae display spectra almost identical to those of the stars embedded in them.

_____ 5. Because of the obscuration of visible light by interstellar dust, we can observe stars only within a few thousand parsecs of Earth.

_____ 6. A typical dark dust cloud is many hundreds of parsecs across.

_____ 7. Because of their low temperatures, dark dust clouds radiate mainly in the radio part of the electromagnetic spectrum.

_____ 8. 21-centimeter radiation provides astronomers with information on interstellar molecular hydrogen gas.

_____ 9. 21-centimeter radiation can pass unimpeded through the entire Milky Way Galaxy.

_____ 10. Given the typical temperatures found in interstellar space, a cloud containing as few as 1000 atoms has sufficient gravity for it to begin to collapse.

_____ 11. The time a solar-type star spends forming is relatively short compared with the time it spends as a main-sequence star.

_____ 12. Most stars form as members of clusters of stars.

_____ 13. A stage 4 object has a luminosity about 1000 times that of our Sun's present luminosity.

_____ 14. The rate of evolution of a stage 5 object is fast compared with the rates at previous stages.

_____ 15. Stages 1 and 2 of star formation can be observed using optical telescopes.

_____ 16. In star formation, more G, K, and M stars form than O and B stars.

Self-Test: Fill in the Blank

1. The interstellar medium is made up of _____ and _____.

2. To scatter a beam of radiation, a particle must be _____ in size to the wavelength of the radiation.

3. Interstellar gas is composed of 90 percent _____ and 9 percent _____.

4. The temperature of a typical emission nebula is about _____ K.

5. Dark dust clouds can have temperatures as low as _____ K.

6. 21-centimeter radiation results from a change in the _____ of the electron in a _____ atom.

7. Molecular clouds typically have temperatures of about _____ K.

8. Emissions from molecular clouds are in the _____ part of the electromagnetic spectrum.

9. The most common constituent of molecular clouds is molecular _____.

10. A molecular cloud complex may contain as much as _____ solar masses of gas.

11. An _____ plots a star's or protostar's changing location on the H–R diagram as the object evolves.

12. During stage 3 of prestellar evolution, as the various pieces of the original interstellar cloud continue to contract, their central densities and temperatures _____.

13. A stage 4 object is plotted in the _____ (upper/lower) _____ (right/left) part of the H–R diagram.

14. At stage 6 the central temperature of the object reaches _____ K.

15. At stage 7, the star has reached the _____.

16. It takes a star like the Sun a total of about _____ million years to form.

17. Astronomers look for emissions at _____ wavelengths to identify interstellar clouds in stages 1 and 2.

18. At stages 4, 5, and 6, objects emit a great deal of radiation in the _____ part of the electromagnetic spectrum.

Review and Discussion

1. What is the composition of interstellar gas? Of interstellar dust?

2. If space is a near-perfect vacuum, how can there be enough dust in it to block light?

3. How is interstellar matter distributed through space?

4. What are some methods that astronomers use to study interstellar dust?

5. What is an emission nebula?

6. Why do emission nebulae appear red in color photographs?

7. Give a brief description of a dark dust cloud.

8. Why can't 21-centimeter radiation be used to probe the interiors of molecular clouds?

9. If our Sun were surrounded by a cloud of gas, would this cloud be an emission nebula? Why or why not?

10. Briefly describe the basic chain of events leading to the formation of a star like the Sun.

11. What is an evolutionary track?

12. Why do stars tend to form in groups?

13. What event must occur in order for a protostar to become a full-fledged star?

14. What are brown dwarfs?

15. Because stars live much longer than we do, how do astronomers test the accuracy of theories of star formation?

16. At what evolutionary stages must astronomers use radio and infrared radiation to study prestellar objects? Why can't they use visible light?

17. Explain the usefulness of the Hertzsprung–Russell diagram in studying the evolution of stars. Why can't evolutionary stages 1–3 be plotted on the diagram?

18. In the formation of a star cluster in which the individual stars have a wide range of masses, is it possible for some stars to die before others have finished forming? If this occurs, do you think it would have any effect on the cluster's formation?

Problems

1. Calculate the total mass of interstellar matter (of density 10^7 hydrogen atoms/m^3, each atom having a mass of 1.7×10^{-27} kg) contained in a volume equal to the volume of Earth.

2. Calculate the frequency of 21-cm radiation.

3. The intensity of a beam of light shining through a dense molecular cloud is diminished by a factor of 2.5 for every 3 pc the beam travels. By what total factor is the intensity reduced if the total thickness of the cloud is 60 pc?

4. Use the radius–luminosity–temperature relation ($L \propto R^2 T^4$; see Section 10.3 to calculate the luminosity, in solar units, of a brown dwarf whose radius is 0.1 solar radii and whose surface temperature is 600 K (0.1 times that of the Sun).

Projects

1. The constellation Orion the Hunter is prominent in the evening sky of winter. Its most noticeable feature is a short, diagonal row of three medium-bright stars: the famous Belt of Orion (Figure 11.20). The line of stars beginning at the middle star of the Belt and extending toward the south represents Orion's sword. Toward the bottom of the sword is the sky's most famous emission nebula, M42, the Orion Nebula. Observe it with your eye, with binoculars, and with a telescope. What is its color? How can you account for this color? With the telescope, try to find the Trapezium,

the grouping of four stars in the center of M42. These are hot, young stars; their energy causes the Orion Nebula to glow.

2. The Trifid Nebula, also known as M20, is a place where new stars are forming. It has been called a "dark-night revelation, even in modest apertures." An 8- to 10-inch telescope is needed to see the triple-lobed structure of the nebula. Ordinary binoculars reveal the Trifid as a hazy patch located in the constellation Sagittarius. This nebula is set against the richest part of the Milky Way, the edgewise projection of our own Galaxy around the sky. It is one of many wonders in this region of the heavens. What are the dark lanes in M20? Why are other parts of the nebula bright? There have been reports of large-scale changes occurring in this nebula in the last century and a half. The reports are based on old drawings that show M20 looking slightly different from how it appears today. Do you think it possible that a cloud in space might undergo a change in appearance on a time scale of years, decades, or centuries?

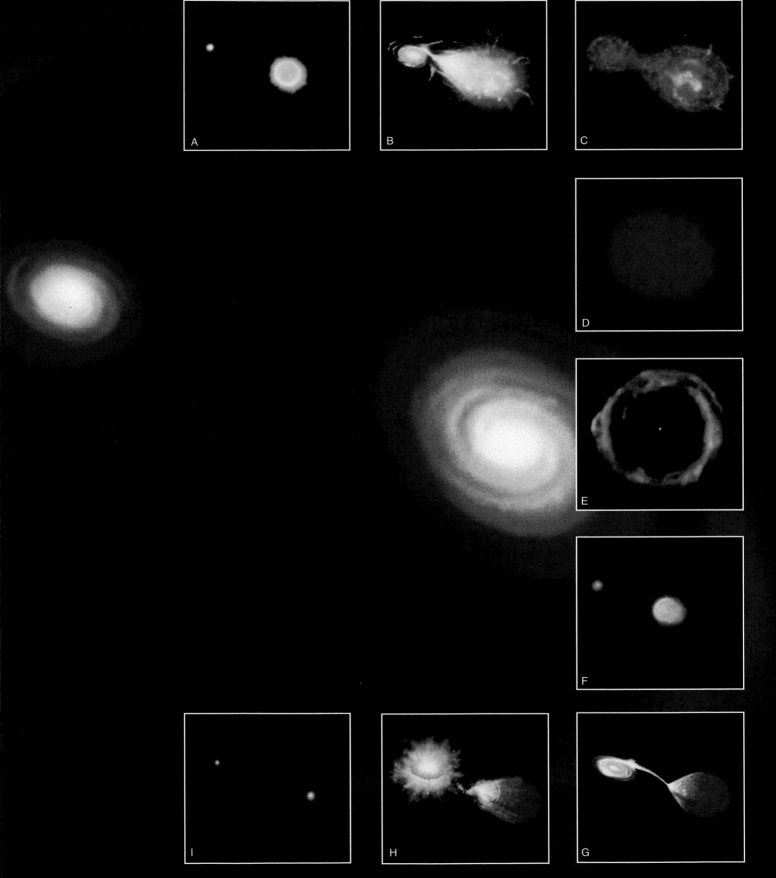

Stellar Evolution

The Lives and Deaths of Stars *www*

◄ (Opposite page, background) These frames are the conceptions of noted space artist Dana Berry. The ten frames depict a sequence of the birth, evolution, and death of a binary-star system. The sequence starts with the large rendering of a 1-solar-mass star and a 4-solar-mass star in the process of formation and then proceeds clockwise from top left. Some highlights: (Inset B) Nearing the end of its life, the 4-solar-mass star swells, spilling gas onto its companion and forming an accretion disk around it. (Inset E) The red giant has reached the point where the gravity of the original two stars cannot contain the gas. The result is a gentle expulsion—forming a planetary nebula rich in oxygen gas (hence the green color). (Inset G) An accretion bridge again joins the two stars. (Inset I) The end point of the system is two white dwarfs of roughly equal mass circling each other forevermore.

LEARNING GOALS

Studying this chapter will enable you to:

1 Explain why stars evolve off the main sequence.

2 Summarize the evolutionary stages followed by a Sun-like star once it leaves the main sequence, and describe the resulting remnant.

3 Explain how white dwarfs in binary systems can become explosively active.

4 Contrast the evolutionary histories of high-mass and low-mass stars.

5 Describe the two types of supernova, and explain how each is produced.

6 Explain the origin of elements heavier than helium, and discuss the significance of these elements for the study of stellar evolution.

7 Discuss the observations that help verify the theory of stellar evolution.

After reaching the main sequence, a newborn star changes little in outward appearance for more than 90 percent of its lifetime. However, at the end of this period, as the star begins to run out of fuel and die, it leaves the main sequence and its properties change greatly. Its ultimate fate depends primarily on its mass. A low-mass star is destined to die quietly, its outer layers eventually escaping into space. However, the potential exists for much more violent activity if a binary companion can provide additional fuel. A high-mass star will die explosively, releasing vast amounts of energy, creating many elements, and scattering debris throughout interstellar space. By continually comparing theoretical calculations with detailed observations of stars of all types, astronomers have refined the theory of stellar evolution into a precise and powerful tool for understanding the universe.

12.1 Leaving the Main Sequence

1 Most stars spend most of their lives on the main sequence. A star like the Sun, for example, after spending a few tens of millions of years in formation (stages 1–6 in Chapter 11), will reside on or near the main sequence (stage 7) for 10 billion years before evolving into something else. ∞ (Sec. 11.4) That "something else" is the topic of this chapter.

Virtually all the low-mass stars that have ever formed still exist as stars. The coolest M stars—red dwarfs—burn so slowly that not one of them has yet left the main sequence. Some of them will burn steadily for a trillion years or more. ∞ (Sec. 10.9) Conversely, the most massive O and B stars evolve away from the main sequence after only a few tens of millions of years. Most of the high-mass stars that have ever existed perished long ago. Between these two extremes, many stars are observed in advanced stages of evolution, their properties quite different from when they formed. By combining these observations with theoretical models, astronomers have built up a comprehensive picture of how stars evolve.

On the main sequence, a star slowly fuses hydrogen into helium in its core. This process is called **core hydrogen burning**. The star's equilibrium during this phase is the result of a balance between gravity and pressure, in which pressure's outward push exactly counteracts gravity's inward pull (Figure 12.1). Eventually, however, as the hydrogen in the core is consumed, the balance starts to shift and both the star's internal structure and its outward appearance begin to change: the star leaves the main sequence. You should keep Figure 12.1 in mind as you study the various stages

of stellar evolution described below. Much of a star's complex behavior can be understood in these simple terms.

Once a star evolves away from the main sequence, its days are numbered. The post–main-sequence stages of stellar evolution—the end of a star's life—depend critically on the star's mass. As a rule of thumb, we can say that low-mass stars die gently, while high-mass stars die catastrophically. The dividing line between these two very different outcomes lies around eight times the mass of the Sun, and in this chapter we will refer to stars of more than 8 solar masses as "high-mass" stars. Realize, however, that within both the

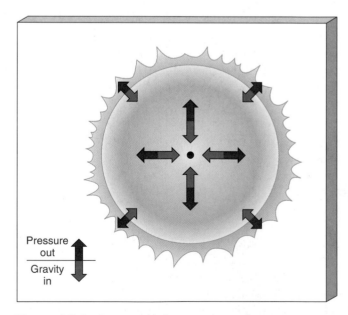

Pressure out

Gravity in

Figure 12.1 In a steadily burning star on the main sequence, the outward pressure exerted by hot gas balances the inward pull of gravity. This is true at every point within the star, guaranteeing its stability.

"high-mass" and the "low-mass" (that is, less than 8 solar masses) categories there are substantial variations. Rather than dwelling on the many details, however, we will concentrate on two representative evolutionary sequences—one specific to a solar-mass star, the other for a "generic" high-mass star much more massive than the Sun.

12.2 Evolution of a Sun-like Star

Formation of a Helium Core

As nuclear burning proceeds, the composition of the star's interior changes. Figure 12.2 illustrates the increase in helium abundance and the corresponding decrease in hydrogen abundance that take place in the stellar core as the star ages. The star's helium content increases fastest at the center, where temperatures are highest and the burning is fastest. The helium content also increases near the edge of the core, but more slowly because the burning rate is less rapid there. The inner, helium-rich region becomes larger and more hydrogen deficient as the star continues to shine. Eventually, hydrogen becomes completely depleted at the center, the nuclear fires there cease, and the location of principal burning moves to higher layers in the core. An inner core of nonburning pure helium starts to grow.

Without nuclear burning to maintain it, the outward-pushing gas pressure weakens in the helium inner core; however, the inward pull of gravity does not. Once the outward push against gravity is relaxed—even a little—structural changes in the star become inevitable. As soon as hydrogen becomes substantially depleted, about 10 billion years after the star arrived on the main sequence, the helium core begins to contract.

If more heat could be generated, then the core might possibly return to equilibrium. For example, if helium in the core were to begin fusing into some heavier element, then all would be well once again. Energy would be created as a by-product of helium burning, and the necessary outward-pushing gas pressure would be reestablished. But the helium at the center cannot burn—not yet, anyway. Despite its high temperature, the core is far too cold to fuse helium into anything heavier.

Recall from Chapter 9 that a temperature of 10^7 K is needed to fuse hydrogen into helium. Only above that temperature do colliding hydrogen nuclei (that is, protons) have enough speed to overwhelm the repulsive electromagnetic force between them. ∞ (Sec. 9.5) Because helium nuclei (with two protons each, compared to one for hydrogen) carry a greater positive charge, their electromagnetic repulsion is larger, and even higher temperatures are needed to cause them

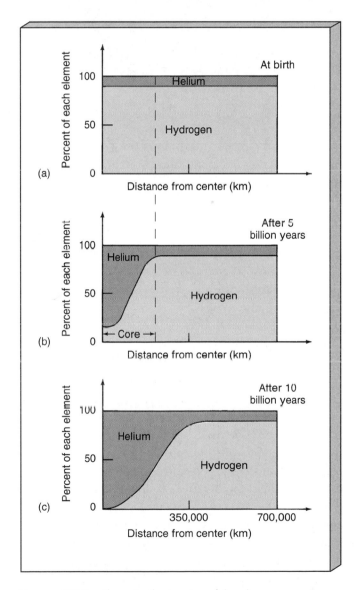

Figure 12.2 Theoretical estimates of the changes in a Sun-like star's composition. Hydrogen and helium abundances are shown (a) at birth, just as the star arrives on the main sequence; (b) after 5 billion years; and (c) after 10 billion years. At stage (b) only about 5 percent of the star's total mass has been converted from hydrogen to helium. The rate at which this change occurs increases as the nuclear burning rate increases with time.

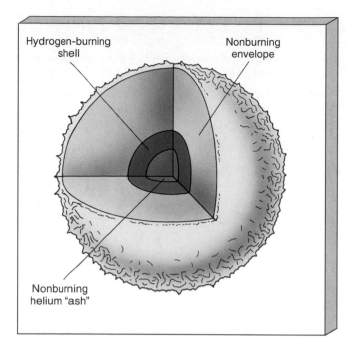

Figure 12.3 As a star's core loses more and more of its hydrogen, the hydrogen in the shell surrounding the nonburning helium ash burns ever more violently.

to fuse—at least 10^8 K. A core composed of helium at 10^7 K thus cannot generate energy through fusion.

The shrinkage of the helium core releases gravitational energy, driving up the central temperature and heating the overlying layers. The higher temperatures—now well over 10^7 K (but still less than 10^8 K)—cause hydrogen nuclei to fuse even more rapidly than before. Figure 12.3 depicts this situation, in which hydrogen is burning at a furious rate in a shell surrounding the nonburning inner core of helium "ash" in the center. This phase is usually known as the **hydrogen-shell-burning** stage. The hydrogen shell generates energy faster than did the original main-sequence star's hydrogen-burning core, and the shell's energy production continues to increase as the helium core continues to shrink. Strange as it may seem, the star's response to the disappearance of the fire at its center is to get brighter!

Red Giants

Conditions in the aging star have clearly changed from the equilibrium that characterized it as a main-sequence object. The helium core is unbalanced and shrinking. The rest of the core is also unbalanced, fus-

ing hydrogen into helium at an ever-increasing rate. The gas pressure exerted by this enhanced hydrogen burning increases, forcing the star's nonburning outer layers to increase in radius. Not even gravity can stop them. Even while the core is shrinking and heating up, the overlying layers are expanding and cooling. The star, aged and unbalanced, is on its way to becoming a red giant. The change from normal main-sequence star to elderly red giant takes about 100 million years.

We can trace these large-scale changes on an H–R diagram. Figure 12.4 shows the star's path away from the main sequence, labeled as stage 7. (Recall from Chapter 11 that stage 7 corresponds to the star's arrival on the main sequence.) ∞ (Sec. 11.4) The star first evolves to the right on the diagram, its surface temperature dropping while its luminosity increases only slightly. The star's roughly horizontal path from its main-sequence location (stage 7) to stage 8 on the

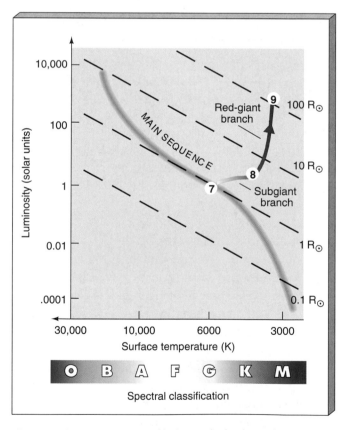

Figure 12.4 As its core of helium ash shrinks and its outer envelope expands, the star leaves the main sequence (stage 7). At stage 8, the star is well on its way to becoming a red giant. The star continues to brighten and grow as it ascends the red giant branch to stage 9. As noted in Chapter 10, the dashed diagonal lines are lines of constant radius, allowing us to gauge the changes in the size of our star.

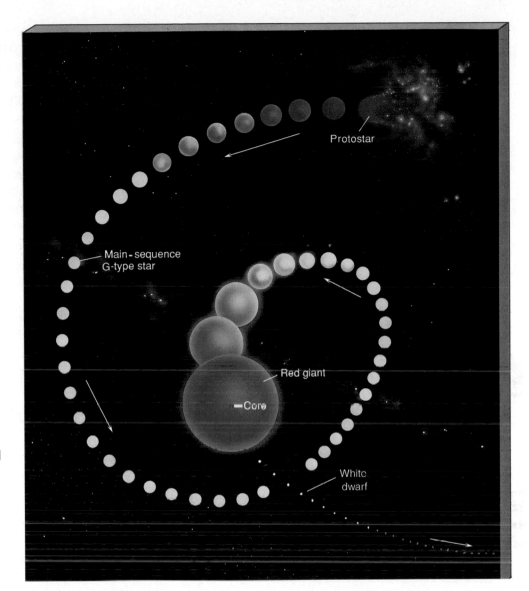

Figure 12.5 Relative sizes and colors of a normal G star (such as our Sun) in its formative stages, on the main sequence, and while passing through the red giant and white dwarf stages. At maximum swelling, the red giant is approximately 70 times the size of its main-sequence parent; the core of the giant is about 1/15 the main-sequence size and would be barely discernible if this figure were drawn to scale. The length of time spent in the various stages—protostar, main-sequence star, red giant, and white dwarf—is roughly proportional to the lengths shown in this imaginary trek through space.

figure is called the **subgiant branch**. By stage 8, the star's radius has increased to about three times the radius of the Sun.

The surface temperature at stage 8 has fallen to the point at which much of the interior is opaque to the radiation from within. Beyond this point, convection carries the core's enormous energy output to the surface. One consequence is that the star's surface temperature remains nearly constant between stages 8 and 9. The nearly vertical path followed by the star between stages 8 and 9 is known as the **red giant branch** of the H–R diagram. By stage 9, the giant's luminosity is many hundreds of times the solar value, and its radius is around 100 solar radii.

Figure 12.5 compares the relative sizes of a G star like our Sun and a stage 9 red giant. It also indicates the stages through which the star will evolve. The red giant is huge—about the size of Mercury's orbit. In contrast, its helium core is surprisingly small—only about 1/1000 the size of the entire star, making that core just a few times larger than Earth.

The density at the center of a red giant is enormous. Continued shrinkage of the red giant's core has compacted its helium gas to approximately 10^8 kg/m^3. Contrast this with the 10^{-3} kg/m^3 in the giant's outermost layers, with the 5000 kg/m^3 average density of Earth, and with the 150,000 kg/m^3 in the present core of the Sun. About 25 percent of the mass of the entire star is packed into its planet-sized core.

Perhaps the most famous red giant is the naked-eye star Betelgeuse in the constellation Orion (shown in Figure 10.4). Despite its great distance from Earth (about 150 pc), its enormous luminosity, 10^4 times that of the Sun, makes it one of the brightest stars in the night sky.

Helium Fusion

Should the unbalanced state of a red giant continue, the core would eventually collapse and the rest of the star would slowly drift into space. However, this simultaneous shrinking and expanding cannot continue indefinitely. A few hundred million years after a solar-mass star leaves the main sequence, something else happens—helium begins to burn in the core. By the time the central density has risen to about 10^8 kg/m^3 (stage 9), the temperature has reached the 10^8 K needed for helium to fuse into carbon, and the central fires reignite.

For stars comparable in mass to the Sun, there is a major complication when helium fusion begins. At the high densities found in the core, the gas has entered a state of matter whose properties are governed by the laws of quantum mechanics rather than by those of classical physics. Up to now, we have been concerned primarily with the nuclei that make up virtually all of the star's mass and participate in the reactions that generate its energy. However, the star contains another important constituent—a vast sea of electrons stripped from their parent nuclei by the ferocious heat in the stellar interior. At this stage in our story, these electrons play a critical role in determining the star's evolution.

Under the conditions found in the stage 9 red giant core, the rules of quantum mechanics prohibit electrons from being squeezed too close together. We can think of electrons as tiny rigid spheres that can be squeezed relatively easily up to the point of contact but become virtually incompressible thereafter. In fact, by stage 9, the pressure in the inner core resisting the force of gravity is supplied almost entirely by tightly packed electrons. Hardly any of the core's support results from "normal" thermal pressure, and this has dramatic consequences once the helium begins to burn.

In a star supported by thermal pressure, the increase in temperature produced by the onset of helium fusion would lead to an increase in pressure. The gas would then expand and cool, reducing the burning rate and reestablishing equilibrium. In the electron-supported core of a solar-mass red giant, however, the pressure is largely *independent* of the temperature. When burning starts and the temperature increases, there is no corresponding rise in pressure, no expansion of the gas, no drop in the temperature, and no stabilization of the core. Instead, the core is unable to respond to the rapidly changing conditions within it. The pressure remains more or less unchanged as the nuclear reaction rates increase, and the temperature rises rapidly in a runaway explosion called the **helium flash**.

For a few hours, the helium burns ferociously, like an uncontrolled bomb. Eventually, the flood of energy released by this period of runaway fusion heats the core to the point where normal thermal pressure once again dominates. Finally able to react to the energy dumped into it by helium burning, the core expands, its density drops, and equilibrium is restored as the inward pull of gravity and the outward push of gas pressure come back into balance. The core, now stable, begins to burn helium into carbon at temperatures well above 10^8 K.

The helium flash terminates the star's ascent of the red giant branch of the H–R diagram at stage 9 in Figure 12.4. Yet despite the explosive detonation of helium in the core, the flash does *not* increase the star's luminosity. On the contrary, the helium flash produces a rearrangement of the core that ultimately results in a *reduction* in the energy output. On the H–R diagram, the star jumps from stage 9 to stage 10, a stable state in which helium burns steadily in the core.

At stage 10 our star is now stably burning helium in its inner core and fusing hydrogen in a shell surrounding that core. It resides in a well-defined region of the H–R diagram known as the **horizontal branch**, where core-helium-burning stars remain for a time before resuming their journey around the H–R diagram. As indicated in Figure 12.6, at this stage the surface temperature is higher than it was on the red giant branch, whereas the luminosity is considerably less than at the helium flash. This adjustment in the star's properties occurs quite quickly—in about 100,000 years.

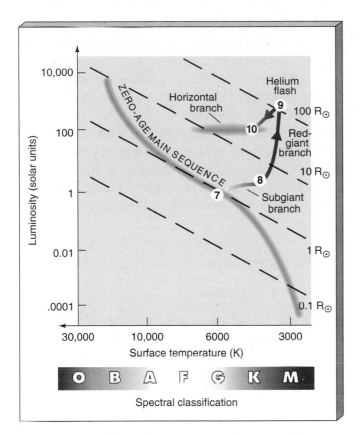

Figure 12.6 After its large increase in luminosity while ascending the red giant branch is terminated by the helium flash, our star settles down into another equilibrium state at stage 10, on the horizontal branch.

The Carbon Core

Nuclear reactions in stars proceed at rates that increase very rapidly with temperature. At the extremely high temperatures found in the core of a horizontal-branch star, the helium fuel doesn't last long—no more than a few tens of million years after the initial flash.

As helium fuses to carbon, a new inner core of carbon ash forms and phenomena similar to the earlier buildup of helium ash begin to occur. Now helium becomes depleted at the very center, and eventually fusion ceases there. In response, the nonburning carbon core shrinks and heats up as gravity pulls it inward, causing the hydrogen- and helium-burning rates in the overlying layers to increase. The star now contains a shrinking carbon-ash inner core surrounded by a helium-burning shell, which is in turn surrounded by a hydrogen-burning shell (Figure 12.7). The outer envelope of the star—the nonburning layers surrounding the core—expands, much as it did earlier in the first

red giant stage. By the time it reaches stage 11 in Figure 12.8, the star has become a swollen red giant for a second time.

The burning rates at the star's center are much fiercer during its second trip to the red giant region, and the radius and luminosity increase to values even greater than those reached during the first visit (stage 9). Our star is now a **red supergiant**. Its carbon core continues to shrink, driving the hydrogen-burning and helium-burning shells to higher and higher temperatures and luminosities.

Table 12.1 summarizes the key stages through which a solar-mass star evolves. It is a continuation of Table 11.2, except that here the density units are the more convenient kilograms per cubic meter and we now express radii in units of the solar radius. The numbers in the "Stage" column refer to the evolutionary stages noted in the figures and discussed in the text.

All the H–R diagrams and evolutionary tracks presented so far are theoretical constructs based largely on computer models. Before continuing our study of stellar evolution, let's take a moment to compare theory with reality. Figure 12.9 shows a real H–R diagram, drawn using the stars of the old globular cluster M3.

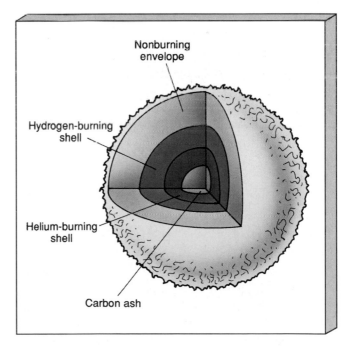

Figure 12.7 Within a few million years after the onset of helium burning (stage 9), carbon ash accumulates in the star's inner core. Above this core, hydrogen and helium are still burning in concentric shells.

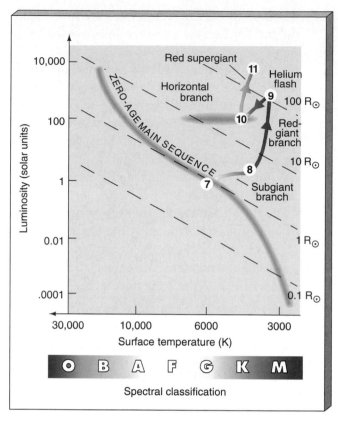

Figure 12.8 A carbon-core star reenters the giant region of the H–R diagram (stage 11) for the same reason it evolved there the first time around: lack of nuclear burning in the inner core causes contraction of the core and expansion of the overlying layers.

The similarity between theory and observation is striking—stars in each of the evolutionary stages 7–11 can be seen, in numbers consistent with the theoretical models. (The points in Figure 12.9 are shifted a little to the left relative to Figure 12.8 because of differences in composition between stars like the Sun and stars in globular clusters; globular cluster stars tend to be slightly hotter than solar-type stars of the same mass.) Astronomers place great confidence in the theory of stellar evolution precisely because its predictions are so often found to be in excellent agreement with plots of real stars.

Planetary Nebulae

As our red supergiant moves from stage 10 to stage 11, its envelope swells while its core, too cool for further nuclear burning, continues to contract. If the central temperature could become high enough for carbon fusion to occur, still heavier elements could be synthesized, and the newly generated energy might again support the star, restoring for a time the equilibrium between gravity and heat. For solar-mass stars, however, this does not occur. The temperature never reaches the 600 million K needed for new nuclear reac-

Stage	Approx. Time to Next Stage (yr)	Central Temperature (K)	Surface Temperature (K)	Central Density (kg/m³)	Radius (km)	Radius (solar radii)	Object
7	10^{10}	1.5×10^7	6000	10^5	7×10^5	1	Main-sequence star
8	10^8	5×10^7	4000	10^7	2×10^6	3	Subgiant
9	10^5	10^8	4000	10^8	7×10^7	100	Helium flash
10	5×10^7	2×10^8	5000	10^7	7×10^6	10	Horizontal branch
11	10^4	2.5×10^8	4000	10^8	4×10^8	500	Red supergiant
12	10^5	3×10^8	100,000	10^{10}	10^4	0.01	Carbon core
	—		3000	10^{-17}	7×10^8	1,000	Planetary nebula[1]
13	—	10^8	50,000	10^{10}	10^4	0.01	White dwarf
14	—	Close to 0	Close to 0	10^{10}	10^4	0.01	Black dwarf

Table 12.1 Evolution of a Sun-like Star

[1]Values in columns 2 through 7 refer to the envelope.

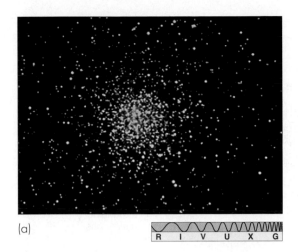

(a)

Figure 12.9 The various evolutionary stages predicted by theory and depicted schematically in Figure 12.8 are clearly visible in this H–R diagram (b) of an old star cluster—the globular cluster M3, shown in (a). The faintest main-sequence stars are not shown here because observational limitations make it difficult to determine the apparent brightness of low-luminosity stars in the cluster.

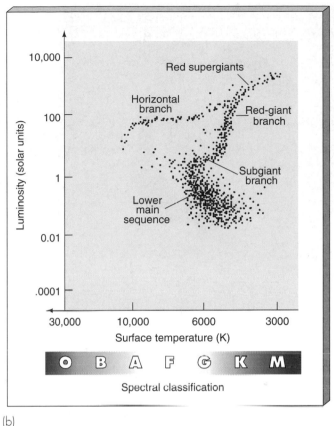

(b)

tions to occur. The red supergiant is now very close to the end of its nuclear-burning lifetime.

Before the carbon core can attain the incredibly high temperatures needed for carbon ignition, its density reaches a point beyond which the core cannot be compressed further. At about 10^{10} kg/m^3, the electrons in the core once again come into contact with one another, stopping the contraction and stabilizing the temperature. This stage (stage 12 in Table 12.1) represents the maximum compression that the star can achieve—there is simply not enough matter in the overlying layers to bear down any harder.

The core density at this stage is extraordinarily high. A single cubic centimeter of core matter would weigh 1000 kg on Earth—a ton of matter compressed into a volume about the size of a grape. Yet despite the extreme compression of the core, the central temperature is "only" about 300 million K. Collisions among nuclei are neither frequent nor violent enough to fuse carbon into any of the heavier elements. The central fires go out once carbon has formed.

Our aged stage 12 star is now in quite a predicament. Its inner carbon core is, for all practical purposes, dead. The outer shells continue to burn hydro-

gen and helium, and, as more and more of the inner core reaches its final, high-density state, the burning increases in intensity. Meanwhile, the envelope continues to expand and cool. Eventually, driven by increasing radiation from within and accelerated by the energy released as electrons recombine with nuclei to form atoms, the envelope becomes unstable and is ejected into space at a speed of a few tens of kilometers per second.

In time, a rather unusual-looking object results. We say unusual because the "star" now has two distinct parts, both of which constitute stage 12. At the center is a small well-defined core of mostly carbon ash. Hot and dense, only the outermost layers of this core still fuse helium to carbon. Well beyond the core, there is a spherical shell of cooler and thinner matter—the ejected envelope of the giant—spread over a volume roughly the size of our solar system. Such an object is called a **planetary nebula** (Figure 12.10).

The term *planetary* here is very misleading, for these objects have no association with planets. The name originated in the eighteenth century when, viewed at poor resolution through small telescopes, these shells of gas looked to some astronomers like the

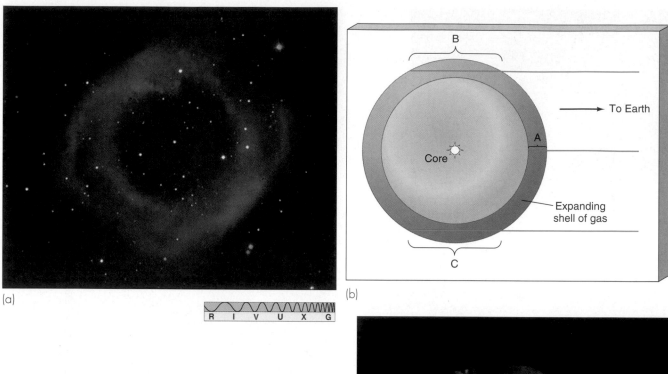

(a)

R   I   V   U   X   G

(b)

To Earth

Core

Expanding shell of gas

A

B

C

(c)

R   I   V   U   X   G

Figure 12.10 A planetary nebula is an object with a small, dense core (central blue-white star) surrounded by an extended shell (or shells) of glowing matter. (a) The Helix Nebula appears to the eye as a small star with a halo around it. About 140 pc from Earth and 0.6 pc across, its apparent size in the sky is roughly half that of the full Moon. (b) The appearance of a planetary nebula can be explained once we realize that the shell of glowing gas around the central core is quite thin. There is very little gas along the line of sight between the observer and the central star (path A), so that part of the shell is invisible. Near the edge of the shell, however, there is more gas along the line of sight (paths B and C), so the observer sees a glowing ring. (c) The Cat's Eye Nebula is an example of a much more complex planetary nebula. It lies about 1000 pc away. It may have been produced by a pair of binary stars (unresolved at the center) that have both shed planetary nebulae.

circular disks of the planets in our solar system. The term *nebula* is also confusing, as it suggests kinship with the emission nebulae studied in Chapter 11). ∞ (Sec. 11.2) In fact, not only are planetary nebulae much smaller than emission nebulae, they are also associated with much older stars. Emission nebulae are the signposts of recent stellar birth. Planetary nebulae indicate impending stellar death.

The "ring" of a planetary nebula is in reality a three-dimensional shell of warm, glowing gas completely surrounding the core. Its halo-shaped appearance is only an illusion. The shell is a complete envelope that has been expelled from around the core, but

we can see it only at the edges, where emitting matter accumulates along our line of sight. As illustrated in Figure 12.10(b), the shell is virtually invisible in the direction of the core. Few planetary nebulae are quite as regular as this simple picture might suggest, however. Figure 12.10(c) shows a system in which the details of the gas-ejection process have apparently played an important role in determining the planetary nebula's shape and appearance.

The planetary nebula continues to spread out with time, becoming more diffuse and cooler, gradually dispersing into interstellar space. In doing so, it enriches the interstellar medium with atoms of helium and carbon dredged up from the depths of the core into the envelope by convection during the star's final years.

White Dwarfs

The carbon core, the stellar remnant at the center of the planetary nebula, continues to evolve. Formerly concealed by the atmosphere of the red giant star, the core becomes visible as the envelope recedes. The core is very small, about the size of Earth, with a mass about half that of the Sun. Shining only by stored heat, not by nuclear reactions, this small star has a white-hot surface when it first becomes visible, although it appears dim because of its small size. The core's heat and size give rise to its new name—*white dwarf*. ∞ (Sec. 10.3) This is stage 13 of Table 12.1. The approximate path followed by the star on the H–R diagram as it evolves from stage 11 red supergiant to stage 13 white dwarf is shown in Figure 12.11.

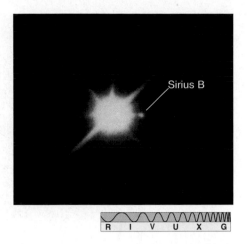

Figure 12.12 Sirius B (the speck of light to the right of the much larger and brighter Sirius A) is a white dwarf star, a companion to Sirius A. (The "spikes" on the image of Sirius A are not real; they are caused by the support struts of the telescope.)

Not all white dwarfs are seen as the cores of planetary nebulae. Several hundred have been discovered "naked," their envelopes expelled to invisibility long ago. Figure 12.12 shows an example of a white dwarf, Sirius B, that happens to lie particularly close to Earth; it is the faint binary companion of the much brighter Sirius A. ∞ (Sec. 10.9) Detailed observations show Sirius B to have the properties listed in Table 12.2. With more than the mass of the Sun packed into a volume smaller than Earth, Sirius B's density is about a million times greater than anything familiar to us in the solar system. Sirius B has an unusually high mass for a white dwarf; it is believed to be the evolutionary product of a star roughly 4 times the mass of the Sun.

Once a star becomes a white dwarf, its evolution, for all practical purposes, is over. It continues to cool

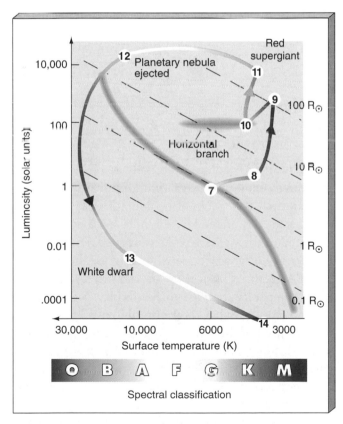

Figure 12.11 A star's passage from the horizontal branch (stage 10) to the white dwarf stage (stage 13) creates an evolutionary path that cuts across the H–R diagram.

Table 12.2 Sirius B—A Nearby White Dwarf Star	
Mass	1.1 solar masses
Radius	0.008 solar radii (5500 km)
Luminosity (total)	0.04 solar luminosities
Surface temperature	24,000 K
Average density	3×10^9 kg/m^3

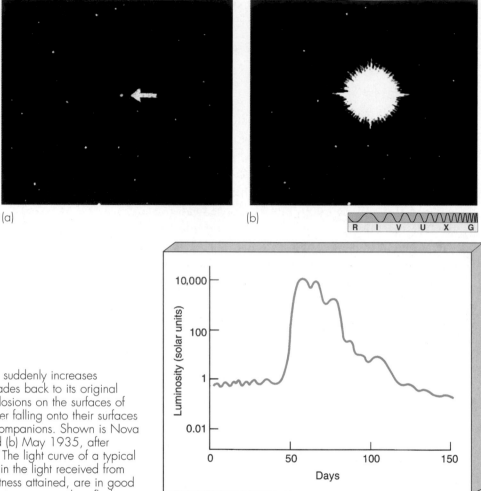

Figure 12.13   A nova is a star that suddenly increases enormously in brightness, then slowly fades back to its original luminosity. Novae are the result of explosions on the surfaces of faint white dwarf stars, caused by matter falling onto their surfaces from the atmosphere of larger binary companions. Shown is Nova Herculis 1934 in (a) March 1935 and (b) May 1935, after brightening by a factor of 60,000. (c) The light curve of a typical nova. The rapid rise and slow decline in the light received from the star, as well as the maximum brightness attained, are in good agreement with the explanation of the nova as a nuclear flash on a white dwarf's surface.

and dim with time, following the white–yellow–red line near the bottom of Figure 12.11, eventually becoming a **black dwarf**—a cold, dense, burned-out ember in space. This is stage 14 of Table 12.1, the graveyard of stars. The cooling dwarf does not shrink much as it fades away, however. Even though its heat is leaking away into space, gravity does not compress it further. At the enormously high densities in the star (from the white dwarf stage on), the resistance of electrons to being squeezed together holds the star up, even as its temperature drops almost to absolute zero. As the dwarf cools, it remains about the size of Earth.

Novae

3   In some cases, the white dwarf stage does not represent the end of the road for a Sun-like star. Given the right circumstances, it is possible for a

white dwarf to become explosively active, in the form of a highly luminous **nova** (plural: novae). The word *nova* means "new" in Latin, and, to early observers, these stars did indeed seem new as they suddenly appeared in the night sky. Astronomers now recognize that a nova is what we see when a white dwarf undergoes a violent explosion on its surface, resulting in a rapid, temporary increase in luminosity. Figures 12.13 (a) and (b) illustrate the brightening of a typical nova over a period of three days. Figure 12.13(c) shows how a nova's luminosity rises dramatically in a matter of days, then fades slowly back to normal over the course of several months. On average, two or three novae are observed each year.

What could cause such an explosion on a faint, dead star? The energy involved is far too great to be explained by flares or other surface activity, and, as we have just seen, there is no nuclear activity in the

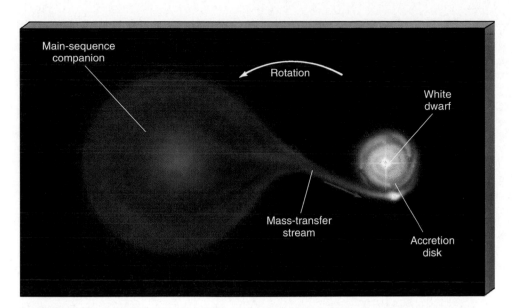

Figure 12.14 If a white dwarf in a binary system is close enough to its companion, its gravitational field can tear matter from the companion's surface. Notice that the matter does not fall directly onto the white dwarf's surface. Instead, it forms an *accretion disk* of gas spiraling down onto the dwarf.

dwarf's interior. The answer to this question lies in the white dwarf's surroundings. If the white dwarf is isolated, then it will indeed cool and ultimately become a black dwarf, as just described. However, should the white dwarf be part of a binary system in which the other star is either a main-sequence star or a giant, an important new possibility exists.

If the distance between the dwarf and the other star is small enough, then the dwarf's gravitational field can pull matter—primarily hydrogen and helium—away from the surface of the companion, as illustrated in Figure 12.14. As it builds up on the white dwarf's surface, the "stolen" gas becomes hotter and denser. Eventually its temperature exceeds 10^7 K, and the hydrogen ignites, fusing into helium at a furious rate. This surface-burning stage is as brief as it is violent. The star suddenly flares up in luminosity, then fades away as some of the fuel is exhausted and the rest is blown off into space. If the event happens to be visible from Earth, we see a nova.

Because of the binary's rotation, material leaving the companion does not fall directly onto the white dwarf. Instead, it "misses" the dwarf, loops around behind it, and goes into orbit around it, forming a swirling, flattened disk of matter called an **accretion disk**. The orbiting matter in the disk then drifts gradually inward, its temperature increasing steadily as it spirals down onto the dwarf's surface. The inner part of the accretion disk becomes so hot that it radiates strongly in the visible, the ultraviolet, even the X-ray portions of the electromagnetic spectrum. Often, the disk outshines the white dwarf itself and is the main source of the light emitted between nova outbursts.

Once the nova explosion is over and the binary has returned to normal, the mass-transfer process can begin again. Astronomers know of many *recurrent novae*—stars that have been observed to "go nova" several times over the course of a few decades; such systems can, in principle, repeat their violent outbursts many dozens, if not hundreds, of times.

12.3 Evolution of Stars More Massive than the Sun

4 Stars leave the main sequence for one basic reason: they run out of hydrogen in their cores. As a result, the early stages of stellar evolution after the main sequence are qualitatively the same in all cases: main-sequence hydrogen burning in the core (stage 7) eventually gives way to the formation of a nonburning, contracting helium core surrounded by a hydrogen-burning shell (stages 8 and 9). All stars leaving the main sequence on the journey toward the red giant region of the H–R diagram have essentially the same

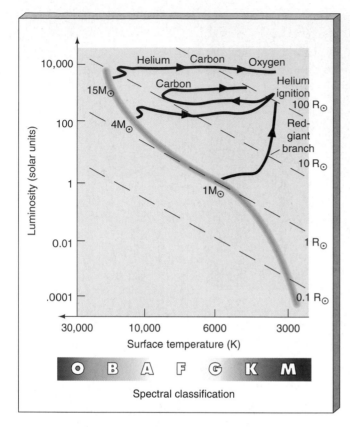

Figure 12.15 Evolutionary tracks for stars of 1, 4, and 15 solar masses (shown only up to helium ignition in the 1-solar-mass case). Stars with masses comparable to the mass of the Sun ascend the giant branch almost vertically, whereas higher-mass stars move roughly horizontally across the H–R diagram from the main sequence into the red giant region. The most massive stars experience smooth transitions into each new burning stage. Some points are labeled with the element that has just started to fuse in the inner core.

internal structure. Thereafter, however, their evolutionary tracks diverge.

Figure 12.15 compares the post–main-sequence evolution of three stars having masses 1, 4, and 15 times the mass of the Sun. Note that, while stars like the Sun ascend the red giant branch almost vertically, higher-mass stars move nearly horizontally across the H–R diagram after leaving the upper main sequence. Their luminosities stay roughly constant as their radii increase and their surface temperatures drop.

In stars having more than about 2.5 times the mass of the Sun, helium burning begins smoothly and stably, *not* explosively—there is no helium flash. Calculations indicate that, the more massive a star, the lower its core density when the temperature reaches the 10^8 K necessary for helium ignition and the smaller

the contribution to the pressure from tightly packed electrons. As a result, above 2.5 solar masses, the unstable core conditions described earlier do not occur. Thus, the 4-solar-mass red giant in Figure 12.15 remains a red giant as helium starts to fuse into carbon. There is no sudden jump to the horizontal branch and no subsequent reascent of the giant branch. Instead, the star loops smoothly back and forth near the top of the H–R diagram.

Fusion of Heavy Elements

A much more important divergence occurs at approximately 8 solar masses—the dividing line between high and low mass mentioned in Section 12.1. A low-mass star never becomes hot enough to burn carbon in its core, and so it ends its life as a carbon white dwarf. A high-mass star, however, can fuse not just hydrogen and helium, but also carbon, oxygen, and even heavier elements as its inner core continues to contract and its central temperature continues to rise. The burning rate accelerates as the core evolves.

Evolution proceeds so rapidly in the 15-solar-mass star whose evolution is shown in Figure 12.15 that the star doesn't even reach the red giant region before helium fusion begins. The star achieves a central temperature of 10^8 K while still quite close to the main sequence, and its evolutionary track continues smoothly across the H–R diagram, seemingly unaffected by each new phase of burning.

Can anything stop this runaway fusion process in a high-mass star? Is there a stable "white dwarf–like" state at the end of the road for such a body? To answer these questions, we must look more carefully at nuclear fusion in massive stars.

Figure 12.16 is a cutaway diagram of a highly evolved high-mass star. Note the numerous layers where various nuclei burn. As the temperature increases with depth, the ash of each burning stage becomes the fuel for the next. At the relatively cool periphery of the core, hydrogen fuses into helium. In the intermediate layers, shells of helium, carbon, and oxygen burn to form heavier nuclei. Deeper down reside neon, magnesium, silicon, and other heavy nuclei, all produced by nuclear fusion in the layers overlying the nonburning inner core. The inner core itself is composed of iron.

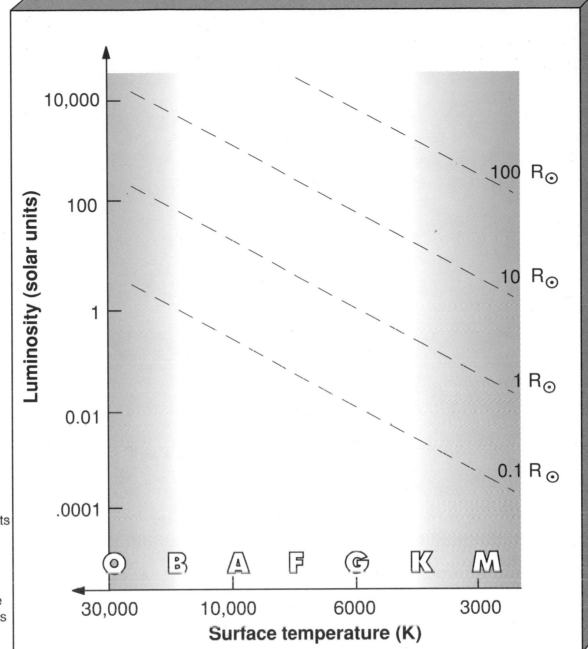

The H-R diagram plots stars by luminosity (vertical axis) and temperature, or spectral class (horizontal axis). The dashed diagonal lines are lines of constant radius.

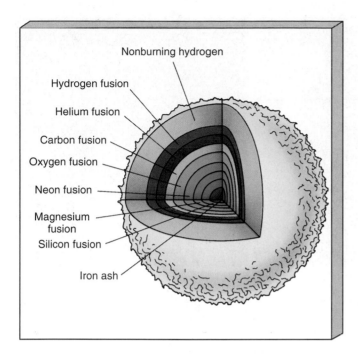

Nonburning hydrogen

Hydrogen fusion

Helium fusion

Carbon fusion

Oxygen fusion

Neon fusion

Magnesium fusion

Silicon fusion

Iron ash

Figure 12.16 Cutaway diagram of the interior of a highly evolved high-mass star. The interior resembles the layers of an onion, with shells of progressively heavier elements burning at smaller and smaller radii and at higher and higher temperatures.

As each element is burned to depletion at the center, the core contracts, heats up, and starts to fuse the ash of the previous burning stage. A new inner core forms, contracts again, heats again, and so on. Through each period of stability and instability, the star's central temperature increases, the nuclear reactions speed up, and the newly released energy supports the star for ever-shorter periods of time. For example, in round numbers, a star 20 times more massive than the Sun burns hydrogen for 10 million years, helium for 1 million years, carbon for 1000 years, oxygen for 1 year, and silicon for a week. Its iron core grows for less than a day.

The Death of a High-Mass Star

Once the inner core begins to change to iron, our high-mass star is in trouble. Nuclear fusion involving iron does not produce energy, because iron nuclei are so compact that energy cannot be extracted by combining them into heavier elements. In effect, iron plays the role of a fire extinguisher, damping the inferno in the stellar core. With the appearance of substantial quan-

tities of iron, the central fires cease for the last time, and the star's internal support begins to dwindle. The star's foundation is destroyed, and its equilibrium is gone forever. Even though the temperature in the iron core has reached several billion kelvins by this stage, the enormous inward gravitational pull of matter ensures catastrophe in the very near future. Gravity overwhelms the pressure of the hot gas, and the star implodes, falling in on itself.

The core temperature rises to nearly 10 billion K. At these temperatures, individual photons are energetic enough to split iron into lighter nuclei and then to break those lighter nuclei apart until only protons and neutrons remain. This process is known as *photo-disintegration*. In less than a second, the collapsing core undoes all the effects of nuclear fusion that occurred during the previous 10 million years! But to split iron and lighter nuclei into smaller pieces requires a lot of energy. After all, this splitting is just the opposite of the fusion reactions that generated the star's energy during earlier times. Photodisintegration *absorbs* some of the core's thermal energy—in other words, it cools the core and so reduces the pressure. As nuclei are destroyed, the core becomes even less able to support itself against its own gravity. The collapse accelerates.

Now the core consists entirely of electrons, protons, neutrons, and photons at enormously high densities, and it is still shrinking. As the core density continues to rise, the protons and electrons are crushed together, combining to form more neutrons and releasing neutrinos. This process is sometimes called the *neutronization* of the core. Recall from our discussion in Chapter 9 that the neutrino is an extremely elusive particle that interacts hardly at all with matter. ∞ (Sec. 9.5) Even though the central density by this time may have reached 10^{12} kg/m^3 or more, most of the neutrinos pass through the core as if it weren't there. They escape into space, carrying away energy as they go.

The disappearance of the electrons and the escape of the neutrinos make matters even worse for the core's stability. There is now nothing to prevent it from collapsing all the way to the point at which the neutrons come into contact with each other, at the incredible density of about 10^{15} kg/m^3. At this density, the neutrons in the shrinking core play a role similar in many ways to that of the electrons in a white dwarf. When far apart, they offer little resistance to compression, but when brought into contact, they produce

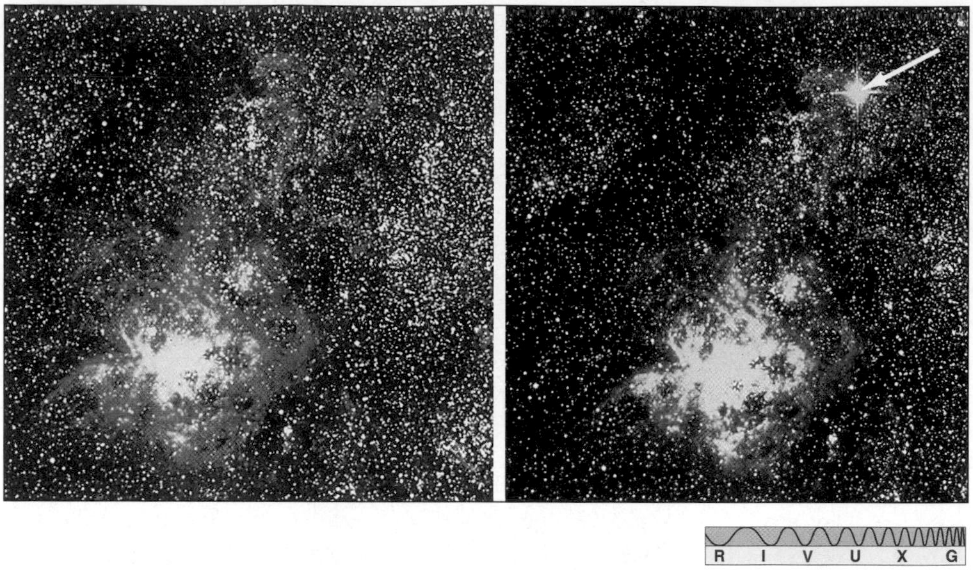

Figure 12.17 A supernova called SN1987A (arrow) was exploding near this nebula (30 Doradus) at the moment the photograph on the right was taken. The photograph on the left is the normal appearance of the star field. (See *Interlude* 12-1.)

enormous pressures that strongly oppose further gravitational collapse. The collapse finally begins to slow. By the time it is actually halted, however, the core has overshot its point of equilibrium and may reach densities as high as 10^{18} kg/m^3 before beginning to reexpand. Like a fast-moving ball hitting a brick wall, the core becomes compressed, stops, then rebounds—with a vengeance!

The events just described do not take long. Only about a second elapses from the start of the collapse to the "bounce" at nuclear densities. Driven by the rebounding core, an enormously energetic shock wave then sweeps outward through the star at high speed, blasting all the overlying layers—including the heavy elements outside the iron inner core—into space. Although the details of how the shock reaches the surface and destroys the star are still uncertain, the end result is not: the star explodes in one of the most energetic events known in the universe (Figure 12.17). This spectacular death rattle of a high-mass star is known as a **core-collapse supernova**.

12.4 Supernova Explosions

Novae and Supernovae

5 Observationally, a **supernova**, like a nova, is a "star" that suddenly increases dramatically in brightness, then slowly dims again, eventually fading from view. However, novae and supernovae are very different phenomena. As we saw in Section 12.2, a nova is a violent explosion on the surface of a white

dwarf that is a member of a binary system.[1] Supernovae are even more energetic—about a million times brighter than novae—and are driven by very different underlying physical processes. A supernova produces a burst of light equaling over a billion solar luminosities, reaching that brightness within just a few hours of the start of the outburst. The total amount of electromagnetic energy radiated by a supernova during the few months it takes to brighten and fade away is roughly the same as the Sun will radiate during its *entire* 10^{10}-year lifetime!

Despite the differences in energy, some supernova light curves appear quite similar to those of novae. For this reason, a distant supernova can look very like a nearby nova—so much so, in fact, that the distinction between the two was not fully appreciated until the 1920s. In addition to the magnitude of the energy involved, another important difference between these two classes of stellar explosion is the fact that the same star may become a nova many times, but a star can become a supernova only once. The nova accretion–explosion cycle described earlier can take place over and over again, but a supernova destroys the star involved, with no possibility of a repeat performance.

There are also important observational differences among supernovae. Some contain very little hydrogen, according to their spectra, while others con-

---

[1] Note that, when discussing novae and supernovae, astronomers tend to blur the distinction between the observed event (the sudden appearance and brightening of an object in the sky) and the process responsible for it (a violent explosion in or on a star). The two terms can have either meaning, depending on context.

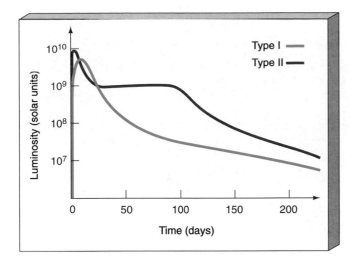

Figure 12.18 The light curves of typical Type I and Type II supernovae. In both cases, the maximum luminosity can sometimes reach that of a billion suns, but there are characteristic differences in the falloff of the luminosity after the initial peak. Type I light curves somewhat resemble those of novae (Figure 12.13c). Type II curves have a characteristic bump in the declining phase.

tain a lot. Also, the light curves of the hydrogen-poor supernovae are qualitatively different from those of the hydrogen-rich ones, as illustrated in Figure 12.18. On the basis of these observations, astronomers divide supernovae into two classes. **Type I supernovae**, the hydrogen-poor kind, have a light curve somewhat similar in shape to that of typical novae—a sharp rise in intensity followed by steady, gradual decline. **Type II supernovae**, whose spectra show lots of hydrogen, usually have a characteristic "plateau" in the light curve a few months after the maximum. Observed supernovae are divided roughly evenly between these two categories.

Type I and Type II Supernovae Explained

What is responsible for the two distinct types of supernova? Is there more than one way in which a supernova explosion can occur? The answer is yes. To understand the alternative supernova mechanism, we must return to the processes that cause novae and reconsider the long-term consequences of their accretion–explosion cycle.

Nova explosions eject matter from a white dwarf's surface, but they do not necessarily expel or burn all the material that has accumulated since the last outburst. In other words, there is a tendency for the

dwarf's mass to increase slowly with each new nova cycle. As its mass grows and the internal pressure required to support its weight rises, the white dwarf can enter into a new period of instability—with disastrous consequences.

Recall that a white dwarf is held up not by thermal pressure (heat) but by the pressure of electrons that have been squeezed so close together that they have effectively come into contact with one another. However, there is a limit to the mass of a white dwarf, beyond which electrons cannot provide the pressure needed to support the star. Detailed calculations show that the maximum mass of a white dwarf is about 1.4 solar masses.

If an accreting white dwarf exceeds this maximum mass, the pressure of electrons in its interior becomes unable to withstand the pull of gravity, and the star immediately starts to collapse. Its internal temperature rapidly rises to the point at which carbon can fuse into heavier elements. Carbon fusion begins everywhere throughout the white dwarf almost simultaneously, and the entire star explodes in a **carbon-detonation supernova**—an event comparable in violence to the core-collapse supernova associated with the death of a high-mass star but very different in cause. In an alternative and (some astronomers think) possibly more common scenario, two white dwarfs in a binary system may collide and merge to form a massive, unstable star. The end result is the same—a carbon-detonation supernova.

We can now understand the differences between Type I and Type II supernovae. The detonation of a carbon white dwarf, the descendent of a *low-mass* star, is a Type I supernova. Because the explosion occurs in a system containing virtually no hydrogen, we can readily see why the spectrum of a Type I supernova shows little evidence of that element. The appearance of the light curve results almost entirely from the radioactive decay of unstable heavy elements produced in the explosion.

The collapse and rebound of the core of a *high-mass* star produces a Type II supernova. The characteristic shape of the Type II light curve is just what is expected from the star's outer envelope expanding and cooling as it is blown into space by the shock wave sweeping up from below. The expanding material consists mainly of unburned hydrogen and helium, so it is not surprising that those elements are strongly represented in the supernova's spectrum.

(a) Type-I Supernova

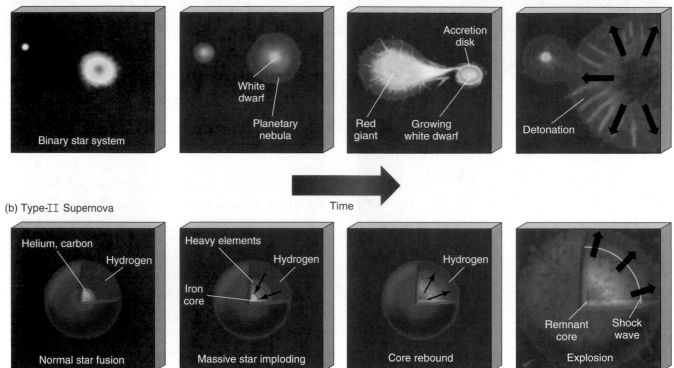

(b) Type-II Supernova

Figure 12.19 Type I and Type II supernovae have different causes. These sequences depict the evolutionary history of each type. (a) A Type I supernova usually results when a carbon-rich white dwarf pulls matter onto itself from a nearby red giant or main-sequence companion. (b) A Type II supernova occurs when the core of a high-mass star collapses and then rebounds in a catastrophic explosion.

Figure 12.19 summarizes the processes responsible for the two types of supernovae. We emphasize that, despite the similarity in the total amounts of energy involved, Type I and Type II supernovae are actually unrelated to one another. They occur in stars of very different types, under very different circumstances. All high-mass stars become Type II (core-collapse) supernovae, but only a tiny fraction of low-mass stars evolve into white dwarfs that ultimately explode as Type I (carbon detonation) supernovae. However, there are far more low-mass stars than high-mass stars, resulting in the remarkable coincidence that the two types of supernova occur at roughly the same rate.

Supernova Remnants *Extension*

We have plenty of evidence that supernovae have occurred in our Galaxy. Occasionally, the supernova explosions are visible from Earth. In other cases, we can detect their glowing remains, or **supernova remnants**. One of the best-studied supernova remnants is the Crab Nebula (Figure 12.20a). Its brightness has greatly dimmed now, but the original explosion in the year A.D. 1054 was so brilliant that it is prominently recorded in the manuscripts of ancient Chinese and Middle Eastern astronomers. For nearly a month, this exploded star reportedly could be seen in broad daylight. Even today, the knots and filaments give a strong indication of past violence. The nebula—the envelope of the high-mass star that exploded to create this Type II supernova—is still expanding into space at speeds of several thousand kilometers per second.

Figure 12.20(b) shows another example (also of Type II). The expansion velocity of the Vela supernova remnant implies that its central star exploded around 9000 B.C. This remnant lies only 500 pc from Earth. Given its proximity, it may have been as bright as the Moon for several months.

Although hundreds of supernovae have been observed in other galaxies during the twentieth century,

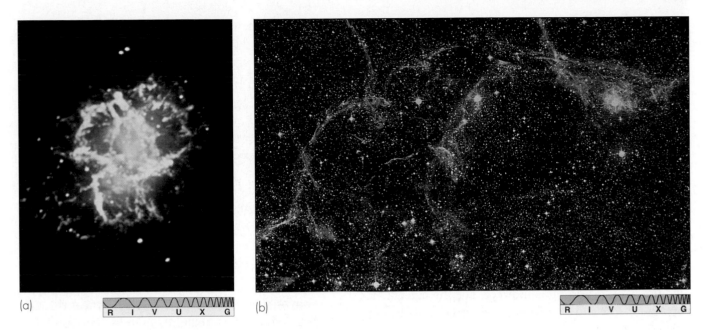

(a)   (b)

Figure 12.20 (a) This remnant of an ancient Type II supernova is called the Crab Nebula (M1 in the Messier catalog). It resides about 1800 pc from Earth and has an angular diameter about one-fifth that of the full Moon. Because its debris is scattered over a region of "only" 2 pc, the Crab is considered to be a young supernova remnant. In A.D. 1054 Chinese astronomers observed the supernova explosion. (b) The glowing gases of the Vela supernova remnant are spread across a large 6° of the sky.

no one using modern equipment has ever observed one in our own Galaxy. A viewable Milky Way star has not exploded since Galileo first turned his telescope to the heavens almost four centuries ago (see, however, *Interlude 12-1* for a discussion of a recent highly publicized supernova in a galaxy very close to our own). Based on stellar evolutionary theory, astronomers estimate that an observable supernova ought to occur in our Galaxy every 100 years or so. Our local neighborhood seems long overdue for a supernova. Unless stars explode much less frequently than predicted by theory, we should be treated to a (relatively) nearby version of nature's most spectacular cosmic event any day now.

Supernovae and the Formation of the Heavy Elements

6 From the point of view of life on Earth, probably the most important aspect of supernovae is their role in creating, and then dispersing, the heavy elements out of which both our planet and our bodies are made.

Since the 1950s, astronomers have come to realize that all of the hydrogen and most of the helium in the universe are *primordial*—that is, they date back to the very earliest times, long before the first stars formed. ∞ (Sec. 17.6) All other elements (and, in particular, virtually everything we see around us on Earth) formed later, through stellar evolution.

We have already seen something of how heavy elements are created from light ones by nuclear fusion. This is the basic process that powers all stars. Hydrogen fuses to helium, then helium to carbon. Subsequently, in high-mass stars, carbon fuses to form still heavier elements. Oxygen, neon, magnesium, sulfur, silicon—in fact, all the known elements up to and including iron—were created in turn by fusion reactions in the cores of the most massive stars.

However, all this stops at iron. The fact that iron will not fuse to create more massive nuclei is the basic underlying cause of Type II supernovae. How then were even heavier elements, such as copper, lead, gold, and uranium, formed? The answer is that they were created during supernova explosions (both Type I and Type II), as neutrons and protons, produced when some nuclei were ripped apart by the almost unimaginable violence of the blast, were crammed into other nuclei, creating heavy elements that could not have formed by any other means. The heaviest elements, then, were formed *after* their parent stars had already died, even as the debris from the explosion signaling the stars' deaths was hurled into interstellar space.

Although no one has ever observed directly the formation of heavy nuclei in stars, astronomers are confident that the chain of events just described really occurs in stars. When nuclear reaction rates (determined from laboratory experiments) are incorporated into detailed computer models of stars and supernovae, the results agree very well, point by point, with the abundances of the elements inferred from spectroscopic studies of planets, stars, and the interstellar medium. The reasoning is indirect, but the agreement between theory and observation is so striking that most astronomers regard it as strong evidence supporting the entire theory of stellar evolution.

The Cycle of Stellar Evolution

We have now seen all the ingredients that make up the complete cycle of star formation and evolution. Let us briefly summarize that process, which is illustrated in Figure 12.21.

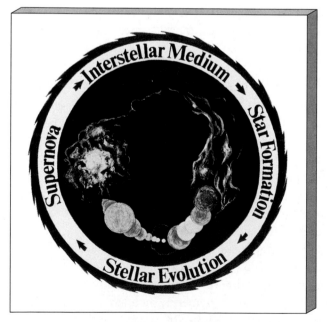

Figure 12.21 The cycle of star formation and evolution continuously replenishes the Galaxy with new heavy elements and provides the driving force for the creation of new generations of stars.

1. Stars form when part of an interstellar cloud is compressed beyond the point at which it can support itself against its own gravity. The cloud collapses and fragments, forming a cluster of stars. ∞ (Sec. 11.4)

2. Within the cluster, stars evolve. The most massive stars evolve fastest, creating heavy elements in their cores and spewing them forth into the interstellar medium in supernova explosions. Low-mass stars create elements up to oxygen and may contribute to the "seeding" of interstellar space when they shed their envelopes as planetary nebulae.

3. The creation and explosive dispersal of new heavy elements is accompanied by further shock waves. The passage of these shock waves through the interstellar medium simultaneously enriches the medium and compresses it into further star formation. Each generation of stars increases the concentration of heavy elements in the interstellar clouds from which the next generation forms. As a result, recently formed stars contain a much greater abundance of heavy elements than stars that formed long ago.

In this way, although some material is used up in each cycle—turned into energy or locked up in low-mass stars—the Galaxy continuously recycles its matter. Each new round of formation creates stars containing more heavy elements than the preceding generation had. From the old globular clusters, which are observed to be deficient in heavy elements relative to the Sun, to the young open clusters, which contain much larger amounts of these elements, we observe this enrichment process in action. Our Sun is the product of many such cycles. We ourselves are another. Without the heavy elements synthesized in the hearts of stars, life on Earth would not exist.

12.5 Observing Stellar Evolution in Star Clusters

7 Star clusters provide excellent test sites for the theory of stellar evolution. Every star in a given cluster formed at the same time, from the same interstellar cloud, with virtually the same composition. Only the mass varies from one star to another. This uniformity allows us to check the accuracy of our theoretical models in a very straightforward way. Having studied in some detail the evolutionary tracks of individual stars, let us now consider how their collective appearance changes in time. The agreement—in detail—between theoretical predictions and observations of real clusters is remarkably good.

We begin our study shortly after the cluster's formation, with the high-mass stars already fully formed and burning steadily on the upper main sequence and lower-mass stars just beginning to arrive on the main sequence, as shown in Figure 12.22(a). The appearance of the cluster at this early stage is dominated by its most massive stars—the bright blue supergiants. Now let's follow the cluster forward in time and ask how its H–R diagram evolves.

Figure 12.22(b) shows the appearance of our cluster's H–R diagram after 10 million years. The most massive O stars have evolved off the main sequence. Most have already exploded and vanished, but one or two may still be visible as supergiants traversing the top of the diagram. Such massive stars evolve so rapidly after leaving the main sequence that the chance of "catching one in the act" is very low. The remaining cluster stars are largely unchanged in appearance—their evolution is slow enough that little happens to them in 10^7 years. The cluster's H–R diagram has a slightly truncated main sequence. Figure 12.23 shows an observed H–R diagram that compares well with the theoretical one shown in Figure 12.22b. The H–R plot in Figure 12.23 is for the twin open clusters h and χ (the Greek letter chi) Persei. Comparing Figure 12.23(b) with H–R diagrams such as those in Figure

Figure 12.22 The changing H–R diagram of a hypothetical star cluster. (a) Initially, stars on the upper main sequence are already burning steadily while the lower main sequence is still forming. (b) At 10^7 years, O stars have already left the main sequence, and a few post–main-sequence supergiants are visible. (c) By 10^8 years, stars of spectral type B have evolved off the main sequence. More supergiants are visible, and the lower main sequence is almost fully formed. (d) At 10^9 years, the main sequence is cut off at about spectral type A. The subgiant and red giant branches are just becoming evident, and the formation of the lower main sequence is complete. A few white dwarfs may be present. (e) At 10^{10} years, only stars less massive than the Sun still remain on the main sequence. The cluster's subgiant, red giant and horizontal branches are all discernible, and many white dwarfs have formed.

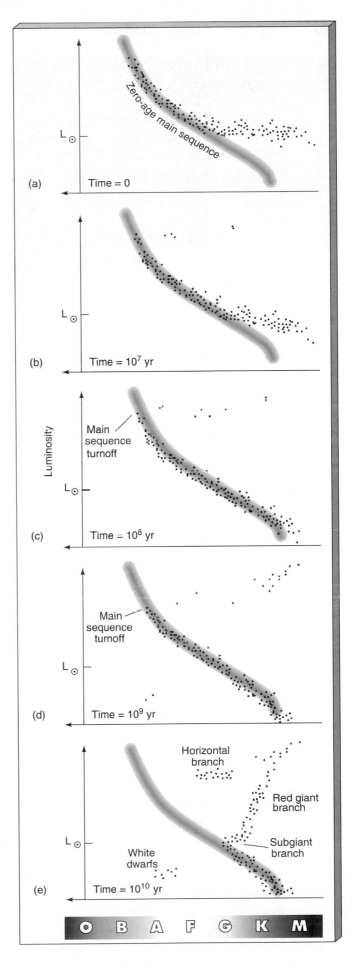

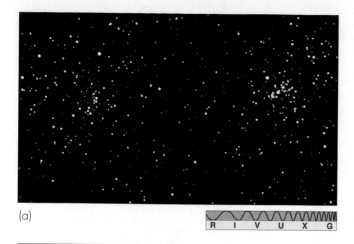

(a)

R I V U X G

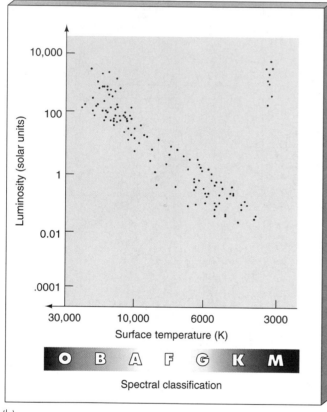

(b)

Figure 12.23 (a) The "double cluster" h and χ Persei. (b) The H–R diagram of the pair indicates that the stars are very young—probably only about 10 million years old.

12.22, astronomers estimate the age of this pair of clusters to be about 10 million years.

After 100 million years (Figure 12.22c), stars brighter than type B5 or so (about 4–5 solar masses) have left the main sequence, and a few more supergiants are visible. By this time, most of the cluster's low-mass stars have finally arrived on the main se-

quence, although the dimmest M stars may still be in their contraction phase. The appearance of the cluster is now dominated by bright B stars and brighter supergiants.

At any time during the evolution, the cluster's original main sequence is intact up to some well-defined stellar mass, corresponding to the stars that are just leaving the main sequence at that instant. We can imagine the main sequence being "peeled away" from the top down, with fainter and fainter stars turning off and heading for the giant branch as time goes on. Astronomers refer to the high-luminosity end of the observed main sequence as the **main-sequence turnoff**. The mass of the star that is just evolving off the main sequence at any moment is known as the *turnoff mass*.

At 1 billion years, the main-sequence turnoff mass is around 2 solar masses, corresponding roughly to spectral type A2. The subgiant and giant branches associated with the evolution of low-mass stars are just becoming apparent, as indicated in Figure 12.22(d). The formation of the lower main sequence is now complete. In addition, the first white dwarfs have just appeared, although they are generally too faint to be observed at the distances of most clusters. Figure 12.24 shows the Hyades open cluster and its H–R diagram. The H–R diagram appears to lie between Figures 20.16(c) and 20.16(d), suggesting that the cluster's age is about 5×10^8 years.

At 10 billion years, the turnoff point has reached solar-mass stars, of spectral type G2. The subgiant and giant branches are now clearly discernible in the H–R diagram (Figure 12.22e), and the horizontal branch appears as a distinct region. Many white dwarfs are also present in the cluster. Although stars in all these evolutionary stages were also present in the 1-billion-year-old cluster shown in Figure 12.22(d), they were few in number—typically only a few percent of the total number of stars in the cluster. Also, because higher-mass stars evolve so rapidly, they spend very little time in these regions. Lower-mass stars are much more numerous and evolve more slowly, so their evolutionary tracks are more easily detected.

Figure 12.25 shows the globular cluster 47 Tucanae. By carefully adjusting their theoretical models until the cluster's main sequence, subgiant, red giant, and horizontal branches are all well matched, astronomers have determined its age to be roughly 11

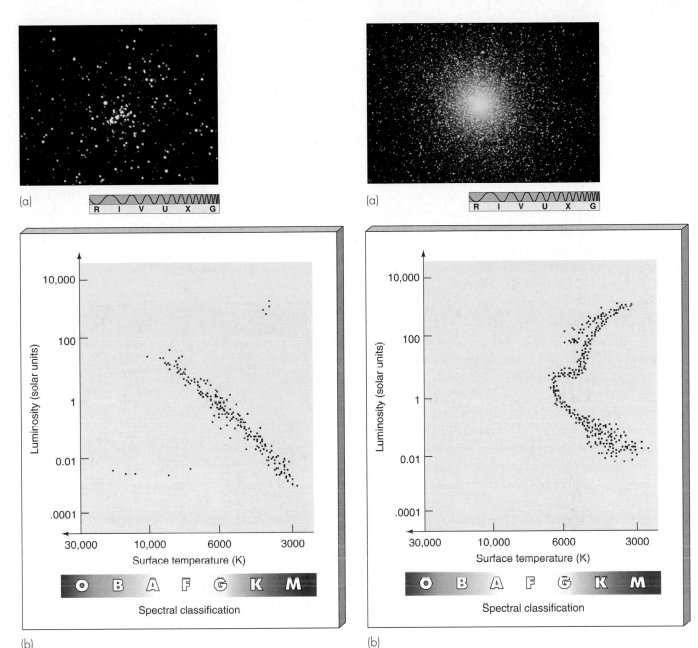

Figure 12.24 (a) The Hyades cluster, a relatively young group of stars visible to the naked eye. (b) The H–R diagram for this cluster is cut off at about spectral type A, implying an age of about 500 million years.

Figure 12.25 (a) The globular cluster 47 Tucanae. (b) Fitting its main-sequence turnoff and its giant and horizontal branches to theoretical models gives 47 Tucanae an age of about 11 billion years, making it one of the oldest known objects in the Milky Way Galaxy.

billion years, a little older than our hypothetical cluster in Figure 12.22(e). In fact, globular cluster ages determined this way show a remarkably small spread. All the globular clusters in our Galaxy appear to have formed between about 10 and 12 billion years ago.

Stellar evolution is one of the great success stories of astrophysics. Like all good scientific theories, it makes definite, testable predictions about the universe, at the same time remaining flexible enough to incorporate new discoveries as they occur. Theory and observation have advanced hand in hand. At the start of the twentieth century, many scientists despaired of ever knowing even the compositions of the stars, let alone why they shine and how they change. Today, the theory of stellar evolution is a cornerstone of modern astronomy.

SUPERNOVA 1987A

In 1987, astronomers were treated to a spectacular supernova in the Large Magellanic Cloud (LMC), a small satellite galaxy orbiting our own. ∞ (Sec. 15.1) Observers in Chile first saw the explosion on February 24, and within a few hours, nearly all Southern Hemisphere telescopes and every available orbiting spacecraft were focused on the object. It was officially named SN1987A. (The SN stands for "supernova," 1987 gives the year, and A identifies it as the first supernova seen that year.) This was one of the most dramatic changes observed in the universe in nearly 400 years. A 15-solar-mass B supergiant star with the catalog name of SK-69°202 exploded, outshining for a few weeks all the other stars in the LMC combined, as shown in the "before" and "after" images of Figure 12.17.

Because the LMC is relatively close to Earth and because the explosion was detected so soon after it occurred, SN1987A has provided astronomers with a wealth of detailed information on supernovae, allowing them to make key comparisons between theoretical models and observational reality. By and large, the theory of stellar evolution described in the text has held up very well. Still, SN1987A did hold some surprises.

According to its hydrogen-rich spectrum, the supernova was of Type II—the core-collapse type—as expected for a high-mass parent star such as SK-69°202. But Figure 12.15 (which was computed for stars in our own Galaxy) shows that, according to theory, the parent star should have been a red supergiant at the time of the explosion—not a blue supergiant, as was actually observed. This unexpected finding caused theorists to scramble in search of an explanation. It now seems that, relative to young stars in the Milky Way, the parent star's envelope was deficient in heavy elements. This deficiency had little effect on the evolution of the core and on the supernova explosion, but it did change the star's evolutionary track on the H–R diagram. Unlike a Milky Way star of the same mass, once helium ignited in the core of SK-69°202, the star shrank and looped back toward the main sequence. The star had just begun to return to the right on the H–R diagram following the ignition of carbon, with a surface temperature of around 20,000 K, when the rapid chain of events leading to the supernova occurred.

The light curve of SN1987A, shown below, also differed somewhat from the "standard" Type II shape (see Figure 12.18). The peak brightness was only about 1/10 the expected value.

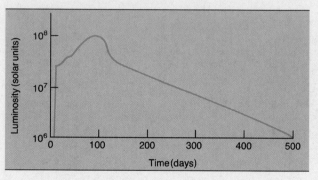

For a few days after its initial detection, the supernova faded as it expanded and cooled rapidly. After about a week, the surface temperature had dropped to about 5000 K, at which point electrons and protons near the expanding surface recombined into atomic hydrogen, making the surface layers less opaque and allowing more radiation from the interior to leak out. As a result, the supernova brightened rapidly as it grew. The temperature of the expanding layers reached a peak in late May, by which point the radius of the expanding photosphere was about 2×10^{10} km—a little larger than our solar system. Subsequently, the photosphere cooled as it expanded, and the luminosity dropped as the internal supply of heat from the explosion leaked away into space.

Much of the preceding description would apply equally well to a Type II supernova in our own Galaxy. The differences between the SN1987A light curve shown above and the Type II light curve in Figure 12.18 are mainly the result of the (relatively) small size of SN1987A's parent star. The peak luminosity of SN1987A was less than that of a "normal" Type II supernova because SK-69°202 was small and quite tightly bound by gravity. A lot of the energy emitted in the form of visible radiation (and evident in Figure 12.18) was used up in expanding SN1987A's stellar envelope, so far less was left over to be radiated into space. Thus, SN1987A's luminosity during the first few months was lower than expected, and the early peak evident in Figure 12.18 did not occur. The peak in the SN1987A light curve at about 80 days actually corresponds to the "plateau" in the Type II light curve in Figure 12.18.

About 20 hours before the supernova was detected optically, a brief (13-second) burst of neutrinos was simultaneously recorded by underground detectors in Japan and the United States. As discussed in the text, the neutrinos are predicted to arise when electrons and protons in the star's collapsing core merge to form neutrons. The neutrinos preceded the light because they escaped during the collapse, whereas the first light of the explosion was emitted only after the supernova shock had plowed through the body of the star to the surface. In fact, theoretical models, consistent with these observations, suggest that vastly more energy was emitted in the form of neutrinos than in any other form. The supernova's neutrino luminosity was many tens of thousands of times greater than its optical energy output.

Despite some unresolved details in SN1987A's behavior, detection of this neutrino pulse is considered to be a brilliant confirmation of theory. This singular event—the detection of neutrinos—may well herald a new age of astronomy. For the first time, astronomers have received information from beyond the solar system by radiation outside the electromagnetic spectrum.

Theory predicts that the expanding remnant of SN1987A will be large enough to be resolvable by optical telescopes in a few years. The accompanying photograph was taken by the *Hubble Space Telescope* in late 1996. It shows the unresolved remnant (at center) surrounded by a much larger

shell of glowing gas (in yellow). Scientists reason that the progenitor star expelled this shell during its red giant phase, some 40,000 years before the explosion. The image we see results from the initial flash of ultraviolet light from the supernova hitting the ring and causing it to glow brightly. In about 10 years, the fastest-moving debris from the remnant will strike the ring, making it a temporary but intense source of X rays.

This overexposed photo also shows the core debris indeed moving outward toward the ring. The four insets, taken over a 24-month interval, directly resolve debris expanding at nearly 3000 km/s. These images also revealed, to everyone's surprise, two additional faint rings (in red) that might be radiation sweeping across the hourglass-shaped bubble of gas. Why the gas should exhibit this odd structure is unclear.

Buoyed by the success of stellar-evolution theory and armed with firm theoretical predictions of what should happen next, astronomers eagerly await future developments in the story of this remarkable object.

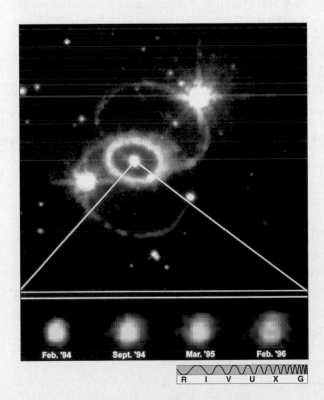

Feb. '94   Sept. '94   Mar. '95   Feb. '96

R   I   V   U   X   G

Chapter Review

Summary

Stars spend most of their lives on the main sequence, in the **core hydrogen burning** (p. 336) phase of stellar evolution, stably fusing hydrogen into helium at their centers. Stars leave the main sequence when the hydrogen in their cores is exhausted. When the central nuclear fires cease, the helium in the star's core is still too cool to fuse into anything heavier. With no internal energy source, the helium core is unable to support itself against its own gravity and begins to shrink. The star at this stage is in the **hydrogen-shell-burning** (p. 338) phase, in which the nonburning helium at the center is surrounded by a layer of burning hydrogen. The energy released by the contracting helium core heats the hydrogen-burning shell, greatly increasing the nuclear reaction rates there. As a result, the star becomes much brighter, while the envelope expands and cools. A low-mass star like the Sun moves off the main sequence on the H–R diagram first along the **subgiant branch** (p. 339), then almost vertically up the **red giant branch** (p. 339).

As the helium core contracts, it heats up. Eventually, it reaches the point at which helium begins to fuse into carbon. In a star like the Sun, conditions at the onset of helium burning are such that electrons in the core present stiff resistance to further compression, making the core unable to "react" to the new energy source. As a result, helium burning begins explosively, in the **helium flash** (p. 340). The flash expands the core and reduces the star's luminosity, sending it onto the **horizontal branch** (p. 340) of the H–R diagram. The star now has a core of burning helium surrounded by a shell of burning hydrogen. As helium burns in the core, it forms an inner core of nonburning carbon. The carbon core shrinks and heats the overlying burning layers, and the star once again becomes a red giant. It reenters the red giant region of the H–R diagram, becoming an extremely luminous **red supergiant** (p. 341) star.

The core of a low-mass star never becomes hot enough to fuse carbon. Such a star continues to brighten and expand until its envelope is ejected into space as a **planetary nebula** (p. 343). At that point the core becomes visible as a hot, faint, and extremely dense white dwarf. The planetary nebula diffuses into space, carrying helium and carbon into the interstellar medium. The white dwarf cools and fades, eventually becoming a cold **black dwarf** (p. 346).

A **nova** (p. 346) is a star that suddenly increases greatly in brightness, then slowly fades back to its normal appearance over a period of months. It is the result of a white dwarf in a binary system drawing hydrogen-rich material from its companion. The gas builds up on the white dwarf's surface, eventually becoming hot and dense enough for the hydrogen to burn explosively, temporarily causing a large increase in the dwarf's luminosity. The matter flowing from the companion star does not fall directly onto the surface of the dwarf. Instead it goes into orbit around it, forming an **accretion disk** (p. 347). The gas spirals slowly inward, heating up and glowing brightly as it nears the dwarf's surface.

Evolutionary changes happen more rapidly for high-mass stars than for low-mass stars because larger mass results in higher central temperatures. Stars more massive than about 8 solar masses do attain central temperatures high enough to fuse carbon. They form heavier and heavier elements in their cores, at a more and more rapid pace. As they evolve, their cores form a layered structure consisting of burning shells of successively heavier elements. A nonburning inner core of iron builds up at the center.

Iron is special in that its nuclei can neither be fused together nor split apart to produce energy. As a result, stellar nuclear burning stops at iron. As a star's iron core grows in mass, it eventually becomes unable to support itself against gravity and begins to collapse. At the enormous densities and temperatures produced during the collapse, iron nuclei are broken down into their constituent particles—protons and neutrons. The protons combine with electrons to form more neutrons. Eventually, when the core becomes so dense that the neutrons are effectively brought into physical contact with one another, their resistance to further squeezing stops the collapse and the core rebounds, sending a violent shock wave out through the rest of the star. The star explodes in a **core-collapse supernova** (p. 350).

Astronomers classify supernovae into two broad categories: Type I and Type II. These classes differ in their light curves and in their composition. **Type I supernovae** (p. 351) are hydrogen poor and have a light curve similar in shape to that of a nova. **Type II supernovae** (p. 351) are hydrogen rich and have a characteristic bump in the light curve a few months after maximum. A Type II supernova is a core-collapse supernova. A Type I supernova occurs when a white dwarf in a binary system exceeds a critical mass of about 1.4 solar masses, collapses and then explodes as its carbon ignites. This type of supernova is called a **carbon-detonation supernova** (p. 351).

Theory predicts that a supernova visible from Earth should occur within our Galaxy about once a century, although none has been observed in the last 400 years. We can see evidence for a past supernova in the form of a **supernova remnant** (p. 352), a shell of exploded debris surrounding the site of the explosion and expanding into space at a speed of thousands of kilometers per second.

All elements heavier than helium formed in stars or in supernova explosions. Comparisons between theoretical predictions of element production and observations of element abundances in stars and supernovae provide strong support for the theory of stellar evolution. The processes of star formation, evolution, and explosion form a cycle that constantly enriches the interstellar medium with heavy elements and sows the seeds of new generations of stars.

The theory of stellar evolution can be tested by observing star clusters, all of whose stars formed at the same time. As time goes by, the most massive stars evolve off the main sequence first, then the intermediate-mass stars, and so on. At any instant, no stars with masses above the cluster's **main-sequence turnoff** (p. 356) mass remain on the main sequence. Stars below this mass have not yet evolved into giants and so still lie on the main sequence. By comparing a particular cluster's main-sequence turnoff mass with theoretical predictions, astronomers can measure the age of the cluster.

Self-Test: True or False?

____ 1. "Low-mass" stars are conventionally taken to have masses less than about 8 solar masses.

____ 2. All the red dwarf stars that ever formed are still on the main sequence today.

____ 3. Once a star is on the main sequence, gravity is no longer important in determining the star's internal structure.

____ 4. The Sun will get brighter as it begins to run out of fuel in its core.

____ 5. As it evolves away from the main sequence, a star gets larger.

____ 6. As it evolves away from the main sequence, a star gets hotter.

____ 7. When the Sun becomes a red giant, its core will be smaller than when the Sun was on the main sequence.

____ 8. When helium starts to fuse inside a solar-mass red giant, it does so very slowly at first; the rate of fusion increases gradually over many years.

____ 9. A planetary nebula is the disk of matter around a star that will eventually form a planetary system.

____ 10. A nova is a sudden outburst of light coming from an old main-sequence star.

____ 11. For a high-mass star, there is no helium flash.

____ 12. High-mass stars can fuse carbon in their cores.

____ 13. It takes less and less time to fuse heavier and heavier elements inside a high-mass star.

____ 14. In a core-collapse supernova, the outer part of the core rebounds from the inner, high-density part of the core, destroying the entire outer part of the star.

____ 15. The spectrum of a Type II supernova shows the presence of lots of hydrogen.

____ 16. Once the process gets under way, the core of a massive star collapses in about 1 s.

_____ 17. Stellar evolution can account for the existence of all elements except hydrogen and helium.

_____ 18. A star cluster 100 million years old still contains many O stars.

Self-Test: Fill in the Blank

1. A main-sequence star doesn't collapse because of the outward _____ produced by hot gases in the stellar interior.
2. The Sun will leave the main sequence in about _____ years.
3. While a star is on the main sequence, _____ is slowly depleted and _____ builds up in the core.
4. A temperature of at least _____ is needed to fuse helium.
5. At the end of its main-sequence lifetime, a star's core starts to _____.
6. As a red supergiant, the Sun will eventually become about _____ times its present size.
7. The various stages of stellar evolution predicted by theory can be tested using observations of stars in _____.
8. By the time the envelope of a red supergiant is ejected, the core has shrunk down to a diameter of about _____ .
9. A typical white dwarf has the following properties: about half a solar mass, fairly _____ surface temperature, small size, and _____ luminosity.
10. As time goes by, the temperature and the luminosity of a white dwarf both _____.
11. In a binary consisting of a white dwarf and a main-sequence or giant companion, matter leaving the companion forms an _____ disk around the dwarf.
12. A nova explosion is due to _____ fusion on the _____ of a white dwarf.
13. When a proton and an electron are forced together, they combine to form a _____ and a _____.
14. A _____ supernova occurs when a white dwarf exceeds 1.4 solar masses.
15. The maximum mass for a white dwarf is roughly _____ solar masses.
16. The two types of supernova can be distinguished observationally by their spectra and by their _____.
17. As a star cluster ages, the luminosity of the main-sequence turnoff _____.

Review and Discussion

1. Why don't stars live forever? Which types of stars live longest?
2. How long can a star like the Sun keep burning hydrogen in its core?
3. Why is the depletion of hydrogen in the core of a star such an important event?
4. What makes an ordinary star become a red giant?
5. How big (in A.U.) will the Sun become when it enters the red giant phase?
6. How long does it take for a star like the Sun to evolve from the main sequence to the top of the red giant branch?
7. What is the helium flash?
8. What are the energy sources of a horizontal-branch star?
9. How do stars of low mass die? How do stars of high mass die?
10. What is a planetary nebula? Why do many planetary nebulae appear as rings?
11. What are white dwarfs? What is their ultimate fate?
12. Under what circumstances will a binary star produce a nova?
13. What occurs in a massive star to cause it to explode?
14. What are the observational differences between Type I and Type II supernovae?
15. How do the mechanisms that cause Type I and Type II supernovae explain their observed differences?

16. What evidence do we have that many supernovae have occurred in our Galaxy?

17. Why do the cores of massive stars evolve into iron and not heavier elements?

Problems

1. Calculate the average density of a red giant core of mass 0.25 solar mass and radius 15,000 km. Compare this with the average density of the giant's envelope, if the mass of the envelope is 0.5 solar mass and its radius is 0.5 A.U. Compare each with the average core density of the Sun.

2. The Crab Nebula is now about 1 pc in radius. If it was observed to explode in A.D. 1054, roughly how fast is it expanding? (Assume constant expansion velocity. Is that a reasonable assumption?)

3. A certain telescope could just detect the Sun at a distance of 10,000 pc. What is the maximum distance at which it could detect a nova having a peak luminosity of 10^5 solar luminosities? Repeat the calculation for a supernova having a peak luminosity 10^{10} times that of the Sun.

4. At what distance would the supernova in question 3 look as bright as the Sun? Would you expect a supernova to occur that close to us?

Projects

1. You can tour the Galaxy without ever leaving Earth, just by looking up. In the winter sky, you'll find the red supergiant Betelgeuse in the constellation Orion. It's easy to see because it's one of the brightest stars visible in our night sky. Betelgeuse is a variable star with a period of about 6.5 years. Its brightness changes as it expands and contracts. At maximum size, Betelgeuse fills a volume of space that would extend from the Sun to beyond the orbit of Jupiter. Betelgeuse is thought to be about 10 to 15 times more massive than our Sun. It is probably between 4 and 10 million years old—and in the final stages of its evolution.

   A similar star can be found shining prominently in midsummer. This is the red supergiant Antares in the constellation Scorpius. Depending on the time of year, can you find one of these stars? Why are they red? What will happen to them in their next stage of evolution?

2. In 1758, the French comet hunter Charles Messier discovered the sky's most legendary supernova remnant, now called M1, or the Crab Nebula. An 8-inch telescope reveals the Crab's oval shape, but it will appear faint. It is located northwest of Zeta Tauri, the star that marks the southern tip of the horns of Taurus the Bull. A 10-inch or larger telescope reveals some of the Nebula's famous filamentary structure.

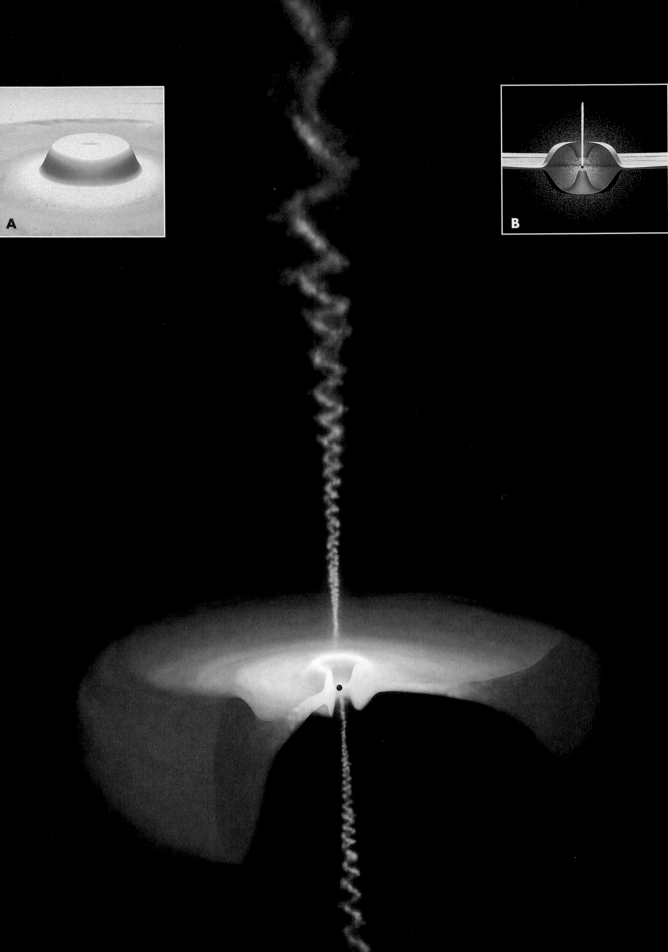

A

B

13 Neutron Stars and Black Holes

Strange States of Matter

LEARNING GOALS

Studying this chapter will enable you to:

1 Describe the properties of neutron stars and explain how these strange objects are formed.

2 Explain the nature and origin of pulsars, and account for their characteristic radiation.

3 List and explain some of the observable properties of neutron-star binary systems.

4 Describe how black holes are formed and discuss their effects on matter and radiation in their vicinity.

5 Relate the phenomena that occur near black holes to the warping of space around them.

6 Discuss some of the ways in which the presence of a black hole might be detected.

Our study of stellar evolution has led us to some unusual and unexpected objects. Red giants, white dwarfs, and supernovae surely represent extreme states of matter completely unfamiliar to us here on Earth, yet stellar evolution can have even more bizarre consequences. The strangest states of all result from the catastrophic implosion-explosion of stars much more massive than our Sun. The incredible violence of a supernova explosion may bring into being objects so extreme in their behavior that they require us to reconsider some of our most hallowed laws of physics. They open up a science fiction writer's dream of fantastic phenomena. They may even one day force scientists to construct a whole new theory of the universe.

13.1 Neutron Stars

1 What remains after a supernova explosion? Is the original star blown to bits and dispersed into interstellar space, or does some portion of it survive? For a Type I supernova (in which a low-mass white dwarf accretes enough additional matter to become unstable and explode), most astronomers regard it as unlikely that anything is left behind after the explosion. The entire star is shattered by the blast. However, for a Type II supernova (involving the collapse and subsequent violent reexpansion of a high-mass star's iron core), theoretical calculations indicate that part of the star might survive. The explosion destroys the parent star, but it may leave a tiny ultra-compressed remnant at its center. Even by the high-density standards of a white dwarf, however, the matter within this severely compressed core is in a very strange state, unlike anything we are ever likely to find (or create) on Earth. ∞ (Sec. 12.2)

Recall from Chapter 12 that during the implosion of a high-mass star—just prior to the supernova explosion—the electrons in the core violently smash into the protons there, forming neutrons and neutrinos. ∞ (Sec. 12.3) The neutrinos leave the scene at (or nearly at) the speed of light, accelerating the collapse of the neutron core, which continues to contract until its particles effectively come into contact with one another. At that point, the central portion of the core rebounds, creating a powerful shock wave that races outward through the star, violently expelling matter into space.

The key point here is that the shock wave does not start at the very center of the collapsing core. The innermost part of the core—the region that rebounds—remains intact as the shock wave it produces destroys the rest of the star. After the violence of the supernova has subsided, this ball of neutrons is all that

is left. Researchers colloquially call this core remnant[1] a **neutron star**, although it is not a star in the true sense of the word—all of its nuclear reactions have ceased forever.

Neutron stars are extremely small and very massive. Composed purely of neutrons packed together in a tight ball about 20 km across, a typical neutron star is not much bigger than a small asteroid or a terrestrial city (Figure 13.1), yet its mass is greater than that of the Sun. With so much mass squeezed into such a small volume, neutron stars are incredibly dense. Their average density can reach 10^{17} or even 10^{18} kg/m^3, nearly a billion times denser than a white dwarf. A single thimbleful of neutron-star material would weigh 100 million tons—about as much as a good-sized terrestrial mountain. Even the density of a normal atomic nucleus is "only" 10^{17} kg/m^3. In a sense, we can think of a neutron star as a single enormous nucleus having an "atomic mass" of around 10^{57}.

Neutron stars are solid objects. Provided that a sufficiently cool one could be found, you might even imagine standing on it. However, this would not be easy, as its gravity is extremely powerful. A 70-kg human would weigh the Earth equivalent of about 1 billion kg (1 million tons). The severe pull of a neutron star's gravity would flatten you much thinner than this piece of paper!

In addition to large mass and small size, newly formed neutron stars have two other very important properties. First, they *rotate* extremely rapidly, with periods measured in fractions of a second. This is a

---

[1]Astronomers commonly use the term *remnant* to mean whatever remains of a star's inner core after evolution has ended. Such objects are small and compact—no larger than Earth. They should not be confused with *supernova remnants*, which are the aftermath of supernova explosions: glowing clouds of debris scattered across many parsecs of interstellar space. ∞ (Sec. 12.4)

Figure 13.1 Neutron stars are not much larger than many of Earth's major cities. In this fanciful comparison, a typical neutron star sits alongside Manhattan Island (outlined).

13.2 Pulsars

2 Can we be sure that objects as strange as neutron stars really exist? The answer is a confident yes. The first observation of a neutron star occurred in 1967, when Jocelyn Bell, a graduate student at Cambridge University, made a surprising discovery. She observed an astronomical object emitting radio radiation in the form of rapid *pulses*. Each pulse consisted of a 0.01-s burst of radiation, after which there was nothing. Then, 1.34 s later, another pulse would arrive. The time interval between pulses was astonishingly uniform—so accurate, in fact, that the repeated emissions could be used as a very precise clock. Figure 13.2 is a recording of the radio radiation from the pulsating object Bell discovered.

Many hundreds of these pulsating objects are now known. They are called **pulsars**. Each has its own characteristic pulse period and duration. The pulse periods of some pulsars are so stable that they are by far the most accurate natural clocks known in the universe—more accurate even than the best atomic clocks on Earth. In some cases, the period is predicted to change by only a few seconds in a million years.

Most pulsars emit their pulses in the form of radio radiation. However, some pulse in the visible, X-ray, and gamma-ray parts of the spectrum as well. Whatever types of radiation are produced, however, all the electromagnetic flashes from a given pulsar are synchronized, occurring at the same regular, repeated time intervals, because they all arise from the same object. The period of most pulsars is usually short, ranging from about 0.03 to 0.3 s, corresponding to a flashing rate of between 3 and 30 times per second. The human eye is insensitive to such rapid flashes, making it impossible to observe the flickering of a pulsar by eye, even using a large telescope. Fortunately, instruments can record pulses that the human eye cannot.

direct result of the law of conservation of angular momentum (Chapter 4), which tells us that any rotating body must spin faster as it shrinks. ∞ (*More Precisely 4-1*) Second, they have very strong *magnetic fields*. The original field of the parent star is amplified by the collapse of the core because the contracting material squeezes the magnetic field lines closer together, creating a magnetic field trillions of times stronger than Earth's field.

In time, theory indicates, a neutron star will spin more and more slowly as it radiates its energy into space, and its magnetic field will diminish. However, for a few million years after its birth, these two properties combine to provide the primary means by which this strange object can be detected and studied.

Figure 13.2 Pulsars emit periodic bursts of radiation. This recording shows the regular change in the intensity of the radio radiation emitted by the first such object known. It was discovered in 1967. Some of the pulses are marked by arrows.

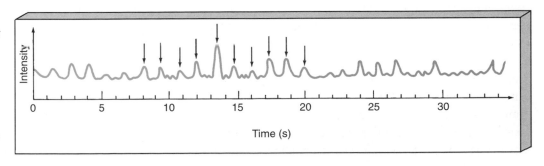

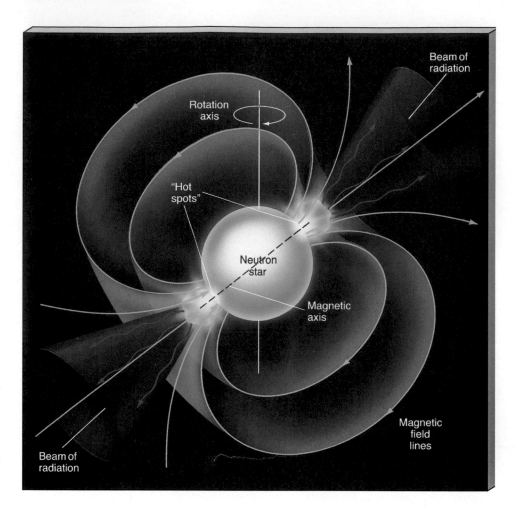

Figure 13.3 This diagram of the lighthouse model of neutron-star emission accounts for many of the observed properties of pulsars. Charged particles, accelerated by the magnetism of the neutron star, flow along the magnetic field lines, producing radio radiation that beams outward.

When Bell made her discovery in 1967, she did not know what she was looking at. No one at the time knew what a pulsar was. The explanation won Bell's thesis advisor, Anthony Hewish, the 1974 Nobel Prize in physics. Hewish reasoned that the only physical mechanism consistent with such precisely timed pulses is a small, rotating source of radiation. Only rotation can cause the high degree of regularity of the observed pulses, and only a small object can account for the sharpness of each pulse. Radiation emitted from different regions of an object larger than a few tens of kilometers across would arrive at Earth at slightly different times, blurring the pulse profile. The best current model describes a pulsar as a compact, spinning neutron star that periodically flashes radiation toward Earth.

Figure 13.3 outlines the important features of this pulsar model. Two "hot spots," either on the surface of a neutron star or in the magnetosphere just above it, continuously emit radiation in a narrow beam. These spots are most likely localized regions near the neutron star's magnetic poles, where charged particles, acceler-

ated to extremely high energies by the star's rotating magnetic field, emit radiation along the star's magnetic axis. The hot spots radiate more or less steadily, and the resulting beams sweep through space, like a revolving lighthouse beacon, as the neutron star rotates. Indeed, this pulsar model is often known as the **lighthouse model**. If the neutron star happens to be positioned such that the beam of its pulses sweeps across Earth, we see the pulses. The period of the pulses is the star's rotation period.

All pulsars are neutron stars, but not all neutron stars are pulsars, for two reasons. First, the two ingredients that make the neutron star pulse—rapid rotation and strong magnetic field—both diminish with time, so the pulses gradually weaken and become less frequent. Theory indicates that, within a few tens of millions of years, the pulsations have all but stopped. Second, even a young, bright pulsar is not necessarily visible from Earth. The pulsar beam depicted in Figure 13.3 is relatively narrow—perhaps only a few degrees across. Only if the neutron star happens to be oriented

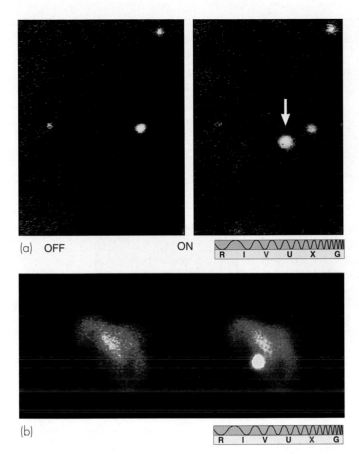

(a) OFF   ON

R I V U X G

(b)

R I V U X G

Figure 13.4 The pulsar in the core of the Crab Nebula blinks on and off about 30 times each second. (a) In this pair of optical images, the pulsing can be seen clearly. (b) The same phenomenon is also detected in X rays.

in just the right way do we see pulses. When we can see the pulses from Earth, we call the body a pulsar.

Given our present knowledge of star formation, stellar evolution, and neutron stars, current pulsar observations are consistent with the idea that *every* high-mass star dies in a supernova explosion, leaving a neutron star behind and that *all* young neutron stars emit beams of radiation, like the pulsars we actually see.

A few pulsars are definitely associated with supernova remnants, clearly establishing those pulsars' explosive origin. (For the reasons just outlined, not all supernova remnants have an observable pulsar within them.) Figure 13.4(a) shows a pair of optical photographs of the Crab pulsar, at the center of the Crab supernova remnant. ∞ (Sec. 12.4) In the left frame, the pulsar is off; in the right frame, it is on. Figure 13.4(b) shows that the Crab also pulses in X rays. By observing the speed and direction in which the Crab's ejected matter is traveling, astronomers can work back-

ward to pinpoint the location where the explosion must have occurred and where the remains of the supernova core should be located. That is precisely the region of the Crab Nebula from which the pulsating signals arise. The Crab pulsar is evidently all that remains of the once-massive star whose supernova was observed in 1054.

13.3 Neutron-Star Binaries

X-ray Bursters

3   The late 1970s saw several important discoveries about neutron stars in binary systems. Numerous X-ray sources were found near the central regions of our Galaxy and also near the centers of a few globular star clusters. Some of these sources, known as **X-ray bursters**, emit much of their energy in violent eruptions, each thousands of times more luminous than our Sun but lasting only a few seconds. A typical burst is shown in Figure 13.5.

This X-ray emission is thought to arise on or near neutron stars that are members of binary systems. Matter torn from the surface of the (main-sequence or giant) companion by the neutron star's strong gravitational pull accumulates on the neutron star's surface. As in the case of white dwarf accretion (Chapter 12), the material does not fall directly onto the surface; instead, it forms an accretion disk. (Figure 12.14 shows the white dwarf equivalent.) ∞ (Sec. 12.2) The gas goes into a tight orbit around the neutron star, then slowly spirals inward. The inner portions of the disk become extremely hot, releasing a steady stream of X rays.

As gas builds up on the neutron star's surface, the gas temperature rises due to the pressure of overlying material. Eventually, it becomes hot enough to fuse hydrogen. The result is a sudden period of rapid nuclear burning that releases a huge amount of energy in a brief but intense flash of X rays—an X-ray burst. After several hours of renewed accumulation, a fresh layer of matter produces the next burst. Thus an X-ray burst is very much like a nova explosion on a white dwarf, but occurring on a far more violent scale because of the neutron star's much stronger surface gravity.

At around the same time as the first X-ray bursts were seen, military satellites detected the first **gamma-ray bursters**—bright, irregular flashes of gamma rays

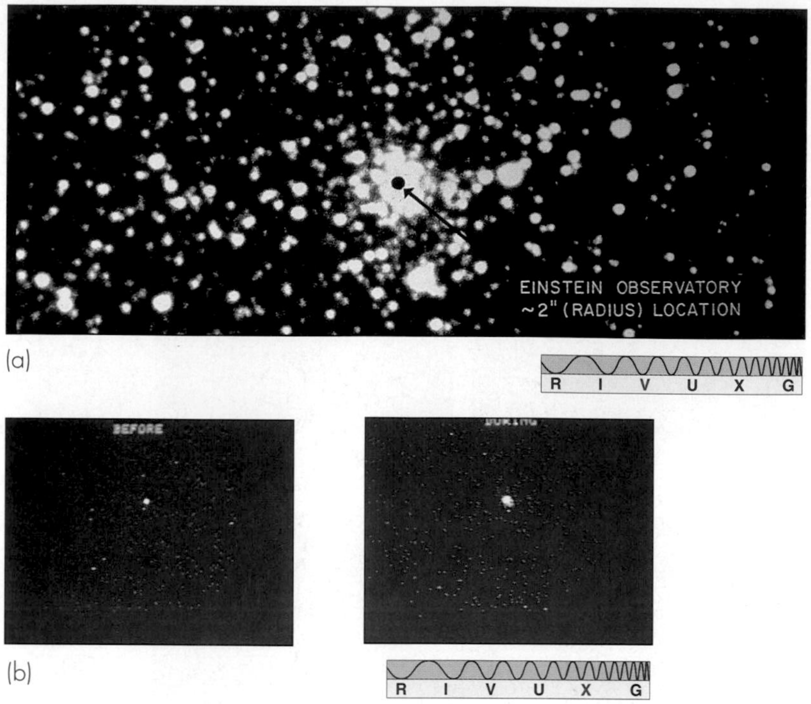

Figure 13.5 An X-ray burster produces a sudden, intense flash of X rays followed by a period of relative inactivity lasting as long as several hours. Then another burst occurs. The bursts are thought to be caused by explosive nuclear burning on the surface of an accreting neutron star, similar to the explosions on a white dwarf that give rise to novae. (a) An optical photograph of the star cluster Terzan 2, showing a 2″ dot at the center where the X-ray bursts originate. (b) X-ray images taken before and during the outburst. The most intense X rays correspond to the position of the dot shown in (a).

lasting only a few seconds. Until fairly recently, it was thought that gamma-ray bursters were basically "scaled-up" versions of X-ray bursters, in which matter accreted from the binary companion experienced even more violent nuclear burning, accompanied by the release of the more energetic gamma rays.

Recently, however, that view has been called into question by observations made using the *Gamma-Ray Observatory*, which has detected many gamma-ray bursters. ∞ (Sec. 3.5) The bursts are found to be distributed roughly uniformly over the entire sky, rather than being confined to the relatively narrow band known as the Milky Way. This widespread distribution suggests to many astronomers that the gamma rays do not originate within our own Galaxy, as had formerly been assumed, but instead are produced at much greater distances.

Strong observational evidence supporting this new picture was obtained in 1997 when a source of visible radiation, thought to be associated with a recently detected gamma-ray burster, was shown to lie at least 1000 Mpc from Earth. In that case, the bursts must be far more energetic than had previously been thought, and a new explanation is required. One possibility under active investigation involves a collision between two neutron stars in a very distant binary system, which may possibly lead to a powerful supernovalike explosion as the two ultracompact remnants merge.

Millisecond Pulsars

In the mid-1980s an important new category of pulsars was found—a class of very rapidly rotating objects called **millisecond pulsars**. These objects spin hundreds of times per second (that is, their pulse period is a few milliseconds). This speed is about as fast as a typical neutron star can spin without flying apart. In some cases, the star's equator is moving at more than 20 percent of the speed of light. This speed suggests a phenomenon bordering on the incredible—a cosmic object of kilometer dimensions, more massive than our Sun, spinning almost at breakup speed, making nearly 1000 complete revolutions *every second*. Yet the observations and their interpretation leave little room for doubt.

The story of these remarkable objects is further complicated by the fact that many of them are found in globular clusters. This is odd because globular clusters are known to be at least 10 billion years old, yet Type II supernovae (the kind that create neutron stars) are associated with massive stars that explode within a few tens of *millions* of years after their formation, and no stars have formed in any globular cluster since the cluster came into being. Thus, no new neutron star has been produced in a globular cluster in a very long time. Furthermore, as mentioned earlier, the pulsar produced in a supernova explosion is expected to slow down and

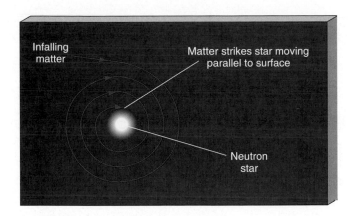

Figure 13.6 Gas from a companion star spirals down onto the surface of a neutron star. As it strikes the star, the infalling matter moves almost parallel to the surface, so it tends to make the star spin faster. Eventually, this process can result in a millisecond pulsar—a neutron star spinning at the incredible rate of hundreds of revolutions per second.

fade away in only a few tens of millions of years—after 10 billion years, its rotation rate should be very slow. The rapid rotation of the pulsars found in globular clusters therefore cannot be a relic of their birth. These pulsars must have been spun up—that is, had their rotation rates increased—by some other, much more recent, mechanism.

The most likely explanation for the high rotation rate of millisecond pulsars is that they have been spun up by drawing in matter from a companion star. As matter spirals down onto the neutron star's surface in an accretion disk, it provides the "push" needed to make the neutron star spin faster (Figure 13.6). Theoretical calculations indicate that this process can spin the star up to breakup speed in about a hundred million years. Subsequently, an encounter with another star may eject the pulsar from the binary, or the pulsar's intense radiation may destroy its companion, and an isolated millisecond pulsar results. This general picture is supported by the finding that, of the 40 or so millisecond pulsars seen in globular clusters, 10 are known currently to be members of binary systems. These numbers are consistent with the rate at which binaries can be broken up by encounters with other cluster members.

Thus, millisecond pulsars are the product of a two-stage process. The neutron star was formed in an ancient supernova, billions of years ago. Only relatively recently, through interaction with a binary companion, has it achieved the rapid spin that we observe

today. Notice that the scenario of accretion onto a neutron star from a binary companion is the same scenario that we just used to explain the existence of X-ray bursters. In fact, the two phenomena are very closely linked. Many X-ray bursters may be on their way to becoming millisecond pulsars.

The way in which a neutron star can become a member of a binary system is the subject of active research, because the violence of a supernova explosion would be expected to blow the binary apart in many cases. Only if the star that exploded lost a lot of mass before the supernova would the binary be likely to survive. Alternatively, by interacting with an existing binary and displacing one of its components, a neutron star may become part of a binary system after it is formed, as depicted in Figure 13.7. Astronomers are eagerly searching the skies for more millisecond pulsars to test their ideas.

Pulsar Planets

Radio astronomers can capitalize on the precision with which pulsar signals repeat themselves to make extremely accurate measurements of pulsar motion. In January 1992, radio astronomers at the Arecibo Observatory found that the pulse period of a recently discovered millisecond pulsar lying some 500 pc from Earth varies in an unexpected but regular way, exhibiting tiny fluctuations (at the level of about 1 part in 10^7) on two distinct time scales—67 and 98 days.

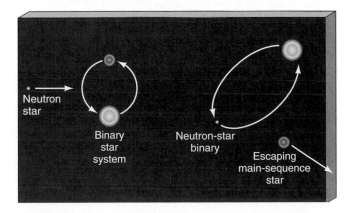

Figure 13.7 A neutron star can encounter a binary made up of two low-mass stars, ejecting one of them and taking its place. This mechanism provides a means of forming a binary system with a neutron-star component (which may later evolve into a millisecond pulsar) without having to explain how the binary survived the supernova explosion that formed the neutron star.

The leading explanation for these fluctuations holds that they are caused by the Doppler effect as the pulsar wobbles back and forth in space. But what causes the wobble? The Arecibo group believes it is the result of the combined gravitational pulls of two small bodies, each about three times the mass of Earth! One orbits the pulsar at a distance of 0.4 A.U. and the other at a distance of 0.5 A.U. Their orbital periods are 67 and 98 days, respectively, matching the timing of the fluctuations. In April 1994, the group announced further observations that not only confirmed their earlier findings, but also revealed the presence of a third body, with a mass comparable to that of Earth's Moon, orbiting only 0.2 A.U. from the pulsar.

These remarkable results constituted the first definite evidence for planet-sized bodies outside our solar system. Since their announcement, a few other millisecond pulsars have been found with similar behavior. However, it is unlikely that any of these planets formed in the same way as our own. Any planetary system orbiting the pulsar's parent star was almost certainly destroyed in the supernova explosion that created the pulsar. As a result, scientists are still uncertain about how these planets came into being. One possibility involves the binary companion that provided the matter necessary to spin the pulsar up to millisecond speeds. Possibly the pulsar's intense radiation and strong gravity destroyed the companion, then spread its matter out into a disk (a little like the solar nebula) in whose cool outer regions the planets might have condensed. These measurements are difficult, however, and their interpretation is still controversial.

Astronomers have been searching for decades for planets orbiting main-sequence stars like our Sun, on the assumption that planets are a natural by-product of star formation. ∞ (Sec. 4.3) Only very recently have these searches begun to yield positive results, although many of these findings remain controversial. ∞ (Interlude 4-2) It is ironic that the first Earth-sized planets to be found outside the solar system orbit a dead star and have little or nothing in common with our own world.

13.4 Black Holes *Animation*

4 Neutron stars are peculiar objects. Nevertheless, theory predicts that they are in equilibrium, like most other stars. For neutron stars, however, equilibrium does not mean a balance between the inward pull of gravity and the outward pressure of hot gas. Instead, as we have seen, the outward force is provided by the pressure of tightly packed neutrons. Squeezed together, the neutrons form a hard ball of matter that not even gravity can compress any further. Or do they? Is it possible that, given enough matter packed into a small enough volume, the collective pull of gravity can eventually crush any opposing pressure? Can gravity continue to compress a massive star into an object the size of a planet, a city, a pinhead—even smaller? The answer, apparently, is yes.

The Final Stage of Stellar Evolution

We have seen that the fate of a star depends critically on its mass. Low-mass stars leave behind a compact object known as a white dwarf. ∞ (Sec. 12.2) High-mass stars may produce an even more compact neutron star. The laws of physics make specific predictions about the masses of these ultracompressed objects. A white dwarf must be less than about 1.4 solar masses, beyond which the resistance of electrons to compression cannot support the core against its own gravity. ∞ (Sec. 12.4) Similarly, a neutron star resulting from a supernova must have a mass between about 1.4 and 3 solar masses. The lower limit of 1.4 solar masses stems from the theory of stellar evolution: the iron core of an evolved star must exceed this mass for core collapse to begin and a supernova to occur. ∞ (Sec. 12.3) The (quite uncertain) upper limit of 3 solar masses is the neutron-star equivalent of the white dwarf mass limit—above this mass, not even tightly packed neutrons can withstand the star's gravitational pull.

In fact, we know of *no* force that can counteract gravity beyond this point. Thus, if enough material is left behind after a supernova explosion, as may happen in the case of an extremely massive star, gravity wins the battle with pressure once and for all, and the central core collapses forever. As the core shrinks, the gravitational pull in its vicinity eventually becomes so great that even light is unable to escape. The resultant object therefore emits no light, no other form of radiation, no information whatsoever. Astronomers call this bizarre endpoint of stellar evolution, in which the core of a very-high-mass star collapses in on itself and vanishes forever, a **black hole**.

Newtonian mechanics—up to now our reliable and indispensable tool in understanding the universe—cannot adequately describe conditions in or near black holes. ∞ (*More Precisely 1-1*) To comprehend these collapsed objects, we must turn instead to the modern theory of gravity, Einstein's **general theory of relativity** (see *More Precisely 13-1*). The Einsteinian description of the universe is equivalent to that of Newton for situations encountered in everyday life, but the two theories diverge radically in circumstances where speeds approach the speed of light and in regions of intense gravitational fields.

Despite the fact that general relativity is necessary for a proper description of black-hole properties, we can still usefully discuss some aspects of these strange bodies in more or less Newtonian terms. Let's reconsider the familiar Newtonian concept of escape speed—the speed needed for one object to escape from the gravitational pull of another—supplemented by two key facts from relativity: (1) nothing can travel faster than the speed of light, and (2) all things, *including light,* are attracted by gravity.

Escape Speed

In Chapter 5, we noted that escape speed is proportional to the square root of a body's mass divided by the square root of its radius. ∞ (*Sec. 5.1*) Earth's radius is 6400 km, and the escape speed from Earth's surface is just over 11 km/s. Now consider a hypothetical experiment in which Earth is squeezed on all sides by a gigantic vise. As our planet shrinks under the pressure, the escape speed increases because the radius is decreasing. Suppose that Earth were compressed to one-fourth its present size. The proportionality just mentioned then predicts that the escape speed would double (because $1/\sqrt{1/4} = 2$). Any object escaping from the surface of this hypothetically compressed Earth would need a speed of at least 22 km/s.

Imagine compressing Earth some more. Squeeze it, for example, by an additional factor of 1000, making its radius hardly more than a kilometer. Now a speed of about 630 km/s—slightly more than the escape speed of the Sun—would be needed to escape. Compress Earth still further, and the escape speed continues to rise. If our hypothetical vise were to squeeze Earth hard enough to crush its radius to about a cen-

timeter, the speed needed to escape its surface would reach 300,000 km/s. But this is no ordinary speed. It is the speed of light, the fastest speed allowed by the laws of physics as we currently know them.

Thus, if by some fantastic means the entire planet Earth could be compressed to less than the size of a grape, the escape speed would exceed the speed of light. And because nothing can exceed that speed, the compelling conclusion is that nothing—absolutely nothing—could escape from the surface of such a compressed body. Even radiation—radio waves, visible light, X rays, photons of all wavelengths—would be unable to escape the intense gravity of our ultracompressed Earth. With no photons leaving, our planet would be invisible and uncommunicative—no signal of any sort could be sent to the universe outside. The origin of the term *black hole* becomes clear. For all practical purposes, such a supercompact Earth could be said to have disappeared from the universe! Only its gravitational field would remain behind, betraying the presence of its mass, now shrunk to a point.[2]

The Event Horizon

Astronomers have a special name for the critical radius at which the escape speed from an object would equal the speed of light and within which the object could no longer be seen. It is the **Schwarzschild radius**, after Karl Schwarzschild, the German scientist who first studied its properties. The Schwarzschild radius of any object is proportional to its mass. For Earth, it is 1 cm; for Jupiter, at about 300 Earth masses, it is about 3 m; for the Sun, at 300,000 Earth masses, it is 3 km. For a 3-solar-mass stellar core, the Schwarzschild radius is about 9 km. As a convenient rule of thumb, the Schwarzschild radius of an object is simply 3 km multiplied by the object's mass measured in solar masses.

The surface of an imaginary sphere having a radius equal to the Schwarzschild radius and centered on a collapsing star is called the **event horizon**. No

[2]In fact, we now know that, regardless of the composition or condition of the object that formed the black hole, only three physical properties can be measured from the outside—the hole's mass, charge, and angular momentum. All other information is lost once matter falls into the black hole. Thus, only three numbers are required to describe completely a black hole's outward appearance. In this chapter, we consider only holes that formed from nonrotating, electrically neutral matter. Such objects are completely specified once their masses are known.

event occurring within this region can ever be seen, heard, or known by anyone outside. Even though there is no matter of any sort associated with it, we can think of the event horizon as the "surface" of a black hole.

A 1.4-solar-mass neutron star has a radius of about 10 km and a Schwarzschild radius of 4.2 km. If we were to keep increasing the neutron star's mass, its Schwarzschild radius would increase, although its physical radius would not; in fact, the physical radius of a neutron star *decreases* slightly with increasing mass. By the time our neutron star's mass exceeded about 3 solar masses, the star would lie within its own event horizon and would collapse of its own accord. It would not stop shrinking at the Schwarzschild radius—the event horizon is not a physical boundary of any kind, just a communications barrier. The neutron star would shrink right past the Schwarzschild radius to ever-diminishing size, on its way to disappearing forever.

Thus, provided that at least 3 solar masses of material remain behind after a supernova explosion, the remnant core will collapse catastrophically, diving below the event horizon in less than a second. The core "winks out," vanishing from view and becoming a small dark region from which nothing can escape—a literal black hole in space. This is the fate of any star whose main-sequence mass exceeds about 20–30 times the mass of the Sun.

13.5 Properties of Black Holes

5 A central concept of general relativity (see *More Precisely 13-1*) is this: matter—all matter—tends to "warp," or curve, space in its vicinity. At the same time, all objects, such as planets and stars, react to this warping by changing their paths. In the Newtonian view of gravity, particles move on curved trajectories because they feel a gravitational force. In Einsteinian relativity, those same particles move on curved trajectories because they are following the curvature of space produced by some nearby massive object. The greater the mass, the greater the warping. Close to a black hole, the gravitational field becomes overwhelming and the curvature of space extreme. At the event horizon, the curvature is so great that space "folds over" on itself, causing objects within to become trapped and disappear.

Rubber Sheets and Curved Space

Some props may help you visualize the curvature of space near a black hole. Bear in mind, however, that these props are in no sense "real"—they are only tools to help you grasp some exceedingly strange concepts.

First, imagine a pool table with the tabletop made of a thin rubber sheet rather than the usual hard felt. As Figure 13.8 suggests, such a rubber sheet becomes distorted when a heavy weight, such as a rock, is placed on it. The otherwise flat sheet sags (becomes distorted), especially near the rock. The heavier the rock, the larger the distortion. Trying to play pool, you would quickly find that balls passing near the rock were deflected by the curvature of the tabletop. In much the same way, both matter *and radiation* are deflected by the curvature of space near a star. For example, Earth's orbital path is governed by the relatively gentle curvature of space created by our Sun. The more massive the object, the more the space surrounding it is curved.

Let's consider another analogy. Imagine a large extended family of people living at various locations on a huge rubber sheet—a sort of gigantic trampoline. Deciding to hold a reunion, they converge on a given place at a given time. As shown in Figure 13.9, one person remains behind, not wishing to attend. He keeps in touch with his relatives by means of "message balls" rolled out to him (and back from him) along

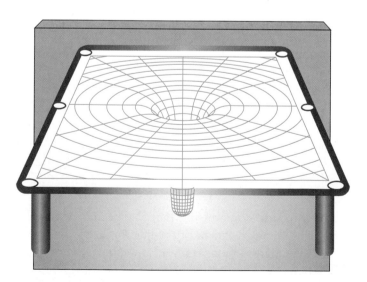

Figure 13.8 A pool table made of a thin rubber sheet sags when a weight is placed on it. Likewise, space is bent, or warped, in the vicinity of any massive object.

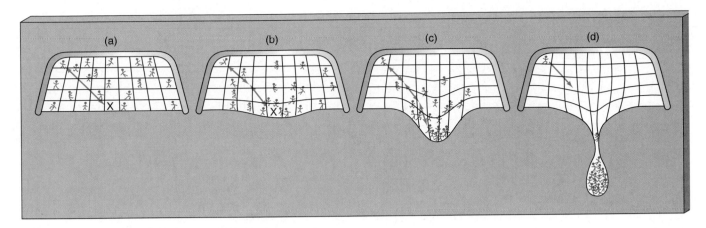

Figure 13.9 Any mass causes a rubber sheet (space) to be curved. (a), (b), (c) As people assemble at the appointed spot on the sheet, the curvature grows progressively larger. The blue arrows represent some directions in which information can be transmitted from place to place. (d) The people are finally sealed inside the bubble, forever trapped and cut off from the outside world.

the surface of the sheet. These message balls are the analog of radiation carrying information through space.

As the people converge, the rubber sheet sags more and more. Their accumulating mass creates an increasing amount of space curvature. The message balls can still reach the lone person far away in nearly flat space, but they arrive less frequently as the sheet becomes more and more warped and stretched and the balls have to climb out of a deeper and deeper well (Figures 13.9b and c). Finally, when enough people have arrived at the appointed spot, the mass becomes too great for the rubber to support. As illustrated in Figure 13.9(d), the sheet pinches off into a "bubble," compressing the people into oblivion and severing their communications with the lone survivor outside. This final stage represents the formation of an event horizon around the party.

Right up to the end—the pinching off of the bubble—two-way communication is possible. Message balls can reach the outside from within (but at a lower and lower rate as the rubber stretches), and messages from outside can get in without difficulty. Once the event horizon (the bubble) forms, balls from the outside can still fall in, but they can no longer be sent back out to the person left behind, no matter how fast they are rolled. They cannot make it past the lip of the bubble in Figure 13.9(d). This analogy (very) roughly depicts how a black hole warps space completely around on itself, isolating its interior from the rest of the universe.

The essential ideas—the slowing down and eventual cessation of outward-going signals and the one-way nature of the event horizon once it forms—all have parallels in the case of stellar black holes.

Cosmic Cleaners? No

Black holes are *not* cosmic vacuum cleaners. They do not cruise around interstellar space, sucking up everything in sight. The orbit of an object near a black hole is essentially the same as its orbit near a star of the same mass as the black hole. Only if the object happens to pass within a few Schwarzschild radii (perhaps 50 or 100 km for a typical black hole formed in a supernova explosion) of the event horizon is there any significant difference between its actual orbit and the one predicted by Newtonian gravity and described by Kepler's laws. From a distance, the main observational difference is that an object orbiting a black hole would appear to orbit a dark, empty region of space.

Black holes, then, do not go out of their way to drag in matter. However, if some matter does happen to fall into one—if a body's orbit happens to take it too close to the event horizon—it will be unable to get out. Black holes are like turnstiles, permitting matter to flow in only one direction—inward. Because a black hole will accrete at least a little material from its surroundings, its mass tends to increase over time. The black hole's size is proportional to its mass, so the radius of the event horizon grows with time.

13-1 MORE PRECISELY

EINSTEIN'S THEORIES OF RELATIVITY

Albert Einstein is probably best known for his two *theories of relativity*, the successors to Newtonian mechanics that form the foundation of twentieth-century physics. The *special theory of relativity* (or just *special relativity*), proposed by Einstein in 1905, deals with the preferred status of the speed of light. We have noted that the speed of light, *c*, is the maximum speed attainable in the universe, but there is more to it than that. In 1887, a fundamental experiment carried out by two American physicists, A. A. Michelson and E. W. Morley, demonstrated a further important and unique aspect of light—that the measured speed of a beam of light is *independent* of the motion of the observer or the source. No matter what our velocity may be relative to the source of the light, we always measure precisely the same value for *c*—299,792.458 km/s.

A moment's thought leads to the realization that this is a decidedly nonintuitive statement. If we were traveling in a car moving at 100 km/h and we fired a bullet forward with a velocity of 1000 km/h relative to the car, an observer standing at the side of the road would see the bullet pass by at 100 + 1000 = 1100 km/h, as illustrated in the accompanying figure. However, if we were traveling in a rocket ship at 1/10 the speed of light, 0.1*c*, and we shone a searchlight beam ahead of us, the Michelson–Morley experiment tells us that an outside observer would measure the speed of the beam not as 1.1*c*, as the preceding example would suggest, but as *c*. The rules that apply to particles moving at or near the speed of light are different from those we are used to in everyday life.

Special relativity is the mathematical framework that allows us to extend the familiar laws of physics from low speeds (that is, speeds much less than *c*, which are often referred to as *nonrelativistic*) to very high (or *relativistic*) speeds comparable to *c*. Relativity is equivalent to Newtonian mechanics when objects move much more slowly than light,

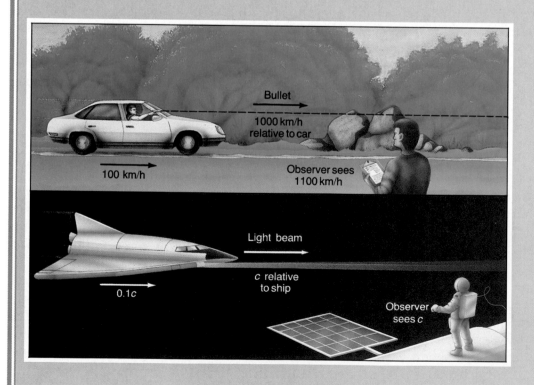

but it differs greatly in its predictions at relativistic speeds. For example, special relativity predicts that a rapidly moving spacecraft will appear to contract in the direction of its motion, its clocks will appear to run slow, and its mass will appear to increase. All the theory's predictions have been repeatedly verified to very high accuracy. Today special relativity lies at the heart of all physical science. No scientist seriously doubts its validity.

General relativity is what results when gravity is incorporated into the framework of special relativity. In 1915, Einstein made the connection between special relativity and gravity with the following famous "thought experiment." Imagine that you are enclosed in an elevator with no windows, so that you cannot directly observe the outside world, and that the elevator is floating in space. You are weightless. Now suppose that you begin to feel the floor press up against your feet. Weight has apparently returned. There are two possible explanations for this, shown in the accompanying diagram. A large mass could have come nearby, and you are feeling its downward gravitational attraction, *or* the elevator has begun to accelerate upward and the force you feel is that exerted by the elevator as it accelerates you, too. The crux of Einstein's argument is this: there is *no* experiment that you can perform within the elevator, without looking outside, that will let you distinguish between these two possibilities.

Thus, Einstein reasoned, there is no way to tell the difference between a gravitational field and an accelerated frame of reference (such as the rising elevator in the thought experiment). Gravity can therefore be incorporated into special relativity as a general acceleration of all particles. However, a major modification must be made to special

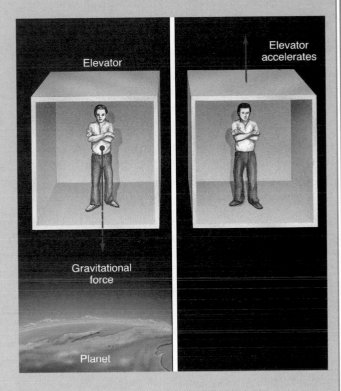

relativity. Central to relativity is the notion that space and time are not separate quantities, but instead must be treated as a single entity—*spacetime*. To incorporate the effects of gravity, the mathematics forces us to the conclusion that spacetime has to be *curved*.

In general relativity, then, gravity is a manifestation of curved spacetime. There is no such thing as a "gravitational field," in the Newtonian sense. Instead, objects move as they do because they follow the curvature of spacetime, and this curvature of spacetime is determined by the amount of matter present. We explore some of the consequences of this view of gravity in more detail in the text.

Cosmic Heaters? Yes

Matter flowing into a black hole is subject to great tidal stress. An unfortunate person falling feet first into a solar-mass black hole would find herself stretched enormously in height and squeezed unmercifully laterally. She would be torn apart even before she reached the event horizon, for the pull of gravity would be much stronger at her feet (which are closer to the hole) than at her head. Anything falling into a black hole—gas, people, space probes—is vertically stretched and horizontally squeezed, while simultaneously being accelerated to extremely high speeds. The net result of all this stretching and squeezing is numerous and violent collisions among the torn-up debris, causing a great deal of frictional heating of the infalling matter. Material is simultaneously torn apart and heated to high temperatures as it plunges into the hole.

The rapid heating of matter by tides and collisions is so efficient that, prior to submersion below the hole's event horizon, matter being sucked into the hole emits radiation of its own accord. For a black hole of solar mass, the energy is expected to be emitted in the form of X rays. In effect, the gravitational energy of matter outside the black hole is converted into heat as that matter falls toward the hole. Once the hot matter falls below the event horizon, its radiation is no longer detectable—it never leaves the hole. Thus, contrary to what we might expect from an object whose defining property is that nothing can escape from it, the region surrounding a black hole is expected to be a *source* of energy.

13.6 Space Travel Near Black Holes

One safe way to study a black hole would be to go into orbit around it, safely beyond the disruptive influence of the hole's strong tidal forces. After all, Earth and the other planets of our solar system all orbit the Sun without falling into it and without being torn apart. The gravity field around a black hole is basically no different. However, even from a stable circular orbit, a close investigation of the hole would be unsafe for humans. Endurance tests suggest that the human body cannot withstand stress greater than about 10 times the pull of gravity on Earth's surface. This breaking point would occur about 3000 km from a 10-solar-mass black hole

(which, recall, would have a 30-km event horizon). Closer than that, the tidal effect of the hole would tear a human body apart.

Approaching the Event Horizon

Let's send an imaginary indestructible astronaut—a mechanical robot, say—in a probe toward the center of the hole, as illustrated in Figure 13.10. Watching from a safe distance in our orbiting spacecraft, we can then examine the nature of space and time, at least down to the event horizon. After that boundary is crossed, there is no way for the robot to return any information about its findings.

Suppose our robot has an accurate clock and a light source of known frequency mounted on it. From our safe vantage point far outside the event horizon, we could use telescopes to read the clock and measure the frequency of the light we receive. What might we discover? We would find that the light from the robot would become more and more redshifted as the robot neared the event horizon. Even if the robot used rocket engines to remain motionless, the redshift would still be detected. The redshift is

Figure 13.10 A hypothetical robot-astronaut can travel toward a black hole while performing experiments that humans, farther away, can monitor in order to learn something about the nature of space near the event horizon.

not caused by motion of the light source—it is not the result of the Doppler effect. Rather, it is a **gravitational redshift** induced by the black hole's gravitational field.

We can explain the gravitational redshift as follows. Photons are attracted by gravity. As a result, in order to escape from a source of gravity, photons must expend some energy—they have to work to get out of the gravitational field. They don't slow down (photons always move at the speed of light); they just lose energy. Because a photon's energy is proportional to the frequency of its radiation, light that loses energy must have its frequency reduced (or, conversely, its wavelength lengthened). As illustrated in Figure 13.11, radiation coming from the vicinity of a gravitating object will be redshifted by an amount depending on the strength of the gravitational field.

As photons traveled from the robot's light source to our orbiting spacecraft, they would become gravitationally redshifted. From our standpoint, a green light, say, would become yellow, then red as the robot astronaut neared the black hole. From the robot's perspective, of course, the light would remain green. As the robot got closer to the event horizon, the radiation from its light source would become undetectable with optical telescopes. The radiation reaching us in the orbiting spacecraft would by then be lengthened so much that infrared and then radio telescopes would be needed to detect it. Closer still to the event horizon, the radiation emitted as visible light from the robot probe would be shifted to wavelengths even longer than conventional radio waves by the time it reached us.

Light emitted *from the event horizon itself* would be gravitationally redshifted to infinitely long wavelengths. In other words, each photon would use all its energy trying to escape from the edge of the hole. What was once light (on the robot) has no energy left upon arrival at the safely orbiting spacecraft. Theoretically, this radiation reaches us—still moving at the speed of light—but with zero energy. The radiation originally emitted has become redshifted beyond our perception.

What about the robot's clock? What time does it tell? Is there any observable change in the rate at which the clock ticks while moving deeper into the hole's gravitational field? We would find that, from the safely orbiting spacecraft, any clock close to the hole would appear to tick more *slowly* than an equivalent

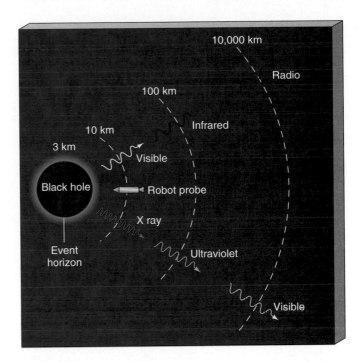

Figure 13.11 As it escapes from the strong gravitational field close to a black hole, a photon must expend energy to overcome the hole's gravity. This energy does not come from a change in the speed at which the photon travels (that speed is always 300,000 km/s, even under these extreme conditions). Rather, the photon "gives up" energy by decreasing its frequency. Thus, the photon's color changes. This figure shows the effect on two beams of radiation, one of visible light and one of X rays, emitted from a space probe as it nears a 1-solar-mass black hole.

clock on board our spacecraft. The closer the clock came to the hole, the slower it would appear to run. Upon reaching the event horizon, the clock would seem to stop altogether. All action would become frozen in time. Consequently, we would *never* actually witness the infalling robot sink below the event horizon. The process would appear to take forever.

This apparent slowing down of the robot's clock is known as **time dilation**. It is another clear prediction of general relativity, and in fact it is closely related to the gravitational redshift. To see this connection, imagine that we use our light source as a clock, with the passage of a wave crest (say) constituting a "tick." The clock thus ticks at the frequency of the radiation. As the wave is redshifted, the frequency drops, and fewer wave crests pass the distant observer each second—the clock appears to slow down. This thought experiment demonstrates that the redshift of the radiation and the slowing of the clock are one and the same thing.

13-2 MORE PRECISELY

TESTS OF GENERAL RELATIVITY

Special relativity is the most thoroughly tested and most accurately verified theory in the history of science. General relativity, however, is on somewhat less firm experimental ground. The problem with verifying general relativity is that its effects on Earth and in the solar system—the places where we can most easily perform tests—are very small. Just as special relativity produces major departures from Newtonian mechanics only when speeds approach the speed of light, general relativity predicts large departures from Newtonian gravity only when extremely strong gravitational fields are involved. In the presence of such intense fields, orbit speeds and escape speeds become relativistic.

Here we consider just two "classical" tests of the theory—solar-system experiments that helped ensure acceptance of Einstein's theory. Bear in mind, however, that there are no known tests of general relativity in the "strong-field" regime—that part of the theory that predicts black holes, for example—so the full theory has never been experimentally tested.

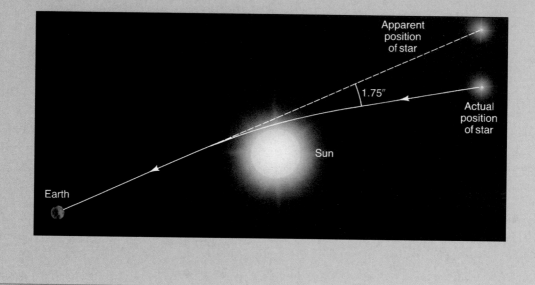

From the point of view of the indestructible robot, however, relativity theory predicts no strange effects at all. The light source hasn't reddened, and the clock keeps perfect time. In the robot's frame of reference, everything is normal. Nothing prohibits the robot from approaching within the Schwarzschild radius of the hole. No law of physics constrains an object from passing through an event horizon. There is no barrier at the event horizon and no sudden lurch as it is crossed; it is only an imaginary boundary in space. Travelers passing through the event horizon of a sufficiently massive hole (such as might lurk in the heart of

our own Galaxy, as we will see) might not even know it—at least until they tried to get out!

Deep Down Inside

No doubt, you are wondering what lies within the event horizon of a black hole. The answer is simple: no one knows. However, the question raises some very fundamental issues that lie at the forefront of modern physics.

General relativity predicts that, without some agent to compete with gravity, the core remnant of a

At the heart of general relativity is the premise that everything, including light, is affected by gravity because of the curvature of spacetime. Shortly after he published his theory in 1915, Einstein noted that light from a star should be deflected by a measurable amount as it passes the Sun. The closer to the Sun the light comes, the more it is deflected. Thus, the maximum deflection should occur for a ray that just grazes the solar surface.

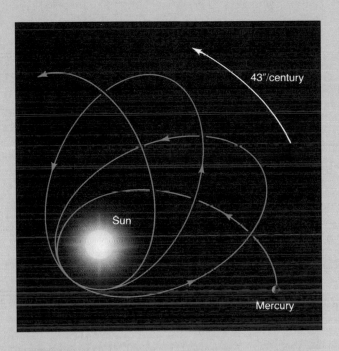

Einstein calculated that the deflection angle should be 1.75″—a small but detectable amount. Of course, it is normally impossible to see stars so close to the Sun. During a solar eclipse, however, when the Moon blocks the Sun's light, the observation becomes possible, as illustrated in the accompanying figure.

In 1919, a team of observers led by the British astronomer Sir Arthur Eddington succeeded in measuring the deflection of starlight during an eclipse. The results were in excellent agreement with the prediction of general relativity. Virtually overnight Einstein became world famous. His previous major accomplishments notwithstanding, this single prediction assured him a permanent position as the best-known scientist on Earth!

Another prediction of general relativity is that planetary orbits should deviate slightly from the perfect ellipses of Kepler's laws. Again, the effect is greatest where gravity is strongest—that is, closest to the Sun. Thus, the largest relativistic effects are found in the orbit of Mercury. Relativity predicts that Mercury's orbit is not exactly an ellipse. Instead, its orbit should rotate slowly, as shown in the (highly exaggerated) diagram below. The amount of rotation is very small—only 43″ per century—but Mercury's orbit is so well charted that even this tiny effect is measurable.

high-mass star will collapse all the way to a point at which both its density and its gravitational field become infinite—a so-called **singularity**. However, we should not take this prediction of infinite density too literally. Singularities always signal the breakdown of the theory producing them. In other words, the present laws of physics are simply inadequate to describe the final moments of a star's collapse.

As it stands today, the theory of gravity is incomplete because it does not incorporate a proper description of matter on very small scales. As our collapsing stellar core shrinks to smaller and smaller radii, we

eventually lose our ability even to describe, let alone predict, its behavior. Perhaps matter trapped in a black hole never actually reaches a singularity. Perhaps it just approaches this bizarre state, in a manner that we will someday understand as the subject of *quantum gravity*—the merger of general relativity with quantum mechanics—develops.

Having said that, we can at least estimate how small the core can get *before* current theory fails. It turns out that by the time that stage is reached, the core is already much smaller than any elementary particle. Thus, although a complete description of the endpoint of stellar collapse may well require a major overhaul of the laws of physics, for all practical purposes the prediction of collapse to a point is valid. Even if a new theory somehow succeeds in doing away with the central singularity, it is very unlikely that the external appearance of the hole, or the existence of its event horizon, will change. Any modifications to general relativity are expected to occur only on submicroscopic scales, not on the macroscopic (kilometer-sized) scale of the Schwarzschild radius.

Singularities are places where the rules break down, and some very strange things may occur near them. Many possibilities have been envisaged—gateways into other universes, time travel, the creation of new states of matter—but none has been proved, and certainly none has been observed. Because these regions are places where science fails, their presence causes serious problems for many of our cherished laws of physics, including causality (the idea that cause should precede effect, which runs into immediate problems if time travel is possible) and energy conservation (which is violated if material can hop from one universe to another through a black hole). It is unclear whether the removal of the singularity by some future, all-encompassing theory would necessarily also eliminate all of these problematic side effects.

Disturbed by the possibility of such chaos in science, some relativists have even proposed a "principle of cosmic censorship": nature always hides *any* singularity inside an event horizon. In that case, even though physics fails, that breakdown cannot affect us outside, so we are safely insulated from any effects the singularity might have. What would happen if we one day found a so-called *naked singularity* somewhere, one uncloaked by an event horizon? Would relativity theory still hold there? For now, we just don't know.

13.7 Observational Evidence for Black Holes *Animation*

6 Theoretical ideas aside, is there any observational evidence for black holes? What proof do we have that these invisible objects really do exist?

Stellar Transits?

One way in which we might think of detecting a black hole would be to observe it transit (pass in front of) a

star. Unfortunately, such an event would be extremely hard to see; the 12,000-km planet Venus is barely noticeable when transiting the Sun, so a 10-km-wide object moving across the image of a faraway star would be completely invisible with either current equipment or any equipment available in the foreseeable future.

Even if we were close enough to the star to resolve the disk of the transiting black hole, the observable effect would not be a black dot superimposed on a bright background. The background starlight would be deflected as it passed the black hole on its way to Earth, as indicated in Figure 13.12. The effect is the same as the bending of distant starlight around the edge of the Sun, a phenomenon that has been repeatedly measured during solar eclipses throughout the last several decades (see *More Precisely 2-2*). With a black hole, much larger deflections would occur. As a result, the image of a black hole in front of a bright companion star would show not a neat, well-defined black dot but rather a fuzzy image virtually impossible to observe, even from nearby.

Black Holes in Binary Systems

A much better way to find black holes is to look for their effects on other objects. Our Galaxy harbors

Figure 13.12 When a black hole transits a star, the gravitational bending of light around the edges of the hole makes it impossible to observe the hole as a black dot superimposed against the bright background of the star.

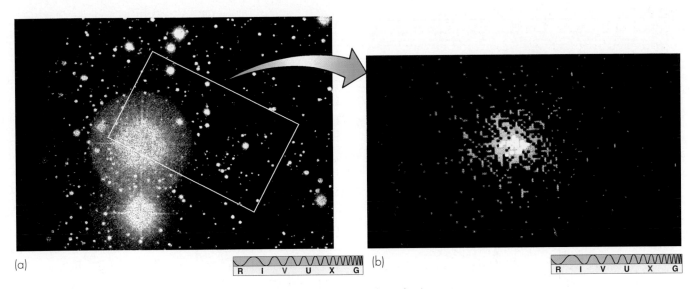

(a)

(b)

Figure 13.13 (a) The largest bright spot in this photograph is a member of a binary system whose unseen companion, called Cygnus X-1, is a leading black-hole candidate. (b) An X-ray image of the field of view outlined by the rectangle in part (a).

many binary-star systems in which only one object can be seen. Recall from our study of binary-star systems in Chapter 10 that we need to observe the motion of only one star to infer the existence of an unseen companion and measure some of its properties. ∞ (Sec. 10.9) In the majority of cases, the invisible companion is simply small and dim, nothing more than an M star hidden in the glare of an O or B partner, or perhaps shrouded by dust or other debris, making it invisible to even the best available equipment. In either case, the invisible object is not a black hole.

A few close binary systems, however, have peculiarities suggesting that one of their members may be a black hole. Some of the most interesting observations, made during the 1970s and 1980s by Earth-orbiting satellites, reveal binary systems in which the invisible member emits large amounts of X rays. The mass of the emitting object is measured as several solar masses, so we know it is not simply a small, dim star, and radiation pressure from the binary members makes circumstellar debris an unlikely explanation for its invisibility.

One particular binary system drawing much attention lies in the constellation Cygnus. Figure 13.13(a) shows the area of the sky where astronomers have reasonably good evidence for a black hole. The rectangle outlines the celestial system of interest, some 2000 pc from Earth. The black-hole candidate is an X-ray source called Cygnus X-1, discovered by the *Uhuru* satellite in the early 1970s. Its visible companion is a blue B supergiant.

Spectroscopic observations indicate that the binary system has an orbital period of 5.6 days. Assuming that the visible component lies on the main sequence, its mass must be around 25 times the mass of the Sun. Combining these pieces of information with further spectroscopic measurements of the visible component's orbital speed, astronomers estimate the total mass of the binary system to be 30–35 solar masses, implying that Cygnus X-1 has a mass between 5 and 10 times the mass of the Sun.

X-ray radiation emitted from the immediate neighborhood of Cygnus X-1 indicates the presence of high-temperature gas, perhaps as hot as several million kelvins. Rapid variability of this X-ray radiation implies that the size of the X-ray-emitting region must be less than a few hundred kilometers across. The reasoning goes as follows. If the emitting region were, say, 300,000 km—1 light second—across, even an instantaneous change in intensity at the source would be smeared out over a time interval of 1 s as seen from Earth, because light from the far side of the object would take 1 s longer to reach us than light from the near side. X rays from Cygnus X-1 have been observed to vary in intensity on time scales as short as a millisecond. For this variation not to be blurred by the travel time of light across the source, Cygnus X-1 cannot be more than 1 light millisecond, or 300 km, in diameter.

13-1 INTERLUDE

GRAVITY WAVES

Electromagnetic waves are common, everyday phenomena. Whether they are radio, infrared, visible, ultraviolet, X-ray, or gamma-ray radiation, all electromagnetic waves involve periodic changes in the strengths of electric and magnetic fields. ∞ (Fig. 2.6) They move through space and transport energy. Any accelerating charged particle, such as an electron in a broadcasting antenna or on the surface of a star, generates electromagnetic waves.

The modern theory of gravity—Einstein's theory of relativity—also predicts waves that move through space. A *gravity wave* is the gravitational counterpart of an electromagnetic wave. *Gravitational radiation* results from changes in the strength of a gravitational field. In principle, any time an object of any mass accelerates, a gravity wave should be emitted at the speed of light. The passage of a gravity wave should produce small distortions in the space through which it passes. Gravity is an exceedingly weak force compared with electromagnetism, so these distortions are expected to be very small—in fact, much smaller than the diameter of an atomic nucleus for the

waves that might be produced by any sources located in the Milky Way Galaxy—yet many researchers believe that these distortions should be measurable. So far, though, no one has succeeded in detecting gravity waves. However, their detection would provide very strong support for the theory of relativity, so scientists are eager to search for them.

Theorists are still debating which kinds of astronomical objects should produce gravity waves detectable on Earth. Leading candidates include (1) the merger of a binary-star system, (2) the collapse of a star into a black hole, and (3) the collision of two black holes or of two neutron stars. Because each of these possibilities involves the acceleration of huge masses, the strength of the gravitational fields should change drastically and rapidly in each case. Other astronomical objects are also expected to emit gravity waves, but only changes involving large masses will produce waves intense enough to be observed.

Of the three candidates, the first one probably presents the best chance to detect gravity waves, at least for the present. Binary-star systems should emit gravitational radiation as the component stars orbit one another. As energy escapes in the form of gravity waves, the two stars slowly spiral toward one another, orbiting more rapidly and emitting even more gravitational radiation. This runaway situation can lead to the

---

These general properties suggest that the invisible X-ray-emitting companion could be a black hole. The X-ray-emitting region is likely an accretion disk formed as matter drawn from the visible star spirals down onto the unseen component. The rapid variability of the X-ray emission indicates that the unseen component must be compact—a neutron star or a black hole. The mass limit on the dark component

argues for the latter because, as mentioned earlier, it is believed that the mass of a neutron star cannot exceed about three times the mass of the Sun. Figure 13.14 is an artist's conception of this intriguing object. As the gas flows toward the black hole, it becomes superheated and emits the X rays we observe, just before both matter and radiation are trapped forever below the event horizon.

decay and eventual merger of close binary systems in a relatively short time (which, in this case, means tens or hundreds of millions of years).

Such a slow but steady decay in the orbit of a binary system has in fact been detected. In 1974, radio astronomer Joseph Taylor and his student Russell Hulse at the University of Massachusetts discovered a very unusual binary system. Both components are neutron stars, and one is observable from Earth as a pulsar. This system has become known as the *binary pulsar*. Measurements of the periodic Doppler shift of the pulsar's radiation prove that its orbit is slowly shrinking. Furthermore, the rate at which the orbit is shrinking is exactly what would be predicted by relativity theory if the energy were being carried off by gravity waves. Even though the gravity waves themselves have not yet been detected, the binary pulsar is regarded by most astronomers as a very strong piece of evidence in favor of general relativity. Taylor and Hulse received the 1993 Nobel Prize in physics for their discovery.

Gravity waves should contain a great deal of information about the physical events in some of the most exotic regions of space. In 1992, funding was approved for an ambitious gravity-wave observatory called LIGO—short for Laser Interferometric Gravity-wave Observatory. Twin detectors, one in the state of Washington, the other in Louisiana, will use laser beams to measure the extremely small distortions of space produced by gravitational radiation. The accompanying figure shows the complex arrangement of test masses, hanging from wires and fitted with mirrors, that lies at the heart of the LIGO detector. The laser beams will be used to measure the tiny motions of these masses that will result should a gravity wave pass by. They should be capable of detecting gravity waves from many Galactic and extragalactic sources. Work began on building the detectors in 1995; they should be operational by 2000. If successful, the discovery of gravity waves could herald a new age in astronomy, in much the same way that invisible electromagnetic waves, unknown a century ago, revolutionized classical astronomy and led to the field of modern astrophysics.

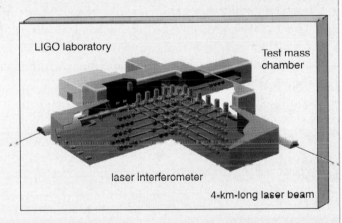

LIGO laboratory

Test mass chamber

laser interferometer

4-km-long laser beam

Have Black Holes Been Detected?

A few other black-hole candidates are known. For example, LMC X-3—the third X-ray source ever discovered in the Large Magellanic Cloud, a small galaxy orbiting our own—is an invisible object that, like Cygnus X-1, orbits a bright companion star. LMC X-3's visible companion seems to be distorted into the shape of an egg by the unseen object's intense gravitational pull. Reasoning similar to that applied to Cygnus X-1 leads to the conclusion that LMC X-3 has a mass nearly 10 times that of the Sun, making it too massive to be anything but a black hole. The Galactic X-ray binary system A0620-00 has been found to contain an invisible compact object of mass 3.8 times the mass of the Sun. In total, there are perhaps half a dozen known

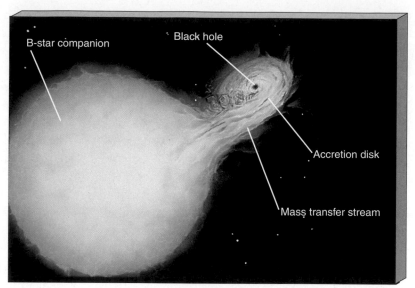

Figure 13.14 Artist's conception of a binary system containing a large, bright, visible star and an invisible, X-ray-emitting black hole.

objects that may turn out to be black holes, although Cygnus X-1, LMC X-3, and A0620-00 have the strongest claims.

So have stellar black holes really been discovered? The answer is probably yes. Skepticism is healthy in science, but only the most stubborn astronomers (and some do exist!) would take serious issue with the reasoning that supports the case for black holes. Can we guarantee that future modifications to the theory of compact objects will not invalidate our arguments? No,

but similar statements could be made in many other areas of astronomy—indeed, about any theory in any area of science. We conclude that, strange as they are, black holes have been detected in our Galaxy. Perhaps some day, future generations of space travelers will visit Cygnus X-1 or LMC X-3 and (carefully!) test these conclusions firsthand. Until then, we must continue to rely on improving theoretical models and observational techniques to guide our discussions of these mysterious objects.

Chapter Review *www*

Summary

A core-collapse supernova may leave behind an ultra-compressed ball of material called a **neutron star** (p. 366). The processes that form neutron stars ensure that these stars are rapidly rotating and strongly magnetized at birth. According to the **lighthouse model** (p. 368), neutron stars, because they are rotating, send regular bursts of electromagnetic energy into space. When we can see the beams from Earth, we call the source neutron star a **pulsar** (p. 367). The pulse period is the rotation period of the neutron star.

A neutron star that is a member of a binary system can draw matter from its companion, forming an accretion disk. The material in the disk heats up even before it reaches the neutron star, and the disk is usually a strong source of X rays. As gas builds up on the

star's surface, it eventually becomes hot enough to fuse hydrogen. As with a nova explosion on a white dwarf, when hydrogen burning starts on a neutron star, it does so explosively. An **X-ray burster** (p. 369) results. **Gamma-ray bursters** (p. 369) may be scaled-up versions of the same basic phenomenon, although it now seems much more likely that they may be much more distant and hence vastly more violent events.

The rapid rotation of the inner part of the accretion disk causes the neutron star to spin faster as new gas arrives on its surface. The eventual result is a very rapidly rotating neutron star—a **millisecond pulsar** (p. 370). Many millisecond pulsars are found in the hearts of old globular clusters. They cannot have formed recently, so they must have been spun up by interactions with other stars. Careful analysis of the radiation received has shown that some millisecond pulsars are orbited by planet-sized objects. The origin of these "pulsar planets" is still uncertain.

The upper limit on the mass of a neutron star is about 3 solar masses. Beyond that mass, the star can no longer support itself against its own gravity, and its central core collapses to form a **black hole** (p. 372), which is a region of space from which no visible light, no other form of radiation, no information whatsoever can escape. Astronomers believe that the most massive stars, after exploding in a supernova, form black holes rather than neutron stars.

Conditions in and near black holes cannot be described by Newtonian mechanics. A proper description involves Einstein's **general theory of relativity** (p. 373), which predicts that no known force can prevent a collapsing star from contracting all the way to a pointlike **singularity** (p. 381), at which point both the density and the gravitational field of the star become infinite. This prediction of relativity theory has yet to be proved, for at singularities the known laws of physics break down.

The radius at which the escape speed from a collapsing star equals the speed of light is called the **Schwarzschild radius** (p. 373). Any collapsing star having a radius smaller than the Schwarzschild radius can never be seen because any radiation it emits cannot escape and travel to our detectors. The surface of an imaginary sphere centered on the collapsing star and having a radius equal to the star's Schwarzschild radius is called the **event horizon** (p. 373).

Relativity theory describes gravity in terms of a warping, or bending, of space by the presence of mass. The more mass, the greater the warping. All particles—including photons—respond to that warping by moving along curved paths. A black hole is a region where the warping is so great that space folds back on itself, cutting off the interior of the hole from the rest of the universe.

To a distant observer, light leaving a spaceship falling into a black hole would be subject to **gravitational redshift** (p. 379) as the light climbed out of the hole's intense gravitational field. Light emitted just at the event horizon would be redshifted to infinite wavelength. At the same time, a clock on the spaceship would show **time dilation** (p. 379)—the clock would appear to slow down as the ship approached the event horizon. The observer would never see the ship reach the surface of the hole.

Once matter falls into a black hole, it can no longer communicate with the outside. However, on its way in, it can form an accretion disk and emit X rays just as in the neutron-star case. The best place to look for a black hole is in a binary system in which one component is a compact X-ray source. Cygnus X-1, a well-studied X-ray source in the constellation Cygnus, is a long-standing black-hole candidate. Studies of orbital motions imply that the compact objects are too massive to be neutron stars, leaving black holes as the only alternative.

Self-Test: True or False?

_____ 1. The density of a neutron star is much greater than the density of an atomic nucleus.

_____ 2. As a result of their high mass and small size, neutron stars have only a weak gravitational pull at their surface.

_____ 3. A millisecond pulsar is a very old neutron star that has been recently spun up by interaction with a neighbor.

_____ 4. Millisecond pulsars are always found in globular clusters, and nowhere else.

_____ 5. Planet-sized bodies will never be found around a pulsar because the supernova that formed the pulsar would have destroyed any planets in the system.

_____ 6. Nothing can travel faster than the speed of light.

_____ 7. All things except light are attracted by gravity.

_____ 8. A black hole is an object whose escape speed equals or exceeds the speed of light.

_____ 9. Although visible light cannot escape from a black hole, high-energy radiation, such as gamma rays, can escape.

_____ 10. If you could touch it, the surface of a black hole, the event horizon, would be very hard.

_____ 11. Thousands of black holes have now been identified.

_____ 12. If the Sun were magically to turn into a black hole, Earth's orbit would immediately change.

Self-Test: Fill in the Blank

1. No neutron star remains after the explosion of a _____ supernova.

2. A typical neutron star is _____ km in diameter.

3. Neutron stars may be characterized as having a _____ rate of rotation and a _____ magnetic field.

4. Pulsars were discovered through observations in the _____ part of the electromagnetic spectrum.

5. Typical pulsar periods range from _____ to _____. (Give numbers and units.)

6. The pulse period of pulsar radiation tells us the _____ of the neutron star emitting the radiation.

7. All millisecond pulsars either are now or once were members of _____ star systems.

8. X-ray bursters result when material from a binary companion accretes onto a _____ star.

9. According to the theory of general relativity, space is warped, or curved, by _____.

10. Photons _____ energy as they escape from a gravitational field.

11. Matter accreting onto a stellar-mass black hole emits radiation in the form of _____.

12. Black holes are believed to have been discovered in _____ systems.

Review and Discussion

1. How does the way in which a neutron star forms determine some of its most basic properties?

2. What would happen to a person standing on the surface of a neutron star?

3. Why aren't all neutron stars seen as pulsars?

4. What are X-ray bursters?

5. What is the favored explanation for the rapid spin rates of millisecond pulsars?

6. Why do you think astronomers were surprised to find a pulsar with planets orbiting it?

7. What does it mean to say that the measured speed of a light beam is independent of the motion of the observer?

8. Use your knowledge of escape speed to explain why black holes are said to be "black."

9. Why is it so difficult to test the predictions of the theory of general relativity? Describe two tests of the theory.

10. What would happen to someone falling into a black hole?

11. What makes Cygnus X-1 a good black hole candidate?

12. Imagine that you had the ability to travel at will through the Milky Way Galaxy. Explain why you would discover many more neutron stars than those known to observers on Earth. Where would you be most likely to find these objects?

13. Do you think that planet-sized objects discovered in orbit around a pulsar should be called planets? Why or why not?

Problems

1. The angular momentum of a spherical body (see *More Precisely 4-1*) is proportional to the body's angular speed times the square of its radius. Using the law of conservation of angular momentum, estimate how fast a collapsed stellar core would spin if its initial spin rate was 1 revolution per day and its radius decreased from 10,000 km to 10 km.

2. Supermassive black holes are believed to exist in the centers of some galaxies. What would be the Schwarzschild radii of black holes of 1 million and 1 billion solar masses? How does the first black hole compare in size with the Sun? How does the second compare in size with the solar system?

3. A 10-km-radius neutron star is spinning 1000 times per second. Calculate the speed of a point on its equator, and compare your answer with the speed of light. (Consider the equator as the circumference of a circle, and recall that circumference = $2\pi \times$ radius.)

Projects

1. Many amateur astronomers enjoy turning their telescopes on the ninth-magnitude companion to Cygnus X-1, the sky's most famous black-hole candidate. Because none of us can see X-rays, no sign of anything unusual can be observed. Still, it's fun to gaze toward this region of the heavens and contemplate Cygnus X-1's powerful energy emission and strange properties. Even without a telescope, it is easy to locate the region of the heavens where Cygnus X-1 resides. The constellation Cygnus contains a recognizable star pattern, or asterism, in the shape of a large cross. This asterism is called the Northern Cross. The star in the center of the crossbar is called Sadr, and the star at the bottom of the cross is called Albireo. Approximately midway along an imaginary line between Sadr and Albireo lies the star Eta Cygni. Cygnus X-1 is located slightly less than 0.5 degree from this star. With or without a telescope, sketch what you see.

2. Set up a demonstration of the densities of various astronomical objects—an interstellar cloud, a star, a terrestrial planet, a white dwarf, and a neutron star. Select a common object that is easily held in your hand, something familiar to anyone—an apple, for example. For the lowest densities, calculate how large a volume would contain the object's equivalent mass. For high densities, calculate how many of the objects would have to be fit into a standard volume, such as 1 cm^3. Present your demonstration to your class or to some other group of students. Tell them about each astronomical object and how it comes by its density.

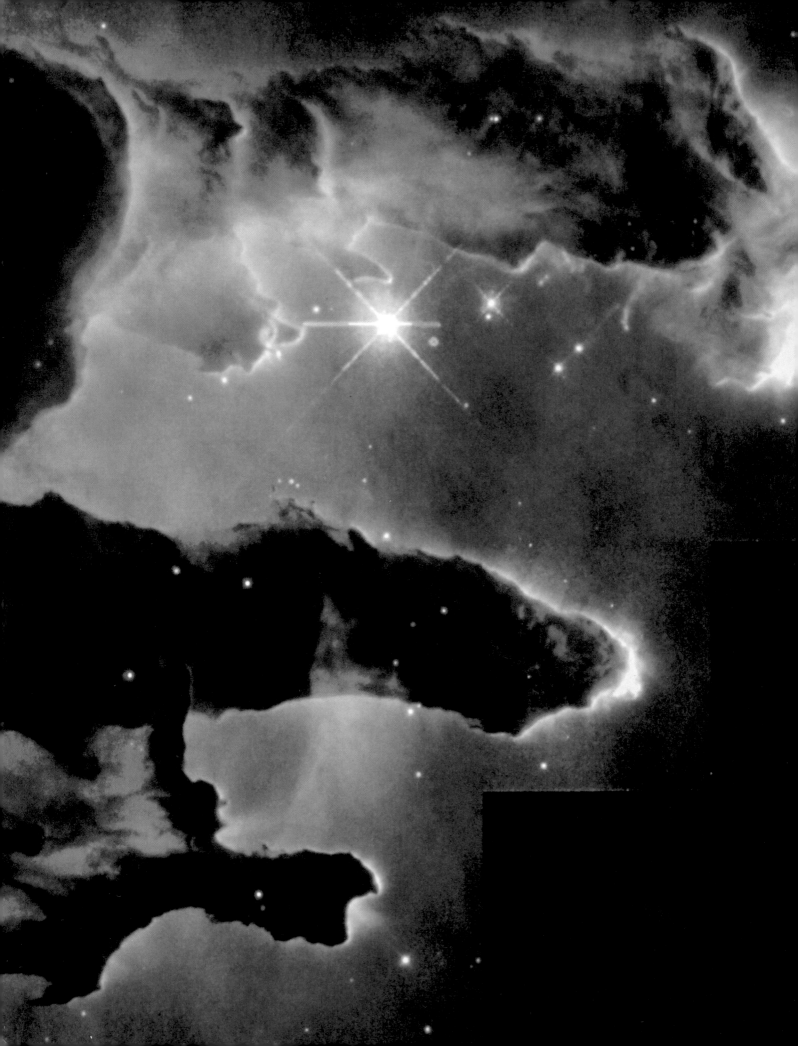

14

The Milky Way Galaxy

A Grand Design ⭐ *www*

The varying interrelationships among the many components of matter in our Milky Way Galaxy comprise a sort of "galactic ecosystem." Its evolutionary balance might be as complex as that of life in a tidepool or a tropical rain forest. Here, stars, gas and dust abound throughout the Eagle Nebula (M16), a rich stellar nursery about 1800 pc from Earth. (See also Figures 11.7 and 11.22.) The largest of these gaseous columns is about 2 light-years long

LEARNING GOALS

Studying this chapter will enable you to:

1 Describe the overall structure of the Milky Way Galaxy, and specify how the various regions differ from one another.

2 Explain the importance of variable stars in determining the size and shape of our Galaxy.

3 Describe the orbital paths of stars in different regions of the Galaxy, and explain how these motions are accounted for by our understanding of how the Galaxy formed.

4 Discuss some possible explanations for the existence of the spiral arms observed in our own and many other galaxies.

5 Explain what studies of galactic rotation reveal about the size and mass of our Galaxy, and discuss the possible nature of dark matter.

6 Describe some of the phenomena observed at the center of our Galaxy.

Looking up on a dark, clear night, we are struck by two aspects of the night sky. The first is a fuzzy band of light—the Milky Way—that stretches across the heavens. From the Northern Hemisphere, this band is most easily visible in the summertime, arcing high above the horizon. Its full extent forms a great circle that encompasses the entire celestial sphere. Away from that glowing band, however, our second impression is that the nighttime sky seems more or less the same in all directions. Bunches of stars cluster here and there, but overall, apart from the band of the Milky Way, the evening sky looks pretty uniform. Yet this is only a local impression. Ours is a rather provincial view. When we consider much larger volumes of space, on scales far, far greater than the distances between neighboring stars, a new level of organization becomes apparent as the large-scale structure of the Milky Way Galaxy is revealed.

14.1 Our Parent Galaxy

1 A **galaxy** is a gargantuan collection of stellar and interstellar matter—stars, gas, dust, neutron stars, black holes—isolated in space and held together by its own gravity. Astronomers are aware of literally millions of galaxies beyond our own. The particular galaxy we happen to inhabit is known as the **Milky Way Galaxy**, or just "the Galaxy," with a capital "G."

Our Sun lies in a part of the Galaxy known as the **Galactic disk**—an immense, circular, flattened region containing most of our Galaxy's luminous stars and interstellar matter (and virtually everything we have studied so far in this book). We do not need sophisti-

cated astronomical equipment to verify this statement. Our own unaided eyes will suffice. Figure 14.1 illustrates how, viewed from within, the Galactic disk appears as a band of light stretching across our night sky, a band known as the *Milky Way*. As indicated in the figure, if we look in a direction away from the Galactic disk (red arrows), relatively few stars lie in our field of view. However, if our line of sight happens to lie within the disk (white and blue arrows), we see so many stars that their light merges into a continuous blur.

Paradoxically, although we can study individual stars and interstellar clouds near the Sun in great detail, our location within the Galactic disk makes deciphering our Galaxy's large-scale structure from Earth a very difficult task—a little like trying to

Galactic disk   Galactic bulge   Galactic center

Earth

R I V U X G

Figure 14.1 Gazing from Earth toward the Galactic Center (white arrow), we see myriad stars stacked up within the thin band of light known as the Milky Way. Looking in the opposite direction (blue arrow), we still see the Milky Way band, but now it is fainter because our position is far from the Galactic center, with the result that we see more stars when looking toward the center than when looking in the opposite direction. Looking perpendicular to the disk (red arrows), we see far fewer stars. The inset is an enhanced satellite view of the sky all around us; the white band is the disk of our Milky Way Galaxy, which can be seen with the naked eye from very dark locations on Earth.

Figure 14.2 (a) The Andromeda Galaxy probably resembles fairly closely the overall layout of our own Milky Way Galaxy. The disk and bulge are clearly visible in this image, which is about 30,000 pc across. The faint halo stars cannot be seen. The white stars sprinkled all across this image are not part of Andromeda's halo; they are foreground stars in our own Galaxy, lying in the same region of the sky as Andromeda but about a thousand times closer. (b) This galaxy, seen nearly face-on, is somewhat similar in its overall structure to our own Milky Way Galaxy and Andromeda. It is known as M83—the eighty-third object in the Messier catalog. (c) The galaxy NGC 891 happens to be oriented in such a way that we see it edge-on, allowing us to see clearly its disk and central bulge.

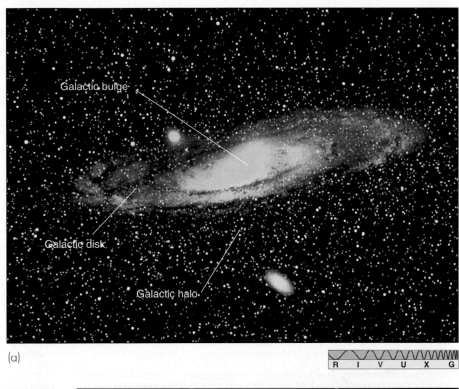

Galactic bulge

Galactic disk

Galactic halo

(a)

R I V U X G

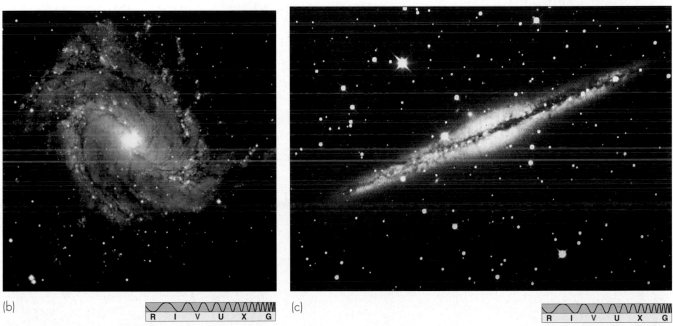

(b)

R I V U X G

(c)

R I V U X G

unravel the layout of paths, bushes, and trees in a city park without being able to leave one particular park bench. In some directions, the interpretation of what we see is ambiguous and inconclusive. In others, foreground objects completely obscure our view of what lies beyond, but we cannot move around them to get a better look. As a result, astronomers who study the Milky Way Galaxy are often guided in their efforts by comparisons with more distant, but much more easily observable, systems.

Figure 14.2 shows three galaxies thought to resemble our own in overall structure. Figure 14.2(a) is the Andromeda Galaxy, the nearest major galaxy to the Milky Way Galaxy, lying about 700 kpc (about 2 million light years) away. Andromeda's apparent elongated shape is a consequence of the angle at which we happen to view it. In fact, this galaxy, like our own, consists of a circular galactic disk of matter that fattens to a **galactic bulge** at the center. The disk and bulge are embedded in a roughly spherical ball of faint old stars known as the **galactic halo**. These three basic galactic regions are indicated on the figure (the halo stars are so faint that they cannot be discerned here; see Figure 14.9). Figures 14.2(b) and (c) show views of two other galaxies—one seen face-on, the other edge-on—that illustrate these points more clearly.

14.2 Measuring the Milky Way

Before the twentieth century, astronomers' conception of the cosmos differed markedly from the modern view. The fact that we live in just one of many enormous "islands" of matter separated by even larger tracts of apparently empty space was completely unknown, and the clear distinction between "our Galaxy" and "the universe" did not exist. The twin ideas that (1) the Sun is not at the center of the Galaxy and (2) the Galaxy is not at the center of the universe required both time and hard observational evidence before they gained widespread acceptance.

Star Counts

In the late eighteenth century, long before the distances to any stars were known, the English astronomer William Herschel tried to estimate the size and shape of our Galaxy simply by counting how many stars he could see in different directions in the sky.

Assuming that all stars were of about equal brightness, he concluded that the Galaxy was a somewhat flattened, roughly disk-shaped collection of stars lying in the plane of the Milky Way, with the Sun at its *center* (Figure 14.3). Subsequent refinements to this approach led to essentially the same picture; early in the twentieth century, some workers went so far as to estimate the dimensions of this "Galaxy" as about 10 kpc in diameter by 2 kpc thick.

Today, the Milky Way Galaxy is known to be several tens of kiloparsecs across, and the Sun lies far from the center. How could the older picture have been so flawed? The answer is that the earlier observations were made in the visible part of the electromagnetic spectrum, and astronomers failed to take into account the (then unknown) absorption of visible light by interstellar gas and dust. ∞ (Sec. 11.1) Only in the 1930s did astronomers begin to realize the true extent and importance of the interstellar medium.

Any objects in the Galactic disk that are more than a few kiloparsecs away from us are hidden from our view by the effects of interstellar absorption. The apparent falloff in the density of stars with distance in the plane of the Milky Way is thus not a real thinning of their numbers in space but simply a consequence of the murky environment in the Galactic disk. The long "fingers" in Herschel's map are directions where the obscuration happens to be a little less severe than in others. However, because some obscuration occurs in all directions in the disk, the falloff is roughly similar

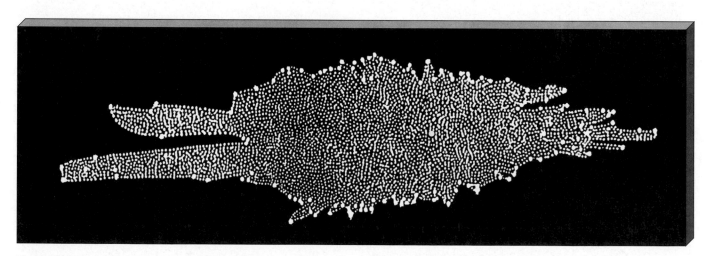

Figure 14.3 Eighteenth-century English astronomer William Herschel constructed this "map" of the Galaxy by counting the numbers of stars he saw in different directions in the sky. He assumed that all stars were of roughly equal luminosity and, within the confines of the Galaxy, were uniformly distributed in space. Our Sun (marked by the large yellow dot) appears to lie near the center of the distribution, and the long axis of the diagram lies in the plane of the Galactic disk.

no matter which way we look, and so the Sun appears to be at the center. In contrast, radiation coming to us from above or below the plane of the Galaxy, where there is less gas and dust along the line of sight, arrives on Earth relatively unscathed. There is still some patchy obscuration, but the Sun happens to lie in a location where the view out of the disk is largely unimpeded by nearby interstellar clouds.

Spiral Nebulae and Globular Clusters

Astronomers' attempts to probe the Galactic disk by optical means are frustrated by the effects of the interstellar medium. However, looking in other directions, out of the Milky Way plane, we can see to much greater distances. During the first quarter of the twentieth century, studies of the large-scale structure of our Galaxy focused on two particularly important classes of objects, both found mainly away from the Milky Way. The first are globular clusters, those tightly bound swarms of old, reddish stars we first met in Chapter 10. ∞ (Sec. 10.10) About 140 are now known. The second were known at the time as *spiral nebulae*. Two examples are shown in Figures 14.2(a) and (b). We know them today as **spiral galaxies**, comparable in size to our own.

At the time, astronomers had no means of determining the distances to any of these objects. They are far too distant to have any observable parallax, and, with the technology of the day, main-sequence stars (after the discovery of the main sequence in 1911) could not be clearly identified and measured. For these reasons, neither of the techniques discussed in Chapter 10 were applicable. ∞ (Sec. 10.1, 10.8) As a result, even the most basic properties—size, mass, and stellar and interstellar content—of globular clusters and spiral nebulae were unknown. It was assumed that the globular clusters lay within our own Galaxy, which was thought to be relatively small (using the size estimates mentioned earlier, based on star counts). The locations of the spiral nebulae were much less clear.

During this period, both the size of our Galaxy and the distances to the spiral nebulae were hotly debated in astronomical circles. One school of thought maintained that the spiral nebulae were relatively small systems contained within our Galaxy and perhaps related in some way to emission nebulae. Other astronomers held that the spirals were much larger objects, lying far outside the Milky Way Galaxy and

comparable to it in size. However, with no firm distance information, both arguments were quite inconclusive. Only with the discovery of a new distance-measurement technique—which we now discuss—was the issue finally settled in favor of the latter view. However, in the process, astronomers' conception of our own Galaxy changed radically and forever.

The growth in our knowledge of the Galaxy and the realization that there are many other distant galaxies similar to our own have gone hand in hand with the development of the cosmic distance scale.

A New Yardstick

2 An important by-product of the laborious effort to catalog stars around the turn of the twentieth century was the systematic study of **variable stars**. These are stars whose luminosity changes with time—some quite erratically, others more regularly. Only a small fraction of stars fall into this category, but those that do are of great astronomical significance.

We have encountered several examples of variable stars in earlier chapters. Often, the variability is the result of membership in a binary system. Eclipsing binaries, novae, and Type I supernovae are cases in point. Novae and supernovae collectively are called *cataclysmic variables* because of their sudden, large changes in brightness. In other instances, the variability is a basic trait of a star and is not dependent on its being a part of a binary system. We call such a star an *intrinsic variable*.

A particularly important class of intrinsic variables is the class known as **pulsating variable stars**,[1] which vary cyclically in luminosity in very characteristic ways. Two types of pulsating variable stars that have played central roles in revealing both the true extent of our Galaxy and the distances to our galactic neighbors are the **RR Lyrae** and **Cepheid** variables. Following long-standing astronomical practice, the names come from the first star of each class to be discovered—in this case the variable star labeled RR in the constellation Lyra and the variable star Delta Cephei, the fourth brightest star in the constellation Cepheus.

RR Lyrae and Cepheid variable stars are recognizable by the characteristic shapes of their light curves. RR Lyrae stars all pulsate in essentially similar

[1]Note, by the way, that these stars have *nothing* whatsoever to do with pulsars.

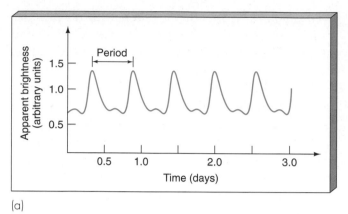

(a)

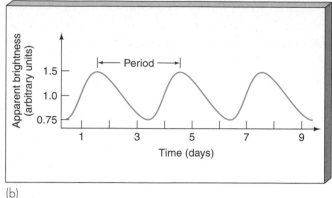

(b)

Figure 14.4 (a) Light curve of the pulsating variable star RR Lyrae. All RR Lyrae-type variables have essentially similar light curves, with periods of less than a day. (b) The light curve of a Cepheid variable star called WW Cygni.

ways (Figure 14.4a), with only small differences in period between one RR Lyrae variable and another. Observed periods range from about 0.5 to 1 day. Cepheid variables also pulsate in distinctive ways (the regular "sawtooth" pattern in Figure 14.4b), but different Cepheids can have very different pulsation periods, ranging from about 1 to 100 days. (Note, incidentally, that the period of any given RR Lyrae or Cepheid variable is, to high accuracy, the same from one cycle to the next.) The key point is that pulsating variable stars can be recognized and identified *just by observing the variations in the light they emit.*

Pulsating variable stars are normal stars experiencing a brief period of instability as a natural part of stellar evolution. The conditions necessary to cause pulsations are not found in main-sequence stars; they occur in evolved post–main-sequence stars as they pass through a region of the Hertzsprung–Russell diagram known as the *instability strip* (Figure 14.5). When a star's temperature and luminosity place it in this strip, the star becomes internally unstable, and both its temperature and its radius vary in a regular way, causing the pulsations we observe: as the star brightens, its surface becomes hotter and its radius shrinks; as its luminosity decreases, the star expands and cools. As we learned in Chapter 12, high-mass stars evolve across the upper part of the H–R diagram; when their evolutionary tracks take them into the instability strip, they become Cepheid variables. RR Lyrae variables are low-mass horizontal-branch stars that lie within the lower portion of the instability strip.

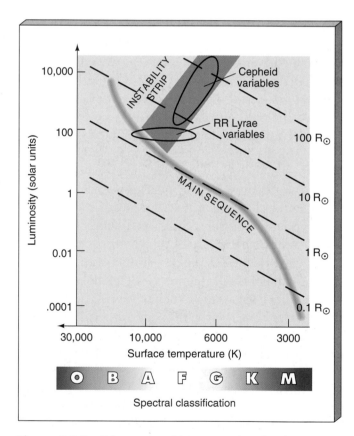

Figure 14.5 Pulsating variable stars are found in the instability strip of the H–R diagram. As a high-mass star evolves through the strip, it becomes a Cepheid variable. Low-mass horizontal-branch stars in the instability strip are RR Lyrae variables.

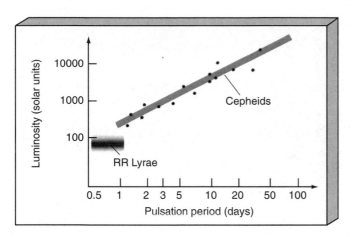

Figure 14.6 A plot of pulsation period versus average absolute brightness (that is, luminosity) for a group of Cepheid variable stars. The two properties are quite tightly correlated. The pulsation periods of some RR Lyrae variables are also shown.

The importance of these stars to Galactic astronomy lies in the fact that, once we recognize a star as being of the RR Lyrae or Cepheid type, we can infer its luminosity, and that in turn allows us to measure its distance. The distance calculation is precisely the same as presented in Chapter 10 during our discussion of spectroscopic parallax. ∞ (Sec. 10.8) Comparing the star's (known) luminosity with its (observed) apparent brightness yields an estimate of its distance, by the inverse-square law: ∞ (Sec. 10.4)

$$\text{apparent brightness} \propto \frac{\text{luminosity}}{\text{distance}^2}.$$

In this way, astronomers can use pulsating variables as a means of determining distances, both within our own Galaxy and far beyond.

How do we infer a variable star's luminosity? For RR Lyrae stars, this is simple. All such stars have basically the same luminosity (averaged over a complete pulsation cycle)—about 100 times that of the Sun—so once a variable star is recognized as being of the RR Lyrae type, its luminosity is immediately known. For Cepheids, we make use of a close correlation between average luminosity and pulsation period, discovered in 1908 by Henrietta Leavitt of Harvard University and known simply as the **period–luminosity relationship**: Cepheids that vary slowly—that is, have long periods—have large luminosities; conversely, short-period Cepheids have low luminosities.

Figure 14.6 illustrates the period–luminosity relationship for Cepheids found within a thousand parsecs or so of Earth. Astronomers can plot such a diagram for relatively nearby stars because they can measure the distances using stellar or spectroscopic parallax. Once distances are known, the luminosities of those stars can be calculated. We know of no exceptions to the period–luminosity relationship, and it is consistent with theoretical calculations of pulsations in evolved stars. Consequently, we assume that it holds for all Cepheids, near and far. Thus, a simple measurement of a Cepheid variable's pulsation period immediately tells us its luminosity—we just read it off the plot in Figure 14.6. (The roughly constant luminosities of the RR Lyrae variables are also indicated in the figure.)

This distance-measurement technique works well, provided the variable star can be clearly identified and its pulsation period measured. With Cepheids, this method allows astronomers to estimate distances out to about 15 million parsecs, enough to take us all the way to the nearest galaxies. The less luminous RR Lyrae stars are not as easily seen as Cepheids, so their useful range is not as great. However, they are much more common, so, within their limited range, they are actually more useful than Cepheids. Figure 14.7 extends our cosmic distance ladder, begun in Chapter 1 with radar ranging in the solar system and expanded in Chapter 10 to include stellar and spectroscopic par-

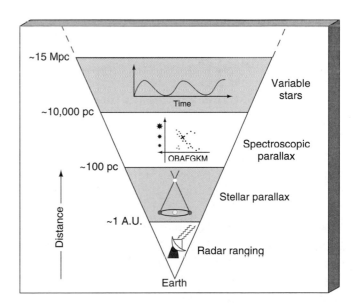

Figure 14.7 Application of the period–luminosity relationship for Cepheid variable stars allows us to determine distances out to about 15 Mpc with reasonable accuracy.

allax, by adding variable stars as a fourth method of determining distance.

The Size and Shape of Our Galaxy

Many RR Lyrae variables are found in globular clusters. Early in the twentieth century, the American astronomer Harlow Shapley used observations of RR Lyrae stars to make two very important discoveries about the Galactic globular cluster system. First, he showed that most globular clusters reside at great distances—many thousands of parsecs—from the Sun. Second, by measuring the direction and distance of each cluster, he was able to determine their three-dimensional distribution in space. In this way, Shapley demonstrated that the globular clusters map out a truly gigantic, and roughly *spherical*, volume of space, about 30 kpc across.[2] However, the center of the distribution lies nowhere near our Sun. It is located nearly 8 kpc away from us, in the direction of the constellation Sagittarius.

In a brilliant intellectual leap, Shapley realized that the distribution of globular clusters maps out the true extent of stars in the Milky Way Galaxy—the region that we now call the Galactic halo. The hub of this vast collection of matter, 8 kpc from the Sun, is the **Galactic center**. As illustrated in Figure 14.8, we live in the suburbs of this huge ensemble, in the Galactic disk—the thin sheet of young stars, gas, and dust that cuts through the center of the halo. Since Shapley's time, astronomers have identified many individual stars—that is, stars not belonging to any globular cluster—within the Galactic halo.

Shapley's bold interpretation of the globular clusters as defining the overall distribution of stars in our Galaxy was an enormous step forward in human understanding of our place in the universe. Five hundred years ago, Earth was considered the center of all things. Copernicus argued otherwise, demoting our planet to an undistinguished place removed from the center of the solar system. In Shapley's time, as we have just seen, the prevailing view was that our Sun was the center not only of the Galaxy but also of the universe. Shapley showed otherwise. With his observations of globular clusters, he simultaneously increased the size of our Galaxy by almost a factor of 10 over earlier esti-

[2]The Galactic globular cluster system and the Galactic halo of which it is a part are somewhat flattened in the direction perpendicular to the disk, but the degree of flattening is quite uncertain. The halo is certainly much less flattened than the disk, however.

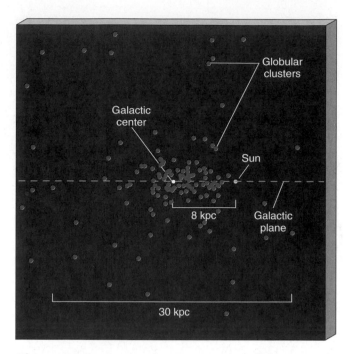

Figure 14.8 Our Sun does not coincide with the center of the very large collection of globular clusters. Instead, more globular clusters are found in one direction than in any other. The Sun resides closer to the edge of the collection, which measures roughly 30 kpc across. We now know that the globular clusters outline the true distribution of stars in the Galactic halo.

mates and banished our parent Sun to its periphery, virtually overnight!

Curiously, Shapley's dramatic revision of the size of the Milky Way Galaxy and our place in it only strengthened his erroneous opinion that the spiral nebulae were part of our Galaxy and that our Galaxy was essentially the entire universe. He regarded as beyond belief the idea that there could be other structures as large as our Galaxy. Only in the late 1920s was the Copernican principle extended to the Galaxy itself, when American astronomer Edwin Hubble observed Cepheids in the Andromeda Galaxy and finally succeeded in measuring its distance.

14.3 The Large-Scale Structure of Our Galaxy

Figure 14.9 illustrates the very different spatial distributions of the disk and halo components of the Milky Way Galaxy. As just mentioned, the Sun lies about 8 kpc from the Galactic center. Based on optical, infrared, and radio observations of stars, gas, and dust found within a thousand or so parsecs of the Sun, astronomers estimate that the disk in the vicinity

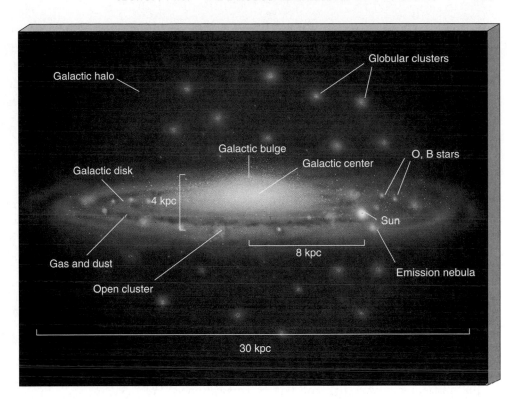

Figure 14.9 Artist's conception of a (nearly) edge-on view of the Milky Way Galaxy, showing the distributions of young blue stars, open clusters, old red stars, and globular clusters.

of the Sun is relatively thin—"only" 300 pc thick, or about 1/100 of the Galactic diameter. Don't be fooled, though. Even if you could travel at the speed of light, it would take you 1000 years to traverse the thickness of the Galactic disk. The disk may be thin compared with the Galactic diameter, but it is huge by human standards.

Also shown in Figure 14.9 is our Galaxy's central bulge, measuring roughly 6 kpc across in the plane of the Galactic disk by 4 kpc perpendicular to the disk plane. Obscuration by interstellar dust makes it difficult to study the detailed structure of the Galactic bulge in optical images of the Milky Way (see, for example, Figure 11.3, which clearly would show a large portion of the bulge were it not for interstellar absorption). However, at longer wavelengths, which are less affected by interstellar matter, a much clearer picture emerges (Figure 14.10; compare Figure 14.2c). Detailed measurements of the motion of gas and stars in and near the bulge imply that it is actually football shaped, about half as wide as it is long, with the long axis of the football lying in the Galactic plane.

Stellar Populations

Aside from their shapes, the three components of the Galaxy—disk, bulge, and halo—have several other properties that distinguish them from one another. For

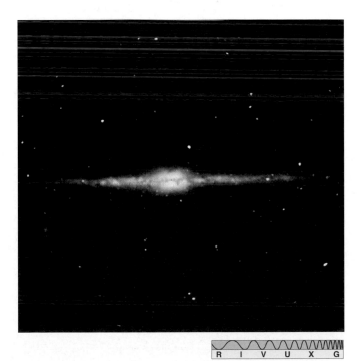

Figure 14.10 A wide-angle infrared image of the disk and bulge of the Milky Way Galaxy, as observed by the *Cosmic Background Explorer (COBE)* satellite.

one thing, the halo contains essentially *no* gas or dust—just the opposite of the disk and bulge, in which interstellar matter is common. For another, there are clear differences in both *appearance* and *composition* between disk, bulge, and halo stars. Stars in the Galactic bulge and halo are found to be distinctly *redder* than stars found in the disk. Observations of other spiral galaxies also show this trend—the blue-white tint of the disk and the yellowish coloration of the bulge are evident in Figures 14.2(a) and (b).

All the bright, blue stars visible in our sky are part of the Galactic disk, as are the young open star clusters and star-forming regions. In contrast, the cooler, redder stars—including those found in the old globular clusters—are more uniformly distributed throughout the disk, bulge, and halo. Galactic disks appear bluish because main-sequence O and B blue supergiants are very much brighter than G, K, and M dwarfs, even though the dwarfs are present in far greater numbers.

The explanation for the marked difference in stellar content between disk and halo is this: whereas the gas-rich Galactic disk is the site of ongoing star formation and so contains stars of all ages, all the stars in the Galactic halo are *old*. The absence of dust and gas in the halo means that no new stars are forming there, and star formation apparently ceased long ago—at least 10 billion years in the past, judging from the types of halo stars we now observe. (Recall from Chapter 12 that most globular clusters are thought to be between 12 and 15 billion years old.) ∞ (Sec. 12.5) The gas density is very high in the inner part of the Galactic bulge, making this region the site of vigorous ongoing star formation, and both very old and very young stars mingle there. The bulge's gas-poor outer regions have properties more similar to those of the halo.

Support for this picture comes from studies of the spectra of halo stars, which indicate that these stars are far less abundant in heavy elements (that is, elements heavier than helium) than are nearby stars in the disk. In Chapter 12 we saw how each successive cycle of star formation and evolution enriches the interstellar medium with the products of stellar nucleosynthesis, leading to a steady increase in heavy elements with time. ∞ (Sec. 12.4) Thus, the scarcity of these elements in halo stars is consistent with the view that the halo formed long ago.

Astronomers often refer to young disk stars as *Population I* stars and old halo stars as *Population II* stars. The idea of two stellar "populations" dates back to the 1930s, when the differences between disk and

halo stars first became clear. It represents something of an oversimplification, as there is actually a continuous variation in stellar ages throughout the Milky Way Galaxy, not a simple division of stars into two distinct "young" and "old" categories. Nevertheless, the terminology is still widely used.

Orbital Motion

3 Now let's turn our attention to the *dynamics* of the Milky Way Galaxy—that is, to the motion of the stars, dust, and gas it contains. Are the internal motions of our Galaxy's members chaotic and random, or are they part of some gigantic "traffic pattern"? The answer depends on our perspective. The motion of stars and clouds we see on small scales (within a few tens of parsecs of the Sun) seems random, but on larger scales (hundreds or thousands of parsecs) the motion is much more orderly.

As we look around the Galactic disk in different directions, a clear pattern of motion emerges. Careful study of the Doppler shifts of radiation received from stars and gas clouds leads to the conclusion that the entire Galactic disk is *rotating* about the Galactic center (Figure 14.11). In the vicinity of the Sun, the

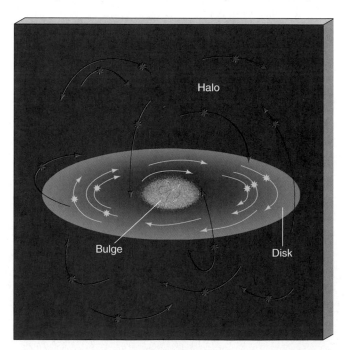

Figure 14.11 Stars in the Galactic disk move in orderly, circular orbits about the Galactic center. In contrast, halo stars have orbits with largely random orientations and eccentricities. The orbit of a typical halo star takes it high above the Galactic disk, then down through the disk plane, then out the other side and far below the disk. The orbital properties of bulge stars are intermediate between those of disk stars and those of halo stars.

Table 14.1 Overall Properties of the Galactic Disk, Halo, and Bulge		
Galactic Disk	**Galactic Halo**	**Galactic Bulge**
highly flattened	roughly spherical—mildly flattened	somewhat flattened and elongated in the plane of the disk ("football shaped")
contains both young and old stars	contains old stars only	contains both young and old stars; more old stars at greater distances from the center
contains gas and dust	contains no gas and dust	contains gas and dust, especially in the inner regions
site of ongoing star formation	no star formation during the last 10 billion years	ongoing star formation in the inner regions
gas and stars move in circular orbits in the Galactic plane	stars have random orbits in three dimensions	stars have largely random orbits, but with some net rotation about the Galactic center
spiral arms	no obvious substructure	ring of gas and dust near center; Galactic nucleus
overall white coloration, with blue spiral arms	reddish in color	yellow-white

orbital speed is about 220 km/s. At the Sun's distance of 8 kpc from the Galactic center, material takes about 225 million years—an interval of time sometimes known as 1 *Galactic year*—to complete one circuit. At other distances from the center, the rotation period is different—shorter closer to the center, longer at greater distances—that is, the Galactic disk rotates not as a solid object, but *differentially*.

This picture of orderly circular orbital motion about the Galactic center applies only to the Galactic disk. Stars in the Galactic halo and bulge are not so well behaved. The old globular clusters in the halo and the faint, reddish individual stars in both the halo and the bulge do *not* share the disk's well-defined rotation. Instead, their orbits are largely random.[3] Although they do orbit the Galactic center, they move in all directions, their paths filling an entire three-dimensional volume rather than a nearly two-dimensional disk. At any given distance from the Galactic center,

bulge or halo stars move at speeds comparable to the disk's rotation speed at that radius but in *all* directions, not just one. Their orbits carry these stars repeatedly through the disk plane and out the other side. Figure 14.11 contrasts the motion of bulge and halo stars with the much more regular orbits of stars in the Galactic disk.

14.4 The Formation of the Milky Way Galaxy

3 Table 14.1 compares some key properties of the three basic components of the Galaxy. Is there some evolutionary scenario that can naturally account for the Galactic structure we see today? The answer is that there is, and it takes us all the way back to the birth of our Galaxy, 10–15 billion years ago. Not all the details are agreed upon by all astronomers, but the overall picture is now fairly widely accepted. For simplicity, we confine our discussion here to the Galactic disk and halo. In many ways, the bulge is intermediate in its properties between these two extremes.

[3]Halo stars do in fact have some net rotation about the Galactic center, but the rotational component of their motion is overwhelmed by the larger random component. The motion of bulge stars also has a rotational component, larger than that of the halo but still smaller than the random component of stellar motion in the bulge.

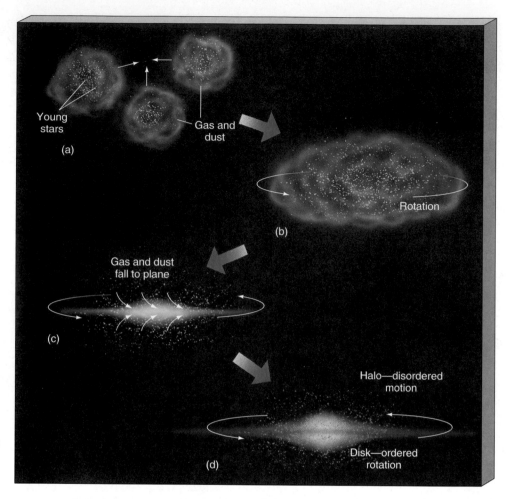

Young
stars

Gas and
dust

(a)

Rotation

(b)

Gas and dust
fall to plane

(c)

Halo—disordered
motion

Disk—ordered
rotation

(d)

Figure 14.12 (a) The Milky Way Galaxy possibly formed via the merger of several smaller systems. (b) Astronomers reason that, early on, our Galaxy was irregularly shaped, with gas distributed throughout its volume. When stars formed during this stage, there was no preferred direction in which they moved and no preferred location in which they were found. (c) In time, rotation caused the gas and dust to fall to the Galactic plane and form a spinning disk. The stars that had already formed were left behind, forming the halo. (d) New stars forming in the disk inherit its overall rotation and so orbit the Galactic center on ordered, circular orbits.

Figure 14.12 illustrates the current view of our Galaxy's evolution, starting (not unlike the star-formation scenario outlined in Chapter 11) from a contracting cloud of pregalactic gas. ∞ (Sec. 11.2) When the first Galactic stars and globular clusters formed, the gas in our Galaxy had not yet accumulated into a thin disk. Instead, it was spread out over an irregular, and quite extended, region of space, spanning many tens of kiloparsecs in all directions. When the first stars formed, they were distributed throughout this volume. Their distribution today (the Galactic halo) reflects that fact—it is an imprint of their birth. Many astronomers believe that the very first stars formed even earlier, in smaller systems that later merged to create our Galaxy; the present-day halo would look the same in either case.

During the past 10–15 billion years, rotation has flattened the gas in our Galaxy into a relatively thin disk. Physically, this process is similar to the flattening of the solar nebula during the formation of the solar system, as described in Chapter 4, except on a vastly larger scale. ∞ (Sec. 4.3) Star formation in the halo ceased billions of years ago when the raw materials fell to the Galactic plane. Ongoing star formation in the disk gives it its bluish tint, but the halo's short-lived blue stars have long since burned out, leaving only the long-lived red stars that give it its characteristic pinkish glow. The Galactic halo is ancient, whereas the disk is full of youthful activity.

The chaotic orbits of the halo stars are also explained by this theory. When the halo developed, the irregularly shaped Galaxy was rotating only very slowly, so there was no strongly preferred direction in which matter tended to move. As a result, halo stars were free to travel along nearly any path once they formed (or when their parent systems merged). As the Galactic disk formed, however, conservation of angular momentum caused it to spin more rapidly. Stars forming from the gas and dust of the disk inherit its rotational motion and so move on well-defined, circular orbits.

14.5 Spiral Structure

Radio Maps of the Milky Way

If we want to look beyond our immediate neighborhood and study the full extent of the Galactic disk, we cannot rely on optical observations, as interstellar absorption severely limits our vision, as mentioned in Section 14.3. In the 1950s, astronomers developed a very important tool to explore the distribution of gas in our Galaxy—spectroscopic radio astronomy.

The keys to observing Galactic interstellar gas are the 21-cm radio emission line produced by atomic hydrogen and the many radio molecular lines formed in molecular cloud complexes. ∞ (Sec. 11.3) Long-wavelength radio waves are largely unaffected by interstellar dust, so they travel more or less unimpeded through the Galactic disk, allowing us to "see" to great distances. Because hydrogen is by far the most abundant element in interstellar space, the 21-cm signals are strong enough that a large portion of the disk can be observed in this way; the molecular lines allow us to study the distribution of the densest interstellar clouds.

Interstellar gas in the Galactic disk exhibits an organized pattern on a grand scale. According to radio studies, the center of the gas distribution coincides roughly with the center of the globular-cluster system, about 8 kpc from the Sun. (In fact, this figure is derived most accurately from radio observations of the distribution of Galactic gas, the center of which is normally taken to define the center of our Galaxy.) Radio-emitting gas has been observed out to at least 50 kpc from the Galactic center. Over much of the inner 20 kpc or so of the disk, the gas is confined within about 100 pc of the Galactic plane. Beyond 20 kpc, the gas distribution spreads out somewhat, to a thickness of several kiloparsecs.

Radio studies provide perhaps the best direct evidence that we live in a spiral galaxy. Figure 14.13 is an artist's conception (based on observational data) of the appearance of our Galaxy as seen from far above the disk, clearly showing our Galaxy's **spiral arms**, pinwheel-like structures originating close to the Galactic bulge and extending outward throughout much of the Galactic disk. One of these arms, as best we can tell, wraps around a large part of the disk and contains our Sun. Notice, incidentally, the scale markers on Figures

Figure 14.13 An artist's conception of our Milky Way Galaxy seen face-on. This illustration is based on data accumulated by legions of astronomers during the past few decades, including radio maps of gas density in the Galactic disk. Painted from the perspective of an observer 100 kpc above the Galactic plane, the spiral arms are at their best-determined positions. All the features are drawn to scale (except for the oversized yellow dot near the top, which represents our Sun). The two small blotches to the left are dwarf galaxies, called the Magellanic Clouds. We study them in Chapter 15.

30 kpc

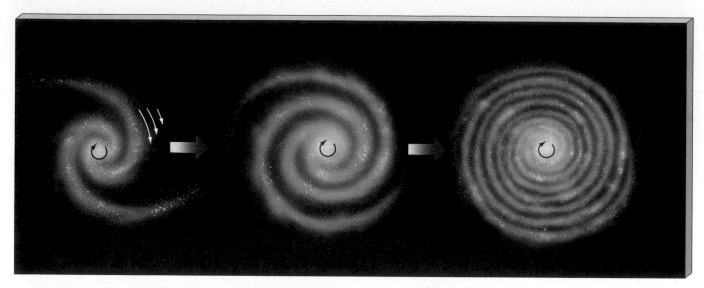

Figure 14.14 The disk of our Galaxy rotates differentially—stars close to the center take less time to orbit the Galactic center than those at greater distances. If spiral arms were somehow tied to the material of the Galactic disk, this differential rotation would cause the spiral pattern to wind up and disappear in a few hundred million years. Spiral arms would be too short-lived to be consistent with the numbers of spiral galaxies we observe today.

14.8, 14.9 and 14.13: the Galactic globular-cluster distribution (shown in Figure 14.8), the luminous stellar component of the disk (Figure 14.9), and the known spiral structure (Figure 14.13) all have roughly the *same* diameter—about 30 kpc.

Persistence of the Spiral Arms

4 The spiral arms in our Galaxy are made up of much more than just interstellar gas and dust. Studies of the Galactic disk within a kiloparsec or so of the Sun indicate that young stellar and prestellar objects—emission nebulae, O and B stars, and recently formed open clusters—are also distributed in a spiral pattern that closely follows the distribution of interstellar clouds. The obvious conclusion is that the spiral arms are the part of the Galactic disk where star formation takes place. The brightness of the young stellar objects just listed is the main reason that the spiral arms of other galaxies are easily seen from afar (Figure 14.2b).

A central problem facing astronomers trying to understand spiral structure is how that structure persists over long periods of time. The basic issue is simple: we know that the inner parts of the Galactic disk rotate more rapidly than do the outer regions. Thus stars in the Galactic disk do not move smoothly together, but ceaselessly change their positions relative to one another as they orbit the Galactic center. This differential rotation makes it impossible for any large-scale structure "tied" to the disk material to survive. Figure 14.14 shows how a spiral pattern consisting always of the same group of stars and gas clouds would necessarily "wind up" and disappear within a few hundred million years. Yet spiral arms clearly do exist in our own galaxy, and their prevalence in other disk galaxies suggests that they last for considerably longer than this. Thus, whatever the spiral arms are, they *cannot* simply be dense star-forming regions orbiting along with the rest of the Galactic disk.

How then do the Galaxy's spiral arms retain their structure over long periods of time in spite of differential rotation? A leading explanation for the existence of spiral arms holds that they are **spiral density waves**—coiled waves of gas compression that move through the Galactic disk, squeezing clouds of interstellar gas and triggering the process of star formation as they go. ∞ (Sec. 11.4) The spiral arms we observe are defined by the denser than normal clouds of gas the density waves create and by the new stars formed as a result of the spiral waves' passage.

This explanation of spiral structure avoids the problem of differential rotation because the wave pattern is not tied to any particular piece of the Galactic disk. The spirals we see are merely patterns moving through the disk, not great masses of matter being transported from place to place. The density wave moves through the collection of stars and gas making up the disk just as a sound wave moves through air or an ocean wave passes through water, compressing different parts of the disk at different times. Even though the rotation rate of the disk material varies with distance from the Galactic center, the wave itself remains intact, defining the Galaxy's spiral arms.

In fact, over much of the visible portion of the Galactic disk (within about 15 kpc of the center), the spiral wave pattern is predicted to rotate *more slowly* than the stars and gas. Thus, Galactic material catches up with the wave, is temporarily slowed down and compressed as it passes through, then continues on its way. (For a more down-to-earth example of an analogous process, see *Interlude 14-1*.)

As shown in Figure 14.15, the slowly moving spiral density wave is outrun by the faster-moving disk material. As gas enters the arm from behind, the gas is compressed and forms stars. Dust lanes mark the regions of highest-density gas. The most prominent stars—the bright O and B blue giants—live for only a short time, so young stellar associations, emission neb-

ulae, and open clusters with long main sequences are found only within the arms, near their birthsites, just ahead of the dust lanes. The brightness of these young systems emphasizes the spiral structure. Further downstream, ahead of the spiral arms, we see mostly older stars and star clusters. These have had enough time since their formation to outdistance the wave and pull away from it. Over millions of years, their random individual motions superimposed on the overall rotation around the Galactic center distort and eventually destroy their original spiral configuration, and they become part of the general disk population.

An alternative possibility is that the formation of stars drives the waves, instead of the other way around. Imagine a row of newly formed massive stars somewhere in the disk. As these stars form, the emission nebulae that appear around them send shock waves through the surrounding gas, possibly triggering new star formation. Similarly, when the stars explode in supernovae, more shocks are formed. ∞ (Sec. 12.4) As illustrated in Figure 14.16(a), the formation of one group of stars thus provides the mechanism for the creation of more stars. Computer simulations suggest that it is possible for the "wave" of star formation thus cre-

Figure 14.15 Density-wave theory holds that the spiral arms seen in our own and many other galaxies are waves of gas compression and star formation moving through the material of the galactic disk. In the painting at right, gas motion is indicated by red arrows, and arm motion is indicated by white arrows. Gas enters an arm from behind, is compressed, and forms stars. The spiral pattern is delineated by dust lanes, regions of high gas density, and newly formed O and B stars. The inset shows the spiral galaxy NGC 1566, which displays many of the features described in the text. Note, incidentally, that although both spirals here have two arms each, astronomers are not completely certain how many arms make up the spiral structure in our own Galaxy (see Figure 14.13). The theory makes no strong predictions on this point.

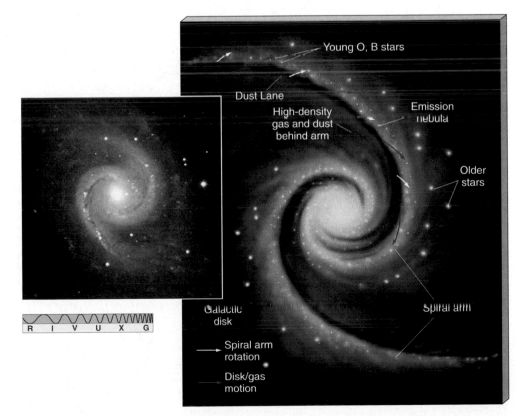

R I V U X G

Young O, B stars

Dust Lane

High-density gas and dust behind arm

Emission nebula

Older stars

Galactic disk

Spiral arm

Spiral arm rotation

Disk/gas motion

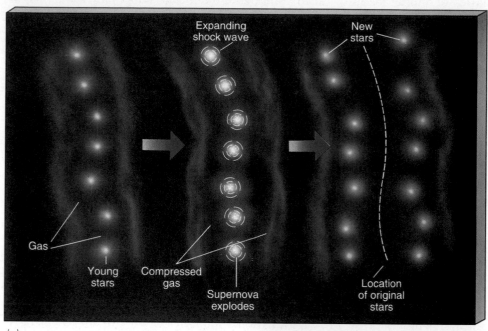

(a)

Figure 14.16 (a) Self-propagating star formation. In this theory of the formation of spiral arms, the shock waves produced by the formation and later evolution of a group of stars provide the trigger for new rounds of star formation. We have used supernova explosions to illustrate the point here, but the formation of emission nebulae and planetary nebulae are also important. (b) This process may well be responsible for the partial spiral arms seen in some galaxies, such as NGC 300, shown here in true color. The distinct blue appearance derives from the vast numbers of young stars that pepper its ill-defined spiral arms.

(b)

R I V U X G

ated to take on the form of a partial spiral and for this pattern to persist for some time. However, this process, sometimes known as *self-propagating star formation*, can produce only pieces of spirals, as are seen in some galaxies (Figure 14.16b). It apparently cannot produce the galaxywide spiral arms seen in other galaxies and present in our own. It may well be that there is more than one process at work in the spectacular spirals we see.

An important question (but one that unfortunately is not answered by either of the two theories just described) is: Where do these spirals come from? What was responsible for generating the density wave in the first place or for creating the line of newborn stars whose evolution drives the advancing spiral arm? Scientists speculate that (1) instabilities in the gas near the Galactic bulge, (2) the gravitational effects of nearby galaxies, or (3) the possible barlike asymmetry within the bulge itself may have had a big enough influence on the disk to get the process going. The fact is that we still don't know for sure how galaxies—including our own—acquire such beautiful spiral arms.

14.6 The Mass of the Milky Way Galaxy

5 We can measure our Galaxy's mass by studying the motions of gas clouds and stars in the Galactic disk. Recall from Chapter 1 that Kepler's third law (as modified by Newton) connects the orbital period, orbit size, and masses of any two objects in orbit around one another: ∞ (Sec. 1.7)

$$\text{total mass (in solar masses)} = \frac{\text{orbit size (in A.U.)}^3}{\text{orbit period (in years)}^2}$$

As we saw in Sections 14.2 and 14.3, the distance from the Sun to the Galactic center is about 8 kpc and the Sun's orbital period is 225 million years. Substituting these numbers into the equation, we obtain a mass of almost 10^{11} solar masses—100 *billion* times the mass of our Sun. The Milky Way Galaxy is truly enormous, in mass as well as in diameter.

But what mass have we just measured? When we performed the analogous calculation in the case of a planet orbiting the Sun (Section 1.7), there was no ambiguity: neglecting the planet's mass, the result of our calculation was the mass of the Sun. However, the Galaxy's matter is not concentrated at the Galactic center (as the Sun's mass is concentrated at the center

of the solar system); instead, Galactic matter is distributed over a large volume of space. Some of it lies inside the Sun's orbit (that is, within 8 kpc of the Galactic center), and some lies outside, at large distances from both the Sun and the center of the Galaxy. What portion of the Galaxy's mass controls the Sun's orbit? Isaac Newton answered this question three centuries ago: the Sun's orbital period is determined by the portion of the Galaxy that lies *within the orbit of the Sun*. This is the mass computed from the foregoing equation.

Dark Matter

To determine the mass of the Galaxy on larger scales—that is, to find how much matter is contained within spheres of progressively larger radii—we must measure the orbital motion of stars and gas farther from the Galactic center than is the Sun. Astronomers have found that the most effective way of doing this is by making radio observations of gas in the Galactic disk because radio waves are relatively unaffected by interstellar absorption and allow us to probe to great distances, far beyond the Sun's orbit. On the basis of these studies, radio astronomers have determined our Galaxy's rotation rate at various distances from the Galactic center. The resultant plot of rotation speed versus distance from the center (Figure 14.17) is called the Galactic **rotation curve**.

Knowing the Galactic rotation curve, we can now repeat our earlier calculation to compute the total mass that lies within any given distance from the Galactic center. We find, for example, that the mass within about 15 kpc from the center—the volume

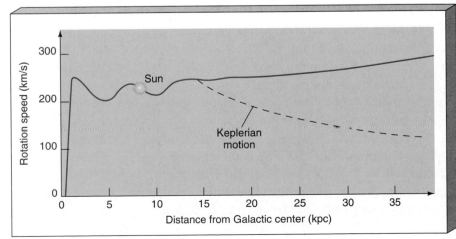

Figure 14.17 The rotation curve for the Milky Way Galaxy plots rotation speed against distance from the Galactic center. We can use this curve to compute the mass of the Galaxy that lies within any given radius. The dashed curve is the rotation curve expected if the Galaxy "ended" abruptly at a radius of 15 kpc, the limit of most of the known spiral structure and the globular cluster distribution. The fact that the red curve does not follow this dashed line, but instead stays well above it, indicates that there must be additional matter beyond that radius.

defined by the globular clusters and the known spiral structure—is roughly 2×10^{11} solar masses, about twice the mass contained within the Sun's orbit. Does most of the matter in the Galaxy "cut off" at this point, where the luminosity drops off sharply? Surprisingly, the answer is no.

If most of the matter in the Galaxy were contained within the edge of the visible structure, Newton's laws of motion predict that the orbital speed of stars and gas beyond 15 kpc would decrease with increasing distance from the Galactic center, just as the orbital speeds of the planets diminish as we move outward from the Sun. The dashed line in Figure 14.17 indicates what the rotation curve would look like in that case. However, the true rotation curve is quite different. Far from falling off at larger distances, it *rises* slightly out to the limits of our measurement capabilities. This implies that the amount of mass contained within successively larger radii continues to grow beyond the orbit of the Sun, apparently out to a distance of at least 40 or 50 kpc. According to the above equation, the amount of mass within 40 kpc is approximately 6×10^{11} solar masses. Since 2×10^{11} solar masses lies within 15 kpc of the Galactic center, we have to conclude that at least twice as much mass lies *outside* the luminous part of our galaxy—the part made up of stars, star clusters, and spiral arms—as lies inside!

Based on these observations of the Galactic rotation curve, astronomers now believe that the luminous portion of the Milky Way Galaxy—the region outlined by the globular clusters and by the spiral arms—is merely the "tip of the Galactic iceberg." Our Galaxy is in reality very much larger. The luminous region is surrounded by an extensive, invisible **dark halo**, which dwarfs the inner halo of stars and globular clusters and extends well beyond the 15-kpc radius once thought to represent the limit of our Galaxy. But what is this dark halo made of? We do not detect enough stars or interstellar matter to account for the mass that our computations tell us must be there. We are inescapably drawn to the conclusion that most of the mass in our Galaxy exists in the form of invisible **dark matter**, which we presently simply do not understand.

The term *dark* here does not refer just to matter undetectable in visible light. The material has (so far) escaped detection at *all* wavelengths, from radio to gamma rays. Only by its gravitational pull do we know of its existence. Dark matter is not hydrogen gas

(atomic or molecular), nor is it made up of ordinary stars. Given the amount of matter that must be accounted for, we would have been able to detect it with present-day equipment if it were in either of those forms. Its nature and its consequences for the evolution of galaxies and the universe are among the most important questions in astronomy today.

Many candidates have been suggested for this dark matter, although none is proven. Among the strongest "stellar" contenders are brown dwarfs—low-mass prestellar objects that never reached the point of core nuclear burning—white dwarfs, and faint, low-mass red dwarfs. ∞ (Sec. 11.5) These objects could in principle exist in great numbers throughout the Galaxy, yet would be exceedingly hard to see. However, recent *Hubble Space Telescope* observations of globular clusters argue against at least the last possibility. The *Hubble* data suggest that there is a cutoff point at about 0.2 solar mass, below which stars rarely form. As a result, very-low-mass stars seem to be unexpectedly scarce in our Galaxy.

A radically different alternative is that the dark matter is made up of exotic *subatomic particles* that pervade the entire universe. Many theoretical astrophysicists believe that these particles could have been produced in abundance during the very earliest moments of our universe. If the particles have survived to the present day, there might be enough of them to account for all the dark matter we believe must be out there. This idea is hard to test, however, because any particles of this nature that might exist would be very hard to detect. Several detection experiments have been attempted, so far without success.

The Search for Stellar Dark Matter

Recently, researchers have obtained insight into the distribution of stellar dark matter by using a key element of Albert Einstein's theory of general relativity (see Chapter 13)—the prediction that a beam of light can be deflected by a gravitational field, which has already been verified in the case of starlight passing close to the Sun. ∞ (*More Precisely 13-2*) Although this effect is small, it has the potential for making otherwise invisible stellar objects observable from Earth. Here's how.

Imagine looking at a distant star as a faint foreground object (such as a brown dwarf or a white dwarf)

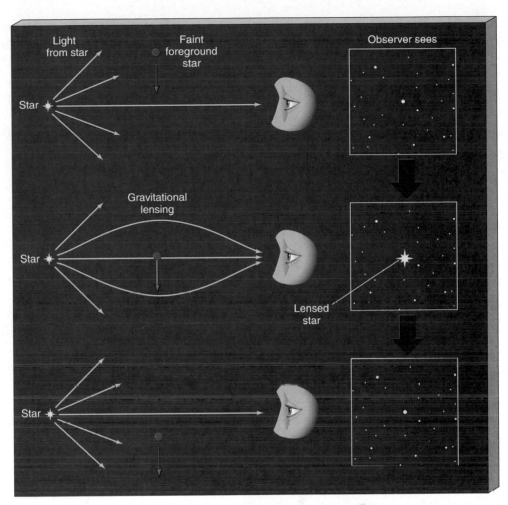

Figure 14.18 Gravitational lensing by a faint foreground object (such as a brown dwarf) can temporarily cause a background star to brighten significantly, providing a means of detecting otherwise invisible stellar dark matter.

happens to cross your line of sight. As illustrated in Figure 14.18, the intervening object deflects a little more starlight than usual toward you, resulting in a temporary, but quite substantial, *brightening* of the distant star. In some ways, the effect is like the focusing of light by a lens, and the process is known as **gravitational lensing**. The foreground object is referred to as a *gravitational lens*. The amount of brightening and the duration of the effect depend on the mass, distance, and speed of the lensing object. Typically, the apparent brightness of the background star increases by a factor of 2 to 5 for a period of several weeks. Thus, even though the foreground object cannot be seen directly, its effect on the light of the background star makes it detectable.

Of course, the probability of one star passing almost directly in front of another, as seen from Earth, is extremely small. But by observing millions of stars every few days over a period of years (using automated telescopes and high-speed computers to reduce the bur-

den of coping with so much data), astronomers have been able to see enough of these events to let them estimate the amount of stellar dark matter in the Galactic halo. The technique represents an exciting new means of probing the structure of our Galaxy. The first lensing events were reported in late 1993. Subsequent observations are consistent with lensing by low-mass white dwarf stars and suggest that such stars could account for at least half of the dark matter inferred from dynamical studies.

Bear in mind, though, that the identity of the dark matter is not necessarily an all-or-nothing proposition. It is perfectly conceivable that more than one type of dark matter exists. For example, it is quite possible that most of the dark matter in the inner (visible) parts of galaxies could be in the form of brown dwarfs and very-low-mass stars, while the dark matter farther out might be primarily in the form of exotic particles. We will return to this perplexing problem in later chapters.

14.7 The Galactic Center

6 Theory predicts that the Galactic bulge, and especially the region close to the Galactic center, should be densely populated with billions of stars. However, we are unable to see this region of our Galaxy—the interstellar medium in the Galactic disk shrouds what otherwise would be a stunning view. Figure 14.19 shows the (optical) view we do have of the region of the Milky Way toward the Galactic center, in the general direction of the constellation Sagittarius.

With the help of infrared and radio techniques, we can peer more deeply into the central regions of our Galaxy than we can with optical techniques. Infrared observations (Figure 14.20a) indicate that the heart of our Galaxy harbors roughly 50,000 stars per cubic par-

Figure 14.19 A photograph of stellar and interstellar matter in the direction of the Galactic center. Because of heavy obscuration, even the largest optical telescopes can see no farther than one-tenth the distance to the center. The M8 nebula (arrow) can be seen at extreme top center. The field is roughly 20°, top to bottom, and is a continuation of the bottom part of Figure 11.4. The circle indicates the location of the center of our Galaxy.

sec. That's a stellar density about a million times greater than in our solar neighborhood, high enough that stars must experience frequent close encounters and even collisions. Infrared radiation has also been detected from what appear to be huge clouds rich in dust. In addition, radio observations indicate a ring of molecular gas nearly 400 pc across, containing some 30,000 solar masses of material and rotating at about 100 km/s. The origin of this ring is unclear, although researchers suspect that the gravitational influence of our Galaxy's elongated, rotating bulge may well be involved. The ring surrounds a bright radio source that marks the Galactic center.

High-resolution radio observations show more structure on small scales. Figure 14.20(b) shows the bright radio source Sagittarius A, which lies at the center of the circle in Figure 14.19 and within the boxed region in Figure 14.20(a), and, we think, at the center of our Galaxy. On a scale of about 100 pc, extended filaments can be seen. Their presence suggests to many astronomers that strong magnetic fields operate in the vicinity of the center, creating structures similar in appearance (but much larger) to those observed on the active Sun. On even smaller scales (Figure 14.20c), the observations indicate a rotating ring or disk of matter only a few parsecs across.

What could cause all this activity? An important clue comes from the Doppler broadening of infrared spectral lines emitted from the central swirling whirlpool of gas. The extent of the broadening indicates that the gas is moving very rapidly. In order to keep this gas in orbit, whatever is at the center must be extremely massive—a million solar masses or more. Given the twin requirements of large mass and small size, a leading contender is a black hole. The hole itself is not the source of the energy, of course. Instead, the vast accretion disk of matter being drawn toward the hole by the enormous gravity emits the energy as it falls in, just as we saw (on a much smaller scale) in Chapter 13 when we discussed X-ray emission from neutron stars and stellar-mass black holes. ∞ (Sec. 13.3, 13.7) The strong magnetic fields are thought to be generated within the accretion disk as matter spirals inward.

Figure 14.21 places these findings into a simplified perspective. Each frame is centered on the Galaxy's core, and each increases in resolution by a factor of 10. Frame (a) renders the Galaxy's overall shape, as painted in Figure 14.13. The scale of this

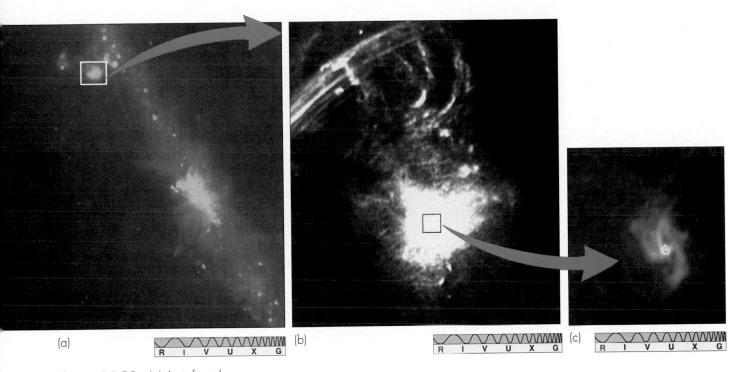

(a)   R I V U X G   (b)   R I V U X G (c)   R I V U X G

Figure 14.20 (a) An infrared image of the region around the center of our Galaxy shows many bright stars packed into a relatively small volume. The average density of matter in this region is estimated to be about a million times that in the solar neighborhood. (b) The central portion of our Galaxy, as seen in the radio part of the spectrum. This image shows a region about 200 pc across surrounding the Galactic center (which lies within the bright blob at bottom right). The long-wavelength radio emission cuts through the Galaxy's dust, providing an image of matter in the immediate vicinity of the Galaxy's center. (c) The spiral pattern of radio emission arising from Sagittarius A, the very center of the Galaxy. The data suggest a rotating ring of matter only 5 pc across.

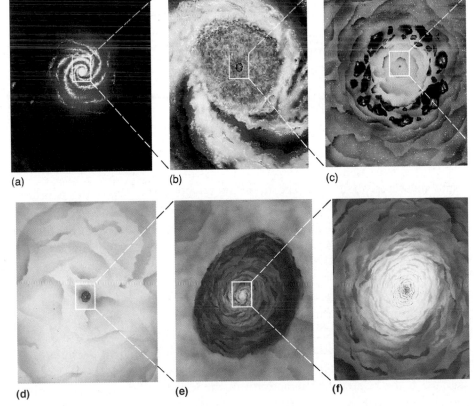

(a)   (b)   (c)

(d)   (e)   (f)

Figure 14.21 Six artist's conceptions, each centered on the Galactic center and each increasing in resolution by a factor of 10. Frame (a) shows the same scene as Figure 14.13. Frame (f) is a rendition of a vast whirlpool within the innermost parsec of our Galaxy. The data imaged in Figure 14.20 do not closely match these artistic renderings because the Figure 14.20 view is parallel to the Galactic disk, whereas these six paintings portray an idealized version perpendicular to that disk.

14-1 INTERLUDE

DENSITY WAVES

In the late 1960s, American astrophysicists C. C. Lin and Frank Shu proposed a way in which spiral arms in the Galaxy could persist for many Galactic rotations. They argued that the arms themselves contain no "permanent" matter. They should not be viewed as assemblages of stars, gas, and dust moving intact through the disk because such structures would quickly be destroyed by differential rotation. Instead, as described in the text, a spiral arm should be envisaged as a *density wave*—a wave of alternating compression and expansion sweeping through the Galaxy.

A wave in water builds up material temporarily in some places (crests) and lets it down in others (troughs). Similarly, as Galactic matter encounters a spiral density wave, the matter is compressed to form a region of slightly higher than normal density. The matter enters the wave, is temporarily slowed down and compressed as it passes through, then continues on its way. This compression triggers the formation of new stars and nebulae. In this way, the spiral arms are formed and reformed repeatedly, without wrapping up. Lin and Shu showed that the process can maintain a spiral pattern for very long periods of time.

The accompanying figure illustrates the formation of a density wave in a much more familiar context—a traffic jam triggered by the presence of a repair crew moving slowly down the road. Cars slow down temporarily as they approach the crew, then speed up again as they pass the worksite and continue on their way. The result observed by a traffic helicopter flying overhead is a region of high traffic density concentrated around the work crew and moving with it. An observer on the side of the road, however, sees that the jam never contains the same cars for very long. Cars constantly catch up to the bottleneck, move slowly through it, then speed up again, only to be replaced by more cars arriving from behind.

The traffic jam is analogous to the region of high stellar density in a Galactic spiral arm. Just as the traffic density wave is not tied to any particular group of cars, the spiral arms are not attached to any particular piece of disk material. Stars and gas enter a spiral arm, slow down for a while, then exit the arm and continue on their way around the Galactic center. The result is a moving region of high stellar and gas density, involving different parts of the disk at different times. Notice also that, just as in our Galaxy, the traffic jam wave moves more slowly than, and independently of, the overall traffic flow.

We can extend our traffic analogy a little further. Most drivers are well aware that the effects of

---

frame measures about 100 kpc from top to bottom. Frame (b) spans a distance of 10 kpc from top to bottom and is nearly filled by the great circular sweep of the innermost spiral arm. Moving in to a 1-kpc span, frame (c) depicts the 400-pc ring of matter mentioned earlier. The dark blobs represent giant molecular clouds, the pink patches emission nebulae associated with star formation within those clouds.

In frame (d), at 100 pc, a pinkish region of ionized gas surrounds the reddish heart of the Galaxy. The source of energy producing this vast ionized cloud is assumed to be related to the activity in the Galactic center. Frame (e), spanning 10 pc, depicts the tilted, spinning whirlpool of hot (10^4 K) gas that marks the center of our Galaxy. The innermost part of this gigantic whirlpool is painted in frame (f), in which a swiftly spinning, white-hot disk of gas with temperatures in the millions of kelvins nearly engulfs a massive black hole too small in size to be pictured (even as a minute dot) on this scale.

If our knowledge of the Galaxy's center seems sketchy, that's because it *is* sketchy. Astronomers are still deciphering the clues hidden within its invisible radiation. We are only beginning to appreciate the full magnitude of this strange new realm deep in the heart of the Milky Way.

a such a tie-up can persist long after the road crew has stopped work and gone home for the night. Similarly, spiral density waves can continue to move through the disk even after the disturbance that originally produced them has long since sub-

sided. According to spiral density wave theory, this is precisely what has happened in the Milky Way Galaxy. Some disturbance in the past produced a wave that has been moving through the Galactic disk ever since.

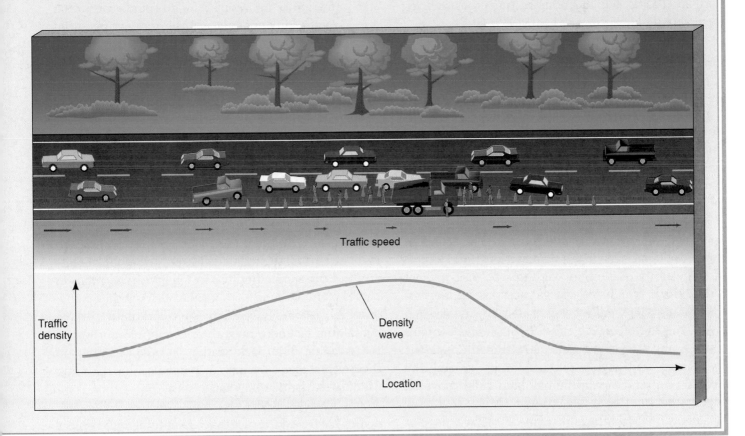

Chapter Review

Summary

A **galaxy** (p. 392) is a huge collection of stellar and interstellar matter isolated in space and bound together by its own gravity. Because we live within it, the **Galactic disk** (p. 392) of our own **Milky Way Galaxy** (p. 392) appears as a broad band of light across the sky, a band called the Milky Way. Near the center, the Galactic disk thickens into the **Galactic bulge** (p. 393). The disk is surrounded by a roughly spherical

Galactic halo (p. 394) of old stars and star clusters. Our Galaxy, like many others visible in the sky, is a **spiral galaxy** (p. 395).

The halo can be studied using **variable stars** (p. 395), whose luminosity changes with time. **Pulsating variable stars** (p. 395) vary in brightness in a repetitive and predictable way. Two types of pulsating variable stars of great importance to astronomers are **RR**

Lyrae variables (p. 395) and **Cepheid** variables (p. 395), whose characteristic light curves make them easily recognizable. All RR Lyrae stars have roughly the same luminosity. For Cepheids, astronomers can determine the luminosity by measuring the pulsation period and using the **period–luminosity relationship** (p. 397), a simple correlation between period and absolute brightness. The brightest Cepheids can be seen at distances of millions of parsecs, extending the cosmic distance ladder well beyond our own Galaxy. RR Lyrae stars are fainter but much more numerous, making them very useful within the Milky Way.

In the early twentieth century, Harlow Shapley used RR Lyrae stars to determine the distances to many of the Galaxy's globular clusters. He found that the clusters have a roughly spherical distribution in space, but the center of the sphere lies far from the Sun. The globular clusters are now known to map out the true extent of the luminous portion of the Milky Way Galaxy. The center of their distribution is close to the **Galactic center** (p. 398), which lies about 8 kpc from the Sun.

Disk and halo stars differ in their spatial distributions, ages, colors, and orbital motion. The luminous portion of our Galaxy has a diameter of about 30 kpc. In the vicinity of the Sun, the Galactic disk is about 300 pc thick. The halo lacks gas and dust, so no stars are forming there. All halo stars are old. The gas-rich disk is the site of current star formation and contains many young stars. Stars and gas within the Galactic disk move on roughly circular orbits around the Galactic center. Stars in the halo and bulge move on largely random three-dimensional orbits that pass repeatedly through the disk plane but have no preferred orientation. Halo stars appeared early on, before the Galactic disk took shape, when there was still no preferred orientation for their orbits. As the gas and dust formed a rotating disk, stars that formed in the disk inherited its overall spin and so moved on circular orbits in the Galactic plane, as they do today.

Attempts to map out the Galactic disk by optical observations are defeated by interstellar absorption. Astronomers use radio observations to explore the Galactic disk because radio waves are largely unaffected by interstellar dust. Regions where most of the hydrogen is in atomic form may be studied using 21-cm radiation. Regions where the gas is mostly molecular are studied through radio molecular emission lines. Gas has been detected in the disk at up to 50 kpc from the Galactic center. Radio observations clearly reveal the extent of our Galaxy's **spiral arms** (p. 403).

The spiral arms in spiral galaxies are regions of the densest interstellar gas and are the places where star formation is taking place. The spirals cannot be "tied" to the disk material, as the disk's differential rotation would have wound them up long ago. Instead, they may be **spiral density waves** (p. 404) that move through the disk, triggering star formation as they pass by. Alternatively, the spirals may arise from self-propagating star formation, when shock waves produced by the formation and evolution of one generation of stars triggers the formation of the next.

The Galactic **rotation curve** (p. 407) plots the orbital speed of matter in the disk against distance from the Galactic center. By applying Newton's laws of motion, astronomers can determine the mass of the Galaxy. They find that the Galactic mass continues to increase beyond the radius defined by the globular clusters and the spiral structure we observe. The rotation curves of our own and other galaxies show that many, if not all, galaxies have an invisible **dark halo** (p. 408) containing far more mass than the visible portion of the galaxies. The **dark matter** (p. 408) making up these dark halos is of unknown composition. Leading candidates include low-mass stars and exotic subatomic particles. Recent attempts to detect stellar dark matter have used the fact that a faint foreground object can occasionally pass in front of a more distant star, deflecting the star's light and causing its apparent brightness to increase temporarily. This deflection is called **gravitational lensing** (p. 409).

Astronomers working at infrared and radio wavelengths have uncovered evidence for energetic activity within a few parsecs of the Galactic center. The leading explanation is that a massive black hole resides at the heart of our Galaxy.

Self-Test: True or False?

_____ **1.** Cepheid variables can be used to determine the distances to the nearest galaxies.

_____ **2.** RR Lyrae stars are a type of cataclysmic variable.

____ 3. The Galactic halo contains about as much gas and dust as the Galactic disk.

____ 4. The Galactic disk contains only old stars.

____ 5. Population I objects are found only in the Galactic halo.

____ 6. Up until the 1930s, the main error made in determining the size of the Galaxy was due to an incorrectly calibrated method of determining stellar distances.

____ 7. Astronomers use 21-cm radiation to study Galactic molecular clouds.

____ 8. Radio techniques are capable of mapping the entire Galactic disk.

____ 9. In the neighborhood of the Sun, the Galaxy's spiral density waves rotate more slowly than the overall Galactic rotation.

____ 10. The mass of the Galaxy is determined by counting stars.

____ 11. Dark matter is now known to be due to large numbers of black holes.

____ 12. A million-solar-mass black hole could account for the unusual properties of the Galactic center.

____ 13. More than 90 percent of the mass of our Galaxy exists in the form of dark matter.

Self-Test: Fill in the Blank

1. One difficulty in studying our Galaxy in its entirety is that we live _____ it.

2. The highly flattened, circular part of the Galaxy is called the Galactic _____.

3. The roughly spherical region of faint old stars in which the rest of the Galaxy is embedded is the Galactic _____.

4. RR Lyrae stars are observed to vary in _____.

5. The pulsation periods for Cepheids range from _____ to _____.

6. According to the period–luminosity relationship, the longer the pulsation period of a Cepheid, the _____ its luminosity.

7. Harlow Shapley used _____ to determine distances to the globular clusters.

8. The Sun lies roughly _____ pc from the Galactic center.

9. The orbital speed of the Sun around the Galactic center is _____.

10. The direction of motion of halo objects at any particular location is _____.

11. The original cloud of gas from which the Galaxy formed probably had a size and shape similar to those of the present Galactic _____.

12. Rotational speeds in the outer part of the Galaxy are _____ than would be expected on the basis of observed stars and gas, indicating the presence of _____.

Review and Discussion

1. What are spiral nebulae? How did they get that name?

2. How are Cepheid variables used in determining distances?

3. Roughly how far out into space can we use Cepheids to measure distance?

4. What important discoveries were made early in this century using RR Lyrae variables?

5. Why can't we study the central regions of the Galaxy using optical telescopes?

6. Of what use is radio astronomy in the study of Galactic structure?

7. Contrast the motions of disk and halo stars.

8. Explain why galactic spiral arms are believed to be regions of recent and ongoing star formation.

9. Describe the motion of interstellar gas as it passes through a spiral density wave.

10. What is self-propagating star formation?

11. What do the red stars in the Galactic halo tell us about the history of the Milky Way Galaxy?

12. What does the rotation curve of our Galaxy tell us about its total mass?

13. What evidence is there for dark matter in the Galaxy?

14. What are some possible explanations for dark matter?

15. Why do astronomers believe that a supermassive black hole lies at the center of the Milky Way Galaxy?

Problems

1. Using the *Hubble Space Telescope* an observer can see a star like the Sun at a distance of 100,000 pc. The brightest Cepheids are 10,000 times the luminosity of the Sun. How far away can *HST* see these Cepheids?

2. Calculate the total mass of the Galaxy within 20 kpc of the center if the rotation speed at that radius is 240 km/s.

3. A density wave made up of two spiral arms is moving through the Galactic disk. At the orbit of the Sun, the wave's speed is 120 km/s. Assuming that the Sun's speed is 220 km/s, calculate how many times the Sun has passed through a spiral arm since the Sun formed 4.5 billion years ago.

Project

If you are far from city lights, look for a hazy band of light arching across the sky. This is our edgewise view of the Milky Way Galaxy. The Galactic center is located in the direction of the constellation Sagittarius, highest in the sky during the summer but visible from spring through fall. Look at the band making up the Milky Way and notice dark regions; these are relatively nearby dust clouds. Sketch what you see. Look for faint fuzzy spots in the Milky Way and note their positions in your sketch. Draw in the major constellations for reference. Compare your sketch with a map of the Milky Way in a star atlas. How many dust clouds did you discover? Can you identify the faint fuzzy spots?

15 Normal Galaxies

The Large-Scale Structure of the Universe www

(Opposite page) This stunning image takes us far out into space, and far back into time. Known as the Hubble Deep Field, it shows thousands of galaxies of many shapes and colors. By "deep," we mean dim and distant—for this is the faintest photograph ever made; it reaches 30th magnitude, or about four billion times fainter than can be seen by the human eye. To create it, the *Hubble* telescope exposed its electronic detectors for about a hundred hours, over the course of ten days. Astronomers have likened this photo to a geologist's "core sample" of the Earth's crust; it covers an area only about one one-hundredth that of the full Moon. The big ground-based Keck telescope is finding that the small blue shards are among the most distant objects ever seen; perhaps galaxies caught in the act of formation. In all, the image implies that there are about 40 billion galaxies in the observable universe— close to what astronomers had estimated before the image was taken.

LEARNING GOALS

Studying this chapter will enable you to:

1 Describe the basic properties of the main types of normal galaxies.

2 Discuss the distance-measurement techniques that enable astronomers to map the universe beyond our Milky Way.

3 Summarize what is known about the large-scale distribution of galaxies in the universe.

4 Describe some of the methods used to determine the masses of distant galaxies.

5 Explain why astronomers think that most of the matter in the universe is invisible.

6 Discuss some theories of how galaxies form and evolve.

7 State Hubble's law and explain how it is used to derive distances to the most remote objects in the observable universe.

Much of our knowledge of the workings of our own Galaxy is based on observations of other galaxies. We know of literally millions of galaxies beyond our own—many smaller than it, some comparable in size, a few much larger. All are vast, gravitationally bound assemblage of stars, gas, dust, dark matter, and radiation, separated from us by almost incomprehensibly large distances. Even a modest-sized galaxy harbors more stars than the number of people who have ever lived on Earth. The light we receive tonight from the most distant galaxies was emitted long before Earth existed. By comparing and classifying the properties of galaxies, astronomers have begun to understand their complex dynamics. By mapping out their distribution in space, astronomers trace out the immense realms of the universe. The galaxies remind us that our position in the universe is no more special than that of a boat adrift at sea.

15.1 Hubble's Galaxy Classification

1 Figure 15.1 shows a vast expanse of space lying about 100 million pc from Earth. Almost every patch or point of light in this figure is a separate galaxy—several hundred can be seen in just this one photograph. Over the years, astronomers have accumulated similar images of many millions of galaxies. We begin our study of these enormous accumulations of matter simply by considering their appearance on the sky.

Seen through even a small telescope, images of galaxies look distinctly nonstellar. They have fuzzy edges, and many are quite elongated—not at all like the sharp, pointlike images normally associated with stars. Although it is difficult to tell from the photograph, some of the blobs of light in Figure 15.1 are actually spiral galaxies like the Milky Way Galaxy and Andromeda. Others, however, are definitely not spirals—no disks or spiral arms can be seen. Even when we take into account their different orientations in space, galaxies do *not* all look the same.

The American astronomer Edwin Hubble was the first to categorize galaxies in a comprehensive way. Working with the then recently completed 2.5-m optical telescope on Mount Wilson in California in 1924, he classified the galaxies he saw into four basic types—*spirals*, *barred spirals*, *ellipticals*, and *irregulars*—solely on the basis of appearance. Many modifications and refinements have been incorporated over the years, but the basic **Hubble classification scheme** is still widely used today.

Spirals

We saw several examples of **spiral galaxies** in Chapter 14—for example, our own Milky Way Galaxy and our neighbor Andromeda. All galaxies of this type contain a flattened galactic disk in which spiral arms are found, a central galactic bulge, and an extended halo of faint, old stars. ∞ (Sec. 14.3) The stellar density (that is, the number of stars per unit volume) is greatest in the galactic nucleus, at the center of the bulge. However, within this general description, spiral galaxies exhibit a wide variety of shapes, as illustrated in Figure 15.2.

In Hubble's scheme, a spiral galaxy is denoted by the letter S and classified as type "a," "b," or "c" according to the size of its central bulge—Type Sa galaxies having the largest bulges, Type Sc the smallest. The size of the bulge is quite well correlated with the tightness of the spiral pattern (although the correspondence is not perfect). Type Sa spiral galaxies, with large central bulges, tend to have tightly wrapped, almost circular, spiral arms. Type Sb galaxies, with smaller bulges, typically have more open spiral arms. Type Sc spirals, with the smallest bulges, often have loose, poorly defined spiral structure. The arms also tend to become more "knotty," or clumped, in appearance as the spiral pattern becomes more open.

Much of our description of the large-scale structure of the Milky Way Galaxy in Chapter 14 applies to spiral galaxies in general. The bulges and halos of spiral galaxies contain large numbers of reddish old stars and globular clusters, similar to those observed in our own Galaxy and in Andromeda. Most of the light from spirals, however, comes from A- through G-type stars in the disk, giving these galaxies an overall whitish

(a)

R I V U X G

(b)

Figure 15.1   (a) A collection of many galaxies, each consisting of hundreds of billions of stars. Called the Coma Cluster, this group of galaxies lies more than 100 million pc from Earth. (The blue spiked object at top right is a nearby star; virtually every other object visible is a galaxy.) (b) A recent *Hubble* image of part of the Coma Cluster.

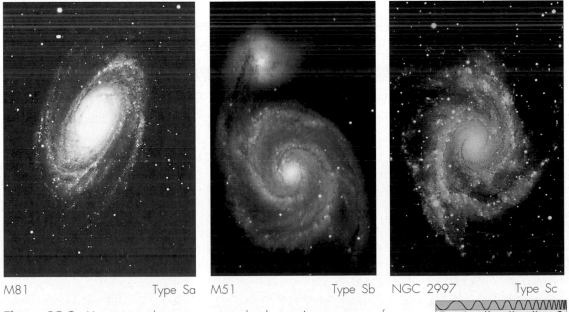

M81   Type Sa   M51   Type Sb   NGC 2997   Type Sc

R I V U X G

Figure 15.2   Variation in shape among spiral galaxies. As we progress from type Sa to Sb to Sc, the bulges become smaller while the spiral arms tend to become less tightly wound.

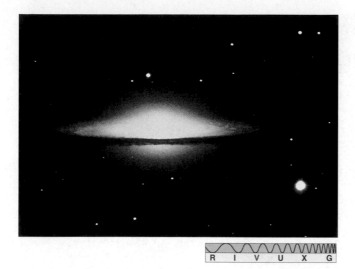

R I V U X G

Figure 15.3 The Sombrero Galaxy, a spiral system seen edge-on. Officially cataloged as M104, this galaxy has a dark band composed of interstellar gas and dust. The large size of this galaxy's central bulge marks it as type Sa, even though its spiral arms cannot be seen.

glow. The spiral arms appear bluish because of the presence of bright blue O and B stars there.

Like the disk of the Milky Way, the flat galactic disks of typical spiral galaxies are rich in gas and dust. The 21-cm radio radiation emitted by spirals betrays the presence of interstellar gas, and detailed photographs clearly show obscuring dust in many systems. Stars are forming within the spiral arms, where the interstellar medium is densest, and the arms contain numerous emission nebulae and newly formed O and B stars, as discussed in Chapter 14. ∞ (Sec. 14.5) Type Sc galaxies contain the most interstellar gas and dust, Sa galaxies the least. The Sc galaxy NGC 2997 shown in Figure 15.2(c) clearly shows the preponderance of interstellar gas, dust, and young blue stars tracing the

spiral pattern. Spirals are not necessarily young galaxies, however. Like our own Galaxy, they are simply rich enough in interstellar gas to provide for continued stellar birth.

Most spirals are not seen face-on as in Figure 15.2. Many are tilted with respect to our line of sight, making their spiral structure hard to discern, However, we do not need to see spiral arms to classify a galaxy as a spiral. The presence of the disk, with its gas, dust, and newborn stars, is sufficient. For example, the galaxy shown in Figure 15.3 is classified as a spiral because of the clear line of obscuring dust seen along its midplane.

A variation of the spiral category in Hubble's classification scheme is the **barred-spiral galaxy**. The barred spirals differ from ordinary spirals mainly by the presence of an elongated "bar" of stellar and interstellar matter passing through the center and extending beyond the bulge, into the disk. The spiral arms project from near the ends of the bar rather than from the bulge (as they do in normal spirals). Barred spirals are designated by the letters SB and are subdivided, like the ordinary spirals, into categories SBa, SBb, and SBc, depending on the size of the bulge. Again like ordinary spirals, the tightness of the spiral pattern is correlated with bulge size. Figure 15.4 shows the variation among barred-spiral galaxies. In the case of the SBc category, it is often hard to tell where the bar ends and the spiral arms begin.

Astronomers often cannot distinguish between spirals and barred spirals, especially when a galaxy happens to be oriented with its galactic plane nearly edge-on toward Earth, as in Figure 15.3. Because of the physical and chemical similarities of spiral and barred-spiral galaxies, some researchers do not even bother to distinguish between them. Other researchers, however,

NGC 3992   Type SBa

NGC1433   Type SBb

NGC1300   Type SBc

R I V U X G

Figure 15.4 Variation in shape among barred-spiral galaxies. The variation from SBa to SBc is similar to that for the spirals in Figure 15.2, except that now the spiral arms begin at either end of a bar through the galactic center.

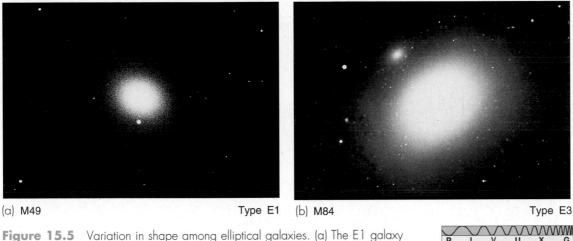

(a) M49   Type E1   (b) M84   Type E3

Figure 15.5 Variation in shape among elliptical galaxies. (a) The E1 galaxy M49 is nearly circular in appearance. (b) M84 is a slightly more elongated elliptical galaxy. It is classified as E3. Both these galaxies lack spiral structure, and neither shows evidence of interstellar matter.

R   I   V   U   X   G

regard the differences in their structures as very important, arguing that these differences suggest basic dissimilarities in the conditions that led to the formation of the galaxies eons ago. The recent findings that the bulge of our own Galaxy is elongated suggest that the Milky Way is a barred spiral, of type SBb or SBc. ∞ (Sec. 14.3)

Ellipticals

The next major category in the Hubble scheme contains the **elliptical galaxies**. Unlike the spirals, ellipticals have no spiral arms and, in most cases, no flattened galactic disk—in fact, they often exhibit little internal structure of any kind. As with spirals, the stellar density increases sharply in the central nucleus. Ellipticals range in shape from highly elongated to nearly circular in appearance. Denoted by the letter E, these systems are subdivided according to how elliptical they are. The most circular are designated E0, slightly flattened systems are labeled E1, and so on, all the way to the most elongated ellipticals, of type E7 (Figure 15.5).

There is a large range in both the size and the number of stars contained in elliptical galaxies. The largest elliptical galaxies are much larger than our own Milky Way Galaxy. These *giant ellipticals* can range up to a few megaparsecs across and contain trillions of stars. At the other extreme, *dwarf ellipticals* may be as small as 1 kpc in diameter and contain fewer than a million stars. The substantial observational differences between giant and dwarf ellipticals have led many

astronomers to conclude that these galaxies are members of separate classes, with quite different formation histories and stellar content. The dwarfs are by far the most common type of ellipticals, outnumbering their brighter counterparts by about 10 to 1. However, most of the *mass* that exists in the form of elliptical galaxies is contained in the larger systems.

Lack of spiral arms is not the only difference between spirals and ellipticals. Most ellipticals also contain little or no gas and dust. The 21-cm radio emission from neutral hydrogen gas is, with few exceptions, completely absent, and no obscuring dust lanes are seen. In most cases, there is no evidence of young stars or of any ongoing star formation. Like the halo of our own Galaxy, ellipticals are made up mostly of old, reddish, low-mass stars. Indeed, having no disk, gas, or dust, elliptical galaxies are, in a sense, "all halo." Again like the halo of our Galaxy, the orbits of stars in ellipticals are disordered, exhibiting little or no overall rotation; objects move in all directions, not in regular, circular paths as in our Galaxy's disk. Apparently all, or nearly all, of the interstellar gas within elliptical galaxies was swept up into stars (or out of the galaxy) long ago, before a disk had a chance to form, leaving stars in random orbits, with no loose gas and dust for the creation of future stellar generations.

Some giant ellipticals are exceptions to many of the foregoing general statements about elliptical galaxies, as they have been found to contain disks of gas and dust in which stars are forming. Astronomers speculate that these galaxies may really be otherwise "normal" ellipticals that have collided and merged with a companion spiral system (although this explanation is not

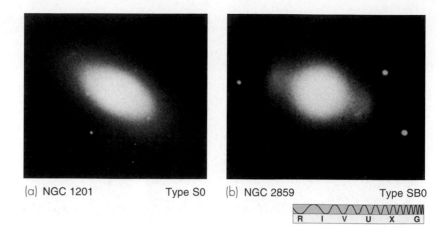

Figure 15.6 (a) S0 galaxies contain a disk and a bulge but no interstellar gas and no spiral arms. They are in many respects intermediate between E7 ellipticals and Sa spirals in their properties. (b) SB0 galaxies are similar to S0 galaxies, except for a bar of stellar material extending beyond the central bulge.

(a) NGC 1201   Type S0   (b) NGC 2859   Type SB0

R I V U X G

yet universally accepted). We will return later in this chapter to the important role played by galactic collisions in determining the appearance of the systems we observe today.

Intermediate between the E7 ellipticals and the Sa spirals in the Hubble classification is a class of galaxies that show evidence of a thin disk and a flattened bulge but that contain no gas and no spiral arms. Two such objects are shown in Figure 15.6. These galaxies are known as **S0 galaxies** if no bar is evident and **SB0 galaxies** if a bar is present. They look a little like spirals whose dust and gas have been stripped away, leaving behind a stellar disk. Observations in recent years have shown that many normal elliptical galaxies have faint disks within them, like the S0 galaxies. As with the S0s, the origin of these disks is uncertain, but some researchers suspect that the S0s and ellipticals actually form a continuous sequence, along which the bulge-to-disk ratio varies smoothly.

Irregulars

The final galaxy class identified by Hubble is a catch-all category—**irregular galaxies**—so named because their visual appearance does not allow us to place them into any of the other categories just discussed. Irregulars tend to be rich in interstellar matter and young, blue stars, but they lack any regular structure, such as well-defined spiral arms or central bulges. They are divided into two subclasses—Irr I galaxies and Irr II galaxies. The Irr I galaxies often look like misshapen spirals. The much rarer Irr II galaxies, in addition to their irregular shape, have other peculiarities, often exhibiting a distinctly explosive or filamentary appearance. Figure 15.7 shows some examples of these

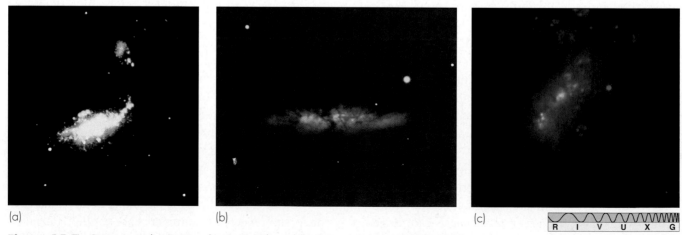

(a)   (b)   (c)

R I V U X G

Figure 15.7 Some irregular (Irr II) galaxies. (a) The oddly shaped galaxies NGC 4485 and NGC 4490 may be close to one another and interacting gravitationally. (b) The galaxy M82 seems to show an explosive appearance, although interpretations remain uncertain. (c) Many irregular galaxies are small and dim, but this one, NGC 4449, is comparable in both size and luminosity to the Milky Way Galaxy.

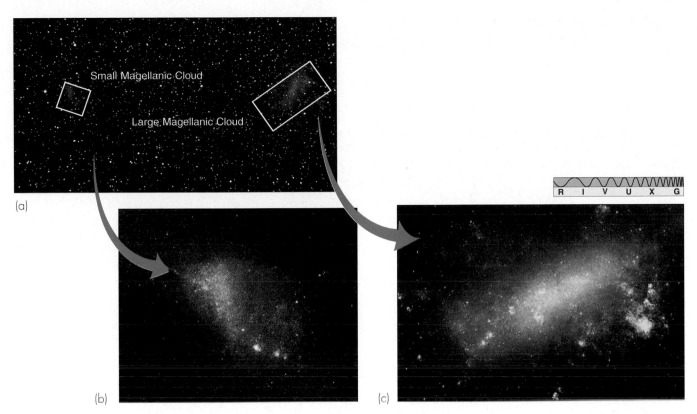

R I V U X G

Figure 15.8 The Magellanic Clouds are prominent members of the sky in the Southern Hemisphere. Named for the sixteenth-century Portuguese explorer Ferdinand Magellan, whose round-the-world expedition first brought word of these fuzzy patches of light to Europe, they are dwarf irregular (Irr I) galaxies, gravitationally bound to our own Milky Way Galaxy. They orbit our Galaxy and accompany it on its trek through the cosmos. (a) The Clouds' relationship to one another in the southern sky. Both the Small (b) and the Large Magellanic Cloud (c) have distorted, irregular shapes, although some observers claim they can discern a single spiral arm in the Large Cloud.

strangely shaped galaxies. Their appearance once led astronomers to suspect that violent events had occurred within them. However, it now seems more likely that in some (but probably not all) cases, we are seeing the result of a close encounter, or even a collision, between two previously "normal" systems.

Irregular galaxies tend to be smaller than spirals but somewhat larger than dwarf ellipticals. They typically contain between 10^8 and 10^{10} stars. The smallest are called *dwarf irregulars*. As with elliptical galaxies, the dwarf type is the most common irregular. Dwarf ellipticals and dwarf irregulars occur in approximately equal numbers and together make up the vast majority of galaxies in the universe. They are often found close to a larger "parent" galaxy.

Figure 15.8 shows the **Magellanic Clouds,** a famous pair of Irr I galaxies that orbit the Milky Way Galaxy. They are shown to proper scale in Figure 14.13. Studies of Cepheid variables within the Clouds show them to be approximately 50 kpc from the center of our Galaxy. The Large Cloud contains about 6 billion solar masses of material and is a few kiloparsecs across. Both Magellanic Clouds contain lots of gas, dust, and blue stars (and the recent, well-documented supernova discussed in *Interlude 12-1*), indicating youthful activity and presumably current star formation. Both also contain many old stars and several old globular clusters, so we know that star formation has been going on there for a very long time.

Table 15.1 summarizes the basic characteristics of the various galaxy types. When he first developed his classification scheme, Hubble arranged the galaxy types into the "tuning fork" diagram shown in Figure 15.9. His aim in doing this was simply to indicate similarities in appearance among galaxies, not to suggest that any connection might exist, but some astronomers have since speculated that perhaps the diagram has deeper significance. Does some sort of evolutionary

Table 15.1 Basic Galaxy Properties by Type

	Spiral/Barred Spiral (S, SB)	Elliptical (E)	Irregular (Irr)
Shape and structural properties	Highly flattened disk of stars and gas, containing spiral arms and thickening to central bulge. Sa and SBa galaxies have largest bulges, the least obvious spiral structure, and roughly spherical stellar halos. SB galaxies have an elongated central "bar" of stars and gas.	No disk. Stars smoothly distributed through an ellipsoidal volume ranging from nearly spherical (E0) to very flattened (E7) in shape. No obvious substructure other than a dense central nucleus.	No obvious structure. Irr II galaxies often have "explosive" appearance.
Stellar content	Disks contain both young and old stars; halos consist of old stars only.	Contain old stars only	Contain both young and old stars.
Gas and dust	Disks contain substantial amounts of gas and dust; halos contain little of either.	Contain little or no gas and dust.	Very abundant in gas and dust.
Star formation	Ongoing star formation in spiral arms.	No significant star formation during the last 10 billion years.	Vigorous ongoing star formation.
Stellar motion	Gas and stars in disk move in circular orbits around the galactic center; halo stars have random orbits in three dimensions.	Star have random orbits in three dimensions.	Stars and gas have very irregular orbits.

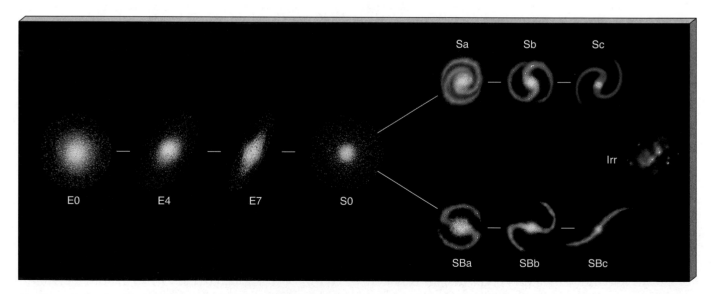

Figure 15.9 Hubble's "tuning fork" diagram, showing his basic galaxy classification scheme. The placement of the four basic galaxy types—ellipticals, spirals, barred spirals, and irregulars—in the diagram is suggestive, but no "evolutionary track" along the sequence (in either direction) is proven.

connection exist among galaxies of different types? The answer, as best we can tell, is no. Isolated normal galaxies do *not* evolve from one type to another. Spirals are not ellipticals with arms, nor are ellipticals spirals that have somehow lost their star-forming disks. In short, astronomers know of no parent-child relationship among normal galaxies. However, the subject of galaxy evolution is still poorly understood.

15.2 The Distribution of Galaxies in Space

3 Now that we have seen some of their basic properties, let us ask how galaxies are spread through the expanse of the universe beyond the Milky Way Galaxy. Are they scattered everywhere throughout intergalactic space all the way out to the very limits of the observable universe, or is there a boundary beyond which galaxies no longer exist? To begin to answer this question, we must first know the *distances* to the galaxies.

Extending the Distance Scale

2 Astronomers estimate that some 100 billion galaxies exist in the observable universe. Although some reside close enough for the Cepheid variable technique to work (within about 15 Mpc; see Figure 15.11), most known galaxies lie much farther away. ∞ (Sec. 14.2) Cepheid variable stars in very distant galaxies simply cannot be observed well enough, even through the world's most sensitive telescopes, to allow us to measure their luminosity and period. To extend our distance-measurement ladder, therefore, we must find some new object to study. What individual objects are bright enough for us to observe at great distances?

One way in which researchers have tackled this problem is through observations of **standard candles**—easily recognizable astronomical objects whose luminosities are confidently known. The basic idea is very simple. Once an object is identified as a standard candle—by its appearance or by the shape of its light curve, say—its luminosity can be estimated. Comparison of this luminosity with the object's apparent brightness then yields the object's distance and hence the distance to the galaxy in which it resides. Note that

the Cepheid variable technique relies on identical reasoning; however, the term *standard candle* tends to be applied only to very bright objects. To be most useful, a standard candle must (1) have a narrowly defined luminosity, so that the uncertainty in estimating its brightness is small and (2) be bright enough to be seen at large distances.

Over the years, astronomers have explored the use of many types of object as standard candles—novae, emission nebulae, planetary nebulae, globular clusters, Type I (carbon-detonation) supernovae, even entire galaxies have been employed. Not all have been equally useful, however; some have larger intrinsic spreads in their luminosities than others, making them less useful for distance-measuring purposes. Planetary nebulae and Type I supernovae have proved particularly reliable. The latter have remarkably consistent peak luminosities and are bright enough to be identified out to distances of hundreds of megaparsecs.

An important alternative to standard candles was discovered in the 1970s when astronomers found a close correlation between the rotational speeds and the luminosities of spiral galaxies within a few tens of megaparsecs of the Milky Way Galaxy. Rotation speed is a measure of a spiral galaxy's total mass, so it is perhaps not surprising that this property is related to luminosity. ∞ (Sec. 14.6) What *is* surprising is how tight the correlation is. The **Tully-Fisher relation**, as it is now known (after its discoverers), allows us to obtain a remarkably accurate estimate of a spiral galaxy's luminosity simply by observing how fast it rotates. As usual, comparing the galaxy's (known) luminosity with its (observed) apparent brightness yields its distance.

To see how the method is used, imagine we are looking edge-on at a distant spiral galaxy and observing one particular emission line, as illustrated in Figure 15.10. Radiation from the side of the galaxy where matter is generally approaching us is blueshifted by the Doppler effect. Radiation from the other side, which is receding from us, is redshifted by a similar amount. The overall effect is that line radiation from the galaxy is "smeared out," or broadened by the galaxy's rotation. The faster the rotation, the greater the amount of broadening. By measuring the amount of broadening, we can determine the galaxy's rotation speed. Once we know that, the Tully–Fisher relation tells us the galaxy's luminosity.

The Tully–Fisher relation can be used to measure distances to spiral galaxies out to about 200 Mpc,

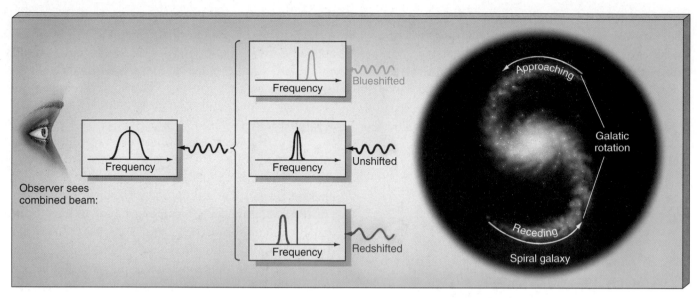

Figure 15.10 A galaxy's rotation causes some of the radiation it emits to be blueshifted and some to be redshifted (relative to what the emission would be from an unmoving source). From a distance, when the radiation from the galaxy is combined into a single beam and analyzed spectroscopically, the redshifted and blueshifted components combine to produce a broadening of the galaxy's spectral lines. The amount of broadening is a direct measure of the rotation speed of the galaxy.

beyond which the line broadening becomes increasingly difficult to measure accurately. A somewhat similar connection exists for elliptical galaxies, linking the broadening of a galaxy's spectral lines (which, in the case of an elliptical, measures the average random velocity of the stars in the galaxy) and the galaxy's size. By measuring the broadening, astronomers can deter-

mine the galaxy's true size, which is then compared with the apparent size to give the distance. ∞ (Sec. P.3) These methods bypass many of the standard candles just described and so provide independent means of determining distances to faraway objects.

Standard candles and Tully–Fisher share the fifth rung of our cosmic distance ladder (Figure 15.11).

Figure 15.11 An inverted pyramid summarizes the distance techniques used to study different realms of the universe. The techniques shown in the bottom four layers—radar ranging, stellar parallax, spectroscopic parallax, and variable stars—take us as far as the nearest galaxies. To go farther, we must use new techniques, the method of standard candles and the Tully–Fisher relation, each based on distances determined by the four lowest techniques.

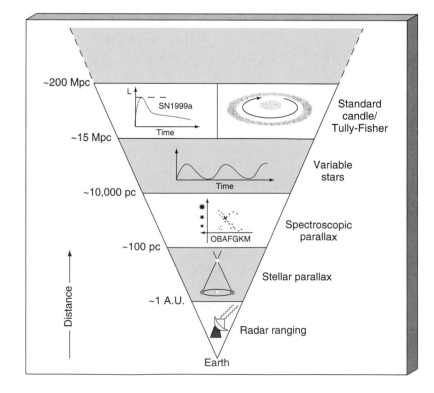

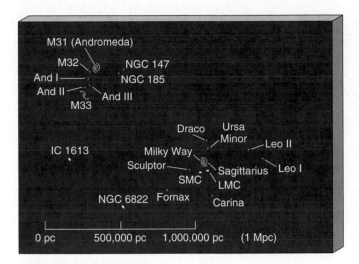

Figure 15.12 Diagram of the Local Group, made up of some 20 galaxies within approximately 1 Mpc of our Milky Way Galaxy. Only a few are spirals; most of the rest are dwarf elliptical or irregular galaxies. Spirals are shown in blue, ellipticals in pink, and irregulars in white.

Just as with the lower rungs, we calibrate the properties of these new techniques using distances measured by more local means. In this way, the distance-measurement process "bootstraps" itself to greater and greater distances. However, at the same time, the errors and uncertainties in each step accumulate, so the distances to the farthest objects are the least well known.

Galaxy Clusters

Figure 15.12 sketches the locations of all the known major astronomical objects within about 1 Mpc of the Milky Way Galaxy. Our Galaxy appears with its three known satellite galaxies—the two Magellanic Clouds and another recently discovered companion (still unnamed, but designated "Sagittarius" in the figure) lying almost within our own Galactic plane. The Andromeda Galaxy, lying 900 kpc from us, is also shown. Two galactic neighbors of Andromeda are shown in Figure 15.13; M33 is a spiral, while M32 is a dwarf elliptical, easily seen in Figure 14.2(a) to the bottom right of Andromeda's central bulge.

All told, some 20 galaxies populate our Galaxy's neighborhood. Three of them (the Milky Way, Andromeda, and M33) are spirals; the remainder are dwarf irregular and ellipticals. Together, these galaxies form the **Local Group**—a new level of structure in the universe above the scale of our Galaxy. As indicated in Figure 15.12, the Local Group's diameter is roughly 1 Mpc. The Milky Way Galaxy and Andromeda are by far its largest members. The combined gravity of the galaxies in the Local Group binds them together, like stars in a star cluster, but on a millionfold larger scale. More generally, a group of galaxies held together by their mutual gravitational attraction is called a **galaxy cluster**.

Moving beyond the Local Group, the next large concentration of galaxies we come to is the Virgo Cluster (Figure 15.14). It lies about 15 Mpc from the Milky Way Galaxy. Like the Local Group, the Virgo Cluster is held together by the mutual gravitational attraction of its member systems. Unlike the Local Group, however, the Virgo Cluster does not contain a mere 20 galaxies. Instead, it houses approximately 2500 galaxies, each containing 100 billion or so indi-

Figure 15.13 Two well-known neighbors of the Andromeda Galaxy (M31): (a) the spiral galaxy M33 and (b) the dwarf elliptical galaxy M32 (also visible in Figure 14.2a, a larger-scale view of the Andromeda system).

R I V U X G

Figure 15.14 The central region of the Virgo Cluster of galaxies, about 15 Mpc from Earth. Several large spiral and elliptical galaxies can be seen. The galaxy near the center is a giant elliptical known as M86.

vidual stars. Those galaxies are bound together in a tightly knit group about 3 Mpc across. Figure 15.15 illustrates the approximate locations of Virgo and several other well-defined clusters in our cosmic neighborhood, lying within about 30 Mpc of the Milky Way Galaxy.

Many thousands of galaxy clusters have now been identified and cataloged, and they come in many shapes and sizes. Large, "rich" clusters, like Virgo,

contain many thousands of individual galaxies distributed fairly smoothly in space. Small clusters, such as the Local Group, contain only a few galaxies and are quite irregular in shape. A small fraction of galaxies are not members of any cluster. They are apparently isolated systems, moving alone through intercluster space.

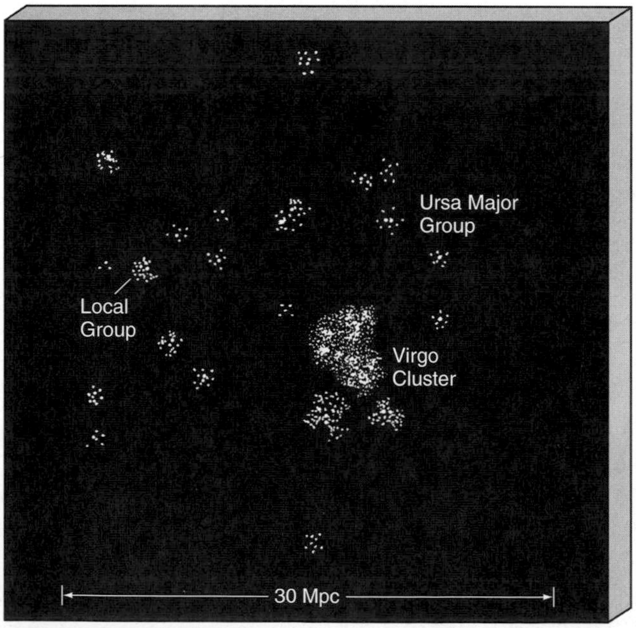

Clusters of Clusters

Does the universe have even greater groupings of matter, or do galaxy clusters top the cosmic hierarchy? Most astronomers believe that the galaxy clusters themselves are clustered, forming titanic agglomerations of matter known as **superclusters**.

Figure 15.16 shows the *Local Supercluster*, containing the Local Group, the Virgo Cluster, and most of the other clusters shown in Figure 15.15. (The direction of view is roughly perpendicular to that shown in the previous figure, so many of the clusters noted there are difficult to discern here.) Each point represents a galaxy, and the diagram is centered on the Milky Way Galaxy. The perspective is such that the disk of our Galaxy is seen edge-on and runs vertically up the page. The two nearly empty V-shaped regions at the top and bottom of the figure are not devoid of galaxies—they are simply obscured from view by the dust in our own Galaxy's plane. The total mass of the entire Local Supercluster is about 10^{15} solar masses. By now it should come as no surprise that the Local Group is not found at its heart, which lies within the Virgo Cluster.

Figure 15.15 Schematic diagram of the location of several galaxy clusters in our part of the universe. Our Milky Way Galaxy is just one of these dots, and our Local Group just one of the clusters.

Ursa Major Group

Local Group

Virgo Cluster

30 Mpc

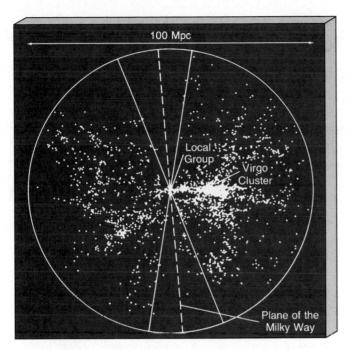

Figure 15.16 The Local Supercluster. Each of the 2200 points shown represents a galaxy, and the Milky Way Galaxy is at the center of the diagram. The Virgo Cluster and the plane of our own Galaxy are marked. Our Galaxy is seen edge-on. Its dust obscures our views to the top and the bottom, resulting in two empty V-shaped regions on the map. The circle shown here is about 100 Mpc across.

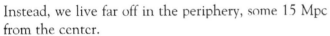

Figure 15.17 The galaxy cluster CL 0939+4713 contains huge numbers of galaxies and resides roughly a billion parsecs from Earth. Every patch of light in this photograph is a separate galaxy. Thanks to the high resolution of the *Hubble Space Telescope*, we can now discern, even at this great distance, spiral structure in some of the galaxies. We also see many galaxies in collision—some tearing matter from one another, others merging into single systems.

Instead, we live far off in the periphery, some 15 Mpc from the center.

The Local Supercluster contains a huge number of individual galaxies—several tens of thousands—yet the great majority of known galaxy clusters and super-clusters lie far beyond its edge. Figure 15.17 is a long-exposure photograph of one such remote cluster. This rich cluster is far outside the limit of the circle shown in Figure 15.16. It is only one of many large and distant groups of galaxies scattered throughout the observable universe. The Coma Cluster, shown in Figure 15.1, is another. On and on, the picture is much the same. The farther we peer into deep space, the more galaxies, clusters of galaxies, and superclusters we see.

Is there structure on even larger scales? As we will see in a moment, the answer is still yes. However, before we make our final leap in distance to the limits of the visible universe, let's pause for a moment to take stock of some important properties of galaxies and galaxy clusters and consider a few current ideas on how galaxies form and evolve.

15.3 Galaxy Masses

4 How can we measure the masses of systems as large as galaxies and galaxy clusters? Surely, we can neither count all their stars nor estimate their interstellar content very well. Galaxies are just too complex to take direct inventory of their material makeup. Instead, we must rely on indirect techniques. As usual, we turn to Newton's law of gravity.

Mass Measurement

Astronomers can calculate the masses of some spiral galaxies by determining their *rotation curves*, which plot rotation speed, obtained by measuring the Doppler shift of various spectral lines, versus distance from the galactic center. ∞ (Sec. 14.6) The mass within any given radius then follows directly from Newton's laws. Some rotation curves for nearby spirals are shown in Figure 15.18. They imply masses ranging from about

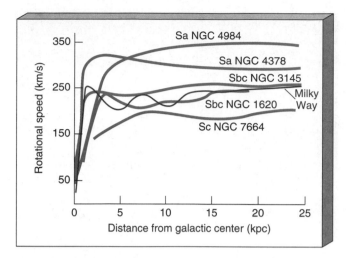

Figure 15.18 Rotation curves for some nearby spiral galaxies indicate masses of a few hundred billion times the mass of the Sun. The corresponding curve for our own Galaxy (Figure 14.17) is marked in red for comparison.

10^{11} to 5×10^{11} solar masses within about 25 kpc of the center—quite comparable to the results obtained for our own Galaxy using the same technique. Distant galaxies are generally too far away for such detailed curves to be drawn. Nevertheless, by observing the broadening of spectral lines—as discussed earlier in the context of the Tully–Fisher relation—we can still measure the overall rotation speed. Estimating the galaxy's size then leads to an estimate of its mass.

This approach is useful only for measuring the mass lying within about 50 kpc of a galaxy's center— the extent of the electromagnetic emission from stellar and interstellar material. To probe farther from the center, galactic astronomers turn to binary and multiple systems of galaxies, much as stellar astronomers study binary stars to determine stellar masses. If we knew the orbit size and period of a galaxy–galaxy binary system, the combined mass of the pair could be inferred from Kepler's third law, just as described in Chapter 10. ∞ (Sec. 10.9) However, there is a problem here. Unlike the stellar case, we cannot watch the galaxies travel even a small fraction of their entire orbit. Instead, the orbital period is usually simply estimated on the basis of two observations: (1) the velocities of the galaxies measured along our line of sight (Figure 15.19a) and (2) the present distance between the two galaxies. Masses obtained in this way, then, are fairly uncertain. However, we can combine many such

measurements to obtain quite reliable *statistical* information about galaxy masses.

From investigations using these methods, we find that most normal spirals (the Milky Way Galaxy in-

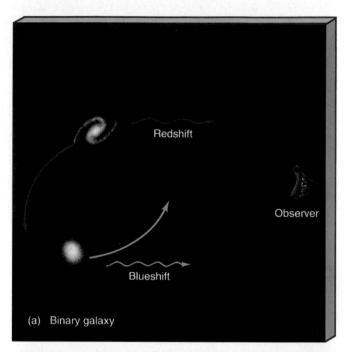

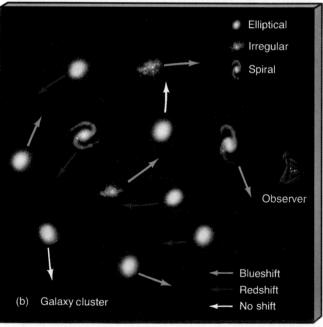

Figure 15.19 (a) In a binary galaxy system, galaxy masses may be estimated by observing the orbit of one galaxy about the other. (b) In a galaxy cluster, cluster masses are measured by observing the motion of many galaxies in the cluster and then estimating how much mass is needed to prevent the cluster from flying apart.

cluded) and large ellipticals contain between 10^{11} and 10^{12} solar masses of material. Irregular galaxies often contain less mass, about 10^8 to 10^{10} times that of the Sun. Dwarf ellipticals and irregulars can contain as little as 10^6 to 10^7 solar masses of material.

We can use another statistical technique to derive the combined mass of all the galaxies within a galaxy cluster. As depicted in Figure 15.19(b), each galaxy within a cluster moves relative to all other cluster members. As with binary galaxies, we cannot watch galaxies move around within a cluster—typical orbit periods are billions of years. But we can estimate the cluster's mass simply by determining, using Newtonian mechanics, how massive it must be in order to bind its galaxies gravitationally. Typical cluster masses obtained in this way lie in the range of 10^{13}–10^{14} solar masses. Notice that this calculation gives us *no* information whatsoever about the masses of individual galaxies. It tells us only about the *total* mass of the entire cluster.

Dark Matter in the Universe

3 5 Radio observations indicate that the rotation curves of many spiral galaxies, such as those shown in Figure 15.18, remain flat (that is, do not decline and may even rise slightly) far beyond the visible image of the galaxy. We conclude that these spiral galaxies—and perhaps all other spiral galaxies as well—must have invisible dark halos similar to that surrounding the Milky Way Galaxy. ∞ (Sec. 14.6) Depending on how far out these halos extend (and observations appear to indicate that they reach at least as far as 50 kpc from the center of a Milky Way–sized galaxy), spiral galaxies may contain 3 to 10 times more mass than can be accounted for in the form of visible matter. Some studies of elliptical galaxies suggest similarly large dark halos surrounding them too, but these observations are still far from conclusive.

Astronomers find an even greater discrepancy when they study galaxy clusters. Calculated cluster masses range from 10 to nearly 100 times the mass suggested by the light emitted by individual cluster galaxies. Stated another way, a lot more mass is needed to bind galaxy clusters than we can see. The problem of dark matter exists, then, not just in our own Milky Way Galaxy, but also in other galaxies and, to an even greater degree, in galaxy clusters as well. It most likely applies to the entire universe. In that case, we must accept the fact that *upward of 90 percent of the universe is composed of dark matter*. As noted in Chapter 14, this matter is not just dark in the visible portion of the spectrum—it is invisible at *all* electromagnetic wavelengths.

As discussed in Chapter 14, many possible explanations for the dark matter have been suggested, ranging from stellar remnants of various sorts to exotic subatomic particles. ∞ (Sec. 14.6) Whatever its nature, however, the dark matter in clusters apparently cannot be simply the accumulation of smaller amounts of dark matter within individual galaxies. Even including the galaxies' dark halos, we still cannot account for all the dark matter in galaxy clusters. As we look on larger and larger scales, we find that a larger and larger fraction of the matter in the universe is dark.

Could the additional dark matter be diffuse intergalactic matter existing among the galaxies within the clusters—that is, intracluster gas? Until the late 1970s, astronomers had no observational evidence for intergalactic matter, either inside or outside galaxy clusters. Then satellites orbiting above Earth's atmosphere detected substantial amounts of X-ray radiation in the direction of many galaxy clusters. Figure 15.20 shows false-color X-ray images of two such clusters. The X-ray-emitting region is centered on, and comparable in size to, the visible image of the cluster. These X-ray observations demonstrated for the first time the existence of large amounts of invisible hot gas—about 1 million K—within galaxy clusters.

How much matter have the X-ray satellites found? The observations suggest that at least as much matter—and in a few cases substantially more—exists within clusters in the form of hot gas as is visible in the form of stars. This is a lot of material, but it still doesn't solve the dark-matter problem. To account for the total masses of galaxy clusters implied by dynamical studies, we would have to find from 10 to 100 times more mass in gas than in stars.

Astronomers have found no evidence for gas beyond the well-defined galaxy clusters. "Extracluster" space is apparently empty of luminous material. Evidently, when the clusters formed eons ago, they did so very efficiently, sweeping up all the matter within any given region of the universe.

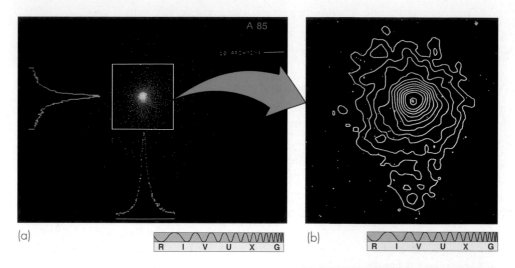

(a)

R I V U X G

(b)

R I V U X G

(c)

R I V U X G

Figure 15.20 (a) X-ray image of Abell 85, an old, distant cluster of galaxies, taken by the *Einstein* X-ray satellite observatory. The cluster's X-ray emission is shown in orange. The green graphs display a smooth, peaked intensity profile centered on the cluster but not associated with individual galaxies. (b) The contour map of X rays is superimposed on an optical photo, showing its X rays peaked on Abell 85's central supergiant galaxy. Images like these demonstrated for the first time that the space between the galaxies within galaxy clusters is filled with superheated gas. (c) A *ROSAT* X-ray image of hot gas within another cluster of galaxies (called Abell 2256). The cluster is nearly a billion parsecs from Earth and measures about 3 Mpc in diameter.

15.4 Galaxy Formation and Evolution

Videos

6 As mentioned earlier, astronomers know of no evolutionary connection among the various types of galaxies in the Hubble classification scheme. How then did those different galaxies come into being? To address this question, we must understand how galaxies formed. Unfortunately, the theory of galaxy formation is still very much in its infancy, and no definitive answers yet exist. Do we understand galaxy formation as well as, say, star formation? The answer is a resounding *no!* The theory is presently unable to answer completely even a basic question such as why spirals and ellipticals exist at all; it certainly cannot predict the different evolutionary tracks that galaxies might take.

There are several good reasons for our lack of understanding of galaxy formation: galaxies are much more complex than stars, they are harder to observe, and the observations are harder to interpret. We have only a partial understanding, and no observations, of the conditions in the universe immediately preceding galaxy formation, quite unlike the corresponding situation for stars. ∞ (Sec. 11.2) Finally, whereas stars almost never collide with one another, with the result that single stars and binaries evolve in isolation, galaxies may suffer many collisions and mergers during their lives (see *Interlude 15-1*), making it much harder to decipher their pasts. Yet despite all these difficulties, some general ideas have begun to gain widespread acceptance, and we can offer a few insights into the processes responsible for the galaxies we see.

Mergers and Acquisitions

Extension

The seeds of galaxy formation were sown in the very early universe, when small density fluctuations in the primordial matter began to grow. For now, let us begin

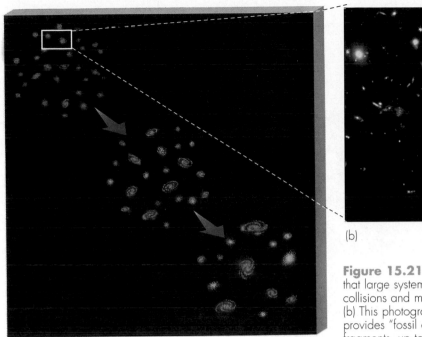

(a)

(b)

R   I   V   U   X   G

Figure 15.21   (a) The present view of galaxy formation holds that large systems were built up from smaller ones through collisions and mergers, as shown schematically in this drawing. (b) This photograph, one of the deepest ever taken of the universe, provides "fossil evidence" for hundreds of galaxy shards and fragments, up to 3000 Mpc distant.

our discussion with these "pregalactic" blobs of gas already formed. The masses of these fragments were quite small—perhaps only a few million solar masses, comparable to the masses of the smallest present-day dwarf galaxies, which may in fact be remnants of this early time. So where did the larger galaxies we find today come from? The key point in our current understanding is the realization that galaxies grow by repeated *merging* of smaller objects, as illustrated in Figure 15.21. Contrast this with the process of star formation, in which a large cloud fragments into smaller pieces that eventually become stars. ∞ (Sec. 11.4)

Theoretical evidence for this scenario is provided by computer simulations of the early universe, which clearly show merging taking place. Further strong support comes from recent observations that indicate that galaxies at large distances from us (well over 1000 Mpc away, meaning that the light we see was emitted more than 4 billion years ago) appear distinctly smaller and more irregular than those found nearby. Figure 15.21(b) and 15.22 show some of these images. The vague bluish patches are separate small galaxies, each containing only a few percent of the mass of the Milky Way Galaxy. Their irregular shape is thought to be the result of galaxy mergers; the bluish coloration comes from young stars that formed during the merger process.

Given that galaxies form by repeated mergers, how might we account for the differences between spirals and ellipticals? The answer is still unclear. One apparently very important factor is just when and where stars first appeared—in the original blobs, during a merger, or later—and how much gas was used up or ejected from the young galaxy in the process. If many stars formed early on and little gas was left over, an elliptical galaxy would be a likely outcome, with many old stars on random orbits and no gas to form a central disk. Alternatively, if a lot of gas remained, it would tend to sink to a central plane and form a rotating disk—in other words, a spiral galaxy would result. However, it is not known what determines the time, the place, or the rate of star formation. For this reason, whether spirals and ellipticals can form in basically the same environment, or if they tend to form in different places, is still an open question.

We do have some important clues to guide us, though. For example, spiral galaxies are relatively rare in regions of high galaxy density, such as the central regions of rich galaxy clusters. Is this because they simply tended not to form there, or is it because their disks are so fragile that they are easily destroyed by collisions and mergers, which are more common in dense galactic environments? Computer simulations suggest that collisions between spiral galaxies can indeed destroy the spirals' disks, ejecting much of the gas into inter-

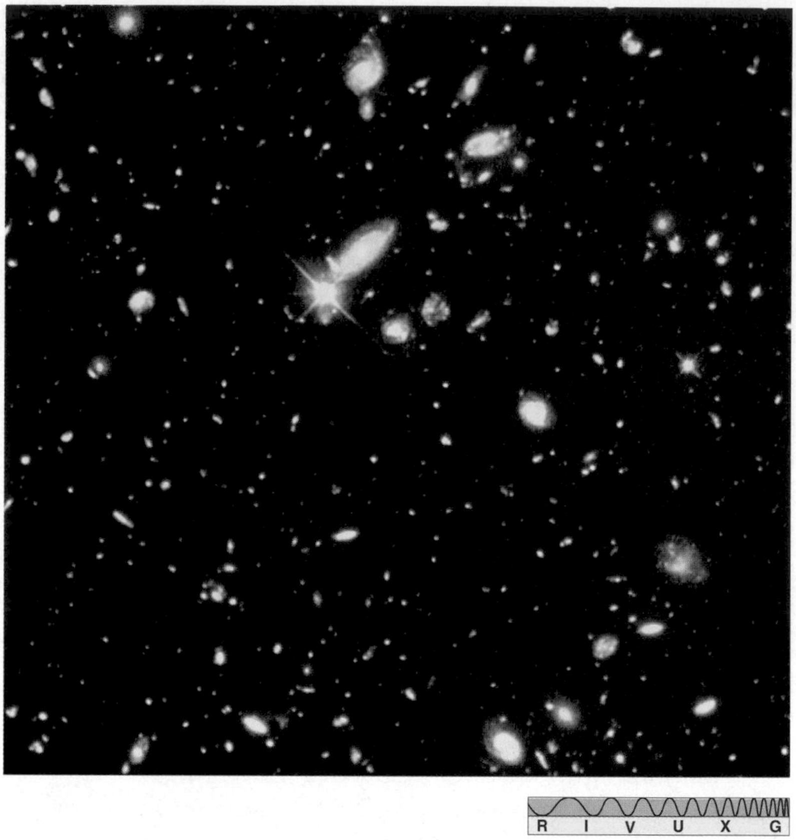

Figure 15.22 Numerous small, irregularly shaped young galaxies can be seen in this very deep optical image. Known as the Hubble Deep Field, this image, made with an exposure of approximately 100 h, captured objects as faint as the 30th magnitude. Redshift measurements indicate that some of the galaxies lie well over 1000 Mpc from Earth. Their size, color, and irregular appearance support the theory that galaxies grew by mergers and were smaller and less regular in the past. The field of view is about 2 arc minutes across. (See also the image at the opening of this chapter.)

galactic space (creating the hot intracluster gas noted in Section 15.3) and leaving behind objects that look very much like ellipticals. Recent observations of interacting galaxies appear to support this scenario. Figure 15.23 shows an example of this phenomenon. Additional evidence comes from the observed fact that spirals are more common at large distances (that is, in the distant past), which implies that their numbers are decreasing with time, presumably as the result of collisions.

However, nothing in this area of astronomy is clear cut. We know of numerous isolated elliptical galaxies in low-density regions of the universe, which are hard to explain as the result of mergers. Apparently some, but *not* all, ellipticals formed in this way.

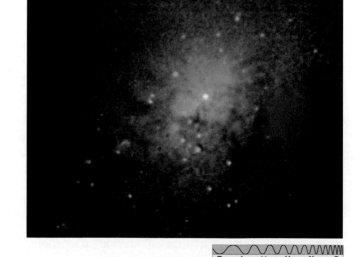

Figure 15.23 The peculiar (Irr II) galaxy NGC 1275 contains a system of long filaments that seem to be exploding outward into space. Its blue blobs, as revealed by the *Hubble Space Telescope*, are probably young globular clusters formed by the collision of two galaxies.

Galaxy Interactions

A related, and very important, question is: When did galaxy formation *stop*? Astronomers are divided on this issue. Some maintain that there was a fairly well-defined time in the past—given by the age of the globular clusters in our own Galaxy, for example—by which most formation was over. Others point out that many galaxies show evidence of mergers and the accumulation of smaller satellite galaxies over an extended period of time—even up to the present day. These astronomers suggest that the many galaxy interactions we observe today (see *Interlude 15-1*) are just part of the same process begun when the first fragments merged. In this view, galaxy formation is still occurring today. In either case, astronomers have ample evidence that galaxies evolve in response to external factors, even long after they first formed.

We now know that spiral galaxies have huge, invisible dark halos surrounding them, and we strongly suspect that all galaxies have similar halos. Consider two galaxies orbiting one another—a binary galaxy system. As they orbit, the galaxies interact with each other's halos, one galaxy stripping halo material from the other by tidal forces. The freed matter is either redistributed within a common envelope or is entirely lost from the binary system. This interaction between the halos changes the orbits of the galaxies, which tend to spiral toward one another, eventually merging. If one galaxy of the pair happens to have a much lower mass than the other, the process is colloquially termed *galactic cannibalism*. Such cannibalism might explain why supermassive galaxies are often found at the cores of rich galaxy clusters. Having dined on their companions, they now lie at the center of the cluster, waiting for more food to arrive. Figure 15.24 is a remarkable

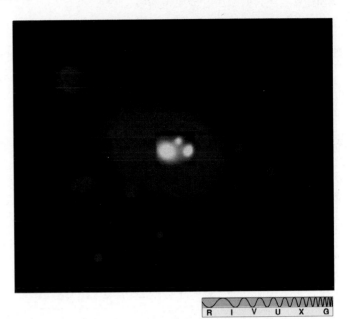

Figure 15.24 This computer-enhanced, false-color composite optical photograph of the galaxy cluster Abell 2199 is thought to show an example of galactic cannibalism. The large central galaxy of the cluster (itself 120 kpc along its long axis) is displayed with a superimposed "window." (This results from a shorter time exposure, which shows only the brightest objects that fall within the frame.) Within the core of the large galaxy are several smaller galaxies (the three bright yellow images at center) apparently already "eaten" and now being "digested" (that is, being torn apart and becoming part of the larger system). Other small galaxies swarm on the outskirts of the swelling galaxy, almost certainly to be eaten, too.

combination of images that has apparently captured this process at work.

Now consider two interacting disk galaxies, one a little smaller than the other but each having a mass comparable to the Milky Way Galaxy. As shown in the computer-generated frames of Figure 15.25, the smaller galaxy can substantially distort the larger one, causing

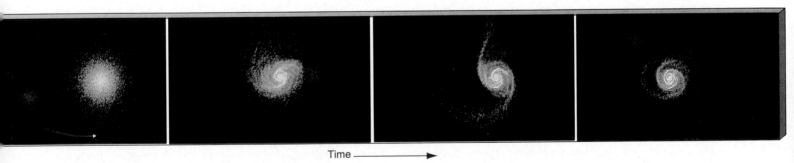

Time ⟶

Figure 15.25 Galaxies can change their shapes long after their formation. In this computer-generated sequence, two galaxies closely interact over several hundred million years. The smaller galaxy, in red, has gravitationally disrupted the larger galaxy, in blue, changing it into a spiral galaxy. Compare the result of this supercomputer simulation with Figure 15.2b, a photograph of M51 and its small companion.

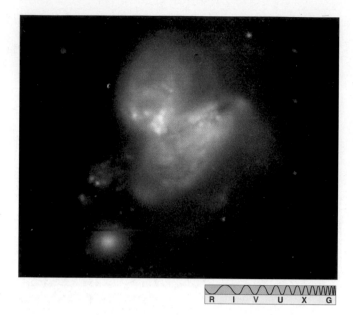

R I V U X G

Figure 15.26 This interacting galaxy pair (IC 694, at left, and NGC 3690) shows starbursts now under way in both galaxies—hence the bluish tint. Such intense, short-lived bursts probably last for no more than a few tens of millions of years—a small fraction of a typical galaxy's lifetime.

spiral arms to appear where none existed before. The entire event requires several hundred million years—a span of evolution that supercomputers can model in minutes.

The final frame of Figure 15.25 looks remarkably similar to the double galaxy shown in Figure 15.2(b). Shown there are two galaxies with sizes, shapes, and velocities corresponding very closely to those in the computer simulation. The magnificent spiral galaxy is M51, popularly known as the Whirlpool Galaxy, about 10 Mpc from Earth. Its smaller companion is an irregular galaxy that may have drifted past M51 millions of years ago. Did this smaller galaxy cause the spiral structure we see in M51? Does the model mirror reality? Perhaps. We need more evidence from other galaxies to confirm the accuracy of these and similar simulations. Still, the computer simulation does demonstrate a plausible way that two galaxies might have interacted millions of years ago and how spiral arms might be created or enhanced as a result.

It now seems that many of the most spectacular changes that occur in galaxies result from interactions with other galaxies. Astronomers have cataloged numerous **starburst galaxies**, such as the one shown in Figure 15.26, where violent events, possibly a near-collision with a neighbor, appear to have rearranged the galaxy's internal structure and triggered a sudden, in-

tense burst of star formation in the recent past. Such close encounters are random events and do not seem to represent any genuine evolutionary sequence linking all spirals to all ellipticals and irregulars. However, it *is* clear that many galaxies and galaxy clusters have evolved greatly since they first formed long ago.

15.5 Hubble's Law

2 7 Now let's turn our attention to the large-scale *motions* of galaxies and galaxy clusters. Within a galaxy cluster, individual galaxies move in essentially random ways. You might expect that, on the largest possible scales, the clusters themselves would also have random, disordered motion—some clusters moving this way, some that. This is not the case at all, however. On the largest scales, galaxies that are not members of a cluster and galaxy clusters alike display very ordered motion.

Universal Recession

In 1912, the American astronomer Vesto M. Slipher, working under the direction of Percival Lowell, discovered that virtually every spiral galaxy he observed had a redshifted spectrum—it was apparently *receding* from our Galaxy. In fact, with the exception of a few nearby systems, *every* known galaxy is part of a general motion away from us in all directions. Individual galaxies that are not part of galaxy clusters are steadily receding. Galaxy clusters too have an overall recessional motion, although their individual member galaxies move randomly with respect to one another. (Consider a jar full of fireflies that has been thrown into the air. The fireflies within the jar, like the galaxies within the cluster, have random motions due to their individual whims, but the jar as a whole, like the galaxy cluster, has some directed motion as well.)

Figure 15.27 shows the optical spectra of several galaxies, arranged in order of increasing distance from the Milky Way Galaxy. These spectra are redshifted, indicating that the galaxies are steadily receding. Furthermore, the extent of the redshift increases progressively from top to bottom in the figure. Because the galaxies' distances from us also increase from top to bottom, we conclude that there is a connection between Doppler shift and distance: the greater the distance, the greater the redshift. This trend holds for

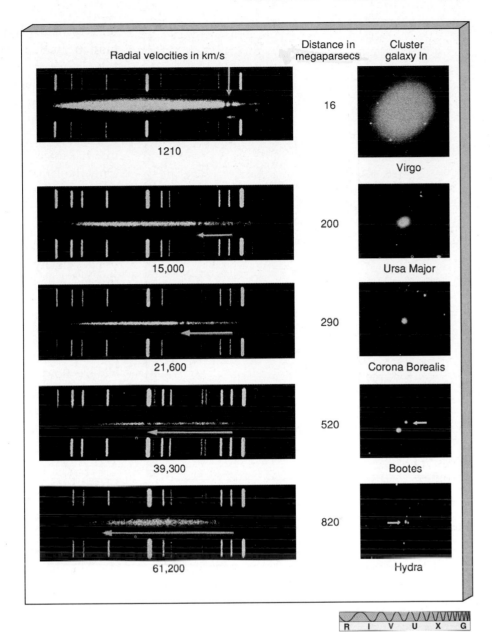

Figure 15.27 Optical spectra (on the left) of several galaxies (on the right). Both the extent of the redshift (denoted by the horizontal yellow arrows) and the distance from the Milky Way Galaxy to each galaxy (numbers in center column) increase from top to bottom. The vertical yellow arrow in the top spectrum highlights a particular spectral feature (a pair of dark absorption lines). The horizontal yellow arrows indicate how this feature shifts to longer wavelengths in spectra of more distant galaxies. The lines at top and bottom of each spectrum are laboratory references.

nearly all galaxies in the universe. (Two galaxies within our Local Group, including Andromeda, and a few galaxies in the Virgo Cluster display blueshifts and so have some motion toward us, but this results from their random motions within their parent clusters. Recall the fireflies in the jar.)

Figure 15.28(a) shows recessional velocity plotted against distance for the galaxies of Figure 15.27. Figure 15.28(b) is a similar plot for some more galaxies within about 1 billion pc of Earth. Plots like these were first made by Edwin Hubble in the 1920s and now bear his name—*Hubble diagrams*. The data points generally fall close to a straight line, indicating that a simple relationship connects recessional velocity and dis-

tance: the rate at which a galaxy recedes is *directly proportional* to its distance from us. This rule is called **Hubble's law**. We could construct such a diagram for any group of galaxies, provided we could determine their distances and velocities. The universal recession described by the Hubble diagram is sometimes called the *Hubble flow*.

To distinguish recessional redshift from redshifts caused by motion *within* an object—for example, galaxy orbits within a cluster or explosive events in a galactic nucleus—the redshift resulting from the Hubble flow is called the **cosmological redshift**. Objects that lie so far away that they exhibit a large cosmological redshift are said to be at *cosmological dis-*

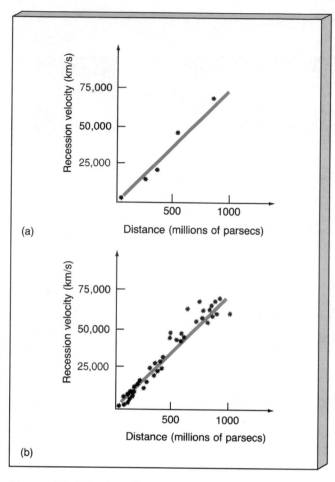

Figure 15.28 Plots of recessional velocity against distance (a) for the galaxies shown in Figure 15.27 and (b) for numerous other galaxies within about 1 billion pc of Earth.

tances—distances comparable to the scale of the universe.

Hubble's law is an *empirical* discovery—that is, a discovery based strictly on observational results. Its central relationship—a statistical correlation between recessional velocity and distance—is well documented as far as galaxy distances can be reliably determined, but no law of nature demands that all galaxies recede, and no law of physics requires that a link exist between velocity and distance. In that sense, Hubble's "law" is not really a law at all. It is simply a convenient way of noting the *observational* fact that any galaxy's recessional velocity is directly proportional to its distance from us.

The recessional motions of the galaxies prove that the cosmos is not steady and unchanging on the largest scales. Its contents are in constant relative motion, and the motion is not random. In fact, the universe is expanding—and expanding in a directed fashion. In short, it is evolving. Before going any further, however, let's be clear on *what* is expanding and what is not. Hubble's law does *not* mean that humans, Earth, the solar system, or even individual galaxies are physically increasing in size. These groups of atoms, rocks, planets, and stars are held together by their own internal forces and are not themselves getting bigger. Only the largest framework of the universe—the ever-increasing distances separating the galaxies and the galaxy clusters—is expanding.

Hubble's law has some fairly obvious and dramatic implications. If nearly all galaxies show recessional velocity according to Hubble's law, then doesn't that mean that they all started their journey from a single point? If we could run time backward, wouldn't all the galaxies fly back to this one point, perhaps the site of some explosion in the remote past? The answer is yes. In Chapter 17 we explore the ramifications of the Hubble flow for the past and future evolution of our universe. For the remainder of this chapter, however, we set aside its cosmic implications and use Hubble's law simply as a convenient distance-measuring tool.

Hubble's Constant

The constant of proportionality between recessional velocity and distance in Hubble's law is known as **Hubble's constant**. It is usually denoted by the symbol H_0. The data shown in Figure 15.28 then obey the equation

$$\text{recessional velocity} = H_0 \times \text{distance}.$$

The value of Hubble's constant is the slope of the straight line—recessional velocity divided by distance—in Figure 15.28(b). Reading the numbers off the graph, this comes to roughly 75,000 km/s divided by 1000 Mpc, or 75 km/s/Mpc (kilometers per second per megaparsec, the most commonly used unit for H_0). Astronomers continually strive to refine the accuracy of the Hubble diagram and the resulting estimate of H_0 because Hubble's constant is one of the most fundamental quantities of nature—it specifies the rate of expansion of the entire cosmos.

The precise value of Hubble's constant is still the subject of considerable debate. In the 1970s, astronomers obtained a value of around 50 km/s/Mpc, using a chain of standard candles to extend their observations to large distances. However, in the early 1980s, when the Tully–Fisher technique had become fairly well established, other researchers used it to obtain a measurement of H_0 that was largely independent of methods relying on standard candles. From observations of galaxies within about 150 Mpc, the latter group deduced a value of $H_0 = 90$ km/s/Mpc, a result inconsistent with the earlier measurements (even allowing for the estimated uncertainties involved). For some reason, the distances obtained using the Tully–Fisher method were only about half those determined by using standard candles, and so the measured value of Hubble's constant nearly doubled using this approach.

Subsequent determinations of H_0 by other researchers, using different galaxies and a variety of distance-measurement techniques, have yielded results mostly within the range 60–90 km/s/Mpc, and most astronomers would be quite surprised if the true value of H_0 turned out to lie outside this range. For now, however, in the absence of any good explanation for the spread, astronomers must live with this uncertainty. Some like to split the difference and use $H_0 = 75$ km/s/Mpc. We will adopt this compromise value as the best current estimate of Hubble's constant for the remainder of the text.

The Cosmic Distance Scale

One very important application of Hubble's law is its use as a means of determining distances. Using Hubble's law, we can derive the distance to a remote object simply by measuring the object's recessional velocity. An astronomer measures the redshift of the object's spectral lines. The extent of the shift is then converted to velocity, using the Doppler relationship discussed in Chapter 2. ∞ (*More Precisely 2-2*) Knowing the object's velocity, the astronomer then finds its distance by using the plot of Figure 15.28(b). Notice, however, that the uncertainty in Hubble's constant translates directly into a similar uncertainty in the distance determined by this method.

Using Hubble's law in this way tops our inverted pyramid of distance-measurement techniques (Figure 15.29). This sixth method simply assumes that Hubble's law holds. If this assumption is correct, Hubble's law enables us to measure great distances in the universe—so long as we can obtain an object's spectrum, we can determine how far away the object is.

Many redshifted objects have recessional motions that are a substantial fraction of the speed of light. The most distant object thus far observed in the universe has the catalog name of QO051-279. Its extremely high redshift implies a recessional velocity 93 percent

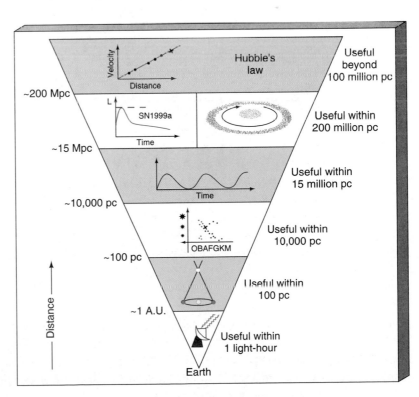

Figure 15.29 Hubble's law tops the hierarchy of distance-measurement techniques. It is used to find the distances of astronomical objects all the way out to the limits of the observable universe.

15-1 INTERLUDE

COLLIDING GALAXIES

Contemplating the congested confines of a rich galaxy cluster (such as Virgo or Coma, with thousands of member galaxies orbiting within a few megaparsecs), we might expect that collisions among galaxies would be common. Gas particles collide in our atmosphere, and hockey players collide in the rink—do galaxies in clusters collide, too? The answer is yes. The first pair of

Whether these galaxies are genuinely colliding or only experiencing a close encounter cannot easily be determined. No human will ever witness an entire collision, for it would last many millions of years. However, computer simulations of these systems display formations remarkably similar to the real thing. The particular simulation shown at left in the second pair of images began with two spiral galaxies, but the details of the original structure have been largely obliterated by the collision. Notice the similarity to the real image of NGC 6240 (right image), an object showing faint tails as well as double galactic centers only a

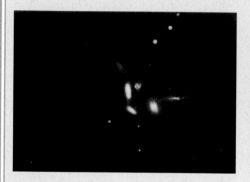

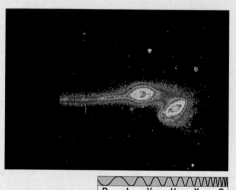

R I V U X G

images show evidence supporting this assertion. The image on the left is an optical photograph of a small group of five galaxies in the constellation Serpens. Connecting clouds seem to link some of them, strongly suggesting that they are (or have just been) interacting with one another. The image on the right is a computer-enhanced view of the pair of galaxies NGC 4676 A and B (also known as "The Mice"), which show streams of gas and stars apparently generated by the encounter between the two.

few hundred parsecs apart. The calculations show that ultimately the two galaxies merge into one.

Direct observational evidence now indicates that galaxies in clusters apparently collide quite often. In the smaller groups, the galaxies' speeds are low enough that interacting galaxies tend to "stick together," and mergers, as shown in the computer simulation, are a very common outcome. In larger groups, galaxies move faster and tend to pass

that of light. At that speed, ultraviolet spectral lines are Doppler shifted all the way into the far infrared! Hubble's law predicts that QO051-279, solely on the basis of its observed redshift, lies more than 4000 Mpc away from us (see *More Precisely 16-1*). This object resides as close to the edge of the observable universe as astronomers have yet been able to probe.

Large-Scale Structure

3 Using Hubble's law, we can complete our census of the large-scale distribution of galaxies. One of the most extensive surveys of the universe yet undertaken has been carried out by astronomers at Harvard University. Using Hubble's law as their dis-

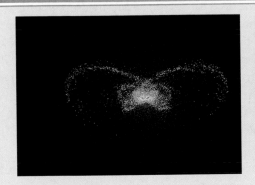

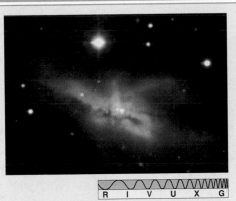

through one another without sticking. Since the early 1980s, it has become increasingly clear, on the basis of both observations and numerical simulations, that collisions can have very large effects on the galaxies involved. The stellar and interstellar contents of each galaxy are rearranged, and the merged interstellar matter very likely experiences shock waves that trigger widespread bursts of star formation (see Figure 15.23). Some researchers go so far as to suggest that most galaxies have been strongly influenced by collisions, in many cases in the relatively recent past.

Curiously, although a collision may wreak havoc on the large-scale structure of the galaxies involved, it has essentially no effect on the individual stars they contain. The stars within each galaxy just glide past one another. Although we have plenty of photographic evidence for galaxy collisions, no one has ever witnessed or photographed a collision between two stars. Stars do collide in other circumstances—in the dense central cores of galactic nuclei and globular clusters, or as a result of stellar evolution in binary systems—but stellar

collisions are a very rare consequence of galaxy interactions.

To understand why individual stars do not collide when galaxies collide, recall that the galaxies within a typical cluster are bunched together fairly tightly. The distance between adjacent galaxies in a given cluster averages a few hundred thousand parsecs, which is only about 10 times greater than the size of a typical galaxy. Galaxies simply do not have that much room to roam around without bumping into one another. By contrast, stars within a galaxy are spread out much more thinly. The average distance between stars within a galaxy is a few parsecs, which is millions of times greater than the size of a typical star. When two galaxies collide, the star population merely doubles for a time, and the stars continue to have so much space that they do not run into each other. The stellar and interstellar contents of each galaxy are certainly rearranged, and the resultant burst of star formation may indeed be spectacular from afar, but from the point of view of the stars, it's clear sailing.

tance indicator, these researchers are compiling a catalog of the positions and redshifts of all galaxies within about 200 Mpc of our Galaxy. This is an extremely painstaking task—even with a large telescope, it takes a long time to obtain a detailed spectrum of a distant galaxy. Rather than trying to cover the entire sky at once, the team elected to map the universe in a series

of wedge-shaped "slices," each 6° thick, starting in the northern sky. The first slice, shown in Figure 15.30, covered a region of the sky containing the Coma Cluster (Figure 15.1), which happens to lie in a direction almost perpendicular to our Galaxy's plane.

The most striking feature of maps such as this is that the distribution of galaxies on very large scales is

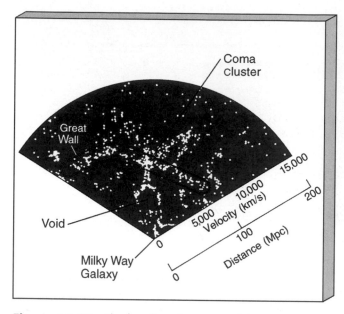

Figure 15.30 The first slice of a survey of the universe, covering 1057 galaxies out to an approximate distance of 200 Mpc, clearly shows that galaxies and clusters are not randomly distributed on large scales. Instead, they appear to have a filamentary structure, surrounding vast, nearly empty voids. The distances shown assume $H_0 = 75$ km/s/Mpc.

36° on the sky perpendicular to the outline in the figure. This extended sheet of galaxies, which has come to be known as the *Great Wall*, measures at least 70 Mpc (out of the plane of the page) by 200 Mpc (across the page). For now at least, it is one of the largest known structures in the universe. Figure 15.31 combines all the available data with results of other surveys of the southern sky. (The large empty regions are either obscured by our Galaxy or not yet mapped.) The Great Wall can be seen arcing around the left side of the figure.

Are there still larger structures in the universe? Only with more extensive surveys will we know for sure. Many theorists now believe that this "frothy" distribution of galaxies, and in fact all structure on scales larger than a few megaparsecs, traces its origin directly to the conditions found in the very earliest stages of the universe (Chapter 17). As such, studies of large-scale structure are vital to our efforts to understand the origin and nature of the cosmos itself.

decidedly nonrandom. The galaxies appear to be arranged in a network of strings, or filaments, surrounding large, relatively empty regions of space known as **voids**. The biggest voids measure some 100 Mpc across. For a time, they were the largest objects in the universe known to astronomers. The most likely explanation for the voids and filamentary structure in Figure 15.30 is that the galaxies and galaxy clusters are spread across the surfaces of vast "bubbles" in space. The voids are the interiors of these gigantic bubbles. The galaxies seem to be distributed like beads on strings only because of the way our slice of the universe cuts through the bubbles. Like suds on soapy water, these bubbles fill the entire universe. The densest clusters and superclusters lie in regions where several bubbles meet.

The notion that the filaments are just the intersection of the survey slice with much larger structures (the bubble surfaces) was confirmed when the next three slices of the survey, lying above and below the first, were completed. The region of Figure 15.30 indicated by the red outline was found to continue through both the other slices, so we know that it covers *at least*

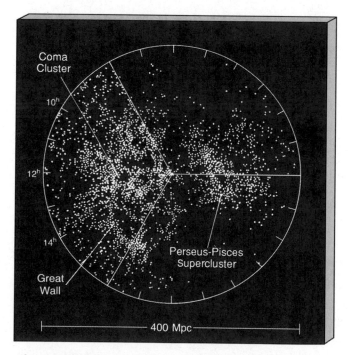

Figure 15.31 Combination of data from several redshift surveys of the universe reveal the extent of large-scale structure within 200–300 Mpc of the Sun. The arc on the left is the Great Wall. The empty regions are mostly areas obscured by our Galaxy. Positions for more than 4500 galaxies are plotted here. We assume $H_0 = 75$ km/s/Mpc.

Chapter Review

Summary

The **Hubble classification scheme** (p. 420) divides galaxies into several classes, depending on their appearance. **Spiral galaxies** (p. 420) have flattened disks, central bulges, and spiral arms. They are further subdivided on the basis of the size of the bulge and the tightness of the spiral structure. The halos of these galaxies consist of old stars, whereas the gas-rich disks are the sites of ongoing star formation. **Barred-spiral galaxies** (p. 422) contain an extended "bar" of material projecting beyond the central bulge. **Elliptical galaxies** (p. 423) have no disk and contain no gas or dust. In most cases, they consist entirely of old stars. They range in size from dwarf ellipticals, which are much less massive than the Milky Way Galaxy, to giant ellipticals, which may contain trillions of stars. **S0 and SB0 galaxies** (p. 424) are intermediate in their properties between ellipticals and spirals. They have extended halos and stellar disks and bulges (and bars, in the SB0 case) but little or no gas and dust. **Irregular galaxies** (p. 424) are galaxies that do not fit into either of the other categories. Some may be the result of galaxy collisions or close encounters. Many irregulars are rich in gas and dust and are the sites of vigorous star formation. The **Magellanic Clouds** (p. 425), two small systems that orbit the Milky Way Galaxy, are examples of this type of galaxy.

Astronomers often use **standard candles** (p. 427) as distance-measuring tools. These are objects that are easily recognizable (by their light curves, spectra, or some other directly observable characteristic) and whose luminosities are known to lie in some reasonably well-defined range. Comparing luminosity and apparent brightness, astronomers determine the distance using the inverse-square law. An alternative is the **Tully–Fisher relation** (p. 427), an empirical correlation between rotational velocity and luminosity in spiral galaxies. By measuring the rotation speed of a spiral and using this relationship, astronomers can determine the galaxy's luminosity and hence its distance.

The Milky Way, Andromeda, and several other smaller galaxies form the **Local Group** (p. 429), a small **galaxy cluster** (p. 429). Galaxy clusters consist of a collection of galaxies orbiting one another, bound together by their own gravity. The nearest large galaxy cluster to the Local Group is known as the Virgo Cluster. Galaxy clusters themselves tend to clump together into **superclusters** (p. 430). The Virgo Cluster, the Local Group, and several other nearby clusters form the Local Supercluster.

The masses of nearby spiral galaxies can be determined by studying their rotation curves. For more distant spirals, masses can be inferred from observations of the broadening of their spectral lines. On larger scales, astronomers use studies of binary galaxies and galaxy clusters to obtain statistical mass estimates of the galaxies involved. As in the Milky Way Galaxy, measurements of the masses of other galaxies and of galaxy clusters reveal the presence of large amounts of dark matter that is presently undetectable at any electromagnetic wavelength. The fraction of dark matter apparently grows as the scale under consideration increases. Large amounts of hot X-ray-emitting gas have been detected among the galaxies in many clusters, but not enough to account for the dark matter inferred from dynamical studies.

Researchers know of no evolutionary sequence that links spiral, elliptical, and irregular galaxies together, and the process of galaxy formation is still only poorly understood. There is growing evidence that large galaxies formed by the merger of smaller ones in a process that may be continuing today. Collisions and mergers of galaxies play very important roles in galactic evolution. Interactions between galaxies appear to be very common. A **starburst galaxy** (p. 438) may result when a galaxy experiences a close encounter with a neighbor. The strong tidal distortions caused by the encounter compress galactic gas, resulting in a widespread burst of star formation.

Distant galaxies are observed to be receding from the Milky Way at rates that increase proportional to their distances from us. This relationship between recessional speed and distance is called **Hubble's law** (p. 439). The constant of proportionality in the law is **Hubble's constant** (p. 440). Its value is believed to lie between 60 and 90 km/s/Mpc. Astronomers use

Hubble's law to determine distances to the most remote objects in the universe. The redshift associated with the Hubble expansion is called the **cosmological redshift** (p. 439).

On very large scales, galaxies and galaxy clusters are not spread randomly throughout space. Instead, they are arranged on the surfaces of enormous "bubbles" of matter surrounding vast low-density regions called **voids** (p. 444). The origin of this structure is thought to be closely related to conditions in the very earliest epochs of the universe.

Self-Test: True or False?

_____ 1. Barred-spiral galaxies have the same properties as normal spirals, except for the "bar" feature.

_____ 2. Elliptical galaxies contain no flattened disk.

_____ 3. Elliptical galaxies contain substantial amounts of interstellar gas but no interstellar dust.

_____ 4. Most ellipticals contain only old stars.

_____ 5. Irregular galaxies, although small, have lots of star formation taking place in them.

_____ 6. Most galaxies are spirals.

_____ 7. Spiral galaxies evolve into ellipticals.

_____ 8. Type I supernovae can be used to determine distances to galaxies.

_____ 9. Every galaxy is a member of some galaxy cluster.

_____ 10. Galaxy collisions can occur, but are extremely rare.

_____ 11. Distant galaxies appear to be much larger than those nearby.

_____ 12. A typical galaxy cluster has a mass of about 10^{11} solar masses.

_____ 13. Most galaxies appear to be receding from the Milky Way Galaxy.

_____ 14. Hubble's law can be used to determine distances to the farthest objects in the universe.

Self-Test: Fill in the Blank

1. Galaxies are categorized by their _____ classification.

2. Spiral galaxies that have tightly wrapped spiral arms tend to have _____ central bulges.

3. Spiral galaxies of type _____ have the least amount of gas; type _____ have the most.

4. The Milky Way Galaxy, the Andromeda Galaxy, and about 20 others form a small cluster known as the _____.

5. In the Tully–Fisher relation, a galaxy's luminosity is found to be related to the _____ of its spectral lines.

6. The diameter of the Local Supercluster is about _____ .

7. When galaxies collide, the star formation rate often _____.

8. Galaxy mass determinations from rotation curves, line broadening, and binary galaxies all make use of _____ law.

9. Dark matter may make up as much as _____ percent of the universe.

10. Intergalactic gas in galaxy clusters emits large amounts of energy in the form of _____.

11. By which process do galaxies form: fragmentation (large objects breaking up into small) or mergers (small objects accumulating into large)? _____

12. Hubble's law is a correlation between the redshifts and the _____ of galaxies.

13. Hubble's constant is believed to lie in the range _____ to _____ km/s/Mpc.

14. The largest known structures in the universe, such as voids and the Great Wall, have sizes on the order of _____.

Review and Discussion

1. In what sense are elliptical galaxies "all halo"?

2. Describe the four rungs in the distance-measurement ladder used to determine the distance to a galaxy lying 5 Mpc away.

3. Describe the contents of the Local Group. How much space does it occupy compared with the volume of the Milky Way Galaxy?

4. How is the Tully–Fisher relation used to measure distances to galaxies?

5. What is the Virgo Cluster?

6. Describe two techniques for measuring the mass of a galaxy.

7. Why do astronomers believe that galaxy clusters contain more mass than we can see?

8. What evidence do we have that galaxies collide with one another?

9. Describe the role of collisions in the formation and evolution of galaxies.

10. Do you think that collisions between galaxies constitute "evolution" in the same sense as the evolution of stars?

11. What is Hubble's law?

12. How is Hubble's law used by astronomers to measure distances to galaxies?

13. What is the most likely range of values for Hubble's constant? Why is the exact value uncertain?

14. What are voids? What is the distribution of galactic matter on very large (more than 100 Mpc) scales?

Problems

1. A supernova of luminosity 1 billion times the luminosity of the Sun is being used as a standard candle to measure the distance to a faraway galaxy. From Earth, the supernova appears as bright as the Sun would appear from a distance of 10 kpc. What is the distance to the galaxy?

2. According to Hubble's law, with $H_0 = 75$ km/s/Mpc, what is the recessional speed of a galaxy at a distance of 200 Mpc? How far away is a galaxy whose recessional speed is 4000 km/s? How do these answers change if $H_0 = 50$ km/s/Mpc?

3. Two galaxies are orbiting one another at a separation of 500 kpc, and the orbital period is estimated to be 30 billion years. Use Kepler's third law (as stated in Chapter 14) to find the total mass of the pair. ∞ (Sec. 14.6)

Projects

1. Look for a copy of the *Atlas of Peculiar Galaxies* by Halton Arp. It is available in book form or on laser disk. Search for examples of interacting galaxies of various types: (1) tidal interactions, (2) starburst galaxies, (3) collisions between two spirals, and (4) collisions between a spiral and an elliptical. For (1), look for galactic material pulled away from a galaxy by a neighboring galaxy. Is the latter galaxy also tidally distorted? In (2), the surest sign of starburst activity are bright knots of star formation. In what type(s) of galaxies do you find starburst activity? For (3) and (4), how do collisions differ depending on the types of galaxies involved?

2. Look for the Virgo Cluster. An 8-inch telescope is a perfect size for this project, although a smaller telescope will also work. The constellation Virgo is visible from the United States during much of fall, winter, and spring. To locate the center of the cluster, first find the constellation Leo. The eastern part of Leo is composed of a distinct triangle of stars — Denebola (β), Chort (θ), and Zosma (δ). Move your eye eastward from Chort to Denebola and then continue along that same line until you have moved a distance equal to the distance between Chort and Denebola. You are now looking at the approximate middle of the Virgo Cluster. Look for the following Messier objects that make up some of the brightest galaxies in the cluster. M49, 58, 59, 60, 84, 86, 87 (which is a giant elliptical thought to have a massive black hole at its center), 89, and 90. Examine each galaxy for unusual features; some have very bright nuclei.

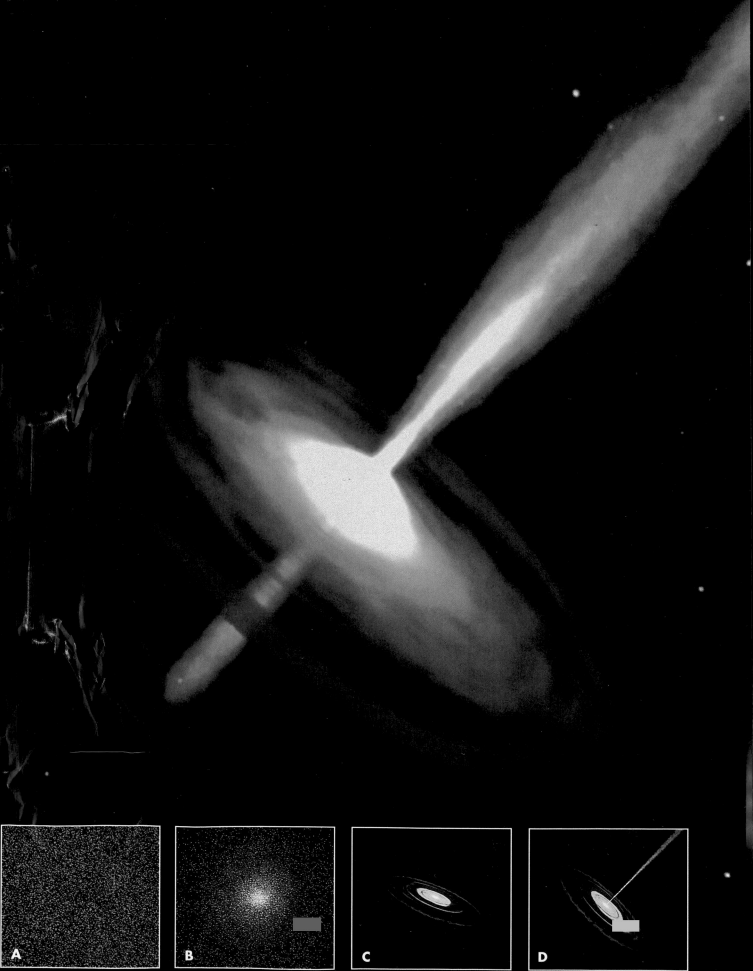

A

B

C

D

16 Active Galaxies and Quasars

Limits of the Observable Universe ✴ *www*

◀ (Opposite page, background) This artist's conception depicts one possible scenario for the "central engine" of an active galaxy. Two jets of matter are shown moving outward perpendicular to a flattened accretion disk surrounding a supermassive black hole.

(Inset A) Numerical simulations—i.e., done on computers—can help us visualize key events. Here a uniform distribution of some thousand stars is spread over a dimension of several hundred parsecs.

(Inset B) As the simulation proceeds, the hypothetical stars congregate preferentially toward the core of the star cluster, forming a "luminosity cusp" of light.

(Inset C) If we could see inside the cusp, an accretion disk would probably be evident, swirling around a giant black hole.

(Inset D) As the accretion disk rotates, tilts, and acquires more matter, jets of matter emerge to cool the environment surrounding the hole.

LEARNING GOALS

Studying this chapter will enable you to:

1 Specify the basic differences between active and normal galaxies.

2 Describe the important features of Seyfert and radio galaxies.

3 Explain what drives the central engine thought to power all active galaxies.

4 Describe the observed properties of quasars and discuss the special properties of the radiation they emit.

5 Discuss the place of active galaxies in current theories of galactic evolution.

Our journey from the Milky Way to the Great Wall in the past two chapters has widened our cosmic field of view by a factor of 10,000, yet the galaxies that make up the structures we see show remarkable consistency in their properties. The overwhelming majority of galaxies fit neatly into the Hubble classification scheme, showing few, if any, unusual characteristics. However, sprinkled through the mix of normal galaxies, even relatively close to the Milky Way Galaxy, are some that are decidedly abnormal in their properties. Although their optical appearance is often quite ordinary, these abnormal galaxies emit huge amounts of energy—far more than normal galaxies—mostly in the nonvisible part of the electromagnetic spectrum. Observing such objects at great distances, we may be seeing some of the formative stages of our own galactic home.

16.1 Beyond the Local Realm

1 Astronomers have recognized and cataloged spiral, elliptical, and irregular galaxies as far away as several hundred megaparsecs from Earth. Beyond this distance, galaxies appear so faint that it is often difficult to discern their shapes. Consequently, their types are largely unknown. Nevertheless, according to their observed redshifts (and Hubble's law), we know that many galaxies lie well beyond this distance, in the farthest reaches of the observable universe. But what kinds of objects are they? Are they normal galaxies—close relatives of the systems that populate our neighborhood—or are they somehow different? The answer is that, while most distant galaxies appear basically normal in their properties, some are distinctly different from galaxies found in our cosmic backyard: they are more luminous—more *active*—than the typical galaxy found close to home. These energetic objects, some of them hundreds, even thousands, of times more luminous than the Milky Way Galaxy, are known collectively as **active galaxies**.

In addition to their greater overall luminosities, active galaxies differ fundamentally from normal galaxies in the *character* of the radiation they emit. Most of a normal galaxy's energy is emitted in or near the visible portion of the electromagnetic spectrum, much like the radiation from ordinary stars. Indeed, to a large extent, the light we see from a normal galaxy *is* just the accumulated light of its many component stars. By contrast, as illustrated schematically in Figure 16.1, the radiation from active galaxies does *not* peak in the visible. Most active galaxies do emit substantial amounts of visible radiation, but far more energy is emitted at longer wavelengths. Put another way, the radiation

from active galaxies is *inconsistent* with what we would expect if it were the combined radiation of myriad stars. Their radiation is said to be *nonstellar*.

The two most important categories of active galaxies are *Seyfert galaxies* and *radio galaxies*. More extreme in their properties are the even more luminous *quasars*. Astronomers conventionally distinguish between active galaxies and quasars based on their appearance, spectra, and distance from us. At visible wavelengths, active galaxies typically *look* like normal galaxies. Indeed, for many purposes, we can think of active galaxies as being otherwise "normal" systems (that is, emitting visible radiation) that happen also to be extremely intense sources of radio and/or infrared

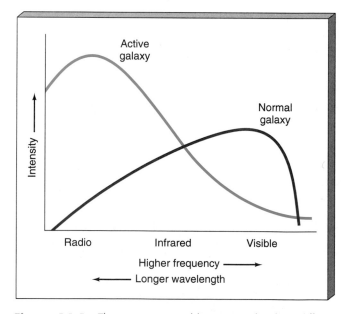

Figure 16.1 The energy emitted by a normal galaxy differs from that emitted by an active galaxy. (This plot illustrates the general run of intensity for all galaxies of a particular type and does not represent any one individual galaxy.)

radiation. Quasars, on the other hand, are mostly so far away that little internal structure can be discerned. However, the distinction between active galaxies and quasars is largely historical, dating back to the days when the connection between them was not understood. As we will see, most astronomers now believe that quasars are simply an early stage of galaxy formation and that there is really no sharp dividing line between quasars and active galaxies. Many researchers now go so far as to include quasars in the "active galaxy" category.

Not all active galaxies are distant, and only a small fraction of all distant galaxies are active. Some active galaxies are found locally, scattered among the normal galaxies that make up most of our cosmic neighborhood, and many faraway normal galaxies are known. As a general rule, though, active galaxies are more common at greater distances, and the most active objects lie farthest from Earth. Physical conditions in the universe at earlier times were undoubtedly different from conditions today, so we should perhaps not be surprised that remote astronomical objects, which emitted long ago the radiation we observe today, differ from nearby objects, which emitted their radiation much more recently. What *is* surprising—in fact, astounding—is the *amount* of energy radiated by some of the most luminous objects. Their tremendous power, nonstellar radiation, and abundance at great distances suggest to many astronomers that the universe was once a much more violent place than it is today.

16.2 Seyfert Galaxies

In 1943 Carl Seyfert, an American optical astronomer studying spiral galaxies from Mount Wilson Observatory, discovered the type of active galaxy that now bears his name. **Seyfert galaxies** are a class of astronomical objects whose properties lie between those of normal galaxies and those of the most violent active galaxies known. This fact suggests to some astronomers that Seyferts represent an evolutionary link between these two extremes. The spectral lines of Seyfert galaxies are usually substantially redshifted, telling us that most Seyferts reside at large distances (hundreds of megaparsecs) from us. ∞ (Sec. 15.5) However, a few lie just 20 or 30 Mpc away.

Figure 16.2 shows two visible-light images of a typical Seyfert galaxy. A casual glance at a long-exposure photograph (Figure 16.2a) reveals nothing

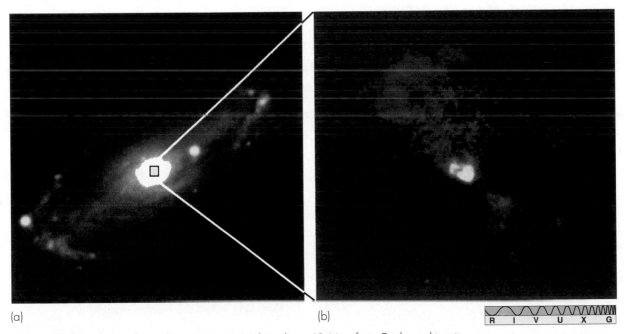

(a)　　　(b)

R I V U X G

Figure 16.2 The Seyfert galaxy NGC 5728 lies about 40 Mpc from Earth, making it one of the closest active galaxies known. (a) A ground-based view. (b) A higher-resolution view from Earth orbit, showing two cone-shaped beams of light and a group of glowing blobs near the galaxy's nucleus, perhaps illuminated by radiation arising from the accretion disk of a black hole.

strange. Superficially, Seyferts resemble normal spiral galaxies. However, closer study reveals some peculiarities not found in normal spirals.

First, maps of Seyfert energy emission show that nearly all of the radiation stems from a small central region known as the **galactic nucleus**. This region lies at the center of the overexposed white patch in Figure 16.2(a); it is shown in more detail in Figure 16.2(b). Astronomers suspect that a Seyfert nucleus may be quite similar to the center of a normal galaxy (the Milky Way or Andromeda, for example), but with one very important difference: the nucleus of a Seyfert is 10,000 times brighter than the center of our Galaxy. Indeed, the brightest Seyfert nuclei are 10 times more energetic than the *entire* Milky Way Galaxy.

Second, Seyfert galaxies emit their radiation in two broad frequency ranges. The stars in a Seyfert's galactic disk and spiral arms produce about the same amount of visible radiation as the stars of a normal spiral galaxy. However, most of the energy from the Seyfert's bright nucleus is emitted in the form of *invisible* radio and infrared radiation, and must be nonstellar in origin.

Third, Seyfert spectral lines bear little or no resemblance to those produced by ordinary stars, although the lines do have many similarities to those observed toward the center of our own Galaxy. ∞ (Sec. 14.7) Seyfert spectra contain strong emission lines of highly ionized heavy elements, especially iron. The lines are very broad, indicating either that the galaxy's gases are tremendously hot (more than 10^8 K) or that they are rotating very rapidly (at about 1000 km/s) around some central object. ∞ (Sec. 2.7) The first possibility can be ruled out because such a high temperature would cause all the gas to be ionized, so that no spectral lines would be produced. Thus, the broadening indicates rapid internal motion in the nucleus.

Finally, long-term monitoring of Seyfert radiation has shown that the energy emission often varies in time (Figure 16.3). These radiative changes are unlike anything found in the Milky Way Galaxy or any other normal galaxy. A Seyfert's luminosity can double or halve within a fraction of a year. These rapid fluctuations in luminosity lead us to conclude that the source of energy emissions must be quite compact. As mentioned in Chapter 13, in order for astronomers to be able to detect a variation in brightness within a certain time interval, the diameter of the source must be less than the distance traveled by light during that interval.

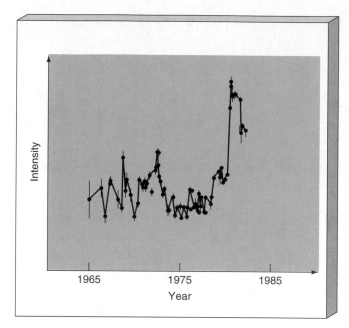

Figure 16.3   The irregular variations of a particular Seyfert galaxy's luminosity over two decades. Because this Seyfert, called 3C 84, emits most strongly in the radio part of the electromagnetic spectrum, these observations were made with large radio telescopes. The optical and X-ray luminosities vary as well.

∞ (Sec. 13.7) Simply put, an object cannot "flicker" in less time than radiation takes to cross it. Because the rise and fall of a Seyfert's radiation usually occurs within one year, we can confidently conclude that the emitting region must be less than one light year across—an extraordinarily small region, considering the amount of energy emanating from it. High-resolution interferometric radio maps of Seyfert nuclei generally confirm this reasoning. ∞ (Sec. 4.3)

The rapid time variability and large radio and infrared luminosities observed in Seyferts together imply violent nonstellar activity in their nuclei. This activity may well be similar in *nature* to processes occurring at the center of our own Galaxy, but its *magnitude* is thousands of times greater than the comparatively mild events within our own Galaxy's heart. ∞ (Sec. 14.7)

16.3 Radio Galaxies

2 As the name suggests, **radio galaxies** are active galaxies that emit most of their energy in the radio portion of the electromagnetic spectrum. They differ from Seyferts not only in the wavelengths at which they radiate most strongly (Seyferts radiate at shorter radio wavelengths and at infrared wavelengths; radio galaxies radiate at longer radio wavelengths) but also in both the appearance and the extent of their emitting regions. Whereas the nonstellar emission from Seyfert galaxies always comes from a small central nucleus, the nonstellar emitting regions of radio galaxies can be much larger—hundreds of kiloparsecs across in some cases.

Core–Halo Radio Galaxies

One common type of radio galaxy is often called a *core–halo* radio galaxy. As illustrated in Figure 16.4, the radio energy from such an object comes mostly from an extremely small central nucleus (which radio astronomers refer to as the *core*) less than 1 pc across, with weaker radio emission coming from an extended region called the *halo* surrounding the nucleus. The halo typically measures about 50 kpc across, comparable in size to the visible galaxy,[1] which is usually elliptical and often quite faint. The radio luminosity from the nucleus can be as great as 10^{37} W—about the same as the total emission from a Seyfert nucleus and comparable to the output from the Milky Way Galaxy at all wavelengths.

[1]The term "visible galaxy" is commonly used to refer to the components of an active galaxy that emit visible "stellar" radiation, as opposed to the nonstellar "active" component of the galaxy's emission.

Figure 16.4 Radio contour map of a typical core–halo radio galaxy. The radio emission from such a galaxy comes from a bright central nucleus, which is surrounded by an extended, less intense radio halo. The radio map is superimposed on an optical image of the galaxy and some of its neighbors, shown previously in Figure 15.14.

Figure 16.5 presents several images of a nearby core–halo radio galaxy. This is a giant elliptical galaxy known as M87—the 87th object in Messier's catalog. (We can be sure that this eighteenth-century Frenchman had no idea what he was really looking at. Nor perhaps do we!) M87 is roughly 15 Mpc distant, a prominent member of the Virgo Cluster, and one of the closest active galaxies. Its nearness and interesting activity have made it one of the most intensely studied of all astronomical objects.

A long time exposure (Figure 16.5a) shows a large, fuzzy ball of light—a fairly normal-looking E1 galaxy whose full extent is about 100 kpc across (considerably larger than the view shown in the figure). A shorter-time exposure of M87 (Figure 16.5b), capturing only the galaxy's bright inner regions, reveals a long thin *jet* of matter ejected from M87's center. The jet is about 2 kpc long and is traveling outward at very high speed, possibly as much as half the speed of light. Computer enhancement shows that the jet is made up of a series of distinct "blobs" more or less evenly spaced along its length. This high-speed jet, which emits energy at the rate of almost 10^{35} W, has been imaged in the radio and infrared regions of the spectrum (Figures 16.5c and d) as well as in the visible region. Jets such as this are a very common feature of active galaxies. As we will see, they play a vital role in our understanding of these energetic objects.

Lobe Radio Galaxies

Many radio galaxies are not of the core–halo type. Very little of their radio emission arises from the compact central nucleus. Instead, most of this emission comes from huge extended regions called **radio lobes**—roundish clouds of gas up to 1 Mpc across, lying well beyond the center of the visible part of the galaxy. For this reason, these objects are known as *lobe radio galaxies*.

The radio lobes of lobe radio galaxies are truly enormous. From end to end, an entire lobe radio galaxy typically is more than 10 times the size of the Milky Way Galaxy, comparable in size to the entire Local Group. The radio luminosity of the lobes can range from 10^{36} to 10^{38} W—between 1/10 and 10 times the total energy emitted by our Galaxy. Several of these strange objects are located relatively nearby, so we can study them at close range. One such system, known as

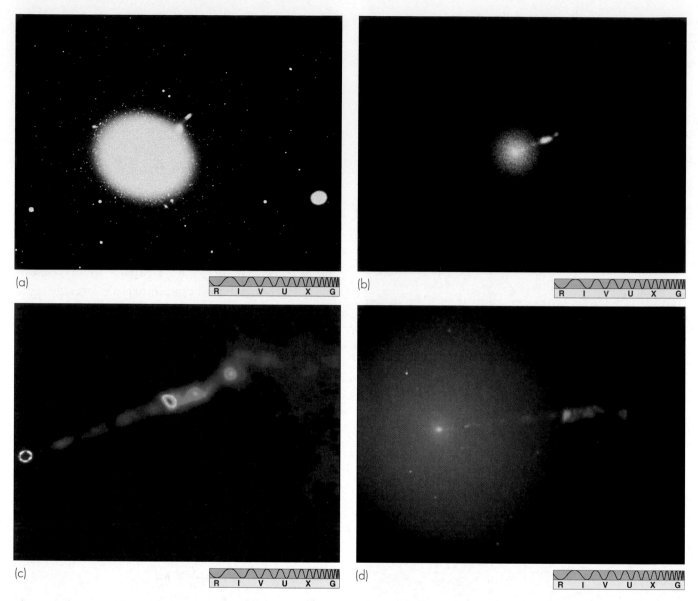

Figure 16.5 A core–halo radio galaxy—the giant elliptical galaxy M87 (also called Virgo A)—is displayed here at several wavelengths. (a) A long optical exposure, showing mainly the galaxy's outer halo. (b) A shorter optical exposure of the galaxy's nucleus and an intriguing jet of matter, on the same scale as (a). (c) A radio image of the jet, on a scale larger than that in (b). The red dot at left marks the bright nucleus of the galaxy; the red and yellow blob near the center of the image corresponds to the bright area visible in the jet in (b). (d) A near-infrared image of the jet, at roughly the same scale as (c).

Centaurus A, is shown in Figure 16.6. It lies only 4 Mpc from Earth.

Figure 16.7 superposes another representation of Figure 16.6 on an optical image of Centaurus A, to show the relationship between the visible and radio emission. In visible light, Centaurus A is a peculiar-looking object, apparently an E2 galaxy bisected by an irregular band of dust. Numerical simulations suggest that this system is probably the result of a merger

between an elliptical galaxy and a smaller spiral galaxy about 500 million years ago. The radio lobes are roughly symmetrically placed, jutting out from the center of the visible galaxy, roughly perpendicular to the dust lane. The elliptical galaxy itself is very large—some 500 kpc in diameter.

The lobes of radio galaxies vary in size and shape from galaxy to galaxy, but they are aligned with the center of the visible galaxy in nearly all cases. This

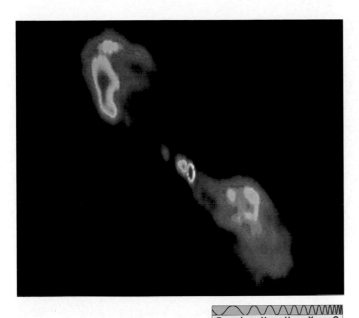

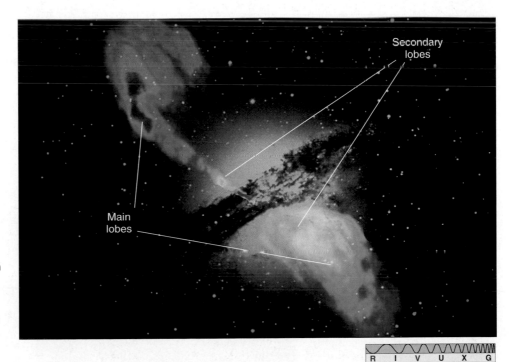

Figure 16.6 Lobe radio galaxies, such as Centaurus A shown here, have huge radio-emitting regions extending a million parsecs or more beyond the center of the galaxy. The lobes emit no visible light and so are observable only with radio telescopes. (The lobes are shown here in false color, with decreasing intensity from red to yellow to green to blue.)

alignment suggests that the lobes consist of material that was somehow ejected in opposite directions by violent events in the galactic nucleus. In the case of Centaurus A, this argument is strengthened by the presence of a pair of secondary lobes, smaller than the

main lobes (about 50 kpc in length, marked in Figure 16.7) and closer to the visible galaxy. Both pairs of lobes share the same high degree of linear alignment with the galactic center. Astronomers believe that the inner lobes were expelled from the nucleus by the same basic process as the outer ones, but more recently, so they have not had time to travel as far. Studies at still higher resolution reveal the presence of a roughly 1-kpc-long jet in the center of Centaurus A, aligned with the lobes.

If material is ejected from the nucleus at close to the speed of light (which seems likely), it follows that Centaurus A's outer lobes were created a few hundred million years ago, quite possibly around the time of the elliptical/spiral merger thought to be responsible for the galaxy's odd optical appearance. Apparently some violent process at the center of Centaurus A—most probably triggered by the merger—started up around that time and has been intermittently firing jets of matter out into intergalactic space ever since.

Further evidence in favor of the interpretation of radio lobes as material ejected from the nucleus of a galaxy is provided by another object, Cygnus A, shown as an optical image in Figure 16.8(a) and as a high-resolution radio map in Figure 16.8(b). The filamentary structure evident in the radio lobes and the thin, radio-emitting line joining the right lobe to the center of the visible galaxy (the dot at the center of the radio image)

Figure 16.7 An optical photograph of Centaurus A, one of the most massive and peculiar galaxies known and believed to be the result of a collision between two galaxies that took place 500 million years ago. The pastel false colors mark the radio emission shown in Figure 16.6, in this case more recently acquired and with higher resolution. Note that the radio lobes emit no visible light.

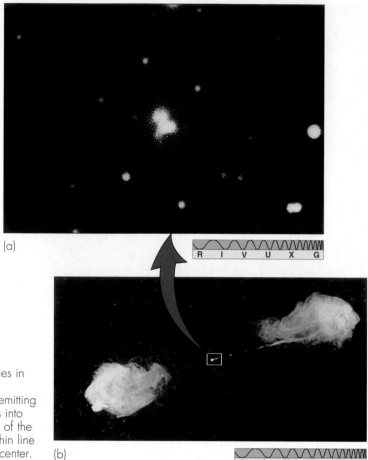

Figure 16.8 (a) Cygnus A also appears to be two galaxies in collision, although it is not completely clear what is really happening. (b) A radio image of Cygnus A shows the radio-emitting lobes on either side of the visible center. To put these images into proper perspective, the optical galaxy in (a) is about the size of the small dot at the center of the radio image in (b). Notice the thin line of radio-emitting material joining the right lobe to the galaxy center.

strongly suggest that we are seeing two oppositely directed, narrow jets of material running into the intracluster gas filling the galaxy cluster of which Cygnus A is a member, then slowing down and spreading out to form the radio lobes. ∞ (Sec. 15.3)

In some systems, known as *head–tail* radio galaxies, the lobes seem to form a "tail" behind the main part of the galaxy. For example, the lobes of radio galaxy NGC 1265, shown in Figure 16.9, appear to be swept back by some onrushing wind, and indeed, this is the most likely

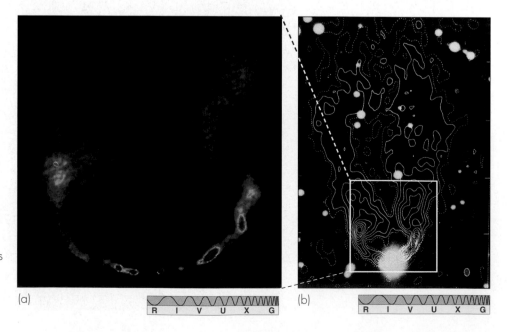

Figure 16.9 (a) Radio image, in false color, of the active "head–tail" galaxy NGC 1265. (b) The same radio data, on a somewhat larger scale and in contour form, superposed on the optical image of the galaxy and its surroundings. Astronomers reason that this object is moving rapidly through space, the motion forcing the lobes to trail behind the central part of the galaxy, forming the tail.

explanation for their appearance. If NGC 1265 were at rest, it would be just another double-lobe source, perhaps looking quite similar to Centaurus A. However, the galaxy is traveling through the intergalactic gas of its parent galaxy cluster (known as the Perseus Cluster), and the outflowing matter forming the lobes tends to be left behind as the galaxy moves.

Radio galaxies share many characteristics with Seyfert galaxies. Both types of active galaxy emit comparably large amounts of energy, and there is good evidence that the energy source in both is a compact region at the center of an otherwise relatively normal-looking galaxy. In lobe radio galaxies, that energy is fired out from the nucleus in the form of narrow, high-speed jets of matter that travel into the intergalactic medium and become extended lobes far from the center of the galaxy. As a result, the energy from a lobe radio galaxy is ultimately emitted (in the form of radio radiation) from a region well outside the visible galaxy. It may be that the differences between core–halo and lobe radio galaxies is largely a matter of perspective (Figure 16.10). If we view the jets and lobes from the side, we see a lobe radio galaxy, but if we view the jet almost head-on—in other words, looking *through* the lobe—we see a core–halo system. If that is so, then in

all cases studied so far the central compact nucleus is the place where the energy is actually produced. Let us now consider the current view of the "engine" that powers all this activity.

16.4 The Central Engine of an Active Galaxy

3 The behavior of active galaxies is contrary to that expected from vast collections of stars. The lobe radio galaxies in particular, with their huge energy emission from far beyond the visible galaxy, are among the most powerful objects in the universe. Can we explain this enormous nonstellar energy output in terms of known physics? Remarkably, the answer is yes. The present consensus among astronomers is that, despite great differences in appearance, Seyferts and radio galaxies may share a common energy-generation mechanism.

As a class, active galaxies (and quasars too, as we will see) have some or all of the following properties:

1. They have *high luminosities*, generally greater than the 10^{37} W characteristic of a fairly bright normal galaxy.

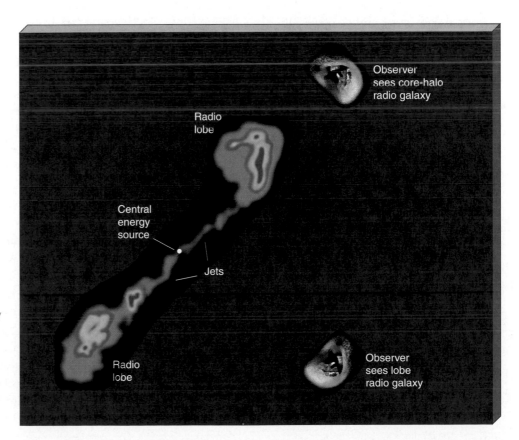

Figure 16.10 A central energy source produces high-speed jets of matter that interact with intergalactic gas to form radio lobes. The system may appear to us as either a lobe or a core–halo radio galaxy, depending on our location with respect to the jets and lobes.

2. Their energy emission is mostly *nonstellar*—it cannot be explained as the combined radiation of even trillions of stars.

3. Their energy output can be highly *variable*, implying that it is emitted from a small central nucleus much less than a parsec across.

4. They often exhibit *jets* and other signs of explosive activity.

5. Their optical spectra may show broad emission lines, indicative of *rapid internal motion* within the energy-producing region.

The principal questions then are: How can such vast quantities of energy arise from these relatively small regions of space? Why is so much of the energy radiated at long wavelengths, in the radio and infrared? And what is the origin of the extended radio-emitting lobes and jets? Let us first consider how the energy is produced.

Energy Production

To develop a feeling for the enormous emissions of active galaxies, consider for a moment an object having a luminosity of 10^{38} W. In and of itself, this energy output is not inconceivably large. The brightest giant ellipticals are comparably powerful. Thus, some 10^{12} stars—a few normal galaxies' worth of material—could *equivalently* power a typical active galaxy. The difficulty arises when we consider that in an active galaxy this energy production is packed into a region much less than a parsec in diameter!

It is difficult to imagine how several Milky Way Galaxies could be compressed into a space no larger than a parsec. Even if we could somehow squeeze that much mass into such a volume, it would immediately collapse to form a huge black hole, and none of the light produced could escape! Thus, even neglecting its nonstellar spectrum, the total energy output of an active galaxy simply cannot be explained as the combined energy of many stars. We must think of something else.

The twin requirements of large energy generation and small physical size bring to mind our discussion of X-ray sources in Chapter 13. ∞ (Sec. 13.3, 13.7) The presence of the jet in M87 and the radio lobes in Centaurus A and Cygnus A strengthen the connection, as such phenomena have also been observed in

some stellar X-ray-emitting systems. Recall that the best current explanation for those "small-scale" phenomena involves the accretion of material onto a compact object—a neutron star or a black hole. Large amounts of energy are produced as matter spirals down onto the central object, and high-speed jets may well be a common by-product of the process. In Chapter 14 we suggested that a similar mechanism, involving a *supermassive black hole*—one having a mass of around a million suns—may also be responsible for the energetic radio and infrared emission observed at the center of our own Galaxy. ∞ (Sec. 14.7)

As illustrated in Figure 16.11, the leading model for the central engine of active galaxies is essentially a scaled-up version of the same accretion process, only now the black holes involved have masses between a few million and a billion times the mass of the Sun. As with this model's smaller-scale counterparts, infalling gas forms an accretion disk and spirals down toward the hole. It is heated to high temperatures by friction within the disk and emits large amounts of radiation as a result. In this case, however, the origin of the accreted gas is not a binary companion, as in stellar X-ray sources, but entire stars and clouds of interstellar gas that come too close to the hole and are torn apart by its strong gravity.

The accretion process is extremely efficient at converting infalling mass (in the form of gas) to energy (in the form of electromagnetic radiation). Detailed calculations indicate that as much as 10 or 20 percent of the total mass-energy of the infalling matter can be radiated away before it crosses the hole's event horizon and is lost forever. ∞ (Sec. 13.4) Since the total mass-energy of a star like the Sun—the mass times the speed of light squared—is about 2×10^{47} J, it follows that the 10^{38} W luminosity of a bright active galaxy can be accounted for by the consumption of only 1 solar mass of gas per decade by a billion-solar-mass black hole. Less luminous active galaxies would require correspondingly less fuel—for example, the central black hole of a 10^{36}-W Seyfert galaxy would devour only one Sun's worth of material every thousand years.

In this picture, the small size of the emitting region is a direct consequence of the compact nature of the central black hole. Even a billion-solar-mass hole has a radius of only 3×10^9 km, or 10^{-4} pc—about 20 A.U.—and theory suggests that the part of the accretion disk responsible for most of the emission would be much less than 1 pc across. Instabilities in the

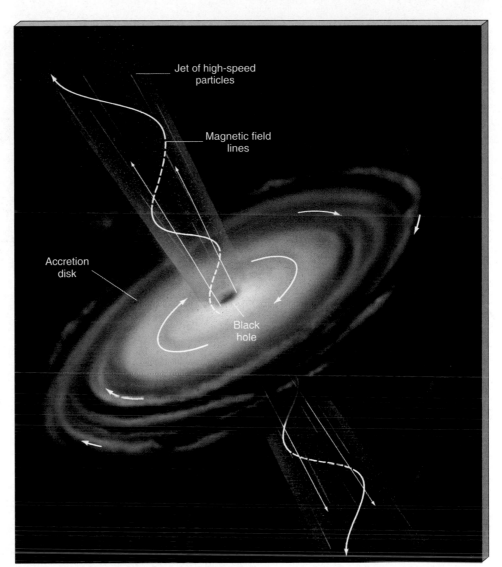

Figure 16.11 The leading theory for the energy source in active galactic nuclei holds that these objects are powered by material accreting onto a supermassive black hole. As matter spirals toward the hole, the matter heats up, producing large amounts of energy. At the same time, high-speed jets of gas may be ejected perpendicular to the accretion disk, giving rise to the jets and lobes seen in many active objects. Magnetic fields generated in the disk are carried by the jets out to the radio lobes, where they play a crucial role in producing the observed radiation.

accretion disk can cause fluctuations in the energy released, leading to the variability observed in many objects. The broadening of the spectral lines observed in the nuclei of many active galaxies results from the rapid orbital motion of the gas in the hole's intense gravity.

Observations of galaxies in the Virgo Cluster by the *Hubble Space Telescope* lend strong support to this general picture. Figure 16.12 shows an image of a disk of gas and dust apparently feeding a possible black hole at the core of a giant elliptical active galaxy. As expected from the theory just described, the disk is perpendicular to the huge jets emanating from the galaxy's center.

Hubble has also allowed astronomers to probe the fine details of the huge core–halo radio galaxy M87 (Figure 16.5), and what they have found is in excellent

agreement with the idea that the energy is produced by accretion onto a large black hole. At M87's distance, *Hubble*'s resolution of 0.05 arc second corresponds to a distance of about 5 pc, so we are still far from seeing the (solar-system–sized) central black hole itself, but the improved "circumstantial" evidence has convinced many doubters of the correctness of the theory. Figure 16.13 shows imaging and spectroscopic data that suggest the existence of a rapidly rotating disk of matter orbiting the galaxy's center, perpendicular to the jet. Measurements of the gas velocity on opposite sides of the disk indicate that the mass within a few parsecs of the center is approximately 3×10^9 solar masses—we assume that this is the mass of the central black hole.

Even more compelling evidence for supermassive black holes has recently come from radio studies. Using the Very Long Baseline Array, a continentwide net-

Figure 16.12 (a) A combined optical/radio image of the giant elliptical galaxy NGC 4261, in the Virgo Cluster, shows a white visible galaxy at center, from which red-orange (false color) radio lobes extend for about 60 kpc. (b) A close-up photograph of the galaxy's nucleus reveals a 100-pc-diameter disk surrounding a bright hub thought to harbor a black hole.

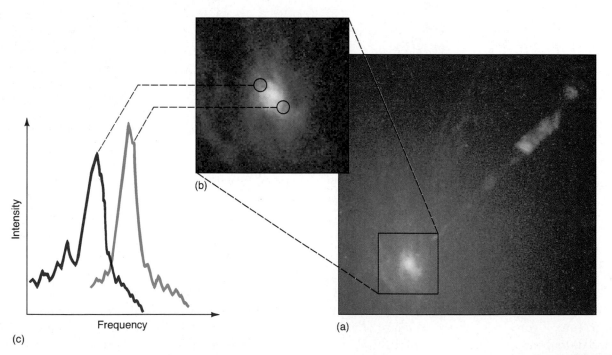

Figure 16.13 Recent imaging and spectroscopic observations of M87 support the idea of a rapidly whirling accretion disk at its heart. (a) An image of the central region of M87, similar to that shown in Figure 16.5(d), shows its bright nucleus and jet. The scale is comparable to the scale of Figure 16.5(c). (b) A magnified view of the nucleus suggests a spiral swarm of stars, gas, and dust. (c) Spectral-line features observed on opposite sides of the nucleus show opposite Doppler shifts, implying that material on one side of the nucleus is coming toward us and material on the other side is moving away from us. The strong implications are that an accretion disk spins perpendicular to the jet and that at its center is a black hole having some 3 billion times the mass of the Sun.

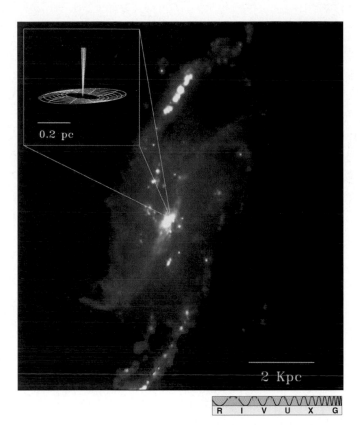

R   I   V   U   X   G

Figure 16.14 A network of radio telescopes has probed the nucleus of the spiral galaxy NGC 4258, shown here in the light of mostly hydrogen emission. Within the innermost 0.2 pc (inset), observations of Doppler-shifted molecular clouds (designated by red, green, and blue dots) show that they obey Kepler's third law perfectly and reveal a slightly warped disk of rotating gas (shown here in artist's conception). At the center of the disk presumably lurks a huge black hole.

work of 10 radio telescopes, a U.S.–Japanese team was able to achieve angular resolution hundreds of times better than that attainable with the *Hubble Space Telescope*. VLBA observations of NGC 4258, a spiral galaxy about 6 Mpc away, have uncovered a group of molecular clouds swirling in an organized fashion about the galaxy's center. Astronomers have detected the red- and blueshifts of molecular spectral lines in those faraway clouds. As depicted in Figure 16.14, the observations reveal a slightly warped, spinning disk centered precisely on the galaxy's heart. The rotation speeds imply the presence of more than 40 million solar masses packed into a region less than 0.2 pc across.

Energy Emission

Having accounted for the source of the energy in active galaxies, let us now turn to the way in which it is eventually emitted into intergalactic space. In gen-

eral, the theoretically predicted radiation spectra for matter accreting onto a black hole do not match well with the spectra observed in active galaxies. In order to account for the details of the spectra of Seyfert galaxies, it is necessary to modify the mechanism just described by adding the assumption that the energy emitted from the accretion disk is "reprocessed"—that is, absorbed and reemitted at infrared and longer wavelengths—by gas and dust surrounding the nucleus.

A different reprocessing mechanism is responsible for the spectra of the jets and lobes seen in many active galaxies, especially radio galaxies. The jets consist of material (mainly protons and electrons) blasted out into space—and out of the visible portion of the galaxy entirely—from the inner regions of the disk. The details of how jets form remain uncertain, but there is a growing consensus among theorists that jets are a common feature of accretion flows, large and small. The jets also contain strong *magnetic fields* (shown in Figure 16.11), possibly generated by the swirling gas motion within the disk, which accompany the gas as it leaves the galaxy.

As sketched in Figure 16.15(a), whenever a charged particle (here an electron) encounters a magnetic field, the particle tends to spiral around the magnetic field lines. We have encountered this idea several times previously, in a variety of contexts (see, for example, the discussion of Earth's magnetosphere in Chapter 5 or solar activity in Chapter 9). As the particles whirl around, they emit electromagnetic radiation, as discussed in Chapter 2. ∞ (Sec. 2.2) The faster the particles move, or the stronger the magnetic field, the greater the amount of energy radiated. In most cases, the electrons are the fastest-moving particles, and they are responsible for essentially all of the radiation we observe.

The electromagnetic radiation produced in this way—called **synchrotron radiation**, after the type of particle accelerator in which it was first observed—is *nonthermal* in nature: there is no link between the emission and the temperature of the radiating object, so the radiation is not described by a black-body curve. Instead, its intensity increases with decreasing frequency, as shown in Figure 16.15(b). This is just what is needed to explain the overall spectrum of radiation recorded from active galaxies (compare Figure 16.15b with Figure 16.1). Observations of the radiation received from the jets and radio lobes of active galaxies are completely consistent with this process.

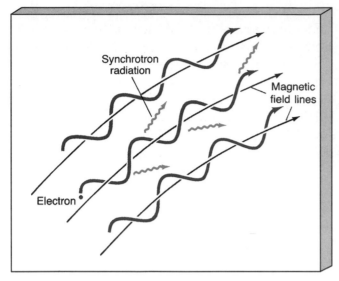

(a)

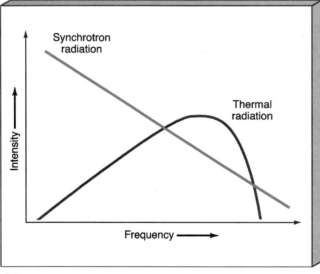

(b)

Figure 16.15 (a) Charged particles, especially fast-moving electrons, emit synchrotron radiation while spiraling in a magnetic field. This process is not confined to active galaxies. It occurs, on smaller scales, when charged particles interact with magnetism in Earth's Van Allen belts (Section 5.6), when charged matter arches above sunspots on the Sun (Section 9.4), in the vicinity of neutron stars (Section 13.2), and at the center of our own Galaxy (Section 14.7). (b) How the intensity of thermal and synchrotron (nonthermal) radiation varies with frequency. Thermal radiation, described by a black-body curve, peaks at some frequency that depends on the temperature of the source. Nonthermal synchrotron radiation, by contrast, is most intense at low frequencies and is independent of the temperature of the emitting object. Compare this figure with Figure 16.1.

Eventually, the jet is slowed and stopped by the intergalactic medium, the flow becomes turbulent, and the magnetic field grows tangled. The result is a gigantic radio lobe, like those pictured in Figures 16.6–16.8, emitting virtually all of its energy in the form of synchrotron radiation. Even though the radio emission comes from an enormously extended volume of space that dwarfs the visible galaxy, the source of the energy is still the (relatively) tiny accretion disk—a billion billion times smaller in volume than the radio lobe—lying at the galactic center. The jets serve merely as a conduit to transport energy from the nucleus, where it is generated, into the lobes, where it is finally radiated into space.

The existence of the inner lobes of Centaurus A and the blobs in M87's jet imply that jet formation may be an intermittent process. There is also evidence to suggest that much, if not all, of the activity observed in nearby active galaxies could have been sparked by recent interaction with a neighbor. Many nearby active galaxies (Centaurus A, for example) appear to have been "caught in the act" of interacting with another galaxy, suggesting that the fuel supply can sometimes be turned on by a companion. Just as tidal forces can trigger star formation in starburst galaxies, they may also divert gas and stars into the galactic nucleus, triggering an outburst that may last for millions or even billions of years. ∞ (Sec. 15.4)

16.5 Quasi-stellar Objects

The Discovery of Quasars *Animation*

In the early days of radio astronomy, many radio sources were detected for which no corresponding visible object was known. By 1960, several hundred such sources were listed in the *Third Cambridge Catalog*, and astronomers were scanning the skies in search of visible counterparts to these radio sources. Their job was made difficult both by the low resolution of the radio observations (which meant that the observers did not know exactly where to look) and by the faintness of these objects at visible wavelengths.

In 1960, astronomers detected what appeared to be a faint blue star at the location of the radio source 3C 48 (the 48th object on the third Cambridge list) and obtained its spectrum. Containing many unknown broad emission lines, the unusual spectrum defied interpretation. 3C 48 remained a unique curiosity until 1962, when another similar-looking, and similarly mysterious, faint blue object with "odd" spectral lines was discovered and identified with the radio source 3C 273. Several more of these peculiar objects are shown in Figure 16.16.

The following year saw a breakthrough when astronomers realized that the strongest unknown lines in 3C 273's spectrum were simply familiar spectral lines of hydrogen redshifted by a very unfamiliar amount—about 16 percent! This large redshift indicated a re-

cessional velocity of about 48,000 km/s. Figure 16.17 shows the spectrum of 3C 273. Some prominent emission lines and the extent of their redshift are marked on the diagram. Once the nature of the strange spectral lines was known, astronomers quickly found a similar explanation for the spectrum of 3C 48. Its 37 percent redshift implied that it is receding from Earth at almost one-third the speed of light.

These huge speeds mean that neither of the two objects can possibly be members of our Galaxy. Applying Hubble's law (with our adopted value of $H_0 = 75$ km/s/Mpc), we obtain distances of 640 Mpc

Figure 16.16 (a) 3C 275, one of the first quasars discovered. Its starlike appearance shows no obvious structure and gives little outward indication of this object's enormous luminosity. However, 3C 275 has a much larger redshift than any of the other stars or galaxies in this image; it is about 2 billion parsecs away. (b) A field of quasars (marked), including QSO 1229+204, one of the most powerful quasars yet discovered, shown enlarged in (c). Like 3C 275, its distance from Earth is about 2000 Mpc.

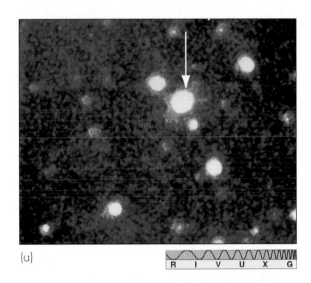

(a)

R   I   V   U   X   G

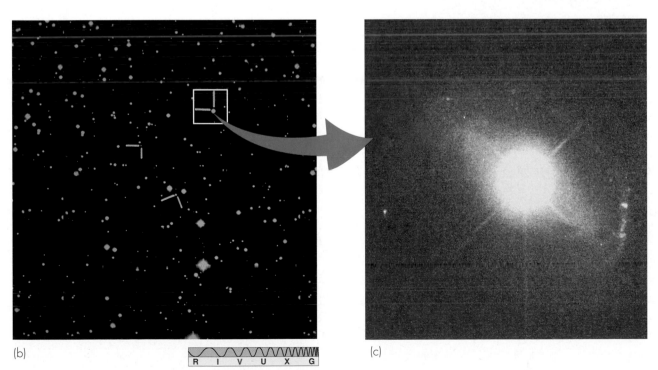

(b)

R   I   V   U   X   G

(c)

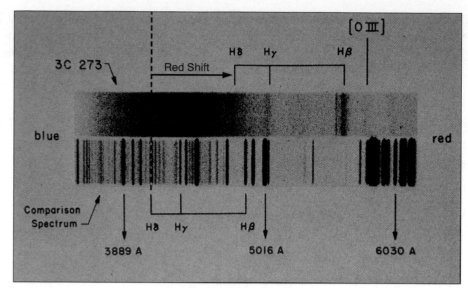

Figure 16.17 Optical spectrum of the distant quasar 3C 273. Notice both the redshift and the widths of the three hydrogen spectral lines marked as Hβ, Hγ, and Hδ. The redshift indicates the quasar's enormous distance. The width of the lines implies rapid internal motion within the quasar. (Note that, in this figure, red is to the right and blue is to the left.)

for 3C 273 and 1300 Mpc for 3C 48. Clearly not stars (with such enormous redshifts), these objects became known as *quasi-stellar radio sources* (quasi-stellar just means "starlike"). The term has been shortened to **quasars**. Because we now know that not all such highly redshifted, starlike objects are strong radio sources, the term **quasi-stellar object** (or QSO) is more common today. However, the name quasar persists, and we will continue to use it here.

Observed Properties of Quasars

The most striking characteristic of the several hundred quasars now known is that their spectra all show large redshifts, ranging from 0.06 (that is, a 6 percent increase in wavelength) up to the current maximum of just under 5. (See *More Precisely 16-1* for an explanation of redshifts greater than 1.) Thus, *all* quasars lie at large distances from us—the closest is 240 Mpc away, the farthest nearly 4700 Mpc (Table 16.1). The majority of quasars lie more than 1000 Mpc from Earth. We therefore see most quasars as they existed long ago— they represent the universe as it was in the distant past.

Thus, despite their unimpressive optical appearance (see, for example, Figure 16.18, which compares a quasar and a spiral galaxy that happen to lie close to one another on the sky), the large distances implied by quasar redshifts mean that these faint "stars" are in fact the brightest known objects in the universe! 3C 273,

for example, has a luminosity of about 10^{40} W. More generally, quasars range in luminosity from around 10^{38} W—about the same as the brightest radio galaxies—up to nearly 10^{42} W. A value of 10^{40} W, comparable to 20 trillion Suns or 1000 Milky Way Galaxies, is fairly typical. Thus quasars outshine the brightest normal and active galaxies by about a factor of 1000.

Quasars have many of the same general properties as active galaxies. Their radiation is nonthermal, and some show evidence of jets and extended emission features (although few quasars are more than a ball of luminous fuzz in visible-light images). Figure 16.19 is an optical photograph of 3C 273. Notice the jet of luminous matter, reminiscent of the jet in M87, extending nearly 3 kpc from the center of the quasar. Often, as shown in Figure 16.20, quasar radio radiation arises from regions lying beyond the bright nucleus, much like the emission from core–halo and lobe radio galaxies. In other cases, the radio emission is confined to the central part of the visible image. Quasars have been observed in the radio, infrared, visible, ultraviolet, and X-ray parts of the electromagnetic spectrum, and some have even been found to emit gamma rays. However, most quasars emit most of their energy in the infrared.

In addition to their own strongly redshifted spectra, many quasars also show additional absorption features that are redshifted by substantially *less* than the lines from the quasar itself. For example, the quasar PHL 938 has an emission-line redshift of 1.955, plac-

Figure 16.18 Although quasars are the most luminous objects in the universe, they are often unimpressive in appearance. In this optical image, a distant quasar (marked by an arrow) is seen close (on the sky) to a nearby spiral galaxy. The quasar's much greater distance makes it appear much fainter than the galaxy.

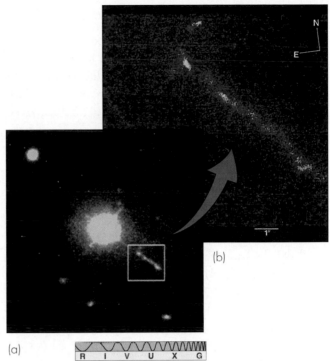

Figure 16.19 (a) The bright quasar 3C 273 displays a luminous jet of matter, but the main body of the quasar is starlike in appearance. (b) The jet extends for about 30 kpc and can be seen better in this high-resolution image.

ing it at a distance of some 3400 Mpc, but it also shows three sets of absorption lines that have redshifts of 1.949, 1.945, and 0.613, respectively. The first two sets may well come from high-speed gas within the quasar (the differences in recession velocities are only a few hundred kilometers per second), but the third is interpreted as arising from intervening gas that is much closer to us (only about 1700 Mpc away), which explains why it has a much smaller redshift than the quasar. The most likely possibility is that this gas is part of an otherwise invisible galaxy lying along the line of sight. Quasar spectra, then, afford astronomers an important means of probing previously undetected parts of the universe.

Many quasars have been observed to vary irregularly in brightness over periods of months, weeks, days, or (in some cases) even hours, in many parts of the electromagnetic spectrum. The same reasoning we used earlier for active galaxies leads to the conclusion that the region generating the energy must be very small—not much larger than our solar system in some cases.

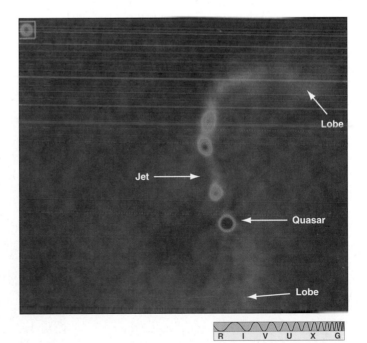

Figure 16.20 Radio image of the quasar 2300-189 showing radio jets feeding faint radio lobes. The bright (red) central object is the quasar, some 400 Mpc away from Earth.

Quasar Energy Generation and Lifetimes

5 Quasars have all of the properties described earlier for active galaxies, so it should come as no surprise that the best current explanation of the quasar engine is basically a scaled-up version of the mechanism powering lower-luminosity active galaxies—accretion onto a central, supermassive black hole.

A black hole having a mass of 10^8 or 10^9 solar masses can emit enough energy to power even the brightest (10^{38} W) radio galaxy by swallowing stars and gas at the relatively modest rate of 1 star every 10 years. To power a 10^{40}-W quasar, which is 100 times brighter, the hole simply consumes 100 times more fuel—10 stars per year. The reprocessing mechanisms that convert the quasar's power to the radiation we detect—namely, the ejection of matter in jets and lobes and the reemission of radiation by surrounding gas and dust—probably operate in much the same manner as described earlier for Seyferts and radio galaxies. The most likely explanation for the large luminosities of quasars is simply that there was more fuel available at very early times, perhaps left over from the formation of the galaxies in which the quasars reside. At the distances of most quasars, the galaxies themselves cannot be seen. Only their intensely bright nuclei are visible from Earth.

In this picture, the brightest known quasars devour about 1000 solar masses of material every year. A simple calculation indicates that if they kept up this rate of energy production for the roughly 10 billion years the universe has been in existence, a total of 10^{13} stars would be destroyed. Unless the galaxies housing quasars are much larger than any other galaxy we know of, most of the quasar's parent galaxy would be completely consumed by now, and the universe should contain many 10^{13}-solar-mass black holes—"burned-out" quasars. We have no evidence for the existence of any such objects. One way around this problem is to suppose that a quasar spends only a fairly short period of time in this highly luminous phase—perhaps a few tens of millions of years. There is theoretical evidence to suggest that black holes tend to eat out "cavities" at the centers of their host galaxies, effectively cutting off their fuel supply through their own greed. Alternatively, as with nearby active galaxies, the high luminosities of quasars may be the result of interactions between galaxies in the early universe. The fact that quasars have been observed in some distant galaxy clusters argues in favor of this latter view.

Quasar "Mirages" *Extension*

In 1979, astronomers were surprised to discover what appeared to be a binary quasar—two quasars with exactly the same redshift and very similar spectra, separated by only a few arc seconds on the sky. Remarkable as the discovery of such a binary would have been, the truth about this pair of quasars turned out to be even more amazing. Closer study of the quasars' radio emission revealed that they were *not* two distinct objects. Instead, they were two separate images of the *same* quasar! Optical views of such a *twin quasar* are shown in Figure 16.21.

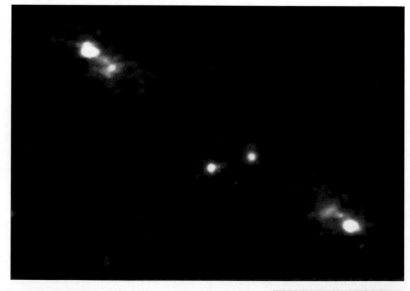

Figure 16.21 This twin quasar (designated AC114 and located about 2 billion parsecs away) is not two separate objects at all. Instead, the two large blobs (at upper left and lower right) are images of the same object, created by a gravitational lens. The lensing galaxy itself is probably not visible in this image—the two objects near the center of the frame are thought to be unrelated galaxies in a foreground cluster.

R I V U X G

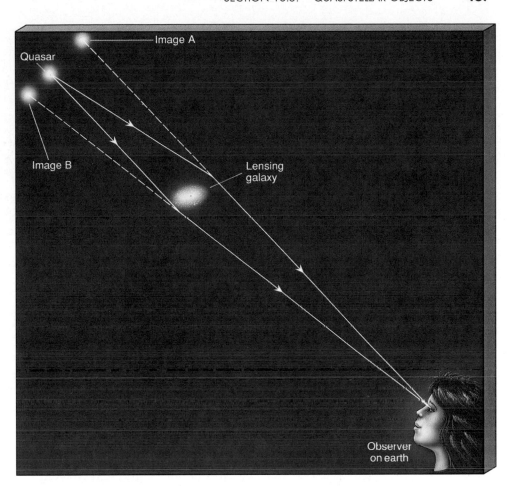

Figure 16.22 When light from a distant object passes close to a galaxy or cluster of galaxies along the line of sight, the image of the background object (here, the quasar) can sometimes be split into two or more separate images (here, A and B). The foreground object is a gravitational lens.

What could produce such a "doubling" of a quasar image? The answer is gravitational lensing—the deflection and focusing of light from a background object by the gravity of some foreground body, as illustrated in Figure 16.22. In Chapter 14 we saw how lensing by white dwarfs or other compact objects in the halo of the Milky Way Galaxy may temporarily cause the light from a distant star to be amplified, allowing

astronomers to detect otherwise invisible stellar dark matter in our Galaxy. ∞ (Sec. 14.6) In the case of quasars, the idea is the same, except that the foreground lensing object is an entire galaxy or galaxy cluster, and the deflection of the light is so great (an arc second or so) that several separate images of the quasar may be formed, as shown in Figure 16.23. About a dozen likely gravitational lenses are known.

Figure 16.23 (a) The Einstein Cross, a multiply imaged quasar. In this *Hubble* view, spanning only a couple of arc seconds, four images of the same quasar have been produced by the galaxy at the center. (b) An artist's conception of what might be occurring here, with Earth at right and the distant quasar at left.

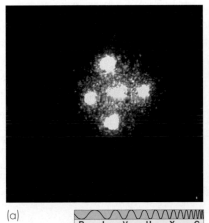

(a)

R I V U X G

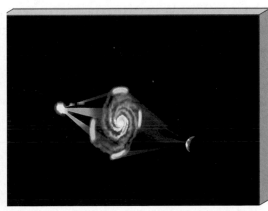

(b)

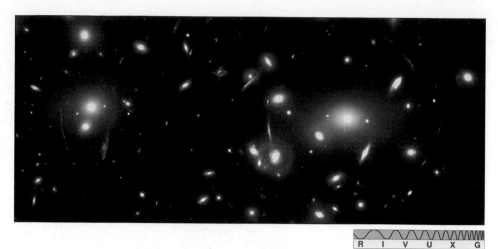

Figure 16.24 A spectacular example of gravitational lensing by a galaxy cluster. This wispy "spider's web" of more than a hundred faint arcs around a foreground galaxy cluster (A2218; still 1 billion parsecs away from us!) is an illusion caused by the cluster's gravitational field, which deflects and distorts the light from very distant background galaxies. By measuring the extent of this distortion, astronomers can estimate the mass of the cluster.

The existence of these multiple images provides astronomers with a number of useful observational tools. First, the lensing tends to amplify the light of the quasar, making it easier to observe. Second, because the light rays forming the images usually follow paths of different lengths, there is often a time delay, ranging from several days to several years, between them. This delay provides advance notice of explosive events, such as sudden flareups in the quasar's brightness—if one image flares up, astronomers know that in time the others will too, so they have a second chance to study the event. The time delay also permits astronomers to determine the distance to the lensing galaxy by carefully timing the measurements. If enough lenses can be found, this method may provide a reliable alternative means of measuring the Hubble constant that is independent of any of the techniques discussed in Chapter 15.

Third, lensing by individual stars in the foreground galaxy can cause large fluctuations in a quasar's brightness, just as discussed in Section 14.6, allowing astronomers to study the stellar content of the lensing galaxy. Finally, by studying the lensing of background quasars and galaxies by foreground galaxy clusters, astronomers can obtain a better understanding of the distribution of dark matter in those clusters, an issue that has great bearing on the large-scale structure of the cosmos. Figure 16.24 shows how the images of faint, background galaxies are bent into arcs by the gravity of a nearby galaxy cluster. The degree of bending allows the total mass of the cluster (*including* the dark matter) to be measured.

16.6 Active Galaxy Evolution

Animation   *Extension*

5 In Chapters 14 and 15 we addressed the issue of evolutionary change in normal galaxies. ∞ (Sec. 14.4, 15.4) Let us now briefly consider the possibility of evolutionary links among active galaxies and between normal and active galaxies. We emphasize that this section is really mostly speculation. Although the consensus is that galaxies began to form about 8 billion years ago (corresponding to a redshift of 5, which is an upper limit on the measured redshifts of known quasars) and that quasars were an early stage of galaxy evolution, the details of the connections among different types of active and normal galaxies are still very uncertain.

Most quasars are very distant, indicating that they were more common in the past than they are today. At the same time, normal galaxies seem to be less common in the distant past. These two pieces of evidence suggest to many astronomers that, when galaxies first formed, they probably were quasars. This view is strengthened by the fact that the same black-hole energy-generation mechanism can account for the luminosity of quasars, active galaxies, and the central regions of normal galaxies. Large black holes do not simply vanish. Thus, the presence of supermassive black holes in the centers of many, if not all, normal galaxies is consistent with the idea that they started off as quasars, then "wound down" to become the rela-

Figure 16.25 A possible evolutionary sequence for galaxies, beginning with the highly luminous quasars, decreasing in violence through the radio and Seyfert galaxies, and ending with normal spirals and ellipticals. The central black holes that powered the early activity are still there at later times; they simply run out of fuel as time goes on.

tively quiescent objects we see today. In this picture, the gradual reduction in violence from a quasar to a Seyfert galaxy to a normal spiral, for example, occurs primarily because the fuel supply is reduced as the galaxy evolves. A similar sequence might connect quasars to radio galaxies to normal ellipticals. These possible evolutionary connections between active and normal galaxies are illustrated in Figure 16.25.

If we accept this appealing (but still unproven) view, we can construct the following possible scenario for the evolution of galaxies in the universe: galaxies began to form about 8 billion years ago. The early round of massive star formation that may have expelled galactic gas and helped determine a galaxy's Hubble type—spiral or elliptical—could also have given rise to many stellar-mass black holes, which sank to the center of the still-forming galaxy and merged into a supermassive black hole there. Alternatively, the supermassive hole may have formed directly by gravitational collapse of the dense central regions of the protogalaxy. Whatever the cause, large black holes appeared at the centers of many galaxies at a time when there was still plenty of fuel available to power them, resulting in many highly luminous quasars. The brightest quasars—the ones we now see from Earth—were those with the greatest fuel supply.

Young galaxies at this early stage were much fainter than their bright quasar cores. As a result, until very recently, astronomers were hard-pressed to discern any galactic structure in quasar images. While computer-enhanced, ground-based observations hinted at wispy "fuzz" in some quasar images, evidence for host galaxies surrounding quasars was inconclusive. In 1996, several groups of astronomers used the *Hubble Space Telescope* to take up the challenge. Early results were ambiguous but, after removing the bright quasar

16-1 MORE PRECISELY

RELATIVISTIC REDSHIFTS AND LOOK-BACK TIME

When discussing very distant objects, astronomers usually talk about their redshifts rather than their distances. Indeed, it is very common for researchers to speak of an event occurring "at" a certain redshift—meaning that the light received today from that event is redshifted by the specified amount. Of course, because of Hubble's law, redshift and distance are equivalent to one another. However, redshift is the preferred quantity because it is a directly observable property of an object, whereas distance is derived from redshift using Hubble's constant, whose value is not accurately known. (In the next chapter, we will see another reason astronomers favor the use of redshift in studies of the cosmos.)

The redshift of a beam of light is, by definition, the *fractional* increase in its wavelength resulting from the recessional motion of the source. ∞ (More Precisely 2-2) Thus, a redshift of 1 corresponds to a *doubling* of the wavelength. Using the formula for the Doppler shift presented in *More Precisely 2-2*, the redshift of radiation received from a source moving away from us with speed v is given by

$$\text{redshift} = \frac{\text{observed wavelength} - \text{true wavelength}}{\text{true wavelength}}$$
$$= \frac{\text{recessional velocity, } v}{\text{speed of light, } c}.$$

Let's illustrate this with two examples, rounding the speed of light, c, to 300,000 km/s. A galaxy at a distance of 100 Mpc has a recessional speed (by Hubble's law) of 75 km/s/Mpc × 100 Mpc = 7,500 km/s. Its redshift therefore is 7,500 km/s ÷ 300,000 km/s = 0.025. Conversely, an object that has a redshift of 0.05 has a recessional velocity of 0.05 × 300,000 km/s = 15,000 km/s and hence a distance of 15,000 km/s ÷ 75 km/s/Mpc = 200 Mpc.

Unfortunately, while it is quite correct for low speeds, the foregoing equation does not take into account the effects of relativity. As we saw in Chapter 13, the rules of everyday physics have to be modified when speeds begin to approach the speed of light, and the formula for the Doppler shift is no exception. ∞ (More Precisely 13-1). In particular, while our formula is valid for speeds much less than the speed of light, when $v = c$, the redshift is not 1, as the equation suggests, but is in fact *infinite*. In other words, radiation received from an object moving away from us at nearly the speed of light would be redshifted to almost infinite wavelength.

Thus do not be alarmed to find that many quasars have redshifts greater than 1. This does not mean that they are receding at speeds faster than that of light! It simply means that their recessional speeds are relativistic—comparable to the speed of light—and the preceding simple formula is not applicable. Table 16.1 presents a conversion chart relating redshift, recession speed, and present distance. The column headed "v/c" gives recession velocities based on the Doppler effect, taking relativity properly into account (but see Section 17.2 for a more correct interpretation of the redshift.) All values are based on reasonable assumptions and usable even for $v \approx c$. As usual, we take Hubble's constant to be 75 km/s/Mpc. The conversions in the table are used consistently throughout this text.

Because the universe is expanding, the "distance" to a galaxy is not very well defined—do we mean the distance when the galaxy emitted the light we see today, or the present distance (as presented in the table, even though we do not see the galaxy as it is today), or is some other measure more appropriate? Largely because of this ambiguity, astronomers prefer to work in terms of a quantity known as the *look-back time* (shown in the last column of Table 16.1), which is simply how long ago an object emitted the radiation we see today. While astronomers talk frequently about redshifts and sometimes about look-back times, they hardly ever talk of the distances to high-redshift objects.

For nearby sources, the look-back time is numerically equal to the distance in light years—the light we receive tonight from a galaxy at a distance of 100 million light years was emitted 100 million years ago. However, for more distant objects, the look-back time and the present distance in light years differ because of the expansion of the universe, and the divergence increases dramatically with increasing redshift. For example, a galaxy now located 15 billion light years from Earth was much closer to us when it emitted the light we now see. Consequently, its light has taken considerably less than 15 billion years—in fact, only about 8 billion years—to reach us.

Table 16.1 Redshift, Distance, and Look-Back Time

Redshift	v/c	Present Distance (Mpc)	Present Distance (10^6 ly)	Look-Back Time (millions of years)
0.000	0.000	0	0	0
0.010	0.010	40	129	129
0.025	0.025	98	320	316
0.05	0.049	193	628	613
0.10	0.095	372	1214	1158
0.25	0.220	844	2753	2473
0.50	0.385	1468	4785	3961
1.00	0.600	2343	7638	5619
2.00	0.800	3381	11021	7019
3.00	0.882	4000	13038	7606
4.00	0.923	4422	14415	7915
5.00	0.946	4733	15431	8101
10.0	0.984	5587	18214	8454
100.0	1.000	7203	23482	8683
∞	1.000	8000	25253	8692

R I V U X G

Figure 16.26 This long-exposure *Hubble* image of a distant quasar clearly shows the young host galaxy in which the quasar resides, lending strong support to the view that quasars represent an early, very luminous phase of galaxy evolution. The quasar has the catalog name PG0052+251 and resides roughly 430 Mpc from Earth.

core from the *Hubble* images and carefully reanalyzing the remnant light, researchers have reported that, in every case studied—a total of about a dozen quasars so far—there appears to be a normal host galaxy enveloping the quasar. Figure 16.26 shows one of the longest exposures ever taken of a quasar. Even without sophisticated computer processing, the host can clearly be seen. Indeed, there is even some resemblance to a normal spiral galaxy, with thin arms extending away from the core.

As a young galaxy developed and its central black hole used up its fuel, the luminosity of the nucleus diminished. While still active, this galaxy no longer completely overwhelmed the emission from the surrounding stars. The result was an active galaxy—either radio or Seyfert—still emitting a lot of energy, but now with a definite "stellar" component in its spectrum.

The central activity continued to decline. Eventually, only the surrounding galaxy could be seen—a normal galaxy, like the majority of those we now see around us. Today, the black holes that generated so much youthful energy lie dormant in galactic nuclei, producing only a relative trickle of radiation. Occasionally, two normal galaxies may interact with one another, causing a flood of new fuel to be directed toward the central black hole of one or both. The engine starts up for a while, giving rise to the nearby active galaxies we observe.

Should this picture be correct—and the confirmation of host galaxies surrounding many quasars lends strong support to this view—then many normal galaxies, including perhaps our own Milky Way Galaxy, were once brilliant quasars. Perhaps some alien astronomer, thousands of megaparsecs away, is at this very moment observing our Galaxy—seeing it as it was billions of years ago—and is commenting on its enormous luminosity, nonstellar spectrum, and high-speed jets, and wondering what exotic physical process could possibly account for its violent activity!

When they were first discovered, active galaxies and quasars seemed to present astronomers with insurmountable problems. For a time, their dual properties of enormous energy output and small size appeared incompatible with the known laws of physics and threatened to overturn our modern view of the universe. Yet the problems were eventually solved, and the laws of physics remain intact. Far from jeopardizing our knowledge of the cosmos, these violent phenomena have become part of the thread of understanding that binds our own Galaxy to the earliest epochs of the universe we live in.

Chapter Review

Summary

Active galaxies (p. 450) are much more luminous than normal galaxies and have spectra that are nonstellar in nature, indicating that the energy emitted by these galaxies is not simply the accumulated light of many stars. Most of the energy from active galaxies is emitted in the radio and infrared parts of the electromagnetic spectrum. The fraction of observed galaxies that display activity increases with increasing distance from us, indicating that galaxies were generally more active in the past than they are today.

A **Seyfert galaxy** (p. 451) looks like a normal spiral galaxy except that the Seyfert has an extremely bright central **galactic nucleus** (p. 452). Spectral lines from Seyfert nuclei are very broad, indicating rapid internal motion. In addition, Seyfert luminosities can vary by large amounts in fractions of a year, implying that the region emitting most of the radiation is much less than 1 light year across. **Radio galaxies** (p. 452) are active galaxies that emit most of their energy in the radio part of the spectrum. They are generally comparable to Seyferts in total energy output. Unlike Seyferts, radio galaxies are usually associated with elliptical galaxies. In a core–halo radio galaxy, most of the energy is emitted from a small central nucleus, as in a Seyfert. In a lobe radio galaxy, the energy comes from enormous **radio lobes** (p. 453) that dwarf the central visible portion of the galaxy and lie far outside it. The lobes are usually symmetrically placed with respect to the center of the visible galaxy.

Many active galaxies have high-speed, narrow jets of matter shooting out from their central nuclei. In lobe radio galaxies, astronomers believe that the jets transport energy from the nucleus, where it is generated, to the lobes, where it is radiated into space. The jets often appear to be made up of distinct "blobs" of gas, suggesting that the process generating the energy is intermittent.

The generally accepted explanation for the observed properties of active galaxies is that the energy is generated by accretion of galactic gas onto a supermassive (billion-solar-mass) black hole lying at the center of the nucleus. As the material spirals down toward the hole, it heats up and releases enormous amounts of energy. The small size of the accretion disk explains the compact extent of the emitting region, and the high-speed orbits of gas in the hole's intense gravity accounts for the rapid motion observed. Typical active galaxy luminosities require the consumption of about one solar mass of material every few years. Some of the infalling matter is blasted out into space, producing magnetized jets that create and feed the extended radio lobes. Charged particles spiraling around the magnetic field lines produce **synchrotron radiation** (p. 461) whose spectrum is consistent with the nonstellar radiation observed in radio galaxies and Seyferts.

Quasars (p. 464), or **quasi-stellar objects** (p. 464), were discovered as starlike radio sources emitting radiation spectra containing unidentified broad spectral lines. In the early 1960s, astronomers realized that the unfamiliar lines are actually those of familiar elements, but redshifted to wavelengths much longer than normal. Even the closest quasars lie at great distances from us. They are the most luminous objects known. Quasars exhibit the same basic features as active galaxies, and astronomers believe that their power source is also basically the same. This source—a black hole—must be much more massive in a quasar and must consume many stars per year. If that is the case, then the brightest quasars consume so much fuel that their energy-emitting lifetimes must be relatively short. Quasars probably represent a brief phase of violent activity early in the life of a galaxy.

Some quasars have been observed to have double or multiple images. These images result from gravitational lensing, where the gravitational field of a foreground galaxy or galaxy cluster bends and focuses the light from the more distant quasar. Analysis of this bending provides a means of determining the masses of galaxy clusters—including the dark matter—far beyond the region defined by the optical images of the galaxies.

Quasars, active galaxies, and normal galaxies may represent an evolutionary sequence. When galaxies began to form, conditions may have been suitable for the formation of large central black holes. If there was a lot of gas available at those early times, a highly luminous quasar would have been the result. As the fuel supply diminished, the quasar dimmed, and the galaxy in which it was embedded became visible as an active galaxy. At even later times, the fuel supply declined to the point where the nucleus became virtually inactive, and a normal galaxy was all that remained.

Self-Test: True or False?

_____ 1. Active galaxies can emit thousands of times more energy than our own Galaxy.

_____ 2. The "extra" radiation emitted by active galaxies is due to the tremendous number of stars they contain.

_____ 3. Active galaxies emit most radiation at visible wavelengths.

____ **4.** Most core–halo radio galaxies are spirals.

____ **5.** The diameter of a billion-solar-mass black hole is about 20 A.U.

____ **6.** Nearby active galaxies are most likely the result of interactions between galaxies.

____ **7.** A redshift greater than 1 means a recessional velocity greater than the speed of light.

____ **8.** All quasars are far away.

____ **9.** Other than a small amount of visible light, quasars emit all of their radiation at radio wavelengths.

____ **10.** Many normal galaxies may once have been quasars.

____ **11.** Quasars emit about as much energy as normal galaxies.

____ **12.** The quasar stage of a galaxy ends because the central black hole uses up all the matter around itself.

Self-Test: Fill in the Blank

1. Active galaxies are more common at _____ distances from Earth.

2. Seyfert galaxies look like normal spirals, but with a very bright galactic _____.

3. In a core–halo radio galaxy, most of the radio radiation is emitted from the _____ .

4. Lobe radio galaxies emit radio radiation from regions that are typically much _____ in size than the visible galaxy.

5. Radio lobes are always found aligned with the _____ of the visible galaxy.

6. For all types of active galaxies, the original source of the tremendous energy emitted is the galactic _____ .

7. The energy source of an active galaxy is unusual in that there is a large amount of energy emitted from a region less than _____ in diameter. (Give size and unit.)

8. The mass of the black hole responsible for energy production in the active galaxy M87 is thought to be approximately equal to _____ solar masses.

9. The amount of mass that must be consumed by a supermassive black hole to provide the energy for an active galaxy is about _____ per _____ .

10. Quasars are also known as quasi-stellar objects because of their _____ appearance.

11. The fact that a typical quasar would consume an entire galaxy's worth of mass in 10 billion years suggests that quasar lifetimes are relatively _____ .

12. The image of a distant quasar can be split into several images by gravitational lensing, produced by a foreground _____ along the line of sight.

Review and Discussion

1. Name two basic differences between normal galaxies and active galaxies.

2. Describe some of the basic properties of Seyfert galaxies.

3. What distinguishes a core–halo radio galaxy from a lobe radio galaxy?

4. What is the evidence that the radio lobes of some active galaxies consist of material ejected from the galaxy's nucleus?

5. What conditions result in a head–tail radio galaxy?

6. Briefly describe the leading model for the central engine of an active galaxy.

7. Which observations of active galaxies lead astronomers to believe that synchrotron radiation is being produced?

8. What was it about the spectra of quasars that was so unexpected and surprising?

9. How do we know that quasars are extremely luminous?

10. How are the spectra of distant quasars used to probe the space between us and them?

11. What evidence do we have that quasars represent an early stage of galaxy evolution?

12. What happened to the energy source at the center of a quasar?

Problems

1. From one radio lobe to the other, Centaurus A spans about 1 Mpc and lies at a distance of 4 Mpc from Earth. What is its angular size? Compare your answer with the angular diameter of the Moon.

2. Assuming the same efficiency as postulated in the text, how much energy would an active galaxy generate if it consumed one Earth mass of material every day? Compare your answer with the luminosity of the Sun.

3. A certain quasar has a recessional speed of 60,000 km/s and the same apparent brightness as the Sun would have if the Sun were 1 kpc from Earth. Assuming $H_0 = 75$ km/s/Mpc, calculate the quasar's luminosity.

4. A Seyfert galaxy is observed to have broadened emission lines indicating a speed of 1000 km/s at a distance of 1 pc from its center. Assuming circular orbits, use Kepler's laws (Section 14.6) to estimate the mass within this 1-pc radius.

Projects

Here are three observing projects that are increasingly challenging.

1. In project 2 of Chapter 15, you were given directions for finding the Virgo Cluster of galaxies. M87, in the central part of this cluster, is the core–halo radio galaxy nearest Earth and has coordinates RA = $12^h 30.8^m$, dec = $+12° 24'$. At magnitude 8.6, it should not be difficult to find in an 8-inch telescope. Its distance is roughly 20 Mpc. Describe its nucleus; compare what you see with the appearance of other nearby ellipticals in the Virgo Cluster.

2. NGC 4151 is the brightest Seyfert galaxy. Its coordinates are RA = $12^h 10.5^m$, dec = $+39° 24'$, and it can be found below the Big Dipper in Canes Venatici. At magnitude 10–12 (it is variable), it should be visible in an 8-inch telescope but will be challenging to find. Its distance from Earth is 13.5 Mpc. Describe its nucleus and compare its appearance with what you have seen for other galaxies.

3. 3C 273 is the nearest and brightest quasar. However, that does not mean it will be easy to find and see! Its coordinates are RA = $12^h 29.2^m$, dec = $+2° 03'$. It is located in the southern part of the Virgo Cluster but is not associated with that cluster. At magnitude 12–13, it may require a 10- or 12-inch telescope to see, but try first with an 8-inch. 3C 273 should appear as a very faint star. The significance of seeing this object is that it is 640 Mpc distant. The light you are seeing left this object over 2 billion years ago! 3C 273 is the most distant object observable with a small telescope.

If you can find the three objects listed here, you have started to become an accomplished observer!

Cosmology
The Big Bang and the Fate of the Universe *www*

(Opposite page, background) By virtually all accounts, the universe began in a fiery explosion some 10–20 billion years ago. Out of this maelstrom emerged all the energy that would later form the galaxies, stars, and planets (depicted here in an artist's rendering). The story of the origin and fate of all these systems—indeed of the universe itself—comprises the subject of cosmology.

The three inset images of spiral galaxies lying at different distances from Earth capture representative views of what the universe was like at much earlier times. (Inset A) About 12 billion years ago, the galaxy's spiral shape is hardly recognizable. (Inset B) About 9 billion years ago, the spiral is still vague, though star burst activity is evident in the outer regions. (Inset C) Some 5 billion years ago, the galaxy's spiral features are more prominent.

LEARNING GOALS

Studying this chapter will enable you to:

1 State the cosmological principle and explain its significance.

2 Explain how the age of the universe is determined and discuss the uncertainties involved.

3 Summarize the leading evolutionary models of the universe and discuss the factors that determine whether or not the universe will expand forever.

4 Describe the cosmic microwave background radiation and explain its importance to our understanding of cosmology.

5 Explain how nuclei and atoms emerged from the primeval fireball.

6 Summarize the horizon and flatness problems, and discuss the theory of cosmic inflation as a possible solution to these problems.

7 Explain the formation of large-scale structure in the cosmos and discuss the observational evidence for our theories of structure formation.

Our field of view now extends for billions of parsecs into space and billions of years back in time. We have asked and answered many questions about the structure and evolution of planets, stars, and galaxies. At last we are in a position to address the central issues of the biggest puzzle of all. How big is the universe? How long has it been around, and how long will it last? What was its origin, and what will be its fate? Is the universe a one-time event, or does it recur and renew itself, in a grand cycle of birth, death, and rebirth? How and when did matter, atoms, our Galaxy form? These are surely basic questions, but they are hard questions. Many cultures have asked them, in one form or another, and have developed their own cosmologies—theories about the nature, origin, and destiny of the universe—to answer them. In this chapter, we see how modern scientific cosmology addresses these important issues and what it has to tell us about the universe we inhabit. After more than 10,000 years of civilization, science may be ready to provide some insight regarding the ultimate origin of all things.

17.1 The Universe on the Largest Scales

The End of Structure

The universe shows structure on every scale we have examined so far. Subatomic particles form nuclei and atoms. Atoms form planets and stars. Stars form star clusters and galaxies. Galaxies form galaxy clusters, superclusters, and even larger structures—voids, filaments, and sheets that stretch across the sky. From protons in an atom's nucleus to the galaxies in the Great Wall, we can trace a hierarchy of "clustering" of matter from the very smallest to the very largest scales. It is natural to ask: Does the clustering ever end? Is there some scale on which the universe may be regarded as more or less smooth and featureless? Perhaps surprisingly, given the trend we have just described, most astronomers think the answer is *yes*.

We saw in Chapter 15 how galaxy surveys have revealed the existence of structures up to about 200 Mpc across. ∞ (Sec. 15.5) Although they cover wide areas of the sky and enormous volumes of space, these studies are still relatively local, in the sense that they span only a few percent of the distance to the farthest quasars (which lie nearly 5000 Mpc from Earth). A major obstacle to extending these wide-angle surveys to much greater distances is the sheer observational effort of measuring the redshifts of all the galaxies within larger and larger volumes of space.

An alternative approach is to narrow the field of view to only a few small patches of the sky and then to study very distant galaxies within those patches. The volume surveyed then becomes a long, thin "pencil beam" extending deep into space, rather than a wide swath through the local universe. Figure 17.1 presents the results of one such survey, showing the distribution of galaxies in two directions perpendicular to the plane of our Galaxy, out to a distance of about 2000 Mpc. Although the numbers of galaxies fall off at large distances—basically because very distant galaxies are very hard to see—a distinctive "on–off" pattern of galaxies, looking a little like a picket fence, can be seen. It is thought that the gaps between the pickets are voids much like those seen closer to home, while the pickets themselves are places where the researchers' field of view intersects sheets like the Great Wall. The data seem to indicate that the largest structures are only 100–200 Mpc across—no voids or clumps of galaxies much larger than that are seen. In short, there is presently no evidence for structure in the universe on scales greater than about 200 Mpc.

On the basis of these rather sketchy data, some theoretical insight, and not a little philosophical preference, *cosmologists* (astronomers who study the large-scale structure and dynamics of the entire universe) assume that the universe is roughly **homogeneous** on scales greater than a few hundred megaparsecs. In other words, if we imagine taking a huge cube—300 Mpc on a side, say—and placing it anywhere in the universe, as illustrated in Figure 17.2, the number of galaxies it enclosed would be pretty much the same—around 100,000, excluding the faint dwarf ellipticals and irregulars. Some of the galaxies would be clustered and clumped into fairly large structures, and some would not be, but the total number would not vary

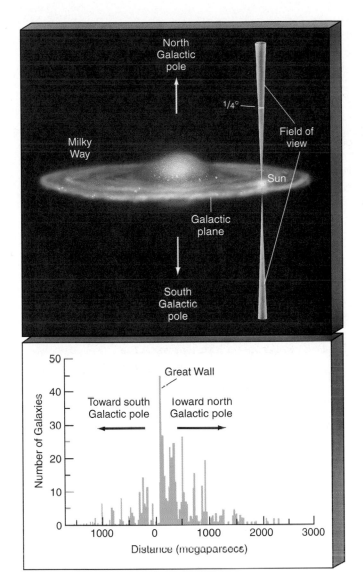

Figure 17.1 The results of a long-range "pencil-beam" survey of two small portions of the sky in opposite directions from Earth perpendicular to the Galactic plane. The graph shows the number of galaxies at different distances from us, out to about 2000 Mpc. The "picket-fence" appearance seems to show voids and sheets of galaxies on scales of 100 or 200 Mpc but provides no indication of any larger structure.

much as the cube was moved from place to place. In short, the universe looks *smooth* on the largest scales.

The Cosmological Principle

1 Cosmic homogeneity is the first of two major assumptions that cosmologists make when studying the large-scale structure of the universe. Observations suggest that it is true, but it is by no

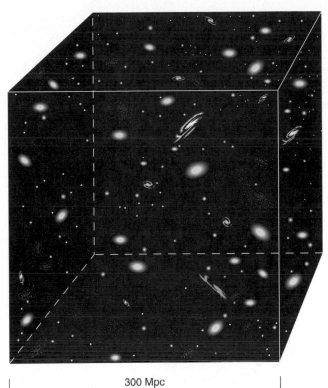

Figure 17.2 Diagram of galaxies contained within an enormous cube, 300 Mpc on a side. Cosmologists believe that the contents of such a cube would look pretty much the same regardless of where in the universe we place the cube.

means proven. The second assumption, also supported by observational evidence and theoretical reasoning, is that the universe is **isotropic**—that is, it looks the same in any direction. Isotropy is on much firmer observational ground than is homogeneity. Apart from regions of the sky that are obscured by our Galaxy, the universe *does* look much the same in all directions, at any wavelength, provided we look far enough. In other words, any deep pencil-beam survey of the sky will count about the same number of galaxies as the study just mentioned, regardless of which patch of the sky is chosen.

The assumptions of homogeneity and isotropy form the foundation of modern **cosmology**—the study of the structure and evolution of the entire universe. Together, these twin pillars of cosmology are known as the **cosmological principle**. No one really knows if this principle is absolutely correct. All we can say is that, so far, it seems consistent with observations. From this point on in this text, we simply assume that it holds.

The cosmological principle has very far-reaching implications. For example, it implies that there can be no *edge* to the universe because the existence of an edge would violate the assumption of homogeneity. Furthermore, it implies that there is no *center* because a center would mean that the universe would not look the same in all directions from any noncentral point, a violation of the assumption of isotropy. Thus, this single principle strongly limits what the overall geometry of the universe can be. The cosmological principle is the ultimate expression of the principle of mediocrity. It states not only that are we not central to the universe, but that *no one* can be central because *the universe has no center!*

17.2 Cosmic Expansion

Olbers's Paradox

Every time you go outside at night and notice that the sky is dark, you are making a profound cosmological observation. Here's why.

According to the cosmological principle, the universe is homogeneous and isotropic. Let's assume that it is also infinite in spatial extent and unchanging in time—precisely the view of the universe that prevailed until the early part of the twentieth century. On average, then, the universe is uniformly populated with galaxies filled with stars. In that case, when you look up at the night sky, your line of sight must *eventually* encounter a star, as illustrated in Figure 17.3. The star may lie at an enormous distance, in some remote galaxy, but the laws of probability dictate that, sooner or later, any line drawn outward from Earth will run into a bright stellar surface. This fact has a dramatic implication: no matter where you look, the sky should be as bright as the surface of a star—the entire night sky should be as brilliant as the surface of the Sun! The obvious difference between this prediction and the actual appearance of the night sky is known as **Olbers's paradox**, after the nineteenth-century German astronomer Heinrich Olbers, who popularized the idea.

What is the resolution of this paradox? Why is the sky dark at night? Having accepted the cosmological principle, we believe that the universe is homogeneous and isotropic. We must conclude, then, that one

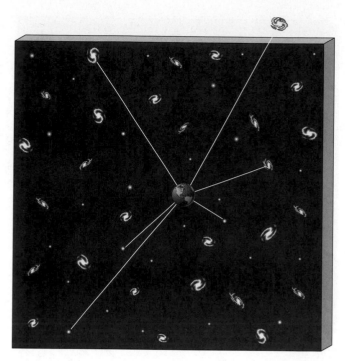

Figure 17.3 If the universe were homogeneous, isotropic, infinite in extent, and unchanging, then any line of sight from Earth should eventually run into a star, and the entire night sky should be bright. This obvious contradiction of the facts is known as Olbers's paradox.

(or both) of the other two assumptions—that the universe is infinite in extent and unchanging in time—is false. Either the universe is finite in extent, or it evolves in time. In fact, the resolution involves a little of each and is intimately tied to the behavior of the universe on the largest scales.

Hubble's Law and the Big Bang

2 We have seen that all the galaxies in the universe are rushing away from us in a manner described by Hubble's law:

$$\text{recession velocity} = H_0 \times \text{distance},$$

where we take Hubble's constant H_0 to be 75 km/s/ Mpc. ∞ (Sec. 17.5) We have used this relationship as a convenient means of determining the distances to galaxies and quasars, but it is much more than that.

Assuming that all velocities have remained constant in time, how long has it taken for any given galaxy to reach its present distance from us? The answer follows from Hubble's law. The time taken is simply the distance traveled divided by the velocity, so we can say

$$\text{time} = \frac{\text{distance}}{\text{velocity}}$$

$$= \frac{\text{distance}}{H_0 \times \text{distance}} \quad \text{(using Hubble's law for the velocity)}$$

$$= \frac{1}{H_0}.$$

For $H_0 = 75$ km/s/Mpc, this time turns out to be about 13 billion years. Notice that the time is *independent* of the distance—galaxies twice as far away are moving twice as fast, so the time they have taken to cross the intervening distance is the same in all cases.

Hubble's law therefore implies that, at some time in the past—13 billion years ago, according to the foregoing simple calculation—*all* the galaxies in the universe lay right on top of one another. In fact, astronomers believe that *everything* in the universe—matter and radiation alike—was confined to a single point at that instant. Then the point exploded, flying apart at high speeds. The present locations and velocities of the galaxies are a direct consequence of that primordial blast. This gargantuan explosion, involving everything in the universe, is known as the **Big Bang**. It marked the beginning of the universe.

Thus, by measuring Hubble's constant, we can estimate the age of the universe to be $1/H_0 \approx 13$ billion years. The range of possible error in this age is considerable, both because Hubble's constant is not known precisely and because the assumption that galaxies moved at constant speed in the past is not a very good one—in fact, galaxies moved faster in the past; their motion has been slowed by the effects of gravity. We will refine our estimate in a moment, but, regardless of the details, the critical fact here is that the age of the universe is *finite*.

This is the explanation of why the sky is dark at night. Olbers's paradox is resolved by the evolution of the universe itself. Whether the universe is actually finite or infinite in extent is irrelevant, at least as far as the appearance of the night sky is concerned. We see only a finite part of the universe—the region lying within roughly 13 billion light years of us. What lies beyond is unknown—its light has not yet had time to reach us.

Where Was the Big Bang?

Let's pause for a moment to take stock of what we have just learned and to shift in a very important way our view of the expansion of the universe as described by Hubble's law. We know *when* the Big Bang occurred. Is there any way of telling *where*? The cosmological principle indicates that the universe is the same everywhere, yet we have just seen that the observed recession of the galaxies described by Hubble's law implies that all the galaxies exploded from a point some time in the past. Wasn't that point, then, different from the rest of the universe, violating the assumption of homogeneity expressed in the cosmological principle? The answer is a definite *no*!

To understand why there is no "center" to the expansion, we must make a great leap in our perception of the universe. If we were to imagine the Big Bang as simply an enormous explosion that spewed matter out into space, ultimately to form the galaxies we see, then the foregoing reasoning would be quite correct—there would be a center and an edge, and the cosmological principle would not apply. But the Big Bang was *not* an explosion in an otherwise featureless, empty universe. The only way that we can have Hubble's law *and* retain the cosmological principle is to realize that the Big Bang involved the entire universe—not just the matter and radiation within it, but the universe *itself*. In other words, the galaxies are not flying apart into the rest of the universe. The universe itself is expanding. Like raisins in a loaf of raisin bread that move apart as the bread expands in an oven, the galaxies are just along for the ride.

Let's reconsider some of our earlier statements in light of this new perspective. We now recognize that Hubble's law describes the expansion of the universe itself. Although galaxies have some small-scale, individual random motions, on average they are not moving with respect to the fabric of space—any such overall motion would pick out a "special" direction in space and violate the assumption of isotropy. On the contrary, the portion of the galaxies' motion that makes up the Hubble flow is really an expansion of space itself. The expanding universe remains homogeneous at all times. There is no "empty space" beyond the galaxies

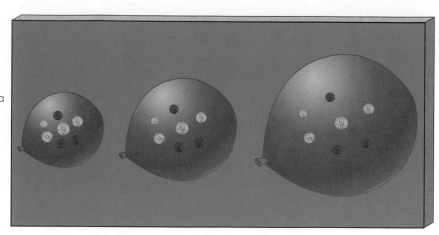

Figure 17.4 Coins taped to the surface of a spherical balloon recede from one another as the balloon inflates (left to right). Similarly, galaxies recede from one another as the universe expands. As the coins recede, the distance between any two of them increases, and the rate of increase of this distance is proportional to the distance between them. Thus, the balloon expands according to Hubble's law.

into which they rush. At the time of the Big Bang, the galaxies did not reside at a point located at some well-defined place within the universe. The *entire universe* was a point. That point was in no way different from the rest of the universe; it *was* the universe. Therefore, there was no one point where the Big Bang "happened"—the Big Bang happened *everywhere* at once.

To illustrate these ideas, imagine an ordinary balloon with coins taped to its surface, as shown in Figure 17.4. (Better yet, do the experiment yourself.) The coins represent galaxies and the two-dimensional surface of the balloon represents the "fabric" of our three-dimensional universe. The cosmological principle applies here because every point on the balloon looks pretty much the same as every other. Imagine yourself as a resident of one of the coin "galaxies" on the leftmost balloon, and note your position relative to your neighbors. As the balloon inflates (that is, as the universe expands), the other galaxies recede from you, and more distant galaxies recede more rapidly. (Notice, incidentally, that the coins do *not* expand along with the balloon, any more than people, planets, stars, or galaxies—all of which are held together by their own internal forces—expand along with the universe.)

Regardless of which galaxy you chose to consider, you would see all the other galaxies receding from you. Nothing is special or peculiar about the fact that all the galaxies are receding from you. Such is the cosmological principle: no observer anywhere in the universe has a privileged position. There is no center to the expansion and no position that can be identified as the location from which the universal expansion began.

Everyone sees an overall expansion described by Hubble's law, with the same value of Hubble's constant in all cases.

Now imagine letting the balloon deflate. This corresponds to running the universe backward from the present time to the moment of the Big Bang. *All* the galaxies (coins) would arrive at the same place at the same time—at the instant the balloon reached zero size. But there is no one point on the balloon that could be said to be *the* place where that occurred. The entire balloon expanded from a point, just as the Big Bang encompassed the entire universe and expanded from a point.

The Cosmological Redshift

This view of the expanding universe requires us to reinterpret the cosmological redshift. Up to now, we have explained the redshift of galaxies as a Doppler shift, a consequence of their motion relative to us. However, we have just argued that the galaxies are *not* in fact moving with respect to the universe, in which case the Doppler interpretation is incorrect. The true explanation is that as a photon moves through space, its wavelength is influenced by the expansion of the universe. In a sense, we can think of the photon as being attached to the expanding fabric of space, so its wavelength expands along with the universe, as illustrated in Figure 17.5. While it is standard practice in astronomy to refer to the cosmological redshift in terms of recessional velocity, bear in mind that, strictly speaking, this is not the right thing to do. The cosmological redshift is a consequence of the

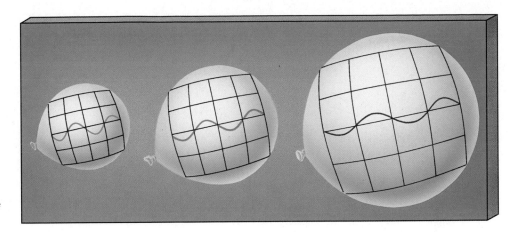

Figure 17.5 As the universe expands, photons of radiation are stretched in wavelength, giving rise to the cosmological redshift.

changing size of the universe and is *not* related to velocity at all.

The redshift of a photon measures the amount by which the universe has expanded since that photon was emitted. For example, when we measure the light from a quasar to have been redshifted by a factor of 5, that means that the light was emitted at a time when the universe was just one-fifth its present size (and that we are observing the quasar as it was at that time). In general, the larger a photon's redshift, the smaller the universe was at the time the photon was emitted, and so the longer ago that emission occurred. Because the universe expands with time and redshift is related to that expansion, cosmologists routinely use redshift as a convenient means of expressing time. ∞ (*More Precisely 16-1*)

17.3 The Fate of the Universe

3 At the present moment, the universe is expanding. Will that expansion continue forever? And if not, what will happen next, and when? These are absolutely fundamental questions concerning the fate of the universe. Yet, remarkably, it turns out that we can address them by considering a simpler and much more familiar problem.

Critical Density

Consider a rocket ship launched from the surface of a planet. What is the likely outcome of that motion? There are basically two possibilities, depending on the initial speed of the ship. If the launch speed is high

enough, it will exceed the planet's escape speed, and the ship will never return to the surface. ∞ (Sec. 5.1) The speed will diminish because of the planet's gravitational pull, but it will never reach zero. The spacecraft leaves the planet on an *unbound* trajectory, as illustrated in Figure 17.6(a). Alternatively, if its launch speed is lower than the escape speed, the ship

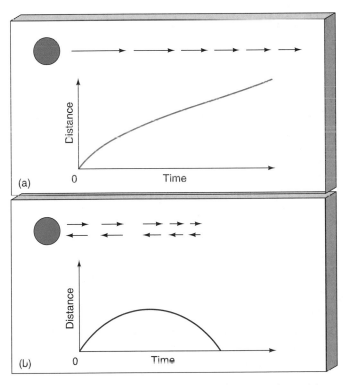

Figure 17.6 (a) A rocket ship (arrow) leaving a planet (blue ball) with a speed greater than the escape speed follows an unbound trajectory. The graph shows the distance between the ship and the planet as a function of time. (b) If the launch speed is less than the escape speed, the ship eventually drops back to the planet. Its distance from the planet first rises, then falls.

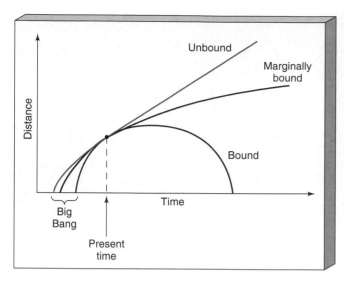

Figure 17.7 Distance between two galaxies as a function of time in each of the possible universes discussed in the text: unbound, bound, and marginally bound. The point where the three curves touch represents the present time.

will reach a maximum distance from the planet, then fall back to the surface. Its *bound* trajectory is shown in Figure 17.6(b).

Similar reasoning applies to the expansion of the universe. Imagine two galaxies at some known distance from one another, their present relative velocity given by Hubble's law. The same two basic possibilities exist for these galaxies as for our rocket ship—the distance between them can increase forever, or it can increase for a while and then start to decrease. What's more, the cosmological principle says that, whatever the outcome, it must be the same for *any* two galaxies—in other words, the same statement applies to the universe *as a whole*. Thus, as illustrated in Figure 17.7, the universe has only two options: it can continue to expand forever—an *unbound* universe—or the present expansion will someday stop and turn around into a contraction—a *bound* universe.

The middle curve on Figure 17.7 marks the dividing line between these two possibilities. It shows a *marginally bound* universe that expands forever, but at an ever-decreasing rate, analogous to our rocket ship's leaving the planet with precisely the escape speed. The three curves are drawn so that they all pass through the same point at the present time. All are possible descriptions of the universe given its present size and expansion rate.

What determines which of these possibilities will actually occur? The answer is the *density* of the uni-

verse. In all cases gravity decelerates the expansion over time. The more matter there is—the denser the universe—the more "pull" there is against the expansion, just as the more mass a planet has, the less likely it is that a rocket ship can escape. In a high-density universe, there is enough mass to stop the expansion and cause a recollapse—the universe is bound. A low-density universe, conversely, is unbound and will expand forever. The dividing line between these two outcomes, which is the density corresponding to a marginally bound universe, is called the **critical density**. For $H_0 = 75$ km/s/Mpc, the present critical density turns out to be about 10^{-26} kg/m^3. That's an extraordinarily low density—about six hydrogen atoms per cubic meter, a volume the size of a typical household closet. In more "cosmological" terms, this density corresponds to about one Milky Way Galaxy (excluding the dark matter) per cubic megaparsec.

Two Futures

If the universe emerged from the Big Bang with a density above the critical value, then it contains enough matter to halt its own expansion. The recession of the galaxies will eventually stop and the universe will start to contract. Figure 17.8(a) illustrates how such a universe will recollapse to a point, requiring just as much time to fall back as it took to reach its maximum size. The entire universe will grow progressively denser and hotter as the end of the contraction is neared. The universe will shrink toward a superdense, superhot singularity, much like the one from which it originated—it will ultimately experience a "heat death," in which all radiation, matter, and life are destined to fry. Some astronomers call the final collapse of this high-density universe the "Big Crunch."

Cosmologists do not know what would happen to the universe upon reaching the point of collapse. We cannot penetrate forward in time beyond the singularity at the Big Crunch any more than we can probe backward past the Big Bang—the laws of physics as we presently understand them are simply inadequate to describe these extreme conditions. However, some theorists speculate that, with both density and temperature increasing as the contraction nears completion, the pressure might somehow be sufficient to overcome gravity, pushing the universe back out into another

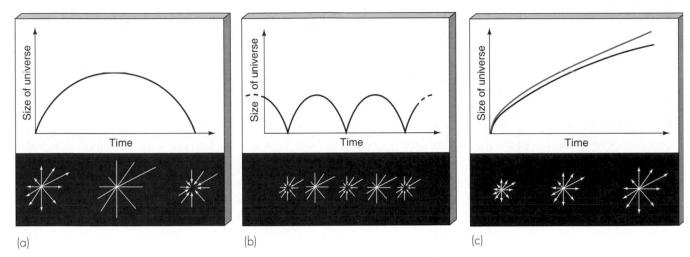

(a) (b) (c)

Figure 17.8 (a) A high-density universe has a beginning, an end, and a finite lifetime. The lower frames illustrate its evolution, from explosion to maximum size to recollapse. (b) An oscillating universe has neither a beginning nor an end. Each expansion–contraction phase ends in a "bounce" that becomes the "Big Bang" of the next expansion. There is currently no information on whether or not this can actually occur. (c) A low-density universe expands forever from its explosive beginning. The upper curve represents a universe with density less than the critical value. The lower curve represents a universe with density exactly equal to the critical value.

cycle of expansion. As depicted in Figure 17.8(b), the universe might not simply end—it might "bounce." A hypothetical universe having many—perhaps infinitely many—cycles of expansion and contraction might be the result. Bear in mind, though, that any discussion of the universe outside of the current cycle is pure speculation.

A quite different fate awaits the universe if its density is below the critical value. In that case, its density always has been, and always will be, too small for gravity to cause it to recontract. Figure 17.8(c) shows how such a low-density universe expands forever. In this scenario, the galaxies will continue to recede, their radiation weakening with increasing distance. In time, an observer on Earth, even with the most powerful telescope, will see no galaxies in the sky beyond the Local Group (which is not itself expanding). The rest of the observable universe will appear dark, the distant galaxies too faint to be seen. Eventually, the Milky Way Galaxy and the Local Group will peter out as their fuel supply is consumed. This universe ultimately experiences a "cold death." All radiation, matter, and life are eventually destined to freeze.

In the intermediate, critical-density case, the universe contains just enough matter eventually to halt the expansion—but only after an infinitely long time. This universe will also expand forever.

Will the Universe Expand Forever?

Is there any way for us to determine which of these models actually describes our universe (that is, apart from just waiting to find out)? The most straightforward approach is to try to estimate the universe's average density because density is the basic quantity that distinguishes one model from the other. As just noted, for $H_0 = 75$ km/s/Mpc, the critical density that separates the two possible futures is about 10^{-26} kg/m^3. Cosmologists conventionally call the ratio of the actual density to the critical value the *cosmic density parameter* and denote it by the symbol Ω_0 ("omega nought"). In terms of this quantity, then, a critical universe has $\Omega_0 = 1$. A universe with Ω_0 less than 1 will expand forever; one with Ω_0 greater than 1 will recollapse.

How can we determine the average density of the universe? On the face of it, it would seem simple—just measure the average mass of the galaxies residing within a large parcel of space, calculate the volume of that space, and compute the total mass density. When astronomers do this, they usually find a little less than 10^{-28} kg/m^3 in the form of luminous matter. Largely independent of whether the chosen region contains only a few galaxies or a rich galaxy cluster, the resulting density is about the same, within a factor of 2 or 3.

Galaxy counts thus yield a value of Ω_0 of about 0.01. If that measure were correct, then the universe would expand forever.

But there is an important additional consideration. We have noted (Chapters 14 and 15) that most of the matter in the universe is *dark*—it exists in the form of invisible material that has been detected only through its gravitational effect in galaxies and galaxy clusters. ∞ (Sec. 14.6, 15.3) We currently do not know what the dark matter is, but we *do* know that it is there. Galaxies may contain as much as 10 times more dark matter than luminous material, and the figure for galaxy clusters is even higher—perhaps as much as 95 percent of the total mass in clusters is invisible. Even though we cannot see it, dark matter contributes to the average density of the universe and plays its part in opposing the expansion. Including all the dark matter that is known to exist in galaxies and galaxy clusters increases the value of Ω_0 to 0.2 or 0.3—still less than 1, implying an unbound universe, but a lot closer to 1 than the 0.01 value calculated using visible matter only.

Unfortunately, the distribution of dark matter on larger scales is not very well known. We can infer its presence in galaxies and galaxy clusters, but we are largely ignorant of its extent in superclusters, voids, or other larger structures. However, there are indications that it accounts for an even greater fraction of the mass on large scales than it does in galaxy clusters. For example, observations of gravitational lensing by galaxy clusters suggest that dark matter may be even more extensive than indicated by the motion of galaxies within the clusters. Furthermore, optical and infrared observations of the overall motion of galaxies in the local supercluster suggest the presence of a nearby huge accumulation of mass known as the *Great Attractor*, which has a total mass of about 10^{17} solar masses and a size of 100–150 Mpc. If the current best estimates of the size and mass of this gargantuan object are correct, then its average density may be quite close to the critical value.

Thus it is quite conceivable that invisible matter may account for as much as 99 percent of the total mass in the universe. In that case, the vast "voids" are not empty at all—they are huge seas of invisible matter, and the visible galaxies are merely insignificant "islands" of brightness within them. If the total amount of dark matter lurking in the darkness beyond the galaxy clusters exceeds the luminous mass by a factor of

100 or more, Ω_0 might even be greater than 1, and the universe will recollapse. This is why it is so important to search for dark matter beyond the galaxies. The measured value of Ω_0 has steadily increased over the past 20 years as larger and larger regions of the universe have been surveyed.

So, what is the ultimate fate of the universe? The answer is still not known with absolute assurance. However, while there is a large uncertainty in the value of Ω_0, most astronomers would probably agree that it lies between 0.1 and 1. Thus, as best we can tell, given the current data, the universe is destined to expand forever.

The Age of the Universe

In Section 17.2, when we estimated the age of the universe from the accepted value of Hubble's constant, we made the assumption that the expansion speeds of the galaxies were constant in the past. However, as we have now found, this is not the case. The effects of gravity have slowed the universe's expansion over time, so, regardless of which evolutionary model turns out to be correct, the universe must have expanded *faster* in the past than it does today. The assumption of a constant expansion rate therefore leads to an overestimate of the universe's age—the universe is actually younger than the 13 billion years we calculated earlier. How much younger depends on how much deceleration has occurred.

Figure 17.9 illustrates this point. It is similar to Figure 17.7, except that here we have added a line corresponding to constant expansion at the present rate—a completely empty universe that is 13 billion years old. Because all the models we have discussed lie below this line, we can see graphically that the true age of the universe is indeed less than the age obtained assuming a constant expansion speed. In the special case of critical density, the age of the universe happens to be particularly easy to calculate—it is *two-thirds* of the foregoing value, or 8.7 billion years. A low-density, unbound universe is older than this (but still less than 13 billion years old); a high-density, bound universe is younger.

Impressive though it is that we can pin down the age of the universe to within less than a factor of 2, these numbers indicate a potentially serious problem. Unless the universe is of very low density, and hence

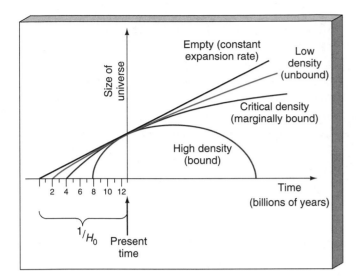

Figure 17.9 As the density of the universe increases, its deceleration increases, too. The universe contains some matter, so whatever the model, its trajectory on this graph will lie below the line for the empty universe. Thus, the age of the universe is always less than $1/H_0$. The true age decreases for larger values of the present-day density.

close to 13 billion years old, the age that we obtain from cosmology is less than the 10–12-billion-year range implied by studies of globular clusters in our own Galaxy (see Chapters 10 and 12). ∞ (Sec. 10.10, 12.5) Because the Galaxy cannot be older than the universe, and because the density of the universe appears to be at least relatively close to the critical value, we are forced to conclude that there may be a glaring contradiction between these two major areas of astronomy! If, as observations improve, Hubble's constant turns out to be close to 60 km/s/Mpc, and if the lower limit on the globular-cluster ages is correct, then the two age estimates might be reconciled. However, if Hubble's constant is more like 80 or 90 km/s/Mpc, then the discrepancy may become a serious embarrassment to astronomers.

17.4 The Geometry of Space

The idea of the entire universe expanding from a point—with *nothing*, not even space and time, outside—takes a lot of getting used to! Nevertheless, it lies at the heart of modern cosmology, and few modern astronomers seriously doubt it. But this description of the universe itself (not just its contents) as a dynamic, evolving object is far beyond the capa-

bilities of Newtonian mechanics, which we introduced in Chapter 1 and which we have used almost everywhere in this book. Instead, the more powerful techniques of Einstein's theory of general relativity (Chapter 13; see especially *More Precisely 13-1*), with its built-in notions of warped space and dynamical spacetime, are needed. ∞ (Sec. 13.1)

The theory of general relativity states that matter curves, or "warps" space in its vicinity. The more matter—that is, the greater the density—the greater the curvature. Furthermore, the curvature must be the same everywhere (because of the cosmological principle), so there are really only three possibilities for the large-scale geometry of the universe. They correspond to the possible futures we have just described. (For more information on the different types of geometry involved, see *More Precisely 17-1*.)

If the average density of the cosmos is above the critical value, space is curved so much that it bends back on itself and closes off, making the universe *finite* in size. Such a universe is known as a **closed universe**. It is difficult to visualize a three-dimensional volume uniformly arching back on itself in this way, but the two-dimensional version is well known: it is just the surface of a sphere, like the balloon we discussed earlier. Figure 17.4, then, is the two-dimensional likeness of a three-dimensional closed universe. Like the surface of a sphere, a closed universe has no boundary, yet it is finite in extent.[1] One remarkable property of a closed universe is illustrated in Figure 17.10: a flashlight beam shone in some direction in space might eventually traverse the entire universe and return from the opposite direction!

The surface of a sphere curves, loosely speaking, "in the same direction" no matter which way we move from a given point. A sphere is said to have *positive curvature*. However, if the average density of the universe is below the critical value, the curvature of space is qualitatively quite different from that of a sphere. The two-dimensional surface corresponding to such a space curves like a saddle and is said to have *negative curvature*. Most people have seen a saddle—it curves "up" in one direction and "down" in another, but no one has ever seen a uniformly negatively curved surface, for the good reason that it cannot be constructed in three-

---

[1]Notice that, for the sphere analogy to work, we must imagine ourselves as two-dimensional "flatlanders" who cannot visualize or experience in any way the third dimension perpendicular to the sphere's surface. Flatlanders are confined to the sphere's surface, just as we are confined to the three-dimensional volume of our universe.

Figure 17.10 In a closed universe, a beam of light launched in one direction might return someday from the opposite direction after circling the universe, just as motion in a "straight line" upon Earth's surface will eventually encircle the globe.

dimensional space! It is just "too big" to fit. A low-density, saddle-curved universe is infinite in extent and is usually called an **open universe**.

The intermediate case, when the density is precisely equal to the critical density, is the easiest to visualize. This **critical universe** has no curvature. It is said to be "flat," and it is infinite in extent. In this case, and only in this case, the geometry of space on large scales is precisely the familiar Euclidean geometry taught in high school. Apart from its overall expansion, this is basically the universe that Newton knew.

Euclidean geometry—the geometry of flat space—is familiar to most of us because it is a good description of space in the vicinity of Earth. It is the geometry of everyday experience. Does this mean that the universe is flat, which would in turn mean that it has exactly the critical density? The answer is no. Just as a flat street map is a good representation of a city, even though we know Earth is really a sphere, Euclidean geometry is a good description of space within the solar system, or even the Galaxy, because the curvature of the universe is negligible on scales smaller than about 1000 Mpc. Only on the very largest scales would the geometrical effects we have just discussed become evident.

17.5 Back to the Big Bang

We now turn from studies of the far future of our universe to the quest to understand its distant past. Just how far back in time can we probe? Is there any way to study the universe beyond the most remote quasar? How close can we come to perceiving directly the edge of time, the very origin of the universe?

The Cosmic Microwave Background

4 A partial answer to these questions was discovered by accident in 1964, during an experiment designed to improve America's telephone system. As part of a project to identify and eliminate unwanted interference in satellite communications, Arno Penzias and Robert Wilson, two scientists at Bell Telephone Laboratories in New Jersey, were studying the Milky Way's emission at microwave (radio) wavelengths. In their data, they noticed a bothersome background "hiss" that just would not go away—a little like the background static on an AM radio station. Regardless of where and when they pointed their antenna, the hiss persisted. Never diminishing or intensifying, the weak signal was detectable at any time of the day, any day of the year, apparently filling all of space.

Eventually, after discussions with colleagues at Bell Labs and theorists at nearby Princeton University, the two experimentalists realized that the origin of the mysterious static was nothing less than the fiery creation of the universe itself. The radio hiss that Penzias and Wilson detected is now known as the **cosmic microwave background**. Their discovery won them the 1978 Nobel Prize in Physics.

In fact, researchers had predicted the existence and general properties of the microwave background well before its discovery. As early as the 1940s, physicists had realized that, in addition to being extremely dense, the early universe must also have been very *hot* and that, shortly after the Big Bang, the universe must have been filled with extremely high-energy thermal radiation—gamma rays of very short wavelength. Researchers at Princeton had extended these ideas, reasoning that the frequency of this primordial radiation would have been redshifted from gamma ray, to X ray, to ultraviolet, eventually all the way into the radio range of the electromagnetic spectrum as the universe expanded and cooled (Figure 17.11). ∞ (Sec. 2.4)

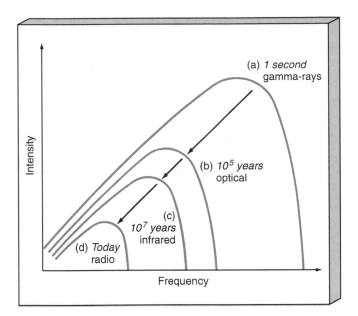

Figure 17.11 Theoretically derived black-body curves for the entire universe (a) 1 second after the Big Bang, (b) 100,000 years after the Big Bang, (c) 10 million years after the Big Bang, and (d) at present, approximately 13 billion years after the Big Bang.

By the present time, they argued, this redshifted "fossil remnant" of the primeval fireball should have a temperature of no more than a few tens of kelvins—peaking in the microwave part of the spectrum. The Princeton group was in the process of constructing a microwave antenna to search for this radiation when Penzias and Wilson announced their discovery.

The Princeton researchers confirmed the existence of the microwave background and estimated its temperature at about 3 K. However, this part of the electromagnetic spectrum happens to be difficult to observe from the ground, and it was 25 years before astronomers could demonstrate conclusively that the radiation was described by a black-body curve. In 1989, the *Cosmic Background Explorer* satellite—COBE for short—measured the intensity of the microwave background at wavelengths straddling the peak, from half a millimeter up to about 10 cm. The results are shown in Figure 17.12. The solid line is the black-body curve that best fits the COBE data. The near-perfect fit corresponds to a universal temperature of about 2.7 K.

A striking aspect of the cosmic microwave background is its high degree of *isotropy*. When we correct for Earth's motion through space (which causes the microwave background to appear a little hotter than average in front of us and slightly cooler behind), the intensity of the radiation is virtually constant (to about

1 part in 10^5) from one direction in the sky to another. This isotropy provides strong support for the cosmological principle.

Remarkably, the cosmic microwave background contains more energy than has been emitted by all the stars and galaxies that have ever existed in the history of the universe. The reason for this is that stars and galaxies, though very intense sources of radiation, occupy only a small fraction of space. When their energy is averaged out over the volume of the entire cosmos, it falls short of the energy in the microwave background by at least a factor of 10. For our present purposes, then, we can ignore much of the rest of this book and regard the cosmic microwave background as the only significant form of radiation in the universe!

Matter and Radiation

The overall density of matter in the universe is not known with certainty, but it is thought to be fairly close to the critical density of 10^{-26} kg/m^3. The universe is apparently open, but barely so. We have just seen, most radiation in the universe exists in the form of the cosmic microwave background. We learned in Section 9.5 that matter is equivalent to energy, so how do these two forms of energy compare? Is matter the dominant constituent of the present universe, or does radiation also play an important role on the largest scales?

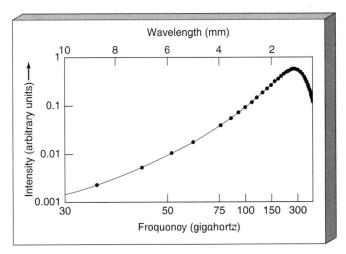

Figure 17.12 The intensity of the cosmic background radiation, as measured by the COBE satellite, agrees very well with theory. The curve is the best fit to the data, corresponding to a temperature of 2.735 K. The experimental errors in this remarkably accurate observation are smaller than the dots representing the data points.

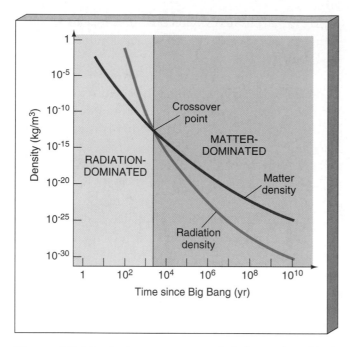

Figure 17.13 As the universe expanded, the number of both matter particles and photons per unit volume decreased. However, the photons were also reduced in energy by the cosmological redshift, reducing their equivalent mass, and hence their density, still further. As a result, the density of radiation fell faster than the density of matter as the universe grew. Tracing the curves back from the densities we observe today, we see that radiation must have dominated matter at early times—that is, at times to the left of the crossover point.

To answer these questions, we must convert matter and radiation to a "common currency"—either mass or energy. Let's compare their masses. We can express the energy in the microwave background as an equivalent density by first calculating the number of photons in any cubic meter of space, then converting the total energy of these photons to a mass using the relation $E = mc^2$. ∞ (Sec. 9.5) When we do this, we arrive at a density for the microwave background of about 5×10^{-30} kg/m³. Thus, *at the present moment* the density of matter in the universe far exceeds the density of radiation. In cosmological terminology, we say that we live in a **matter-dominated universe**.

Was the universe always matter dominated? To address this question, we must ask how the densities of both matter and radiation change as the universe expands. Both decrease, as the expansion dilutes the numbers of atoms and photons alike. But the radiation is also diminished in energy by the cosmological redshift, so its equivalent density falls faster than the density of matter as the universe grows (Figure 17.13).

Conversely, as we look back in time, closer and closer to the Big Bang, the radiation density increases faster than that of matter. Accordingly, even though today the radiation density is much less than the matter density, there must have been a time in the past when they were equal. Before that time, radiation was the main constituent of the cosmos. The universe is said to have been **radiation-dominated** then.

Throughout this book, we have been concerned exclusively with the history of the universe long after it became matter dominated—the formation and evolution of galaxies, stars, and planets as the universe thinned and cooled toward the state we see today. We now consider some important events in the early universe, long before any star or galaxy existed, that played no less a role in determining the present condition of the cosmos.

17.6 The Formation of Nuclei and Atoms

5 At the moment of the Big Bang, the universe was unimaginably hot and dense. It has been expanding and cooling ever since. At the very earliest times, theorists believe, the cosmos consisted entirely of radiation. However, during the first minute or so of the universe's existence, temperatures were high enough that individual photons of radiation had sufficient energy to transform themselves into matter, in the form of elementary particles. This period saw the creation of all the basic "building blocks" of matter we know today—protons and neutrons, electrons, dark matter. Since then, matter has evolved, clumping together into more and more complex structures, forming the nuclei, atoms, planets, stars, galaxies and large-scale structure we know today, but no *new* matter has been created. *Everything we see around us was created out of radiation as the early universe expanded and cooled.*

Helium Formation

Let's complete our story of the creation of the elements by tying up a loose end left over from Chapter 12. Simply put, there is far more helium in the universe than can be explained by nuclear fusion in stars. No matter where they look, and no matter how low a star's abundance of heavy elements may be, astronomers find that there is a minimum amount of helium—a little less than 25 percent by mass—in all stars. The accepted explanation is that this base level of helium is *primordial*—that is, it was created during the early, hot epochs of the universe, before any stars had formed. The production of elements heavier than hydrogen by nuclear fusion shortly after the Big Bang is called **primordial nucleosynthesis**.

By about 100 s after the Big Bang, the temperature had fallen to about 1 billion kelvins and, apart from "exotic" dark matter particles, matter in the universe consisted of electrons, protons, and neutrons, the protons outnumbering the neutrons by about 5 to 1. Fusion reactions between protons and neutrons to form deuterium were frequent.[2] However, the matter was embedded in a billion-kelvin "sea" of radiation, and the deuterium nuclei were broken apart by high-energy gamma rays as quickly as they were created. Only when the temperature of the universe fell below about 900 million K, roughly 2 minutes after the Big Bang, was deuterium at last able to endure. Once that occurred, numerous other fusion reactions quickly converted deuterium into heavier elements, especially helium-4. In just a few minutes, most of the available neutrons were consumed, leaving a universe whose matter content was primarily hydrogen and helium.

We might imagine that the fusion chain would have continued to create heavier and heavier elements, just as in the cores of stars, but this did not occur. In stars, the density and the temperature both *increase* slowly with time, allowing more and more massive nuclei to form, but in the early universe, exactly the opposite was true. The temperature and density were both *decreasing* rapidly, making conditions less and less favorable for fusion as time went by. For all practical purposes the expansion of the universe caused primordial nucleosynthesis to stop at helium-4. By about 15 minutes after the Big Bang, the cosmic elemental abundance was set, and helium accounted for approximately one quarter of the total mass of matter in the universe. The remaining 75 percent of the matter in the universe was hydrogen. It would be almost a billion years before nuclear reactions in stars would change these figures.

Nucleosynthesis and the Density of the Universe

During the early epoch of helium formation, while most deuterium was quickly burned into helium as soon as it formed, a small amount was left over when the primordial nuclear reactions ceased. This fact is of great importance to astronomers, because detailed calculations indicate that the amount of deuterium produced is a very sensitive indicator of the present-day density of matter in the universe. The *denser* the universe is today, the more matter (protons and neutrons) there was at those early times to react with the deuterium, and the *less* deuterium was left over when nucleosynthesis ended. Unlike helium, deuterium is not produced to any significant degree in stars, so any deuterium we see today *must* be primordial. Comparison between theoretical calculations of the production of deuterium in the early universe and the observed abundance of deuterium in stars and the interstellar medium implies a present-day cosmic density of *at most* 3×10^{-28} kg/m^3—about 3 percent of the critical value.

But before we jump to the conclusion that the universe is open and will expand forever, we must make a very important qualification. Primordial nucleosynthesis as just described depends *only* on the presence of protons and neutrons in the early universe. Thus, measurements of the abundance of helium and deuterium tell us only about the density of "normal" matter—matter made up of protons and neutrons—in the cosmos.

This finding has a momentous implication for the overall composition of the universe. As we saw earlier, astronomers have concluded, on the basis of studies of the motions of galaxies in clusters and superclusters, that Ω_0, the ratio of the total cosmic mass density to the critical density, is at least 0.2 or 0.3 and may possibly be considerably more. If this reasoning turns out to be correct, and if the density of matter in the form of protons and neutrons is at most 0.03 of the critical value, then we are forced to admit that not only is most of the matter in the universe dark, but most of the dark matter *is not* composed of protons and neutrons.

Thus, the bulk of the matter in the universe apparently exists in the form of elusive subatomic particles (discussed as dark-matter candidates in Chapter 14), whose nature we do not fully understand and whose very existence has yet to be conclusively

---

[2]Recall from Chapter 9 that deuterium is simply a heavy form of hydrogen. Its nucleus contains one proton and one neutron. ∞ (Sec. 9.5)

17-1 MORE PRECISELY

CURVED SPACE

*E*uclidean geometry is the geometry of flat space—the geometry taught in high schools everywhere. Set forth by one of the most famous of the ancient Greek mathematicians, Euclid, who lived around 300 BC, it is the geometry of everyday experience. Houses are usually built with flat floors. Writing tablets and blackboards are also flat. We work easily with flat, straight objects because a straight line is the shortest distance between any two points.

In constructing houses or any other straight-walled buildings on the surface of Earth, the other basic axioms of Euclid's geometry also apply: parallel lines never meet even when extended to infinity; the angles of any triangle always sum to 180°; the circumference of a circle equals π times its diameter. If these rules were not obeyed, walls and roof would never meet to form a house!

In reality, though, the geometry of Earth's surface is not really flat. It is curved. We live on the surface of a sphere, and on that *surface*, Euclidean geometry breaks down. Instead, the rules for the surface of a sphere are those of *Riemannian geometry*, named after the nineteenth-century German mathematician Georg Friedrich Riemann. For example, there are no parallel lines (or curves) on a sphere's surface; any lines drawn on the surface and around the full circumference will eventually intersect. The sum of a triangle's angles, when drawn on the surface of a sphere, exceeds 180°—in the 90°-90°-90° triangle shown in the figure, the sum is 270°. And the circumference of a circle is less than π times its diameter.

We see that the curved surface of a sphere, governed by the spherical geometry of Riemann, differs greatly from the flat-space geometry of Euclid. These two geometries are approximately the same only if we confine ourselves to a small patch on the surface. If the patch is much smaller than the sphere's radius, the surface surrounding any point looks flat, and Euclidean geometry is approximately valid. This is why we can draw a usable map of our home, our city, even our state, on a flat sheet of paper, but an accurate map of the entire Earth must be drawn on a globe.

When we work with larger parts of Earth, we must abandon Euclidean geometry. World navigators are fully aware of this. Aircraft do not fly along what you might regard as a straight-line path from one point to another. Instead they follow a *great circle*, which, on the curved surface of a sphere, is the shortest distance between two points. For example, a flight from Los Angeles to London does not proceed directly across the United States and the Atlantic Ocean as you might expect from looking at a flat map. Instead, it goes far to the north, over Canada and Greenland, above the Arctic Circle, finally coming in over Scotland for a landing at London. This is the great circle—the shortest path—between the two cities. Look at a globe of Earth (or see the accompanying figure) to verify that this is indeed the case.

The *positively curved* space of Riemann is not the only possible departure from flat space. Another is the *negatively curved* space first studied by Nikolai Ivanovich Lobachevsky, a nineteenth-century Russian mathematician. In this geometry, there are an *infinite* number of lines through any given point parallel to another line, the sum of a triangle's angles is *less* than 180° (see the

---

demonstrated in laboratory experiments. ∞ (Sec. 14.6) For the sake of brevity, we will adopt from here on the convention that the term "dark matter" refers only to these unknown particles, and not to "stellar" dark matter, such as black holes and white dwarfs (also discussed in Chapter 14), which are made of normal matter.

The Formation of Atoms

When the universe was a few thousand years old, matter began to dominate over radiation. At that time, matter consisted of electrons, protons, helium nuclei (formed by primordial nucleosynthesis), and dark matter. The temperature was several tens of thousands of

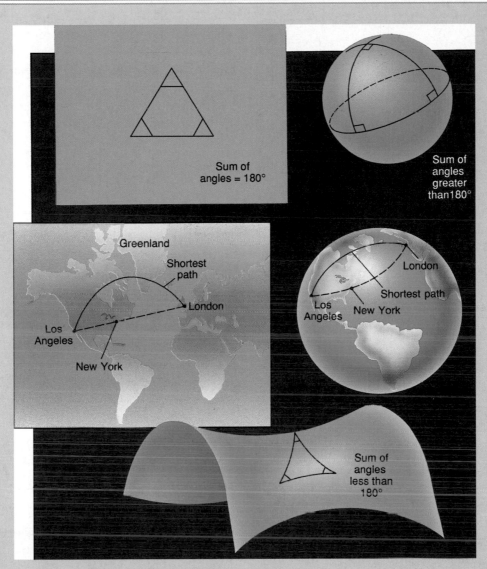

figure), and the circumference of a circle is *greater* than π times its diameter. Instead of the flat surface of a plane or the curved surface of a sphere, this type of space is described by the *surface* of a curved saddle. It is a hard geometry to visualize!

Even though the universe on the largest scales may be curved as we have just described, most of the local realm of the *three*-dimensional universe (including the solar system, the neighboring stars, and even our Milky Way Galaxy) is correctly described by Euclidean geometry.

kelvins—far too hot for atoms of hydrogen to exist, although some helium ions may already have formed. During the next few hundred thousand years, a major change occurred. The universe expanded by another factor of 10, the temperature dropped to a few thousand kelvins, and electrons and nuclei combined to form neutral atoms. By the time the temperature had

fallen to 4500 K, the universe consisted of atoms, photons, and dark matter.

The period during which nuclei and electrons combined to form atoms is often called the epoch of **decoupling**, for it was during this period that the radiation background parted company with normal matter. At early times, when matter was ionized, the universe

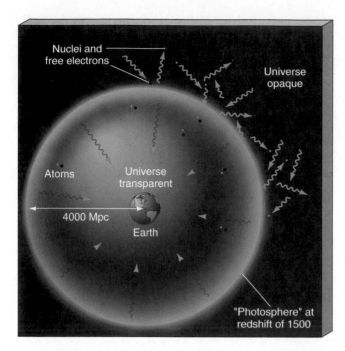

Figure 17.14 When atoms formed, the universe became virtually transparent to radiation. Thus, observations of the cosmic background radiation allow us to study conditions in the universe around a time at a redshift of 1500, when the temperature dropped below about 4500 K.

was filled with large numbers of free electrons, which interacted frequently with electromagnetic radiation of all wavelengths. As a result, a photon could not travel far before encountering an electron and scattering off it. In effect, the universe was opaque to radiation (rather like the deep interior of a star like the Sun). Matter and radiation were strongly "tied," or *coupled*, to one another by these interactions. When the electrons combined with nuclei to form atoms of hydrogen and helium, however, only certain wavelengths of radiation—the ones corresponding to the spectral lines of those atoms—could interact with matter. Radiation of other wavelengths could travel virtually forever without being absorbed. The universe became nearly *transparent*. From that time on, the photons passed generally unhindered through space. As the universe expanded, the radiation simply cooled, eventually to become the microwave background we see today.

The microwave photons now detected on Earth have been traveling through the universe ever since they decoupled. Their last interaction with matter (at the epoch of decoupling) occurred when the universe was a few hundred thousand years old and roughly 1500 times smaller (and hotter) than it is today—that is, at a redshift of 1500. As illustrated in Figure 17.14, the epoch of atom formation created a kind of "photo-

sphere" in the universe, completely surrounding Earth at a distance of approximately 4000 Mpc, the distance the photons have traveled since they decoupled. On our side of the photosphere—that is, since decoupling—the universe is transparent. On the far side—before decoupling—it was opaque. Thus, by observing the microwave background, we are probing conditions in the universe almost all the way back in time to the Big Bang, in much the same way as the study of sunlight tells us about the surface layers of the Sun.

17.7 Cosmic Inflation

The Horizon and Flatness Problems

6 In the late 1970s, cosmologists trying to piece together the evolution of the universe were confronted with two nagging problems that had no easy explanation within the standard Big Bang model. The first is known as the **horizon problem**. Imagine observing the microwave background in two opposite directions on the sky, as illustrated in Figure 17.15. As we have just seen, in doing so we are actually observing two distant regions of the universe, marked A and B on the figure, where the radiation background last

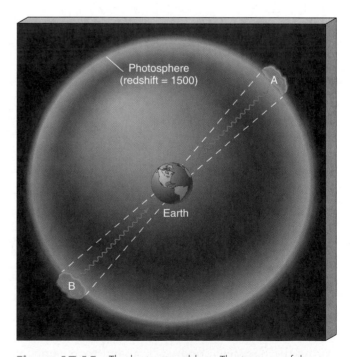

Figure 17.15 The horizon problem. The isotropy of the microwave background indicates that regions A and B in the universe were very similar to one another when the radiation we observe left them, but there has not been enough time since the Big Bang for them ever to have interacted with one another. Why, then, should they look the same?

interacted with matter. The fact that the background radiation is known to be *isotropic* to high accuracy means that regions A and B must have had very similar densities and temperatures at the time the radiation we see left them. The problem is, within the Big Bang theory as just described, there is no particular reason *why* these regions should be so similar to one another.

To take an everyday example, we all know that heat flows from regions of high temperature to regions of low temperature, but it takes time for this to occur. If we light a fire in one corner of a room, we have to wait a while for the other corners to warm up. Eventually, the room will reach a more or less uniform temperature, but only after the heat from the fire—or, more generally, the *information* that the fire is there—has had time to spread. Similar reasoning applies to regions A and B in Figure 17.15. They are separated by many megaparsecs, and there has not been time for information, which can travel no faster than the speed of light, to travel from one to the other. In cosmological parlance, the two regions are said to be outside each others' *horizon*. But if that is so, then how do they "know" that they are supposed to look the same? With no possibility of communication between them, the only alternative is that regions A and B simply started off looking alike—an assumption that cosmologists are very unwilling to make.

The second problem with the standard Big Bang model is called the **flatness problem**. Whatever the exact value of Ω_0, it appears to be quite close to 1—the density of the universe is fairly near the critical value needed for the expansion to barely continue forever. In terms of spacetime curvature, we can say that the universe is remarkably close to being flat. We say "remarkably" here because again there is no good reason *why* the universe should have formed with a density very close to critical. Why not a millionth or a million times that value? Furthermore, as can be seen in Figure 17.16, a universe that starts off close to, but not exactly on, the critical curve soon deviates greatly from it, so if the universe is close to critical now, it must have been *extremely* close to critical in the past. For example, if $\Omega_0 - 0.1$ today, the departure from the critical density at the time of nucleosynthesis would have been only 1 part in 10^{15} (a thousand trillion).

These observations constitute "problems" because cosmologists want to be able to *explain* the present condition of the universe, not just accept it "as is." They would prefer to resolve the horizon and flatness problems in terms of physical processes that could

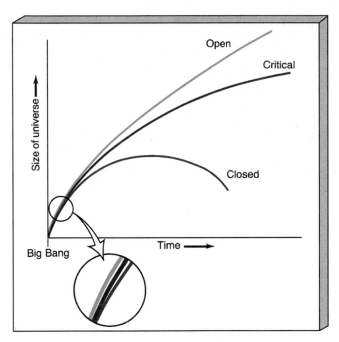

Figure 17.16 The flatness problem. If the universe deviates even slightly from the critical case, that deviation grows rapidly in time. For the universe to be as close to critical as it is today, it must have differed from the critical density in the past by only a tiny amount.

have taken a universe with no special properties and caused it to evolve into the cosmos we now see. The resolution of both problems takes us back in time even earlier than nucleosynthesis or the formation of any of the elementary particles we know today—back, in fact, almost to the instant of the Big Bang itself.

The Epoch of Inflation

In the 1970s and 1980s, theoretical physicists succeeded in combining, or *unifying*, the three nongravitational forces in the universe—electromagnetism, the strong force (which binds protons and neutrons together to form nuclei), and the weak force (which plays a role in many radioactive decays)—into a single all-encompassing "superforce." A general prediction of the **Grand Unified Theories** (GUTs) that describe this superforce is that the three forces are unified and indistinguishable from one another *only* at enormously high energies—corresponding to temperatures in excess of 10^{28} K. At lower temperatures, the superforce splits into three, revealing its separate electromagnetic, strong, and weak characters.

In the early 1980s, cosmologists discovered that Grand Unified Theories had some remarkable implica-

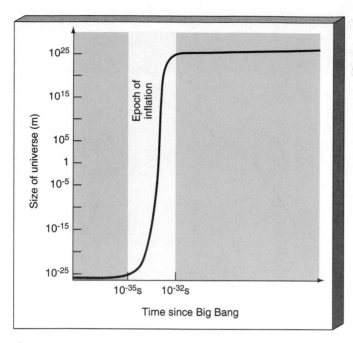

Figure 17.17 During the period of inflation, the universe expanded enormously in a very short time. Afterward, it resumed its earlier "normal" expansion rate, except that now the size of the cosmos was about 10^{50} times bigger than it was before inflation.

tions for the very early universe. About 10^{-34} s after the Big Bang, as temperatures fell below 10^{28} K and the basic forces of nature became reorganized—a little like a gas liquefying or water freezing as the temperature drops—the universe briefly entered a very odd, and unstable, high-energy state that physicists call the "false vacuum." In essence, the universe remained in the "unified" condition a little too long, like water that has been cooled below freezing but has not yet turned to ice. This had dramatic consequences. For a short while, empty space acquired an enormous *pressure*, which temporarily overcame the pull of gravity and accelerated the expansion of the universe at an enormous rate. The pressure remained constant as the cosmos expanded, and the acceleration grew more and more rapid with time—in fact, the size of the universe *doubled* every 10^{-34} s or so! This period of unchecked cosmic expansion, illustrated in Figure 17.17, is generally known as the epoch of **inflation**.

Eventually, the universe returned to the lower-energy "true vacuum" state. Regions of normal space began to appear within the false vacuum and rapidly spread to include the entire cosmos. With the return of the true vacuum, inflation stopped. The whole episode lasted a mere 10^{-32} s, but during that time the universe swelled in size by the incredible factor of about 10^{50}.

With the normal vacuum restored, the universe once again resumed its (relatively) leisurely expansion, slowly decelerated by the effect of gravity. However, some important changes had occurred that would have far-reaching ramifications for the evolution of the cosmos.

Implications for the Universe

The inflationary epoch provides a natural solution for the horizon and flatness problems. The horizon problem is solved because inflation took regions of the universe that had already had time to communicate with one another—and so had established similar physical properties—and then dragged them far apart, well out of communications range of one another. In effect, the universe expanded much faster than the speed of light during the inflationary epoch, so what was once well within the horizon now lies far beyond it. (Relativity restricts matter and energy to speeds less than the speed of light, but imposes no such limit on the universe as a whole.) Regions A and B in Figure 17.15 have been out of contact with one another since 10^{-32} s after creation, but they *were* in contact before then. Their properties are the same today because they were the same long ago, before inflation separated them.

To see how inflation solves the flatness problem, let's return once more to our earlier balloon analogy. Imagine that you are a 1-mm-long ant sitting on the surface of the balloon as it expands, as illustrated in Figure 17.18. When the balloon is just a few centimeters across, you can easily perceive the surface to be curved because its circumference is only a few times your own size. When the balloon expands to, say, a few meters in diameter, the curvature of the surface will be less pronounced but still perceptible. However, by the time the balloon has expanded to a few *kilometers* across, your "ant-sized" patch of the surface will look quite flat, just as the surface of Earth looks flat to us.

Now imagine that the balloon expands 100 trillion trillion trillion trillion times, as the universe did during the period of inflation. Your local patch of the surface would be completely indistinguishable from a perfectly flat plane. Exactly the same argument applies to the universe. Any curvature the universe may have had before inflation has been expanded so much that space is now perfectly flat, at least on the scale of the observable universe (and, in fact, on much larger scales too).

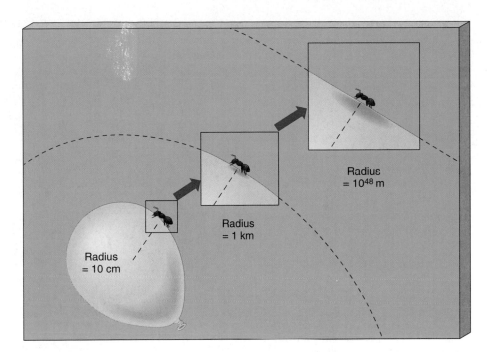

Figure 17.18 Inflation solves the flatness problem by taking a curved surface, here represented by the surface of the expanding balloon, and expanding it enormously. To an ant on the surface, the balloon looks perfectly flat when the expansion is over.

This resolution to the flatness problem—observations indicate that the universe is close to being flat because the universe *is* flat, to very high accuracy—has a very important consequence. Because the universe is flat, the density of matter must be exactly critical: $\Omega_0 = 1$. That means that there must be a lot of invisible matter in the universe beyond the clusters and the superclusters, filling the huge voids on the largest scales. And because we just saw that primordial nucleosynthesis implies that the density of normal matter is at most 0.03 the critical value, it follows that the rest of the mass—97 percent of all the matter in the universe—must be in the form of dark matter (whatever it may be).

When the idea of inflation was first suggested, many astronomers were very skeptical. Some still are. They point out that there is *no* direct observational evidence for a cosmic density as high as the critical value, and in that contention they are correct. However, as we saw earlier, there is ample evidence that the universe contains dark matter equivalent to as much as 20 or 30 percent of the critical density, and estimates of Ω_0 seem to increase as the scale under consideration increases. Furthermore, inflation is a clear prediction of a group of theories—the GUTs—that are becoming more and more firmly established as the standard description of matter at high energies. If the GUTs are indeed correct, then inflation *must* have occurred. Finally, inflation provides a neat solution to two serious difficulties in the standard noninflationary Big Bang model.

17.8 The Formation of Large-Scale Structure

In Chapter 15 we saw a little of how pregalactic fragments may have interacted and merged to form the galaxies we see today. ∞ (Sec. 15.4) Here we concern ourselves with the formation of structure on much larger scales.

Dark Matter and the Growth of Inhomogeneities

7 As best we can tell, all of the present large-scale structure in the universe grew from small *inhomogeneities*—slight deviations from perfectly uniform density—that existed in the matter of the early universe. Denser than average clumps of matter contracted under the influence of gravity, eventually reaching the point where stars began to form, and luminous galaxies appeared. However, by the early 1980s, cosmologists had come to realize that galaxies could not have grown from inhomogeneities involving only *normal* matter. Three lines of reasoning led to this conclusion:

1. Calculations show that, before decoupling (which occurred at a redshift of 1500), the intense background radiation would have prevented clumps of normal matter from contracting. Thus, any such clumps would have had to

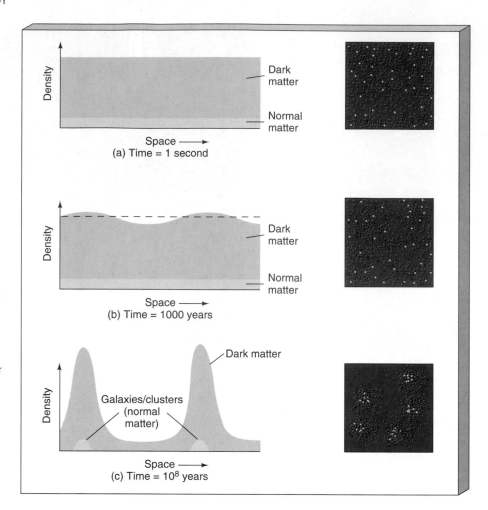

Figure 17.19 The formation of structure in the cosmos. (a) The universe started out as a mixture of (mostly) dark and normal matter. (b) A few thousand years after the Big Bang, the dark matter began to clump. (c) Eventually the dark matter formed large structures (represented here by the two high-density peaks) into which normal matter flowed, ultimately to form the galaxies we see today. The three frames at right represent the densities of dark matter (red) and normal matter (yellow) graphed at left.

wait until *after* decoupling before their densities could increase.

2. Because the cosmic background radiation was "tied" to normal matter up until decoupling, any variations in the matter density would have led to temperature variations in the cosmic background radiation—denser regions would have been a little hotter than less dense ones. The high degree of isotropy observed in the microwave background indicates that any density variations from one region of space to another must have been small—only a few parts in 10^5.

3. Galaxies—or, at least, quasars—are known to have formed by a redshift of 5. ∞ (Sec. 16.6) Because the contracting matter had to "fight" the general expansion of the universe, the small inhomogeneities permitted by observations of the microwave background could not have grown into galaxies in the time available.

Fortunately for cosmology, the fact that most of the universe is made of *dark* matter, which has proper-

ties very different from those of normal matter, provides an alternative explanation. Whatever its true nature, dark matter interacts only very weakly with normal matter and radiation. As a result, the density of initially high-density regions has been increasing ever since matter first began to dominate the universe (which occurred at a redshift of about 20,000). Thus, point (1) above does not apply because the dark matter started clumping well before decoupling (redshift 1500). Furthermore, because the dark matter is not directly tied to the radiation, dark-matter density inhomogeneities could have been quite large at the time of decoupling without having a correspondingly large effect on the microwave background, so the constraints of point (2) are also avoided. In short, dark matter could clump to form large-scale structure in the universe without running into the problems just described for normal matter.

In this picture (Figure 17.19), dark matter determines the overall distribution of mass in the universe and clumps to form the observed large-scale structure without violating any observational constraints on the

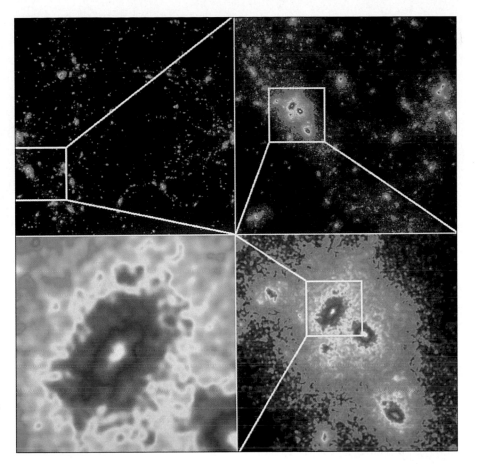

Figure 17.20 Successively magnified views of a 100 × 100 × 100 Mpc cube in a simulated dark-matter universe in which $\Omega_0=1$, showing the present-day structure that results from the growth of small density inhomogeneities in the very early universe. Colors represent mass density, ranging from the cosmic mean (dark blue), through green, yellow, and red, to 100 times the mean (white). The enlargements zoom in on one particular galaxy in one particular small group of galaxies. The last frame (at bottom left) is roughly 1.5 Mpc across. Notice both the large-scale filamentary structure evident in the top two frames and the extensive dark-matter halos surrounding individual galaxies (the galaxies are roughly the white regions in the bottom two frames).

microwave background. Then, at later times, normal matter is drawn by gravity into the regions of highest density, eventually forming galaxies and galaxy clusters. This picture explains why so much dark matter is found outside the visible galaxies. The luminous material is strongly concentrated near the density peaks and dominates the dark matter there, but the rest of the universe is essentially devoid of normal matter. Like foam on the crest of an ocean wave, the universe we see is only a tiny fraction of the total.

Figure 17.20 shows the results of a recent supercomputer simulation of a (mainly) dark-matter universe. Compare these images with the real observations of nearby structure shown in Figures 15.30 and 15.31. Although calculations like this do not prove that dark-matter models are the correct description of the universe, the similarities between the models and reality are certainly very striking.

The Microwave Background

Because dark matter does not interact directly with photons, its density variations do not cause large tem-

perature variations in the microwave background. However, radiation is influenced slightly by the *gravity* of the growing dark clumps, experiencing a slight gravitational redshift that varies from place to place depending on the dark-matter density. As a result, dark-matter models predict that there should be tiny "ripples" in the microwave background—temperature variations of only a few parts per million from place to place on the sky.

Until the late 1980s, these ripples were too small to be accurately measured, but cosmologists were confident that they would be found. In 1992, after almost two years of careful observation, the COBE team announced that the expected ripples had been detected. The temperature variations are tiny—only 30–40 millionths of a kelvin from place to place in the sky—but they are there. The COBE results are displayed as a temperature map of the microwave sky in Figure 17.21.

Initially, it seemed that the inhomogeneities in the microwave background were not consistent with the "standard" dark-matter models that provided the best agreement with actual observations of structure in the present-day universe. The ripples seen by COBE,

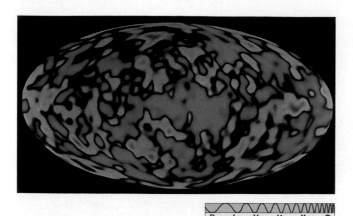

Figure 17.21 COBE map of temperature inhomogeneities in the cosmic microwave background over the entire sky. Hotter than average regions are shown in red, cooler than average regions in blue. The total temperature range shown is ±200 millionths of a kelvin. The temperature variation due to Earth's motion has been subtracted out, as has been the radio emission from the Milky Way Galaxy, and temperature deviations from the average are displayed.

taken in conjunction with the models, appeared to imply too little structure on large scales—that is, the computer simulations predicted fewer superclusters, voids, Great Walls, and so on, than are actually seen. However, with some modifications to the details of the models, it now looks as though the disagreement is not as serious as it first seemed, and a growing number of cosmologists are coming to regard the COBE observations as confirmation of a central prediction of dark-matter theory. The simulation shown in Figure 17.21 implies temperature variations in the microwave background that agree very well with the present COBE observations.

If the COBE results hold up—as they are checked and rechecked by collaborators and competitors alike—they may one day come to rank alongside the discovery of the microwave background itself in terms of their importance to the field of cosmology.

Chapter Review

Summary

On scales larger than a few hundred megaparsecs, the universe appears roughly **homogeneous** (p. 478) (the same everywhere) and **isotropic** (p. 479) (the same in all directions). In **cosmology** (p. 479)—the study of the universe as a whole—researchers usually just assume that the universe is homogeneous and isotropic. This assumption is known as the **cosmological principle** (p. 479).

If the universe were homogeneous, isotropic, infinite, and unchanging, then the night sky would be bright because any line of sight would eventually intercept a star. The fact that the night sky is dark instead of being uniformly bright is called **Olbers's paradox**. (p. 480) Its resolution lies in the fact that, regardless of whether or not the universe is infinite, we see only a finite part of it from Earth—the region within 13 billion light years, beyond which light has not yet had time to reach us.

Tracing the observed motions of galaxies back in time implies that, about 13 billion years ago, the universe consisted of a single point that then began to expand rapidly at the time of the **Big Bang** (p. 481). The Big Bang did not happen at any particular location in space, because space itself was compressed to a point at that instant—the Big Bang happened everywhere at once. The cosmological redshift occurs as a photon's wavelength is "stretched" by cosmic expansion. The extent of the observed redshift is a direct measure of the expansion of the universe since the photon was emitted.

If the density of the universe is greater than the **critical density** (p. 484), then there is enough cosmic matter to stop the expansion and cause a recollapse. If the density of the universe is less than this value, the universe will expand forever. Luminous matter by itself contributes only about 1 percent of the critical density. When dark matter in galaxies and clusters is taken into account, the figure rises to 20 or 30 percent. The fraction of dark matter on larger scales is uncertain, but it may be even greater than is found in clusters. Most astronomers believe that the present density lies somewhere between 10 and 100 percent of the critical value.

General relativity provides a description of the geometry of the universe on the largest scales. The curvature of spacetime in a high-density **closed universe**

(p. 487) is sufficiently large that the universe "bends back" on itself and is finite in extent, somewhat like the surface of a sphere. A low-density **open universe** (p. 488) has curvature a little like that of a two-dimensional saddle surface and is infinite in extent. The intermediate-density **critical universe** (p. 488) will expand forever, but it is spatially flat.

The **cosmic microwave background** (p. 488) is isotropic black-body radiation that fills the entire universe. Its present temperature is about 3 K. The existence of the microwave background is direct evidence that the universe expanded from a hot, dense state. As the universe has expanded, the initially high energy radiation has been redshifted to lower and lower temperatures.

At the present time, the density of matter in the universe greatly exceeds the equivalent mass density of radiation. We live in a **matter-dominated** (p. 490) universe. The density of matter was much greater in the past, when the universe was smaller. However, because radiation is redshifted as the universe expands, the density of radiation was greater still. The early universe was **radiation dominated** (p. 490).

All the matter in the universe was created during the first few minutes after the Big Bang. Most of the helium observed in the universe today is primordial, created by fusion between protons and neutrons in the early universe. This process is known as **primordial nucleosynthesis** (p. 491). Detailed studies of primordial deuterium formation indicate that "normal" matter can account for at most 3 percent of the critical density. All other matter in the universe is dark matter made up of unknown particles. By the time the uni-

verse was about 1500 times smaller than it is today, the temperature had become low enough for atoms to form. At that time, the (then visible) radiation background **decoupled** (p. 493) from the matter. The photons that now make up the microwave background have been traveling freely through space ever since.

According to modern **Grand Unified Theories** (p. 495), the three nongravitational forces of nature first began to display their separate characters about 10^{-34} s after the Big Bang. A brief period of rapid cosmic expansion called the epoch of **inflation** (p. 496) ensued, during which the size of the universe increased by a factor of about 10^{50}. The **horizon problem** (p. 494) is the fact that, according to the standard (that is, noninflationary) Big Bang model, there is no good reason for widely separated parts of the universe to be as similar as they are. Inflation solves the horizon problem by taking a small homogeneous patch of the early universe and expanding it enormously. The patch is still homogeneous, but it is now much larger than the portion of the universe we can see today. Inflation also solves the **flatness problem** (p. 495), which is the fact that there is no obvious reason why the density of the universe is so close to critical. Inflation implies that the cosmic density is exactly critical.

The large-scale structure observed in the universe today could not have formed out of density inhomogeneities in normal gaseous matter. Instead, dark matter clumped and grew to form the "skeleton" of the structure now observed. Normal matter then flowed into the densest regions of space, eventually forming the galaxies we now see.

Self-Test: True or False?

_____ 1. Cosmic homogeneity can be tested observationally, but cosmic isotropy has no observational test.

_____ 2. Olbers's paradox asks, "Why is the night sky dark?"

_____ 3. Hubble's law implies that the universe will expand forever.

_____ 4. The Big Bang is an expansion only of matter, not of space.

_____ 5. All points in space appear to be at the center of the expanding universe.

_____ 6. The cosmological redshift is a direct measure of the expansion of the universe.

_____ 7. As it travels through space, a photon's wavelength expands at the same rate as the universe expands.

_____ 8. According to the standard Big Bang model of the cosmos, the universe has only two possible futures: either it will continue to expand forever, or it will someday stop expanding and start to contract.

_____ 9. The cosmic microwave background is the highly redshifted radiation of the early Big Bang.

_____ 10. The light emitted from all the stars in the universe now far outshines the cosmic microwave background.

_____ 11. Most of the helium in the universe is primordial.

_____ 12. The present-day microwave background radiation was created at the time of decoupling.

_____ 13. After decoupling, neutral atoms could no longer exist.

_____ 14. The horizon problem has to do with the isotropy found in the microwave background radiation.

_____ 15. The flatness problem is the fact that the observed density of matter is unexpectedly different from the critical density.

_____ 16. If the inflation model is correct, then the density of the universe is exactly equal to the critical density.

Self-Test: Fill in the Blank

1. Pencil-beam surveys suggest that the largest structures in space are no larger than about _____ Mpc across.

2. Homogeneous means "the same _____."

3. Isotropic means "the same in all _____."

4. Together, the assumptions of cosmic homogeneity and isotropy are known as the _____.

5. If the universe had an edge, this would violate the assumption of _____.

6. If the universe had a center, this would violate the assumption of _____.

7. A value of 75 km/s/Mpc for Hubble's constant gives a maximum age for the universe of _____ billion years.

8. _____ is slowing down the expansion of the universe.

9. Luminous matter makes up at most _____ percent of the critical density.

10. The surface of a sphere is a two-dimensional example of a _____ universe.

11. The present average temperature of the universe, as measured by the cosmic microwave background, is _____ K.

12. Comparing the mass density of radiation and matter, we find that, at the present time, _____ dominates.

13. When the universe was a few minutes old, nuclear fusion produced _____ and _____.

14. Elements heavier than helium were not formed primordially because the density and temperature of the universe became too _____ after helium formed.

15. If the inflation model is correct, then the density of the universe is _____ (greater than, equal to, less than) the critical density.

16. Theory predicted tiny inhomogeneities in the _____ of the microwave background; the _____ satellite found them.

Review and Discussion

1. What evidence do we have that there is no structure in the universe on very large scales? How large is "very large"?

2. What is the cosmological principle?

3. What is Olbers's paradox? How is it resolved?

4. Explain how an accurate measure of Hubble's constant can lead to an estimate of the age of the universe.

5. Why isn't it correct to say that the expansion of the universe involves galaxies flying outward into empty space?

6. Where and when did the Big Bang occur?

7. How does the cosmological redshift relate to the expansion of the universe?

8. What measurable property of the universe determines whether or not it will expand forever?

9. Is there enough matter in the universe to halt the current cosmic expansion?

10. What is the cosmic microwave background, and why is it so significant?

11. Why do all stars, regardless of their abundance of heavy elements, contain at least 25 percent helium by mass?

12. What is Ω_0? How do measurements of the cosmic deuterium abundance provide an estimate of Ω_0?

13. When did the universe become transparent to radiation?

14. What is cosmic inflation? How does inflation solve the horizon problem? The flatness problem?

15. What is the connection between dark matter and the formation of large-scale structure in the universe?

16. What did dark-matter models predict that was later found by the COBE satellite?

17. Many cultures throughout history have developed their own cosmologies. Do you think the modern scientific cosmology is more likely to endure than any other? Why or why not?

18. Estimates of the age of the universe based on Hubble's constant are glaringly different from estimates of the ages of globular clusters in our own Galaxy. How do you think astronomers should proceed in resolving the controversy?

Problems

1. According to the Big Bang theory described in this chapter, what is the maximum possible age of the universe if $H_0 = 50$ km/s/Mpc? 75 km/s/Mpc? 100 km/s/Mpc?

2. For $H_0 = 75$ km/s/Mpc, the critical density is 10^{-26} kg/m^3. (a) How much mass does that correspond to in a volume of 1 A.U.^{3}? (b) How large a cube would be required to enclose 1 Earth mass of material?

3. Assuming critical density and using the distances presented in Table 16.1, estimate the total amount of matter in the observable universe. Express your answer (a) in kilograms, (b) in solar masses, and (c) in "galaxies," where 1 galaxy = 10^{11} solar masses.

4. How many times did the universe double in size during the inflationary period if its final size was 10^{50} larger than when it started?

Projects

1. Make a model of a two-dimensional universe and examine Hubble's law on it. Find a balloon that will blow up into a nice large sphere. Blow it up about halfway and mark dots all over its surface; the dots represent galaxies. Arbitrarily choose one dot as your home galaxy. Using a cloth measuring tape, measure the distances from your home galaxy to various other galaxies, numbering the dots so you do not confuse them later. Now blow the balloon up to full size and measure the distances again. Calculate the change in the distances for each galaxy; this is a measure of their speed (= change in position/change in time; the time is the same for each and is arbitrary). Plot their speeds against their new distances as in Figure 15.28. Do you get a straight-line correlation (that is, a plot that obeys Hubble's law)?

Try this again using a different dot for your home galaxy. Do you still get a plot that obeys Hubble's law? Does it matter which dot you choose as home?

2. Write a paper on the philosophical differences between living in an open, closed, or flat universe. It is quite possible that astronomers may finally determine which of these is correct within your lifetime. Does it really matter? Are there aspects of any of these three possibilities that are uncomfortable to accept? Do you have a preference?

3. Go to your library and read about the model called the *steady-state universe*, which enjoyed some measure of popularity in the 1950s and 1960s. How does it differ from the standard Big Bang model? Why do you think the steady-state model is not accepted by most cosmologists today?

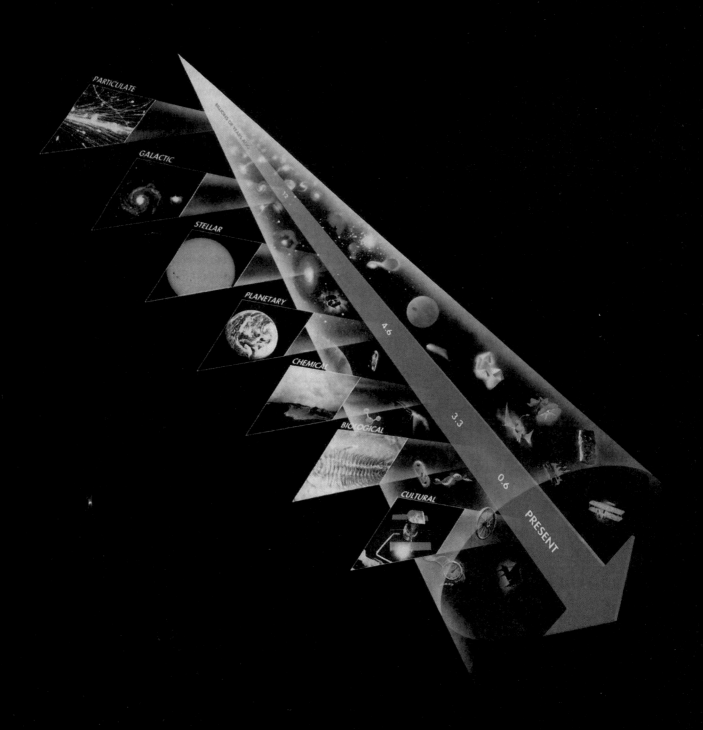

18 Life in the Universe

Are We Alone?

(Opposite page) The arrow of time, from the origin of the universe to the present and beyond. Cosmic evolution is the study of the many varied changes among energy, matter, and life in the thinning and cooling universe.

Inserted along the arrow of time are seven "windows" outlining some of the key events in the history of the universe:

- evolution of primal energy into elementary particles,
- evolution of atoms into galaxies and stars,
- evolution of stars into heavy elements,
- evolution of elements into rocky planets,
- evolution of the molecular building blocks of life,
- evolution of those molecules into life itself
- evolution of advanced life forms into intelligence, culture, and a technological civilization.

LEARNING GOALS

Studying this chapter will enable you to:

1 Summarize the process of cosmic evolution as it is currently understood.

2 Evaluate the chances of finding life in the solar system.

3 Summarize the various probabilities used to estimate the number of advanced civilizations that might exist in our Galaxy.

4 Discuss some of the techniques we might use to search for extraterrestrials and to communicate with them.

Are we unique? Is life on our planet the only example of life in the universe? These are difficult questions, for the subject of extraterrestrial life is one on which we have no data, but they are important questions, with profound implications for the human species. Earth is the only place in the universe where we know for certain that life exists. In this chapter we take a look at how humans evolved on Earth and then consider whether those evolutionary steps might have happened elsewhere. Having done that, we assess the likelihood of our having Galactic neighbors and consider how we might learn about them if they exist.

18.1 Cosmic Evolution

1 Figure 18.1 (see also the chapter opening art) identifies seven major phases in the history of the universe: *particulate*, *galactic*, *stellar*, *planetary*, *chemical*, *biological*, and *cultural* evolution. Together, these evolutionary stages make up the grand sweep of **cosmic evolution**—the continuous transformation of matter and energy that has led to the appearance of life and civilization on Earth. The first four phases represent, in reverse order, the contents of this book. We now expand our field of view beyond astronomy to include the other three.

From the Big Bang to the evolution of intelligence and culture, the universe has evolved from simplicity to complexity. We are the result of an incredibly complex chain of events that spanned billions of years. Were those events random, making us unique, or are they in some sense *natural* and technological civilization inevitable? Put another way, are we alone in the universe, or are we just one among countless other intelligent life forms?

Life in the Universe

Before embarking on our study, we need a working definition of *life*. This seemingly simple task is not an easy one—the distinction between the living and the nonliving is not as obvious as we might at first think. Though most physicists would agree on the definitions of matter and energy, biologists have not arrived at a clear-cut definition of life. Generally speaking, scientists regard the following as characteristics of living organisms: (1) they can *react* to their environment and can often heal themselves when damaged; (2) they can *grow* by taking in nourishment from their surroundings; (3) they can *reproduce*, passing along some of their own characteristics to their offspring; and (4) they have the

capacity for genetic change and can therefore *evolve* from generation to generation so as to adapt to a changing environment.

These rules are not hard and fast, and there is great leeway in interpreting them. Stars, for example, react to the gravity of their neighbors, grow by accretion, generate energy, and "reproduce" by triggering the formation of new stars, but no one would suggest that they are alive. A virus is crystalline and inert when isolated from living organisms, but, once inside a living system, it exhibits all the properties of life, seizing control of the cell and using the cell's own genetic machinery to grow and reproduce. Most researchers now believe that the distinction between living and nonliving is more one of structure and complexity than a simple checklist of rules.

The general case in favor of extraterrestrial life is summed up in what are sometimes called the *assumptions of mediocrity*: (1) because life on Earth depends on just a few basic molecules, (2) because the elements that make up these molecules are (to a greater or lesser extent) common to all stars, and (3) if the laws of science we know apply to the entire universe (as we have supposed throughout this book), then—given sufficient time—life must have originated elsewhere in the cosmos. The opposing view maintains that intelligent life on Earth is the product of a series of extremely fortunate accidents—astronomical, geological, chemical, and biological events unlikely to have occurred anywhere else in the universe. The purpose of this chapter is to examine some of the arguments for and against these viewpoints.

Chemical Evolution

What information do we have about the earliest stages of planet Earth? Unfortunately, not very much. Geological hints about the first billion years or so were

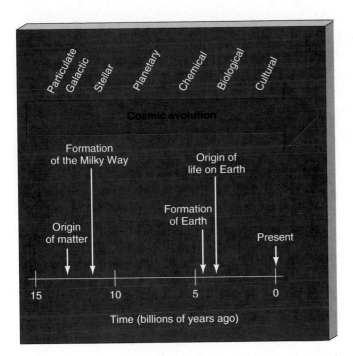

Figure 18.1   Some highlights of cosmic history are indicated along this arrow of time, reaching from the beginning of the universe to the present. Noted along the top of the arrow are the major phases of cosmic evolution.

largely erased by violent surface activity as volcanoes erupted and meteorites bombarded our planet; subsequent erosion by wind and water has seen to it that little evidence has survived to the present day. Scientists believe, however, that the early Earth was barren, with shallow, lifeless seas washing upon grassless, treeless continents. Outgassing from our planet's interior through volcanoes, fissures, and geysers produced an atmosphere rich in hydrogen, nitrogen, and carbon compounds and poor in free oxygen. As Earth cooled, ammonia, methane, carbon dioxide, and water formed. The stage was set for the appearance of life.

The surface of the young Earth was a very violent place. Natural radioactivity, lightning, volcanism, solar ultraviolet radiation and meteoritic impacts all provided large amounts of energy that eventually shaped the ammonia, methane, carbon dioxide and water into more complex molecules known as **amino acids** and **nucleotide bases**—organic (carbon-based) molecules that are the building blocks of life as we know it. Amino acids build *proteins*, and proteins control *metabolism*, the daily utilization of food and energy by means of which organisms stay alive and carry out their vital activities. Sequences of nucleotide bases form *genes*—parts of the DNA molecule—which direct the synthesis of proteins and thus determine the characteristics of

organisms. These same genes also carry an organism's hereditary characteristics from one generation to the next. In all living creatures on Earth—from bacteria to amoebas to humans—genes mastermind life, while proteins maintain it.

The idea that complex molecules could have evolved naturally from simpler ingredients found on the primitive Earth has been around since the 1920s. The first experimental verification was provided in 1953 when scientists Harold Urey and Stanley Miller, using laboratory equipment somewhat similar to that shown in Figure 18.2, took a mixture of the materials thought to be present on Earth long ago—a "primordial soup" of water, methane, carbon dioxide, and ammonia—and energized it by passing an electrical discharge ("lightning") through it. After a few days, they analyzed their mixture and found that it contained many of the same amino acids found today in all living things on Earth. About a decade later, scientists succeeded in constructing nucleotide bases in a similar manner. These experiments have been repeated in

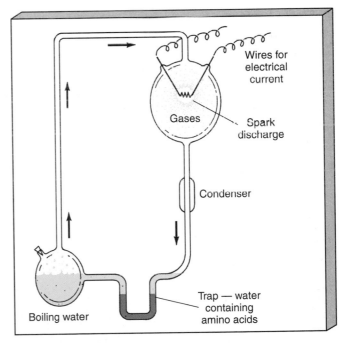

Figure 18.2   This chemical apparatus is designed to synthesize complex biochemical molecules by energizing a mixture of simple chemicals. A mixture of gases (ammonia, methane, carbon dioxide, water vapor) is placed in the upper bulb to simulate the primordial Earth atmosphere and then energized by spark-discharge electrodes. After about a week, amino acids and other complex molecules are found in the trap at the bottom, which simulates the primordial oceans into which heavy molecules produced in the overlying atmosphere would have fallen.

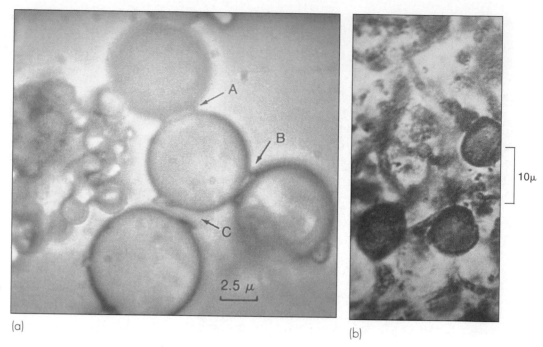

Figure 18.3 (a) These carbon-rich, proteinlike droplets display the clustering of as many as a billion amino acid molecules in a liquid. Droplets can "grow," and parts of droplets can separate from the "parent" to become new individual droplets (such as at A, B, C). The scale of 2.5 microns noted here is 1/4000 of a centimeter. (b) This photograph, taken through a microscope, shows a fossilized organism found in sediments radioactively dated as being 2 billion years old. This primitive system possesses concentric spheres or walls connected by smaller spheroids. The roundish fossils here measure about a thousandth of a centimeter in diameter.

many different forms, with more realistic mixtures of gases and a variety of energy sources, but always with the same basic outcomes.

While none of these experiments has ever produced a living organism, or even a single strand of DNA, they do demonstrate conclusively that "biological" molecules can be synthesized by strictly nonbiological means, using raw materials available on the early Earth. More advanced experiments, in which amino acids are united under the influence of heat, have fashioned proteinlike blobs (Figure 18.3a) that behave to some extent like true biological cells. Such near-protein material resists dissolution in water (so it would remain intact when it fell from the primitive atmosphere into the ocean) and tends to cluster into small droplets called *microspheres*—a little like oil globules floating on the surface of water. The walls of these laboratory-made droplets permit the inward passage of small molecules, which then combine within the droplet to construct more complex molecules too large to pass back out through the walls. As the droplets "grow," they tend to "reproduce," forming smaller droplets.

Can we consider these proteinlike microspheres to be alive? Almost certainly not. Most biochemists would say that the microspheres are not life itself but contain many of the basic ingredients needed to form life. The microspheres lack the hereditary molecule DNA. However, they do have similarities to ancient cells found in the fossil record (Figure 18.3b). Thus, while no actual living cells have yet been created "from scratch" in any laboratory, many biochemists feel that the chain of events leading from simple nonbiological molecules almost to the point of life itself has been amply demonstrated.

Recently, however, a dissenting view has emerged. Some scientists have argued that Earth's primitive atmosphere may *not* have been a particularly suitable environment for the production of complex molecules. There might not have been sufficient energy available to power the chemical reactions, and the early atmosphere may not have contained enough raw material for the reactions to have become important in any case. These researchers suggest that much, if not all, of the organic material that combined to form the first living cells was produced in *interstellar space* and subsequently arrived on Earth in the form of interplanetary dust and meteors that did not burn up during their descent through the atmosphere.

Interstellar molecular clouds are known to contain very complex molecules, and large amounts of organic material were detected on comet Halley by space probes when Halley last visited the inner solar system (see Section 4.2). More recently, similarly complex molecules have been observed on comet Hale-Bopp. ∞ (*Interlude 4-1*) Thus, the idea that organic matter is constantly raining down on Earth from space in the form of interplanetary debris is quite plausible.

Whether or not this was the *primary* means by which complex molecules first appeared in Earth's oceans remains unclear. For now, the issue is unresolved.

Diversity and Culture

However the basic materials appeared on Earth, we know that life *did* appear. The fossil record chronicles how life on Earth became widespread and diversified over time. The study of fossil remains shows the initial appearance about 3.5 billion years ago of simple one-celled organisms. These were followed about 2 billion years ago by more complex one-celled creatures. Multi-cellular organisms did not appear until about 1 billion years ago, after which there flourished a wide variety of increasingly complex organisms—insects, reptiles, mammals, and humans.

The fossil record leaves no doubt that biological organisms have changed over time—all scientists accept the reality of *biological evolution*. As conditions on Earth have shifted and Earth's surface has evolved, those organisms that could best take advantage of their new surroundings succeeded and thrived—often at the expense of those organisms that could not make the necessary adjustments and consequently became extinct.

Many anthropologists believe that, like any other highly advantageous trait, intelligence is strongly favored by natural selection. The social cooperation that went with coordinated hunting efforts was an important competitive advantage that developed as brain size increased. Perhaps most important of all was the development of language. Now our ancestors could share ideas as well as more tangible necessities, such as food and shelter. Experience, stored in the brain as memory, could be passed down from generation to generation. A new kind of evolution had begun, namely, *cultural evolution*—the changes in the ideas and behavior of society. Our more recent ancestors created, within about 10,000 years, the entirety of human civilization.

18.2 Life in the Solar System

2 Simple one-celled life forms reigned supreme on Earth for most of our planet's history. It took time—a great deal of time—for life to emerge from the oceans, to evolve into simple plants, to con-tinue to evolve into complex animals, and to develop intelligence, culture, and technology. Have those (or similar) events occurred elsewhere in the universe? Let's try to assess what little evidence we have on the subject.

Life as We Know It

"Life as we know it" is generally taken to mean carbon-based life that originated in a liquid-water environment—in other words, life on Earth. Is there any reason to suppose that such life might exist elsewhere in our solar system? The answer appears to be no. It seems that no environment in the solar system besides Earth's is particularly well suited for sustaining Earth-like life.

The Moon and Mercury lack liquid water, protective atmospheres, and magnetic fields, and so these two bodies are subjected to fierce bombardment by solar ultraviolet radiation, the solar wind, meteoroids, and cosmic rays. Simple molecules could not possibly survive in such hostile environments. Venus has far too much protective atmosphere! Its dense, dry, scorchingly hot atmospheric blanket effectively rules it out as a possible abode for life. The jovian planets have no solid surfaces, while Pluto and most of the moons of the outer planets are too cold. However, the possibility of liquid water below Europa's icy surface has refueled speculation about the development of life there. ∞ (Sec. 8.2) Saturn's moon Titan, with its atmosphere of methane, ammonia, and nitrogen gases, and possibly with some liquid on its surface, is conceivably another site where life might have arisen, although the results of the 1980 *Voyager 1* flyby suggest that Titan's frigid surface conditions are inhospitable for anything familiar to us. ∞ (Sec. 8.3)

What about the cometary and meteoritic debris that orbits within our solar system? Comets contain many of the basic ingredients for life—ammonia, methane, and water vapor, and many more complex molecules have been observed—and although comets are frozen, their icy matter warms while nearing the Sun. In addition, a small fraction of the meteorites that survive the plunge to Earth's surface do contain organic compounds. The Murchison meteorite, which fell near Murchison, Australia, in 1969, is a well-studied example. Located soon after crashing to the ground, this meteorite contains many of the well-known amino acids normally found in living cells (Figure 18.4). The moderately large molecules found in

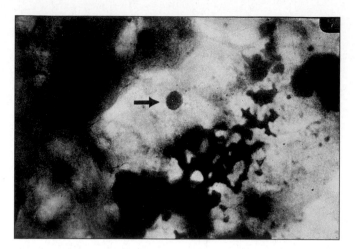

Figure 18.4 The Murchison meteorite contains relatively large amounts of amino acids and other organic material, indicating that chemical evolution of some sort has occurred beyond our own planet. In this magnified view of a fragment from the meteorite, the arrow points to a microscopic sphere of organic matter.

meteorites and in interstellar clouds are our only evidence that chemical evolution has occurred elsewhere in the universe. Most researchers regard this organic matter as prebiotic—that is, matter that could eventually lead to life but that has not yet done so.

The planet most likely to harbor life (or to have harbored it in the past) seems to be Mars. This planet seems harsh by Earth standards—liquid water is scarce, the atmosphere is thin, and the lack of magnetism and of an ozone layer allows solar high-energy particles and ultraviolet radiation to reach the surface unabated. But, as we have seen, the Martian atmosphere was thicker, and the surface warmer and much wetter, in the past. ∞ (Sec. 6.6) As discussed in *Interlude 6.1*, the *Viking* landers found no evidence of life, but they landed on the safest Martian terrain, not in the most interesting regions, such as near the moist polar caps.

Some scientists think that a different type of biology might be operating, or might have operated, on the Martian surface. They suggest that Martian microbes capable of eating and digesting oxygen-rich compounds in the Martian soil could explain the *Viking* results. This speculation will be greatly strengthened if recent announcements of fossilized bacteria in meteorites originating on Mars are confirmed. ∞ (*Interlude* 6-1) The consensus among biologists and chemists today is that Mars does not house any life similar to that on Earth, but a solid verdict regarding life on Mars will likely not be reached until we have thoroughly explored our intriguing neighbor.

Alternative Biochemistries

Conceivably, some types of biology may be so different from life on Earth that we cannot recognize them and do not know how to test for them. Some researchers have pointed out that the abundant element silicon has chemical properties somewhat similar to those of carbon and have suggested silicon as a possible alternative to carbon as the basis for living organisms. Ammonia (made of the common elements hydrogen and nitrogen) is sometimes put forward as a possible liquid medium in which life might develop, at least on a planet cold enough for ammonia to exist in the liquid state. Together or separately, these alternatives would surely give rise to organisms having biochemistries radically different from those we know on Earth. Conceivably, we might have difficulty even recognizing these organisms as being alive.

While the possibility of such alien life forms is a fascinating scientific problem, most biologists would argue that chemistry based on carbon and water is the one most likely to give rise to life. Silicon's chemical bonds are weaker than those of carbon and may not be able to form complex molecules—an apparently essential aspect of carbon-based life. Also, the low temperatures necessary for ammonia to be liquid might inhibit or even prevent completely the chemical reactions leading to the equivalent of amino acids and nucleotide bases. Still, we must admit that we know next to nothing about noncarbon, nonwater biochemistries, for the good reason that there are no examples of them to study experimentally. We can speculate about alien life forms and try to make general statements about their properties, but we can say little of substance about them.

18.3 Intelligent Life in the Galaxy

3 With humans apparently the only intelligent life in the solar system, we must broaden our search for extraterrestrial intelligence to other stars, perhaps even other galaxies. At such distances, though, we have little hope of detecting life with current equipment. Instead, we must ask: "How likely is it that life in any form—carbon based, silicon based, water based, ammonia based, or something we cannot even dream of—exists?" The word *likely* in the last sentence speaks of probabilities, so let's look at some numbers to develop statistical estimates of the probability of life elsewhere in the universe.

The Drake Equation

An early approach to this statistical problem is known as the **Drake equation**, after the U.S. astronomer who pioneered this analysis:

$$
\begin{array}{c}
\text{number of}\\
\text{technological,}\\
\text{intelligent}\\
\text{civilizations}\\
\text{now present}\\
\text{in the Galaxy}
\end{array}
=
\begin{array}{c}
\text{rate of star}\\
\text{formation,}\\
\text{averaged over}\\
\text{the lifetime}\\
\text{of the Galaxy}
\end{array}
$$

$$
\times
\begin{array}{c}
\text{fraction of stars}\\
\text{having}\\
\text{planetary}\\
\text{systems}
\end{array}
\times
\begin{array}{c}
\text{average number}\\
\text{of habitable planets}\\
\text{within those}\\
\text{planetary systems}
\end{array}
$$

$$
\times
\begin{array}{c}
\text{fraction of those}\\
\text{habitable planets}\\
\text{on which life}\\
\text{arises}
\end{array}
\times
\begin{array}{c}
\text{fraction of those}\\
\text{life-bearing planets}\\
\text{on which intelli-}\\
\text{gence evolves}
\end{array}
$$

$$
\times
\begin{array}{c}
\text{fraction of those}\\
\text{intelligent-life}\\
\text{planets that}\\
\text{develop tech-}\\
\text{nological society}
\end{array}
\times
\begin{array}{c}
\text{average lifetime of}\\
\text{a technologically}\\
\text{competent}\\
\text{civilization.}
\end{array}
$$

Several of the terms in this equation are largely a matter of opinion. We do not have nearly enough information to determine—even approximately—all of the terms, so the Drake equation cannot give us a hard-and-fast answer. Its real value is that it subdivides a large and very difficult question into smaller pieces that we can attempt to answer separately. Figure 18.5 illustrates how, as our requirements become more and more stringent, only a small fraction of star systems in the Milky Way Galaxy are likely to generate the advanced qualities specified by the combination of terms on the right-hand side of the equation.

Let's examine the terms in the equation one by one and make some educated guesses about their values. Bear in mind, though, that if you ask two scientists for their best estimates of any given term, you will likely get two very different answers!

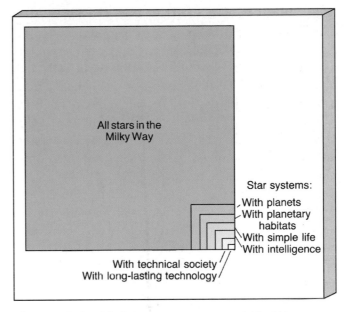

Figure 18.5 Of all the star systems in our Milky Way Galaxy (represented by the largest box), progressively fewer and fewer have each of the qualities typical of a long-lasting technological society (represented by the smallest box at the lower right corner).

Rate of Star Formation

We can estimate the average number of stars forming each year in the Galaxy simply by noting that at least 100 billion stars now shine in it. Dividing this number by the roughly 10-billion-year lifetime of the Galaxy, we obtain a formation rate of 10 stars per year. This may be an overestimate because astronomers believe that fewer stars are forming now than formed during earlier epochs, when more interstellar gas was available. However, we do know that stars are forming today, and our estimate does not include stars that formed in the past and have since died, so our value of 10 stars per year is probably reasonable when averaged over the lifetime of the Milky Way Galaxy.

Fraction of Stars Having Planetary Systems

Many astronomers believe planet formation to be a natural result of the star-formation process. If the condensation theory (see Chapter 4) is correct, and if there is nothing special about our Sun, as we have argued throughout this book, we would expect many stars to have at least one planet. ∞ (Sec. 4.3) Indeed, as we have seen, increasingly sophisticated observations suggest the presence of disks around other young stars. Could these disks be protosolar systems? The condensation theory suggests they are.

No planets like our own have yet been *seen* orbiting any other star. The light reflected by an Earth-like planet circling even the closest star would be too faint to detect even with the very best equipment. The light would be lost in the glare of the parent star. Large orbiting telescopes may soon be able to detect Jupiter-sized planets orbiting the nearest stars, but even those huge planets would be barely visible. However, as described in *Interlude 4-2* there is mounting *indirect* evidence for planets orbiting other stars. Several groups of observers have reported detection of the back-and-forth or side-to-side motions produced in a parent star by an orbiting planet's gravitational pull. The planets found so far are Jupiter-sized rather than Earth-sized, but astronomers are confident that an Earth-like planet will one day be detected by these means.

Accepting the condensation theory and its consequences, and without being either too conservative or naively optimistic, we assign a value near 1 to this term—that is, we believe that essentially all stars have planetary systems.

Number of Habitable Planets per Planetary System

Temperature, more than any other single quantity, determines the feasibility of life on a given planet. The surface temperature of a planet depends on two things: the planet's distance from its parent star and the thickness of its atmosphere. Planets with a nearby parent star (but not too close) and some atmosphere (though not too thick) should be reasonably warm, like Earth or Mars. Planets far from the star and with no atmosphere, like Pluto, will surely be cold by our standards. And planets too close to the star and with a thick atmosphere, like Venus, will be very hot indeed.

Figure 18.6 illustrates that a three-dimensional zone of "comfortable" temperatures—often called a *habitable zone*—surrounds every star. It represents the range of distances within which a planet of mass and composition similar to those of Earth would have a surface temperature between the freezing and boiling points of water. (Our Earth-based bias is plainly evident here!) The hotter the star, the larger this zone. For example, an A or an F star has a rather large hab-

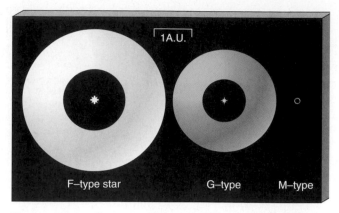

Figure 18.6 The extent of the habitable zone is much larger around a hot star than around a cool one. For a star like the Sun (a G star), the zone extends from about 0.85 A.U. to 2.0 A.U. For an F star, the range is 1.2 to 2.8 A.U. For a faint M star, only planets orbiting between about 0.02 and 0.06 A.U. would be habitable.

itable zone. G, K, and M stars have successively smaller zones. O and B stars are not considered here because they are not expected to last long enough for life to develop, even if they do have planets.

Three planets—Venus, Earth, and Mars—reside within the habitable zone surrounding our Sun. Venus is too hot because of its thick atmosphere and proximity to the Sun. Mars is a little too cold because its atmosphere is too thin and it is too far from the Sun. But if Venus had Mars's thin atmosphere and if Mars had Venus's thick atmosphere, both of these nearby planets might conceivably have surface conditions resembling those on Earth.

To estimate the number of habitable planets per planetary system, we first take inventory of how many stars of each type shine in our Galaxy and calculate the sizes of their habitable zones. Then we eliminate binary-star systems because a planet's orbit within the habitable zone of a binary would be unstable in many cases, as illustrated in Figure 18.7. Given the known properties of binaries in our Galaxy, habitable planetary orbits would probably be unstable in most cases, so there would not be time for life to develop.

Taking all these factors into account, we assign a value of 1/10 to this term in our equation. In other words, we believe that, on average, there is 1 potentially habitable planet for every 10 planetary systems that might exist in our Galaxy. Single F, G, and K stars are the best candidates.

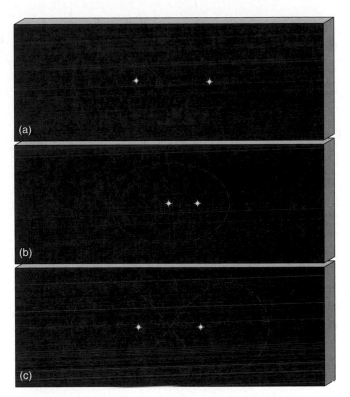

Figure 18.7 In binary-star systems, planets are restricted to only a few kinds of orbits that are gravitationally stable. (a) This orbit is stable only if the planet lies very close to its parent star, so the gravity of the other star is negligible. (b) A planet circulating at a great distance about both stars in an elliptical orbit. This orbit is stable only if it lies far from both stars. (c) Another possible but unstable path interweaves between the two stars in a figure-8 pattern.

Fraction of Habitable Planets on Which Life Arises

The number of possible combinations of atoms is incredibly large. If the chemical reactions that led to the complex molecules that make up living organisms occurred completely at random, then it is extremely unlikely that those molecules could have formed at all. In that case, life is extraordinarily rare, this term is close to zero, and we are probably alone in the Galaxy, perhaps even in the entire universe.

However, laboratory experiments (like the Urey–Miller experiment described earlier) seem to suggest that certain chemical combinations are strongly favored over others—that is, the reactions are not random. Of the billions upon billions of basic organic groupings that could possibly occur on Earth from the random combination of all sorts of simple atoms and molecules, only about 1500 actually do occur. Furthermore, these 1500 organic groups of terrestrial

biology are made from only about 50 simple "building blocks" (including the amino acids and nucleotide bases mentioned earlier). This suggests that molecules critical to life may not be assembled by pure chance. Apparently, additional factors are at work at the microscopic level. If a relatively small number of chemical "evolutionary tracks" is likely to exist, then the formation of complex molecules—and hence, we assume, life—becomes much more likely, given sufficient time.

To assign a very low value to this term in the equation is to believe that life arises randomly and rarely. To assign a value close to 1 is to believe that life is inevitable, given the proper ingredients, a suitable environment, and a long enough period of time. No easy experiment can distinguish between these extreme alternatives, and there is little or no middle ground. To many researchers, the discovery of life—past or present—on Mars, Europa, Titan, or any other object in our solar system would convert the appearance of life from an unlikely miracle to a virtual certainty throughout the Galaxy. We will take the optimistic view and adopt a value of 1.

Fraction of Life-Bearing Planets on Which Intelligence Arises

As with the evolution of life, the appearance of a well-developed brain is a very unlikely event if only random chance is involved. However, biological evolution through natural selection is a mechanism that generates apparently highly improbable results by singling out and refining useful characteristics. Organisms that profitably use adaptations can develop more complex behavior, and complex behavior provides organisms with the *variety* of choices needed for more advanced development.

One school of thought maintains that, given enough time, intelligence is inevitable. In this view, assuming that natural selection is a universal phenomenon, at least one organism on a planet will always rise to the level of "intelligent life." If this is correct, then the fifth term in the Drake equation equals or nearly equals 1. Others argue that there is only one known case of intelligence, and that case is life on Earth. For 2.5 billion years—from the start of life about 3.5 billion years ago to the first appearance of multicellular organisms about 1 billion years ago—life did not advance beyond the one-celled stage. Life remained simple and dumb, but it survived. If this latter view is correct, then the fifth term in our equation is very small, and we are faced with the depressing prospect that humans may be the smartest form of life anywhere in the Galaxy. As with the previous term, we will be optimistic and adopt a value of 1 here.

Fraction of Planets on Which Intelligent Life Develops and Uses Technology

To evaluate the sixth term of our equation, we need to estimate the probability that intelligent life eventually develops technological competence. Should the rise of technology be inevitable, this term is close to 1, given long enough periods of time. If it is not inevitable—if intelligent life can somehow "avoid" developing technology—then this term could be much less than 1. The latter possibility envisions a universe possibly teeming with intelligent civilizations, but very few among them ever becoming technologically competent. Perhaps only one managed it—ours.

Again, it is difficult to decide conclusively between these two views. We don't know how many prehistoric Earth cultures either failed to develop technology or rejected its use. We do know that tool-using societies arose independently at several places on Earth, including Mesopotamia, India, China, Egypt, Mexico, and Peru. Because so many of these ancient "technological" cultures originated *independently* at about the same time, it is tempting to conclude that the chances are good that some sort of technological society will inevitably develop, given some basic intelligence and enough time.

If technology is inevitable, then why haven't other life forms on Earth also found it useful? Possibly the competitive edge given by intellectual and technological skills to humans, the first species to develop them, allowed us to dominate so rapidly that other species—gorillas and chimpanzees, for example—simply haven't had time to catch up. The fact that only one technological society exists on Earth does not imply that the sixth term in the Drake equation must be very much less than 1. On the contrary, it is precisely because some species will probably always fill the niche of technological intelligence that we will take this term to be close to 1.

Average Lifetime of a Technological Civilization

The reliability of the estimate of each term on the right-hand side of the Drake equation declines markedly from left to right. Our knowledge of astronomy allows us to make a reasonably good stab at the first term, but it is much harder to evaluate some of the later terms, and the term is totally unknown. There is only one known example of such a civilization—humans on planet Earth. Our own civilization has presently survived in its technological state for only about 100 years, and how long we will be around before a human-made catastrophe or a planetwide natural disaster (see *Interlude 18-1*) ends it all is impossible to tell.

One thing is certain: if the correct value for *any one term* in the equation is very small, then few technological civilizations now exist in the Galaxy. If the pessimistic view of the development of life or of intelligence is correct, then we are unique, and that is the end of our story. However, if both life and intelligence are inevitable consequences of chemical and biological evolution, as many scientists believe, and if intelligent life always becomes technological, then we can plug the higher, more optimistic values into the Drake equation. In that case, combining our estimates for the other six terms (and noting that $10 \times 1 \times 1/10 \times 1 \times 1 \times 1 = 1$), we can say

number of technological, intelligent civilizations now present in the Milky Way Galaxy	$=$	average lifetime of a technologically competent civilization, *in years*.

Thus, if advanced civilizations typically survive for 1000 years, there should be 1000 of them currently in existence scattered throughout the Galaxy. If they survive for a million years, on average, we would expect there to be a million of them in the Galaxy, and so on.

18.4 The Search for Extraterrestrial Intelligence

4 Let us continue our optimistic assessment of the prospects for life and assume that civilizations enjoy a long stay on their parent planet once their initial technological "teething problems" are past.

In that case, they are likely to be plentiful in the Galaxy. How might we become aware of their existence?

Meeting Our Neighbors

For definiteness, let us assume that the average lifetime of a technological civilization is 1 million years—only 1 percent of the reign of the dinosaurs, but 100 times longer than human civilization has survived thus far. Given the size and shape of our Galaxy, we can then estimate the average *distance* between these civilizations to be some 50 pc, or about 150 light years. Thus, any two-way communication with our neighbors—using signals traveling at or below the speed of light—will take at least 300 years (150 years for the message to reach the planet and another 150 years for the reply to travel back to us).

One obvious way to search for extraterrestrial life would be to develop the capability to travel far outside our solar system. However, this may never be a practical possibility. At a speed of 50 km/s, the speed of the current fastest space probes, the trip to our nearest neighbor, Alpha Centauri, would take about 25,000 years. The journey to the nearest technological civilization (assuming a distance of 150 pc) would take almost 1 million years. Interstellar travel at these speeds is clearly not feasible. Speeding up our ships to near the speed of light would reduce the travel time, but this is far beyond our present technology.

Actually, our civilization has already launched some interstellar probes, although they have no specific stellar destination. Figure 18.8 is a reproduction of a plaque mounted on board the *Pioneer 10* spacecraft launched in the mid-1970s and now well beyond the orbit of Pluto, on its way out of the solar system. Similar information was also included aboard the *Voyager* probes launched in 1978. While these spacecraft would be incapable of reporting back to Earth the news that they had encountered an alien culture, scientists hope that the civilization on the other end would be able to unravel most of its contents using the universal language of mathematics. The caption to Figure 18.8 notes how the aliens might discover from where and when the *Pioneer* and *Voyager* probes were launched.

Setting aside the many practical problems of establishing direct contact with extraterrestrials, some scientists have argued that it might not even be a par-

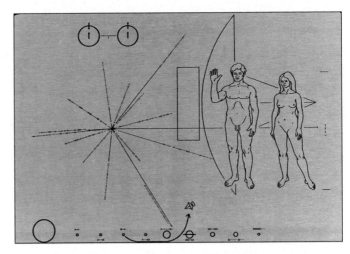

Figure 18.8 A replica of a plaque mounted on board the *Pioneer 10* spacecraft. The important features of the plaque include a scale drawing of the spacecraft, a man, and a woman; a diagram of the hydrogen atom undergoing a change in energy (top left); a starburst pattern representing various pulsars and the frequencies of their radio waves that can be used to estimate when the craft was launched (middle left); and a depiction of the solar system, showing that the spacecraft departed the third planet from the Sun and passed the fifth planet on its way into outer space (bottom). All the drawings have computer- (binary) coded markings from which actual sizes, distances, and times can be derived.

ticularly good idea. Our recent emergence as a technological civilization implies that we must be one of the least advanced technological intelligences in the entire Galaxy. Any other civilization that we discover, or that discovers us, will almost surely be more advanced than us. Consequently, a healthy degree of caution is warranted. If extraterrestrials behave even remotely like humans, then the most advanced aliens might naturally try to dominate all others. The behavior of the "advanced" European cultures toward the "primitive" races they encountered on their voyages of discovery in the seventeenth, eighteenth, and nineteenth centuries should serve as a clear warning of the possible undesirable consequences of contact. Of course, the aggressiveness of Earthlings may not exist in extraterrestrials, but, given the history of the one intelligent species we know, the cautious approach may be advisable.

Radio Communication

A cheaper, and much more practical, alternative is to try to make contact with extraterrestrials using electro-

18-1 INTERLUDE

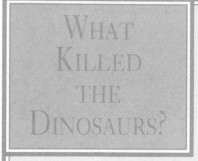

WHAT KILLED THE DINOSAURS?

The name *dinosaur* derives from the Greek words *deinos* (terrible) and *sauros* (lizard). Dinosaurs were no ordinary reptiles. In their prime, roughly 100 million years ago, the dinosaurs were the all-powerful rulers of Earth. Their fossilized remains have been uncovered on all the world's continents. Despite their dominance, according to the fossil record, these creatures vanished from Earth quite suddenly about 65 million years ago. What happened to them?

Until fairly recently, the prevailing view among paleontologists—scientists who study prehistoric life—was that dinosaurs were rather small-brained, cold-blooded creatures. In chilly climates, or even at night, the metabolisms of these huge reptiles would have become sluggish, making it difficult for them to move around and secure food. The suggestion was that they were poorly equipped to adapt to sudden changes in Earth's climate, so they eventually died out. However, a competing view of dinosaurs has emerged in recent years. Fossil evidence suggests that many of these monsters may in fact have been warm-blooded and relatively fast-moving creatures—not at all the dull-witted, slow-moving giants of earlier conception.

If the dinosaurs didn't die out simply because of stupidity and inflexibility, then what happened to cause their sudden and complete disappearance? Many explanations have been offered: devastating plagues, magnetic field reversals, increased geological activity, severe climate changes, supernova explosions. In the 1980s, it was suggested that a huge extraterrestrial object collided with Earth 65 million years ago. According to this idea, illustrated in the accompanying figure, a 10- to 15-km-wide asteroid or comet struck Earth, releasing as much energy as 10 million or more of the largest hydrogen bombs humans have ever constructed and kicking huge quantities of dust (including the pulverized remnants of the object itself) high into the atmosphere. The dust may have shrouded our planet for many years, virtually extinguishing the Sun's rays during this

magnetic radiation, the fastest known means of transferring information from one place to another. Because light and other short-wavelength radiation are heavily scattered while moving through dusty interstellar space, long-wavelength radio radiation seems to be the best choice. We would not attempt to broadcast to all nearby candidate stars, however—that would be far too expensive and inefficient. Instead, radio telescopes on Earth listen *passively* for radio signals emitted by other civilizations. Some preliminary searches of selected nearby stars are now under way, thus far without success.

In what direction should we aim our radio telescopes? The answer to this question at least is fairly easy. On the basis of our earlier reasoning, we should target all F, G, and K stars in our vicinity. But are extraterrestrials broadcasting radio signals? If they are not, this search technique will obviously fail. If they are, how do we distinguish their artificially generated radio signals from signals naturally emitted by inter-time. On the darkened surface, plants could not survive. The entire food chain was disrupted, and the dinosaurs, at the top of that chain, eventually died out.

Although we have no direct astronomical evidence to confirm or refute this idea, we can estimate the chances of a large asteroid or comet striking the Earth today, on the basis of observations of the number of objects presently on Earth-crossing orbits. The accompanying figure shows the likelihood of an impact as a function of the energy released by the collision, measured in *megatons* of TNT. (The megaton—4.2×10^{16} joules, the explosive yield of a large nuclear warhead—is the

only common terrestrial measure of energy adequate to describe the violence of these occurrences.) As we saw in Chapter 4, large meteoroids or asteroids are relatively rare, so 100-million-megaton events like the planetwide catastrophe that supposedly wiped out the dinosaurs are very infrequent, occurring only once every 10 million years or so. ∞ (Sec. 4.2) However, smaller impacts, equivalent to "only" a few tens of kilotons of TNT, could happen every few years—we may be long overdue for one. The most recent large impact was the Tunguska explosion in Siberia, in 1908, which packed a roughly 1-megaton punch (Figure 4.16).

While this theory is (arguably) the leading explanation for the demise of the dinosaurs, it is by no means universally accepted. Perhaps predictably, geologists tend to be less convinced than astronomers, citing numerous pieces of geological data for which the impact theory offers only partial explanation and presenting alternative explanations for many of the facts used to support the impact theory. Whatever killed the dinosaurs, however, dramatic environmental change of some sort was almost surely responsible. It is important that we continue the search for the cause of their extinction, for there is no telling if and when that sudden change might strike again. As the dominant species on Earth, we are the ones who now stand to lose the most.

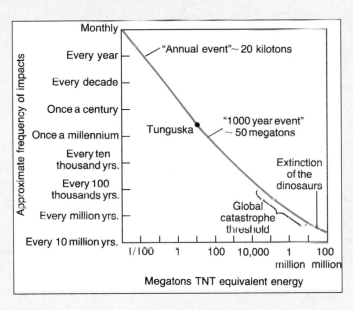

stellar gas clouds? At what frequency should we tune our receivers? This depends on whether the signals are produced deliberately or are simply "waste radiation" escaping from a planet.

Consider how radio wavelengths originating on Earth would look to extraterrestrials. Figure 18.9 shows the pattern of radio signals we emit into space. From the viewpoint of a distant observer, the spinning Earth emits a bright flash of radio radiation every few hours as the most technological regions of our planet rise or

set. In fact, Earth is now a more intense radio emitter than the Sun. Our radio and television broadcasts race out into space, and have been doing so since the invention of these technologies more than six decades ago. Another civilization as advanced as ours might have constructed devices capable of detecting this radiation. If any sufficiently advanced (and sufficiently interested) civilization resides within about 65 light years (20 pc) of Earth, then we have already broadcast our presence to them.

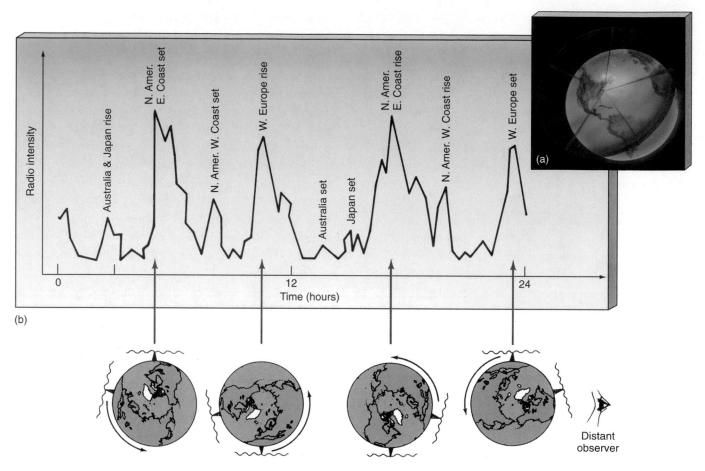

Figure 18.9 Radio radiation now leaks from Earth into space because of the daily activities of our technological civilization. (a) FM radio and television transmitters broadcast their energy parallel to Earth's surface, so they send a great "sheet" of electromagnetic radiation into interstellar space, producing the strongest signal in any given direction when they happen to lie on Earth's horizon, as seen from that direction. (The more common AM broadcasts are trapped below our ionosphere, so those signals never leave Earth.) (b) Because the great majority of transmitters are clustered in the eastern United States and western Europe, a distant observer would detect blasts of radiation from Earth as our planet rotates each day.

The Water Hole

Now let us suppose that a civilization has decided to assist searchers by actively broadcasting its presence to the rest of the Galaxy. At what frequency should we listen for such an extraterrestrial beacon? The electromagnetic spectrum is enormous; the radio domain alone is vast. To hope to detect a signal at some unknown radio frequency is like searching for a needle in a haystack. Are some frequencies more likely than others to carry alien transmissions?

Some basic arguments suggest that civilizations might well communicate at a wavelength near 20 cm. ∞ (Sec. 11.3) As we saw in Chapter 11, the basic building blocks of the universe, namely, hydrogen atoms, naturally radiate at a wavelength of 21 cm. Also, one of the simplest molecules, hydroxyl (OH), radiates near 18 cm. Together, these two substances form water (H_2O). Arguing that water is likely to be the interaction medium for life anywhere, and that radio radiation travels through the disk of our Galaxy with the least absorption by interstellar gas and dust, some researchers have proposed that the interval between 18 and 21 cm is the best wavelength range for civilizations to transmit or listen. Called the **water hole**, this radio interval might serve as an "oasis" where all advanced Galactic civilizations would gather to conduct their electromagnetic business.

This water-hole frequency interval is only a guess, of course, but it is supported by other arguments. Figure

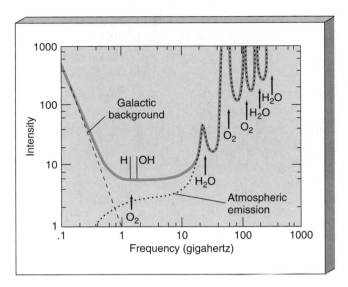

Figure 18.10 The "water hole" is bounded by the natural emission frequencies of the hydrogen (H) atom (21-cm wavelength) and the hydroxyl (OH) molecule (18-cm wavelength). The topmost solid (blue) curve sums the natural emissions of our Galaxy (dashed line on left side of diagram labeled "Galactic background") and Earth's atmosphere (dotted line on right side of diagram denoted by various chemical symbols). This sum is minimized near the water hole frequencies. Perhaps all intelligent civilizations conduct their interstellar communications within this quiet "electromagnetic oasis."

18.10 shows the water hole's location in the electromagnetic spectrum and plots the amount of natural emission from our Galaxy and from Earth's atmosphere. The 18- to 21-cm range lies within the quietest part of the spectrum, where the galactic "static" from stars and interstellar clouds happens to be minimized. Furthermore, the atmospheres of typical planets are also expected to interfere least at these wavelengths. Thus, the water hole seems like a good choice for the frequency of an interstellar beacon, although we cannot be sure of this reasoning until contact is actually achieved. A few radio searches are now in progress at frequencies in and around the water hole. So far, however, nothing resembling an extraterrestrial signal has been detected.

The space surrounding all of us could be, right now, flooded with radio signals from extraterrestrial civilizations. If only we knew the proper direction and frequency, we might be able to make one of the most startling discoveries of all time. The result would provide whole new opportunities to study the cosmic evolution of energy, matter, and life throughout the universe.

Chapter Review

Summary

Major phases in the history of the universe are particulate, galactic, stellar, planetary, chemical, biological, and cultural evolution. **Cosmic evolution** (p. 506) is the continuous process that has led to the appearance of galaxies, stars, planets, and life on Earth.

Living organisms may be characterized by their ability to react to their environment, to grow by taking in nutrition from their surroundings, to reproduce, passing along some of their own characteristics to their offspring, and to evolve in response to a changing environment.

Powered by natural energy sources, reactions between simple molecules in the oceans of the primitive Earth are believed to have led to the formation of **amino acids** (p. 507) and **nucleotide bases** (p. 507), the basic molecules of life. Alternatively, some complex molecules may have been formed in interstellar space and then delivered to Earth by meteors or comets.

Organisms that can best take advantage of their new surroundings succeed at the expense of those organisms that cannot make the necessary adjustments. Intelligence is strongly favored by natural selection.

The best hope for life beyond Earth in the solar system is the planet Mars, although no evidence for living organisms has been found. Jupiter's Europa and Saturn's Titan may also be possibilities, but conditions on both those bodies are harsh by terrestrial standards.

The **Drake equation** (p. 511) provides a means of estimating the probability of other intelligent life in the Galaxy. The astronomical terms in the equation are the Galactic star-formation rate, the likelihood of planets, and the number of habitable planets. Chemical and biological terms are the probability of life appearing and the probability that it subsequently develops intelligence. Cultural and political terms are the probability that intelligence leads to technology and the lifetime of a technological civilization. Taking an optimistic view of the development of life and intelligence leads to the conclusion that the total number of technologically competent civilizations in the Galaxy is approximately equal to the lifetime of a typical civilization, expressed in years.

Even with optimistic assumptions, the distance to our nearest intelligent neighbor is likely to be many hundreds of parsecs. Space travel is not presently a feasible means of searching for intelligent life. Current programs to discover extraterrestrial intelligence involve scanning the electromagnetic spectrum for signals. So far, no intelligent broadcasts have been received. A technological civilization would probably "announce" itself to the universe by the radio and television signals it emits into space. Observed from afar, our planet would appear as a radio source with a 24-hour period, as different regions of the planet rise and set. The **water hole** (p. 518) is a region in the radio range of the electromagnetic spectrum, near the 21-cm line of hydrogen and the 18-cm line of hydroxyl, where natural emissions from the Galaxy happen to be minimal. Many researchers regard this as the best part of the spectrum for communications purposes.

Self-Test: True or False?

_____ 1. The assumptions of mediocrity favor the existence of extraterrestrial life.

_____ 2. The definition of life requires only that, to be considered "alive," you must be able to reproduce.

_____ 3. The Urey–Miller experiment produced biological molecules from nonbiological molecules.

_____ 4. Organic molecules important for life could have reached Earth's surface via comets.

_____ 5. Organic molecules exist only on Earth.

_____ 6. Laboratory experiments have created living cells from nonbiological molecules.

_____ 7. For most of the history of Earth, life consisted of only single-celled life forms.

_____ 8. The *Viking* landers on Mars discovered microscopic evidence of life but found no large fossil evidence.

_____ 9. The rate of star formation in the Galaxy is reasonably well known.

_____ 10. As yet there is no direct evidence that Earth-like planets orbit other stars.

_____ 11. In estimating whether intelligence arises and develops technology, we have only life on Earth as an example.

_____ 12. Dinosaurs existed on Earth for 10,000 times longer than human civilization has existed to date.

_____ 13. Our civilization has launched probes into interstellar space.

_____ 14. One disadvantage of interstellar radio communication is that we can do it only with another civilization that has a technology equal to or greater than our own.

Self-Test: Fill in the Blank

1. Amino acids are the building blocks of _____.
2. The naturally occurring molecules present on the young Earth included water, carbon dioxide, _____, and _____.
3. Two sources of energy for chemical reactions on the young Earth are _____ and _____.
4. The fossil record clearly shows evidence of life dating back _____ years.
5. Multicellular organisms did not appear on Earth until about _____ years ago.
6. The Murchison meteorite was discovered to contain relatively large amounts of _____.
7. The Drake equation estimates the number of _____ in the Milky Way Galaxy.
8. Planets in binary star systems are not considered habitable because the planetary orbits are usually _____.
9. Direct contact between extraterrestrial lifeforms may be impractical because of the large _____ between civilizations.
10. Radio communication over interstellar distances is practical because the signals travel at the speed of _____.

11. Radio waves can travel throughout the Galaxy because they are not blocked by interstellar _____.

12. Radio waves leaking away from Earth have now traveled out a distance of _____ light-years.

13. Radio wavelengths between 18 and 21 cm are referred to as the _____.

14. A two-way communication with another civilization at a distance of 100 light-years will require _____ years.

Review and Discussion

1. Why is life difficult to define?

2. What is chemical evolution?

3. What is the Urey–Miller experiment? What important organic molecules were produced in this experiment?

4. What are the basic ingredients from which biological molecules formed on Earth?

5. How do we know anything at all about the early episodes of life on Earth?

6. What is the role of language in cultural evolution?

7. Where else, besides Earth, have organic molecules been found?

8. Where—besides the planet Mars—might we find signs of life in our solar system?

9. Do we know whether Mars ever had life at any time during its past? What argues in favor of the position that it may once have harbored life?

10. What is generally meant by "life as we know it"? What other forms of life might be possible?

11. How many of the terms in the Drake equation are known with any degree of certainty? Which factor is least well known?

12. What is the relationship between the average lifetime of Galactic civilizations and the possibility of our someday communicating with them?

13. How would Earth appear at radio wavelengths to extraterrestrial astronomers?

14. What are the advantages in using radio waves for communication over interstellar distances?

15. What is the water hole? What advantage does it have over other parts of the radio spectrum?

Problems

1. Suppose that each of the "fraction" terms in the Drake equation turns out to have a value of 1/10, that stars form at an average rate of 20 per year, and that each star has exactly 1 habitable planet orbiting it. Estimate the present number of technological civilizations in the Milky Way Galaxy if the average lifetime of a civilization is (a) 100 years, (b) 10,000 years, (c) 1,000,000 years.

2. Assuming that there are 10,000 FM radio stations on Earth, each transmitting at a power level of 50 kW, calculate the total radio luminosity of Earth in the FM band. Compare this with the Sun's emission in the same frequency range.

3. There are 20,000 stars within 100 light-years of Earth to be searched for radio communications. How long will the search take if one hour is spent looking at each star? If one day is spent per star?

Projects

1. Some people suggest that if extraterrestrial life is discovered, it will have a profound effect on people. Interview as many people as you can and ask the following two questions: (1) Do you believe that extraterrestrial life exists? (2) Why? From your results, try to decide whether there will be a profound effect on people if extraterrestrial life is discovered.

2. Conduct another poll, or do it at the same time you do the first one. Ask the following question: What one question would you like to ask an extraterrestrial life form in a radio communication? How many responses do you receive that indicate the person is very "Earth centered" in thinking? How many responses suggest a lack of understanding of how alien an extraterrestrial life form might be?

Onward and upward. Photograph of an astronaut during the 1997 servicing mission to the *Hubble Space Telescope* in low-Earth orbit.

Appendix 1 Scientific Notation

The objects studied by astronomers range in size from the smaller particles to the largest expanse of matter we know—the entire universe. Subatomic particles have sizes of about 0.000000000000001 meter, while galaxies (like that shown in Figure P.3) typically measure some 1,000,000,000,000,000,000,000 meters across. The most distant known objects in the universe lie on the order of 100,000,000,000,000,000,000,000,000 meters from Earth.

Obviously, writing all those zeros is both cumbersome and inconvenient. More important, it is also very easy to make an error—write down one zero too many or too few and your calculations become hopelessly wrong! To avoid this, scientists always write large numbers using a shorthand notation in which the number of zeros following or preceding the decimal point is denoted by a superscript power, or *exponent*, of 10. The exponent is simply the number of places between the first significant (nonzero) digit in the number (reading from left to right) and the decimal point. Thus 1 is 10^0, 10 is 10^1, 100 is 10^2, 1000 is 10^3, and so on. For numbers less than 1, with zeros between the decimal point and the first significant digit, the exponent is negative: 0.1 is 10^{-1}, 0.01 is 10^{-2}, 0.001 is 10^{-3}, and so on. Using this notation we can shorten the number describing subatomic particles to 10^{-15} meter and write the number describing the size of a galaxy as 10^{21} meters.

More complicated numbers are expressed as a combination of a power of 10 and a multiplying factor. This factor is conventionally chosen to be a number between 1 and 10, starting with the first significant digit in the original number. For example, 150,000,000,000 meters (the distance from Earth to the Sun, in round numbers) can be more concisely written at 1.5×10^{11} meters, 0.000000025 meters as 2.5×10^{-8} meter, and so on. The exponent is simply the number of places the decimal point must be moved *to the left* to obtain the multiplying factor.

Some other examples of scientific notation are:

- the approximate distance to the Andromeda Galaxy − 3,000,000 light years = 3×10^6 light years
- the size of a hydrogen atom = 0.00000000005 meter = 5×10^{-11} meter

- the diameter of the Sun = 1,392,000 kilometers = 1.392×10^6 kilometers
- the U.S. national debt (as of midnight, October 1, 1997) = \$5,411,051,000,000.00 = \$5.411051 trillion = 5.411051×10^{12} dollars.

In addition to providing a simpler way of expressing very large or very small numbers, this notation also makes it easier to do basic arithmetic. The rule for multiplication of numbers expressed in this way is simple: Just multiply the factors and add the exponents. Similarly for division: Divide the factors and subtract the exponents. Thus, 3.5×10^{-2} multiplied by 2.0×10^3 is simply $(3.5 \times 2.0) \times 10^{-2+3} = 7.0 \times 10^1$—that is, 70. Again, 5×10^6 divided by 2×10^4 is just $(5/2) \times 10^{6-4}$, or 2.5×10^2 ($= 250$). Applying these rules to unit conversions, 200,000 nanometers is $200,000 \times 10^{-9}$ meter (since 1 nanometer $= 10^{-9}$ meter; see Appendix 2), or $2 \times 10^5 \times 10^{-9}$ meter, or $2 \times 10^{5-9} = 2 \times 10^{-4}$ meter. Verify these rules yourself with a few examples of your own. The advantages of this notation when considering astronomical objects will soon become obvious.

Scientists often use "rounded-off" versions of numbers, both for simplicity and for ease of calculation. For example, we will usually write the diameter of the Sun as 1.4×10^6 kilometers, instead of the more precise number given earlier. Similarly, the diameter of the Earth is 12,756 kilometers, or 1.2756×10^4 kilometers, but for "ballpark" estimates, we really don't need so many digits, and the more approximate number 1.3×10^4 kilometers will suffice. Very often, we perform rough calculations using only the first one or two significant digits in a number, and that may be all that is necessary to make a particular point. For example, to support the statement, "The Sun is much larger than the Earth," we need only say that the ratio of the two diameters is roughly 1.4×10^6 divided by 1.3×10^4. Since 1.4/1.3 is close to 1, the ratio is approximately $10^6/10^4 = 10^2$, of 100. The essential fact here is that the ratio is much larger than 1; calculating it to greater accuracy (to get 109.13) would give us no additional *useful* information. This technique of stripping away the arithmetic details to get to the essence of a calculation is very common in astronomy, and we will use it frequently throughout this text.

Appendix 2 Astronomical Measurement

Astronomers use many different kinds of units in their work, simply because no single system of units will do. Rather than the *Système Internationale* (SI), or meter-kilogram-second (MKS), metric system used in most high school and college science classes, many professional astronomers still prefer the older centimeter-gram-second (CGS) system. However, astronomers also commonly introduce new units when convenient. For example, when discussing stars, the mass and radius of the Sun are often used as reference points. The solar mass, written as $M_\odot$, is equal to 2.0×10^{33} g, or 2.0×10^{30} kg (since 1 kg = 1000 g). The solar radius, $R_\odot$, is equal to 700,000 km, or 7.0×10^8 m (1 km = 1000 m). The subscript $\odot$ always stands for Sun. Similarly, the subscript $\oplus$ always stands for Earth. In this book, we will use the units that astronomers commonly use in any given context, but we will also give the "standard" SI equivalents where appropriate.

Of particular importance are the units of length astronomers use. On small scales, the *angstrom* (1 Å = $10^{-10} = 10^{-8}$ cm), the *nanometer* (1 nm = 10^{-9} m = 10^{-7} cm), and the *micron* (1 μm = 10^{-6} m = 10^{-4} cm) are used. Distances within the solar system are usually expressed in terms of the *astronomical unit* (A.U.), the mean distance between Earth and the Sun. One A.U.

is approximately equal to 150,000,000 km, or 1.5×10^{11} m. On larger scales, the *light year* (1 ly = 9.5×10^{15} m = 9.5×10^{12} km) and the parsec (1 pc = 3.1×10^{16} m = 3.1×10^{13} km = 3.3 ly) are commonly used. Still larger distances use the regular prefixes of the metric system: *kilo* for one thousand and *mega* for one million. Thus, 1 *kiloparsec* (kpc) = 10^3 pc = 3.1×10^{19} m, 10 megaparsecs (Mpc) = 10^7 pc = 3.1×10^{23} m, and so on.

Astronomers use units that make sense within a context, and as contexts change, so do the units. For example, we might measure densities in grams per cubic centimeter (g/cm^3), in atoms per cubic meter (atoms/m^3), or even in solar masses per cubic megaparsec ($M_\odot$/Mpc^3), depending on the circumstances. The important thing to know is that once you understand the units, you can convert freely from one set to another. For example, the radius of the Sun could equally well be written as $R_\odot = 6.96 \times 10^8$ m, or 6.96×10^{10} cm, or 109 $R_\oplus$, or 4.65×10^{-3} A.U., or even 7.36×10^{-8} ly—whichever happens to be most useful. Some of the more common units used in astronomy, and the contexts in which they are most likely to be encountered, follow.

Length:

1 angstrom (Å)	= 10^{-10} m	atomic physics,
1 nanometer (nm)	= 10^{-9} m	spectroscopy
1 micron (μm)	= 10^{-6} m	interstellar dust and gas
1 centimeter (cm)	= 0.01 m	in widespread use
1 meter (m)	= 100 cm	throughout all
1 kilometer (km)	= 1000 m = 10^5 cm	astronomy
Earth radius ($R_\oplus$)	= 6378 km	planetary astronomy
Solar radius ($R_\odot$)	= 6.96×10^8 m	solar system,
1 astronomical unit (A.U.)	= 1.496×10^{11} m	stellar evolution
1 light year (ly)	= 9.46×10^{15} m = 63,200 A.U.	galactic astronomy,
1 parsec (pc)	= 3.09×10^{16} m = 3.26 ly	stars and star clusters
1 kiloparsec (kpc)	= 1000 pc	galaxies, galaxy clusters,
1 megaparsec (Mpc)	= 1000 kpc	cosmology

Mass:

1 gram (g)		in widespread use in
1 kilogram (kg)	= 1000 g	many different areas
Earth mass ($M_\oplus$)	= 5.98×10^{24} kg	planetary astronomy
Solar mass ($M_\odot$)	= 1.99×10^{30} kg	"standard" unit for all mass scales larger than Earth

Time:

1 second (s)		in widespread use throughout astronomy
1 hour (h)	= 3600 s	planetary and stellar
1 day (d)	= 86,400 s	scales
1 year (yr)	= 3.16×10^7 s	virtually all processes occurring on scales larger than a star

Appendix 3 Tables

Table 1 Some Useful Constants and Physical Measurements[1]

1 astronomical units
A.U. $= 1.496 \times 10^8$ km $(1.5 \times 10^8$ km)

1 light-year
ly $= 9.46 \times 10^{12}$ km $(10^{13}$ km; 6 trillion miles)

1 parsec
pc $= 3.09 \times 10^{13}$ km $= 3.3$ ly

speed of light
$c = 299{,}792.458$ km/s $(3 \times 10^5$ km/s)

gravitational constant
$G = 6.67 \times 10^{-11}$ m^3/kg/s^2

mass of the Earth
$M_\oplus = 5.97 \times 10^{24}$ kg $(6 \times 10^{24}$ kg; about 6000 billion billion tons)

radius of the Earth
$R_\oplus = 6378$ km (6500 km)

mass of the Sun
$M_\odot = 1.99 \times 10^{30}$ kg $(2 \times 10^{30}$ kg)

radius of the Sun
$R_\odot = 6.96 \times 10^5$ km $(7 \times 10^5$ km)

luminosity of the Sun
$L_\odot = 3.90 \times 10^{26}$ W $(4 \times 10^{26}$ W)

effective temperature of the Sun
$T_\odot = 5778$ K (5800 K)

Hubble constant
$H_0 \simeq 75$ km/s/Mpc

mass of the electron
$m_e = 9.11 \times 10^{-31}$ kg

mass of the proton
$m_p = 1.67 \times 10^{-27}$ kg

[1] The rounded-off values used in the text are shown above in parentheses.

CONVERSIONS BETWEEN COMMON ENGLISH AND METRIC UNITS

ENGLISH	METRIC
1 inch	= 2.54 centimeters (cm)
1 foot (ft)	= 0.3048 meters (m)
1 mile	= 1.609 kilometers (km)
1 pound (lb)	= 453.6 grams (g) or 0.4536 kilograms (kg) (on Earth)

TABLE 2A   Planetary Data: Orbital Properties

Planet	Semi-major Axis (A.U.)	(10^6 km)	Sidereal Period (tropical years)	Mean Orbital Speed (km/s)	Orbital Eccentricity e	Inclination to the Ecliptic (degrees)
Mercury	0.39	57.9	0.241	47.9	0.206	7.00
Venus	0.72	108.2	0.615	35.0	0.007	3.39
Earth	1.00	149.6	1.00	29.8	0.017	0.01
Mars	1.52	227.9	1.88	24.1	0.093	1.85
Jupiter	5.20	778.3	11.86	13.1	0.048	1.31
Saturn	9.54	1427	29.46	9.64	0.056	2.49
Uranus	19.19	2870	84.01	6.81	0.046	0.77
Neptune	30.06	4497	164.8	5.43	0.010	1.77
Pluto	39.53	5914	248.6	4.74	0.248	17.15

TABLE 2B Planetary Data: Physical Properties

Planet	Equatorial Radius (km)	Equatorial Radius (Earth = 1)	Mass (kg)	Mass (Earth = 1)	Mean Density (kg/m³)	Sidereal Rotation Period (solar days)[1]
Mercury	2,439	0.38	3.30×10^{23}	0.06	5,430	58.6
Venus	6,051	0.95	4.87×10^{24}	0.81	5,250	-243.0
Earth	6,378	1.00	5.97×10^{24}	1.00	5,520	0.9973
Mars	3,397	0.53	6.42×10^{23}	0.11	3,930	1.026
Jupiter	71,492	11.21	1.90×10^{27}	317.9	1,330	0.41
Saturn	60,268	9.45	5.69×10^{26}	95.18	710	0.43
Uranus	25,559	4.01	8.68×10^{25}	14.54	1,240	-0.69
Neptune	24,764	3.88	1.02×10^{26}	17.13	1,670	0.72
Pluto	1,123	0.18	1.46×10^{22}	0.025	2,290	-6.387

Planet	Axial Tilt (degrees)	Surface Gravity (Earth = 1)	Escape Speed (km/s)	Surface Temperature[2] (k)	Number of Moons
Mercury	7.0	0.38	4.3	100 to 700	0
Venus	177.4	0.90	10.4	730	0
Earth	23.5	1.00	11.2	290	1
Mars	24.0	0.38	5.0	180 to 270	2
Jupiter	3.1	2.53	60	124	16
Saturn	26.7	1.07	36	95	18
Uranus	97.9	0.90	21	58	15
Neptune	29.6	1.14	24	59	8
Pluto	118	0.07	1.3	40 to 60	1

[1]A negative sign indicates retrograde rotation.
[2]Temperature is effective temperature for jovian planets.

TABLE 3 The Twenty Brightest Stars

Name	Star	Spectral Type[1] A	Spectral Type[1] B	Parallax (arc seconds)	Distance (pc)	Apparent Visual Magnitude[1] A	Apparent Visual Magnitude[1] B
Sirius	α CMa	A1V	wd[2]	0.37	2.7	−1.46	+8.7
Canopus	α Car	F01b-II		0.033	30	−0.72	
Rigel Kentaurus	α Cen	G2V	K0V	0.77	1.3	−0.01	+1.3
Arcturus	α Boo	K2IIIp		0.091	11	−0.06	
Vega	α Lyr	A0V		0.13	8.0	+0.04	
Capella	α Aur	GIII	M1V	0.071	14	+0.05	+10.2
Rigel	β Ori	B8 Ia	B9	—	250	+0.14	+6.6
Procyon	α CMi	F5IV-V	wd	0.29	3.5	+0.37	+10.7
Betelgeuse	α Ori	M2Iab		—	150	+0.41	
Achernar	α Eri	B5V		0.050	20	+0.51	
Hadar	β Cen	B1III	?	0.011	90	+0.63	+4
Altair	α Aql	A71V-V		0.20	5.1	+0.77	
Acrux	α Cru	B1IV	B3	0.008	120	+1.39	+1.9
Aldebaran	α Tau	K5III	M2V	0.063	16	+0.86	+13
Spica	α Vir	B1V		0.013	80	+0.91	
Antares	α Sco	MIIb	B4V	0.008	120	+0.92	+5.1
Pollux	β Gem	K0III		0.083	12	+1.16	
Fomalhaut	α PsA	A3V		0.14	7.0	+1.19	+6.5
Deneb	α Cyg	A2Ia		—	430	+1.26	
Mimosa	β Cru	B1IV		—	150	+1.28	

Name	Luminosity (Sun = 1) A	Luminosity (Sun = 1) B	Absolute Visual Magnitude[1] A	Absolute Visual Magnitude[1] B	Proper Motion (arc seconds/yr)	Transverse Velocity (km/s)	Radial Velocity (km/s)
Sirius	23.5	0.003	+1.4	+11.6	1.33	17.0	−7.6[3]
Canopus	1510		−3.1		0.02	2.8	+20.5
Rigel Kentaurus	1.56	0.46	+4.4	+5.7	3.68	22.7	−24.6
Arcturus	115		−0.3		2.28	119	−5.2
Vega	55.0		+0.5		0.34	12.9	−13.9
Capella	166	0.01	−0.7	+9.5	0.44	29	+30.2[3]
Rigel	4.6×10^4	126	−6.8	−0.4	0.00	1.2	+20.7[3]
Procyon	7.7	0.0006	+2.6	+13.0	1.25	20.7	−3.2[3]
Betelgeuse	1.4×10^4		−5.5		0.03	21	+21.0[3]
Achernar	219		−1.0		0.10	9.5	+19
Hadar	3800	182	−4.1	−0.8	0.04	17	−12[3]
Altair	11.5		+2.2		0.66	16	−26.3
Acrux	3470	2190	−4.0	−3.5	0.04	24	−11.2
Aldebaran	105	0.0014	−0.2	+12	0.20	15	+54.1
Spica	2400		−3.6		0.05	19	+1.0[3]
Antares	5500	115	−4.5	−0.3	0.03	17	−3.2
Pollux	41.7		+0.8		0.62	35	+3.3
Fomalhaut	13.8	0.10	+2.0	+7.3	0.37	12	+6.5
Deneb	5.0×10^4		−6.9		0.003	6	−4.6[3]
Mimosa	6030		−4.6		0.05	36	

[1]A and B columns identify individual components of binary systems.
[2]"wd" stands for "white dwarf."
[3]Average value of variable velocity.

TABLE 4 The Twenty Nearest Stars

Name	Spectral Type[1] A	Spectral Type[1] B	Parallax (arc seconds)	Distance (pc)	Apparent Visual Magnitude[1] A	Apparent Visual Magnitude[1] B
Sun	G2V				−26.72	
Proxima Cen	M5e		0.772	1.30	+11.05	
Alpha Centauri	G2V	K0V	0.750	1.33	−0.01	+1.33
Barnard's Star	M5V		0.545	1.83	+9.54	
Wolf 359	M8V		0.421	2.38	+13.53	
BD 1 36°2147	M2V		0.397	2.52	+7.50	
Luyten 726-8	M5.5V	M5.5V	0.387	2.58	+12.52	+13.02
Sirius	A1V	wd[2]	0.377	2.65	−1.46	+8.3
Ross 154	M4.5V		0.345	2.90	+10.45	
Ross 248	M6V		0.314	3.18	+12.29	
ε Eridani	K2V		0.303	3.30	+3.73	
Ross 128	M5V		0.298	3.36	+11.10	
61 Cygni	K5V	K7V	0.294	3.40	+5.22	+6.03
ε Indi	K5V		0.291	3.44	+4.68	
BD 1 43°44	M1V	M6V	0.290	3.45	+8.08	+11.06
Luyten 789-6	M6V		0.290	3.45	+12.18	
Procyon	F5IV-V	wd	0.285	3.51	+0.37	+10.7
BD 1 59°1915	M4V	M5V	0.285	3.55	+8.90	+9.69
CD 2 36°15693	M2V		0.279	3.58	+7.35	
G51-15	MV		0.278	3.60	+14.81	

Name	Luminosity (Sun = 1) A	Luminosity (Sun = 1) B	Absolute Visual Magnitude[1] A	Absolute Visual Magnitude[1] B	Proper Motion (arc seconds/yr)	Transverse Velocity (km/s)	Radial Velocity (km/s)
Sun	1.0		+4.85				
Proxima Cen	0.00006		+15.5		3.86	23.8	−16
Alpha Centauri	1.6	0.45	+4.4	+5.7	3.68	23.2	22
Barnard's Star	0.00045		+13.2		10.34	89.7	−108
Wolf 359	0.00002		+16.7		4.70	53.0	+13
BD 1 36°2147	0.0055		+10.5		4.78	57.1	−84
Luyten 726-8	0.00006	0.00004	+15.5	16.0	3.36	41.1	+30
Sirius	23.5	0.003	+1.4	11.2	1.33	16.7	−8
Ross 154	0.00048		+13.3		0.72	9.9	−4
Ross 248	0.00011		+14.8		1.58	23.8	−81
ε Eridani	0.30		+6.1		0.98	15.3	+16
Ross 128	0.00036		+13.5		1.37	21.8	−13
61 Cygni	0.082	0.039	+7.6	+8.4	5.22	84.1	−64
ε Indi	0.14		+7.0		4.69	76.5	−40
BD 1 43°44	0.0061	0.00039	+10.4	+13.4	2.89	47.3	+17
Luyten 789-6	0.00014		+14.6		3.26	53.3	−60
Procyon	7.65	0.00055	+2.6	+13.0	1.25	2.82.8	−3
BD 1 59°1915	0.0030	0.0015	+11.2	+11.9	2.28	38.4	+5
CD 2 36°15693	0.013		+9.6		6.90	117	+10
G51−15	0.00001		+17.0		1.26	21.5	—

[1]A and B columns identify individual components of binary systems.
[2]"wd" stands for "white dwarf."

Glossary

A

A ring One of three Saturnian rings visible from Earth. The A ring is farthest from the planet and is separated from the B ring by the Cassini division. (p. 240)

absolute brightness The apparent brightness a star would have if it were placed at a standard distance of 10 parsecs from Earth. (p. 284)

absolute magnitude The apparent magnitude a star would have if it were placed at a standard distance of 10 parsecs from Earth. (p. 286)

absorption line Dark line in an otherwise continuous bright spectrum, where light within one narrow frequency range has been removed. (p. 58)

acceleration The rate of change of velocity of a moving object. (p. 34)

accretion Gradual growth of bodies, such as stars or planets, by the accumulation of gas or other, smaller, bodies. (p. 127)

accretion disk Flat disk of matter spiraling down onto the surface of a star or black hole. Often, the matter originated on the surface of a companion star in a binary system. (p. 347)

active galaxies The most energetic galaxies, which can emit hundreds or thousands of times more energy per second than the Milky Way. (p. 450)

active optics Collection of techniques now being used to increase the resolution of ground-based telescopes. Minute modifications are made to the overall configuration of an instrument as its temperature and orientation change, to maintain the best possible focus at all times. (p. 87)

active regions Region of the photosphere of the Sun surrounding a sunspot group, which can erupt violently and unpredictably. During sunspot maximum, the number of active regions is also a maximum. (p. 268)

active Sun The unpredictable aspects of the Sun's behavior, such as sudden explosive outbursts of radiation in the form of prominences and flares. (p. 265)

adaptive optics Technique used to increase the resolution of a telescope by deforming the shape of the mirror's surface under computer control while a measurement is being taken, to undo the effects of atmospheric turbulence. (p. 87)

amino acids Organic molecules which form the basis for building the proteins that direct metabolism in living creatures. (p. 507)

amplitude The maximum deviation of a wave above or below the zero point. (p. 46)

angular resolution The ability of a telescope to distinguish between adjacent objects in the sky. (p. 79)

annular eclipse Solar eclipse occurring at a time when the Moon is far enough from Earth that it fails to cover the disk of the Sun completely, leaving a ring of sunlight visible around its edge. (p. 22)

apparent brightness The brightness that a star appears to have, as measured by an observer on Earth. (p. 284)

apparent magnitude The apparent brightness of a star, expressed using the magnitude scale. (p. 286)

association Small grouping of (typically 100 or less) stars, spanning up to a few tens of parsecs across, usually rich in very young stars. (p. 328)

asteroid belt A region of the solar system, between the orbits of Mars and Jupiter, in which most asteroids are found. (pp. 108, 228)

asteroid One of thousands of very small members of the solar system orbiting the Sun between the orbits of Mars and Jupiter. Asteroids are often referred to as "minor planets." (pp. 108, 228)

astronomical unit The average distance of the Earth from the Sun. Precise radar measurements yield a value for the A.U. of 149,603,500 km. (p. 31)

astronomy Branch of science dedicated to the study of everything in the universe that lies above Earth's atmosphere. (p. 2)

atmosphere Layer of gas confined close to a planet's surface by the force of gravity. (p. 139)

atom Building block of matter, composed of positively charged protons and neutral neutrons in the nucleus, surrounded by negatively charged neutrons. (p. 60)

aurora Event which occurs when atmospheric molecules are excited by incoming charged particles from the solar wind, then emit energy as they fall back to their ground states. Aurorae generally occur at high latitudes, near the north and south magnetic poles. (p. 156)

autumnal equinox Date on which the Sun crosses the celestial equator moving southward, occurring on or near September 22. (p. 16)

B

B ring One of three Saturnian rings visible from Earth. The B ring is the brightest of the three, and lies just within the Cassini division, closer to the planet than the A ring. (p. 240)

barred-spiral galaxy Spiral galaxy containing a central bar of stars and gas, with the spiral arms beginning near the ends of the bar. (p. 422)

baseline The distance between two observing locations used for the purposes of triangulation measurements. The larger the baseline, the better the resolution attainable. (p. 8)

belt Dark, low-pressure region, where gas flows downward in the atmosphere of a jovian planet. (p. 207)

Big Bang Event that cosmologists consider the beginning of the universe, in which all matter and radiation in the entire universe came into being. (p. 481)

binary-star system A system which consists of two stars in orbit about their common center of mass, held together by their mutual gravitational attraction. Most stars are found in binary-star systems. (p. 295)

black-body curve The characteristic way in which the intensity of radiation emitted by a hot object depends on frequency. The frequency at which the emitted intensity is highest is an indication of the temperature of the radiating object. Also referred to as the Planck curve. (p. 52)

black dwarf The end-point of the evolution of an isolated, low mass star. After the white dwarf stage, the star cools to the point where it is a dark "clinker" in interstellar space. (p. 346)

black hole A region of space where the pull of gravity is so great that nothing—not even light—can escape. A possible outcome of the evolution of a very massive star. (p. 372)

blue giant Large, hot, bright star at the upper left end of the main sequence on the Hertzsprung–Russell diagram. Its name comes from its color and size. (p. 292)

blue supergiant The very largest of the large, hot, bright stars at the uppermost left end of the main sequence on the Hertzsprung–Russell diagram. (p. 292)

Bohr model First theory of the hydrogen atom to explain the observed spectral lines. This model rests on three ideas: that there is a state of lowest energy for the electron, that there is a maximum energy, beyond which the electron is no longer bound to the nucleus, and that within these two energies the electron can only exist in certain energy levels. (p. 61)

brown dwarf Remnant of a fragment of collapsing gas and dust that did not contain enough mass to initiate core nuclear fusion. Such objects are frozen somewhere along their pre-main-sequence contraction phase, continually cooling into compact dark objects. Because of their small sizes and low temperatures they are extremely difficult to detect observationally. (p. 324)

brown oval Feature of Jupiter's atmosphere that appears only at latitudes near 20 degrees N, this structure is a long-lived hole in the clouds that allows us to look down into Jupiter's lower atmosphere. (p. 211)

C

C ring One of three Saturnian rings visible from Earth. The C ring is the faintest of the three and lies between the B ring and the planet. (p. 240)

carbon-detonation supernova See type-I supernova. (p. 351)

Cassegrain telescope A type of reflecting telescope in which incoming light hits the primary mirror and is then reflected upward toward the prime focus, where a secondary mirror reflects the light back down through a small hole in the main mirror, into a detector or eyepiece. (p. 77)

Cassini Division A relatively empty gap in Saturn's ring system between the A and B rings, discovered in 1675 by Giovanni Cassini. It is now known to contain a number of thin ringlets. (p. 240)

celestial coordinates Pair of quantities—right ascension and declination—similar to longitude and latitude on Earth, used to pinpoint locations of objects on the celestial sphere. (p. 7)

celestial equator The projection of Earth's equator onto the celestial sphere. (p. 7)

celestial sphere Imaginary sphere surrounding the Earth, to which all objects in the sky were once considered to be attached. (p. 6)

center of mass The "average" position in space of a collection of massive bodies, taking their masses into account. In an isolated system this point moves with constant velocity, according to Newtonian mechanics. (p. 37)

Cepheid variable Star whose luminosity varies in a characteristic way, with a rapid rise in brightness followed by a slower decline. The period of a Cepheid variable star is related to its luminosity, so a determination of this period can be used to obtain an estimate of the star's distance. (p. 395)

charge-coupled device (CCD) Electronic device used for data acquisition, composed of many tiny pixels, each of which records a buildup of charge to measure the amount of light striking it. (p. 86)

chromosphere The Sun's lower atmosphere, lying just above the visible photosphere. (p. 256)

closed universe Geometry that the universe as a whole would have if the density of matter is above the critical value. A closed universe is finite in extent, and has no edge, like the surface of a sphere. It has enough mass to stop the present expansion, and will eventually recollapse. (p. 487)

collecting area The total area of a telescope that is capable of capturing incoming radiation. The larger the telescope, the greater its collecting area, and the fainter the objects it can detect. (p. 79)

coma The brightest part of a comet, often referred to as the "head." (p. 111)

comet A small body, composed mainly of ice and dust, in an elliptical orbit about the Sun. As it comes close to the Sun, some of its material is vaporized to form a gaseous head and extended tail. (pp. 108, 232)

condensation nuclei Dust grains in the interstellar medium which act as seeds around which other material can coagulate. The presence of dust was very important in causing matter to clump during the formation of the solar system. (p. 125)

condensation theory Currently favored model of solar system formation which combines features of the old nebular theory with new information about interstellar dust grains, which acted as condensation nuclei. (p. 125)

constellation A human grouping of stars in the night sky into a recognizable pattern. (p. 4)

continuous spectrum Spectrum in which the radiation is distributed over all frequencies, not just a few specific frequency ranges. A prime example is the black-body radiation emitted by a hot, dense body. (p. 55)

convection Churning motion resulting from the constant upwelling of warm fluid and the concurrent downward flow of cooler material to take its place. (p. 143)

convection zone Region of the Sun's interior, lying just below the surface, where the material of the Sun is in constant convective motion. This region extends into the solar interior to a depth of about 200,000 km. (p. 256)

Copernican revolution The realization toward the end of the sixteenth century that Earth is not at the center of the universe. (p. 35)

core The central region of Earth, surrounded by the mantle. (p. 139) The central region of the Sun. (p. 257)

core hydrogen burning The energy burning stage for main-sequence stars, in which helium is produced by hydrogen fusion in the central region of the star. A typical star spends up to 90 percent of its lifetime in a state of equilibrium brought about by the balance between gravity and the energy generated by core hydrogen burning. (p. 336)

core-collapse supernova See type-II supernova. (p. 350)

corona One of numerous large, roughly circular regions on the surface of Venus, thought to have been caused by upwelling mantle material causing the planet's crust to bulge outward. (p. 178)

corona The tenuous outer atmosphere of the Sun, which lies just above the chromosphere, and at great distances turns into the solar wind. (p. 256)

coronal hole Vast regions of the Sun's atmosphere where the density of matter is about 10 times lower than average. The gas there streams freely into space at high speeds, escaping the Sun completely. (p. 264)

cosmic distance scale Collection of direct and indirect distance-measurement techniques that astronomers use to measure the scale of the universe. (p. 7)

cosmic evolution The collection of the seven major phases of the history of the universe, namely galactic, stellar, planetary, chemical, biological, cultural, and future evolution. (p. 506)

cosmic microwave background The almost perfectly isotropic radio signal that is the remnant of the Big Bang explosion. (p. 488)

cosmological principle Two assumptions which form the foundation of modern cosmology, namely that the universe is homogeneous and isotropic on sufficiently large scales. (p. 479)

cosmological redshift The component of the redshift of an object which is due to the expansion of the universe. (p. 439)

cosmology The study of the structure and evolution of the entire universe. (p. 479)

crater Bowl-shaped depression on the surface of a planet or moon, resulting from a collision with interplanetary debris. (p. 148)

critical density The cosmic density corresponding to the dividing line between a universe that recollapses and one that expands forever. (p. 484)

critical universe Geometry that the universe would have if the density of matter is exactly the critical density. The universe is infinite in extent, and has zero curvature. The expansion will continue forever, but approach an expansion speed of zero. (p. 488)

crust Layer of the Earth which contains the solid continents and the seafloor. (p. 139) The solid surface of any planet or moon.

D

D ring Collection of very faint, thin rings, extending from the inner edge of the C ring down nearly to the cloud tops of Saturn. This region contains so few particles that it is completely invisible from Earth. (p. 244)

dark dust cloud A large cloud, often many parsecs across, which contains gas and dust in a ratio of about 10^{12} gas atoms for every dust particle. Typical densities are a few tens or hundreds of millions of particles per cubic meter. (p. 314)

dark halo Region of a galaxy beyond the visible halo where dark matter predominates. (p. 408)

dark matter Term used to describe the mass in galaxies and clusters whose existence we infer from rotation curves and other techniques, but which has not been confirmed by observations at any electromagnetic wavelength. (p. 408)

declination Celestial coordinate used to measure latitude above or below the celestial equator on the celestial sphere. (p. 7)

decoupling Event in the early universe when atoms first formed, and after which photons could propagate freely through space. (p. 493)

deferent A construct of the geocentric model of the solar system which was needed to explain observed planetary motions. A deferent is a large circle encircling the Earth, on which an epicycle moves. (p. 24)

density A measure of the compactness of the matter within an object, computed by dividing the mass by the volume of the object. Units are kilograms per cubic meter (kg/m^3), or grams per cubic centimeter (g/cm^3). (p. 105)

differential rotation The tendency for a gaseous sphere, such as a jovian planet or the Sun, to rotate at a different rate at the equator than at the poles or for the rotation rate to vary with depth. For a galaxy or other object, a condition where the angular speed varies with location within the object. (p. 205)

differentiation Variation with depth in the density and composition of a body, such as Earth, with low-density material on the surface and higher density material in the core. (p. 154)

diffraction The tendency of waves to bend around corners. The diffraction of light establishes its nature as a wave. (p. 81)

Drake equation Expression which gives an estimate of the probability that intelligence exists elsewhere in the galaxy, based on a number of supposedly necessary conditions for intelligent life to develop. (p. 511)

dust grain An interstellar dust particle, roughly 10^{-8} m in size, comparable to the wavelength of visible light. (p. 309)

dust lane A lane of dark, obscuring interstellar dust in an emission nebula or galaxy. (p. 311)

dust tail The component of a comet's tail that is composed of dust particles. (p. 113)

dwarf Any star with radius comparable to, or smaller than, that of the Sun (including the Sun itself). (p. 284)

E

E ring A faint ring, well outside the main ring system of Saturn, which was discovered by *Voyager* and is believed to be associated with volcanism on the moon Enceladus. (p. 244)

Earth-crossing asteroid An asteroid whose orbit crosses that of the Earth. Earth-crossing asteroids are also called Apollo asteroids, after the first of the type discovered. (p. 109)

earthquake A sudden dislocation of rocky material near the Earth's surface. (p. 152)

eccentricity A measure of the flatness of an ellipse, equal to the distance between the two foci divided by the length of the major axis. (p. 29)

eclipse Event during which one body passes in front of another, so that the light from the occulted body is blocked. (p. 19)

eclipsing binary Rare binary-star system that is aligned in such a way that from Earth we periodically observe one star pass in front of the other, eclipsing the other star. (p. 296)

ecliptic The apparent path of the Sun, relative to the stars on the celestial sphere, over the course of a year. (p. 10)

electric field A field extending outward in all directions from a charged particle, such as a proton or an electron. The electric field determines the electric force exerted by the particle on all other charged particles in the universe; the strength of the electric field decreases with increasing distance from the charge according to an inverse-square law. (p. 47)

electromagnetic radiation Another term for light, electromagnetic radiation transfers energy and information from one place to another, even through the vacuum of empty space. (p. 45)

electromagnetic spectrum The complete range of electromagnetic radiation, from radio waves to gamma rays, including the visible spectrum. All types of electromagnetic radiation are basically the same phenomenon, differing only by wavelength, and all move at the speed of light. (p. 50)

electromagnetism The union of electricity and magnetism, which do not exist as independent quantities, but are in reality two aspects of a single physical phenomenon. (p. 48)

electron An elementary particle with a negative electric charge, one of the components of the atom. (p. 47)

element Matter made up of one particular atom. The number of protons in the nucleus of the atom determines which element it represents. (p. 64)

ellipse Geometric figure resembling an elongated circle. An ellipse is characterized by its degree of flatness, or eccentricity, and the length of its long axis. In general, bound orbits of objects moving under gravity are elliptical. (p. 29)

elliptical galaxy Category of galaxy in which the stars are distributed in an elliptical shape on the sky, ranging from highly elongated to nearly circular in appearance. (p. 423)

emission line Bright line in a specific location of the spectrum of radiating material, corresponding to emission of light at a certain frequency. A heated gas in a glass container produces emission lines in its spectrum. (p. 57)

emission nebula A glowing cloud of hot interstellar gas. The gas glows as a result of a nearby young star which is ionizing the gas. Since this gas is mostly hydrogen, the emitted radiation falls predominantly in the red region of the spectrum, because of a dominant hydrogen emission line. (p. 311)

emission spectrum The pattern of spectral emission lines produced by an element. Each element has its own unique emission spectrum. (p. 58)

Encke Division A small gap in Saturn's A ring. (p. 240)

epicycle A construct of the geocentric model of the solar system which was necessary to explain observed planetary motions. Each planet rides on a small epicycle whose center in turn rides on a larger circle (the deferent). (p. 24)

equinox See autumnal equinox and vernal equinox. (p. 16)

escape speed The speed necessary for an object to escape the gravitational pull of an object. Anything that moves away from the object with more than the escape speed will never return.

event horizon Imaginary spherical surface surrounding a collapsing star, with radius equal to the Schwarzschild radius, within which no event can be seen, heard, or known about by an outside observer. (p. 373)

evolutionary theory A theory which explains observations in a series of gradual steps, explainable in terms of well-established physical principles. (p. 25)

evolutionary track A graphical representation of a star's life, as a path on the Hertzsprung–Russell diagram. (p. 321)

excited state State of an atom when one of its electrons is in a higher energy orbital than the ground state. Atoms can become excited by absorbing a photon of a specific energy, or by colliding with a nearby atom. (p. 61)

F

F ring Faint narrow outer ring of Saturn, discovered by *Pioneer* in 1979. The F ring lies just inside the Roche limit of Saturn, and was shown by *Voyager* to be made up of several ring strands apparently braided together. (p. 243)

flare Explosive event occurring in or near an active region on the Sun. (p. 269)

flatness problem One of two conceptual problems with the Standard Big Bang model, which is that there is no natural way to explain why the density of the universe is so close to the critical density. (p. 495)

focus One of two special points within an ellipse, whose separation from each other indicates the eccentricity. In a bound orbit, objects move in ellipses about one focus. (p. 29)

force Action on an object that causes its momentum to change. The rate at which the momentum changes is numerically equal to the force. (p. 34)

fragmentation The breaking up of a large object into many smaller pieces (for example, as the result of high-speed collisions between planetesimals and protoplanets in the early solar system). (p. 128)

frequency The number of wave crests passing any given point per unit of time. (p. 46)

full Moon Phase of the Moon in which it appears as a complete circular disk in the sky. (p. 17)

G

galactic bulge Thick distribution of warm gas and stars around the galactic center. (p. 393)

galactic center The center of the Milky Way or any other galaxy. The point about which the disk of a spiral galaxy rotates. (p. 398)

galactic disk Flattened region of gas and dust that bisects the galactic halo in a spiral galaxy. This is the region of active star formation. (p. 392)

galactic halo Region of a galaxy extending far above and below the galactic disk, where globular clusters and other old stars reside. (p. 394)

galactic nucleus Small central high-density region of a galaxy. Nearly all of the radiation from an active galaxy is emitted from the nucleus. (p. 452)

galaxy Gravitationally bound collection of a large number of stars. The Sun is a star in the Milky Way Galaxy. (p. 392)

galaxy cluster A collection of galaxies held together by their mutual gravitational attraction. (p. 429)

gamma ray Region of the electromagnetic spectrum, far beyond the visible spectrum, corresponding to radiation of very high frequency and very short wavelength. (p. 45)

gamma-ray burster Object that emits a large amount of energy in the form of a brief burst of gamma rays. (p. 369)

general theory of relativity Einstein's theory of gravity, in which the force of gravity is reinterpreted as a curvature of spacetime in the vicinity of a massive object. (p. 373)

geocentric model A model of the solar system which holds that the Earth is at the center of the universe and all other bodies are in orbit around it. The earliest theories of the solar system were geocentric. (p. 24)

giant A star with a radius between 10 and 100 times that of the Sun. (p. 283)

globular cluster Tightly bound, roughly spherical collection of hundreds of thousands, and sometimes millions, of stars, spanning about 50 parsecs. Globular clusters are distributed in the halos around the Milky Way and other galaxies. (p. 301)

Grand Unified Theories Theories which describe the behavior of the single force that results from unification of the strong, weak, and electromagnetic forces in the early universe. (p. 495)

granulation Mottled appearance of the solar surface, caused by rising (hot) and falling (cool) material in convective cells just below the photosphere. (p. 260)

gravitational field Field created by any object with mass, extending outward in all directions, which determines the influence of that object on all others. The strength of the gravitational field decreases as the square of the distance. (p. 36)

gravitational lensing The effect induced on the image of a distant object by a massive foreground object. Light from the distant object is bent into two or more separate images. (p. 409)

gravitational redshift A prediction of Einstein's general theory of relativity. Photons lose energy as they escape the gravitational field of a massive object. Because a photon's energy is proportional to its frequency, a photon that loses energy suffers a decrease in frequency, which corresponds to an increase, or redshift, in wavelength. (p. 379)

gravity The attractive effect that any massive object has on all other massive objects. The greater the mass of the object, the stronger its gravitational pull. (p. 35)

Great Dark Spot Prominent storm system in the atmosphere of Neptune, located near the equator of the planet. The system is comparable in size to the Earth. (p. 215)

Great Red Spot A large, high-pressure, long-lived storm system visible in the atmosphere of Jupiter. The Red Spot is roughly twice the size of the Earth. (p. 206)

greenhouse effect The partial trapping of solar radiation by a planetary atmosphere, similar to the trapping of heat in a greenhouse. (p. 146)

ground state The lowest energy state that an electron can have within an atom. (p. 61)

H

heliocentric model A model of the solar system which is centered on the Sun, with the Earth in motion about the Sun. (p. 25)

helioseismology The study of conditions far below the Sun's surface through the analysis of internal "sound" waves that repeatedly cross the solar interior. (p. 258)

helium capture The formation of heavy elements by the capture of a helium nucleus. For example, carbon can form heavier elements by fusion with other carbon nuclei, but it is much more likely to occur by helium capture, which requires less energy. (p. 356)

helium flash An explosive event in the post-main-sequence evolution of a low-mass star. When helium fusion begins in a dense stellar core, the burning is explosive in nature. It continues until the energy released is enough to expand the core, at which point the star achieves stable equilibrium again. (p. 340)

Hertzsprung–Russell (H–R) diagram A plot of luminosity against temperature (or spectral class) for a group of stars. (p. 291)

high-energy telescope Telescope designed to detect radiation in X-rays and gamma rays. (p. 95)

highlands Relatively light-colored regions on the surface of the Moon which are elevated several kilometers above the maria. Also called terrae. (p. 147)

homogeneous The same everywhere. In a homogeneous universe, the number of galaxies in an imaginary large cube is the same no matter where in the universe the cube is placed. (p. 478)

horizon problem One of two conceptual problems with the standard Big Bang model, which is that some regions of the universe which have very similar properties are too far apart to have exchanged information in the age of the universe. (p. 494)

horizontal branch Region of the Hertzsprung–Russell diagram where post-main sequence stars again reach hydrostatic equilibrium. At this point, the star is burning helium in its core, and hydrogen in a shell surrounding the core. (p. 340)

Hubble classification Method of classifying galaxies according to their appearance, developed by Edwin Hubble. (p. 420)

Hubble's constant The constant of proportionality which gives the relation between recessional velocity and distance in Hubble's law. (p. 440)

Hubble's law Law that relates the observed velocity of recession of a galaxy to its distance from us. The velocity of recession of a galaxy is proportional to its distance. (p. 439)

hydrogen envelope An invisible region engulfing the coma of a comet, usually distorted by the solar wind, and extending across millions of kilometers of space. (p. 112)

hydrogen shell burning Fusion of hydrogen in a shell that is driven by contraction and heating of the helium core. Once hydrogen is depleted in the core of a star, hydrogen burning stops and the core contracts due to gravity, causing the temperature to rise, heating the surrounding layers of hydrogen in the star, and increasing the burning rate there. (p. 338)

hydrosphere Layer of the Earth which contains the liquid oceans and accounts for roughly 70 percent of Earth's total surface area. (p. 139)

I

inertia The tendency of an object to continue in motion at the same speed and in the same direction, unless acted upon by a force. (p. 34)

inflation Short period of unchecked cosmic expansion early in the history of the universe. During inflation, the universe swelled in size by a factor of about 10^{50}. (p. 496)

infrared Region of the electromagnetic spectrum just outside the visible range, corresponding to light of a slightly longer wavelength than red light. (p. 45)

infrared telescope Telescope designed to detect infrared radiation. Infrared telescopes are designed to be lightweight so that they can be carried above most of Earth's atmosphere by balloons, airplanes, or satellites. (p. 93)

inner core The central part of Earth's core, believed to be solid, and composed mainly of nickel and iron. (p. 153)

intensity A basic property of electromagnetic radiation that specifies the amount or strength of the radiation. (p. 52)

intercrater plains Regions on the surface of Mercury that do not show extensive cratering, but are relatively smooth. (p. 175)

interferometer Collection of two or more telescopes working together as a team, observing the same object at the same time and at the same wavelength. The effective diameter of an interferometer is equal to the distance between its outermost telescopes. (p. 91)

interferometry Technique in widespread use to dramatically improve the resolution of radio and other telescopes. Several telescopes observe an object simultaneously, and a computer analyzes how the signals interfere with one another to reconstruct a detailed image of the field of view. (p. 91)

interstellar medium The matter between stars, composed of two components, gas and dust, intermixed throughout all of space. (p. 308)

inverse-square law The law that a field follows if its strength decreases with the square of the distance. Fields that follow the inverse square law rapidly decrease in strength as the distance increases, but never quite reach zero. (p. 35)

ion An atom that has lost one or more electrons. (p. 61)

ion tail Thin stream of ionized gas that is pushed away from the head of a comet by the solar wind. It extends directly away from the Sun. Often referred to as a plasma tail. (p. 112)

irregular galaxy A galaxy which does not fit into any of the other major categories in the Hubble classification scheme. (p. 424)

isotropic Assumed property of the universe such that the universe looks the same in every direction. (p. 479)

J

jovian planet One of the four giant outer planets of the solar system, which resembles Jupiter in physical and chemical composition. (p. 107)

K

Kirchhoff's laws Three rules governing the formation of different types of spectra. (p. 59)

Kuiper belt A region in the plane of the solar system outside the orbit of Neptune where most short-period comets are thought to originate. (p. 114)

L

law of conservation of mass and energy A fundamental law of modern physics which states that the sum of mass and energy must always remain constant in any physical process. In fusion reactions, the lost mass is converted into energy, primarily in the form of electromagnetic radiation. (p. 271)

laws of planetary motion Three laws, based on precise observations of the motions of the planets by Tycho Brahe, which summarize the motions of the planets about the Sun. (p. 29)

light See electromagnetic radiation. (p. 52)

light curve A plot of the variation in brightness of a star with time. (p. 297)

light year The distance that light, moving at a constant speed of 300,000 km/s, travels in one year. One light year is about 10 trillion kilometers. (p. 2)

lighthouse model The leading explanation for pulsars. A small region of the neutron star, near one of the magnetic poles, emits a steady stream of radiation which sweeps past Earth each time the star rotates. Thus the period of the pulses is just the star's rotation period. (p. 368)

Local Group The small galaxy cluster that includes the Milky Way Galaxy. (p. 429)

luminosity One of the basic properties used to characterize stars, luminosity is defined as the total energy radiated by a star each second, at all wavelengths. (p. 257)

luminosity class A classification scheme which groups stars according to the width of their spectral lines. For a group of stars with the same temperature, luminosity class differentiates between supergiants, giants, main-sequence stars, and subdwarfs. (p. 294)

lunar eclipse Celestial event during which the Moon passes through the shadow of the Earth, temporarily darkening its surface. (p. 19)

M

Magellanic Clouds Two small irregular galaxies that are gravitationally bound to the Milky Way Galaxy. (p. 425)

magnetic field Field which accompanies any changing electric field, and governs the influence of magnetized objects on one another. (p. 47)

magnetosphere A zone of charged particles trapped by a planet's magnetic field, lying above the atmosphere. (p. 139)

magnitude scale A system of ranking stars by apparent brightness, developed by the Greek astronomer Hipparchus. Originally, the brightest stars in the sky were categorized as being of first magnitude, while the faintest stars visible to the naked eye were classified as sixth magnitude. The scheme has since been extended to cover stars and galaxies too faint to be seen by the unaided eye. Increasing magnitude means fainter stars, and a difference of 5 magnitudes corresponds to a factor of 100 in apparent brightness. (p. 285)

main sequence A well-defined band on a Hertzsprung–Russell diagram, on which most stars tend to be found, running from the top left of the diagram to the bottom right. (p. 291)

main-sequence turnoff Special point on a Hertzsprung–Russell diagram for a cluster. If all the stars in a particular cluster are plotted, the lower mass stars will trace out the main sequence up to the point where stars begin to evolve off the main sequence toward the red giant branch. The point where stars are just beginning to evolve off is the main-sequence turnoff. (p. 356)

mantle Layer of the Earth just interior to the crust. (p. 139)

mare Relatively dark-colored and smooth region on the surface of the Moon. (plural: maria) (p. 147)

mass A measure of the total amount of matter contained within an object. (p. 34)

matter-dominated universe A universe in which the density of matter exceeds the density of radiation. The present-day universe is matter-dominated. (p. 490)

meteor Bright streak in the sky, often referred to as a "shooting star," resulting from a small piece of interplanetary debris entering Earth's atmosphere and heating air molecules, which emit light as they return to their ground states. (p. 115)

meteor shower Event during which many meteors can be seen each hour, caused by the yearly passage of the Earth through the debris spread along the orbit of a comet. (p. 237)

meteorite Any part of a meteoroid that survives passage through the atmosphere and lands on the surface of Earth. (p. 116)

meteoroid Chunk of interplanetary debris prior to encountering Earth's atmosphere. (p. 108)

meteoroid swarm Pebble-sized cometary fragments dislodged from the main body, moving in nearly the same orbit as the parent comet. (p. 116)

micrometeoroids Relatively small chunks of interplanetary debris ranging from dust particle size to pebble-sized fragments. (p. 116)

Milky Way Galaxy The spiral galaxy in which the Sun resides. The disk of our Galaxy is visible in the night sky as the faint band of light known as the Milky Way. (p. 392)

millisecond pulsar A pulsar whose period indicates that the neutron star is rotating nearly 1000 times each second. The most likely explanation for these rapid rotators is that the neutron star has been spun up by drawing in matter from a companion star. (p. 370)

molecular cloud A cold, dense interstellar cloud which contains a high fraction of molecules. It is widely believed that the relatively high density of dust particles in these clouds plays an important role in the formation and protection of the molecules. (p. 317)

molecular cloud complex Collection of molecular clouds that spans as much as 50 parsecs and may contain enough material to make millions of Sun-sized stars. (p. 317)

molecule A tightly bound collection of atoms held together by the electromagnetic fields of the atoms. Molecules, like atoms, emit and absorb photons at specific wavelengths. (p. 64)

N

nebula General term used for any "fuzzy" patch on the sky, either light or dark. (p. 124)

nebular theory One of the earliest models of solar system formation, dating back to Descartes, in which a large cloud of gas began to collapse under its own gravity to form the Sun and planets. (p. 124)

neutrino Virtually massless and chargeless particle that is one of the products of fusion reactions in the Sun. Neutrinos move at close to the speed of light, and interact with matter hardly at all. (p. 271)

neutrino oscillations Possible solution to the solar neutrino problem, in which the neutrino has a very tiny mass. In this case, the correct number of neutrinos can be produced in the solar core, but on their way to Earth, some can "oscillate," or become transformed into other particles, and thus go undetected. (p. 273)

neutron An elementary particle with roughly the same mass as a proton, but which is electrically neutral. Along with protons, neutrons form the nuclei of atoms. (p. 64)

neutron star A dense ball of neutrons that remains at the core of a star after a supernova explosion has destroyed the rest of the star. Typical neutron stars are about 20 km across, and contain more mass than the Sun. (p. 366)

new Moon Phase of the moon during which none of the lunar disk is visible. (p. 17)

Newtonian mechanics The basic laws of motion, postulated by Newton, which are sufficient to explain and quantify virtually all of the complex dynamical behavior found on Earth and elsewhere in the universe. (p. 33)

Newtonian telescope A reflecting telescope in which incoming light is intercepted before it reaches the prime focus and is deflected into an eyepiece at the side of the instrument. (p. 77)

north celestial pole Point on the celestial sphere directly above Earth's north pole. (pp. 6, 7)

nova A star that suddenly increases in brightness, often by a factor of as much as 10,000, then slowly fades back to its original luminosity. A nova is the result of an explosion on the surface of a white dwarf star, caused by matter falling onto its surface from the atmosphere of a binary companion. (p. 346)

nuclear fusion Mechanism of energy generation in the core of the Sun, in which light nuclei are combined, or fused, into heavier ones, releasing energy in the process. (p. 270)

nucleotide base An organic molecule, the building block of genes that pass on hereditary characteristics from one generation of living creatures to the next. (p. 507)

nucleus Dense, central region of an atom, containing both protons and neutrons, and orbited by one or more electrons. (p. 60)

nucleus The solid region of ice and dust that composes the central region of the head of a comet. (p. 111)

O

Olbers's paradox A thought experiment suggesting that if the universe were homogeneous, infinite, and unchanging, the entire night sky would be as bright as the surface of the Sun. (p. 480)

Oort Cloud Spherical halo of material surrounding the solar system, out to a distance of about 50,000 A.U., where most comets originate. (p. 114)

opacity A quantity that measures a material's ability to block electromagnetic radiation. Opacity is the opposite of transparency. (p. 56)

open cluster Loosely bound collection of tens to hundreds of stars, a few parsecs across, generally found in the plane of the Milky Way. (p. 299)

open universe Geometry that the universe would have if the density of matter were less than the critical value. In an open universe there is not enough matter to halt the expansion of the universe. An open universe is infinite in extent. (p. 488)

outer core The outermost part of Earth's core, believed to be liquid, and composed mainly of nickel and iron. (p. 153)

outflow channel Surface features on Mars, evidence that liquid water once existed there in great quantity, believed to be the relics of catastrophic flooding about 3 billion years ago. Found only in the equatorial regions of the planet. (p. 184)

ozone layer Layer of the Earth's atmosphere at an altitude of 20 to 50 km where incoming ultraviolet solar radiation is absorbed by oxygen, ozone, and nitrogen in the atmosphere. (p. 144)

P

parallax The apparent motion of a relatively close object with respect to a more distant background as the location of the observer changes. (p. 10)

parsec The distance at which a star must lie in order that its measured parallax is exactly 1 arc second, equal to 206,000 A.U. (p. 280)

partial eclipse Celestial event during which only a part of the occulted body is blocked from view. (p. 20)

penumbra Portion of the shadow cast by an eclipsing object in which the eclipse is seen as partial. (p. 20)

penumbra The outer region of a sunspot surrounding the umbra, which is not as dark and not as cool as the central region. (p. 265)

period The time needed for an orbiting body to complete one revolution about another body. (p. 30)

period-luminosity relation A relation between the pulsation period of a Cepheid variable and its absolute brightness. Measurement of the pulsation period allows the distance of the star to be determined. (p. 397)

permafrost Layer of permanently frozen water ice believed to lie just under the surface of Mars. (p. 184)

phase See lunar phase. (p. 17)

photon Individual packet of electromagnetic energy that makes up electromagnetic radiation. (p. 62)

photosphere The visible surface of the Sun, lying just above the uppermost layer of the Sun's interior, and just below the chromosphere. (p. 256)

planetary nebula The ejected envelope of a red giant star, spread over a volume roughly the size of our solar system. (p. 343)

planetesimal Term given to objects in the early solar system that had reached the size of small moons, at which point their gravitational fields were strong enough to begin to influence their neighbors. (p. 128)

plate tectonics The motions of regions of Earth's crust, which drift with respect to one another. Also known as continental drift. (p. 159)

positron Atomic particle with properties identical to those of a negatively charged electron, except for its positive charge. The positron is the antiparticle of the electron. Positrons and electrons annihilate one another when they meet, producing pure energy in the form of gamma rays. (p. 271)

precession The slow change in the direction of the axis of a spinning object, caused by some external influence. (p. 17)

primary atmosphere The chemical components that would have surrounded Earth just after it formed. (p. 187)

prime focus The point in a reflecting telescope where the mirror focuses incoming light to a point. (p. 74)

primordial nucleosynthesis The production of elements heavier than hydrogen by nuclear fusion in the high temperatures and densities which existed in the early universe. (p. 491)

prominence Loop or sheet of glowing gas ejected from an active region on the solar surface, which then moves through the inner parts of the corona under the influence of the Sun's magnetic field. (p. 268)

proper motion The angular movement of a star across the sky, as seen from Earth, measured in seconds of arc per year. This movement is a result of the star's actual motion through space. (p. 281)

proton An elementary particle carrying a positive electric charge, a component of all atomic nuclei. The number of protons in the nucleus of an atom dictates what type of atom it is. (p. 47)

proton-proton chain The chain of fusion reactions, leading from hydrogen to helium, that powers most main-sequence stars. (p. 271)

protoplanet Clump of material, formed in the early stages of solar system formation, that was the forerunner of the planets we see today. (p. 124)

protostar Stage in star formation when the interior of a collapsing fragment of gas is sufficiently hot and dense that it becomes opaque to its own radiation. The protostar is the dense region at the center of the fragment. (p. 320)

protosun The central accumulation of material in the early stages of solar system formations, the forerunner of the present-day Sun. (p. 124)

Ptolemaic model Solar system model, developed by the second century astronomer Claudius Ptolemy, perhaps the best geocentric model to be proposed. It predicted with great accuracy the positions of the known planets, using more than 80 circles to model the (then known) planets. (p. 24)

pulsar Object that emits radiation in the form of rapid pulses with a characteristic pulse period and duration. Charged particles, accelerated by the magnetic field of a rapidly rotating neutron star, flow along the magnetic field lines, producing radiation that beams outward as the star spins on its axis. (p. 367)

pulsating variable star A star whose luminosity varies in a predictable, periodic way. (p. 395)

Q

quantized The fact that light and matter on small scales behave in a discontinuous manner, and manifest themselves in the form of tiny "packets" of energy, called quanta. (p. 61)

quarter Moon Lunar phase in which the moon appears as a half disk. (p. 17)

quasar Star-like radio source with an observed redshift that indicates extremely large distances from Earth. A member of the most energetic class of active galaxies. (p. 464)

quasi-stellar object (QSO) See quasar. (p. 464)

quiet Sun The underlying predictable elements of the Sun's behavior, such as its average photospheric temperature, which do not change in time. (p. 265)

R

radar Acronym for RAdio Detection And Ranging. Radio waves are bounced off an object, and the time at which the echo is received indicates its distance. (p. 32)

radiation-dominated universe Early epoch in the universe, when the density of radiation in the cosmos exceeded the density of matter. (p. 490)

radiation zone Region of the Sun's interior where extremely high temperatures guarantee that the gas is completely ionized. Photons are only occasionally diverted by electrons, and travel through this region with relative ease. (p. 256)

radio Region of the electromagnetic spectrum corresponding to radiation of the longest wavelengths. (p. 45)

radio galaxy Type of active galaxy that emits most of its energy in the form of long-wavelength radiation. (p. 452)

radio lobe Roundish region of radio-emitting gas lying well beyond the center of a radio galaxy. (p. 453)

radio telescope Large instrument designed to detect radiation from space in radio wavelengths. (p. 89)

radioactivity The release of energy by rare, heavy elements when their nuclei decay into lighter nuclei. (p. 154)

radius-luminosity-temperature relation A mathematical proportionality, arising from simple geometry and Stefan's law, which allows astronomers to indirectly determine the radius of a star once its luminosity and temperature are known. (p. 283)

red dwarf Small, cool faint star at the lower-right end of the main sequence on the Hertzsprung–Russell diagram, whose color and size give its name. (p. 292)

red giant A giant star whose surface temperature is relatively low, so that it glows with a red color. (p. 283)

red giant branch The section of the evolutionary track of a star that corresponds to continued heating from rapid hydrogen shell burning, which drives a steady expansion and cooling of the outer envelope of the star. As the star gets larger in radius and its surface temperature cools, it becomes a red giant. (p. 339)

red giant region The upper-right-hand corner of the Hertzsprung–Russell diagram, where red-giant stars are found. (p. 293)

red supergiant An extremely luminous red star. Often found on the asymptotic giant branch of the Hertzsprung–Russell diagram. (p. 341)

reddening Dimming of starlight by interstellar matter, which tends to scatter higher-frequency (blue) components of the radiation more efficiently than the lower-frequency (red) components. (p. 309)

reflecting telescope A telescope which uses a mirror to gather and focus light from a distant object. (p. 74)

refracting telescope A telescope which uses a lens to gather and focus light from a distant object. (p. 74)

refraction The tendency of a wave to bend as it passes from one transparent medium to another. (p. 74)

retrograde motion Backward, westward loop traced out by a planet with respect to the fixed stars. (p. 23)

revolution Orbital motion of one body about another, such as Earth about the Sun. (p. 14)

right ascension Celestial coordinate used to measure longitude on the celestial sphere. The zero point is the position of the Sun on the vernal equinox. (p. 7)

ringlet Narrow region in Saturn's planetary ring system where the density of ring particles is high. *Voyager* discovered that the rings visible from Earth are actually composed of tens of thousands of ringlets. (p. 242)

Roche limit Often called the tidal stability limit, the Roche limit gives the distance from a planet at which the tidal force, due to the planet, between adjacent objects exceeds their mutual attraction. Objects within this limit are unlikely to accumulate into larger objects. The rings of Saturn occupy the region within Saturn's Roche limit. (p. 242)

rotation Spinning motion of a body about an axis. (p. 6)

rotation curve Plot of the orbital speed of disk material in a galaxy against its distance from the galactic center. Analysis of rotation curves of spiral galaxies indicates the existence of dark matter. (p. 407)

RR Lyrae variable Variable star whose luminosity changes in a characteristic way. All RR Lyrae stars have more or less the same period. (p. 395)

runaway greenhouse effect A process in which the heating of a planet leads to an increase in its atmosphere's ability to retain heat and thus to further heating, quickly causing extreme changes in the temperature of the surface and the composition of the atmosphere. (p. 189)

runoff channel River-like surface feature on Mars, evidence that liquid water once existed there in great quantities. Runoff channels are found in the southern highlands, and are thought to have been formed by water that flowed nearly 4 billion years ago. (p. 184)

S

S0 galaxy Galaxy which shows evidence of a thin disk and a bulge, but which has no spiral arms and contains little or no gas. (p. 424)

SB0 galaxy S0-type galaxy whose disk shows evidence of a bar. (p. 424)

scarp Surface feature on Mercury believed to be the result of cooling and shrinking of the crust, forming a wrinkle on the face of the planet. (p. 176)

Schwarzschild radius The distance from the center of an object such that, if all the mass was compressed within that region, the escape velocity would equal the speed of light. Once a stellar remnant collapses within this radius, light cannot escape and the object is no longer visible. (p. 373)

seasons Changes in average temperature and length of day that result from the tilt of Earth's (or any planet's) axis with respect to the plane of its orbit. (p. 11)

secondary atmosphere The chemicals that composed Earth's atmosphere after the planet's formation, once volcanic activity outgassed chemicals from the interior. (p. 188)

seeing   A term used to describe the ease with which good telescopic observations can be made from Earth's surface, given the blurring effects of atmospheric turbulence. (p. 82)

seeing disk   Roughly circular region on a detector over which a star's pointlike images is spread, due to atmospheric turbulence. (p. 82)

seismic wave   A wave that travels outward from the site of an earthquake through the Earth. (p. 152)

semi-major axis   One half of the major axis of an ellipse. The semi-major axis is the way in which the size of an ellipse is usually quantified. (p. 29)

Seyfert galaxy   Type of active galaxy whose emission comes from a very small region within the nucleus of an otherwise normal-looking spiral system. (p. 451)

shepherd satellite   Satellite whose gravitational effects on a ring preserve its shape, such as the two satellites of Saturn, Prometheus and Pandora, whose orbits lie on either side of the F ring. (p. 244)

shield volcano   A volcano produced by repeated nonexplosive eruptions of lava, creating a gradually sloping, shield-shaped low dome. Often contains a caldera at its summit. (p. 178)

sidereal day   The time required for Earth to rotate exactly once, relative to the stars. (p. 14)

sidereal month   Time required for the Moon to complete one orbit around Earth. (p. 19)

sidereal year   The time required for the constellations to complete one cycle around the sky and return to their starting points, as seen from a given point on Earth. (p. 17)

singularity   A point in the universe where the density of matter and the gravitational field are infinite, such as at the center of a black hole. (p. 381)

solar constant   The amount of solar energy reaching Earth per unit area per unit time, approximately 1400 W/m². (p. 257)

solar cycle   The 22-year period that is needed for both the average number of spots and the Sun's magnetic polarity to repeat themselves. The Sun's polarity reverses on each new 11-year sunspot cycle. (p. 267)

solar day   The period of time between the instant when the Sun is directly overhead (i.e. at noon) to the next time it is directly overhead. (p. 14)

solar eclipse   Celestial event during which the new Moon passes directly between the Earth and Sun, temporarily blocking the Sun's light. (p. 20)

solar nebula   The swirling gas surrounding the early Sun during the epoch of solar system formation, also referred to as the primitive solar system. (p. 124)

solar neutrino problem   The discrepancy between the theoretically predicted numbers of neutrinos streaming from the Sun as a result of fusion reactions in the core and the numbers actually observed. The observed number of neutrinos is only about half the predicted number. (p. 273)

solar system   The Sun, and all the planets that orbit the Sun—Mercury, Venus, Earth, Mars, Jupiter, Saturn, Uranus, Neptune, and Pluto. (p. 104)

solar wind   An outward flow of fast-moving charged particles from the Sun. (p. 113)

south celestial pole   Point on the celestial sphere directly above the Earth's south pole. (pp. 6, 7)

spectral class   Classification scheme, based on the strength of stellar spectral lines, which is an indication of the temperature of a star. (p. 290)

spectrometer   Instrument used to produce detailed spectra of stars. Usually, a spectrograph records a spectrum on a photographic plate, or more recently, in electronic form on a computer. (p. 84)

spectroscope   Instrument used to view a light source so that it is split into its component colors. (p. 55)

spectroscopic binary   A binary-star system which from Earth appears as a single star, but whose spectral lines show back-and-forth Doppler shifts as two stars orbit one another. (p. 296)

spectroscopic parallax   Method of determining the distance to a star by measuring its temperature and then determining its absolute brightness by comparing with a standard H–R diagram. The absolute and apparent brightnesses of the star give the star's distance from Earth. (p. 293)

spectroscopy   The study of the way in which atoms absorb and emit electromagnetic radiation. Spectroscopy allows astronomers to determine the chemical composition of stars. (p. 59)

speed of light   The fastest possible speed, according to the currently known laws of physics. Electromagnetic radiation exists in the form of waves or photons moving at the speed of light. (p. 54)

spiral arm   Distribution of material in a galaxy in a pinwheel-shaped design apparently emanating from near the galactic center. (p. 403)

spiral density wave   A proposed explanation for the existence of galactic spiral arms, in which coiled waves of gas compression move through the galactic disk, triggering star formation. (p. 404)

spiral galaxy   Galaxy composed of a flattened, star-forming disk component which may have prominent spiral arms and a large central galactic bulge. (p. 395)

standard candle   Any object with an easily recognizable appearance and known luminosity, which can be used in estimating distances. Supernovae, which all have the same peak luminosity (depending on type), are good examples of standard candles and are used to determine distances to other galaxies. (p. 427)

Standard Solar Model   A self-consistent picture of the Sun, developed by incorporating the important physical processes that are believed to be important in determining the Sun's internal structure, into a computer program. The results of the program are then compared with observations of the Sun, and modifications are made to the model. The Standard Solar Model, which enjoys widespread acceptance, is the result of this process. (p. 258)

star   A glowing ball of gas held together by its own gravity and powered by nuclear fusion in its core. (p. 256)

star cluster   A grouping of anywhere from a dozen to a million stars which formed at the same time from the same cloud of interstellar gas. Stars in clusters are useful to aid our understanding of stellar evolution because they are all roughly the same age and chemical composition, and lie at roughly the same distance from Earth. (p. 299)

starburst galaxy   Galaxy in which a violent event, such as near-collision, has caused a sudden, intense burst of star formation in the recent past. (p. 438)

Stefan's law   Relation that gives the total energy emitted per square centimeter of its surface per second by an object of a given temperature. Stefan's law shows that the energy emitted increases rapidly with an increase in temperature, proportional to the temperature raised to the fourth power. (p. 54)

stellar occultation   The dimming of starlight produced when a solar system object such as a planet, moon, or ring, passes directly in front of a star. (p. 244)

subgiant branch   The section of the evolutionary track of a star that corresponds to changes that occur just after hydrogen is depleted in the core, and core hydrogen burning ceases. Shell hydrogen burning heats the outer layers of the star, which causes a general expansion of the stellar envelope. (p. 339)

summer solstice   Point on the ecliptic where the Sun is at its northernmost point above the celestial equator, occurring on or near June 21. (p. 15)

sunspot   An Earth-sized dark blemish found on the surface of the Sun. The dark color of the sunspot indicates that it is a region of lower temperature than its surroundings. (p. 265)

sunspot cycle   The fairly regular pattern that the number and distribution of sunspots follows, in which the average number of spots reaches a maximum every 11 or so years, then falls off to almost zero. (p. 267)

supercluster   Grouping of several clusters of galaxies into a larger, but not necessarily gravitationally bound, unit. (p. 430)

supergiant   A star with a radius between 100 and 1000 times that of the Sun. (p. 283)

supergranulation   Large-scale flow pattern on the surface of the Sun, consisting of cells measuring up to 30,000 km across, believed to be the imprint of large convective cells deep in the solar interior. (p. 261)

supernova   Explosive death of a star, caused by the sudden onset of nuclear burning (type I), or an enormously energetic shock wave (type II). One of the most energetic events of the universe, a supernova may temporarily outshine the rest of the galaxy in which it resides. (p. 399)

supernova remnant The scattered glowing remains from a supernova that occurred in the past. The Crab Nebula is one of the best-studied supernova remnants. (p. 352)

synchronous orbit State of an object when its period of rotation is exactly equal to its average orbital period. The Moon is in a synchronous orbit, and so presents the same face toward Earth at all times. (p. 138)

synchrotron radiation Type of nonthermal radiation caused by high-speed charged particles, such as electrons, emitting radiation as they are accelerated in a strong magnetic field. (p. 461)

synodic month Time required for the Moon to complete a full cycle of phases. (p. 19)

T

T Tauri star Protostar in the late stages of formation, often exhibiting violent surface activity. T Tauri stars have been observed to brighten noticeably in a short period of time, consistent with the idea of rapid evolution during this final phase of stellar formation. (p. 322)

tail Component of a comet that consists of material streaming away from the main body, sometimes spanning hundreds of millions of kilometers. May be composed of dust or ionized gases. (p. 111)

telescope Instrument used to capture as many photons as possible from a given region of the sky and concentrate them into a focused beam for analysis. (p. 74)

temperature A measure of the amount of heat in an object, and an indication of the speed of the particles that comprise it. (p. 58)

terrestrial planet The four innermost planets of the solar system, resembling the Earth in general physical and chemical properties. (p. 107)

tidal bulge Elongation of the Earth caused by the difference between the gravitational force on the side nearest the Moon and the force on the side farthest from the Moon. The long axis of the tidal bulge points toward the Moon. (p. 140) More generally, the deformation of any body produced by the tidal effect of a nearby gravitating object.

tidal force The variation in one body's gravitational force from place to place across another body—for example, the variation of the Moon's gravity across the Earth. (p. 141)

tides Rising and falling motion that bodies of water follow, exhibiting daily, monthly, and yearly cycles. Ocean tides on Earth are caused by the competing gravitational pull of the Moon and Sun on different regions of the Earth. (p. 140)

time dilation A prediction of the theory of relativity, closely related to the gravitational redshift. To an outside observer, a clock lowered into a strong gravitational field will appear to run slow. (p. 379)

total eclipse Celestial event during which one body is completely blocked from view by another. (p. 20)

transition zone The region of rapid temperature increase that separates the Sun's chromosphere from the corona. (p. 256)

triangulation Method of determining distance based on the principles of geometry. A distant object is sighted from two well-separated locations. The distance between the two locations and the angle between the line joining them and the line to the distant object are all that are necessary to ascertain the object's distance. (p. 7)

Trojan asteroids One of two groups of asteroids which orbit at the same distance from the Sun as Jupiter, 60 degrees ahead and behind the planet. (p. 108)

tropical year The time interval between one vernal equinox and the next. (p. 16)

Tully-Fisher relation A relation used to determine the absolute luminosity of a spiral galaxy. The rotational velocity, measured from the broadening of spectral lines, is related to the total mass, and hence the total luminosity. (p. 427)

21-centimeter radiation Radio radiation emitted when an electron in the ground state of a hydrogen atom flips its spin to become parallel to the spin of the proton in the nucleus. (p. 316)

Type I supernova One possible explosive death of a star, in which a white dwarf in a binary system accretes so much mass that it cannot support its own weight. The star collapses and temperatures become high enough for carbon fusion to occur. Fusion begins throughout the white dwarf almost simultaneously and an explosion results. (p. 351)

Type II supernova One possible explosive death of a star, in which the highly evolved stellar core rapidly implodes and then explodes, destroying the surrounding star. (p. 351)

U

ultraviolet Region of the electromagnetic spectrum, just outside the visible range, corresponding to wavelengths slightly shorter than blue light. (p. 45)

ultraviolet telescope A telescope that is designed to collect radiation in the ultraviolet part of the spectrum. The Earth's atmosphere is partially opaque to these wavelengths, so ultraviolet telescopes are put on rockets, balloons, or satellites to get high above most or all of the atmosphere. (p. 94)

umbra Central region of the shadow cast by an eclipsing body. (p. 20)

umbra The central region of a sunspot, which is its darker and cooler part. (p. 265)

universe The totality of all space, time, matter, and energy. (p. 2)

V

Van Allen belts At least two doughnut-shaped regions of magnetically trapped charged particles high above Earth's atmosphere. (p. 156)

variable star A star whose luminosity changes with time. (p. 395)

vernal equinox Date on which the Sun crosses the celestial equator moving northward, occurring on or near March 21. (p. 16)

visible The small range of the electromagnetic spectrum that human eyes perceive as light. The visible spectrum ranges from about 400 to 700 nm, corresponding to blue through red light. (p. 45)

visual binary A binary star system in which both members are resolvable from Earth. (p. 296)

void Large, relatively empty region of the universe around which superclusters of galaxies are organized. (p. 444)

volcano Upwelling of hot lava from below Earth's crust to the planet's surface. (p. 153)

W

water hole The radio interval between 18 cm and 21 cm, the wavelengths at which hydroxyl (OH) and hydrogen (H) radiate, respectively, in which intelligent civilizations might conceivably send their communication signals. (p. 518)

wave A pattern that repeats itself cyclically in both time and space. Waves are characterized by the velocity with which they move, their frequency, and their wavelength. (p. 45)

wave period The amount of time required for a wave to repeat itself at a specific point in space. (p. 46)

wavelength The length from one point on a wave to the point where it is repeated exactly in space, at a given time. (p. 46)

weird terrain A region on the surface of Mercury of oddly rippled features. This feature is thought to be the result of a strong impact which occurred on the other side of the planet, and sent seismic waves traveling around the planet, converging in the weird region. (p. 177)

white dwarf A dwarf star with a surface temperature that is hot, so that the object glows white. (p. 284)

white dwarf region The bottom left-hand corner of the Hertzsprung–Russell diagram, where white dwarf stars are found. (p. 293)

white oval Light-colored region near the Great Red Spot in Jupiter's atmosphere. Like the red spot, such regions are apparently rotating storm systems. (p. 211)

Wien's law Relation which gives the connection between the wavelength at which a black-body curve peaks and the temperature of the emitter. The temperature is inversely proportional to the peak wavelength, so the hotter the object, the bluer its radiation. (p. 53)

winter solstice   Point on the ecliptic where the Sun is at its southernmost point below the celestial equator, occurring on or near December 21. (p. 16)

X

X ray   Region of the electromagnetic spectrum corresponding to radiation of high frequency and short wavelengths, far outside the visible spectrum. (p. 45)

X-ray burster   X-ray source that radiates thousands of times more energy than our Sun, in short bursts that last only a few seconds. A neutron star in a binary system accretes matter onto its surface until temperatures reach the level needed for hydrogen fusion to occur. The result is a sudden period of rapid nuclear burning and release of energy. (p. 369)

Z

zero-age main sequence   The region on the Hertzsprung–Russell diagram, as predicted by theoretical models, where stars are located at the onset of nuclear burning in their cores. (p. 323)

zodiac   The twelve constellations through which the Sun moves as it follows its path on the ecliptic. (p. 17)

zonal flow   Alternating regions of westward and eastward flow, roughly symmetrical about the equator, associated with the belts and zones in the atmosphere of a jovian planet. (p. 207)

zone   Bright, high-pressure region, where gas flows upward, in the atmosphere of a jovian planet. (p. 185)

Answers to Self-Test Questions

Chapter 1
True or False? 1. T 2. F 3. F 4. F 5. F 6. T 7. F 8. F 9. F 10. T 11. F 12. F 13. F 14. T 15. T 16. F 17. F 18. T 19. T 20. T
Fill in the Blank 1. stars 2. ecliptic 3. winter solstice, lowest 4. quarter 5. lunar 6. retrograde 7. Copernicus 8. Earth's 9. moons, phases, sunspots 10. ellipse, circle 11. square, cube 12. radar 13. force 14. product, square 15. masses

Chapter 2
True or False? 1. T 2. F 3. F 4. T 5. F 6. F 7. F 8. T 9. T 10. F 11. T 12. T 13. F 14. T 15. T 16. T 17. F 18. T 19. F 20. F
Fill in the Blank 1. 300,000 2. wavelength 3. frequency 4. electric, magnetic 5. 400,700 6. red 7. radio, infrared, visible 8. temperature 9. the 1200 K object 10. hot 11. prism 12. continuous 13. the Sun 14. dense 15. cool 16. particle 17. positive, negative 18. absorbs 19. emits 20. difference

Chapter 3
True or False? 1. F 2. F 3. T 4. F 5. F 6. T 7. T 8. F 9. T 10. T 11. F 12. F 13. F
Fill in the Blank 1. refracting 2. reflecting 3. reflecting 4. area or diameter 5. diameter, wavelength 6. atmosphere 7. one 8. digital 9. resolution 10. reflecting 11. interferometer 12. infrared

Chapter 4
True or False? 1. T 2. F 3. F 4. F 5. F 6. T 7. F 8. F 9. F 10. F 11. T 12. T 13. F 14. F 15. F 16. F 17. T 18. F
Fill in the Blank 1. Mercury, Pluto 2. rocky 3. Mars, Jupiter 4. hundreds, 100 5. Jupiter 6. eccentric 7. a few, 1 A.U., or hundreds of millions of kilometers 8. meteoroid swarm 9. comets 10. meteor 11. 4.6 billion 12. interstellar dust 13. gravity 14. gas 15. icy 16. fragmentation 17. comets 18. comets 19. Jupiter 20. size

Chapter 5
True or False? 1. T 2. F 3. F 4. T 5. T 6. F 7. T 8. F 9. T 10. F 11. T 12. F 13. F 14. T 15. F 16. F 17. F
Fill in the Blank 1. 6500 2. 1/4 3. radar 4. crust 5. liquid water 6. difference 7. maria 8. impacts 9. mantle 10. nitrogen, oxygen 11. convection 12. infrared 13. raise or increase 14. aurora 15. solid, liquid 16. molten 17. heavy elements 18. plate tectonics

Chapter 6
True or False? 1. T 2. T 3. T 4. F 5. F 6. F 7. F 8. T 9. F 10. T 11. T 12. F 13. F 14. T 15. F 16. F 17. F 18. F 19. T 20. F
Fill in the Blank 1. larger 2. nature 3. poles 4. magnetic field 5. density 6. slow or retrograde 7. carbon dioxide 8. closer 9. radar 10. coronae 11. volcanism 12. temperature, pressure

13. cratered 14. bulge 15. volcanoes 16. gravity 17. permafrost or frozen water 18. water 19. flooding 20. billion

Chapter 7
True or False? 1. F 2. F 3. T 4. F 5. F 6. F 7. F 8. T 9. F 10. T 11. F 12. F 13. T 14. T 15. F 16. F 17. F 18. F 19. T
Fill in the Blank 1. density 2. Uranus 3. Neptune 4. methane 5. hydrogen, helium 6. zones, bands 7. hurricanes 8. two 9. ammonia 10. gravity 11. Neptune 12. rotation 13. perpendicular 14. liquid 15. two 16. hydrogen 17. helium precipitation

Chapter 8
True or False? 1. F 2. T 3. T 4. T 5. T 6. F 7. F 8. T 9. F 10. T 11. T 12. F 13. T 14. T 15. T 16. T 17. F 18. F
Fill in the Blank 1. Ganymede 2. water 3. volcanoes 4. 3 5. A, B 6. Roche limit 7. shepherd 8. nitrogen 9. water 10. Miranda 11. retrograde 12. nitrogen 13. Triton 14. Triton

Chapter 9
True or False? 1. T 2. F 3. F 4. F 5. T 6. T 7. F 8. F 9. F 10. F 11. T 12. F 13. T 14. T 15. F 16. F 17. F 18. T
Fill in the Blank 1. photosphere 2. photosphere, chromosphere, corona 3. convective, radiative, core 4. granules 5. photosphere 6. hydrogen 7. helium 8. 98 or 99 9. ionized 10. solar wind 11. cooler 12. 11, 22 13. flare 14. core 15. 4, helium, gamma rays 16. few

Chapter 10
True or False? 1. T 2. T 3. F 4. T 5. F 6. T 7. F 8. F 9. F 10. T 11. F 12. F 13. T 14. T 15. F
Fill in the Blank 1. 2 A.U. 2. spectrum, Doppler 3. proper motion, distance 4. luminosity, temperature 5. white dwarfs 6. temperature 7. ionized 8. neutral 9. G2 10. spectral type, luminosity 11. main sequence 12. red giants 13. white dwarfs 14. binary 15. decrease 16. cluster

Chapter 11
True or False? 1. F 2. T 3. T 4. F 5. T 6. F 7. F 8. F 9. T 10. F 11. T 12. T 13. T 14. F 15. F 16. T
Fill in the Blank 1. gas, dust 2. similar or larger 3. hydrogen, helium 4. 8,000 5. 10 or 20 6. spin, hydrogen 7. 20 8. radio 9. hydrogen 10. millions of 11. evolutionary track 12. increase 13. upper, right 14. 10 million 15. main sequence 16. 50 17. radio 18. infrared

Chapter 12
True or False? 1. T 2. T 3. F 4. T 5. T 6. F 7. T 8. F 9. F 10. F 11. T 12. T 13. T 14. T 15. T 16. T 17. T 18. F

Fill in the Blank 1. pressure 2. 5 billion 3. hydrogen, helium 4. 100 million K 5. contract 6. 100 7. clusters 8. the Earth 9. high, low 10. decrease 11. accretion 12. hydrogen, surface 13. neutron, neutrino 14. Type I 15. 1.4 16. light curves 17. decreases

Chapter 13
True or False? 1. F 2. F 3. T 4. F 5. F 6. T 7. F 8. T 9. F 10. F 11. F 12. F
Fill in the Blank 1. Type I 2. 20 3. high, strong 4. radio 5. 0.03 to 0.3 seconds 6. rate of rotation 7. binary 8. neutron 9. gravity 10. lose 11. X-rays 12. binary

Chapter 14
True or False? 1. T 2. F 3. F 4. F 5. F 6. F 7. F 8. T 9. T 10. F 11. F 12. T 13. T
Fill in the Blank 1. inside 2. disk 3. halo 4. brightness 5. 1, 100 6. higher or brighter 7. RR Lyrae stars 8. 8,000 9. 220 km/s 10. random 11. halo 12. higher, dark matter

Chapter 15
True or False? 1. T 2. T 3. F 4. T 5. T 6. F 7. F 8. T 9. F 10. F 11. F 12. F 13. T 14. T
Fill in the Blank 1. Hubble 2. large 3. Sa, Sc 4. Local Group 5. broadening 6. 100 Mpc 7. increases 8. Kepler's Third 9. 90 10. X-rays 11. mergers 12. distance 13. 60 and 90 14. 100 to 200

Chapter 16
True or False? 1. T 2. F 3. F 4. F 5. T 6. T 7. F 8. T 9. F 10. T 11. F 12. T
Fill in the Blank 1. large 2. nucleus 3. core 4. larger 5. nucleus 6. nucleus 7. one parsec 8. 1 to 3 billion 9. one solar mass, decade 10. star-like 11. short 12. galaxy or cluster of galaxies

Chapter 17
True or False? 1. F 2. T 3. T 4. F 5. T 6. T 7. T 8. T 9. T 10. F 11. T 12. T 13. F 14. T 15. F 16. T
Fill in the Blank 1. 200 2. everywhere 3. directions 4. Cosmological Principle 5. homogeneity 6. isotropy 7. 15 8. gravity 9. one 10. closed 11. 2.7 12. matter 13. deuterium, helium 14. low 15. equal to 16. temperature, COBE

Chapter 18
True or False? 1. T 2. F 3. T 4. T 5. F 6. F 7. T 8. F 9. T 10. T 11. T 12. T 13. T 14. T
Fill in the Blank 1. proteins 2. ammonia, methane 3. any two of lightning, ultraviolet light, radioactivity, volcanism, meteoritic impact 4. 3.5 billion 5. 1 billion 6. amino acids 7. technical civilizations 8. unstable 9. distances 10. light 11. dust 12. 65 13. Water Hole 14. 200

Photo Credits

Prologue
COP J. Sanford/Astrostock; AURA; NASA; AURA; Harvard Medical School   **P.1** NASA
P.2 AURA   **P.3** R.J. Dufour/Hansen Planetarium   **P.4** NASA   **P.5** D. Berry   **P.6a** S. Westphal
P.8 Astronomical Society of the Pacific

Chapter 1
CO1 Harvard College Observatory; Harvard College Observatory; Harvard College Observatory; Vassar College; Harvard College Observatory   **1.5** Lick Observatory Publications Office   **1.8** G. Schneider
1.9 NOAA   **1.10** G. Schneider   **1.11** G. Schneider   **1.13b** Boston Museum of Science
1.16 Erich Lessing/Art Resource   **1.18** Art Resource   **1.19** New Mexico State University
1.20 Erich Lessing/Art Resource   **1.24** The Granger Collection   **Int. 1-1** G. Schneider

Chapter 2
CO2 AURA; Smithsonian Astronomical Observatory   **2.1** T. Hallas   **2.11** Harvard-Smithsonian Center for Astrophysics; J. Moran; AURA; NASA   **2.14** Bausch & Lomb
2.15 AURA   **2.20b** Bausch & Lomb

Chapter 3
CO3 D. Berry   **3.6** Palomar Observatory   **3.7** AURA   **3.8** R. Wainscoat/Inst. for Astronomy; Keck Observatory   **3.10** AURA   **3.13** European Southern Observatory   **3.14** AURA; R. Wainscoat/Peter Arnold, Inc.   **3.15** NASA; NASA; NASA; AURA   **3.16** B. Brandl/Sky Publishing Corporation   **3.17** R. Ressmeyer/Starlight Collection, A Division of Corbis; MIT Lincoln Laboratory   **3.18** NRAO   **3.19** Astronomical Society of the Pacific   **3.20** AURA
3.21 NRAO   **3.22** NRAO; AURA   **3.23** Harvard College Observatory   **3.24** Smithsonian Astrophysical Observatory; NASA   **3.25** NASA; J. Sanford/Astrostock   **3.26** NASA   **3.28** Smithsonian Astrophysical Observatory
3.29 NASA   **3.30** NASA   **3.31** NRAO; NASA; Lund Observatory; NASA   **Int. 3-1** NASA; D. Malin; Hubble Space Telescope/NASA

Chapter 4
CO4 J. Lodriguss   **4.3** S. Westphal   **4.5** Palomar Observatory/California Institute of Technology
4.6 NASA   **4.7b** NOAO   **4.8** U.S. Naval Observatory; NASA   **4.10** European Space Agency
4.12 P. Parviainen/Science Photo Library/Photo Researchers, Inc.   **4.14** U.S. Geological Survey, U.S. Dept. of the Interior   **4.15** NASA
4.16 Sovfoto/Eastfoto   **4.17** Science Graphics
4.19 NASA; D. Berry   **4.20** D. Malin/Anglo-Australian Observatory   **Int. 4-1** NASA; H. Weaver/NASA   **MP4-1** AP/Wide World Photos

Chapter 5
CO5 NASA   **5.7** Lick Observatory Publications Office   **5.8** Defense Department Photo   **5.9** Lick Observatory Publications Office; California Institute of Technology   **5.10** NASA   **5.12** NASA
5.13 NASA   **5.20** NASA   **5.22** Earth Satellite Corp./Science Photo Library   **5.23b** P. Menzel/P. Menzel Photography   **5.25** W. Benz, W. Slatery & A. Cameron   **5.26** NASA

Chapter 6
CO6 NASA   **6.1** Palomar Observatory/California Institute of Technology   **6.4** Lick Observatory Publications Office   **6.5** Pic-du-Midi Observatory
6.7 NASA   **6.8** NASA   **6.9** NASA
6.10 NASA   **6.11** NASA   **6.12** NASA
6.13 Russian Space Agency   **6.14** NASA

6.15 NASA   **6.16** Russian Space Agency
6.17 NASA   **6.18** Hubble Space Telescope/NASA; NASA   **6.19** NASA   **6.20** NASA
6.21 NASA   **6.22** NASA   **6.23** NASA; Apollo/NASA   **6.24** NASA   **6.25** Lick Observatory Publications Office   **6.28** NASA
Int. 6-1 NASA, and the Natural History Museum, London; NASA

Chapter 7
CO7 NASA   **7.1** AURA; NASA   **7.2** Palomar Observatory/California Institute of Technology; NASA   **7.3** Lick Observatory Publications Office; NASA   **7.4** Lick Observatory Publications Office; NASA   **7.10** NASA   **7.11** NASA
7.12 NASA   **7.13** NASA   **7.15** NASA
7.16 J. Westphal/NASA   **7.17** NASA
7.18 NASA   **Int. 7-1** NASA

Chapter 8
CO8 NASA   **8.1** NASA   **8.2** NASA
8.3 NASA   **8.4** NASA   **8.5** NASA
8.6 NASA   **8.7** NASA; NASA; Hubble Space Telescope/NASA   **8.8** NASA   **8.9** NASA
8.10 NASA   **8.11** NASA   **8.12** NASA
8.13 NASA   **8.16** NASA   **8.17** NASA
8.18 NASA   **8.20** NASA   **8.21** NASA
8.22 NASA   **8.23** Lick Observatory Publications Office   **8.24** NASA/ European Space Agency
8.25 U.S. Naval Observatory; NASA

Chapter 9
CO9 NASA   **9.1** NOAO   **9.4** National Solar Observatory   **9.7** Palomar Observatory/California Institute of Technology   **9.8** Palomar Observatory/California Institute of Technology   **9.9** G. Schneider   **9.10** NOAO   **9.11** Sky Publishing Corporation   **9.13** NASA   **9.14** Palomar Observatory/California Institute of Technology; M. Penn/National Solar Observatory
9.15 Palomar Observatory/California Institute of Technology; M. Penn/National Solar Observatory
9.20 National Solar Observatory; NASA
9.21 National Solar Observatory   **9.22** National Solar Observatory   **9.24** R. Davis

Chapter 10
CO10 J. Sanford/Astrostock-Sanford
10.2 Harvard College Observatory
10.4 NOAO   **10.9** J. Sanford/Astrostock-Sanford   **10.17** Harvard College Observatory
10.22 NOAO   **10.23** NOAO

Chapter 11
CO11 NASA   **11.1** Palomar Observatory/California Institute of Technology   **11.3** Palomar Observatory/California Institute of Technology
11.4 Harvard College Observatory   **11.5** Royal Observatory, Edinburgh/Science Photo Library/Photo Researchers, Inc.   **11.6** NOAO; D. Malin/Anglo-Australian Observatory   **11.7** AURA; AURA; AURA; NASA   **11.8** AURA   **11.9** Harvard-Smithsonian Center for Astrophysics; NASA
11.10 Royal Observatory of Belgium; D. Malin/Anglo-Australian Observatory   **11.12** NASA
11.19 NASA   **11.20** J. Sanford/ Astrostock-Sanford; AURA; Harvard-Smithsonian Center for Astrophysics; NASA   **11.21** NASA
11.22 J. Hester/NASA   **11.23** Link Observatory Publications Office NASA; NASA   **11.24** NOAO

Chapter 12
CO12 D. Berry   **12.9** Anglo-Australian Telescope Board/Anglo-Australian Observatory   **12.10** D. Malin/Anglo-Australian Observatory; J. P. Harrington and K. J. Borkowski/NASA   **12.12** Palomar Observatory/California Institute of Technology

12.13 Lick Observatory Publications Office
12.17 European Space Agency   **12.20** NRAO
12.21 L. Chaisson   **12.23** NOAO
12.24 NASA   **12.25** European Space Agency
Int. 12-1 C. Burrows/NASA; NASA

Chapter 13
CO13 D. Berry   **13.1** NASA   **13.4** Lick Observatory Publications Office; Lick Observatory Publications Office; Max-Planck Institut fur Physik
13.5 Harvard-Smithsonian Center for Astrophysics; NASA   **13.13** Harvard-Smithsonian Center for Astrophysics; NASA   **13.14** L. Chaisson
Int. 13-1 California Institute of Technology

Chapter 14
CO14 NASA   **14.1** NASA   **14.2** T. Hallas; D. Malin; Palomar Observatory/California Institute of Technology   **14.10** NOAO   **14.13** L. Chaisson
14.15 AURA   **14.16** European Southern Observatory   **14.19** European Space Agency
14.20 NASA; NRAO; NRAO   **14.21** L. Chaisson

Chapter 15
CO15 R. Williams/NASA   **15.1** Palomar Observatory/California Institute of Technology; NASA   **15.2** NOAA; D. Malin; D. Malin
15.3 AURA   **15.4** Smithsonian Institution; M. J. Geller; P. Huchra/Smithsonian Astrophysical Observatory; University of Hawaii   **15.5** AURA
15.6 Palomar Observatory/California Institute of Technology   **15.7** AURA; Palomar Observatory/California Institute of Technology; D. Malin
15.8 NOAO; Royal Observatory, Edinburgh/Science Photo Library/Photo Researchers, Inc.; Royal Observatory, Edinburgh/Science Photo Library   **15.13** NASA; Palomar Observatory/California Institute of Technology   **15.14** Royal Observatory, Edinburgh/Science Photo Library/Photo Researchers, Inc.   **15.17** NASA   **15.20** NASA; AURA; Max-Planck Institut fur Physik
15.21 NASA   **15.22** NASA   **15.23** NASA
15.24 Harvard College Observatory
15.25 J. Barnes & L. Hernquist   **15.26** D. Malin
15.27 Palomar Observatory/California Institute of Technology   **Int. 15-1** Palomar Observatory/California Institute of Technology; W. Keel; AURA; D. Malin

Chapter 16
CO16 D. Berry   **16.2** AURA; NASA
16.4 Harvard-Smithsonian Center for Astrophysics
16.5 AURA; AURA; NRAO; NASA
16.6 NRAO   **16.7** J. Burns   **16.8** NOAO; NRAO   **16.9** NRAO; Palomar Observatory/California Institute of Technology   **16.12** Hubble Space Telescope/ NASA   **16.13** NASA
16.14 J. Moran   **16.16** NOAO; Palomar Observatory/California Institute of Technology; NASA   **16.17** Palomar Observatory/California Institute of Technology   **16.18** Astronomical Society of the Pacific   **16.19** AURA
16.20 NRAO   **16.21** NASA   **16.23** NASA; D. Berry   **16.24** NASA   **16.26** NASA

Chapter 17
CO17 D. Berry; NASA   **17.20** Dr. Edward Bertschinger, MIT   **17.21** NASA

Chapter 18
CO18 D. Berry   **18.3** M. & S. Fox; E. Barghoorn
18.4 Harvard-Smithsonian Center for Astrophysics
18.8 NASA   **Int. 18-1** C. Butler/Astrostock-Sanford

Index

These star maps show the brighter stars and the prominent constellations as they appear on the dates and at the times indicated. To use these maps, face the south and hold the book overhead with top of the map toward the north and the right-hand edge toward the west. The brightest stars are indicated by the star symbol (☆) and the names are indicated. (Star maps courtesy of Robert Dixon, *Dynamic Astronomy,* 6th ed., Prentice Hall, 1992.)

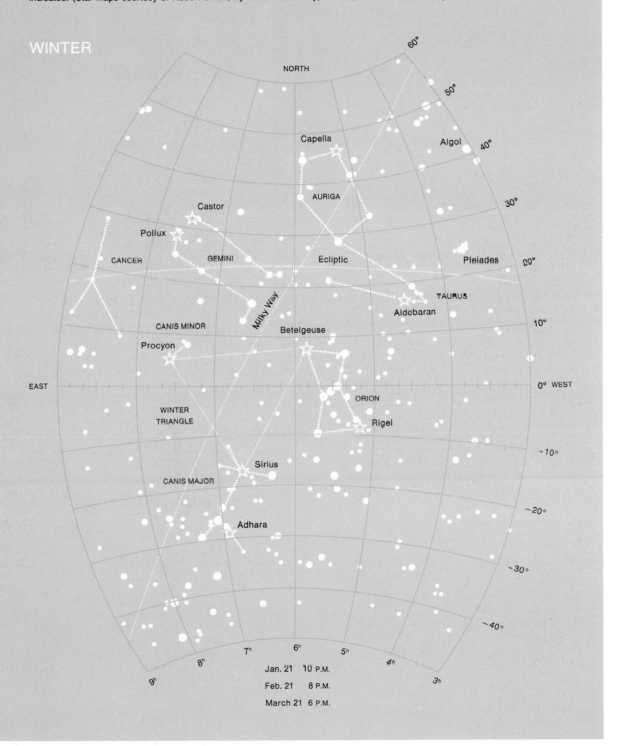

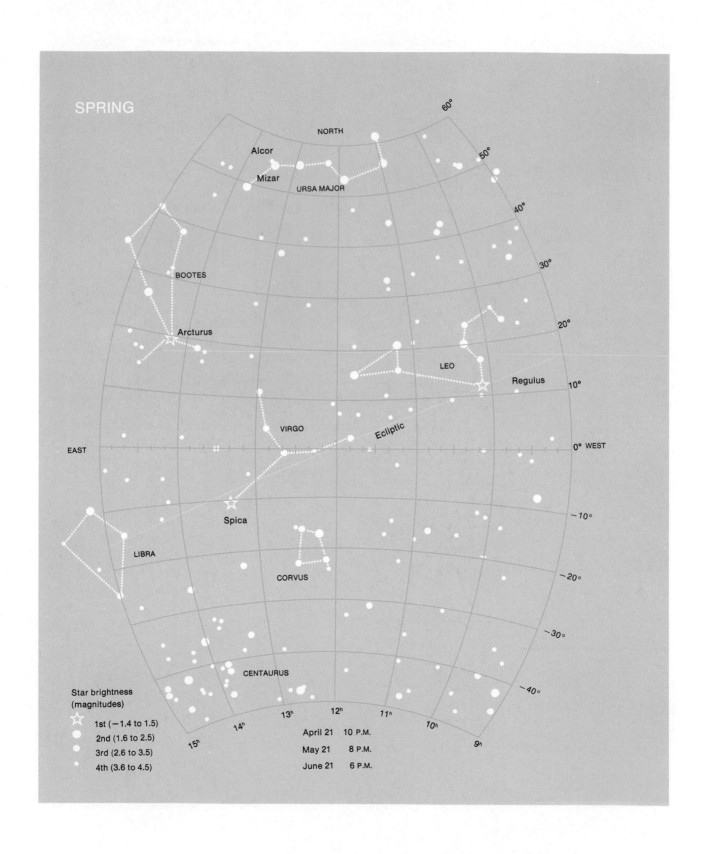

SPRING

60°

NORTH

50°

Alcor

Mizar

URSA MAJOR

40°

30°

BOOTES

20°

Arcturus

LEO

Regulus   10°

VIRGO

Ecliptic

EAST

0° WEST

−10°

Spica

LIBRA

CORVUS

−20°

−30°

CENTAURUS

−40°

Star brightness
(magnitudes)

1st (−1.4 to 1.5)
2nd (1.6 to 2.5)
3rd (2.6 to 3.5)
4th (3.6 to 4.5)

15ʰ   14ʰ   13ʰ   12ʰ   11ʰ   10ʰ   9ʰ

April 21   10 P.M.

May 21   8 P.M.

June 21   6 P.M.

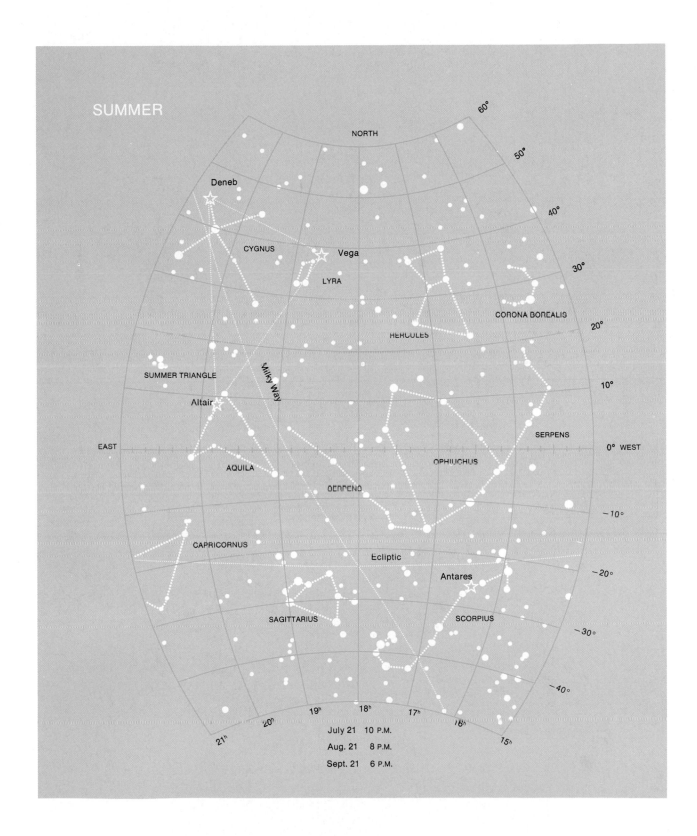

SUMMER

NORTH

60°

50°

40°

Deneb

30°

CYGNUS   Vega

LYRA   CORONA BOREALIS

20°

HERCULES

SUMMER TRIANGLE

10°

Milky Way

Altair   SERPENS

EAST   0° WEST

AQUILA   OPHIUCHUS

SERPENS

−10°

CAPRICORNUS

Ecliptic

Antares

−20°

SAGITTARIUS   SCORPIUS

−30°

−40°

21ʰ   20ʰ   19ʰ   18ʰ   17ʰ   16ʰ   15ʰ

July 21   10 P.M.

Aug. 21   8 P.M.

Sept. 21   6 P.M.

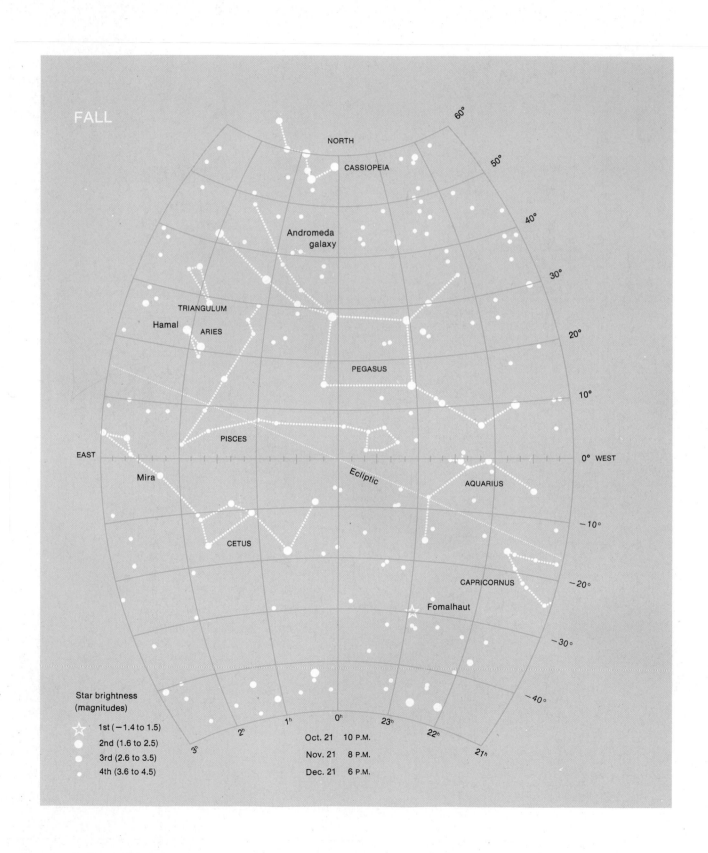

FALL

NORTH

CASSIOPEIA

60°
50°
40°
30°
20°
10°
0° WEST
−10°
−20°
−30°
−40°

Andromeda galaxy

TRIANGULUM

Hamal   ARIES

PEGASUS

PISCES

EAST

Mira

Ecliptic

AQUARIUS

CETUS

CAPRICORNUS

Fomalhaut

Star brightness
(magnitudes)

1st (−1.4 to 1.5)
2nd (1.6 to 2.5)
3rd (2.6 to 3.5)
4th (3.6 to 4.5)

3ʰ   2ʰ   1ʰ   0ʰ   23ʰ   22ʰ   21ʰ

Oct. 21   10 P.M.

Nov. 21   8 P.M.

Dec. 21   6 P.M.

Main-Sequence Stellar Properties by Spectral Class

Spectral Class	Typical Surface Temperature (K)	Color	Mass* ($M_\odot$)	Luminosity* ($L_\odot$)	Lifetime* (10^6 yr)	Familiar Examples
O	>30,000	Electric Blue	>20	>100,000	<2	
B	20,000	Blue	8	3000	30	Spica (B1)
A	10,000	White	3	75	400	Vega (A0) Sirius (A1)
F	7,000	Yellow - white	1.5	4	4000	Procyon (F5)
G	6,000	Yellow	1.0	1.5	9000	Sun (G2) Alpha Centauri (G2)
K	4,000	Orange	0.5	0.1	60,000	Epsilon Eridani (K2)
M	3,000	Red	0.1	0.005	200,000	Proxima Centauri (M5) Barnard's Star (M5)

*Approximate values for stars of solar composition